Thermal Fluctuations and Relaxation Processes in Nanomagnets

Thermal Fluctuations and Relaxation Processes in Nanomagnets

William T Coffey
Trinity College Dublin, Ireland

Yuri P Kalmykov
Université de Perpignan via Domitia, France

Sergei V Titov
Russian Academy of Sciences, Russia

World Scientific

NEW JERSEY • LONDON • SINGAPORE • BEIJING • SHANGHAI • HONG KONG • TAIPEI • CHENNAI • TOKYO

Published by

World Scientific Publishing Co. Pte. Ltd.
5 Toh Tuck Link, Singapore 596224
USA office: 27 Warren Street, Suite 401-402, Hackensack, NJ 07601
UK office: 57 Shelton Street, Covent Garden, London WC2H 9HE

Library of Congress Control Number: 2020938328

British Library Cataloguing-in-Publication Data
A catalogue record for this book is available from the British Library.

THERMAL FLUCTUATIONS AND RELAXATION PROCESSES IN NANOMAGNETS

ISBN 978-981-121-727-2 (hardcover)
ISBN 978-981-121-728-9 (ebook for institutions)
ISBN 978-981-121-729-6 (ebook for individuals)

For any available supplementary material, please visit
https://www.worldscientific.com/worldscibooks/10.1142/11747#t=suppl

«Telle est la faiblesse de l'esprit humain, qu'il a souvent besoin de s'aider d'hypothèses, pour lier les faits entre eux.»

P.S. Laplace, Traité de Mécanique Céleste, vol. 3, De L'Imprimerie de Crapelet, Paris, 1802.

Preface

Fine single-domain ferromagnetic nanoparticles exhibit *unstable behavior* of the magnetization caused by thermal agitation and resulting in superparamagnetism (so-called) because each particle effectively behaves as an *enormous paramagnet* of magnetic moment (~10^4–10^5 Bohr magnetons μ_B). At temperatures above a certain critical temperature called the *blocking temperature*, the magnetization may spontaneously reverse its direction due to thermal fluctuations so that the stable magnetic behavior so characteristic of a ferromagnet is destroyed. It follows that the onset of *thermal instability* characterized by a time-dependent magnetization in the magnetic nanoparticles used is of profound significance in magnetic recording as these are continually being reduced in size to provide both increased signal-to-noise ratio and greater storage density. Furthermore, the thermal instability is also an issue fundamental to rock magnetism as the magnetic record keeping ability of igneous rocks depends on the fact that the fine particles preserve the direction of the earth's magnetic field from the epoch in which the temperature of the environment has fallen below (e.g., due to cooling of the primeval earth) the blocking temperature of the particles. Thus, they constitute *magnetic fossils* and as such are indispensable in the study of paleomagnetism.

In grappling with the magnetization relaxation of fine ferromagnetic nanoparticles over many years, it became increasingly obvious to us that a reasonably comprehensive account of the theory of thermal fluctuations and relaxation processes in nanomagnets was imperative for the accurate interpretation of experimental and computer simulation data using rigorous mathematical models rooted in the principles of nonequilibrium statistical mechanics. Now as may or may not be widely appreciated Louis Néel in 1949 [15,16] formulated his theory of magnetization reversal over a potential barrier inside a nanoparticle by *implicitly* adapting an *Ansatz* proposed by Debye, v. his famous 1929 book *Polar Molecules* [111], to explain the dielectric relaxation of solids which after melting yield polar liquids. Debye considered an ensemble of noninteracting polar molecules each having the same permanent dipole moment and situated for example at the intersection points (sites) of a space lattice. He further supposed that if a uniform field acts on a dipole then it can point only in one of two *definite* directions, namely, that of the field and its opposite. Hence this model later, greatly elaborated upon by Fröhlich in 1949 in another famous book *Theory of Dielectrics* [110] and also by Brown in 1956 in his *Handbuch* article on Dielectrics [112], is known as *discrete orientation* in contrast to the better-known *continuous* distribution of orientations model of noninteracting polar molecules representing a polar liquid also due to Debye. The second model initially based on a fixed-axis rotator version of Einstein's theory of the Brownian motion (during his *Annus Mirabilis,* 1905 [97]) was then extended to rotation in space. In the discrete orientation model, where the transition from one orientation to another occurs in a *single* big jump, as applied to polar dielectrics the essential difference between it and the free rotational diffusion one is that the latter, where the transition from one orientation at time t to another at $t+dt$ occurs by a succession of small jumps, predicts dispersion and absorption in the microwave region while typically the former roughly explains the dispersion and absorption in ice occurring at km wavelengths. Thus, a solid such as ice behaves as if it were a polar liquid with very high internal friction, hence a very long relaxation time. Indeed, *exponentially* long as emphasized by Fröhlich because the probability of jumping of a dipole over a potential barrier is proportional to the appropriate Boltzmann factor $e^{-\Delta E}$ resulting in Arrhenius behaviour of the escape rate $\Gamma = Ae^{-\Delta E}$, which is inversely proportional to the overbarrier (longest) relaxation time τ (ΔE is a dimensionless potential barrier and A is a prefactor).

However, an essentially similar result was obtained for the longitudinal magnetization relaxation of a single-domain ferromagnetic nanoparticle over an internal magnetocrystalline-anisotropy-Zeeman energy potential barrier by Néel [15,16]. This approach to the origin of the magnetization reversal time mechanism based on statistical mechanics was initiated via his hypothesis that the reversal was governed by an activation process pertaining to escape over a barrier in the form of the discrete orientation model. However, the vital difference from standard reaction rate theory is since we treat a magnetization *energy density potential*, that the longest relaxation time of the magnetization, now *depending exponentially* on the volume of the particle, can vary from nanoseconds to geological epochs. Néel's ideas were later refined by Brown in 1963 [17] essentially building on his lucid *Handbuch* article [112] by setting the entire problem within the general theory of stochastic processes so involving a form of Boltzmann's *Stosszahlansatz* [97] (of which the Brownian motion is a particular case) just as achieved by Debye for dielectric relaxation. Brown's criticism of Néel's results for the reversal time of the magnetization rests on two indisputable facts (i) the relaxing system is not explicitly treated as a gyromagnetic one; (ii) it relies on the discrete orientation *Ansatz*. Hence despite the disturbance to the orientational Boltzmann distribution in a potential well due to the loss of magnetization (i.e., representative points) at the barrier that distribution *still prevails everywhere* in accordance with transition state theory. The latter assumption is an innate flaw of the discrete orientation *Ansatz* with the very important consequence that it is impossible to accurately calculate the escape rate and hence the relaxation time, because the effect of the energy dissipation on the escape rate is completely ignored. Brown, however, neatly circumvented both difficulties by formulating from the magnetic Langevin equation, i.e., the Landau–Lifshitz–Gilbert equation for the magnetization evolution as supplemented by stochastic terms, the magnetic Fokker–Planck equation for the surface probability density $W(\vartheta, \varphi, t)$ of the magnetic moment orientations on the unit sphere (ϑ and φ are, respectively, the polar and azimuthal angles of the spherical polar coordinate system). Thus, in a single insightful action he achieved the twin goals of setting the magnetization stochastic process within the framework of the general theory of the Brownian motion and simultaneously removing the difficulties associated with the discrete orientation *Ansatz* just as Debye achieved earlier with his continuous distribution of orientations model with an external applied field, an achievement which with the addition of a mean field potential was later used by Maier and Saupe in the study of the dielectric relaxation of nematic liquid crystals [97].

Furthermore, Brown was able to unambiguously identify the overbarrier relaxation (or magnetization reversal) time with the inverse of the smallest nonvanishing eigenvalue of the magnetic Fokker–Planck equation governing precession-aided rotational diffusion of the magnetic moment in the anisotropy-Zeeman energy potential thereby posing for a given potential a Sturm–Liouville problem. Brown used, *as an indicative example only*, the simplest uniaxial potential of the magnetocrystalline anisotropy-Zeeman energy. Hence the gyromagnetic terms are not explicitly involved in the longitudinal relaxation, because they immediately drop out of the Fokker–Planck equation. Thus, in this effectively one-dimensional problem by suitably adapting the Kramers theory [84] of escape of particles over potential barriers to classical macrospins, he found an asymptotic result which agrees with that of Néel [15,16] as far as the exponential dependence of the magnetization reversal time on the potential barrier is concerned. However, Brown's general formulation of the problem in terms of the magnetic Langevin and Fokker–Planck equations always enables one in principle to rigorously calculate the reversal time including both the gyromagnetic term and the fluctuation-dissipation process due to the bath as each is manifested in the prefactor of that time. This is quite unlike the situation prevailing in the discrete orientation *Ansatz*, where it is merely assumed that the prefactor τ_0 in Néel's formula for the relaxation time $\tau = \tau_0 e^{\Delta E}$ is the inverse of the precession frequency of the magnetization in the shallowest well of the anisotropy potential. We emphasize that Brown did not solve the magnetic Fokker–Planck equation at all rather he only gave an asymptotic expression for the smallest nonvanishing eigenvalue for the *very restricted* case of uniaxial anisotropy since just one canonical variable, namely, the colatitude ϑ of the magnetic moment is explicitly involved because then the azimuth just gives rise to a steady precession. His asymptotic result was later verified by Aharoni [334], who numerically solved the Fokker–Planck equation for the uniaxial anisotropy-Zeeman energy so demonstrating that Brown's result for the smallest nonvanishing eigenvalue is correct. Nevertheless, the magnetization reversal time problem in noncircularly symmetric potentials has generally not been well understood to this day. With such potentials, the conceptual difficulties arise because the three damping regimes of escape-rate theory originally identified for Newtonian particles by Kramers in 1940 [84] also appear for classical spins governed by noncircular symmetry, namely, very low damping, the Kramers turnover region and intermediate to high damping [166]. The reason being that *two canonical variables*, namely, the polar and the azimuthal angles of the spherical polar coordinate system specifying the magnetization orientation are now involved, i.e., both degrees of freedom explicitly contribute to the magnetization relaxation. Moreover, the spin Hamiltonian is *no longer separable and additive* unlike that of point particles or rigid rotators, the sole exception being uniaxial anisotropy where the azimuthal angle φ does not explicitly enter in order to describe the longitudinal relaxation. Hence to verify escape-rate formulas for noncircularly symmetric potentials which intrinsically involve interesting phenomena such as the role played by *longitudinal and precessional mode coupling* in the magnetization reversal time and dynamic susceptibility, one must be able to exactly solve the relevant Fokker–Planck or Langevin equation.

Our terminology is as follows, by the Fokker–Planck method, we mean expansion of the solution of that equation as a Fourier–Laplace series in the spherical harmonics $Y_{nm}(\vartheta, \varphi)$, which are the appropriate basis set for the evaluation of observables such as the magnetization yielding a multiterm differential-recurrence relation for the statistical moments (averages of the spherical harmonics). However, the completely equivalent but more easily visualized procedure based on the magnetic Langevin equation at first involves rewriting that equation as a stochastic evolution equation for a spherical harmonic of arbitrary rank with the aid of the theory of angular momentum. One then averages the resulting equation during an infinitesimal time as is usual in the theory of the Brownian motion over an ensemble of macrospins [97], which all had at some initial time the same *sharp* colatitude and azimuth (ϑ and φ), i.e., the same orientation. This procedure leads to the equation of motion of a spherical harmonic in terms of the sharp values ϑ and φ. However, since the latter are *themselves random variables*, then a second (spatial) average over their distribution yields a time-dependent multiterm ordinary differential-recurrence relation for the statistical moments — averages of the spherical harmonics. This relation concurs with that of the Fokker–Planck operator proving the equivalence of the two methods. Having determined the hierarchy of recurrence relations for an arbitrary anisotropy potential it is then possible to transform it into a tridiagonal vector-valued differential recurrence relation. This relation admits of a formally exact solution in terms of a sum of products of matrix continued fractions (*Kettenbrüche*) in the frequency domain ultimately yielding the observables (reversal time of the magnetization, longitudinal and transverse components of the dynamic susceptibility tensor, etc.) for any given potential as illustrated by many examples.

Brown's magnetic Langevin–Fokker–Planck analytic approach may also be extended to thermal fluctuations in spin torque transfer (STT) devices [95,96]. Because of the STT effect [114,115], the magnetization of a nanoscale ferromagnet may be altered by spin-polarized currents because an electric current with spin polarization in a ferromagnet has an associated flow of angular momentum thereby exerting a *macroscopic spin-torque*. The phenomenon underpins the novel subject of spintronics, i.e., *current-induced control over magnetic nanostructures*. In STT devices, the thermal fluctuations cannot be ignored at *finite* temperatures because they lead to mainly noise-induced switching at currents far less than the critical switching current without noise as well as introducing randomness into the precessional orbits These fluctuations represent a very significant feature of the operation of such devices and are usually treated via an adaptation of the Néel–Brown model including STT. The relevant magnetic Langevin equation is then the Landau-Lifshitz-Gilbert equation for the evolution of the magnetization as modified by Slonczewski [114] and Berger [115] and including noise terms. This equation may then be recast as before as a tridiagonal vector-differential recurrence relation for the observables for an arbitrary magnetocrystaline anisotropy-Zeeman energy potential along with the nonconservative STT potential.

Yet another advantage of the analytic approach is that it allows one in tandem with the representation-space method (described below) to treat quantum effects in the magnetization relaxation processes. This method postulates following Wigner in 1932 [6,278,279] that quantum mechanics may be formally regarded as a statistical theory on a representation or phase space which has a classical meaning. Referring first to particles both position and momentum are placed on an equal footing as in Hamiltonian mechanics unlike in the Schrödinger picture which uses either position or momentum representations. Hence, the quantum state of a particle is described by a quasiprobability distribution function (because the uncertainty principle forbids *precise location* within phase-space regions smaller than $\hbar$), commonly called the Wigner function, in momentum p and position x and not a wave function, state vector or density matrix. As demonstrated by Wigner in 1932 (v. for an illuminating discussion Wyatt [216]), the marginal distributions of the Wigner function which unlike the latter are true probability distributions yield the correct quantum mechanical probability for either the position or the momentum which are conjugate variables. These results follow from the definition of the Wigner function, assuming a mixed state, as the Fourier transform of the density operator with the argument written in overlap form. In addition, because the phase-space formulation does not *explicitly* involve operators and wave functions, observables and matrix elements may be calculated from phase-space integrals of *c*-number functions (classical kernels) via the Weyl correspondence. This is the correspondence of phase-space functions to quantum mechanical operators in Hilbert space as envisaged by Weyl in 1927 [493] which are now called Weyl-ordered operators. The Wigner function and its associated evolution equation in phase space called the Wigner–Moyal equation (analogous to the Liouville equation of classical mechanics) allows one via perturbation theory in a quantum parameter to calculate from its stationary solution relatively high temperature quantum corrections to the classical Maxwell–Boltzmann distribution. Indeed, as first realized by Groenewold in 1946 [494] and Moyal in 1949 [286], the Wigner function has a rather deep meaning, because it is in effect the inverse of Weyl's rule for associating quantum mechanical operators with corresponding classical quantities. Thus, the Weyl correspondence should be regarded as an *invertible transform* rather than as a consistent quantization rule, amounting to a change of representation being only one of many such maps. In the context of the quantum statistical mechanics, the Wigner function in representation space is the Wigner transform of the density matrix in Hilbert space, while the density matrix is the inverse transform of the Wigner function. One may further infer that the Weyl association rule mapping invertibly *c*-number phase-space functions or phase-space kernels to operators in a *given ordering prescription* is but one of a multitude of alternate equivalent association rules which specify corresponding representations.

In the spin context these representations are usually based on the coherent (minimum uncertainty) state representation of the density matrix as widely used in quantum optics [184,246,275] and which is particularly appropriate when seeking semiclassical corrections for quantum effects. Such a representation when applied to spin systems allows one to analyze quantum spin relaxation using a master equation for a quasiprobability distribution $W(\vartheta, \varphi, t)$ of spin orientations in a phase (here configuration) space (ϑ, φ); ϑ and φ are as before the polar and azimuthal angles constituting the canonical variables. The quasiprobability function $W(\vartheta, \varphi, t)$ underpinned by a bijective mapping $\hat{w}$ in Hilbert space onto (ϑ, φ) and vice versa was originally introduced by Stratonovich in 1956 [273] for closed systems (zero dissipation) as part of a general discussion of *c*-number quasiprobability distributions for quantum systems in a representation space based on the symmetry properties of the underlying group and comprises a generalization of the Weyl correspondence. Examples are the Heisenberg–Weyl group for particles and the SU(2) group for rotations [275]. The function $W(\vartheta, \varphi, t)$ was further developed both for closed and open systems and is entirely analogous to the translational Wigner distribution $W(x, p, t)$ in the phase space (x, p) which is the quasiprobability representation of the density operator except that certain *intrinsic differences* arise because of the angular momentum commutation relations. The Wigner function $W(\vartheta, \varphi, t)$ of spin orientations in a configuration space (ϑ, φ) just as the Wigner function $W(x, p, t)$ for translational motion of a particle in a phase space (x, p) enables the expected value $\langle \hat{A} \rangle (t)$ of a quantum spin operator $\hat{A}$ in Hilbert space to be calculated via the corresponding *c*-number (Weyl) function $A(\vartheta, \varphi)$. Thus, quantum mechanical averages involving the density matrix may all be calculated in the manner of classical ones which *sui generis* is naturally suited to the determination of quantum corrections to them. In addition, the formalism is easy to implement because semiclassical master equations in phase space also enable powerful techniques originally developed to treat classical Fokker–Planck equations, e.g., continued fractions, mean first-passage times etc. to be seamlessly carried over into the quantum domain.

Chapter 1, comprises several sections, where the foundations of spin Brownian motion are laid down starting with the classical treatment of the Larmor precession of magnetic moments then proceeding to the Langevin equation of motion of the magnetization, i.e., the random variable, and its accompanying probability density diffusion equation or Fokker–Planck equation and how both may be formally solved by the matrix continued fraction method. A detailed account of how approximate closed expressions for relaxation times etc. may be obtained via the Kramers escape-rate theory as adapted to classical spins is then given. It is also shown how the overall theory may be applied to switching-field curves and surfaces as well as to the Stoner–Wohlfarth astroids so factoring in the effect of thermal agitation. The principles of the quasiprobability treatment of the quantum Brownian motion of spins are then extensively discussed starting with a derivation of the reduced density matrix evolution equation for a spin system in contact with a thermal bath and the assumptions that are made. Then how one may formally solve the reduced equation using the method of statistical moments as expressed in terms of averages of the polarization operators. These operators are the Hilbert space correspondents of the spherical harmonics used in the classical solution and transform under rotation in exactly the same way [275]. Next, we demonstrate the phase-space formulation of the magnetization evolution starting for purposes of illustration with the Wigner distribution for particles with separable and additive Hamiltonians as outlined earlier and how it may be applied via a phase-space master equation, a generalization of the Wigner–Moyal equation, to the quantum translational Brownian motion. Then beginning with Stratonovich's group-theoretic approach based on his discussion of c-number quasiprobability distributions for quantum systems (in particular, spins), it is shown how his bijective operator $\hat{w}$ formulation leads to these quasidistributions expressed in the classically meaningful form of finite Fourier–Laplace series spanning the spin number. As a sequel to this both the Weyl symbols of some common spin operators are derived as well as the general statistical moment equations for spin relaxation in phase space in terms of averages of polarization operators.

Chapter 2 is concerned with applications of classical spin relaxation theory developed in the first half of Chapter 1. We present both exact (in terms of continued fractions) and approximate analytical solutions of various problems appearing in the magnetization relaxation of nanomagnets. These include the longitudinal and transverse responses, their characteristic times, linear and nonlinear dynamic susceptibilities, the magnetic stochastic resonance, and dynamic hysteresis. We also evaluate the magnetization reversal time using the Kramers escape-rate theory and its generalization to classical spin systems. As particular examples, we evaluate the reversal time for cubic, biaxial, and mixed (uniaxial and cubic) anisotropy potentials while STT effects in the magnetization reversal and nonlinear ac stationary response of nanomagnets are treated via the generic nanopillar model introduced in Chapter 1, Section 1.3.3. All results are presented in a form suitable for comparison with experimental data.

Chapter 3 on the other hand is concerned with practical applications of the representation of quantum spin dynamics as a statistical theory on configuration space (ϑ, φ). Indeed, in conjunction with [271] and the second half of Chapter 1, it could also be seen as an introduction to representation-space methods. The list of contents taken together with the summary at the end is once more the best possible guide to the material presented. However, we single out Section 3.2.6, where the effect of quantum tunnelling on the TST reversal time is particularly obvious and also Section 3.3.10, where we see how quantum characteristic times go over to the corresponding classical ones as the spin number goes to infinity. We also show how calculations of statistical moments from the master equation in phase space for noncircularly symmetric problems may be fruitfully described using the polarization operators.

The mathematical details of many of the calculations in Chapters 2 and 3 have been relegated to the Appendixes so as not to overwhelm the reader with them. Some properties of the special mathematical functions used throughout the book are also described there.

We would like to thank many individuals and organizations who have greatly helped us, directly or indirectly, in the thirty or so years since we began this work. In particular, Declan Byrne, Roy Chantrell, Liam Cleary, Patrick Cregg, Derrick Crothers, the late Jean–Louis Dormann, Liam Dowling, Paul Fannin, Dmitry Garanin, Lawrence Geoghegan, Elaine Kennedy, Dave McCarthy, Bernard Mulligan, Bachir Ouari, Stuart Rice, Brendan Scaife, John Waldron, Jean–Eric Wegrowe, Nijun Wei, Wolfgang Wernsdorfer, Tony Wickstead, and Marios Zarifakis. We are indebted to Pierre–Michel Déjardin, Yuri Raikher, Igor Poperechny, Victor Stepanov, and André Thiaville for useful comments and suggestions as well as Michael O'Riordan for clerical assistance. We would also like to thank the Trinity College Dublin Trust, the Royal Irish Academy, the Royal Society, the Enterprise Ireland International Collaboration Fund, INTAS, EGIDE, the Soros Foundation, the United States Air Force Research Laboratory, European Office of Aerospace Research and Development, Agence Nationale de la Recherche (ANR), and Science Foundation Ireland for financial support.

Finally, we thank our families who have borne with us during the heavy work involved in preparing this book.

Dublin, Ireland	William T. Coffey
Perpignan, France	Yuri P. Kalmykov
Fryazino, Russia	Sergei V. Titov

November 2019

Contents

Preface vii

Contents xvii

List of Abbreviations xxiii

Chapter 1 Introductory Concepts 1

1.1 Introduction 1
1.2 Motion of magnetic moments: classical treatment 5
1.2.1 Larmor equation 5
1.2.2 Magnetic properties of solids 8
1.2.3 Magnetocrystalline anisotropy 12
1.2.4 Landau–Lifshitz and Gilbert equations 14
1.2.5 Bloch equation 21
1.3 Stochastic motion of the magnetization 24
1.3.1 Langevin equation for magnetic moment 24
1.3.2 Fokker–Planck equation for magnetic moments in configuration space 26
1.3.3 Spin-transfer torque (STT) effects in the magnetization dynamics 38
1.3.4 Method of statistical moments 45
1.3.5 Methods of solution of infinite sets of differential-recurrence equations 54
1.3.6 Calculation of observables 58
1.3.7 Langevin and Fokker–Planck equations for magnetic moments in energy space 62
1.3.8 Energy-controlled diffusion equation in the presence of STT 69
1.4 Magnetization relaxation times 71
1.4.1 Relaxation processes 71
1.4.2 Discrete orientation model 76
1.4.3 Mean first-passage time (MFPT) 80
1.4.4 MFPT for circularly symmetric free energy 84

1.4.5 MFPT in the VLD regime ... 87
1.4.6 MFPT in the VLD regime in the presence of STT ... 89
1.5 Escape-rate theory for classical spins ... 91
1.5.1 Transition state theory for spins ... 92
1.5.2 Kramers' escape rate for Brownian particles ... 94
1.5.3 Langer's treatment of the IHD regime ... 99
1.5.4 IHD escape rate for spins ... 105
1.5.5 VLD escape rate for spins by the Kramers method ... 110
1.5.6 Turnover escape rate for spins ... 112
1.6 Switching-field curves and surfaces ... 114
1.6.1 Stoner–Wolhfarth astroids ... 114
1.6.2 Effect of thermal agitation ... 119
1.7 Ferrofluids ... 122
1.8 Quantum treatment of the motion of spins ... 137
1.8.1 Early theories ... 138
1.8.2 Density matrix formulation of spin relaxation ... 142
1.8.3 Collision kernel in the Markov approximation ... 151
1.8.4 Method of statistical moments for quantum spins ... 162
1.9 Phase-space formulation of magnetization relaxation ... 166
1.9.1 Quasiprobability distribution functions for particles ... 168
1.9.2 The Wigner distribution function for particles ... 172
1.9.3 Application to transition state theory ... 182
1.9.4 Application to quantum Brownian motion ... 187
1.9.5 Quasiprobability distribution functions for spins ... 193
1.9.6 Spin phase-space distribution functions ... 197
1.9.7 Weyl symbols of some spin operators ... 211
1.9.8 Master equation and statistical moment equations for spin relaxation in phase space ... 214
1.10 Conclusion and Summary ... 217

Chapter 2 Magnetization Relaxation and Resonance in Nanomagnets ... 221

2.1 Introduction ... 221
2.2 Reversal time of the magnetization at VLD: the MFPT approach ... 226
2.2.1 Reversal time for circularly symmetric potentials ... 226
2.2.2 VLD reversal time for uniaxial anisotropy with an oblique field ... 228
2.2.3 VLD reversal time for biaxial anisotropy ... 236
2.2.4 Comparison of analytical, asymptotic, and numerical results ... 242
2.2.5 STT effects on the magnetization reversal time at VLD ... 245
2.3 Magnetization reversal time in nanomagnets with noncircularly symmetric potentials: Kramers' escape-rate theory approach ... 252
2.3.1 Basic escape-rate equations for nanomagnets ... 254

2.3.2 Uniaxial nanomagnet in an oblique dc magnetic field..... 257
2.3.3 Biaxial nanomagnet in a uniform dc magnetic field applied along the easy axis 263
2.3.4 Reversal time for cubic anisotropy 266
2.3.5 Mixed anisotropy: breakdown of the paraboloid approximation 270
2.3.6 STT effects in the thermally assisted magnetization reversal 281
2.4 Magnetization relaxation in uniaxial nanomagnets 287
2.4.1 Longitudinal relaxation and dc field effect 293
2.4.1.1 Matrix solution 294
2.4.1.2 Matrix continued fraction solution 296
2.4.2 Characteristic times and magnetic susceptibility 297
2.4.3 Depletion effect in a biased bistable potential 301
2.4.4 Magnetic stochastic resonance in uniaxial nanomagnets. 306
2.4.5 Transverse response: ferromagnetic resonance 313
2.5 Magnetization relaxation in nanomagnets with noncircularly symmetric anisotropy 319
2.5.1 Uniaxial nanomagnet in an oblique field 319
2.5.2 Cubic anisotropy 329
2.5.2.1 Complex susceptibility and relaxation times....... 334
2.5.2.2 Magnetic stochastic resonance in nanomagnets with cubic anisotropy 336
2.5.2.3 Effect of a dc bias field 339
2.6 Nonlinear stationary ac response of nanomagnets 345
2.6.1 Uniaxial nanomagnets 345
2.6.2 Biaxial nanomagnets 358
2.7 Dynamic magnetic hysteresis 364
2.7.1 DMH in uniaxial nanomagnets 364
2.7.2 DMH in biaxial nanomagnets 375
2.8 Spin-transfer torque effects 380
2.8.1 STT effects in the magnetization reversal 380
2.8.2 Nonlinear ac stationary response 391
2.8.3 Linear dynamic susceptibility 396
2.8.4 Nonlinear response 398
2.8.5 Dynamic magnetic hysteresis 401
2.9 Antiferromagnetic nanoparticles 405
2.10 Finite barrier correction for the ferromagnetic resonance frequency 425
2.10.1 FMR frequency for uniaxial anisotropy with a transverse magnetic field 428
2.10.2 FMR frequency for biaxial anisotropy 431
2.10.3 Mixed anisotropy: breakdown of the paraboloid approximation 435
2.11 Magnetization dynamics of two interacting spins in an external magnetic field 439

2.11.1 Exchange interaction 445
2.11.2 Dipole-dipole coupling 459
2.12 Concluding remarks 465

Chapter 3 Quantum Effects in the Magnetization Relaxation of Nanomagnets 467

3.1 Introduction 467
3.2 Equilibrium phase-space distribution functions for spins 473
3.2.1 Spins in a uniform external field 477
3.2.2 Uniaxial nanomagnet in an external field 483
3.2.3 Uniaxial nanomagnet in a transverse field 489
3.2.4 Biaxial anisotropy 490
3.2.5 Cubic anisotropy 492
3.2.6 TST reversal time 494
3.2.7 Switching-field curves 500
3.3 Master equation in phase space for circularly symmetric systems 505
3.3.1 Master equation for a uniaxial nanomagnet subjected to a dc magnetic field 505
3.3.2 Differential-recurrence relations for the statistical moments 510
3.3.3 Spin relaxation in a dc magnetic field 516
3.3.4 Quantum analog of the magnetic Langevin equation 519
3.3.5 Exact solution of the master equation for longitudinal relaxation 523
3.3.6 Nonlinear longitudinal relaxation time 529
3.3.7 Linear response 532
3.3.8 Single-mode approximation 534
3.3.9 Longitudinal relaxation of uniaxial nanomagnets 538
3.3.10 Analytic equations for the characteristic relaxation times and dynamic susceptibility 544
3.3.11 Nonlinear longitudinal relaxation in superimposed ac and dc magnetic fields 562
3.3.12 Dynamic magnetic hysteresis 569
3.3.13 Quantum effects in stochastic resonance 575
3.4 Master equation in phase space for noncircularly symmetric systems 582
3.4.1 Uniaxial nanomagnet subjected to a dc bias field of arbitrary orientation 582
3.4.2 Characteristic relaxation times and dynamic susceptibility 586

General Conclusions and Summary 595

Appendixes: Special Functions, Operators, etc. 605

A.1. Relevant special functions and associated formulas 605
A.1.1. Gamma and related functions 605
A.1.2. Confluent hypergeometric (Kummer) functions 605
A.1.3. Error function 607
A.1.4. Hermite and Laguerre polynomials 608
A.1.5. Legendre polynomials 609
A.1.6. Modified Bessel functions of the first and second kinds 611
A.1.7. Spherical harmonics 614
A.1.8. Wigner's *D*-functions 618
A.1.9. Elliptic integrals and functions 620
A.1.10. Dirac-delta and Heaviside functions 622
A.2. Spin and polarization operators 623
A.3. Derivation of the master equation for a uniaxial paramagnet subjected to a dc magnetic field 628
A.4. Characteristic times of relaxation and correlation functions 641

References 649

Index 679

List of Abbreviations

ACF	autocorrelation function
DMH	dynamic magnetic hysteresis
FMR	ferromagnetic resonance
IHD	intermediate-to-high damping
MFPT	mean first-passage time
PDF	probability density function
STT	spin torque transfer
TST	transition state theory
VHD	very high damping
VLD	very low damping

Chapter 1

Introductory Concepts

1.1 Introduction

De Magnete so runs [1] the title of the renowned late Elizabethan era treatise of William Gilbert, published in the year 1600 which more or less represented the first attempt to describe magnetism theoretically in the early modern age. The word *magnet* itself (cf. the Magnetes from Thessaly) appears to have originated in classical times from the Ionian Greek city of Magnesia on the river Maeander in south western Asia Minor, where lodestones may be mined. Indeed, Magnesia has ruins which are still extant today (near the modern village of Tekin, Germencik district of Aydin Province, Turkey). Admirable accounts of these and more modern developments and indeed the subject since classical times have been given by Mattis [1] and Coey [2] with the attraction of many amusing quotations from the original works.

Fast forward now to the beginning of the last century when about the year 1900 (i.e., almost contemporaneously with the advent of quantum mechanics) physicists began to develop a theory of magnetism capable of describing magnetic interactions on a microscopic scale. In turn their efforts were given renewed impetus by the rapid development of the modern quantum theory in the years 1925–1928 at the hands of Werner Heisenberg, Erwin Schrödinger, Max Born, Pascual Jordan, Paul Dirac and Wolfgang Pauli [3–5].

Now for the past sixty or so years, i.e., roughly the period covered in this book (dealing with the relatively specialized topic of the role of stochastic spin dynamics in the theory of fine particle magnetization including quantum effects which we shall mainly treat by the representation space methods initiated by Eugene Wigner from 1930 [6] onwards) much scientific and technological effort has been directed towards producing progressively smaller magnetic structures. These are used for the long-term storage of data in hard disk drives of computers, magnetoresistive random access memory, mobile phones and in other novel devices involving information systems as well as in biomedical applications so that magnetic micro-systems are now all-pervasive in daily life. As far as many technological applications are concerned hard disk drive manufacturers are constantly seeking ways to improve areal density, as this parameter equates to a greater storage capacity. Indeed, *areal density* is widely regarded as the crucial metric driving that industry. The trend towards more and more miniaturization cannot, however, continue indefinitely, since present methods of hard disk production are fast approaching physical limits, and the areal density will no longer be able to increase beyond these fundamental limits. To overcome these physical obstacles, one must therefore study the behavior of magnets on the microscopic scales used in hard disks to find potential solutions to the technological problems posed by them.

Experimental investigations allow certain observations to be made on actual systems, however they are severely limited by the detail which can be extracted from these measurements. Nevertheless, if appropriate theoretical treatments were available, the resulting analytical and numerical models would provide an all-important link between experiment and theory. Hence it would be possible to systematically interpret experimental results. Indeed, new designs for experiments could emerge and moreover be subsequently developed thereby reducing costs.

From a theoretical aspect the orientation of magnetic moments on a *small* scale is governed by four competing energy terms, viz., the *dipolar* energy causing magnets to align from north pole to south pole; the *exchange* energy in ferromagnetic materials compelling the magnetic moments in the immediately surrounding space to lie parallel to each other; the *anisotropy* energy which is least if the moments are aligned along a particular crystalline direction, and the *Zeeman* energy which is least when the magnetic moments lie in the same direction as an external applied magnetic field. Since the most efficient magnetic alignment, or *configuration*, is that where the energy is lowest, these four energy terms try to minimize the overall energy at the expense of one another resulting in rather complex physics. Clearly the competition between these interactions under various conditions is responsible for the overall behavior of a magnetic system, and the ability to predict this will permit a greater understanding of the latter. Furthermore, in tandem with the rapid growth of quantum mechanics, all kinds of magnetic interactions have become better understood on a microscopic level.

Nanomagnetism as a separate subfield, dealing specifically with the behavior of ferromagnetic materials at fine (nanometer) length scales, effectively began in 1963 when William Fuller Brown Jr. published his paper on antiparallel domain wall structure [7]. Consequently, our overall purpose in writing this book is to describe some of the theories underpinning nanomagnetism, i.e., the theoretical basis for studying the foundations of the relaxation and resonance phenomena in nanomagnets which may be briefly reviewed as follows.

Fine single-domain ferromagnetic particles exhibit unstable behavior of the magnetization due to thermal agitation causing superparamagnetism [8] (so-called) because each nanoparticle effectively behaves as an *enormous* paramagnet of magnetic moment ~10^4–10^5 Bohr magnetons. The thermal fluctuations, causing relaxation of the magnetization of such nanoparticles, play a major role both in information storage [9–11] and biomedical applications [12,13]. In particular, all magnetic recording media rely on the properties of these fine (~10 nm) ferromagnetic nanoparticles because [8,14–18] the ferromagnetic state intrinsically associated with a given orientation of the magnetic moment of a single-domain nanoparticle has a *remanent* magnetization, a property which has led directly to the establishment of the modern magnetic recording industry. However, the *apparently stable* ferromagnetic state of a tape or a magnet is *only one of many possible local minima* of the free energy hence thermal agitation may cause spontaneous jumps of the magnetic moment from one such metastable state to another. Therefore, at temperatures exceeding a certain critical temperature called the *blocking temperature*, the magnetization may reverse its direction due to thermal agitation so that it exhibits *thermal instability*. In this manner the stable magnetic behavior, so characteristic of a ferromagnet, is destroyed due to the loss of the elementary representative points comprising the magnetization from the well in question, to another one due to the shuttling action of the thermal agitation. This is represented by the Brownian motion and ultimately results in the complete loss of the recording, (e.g., witness the deterioration of old video tapes). Now the onset of thermal instability (characterized by a time-dependent magnetization) in the fine magnetic particles used in magnetic recording is of marked technological significance as these are continually being reduced in size to provide both increased signal-to-noise ratio and greater storage density for the recording and information industries. The thermal instability is also of profound interest in rock magnetism [8,19] as the magnetic record keeping ability of igneous rocks occurs because the fine particles preserve the direction of the earth's magnetic field from the epoch, in which the temperature of the environment has fallen below (e.g., due to cooling of the primeval earth) the blocking temperature of the particles. Thus, they constitute *magnetic fossils* and, as such, are indispensable in the study of paleomagnetism. Yet another consideration is that with recent progress in the development of magnetic nanotechnologies experimental studies of relaxation processes in *individual* particles have now become possible [9,20]. Clearly the main parameter characterizing the thermal stability is the *magnetization reversal time* (superparamagnetic relaxation time) of the nanoparticles. This time is crucially affected by thermal interactions of the particles with their surrounding heat bath resulting in fluctuations and dissipation, ultimately leading to a complete loss of the stored information. Thus, it is vital for information storage purposes to determine the dependence of the reversal time on the dissipative coupling to the bath at a given temperature. Moreover, estimates of that time over extensive ranges of temperature and damping are required in numerous other physical applications, e.g., in the determination of linear and nonlinear dynamic susceptibilities (e.g., Refs. [21–40]), the loop shape, coercive force and specific power loss in dynamic magnetic hysteresis (e.g., Refs. [41–54]), the signal-to-noise ratio in stochastic resonance (e.g., Refs. [55–69]), the switching-field curves and surfaces at finite temperatures (e.g., Refs. [14,70–75]), Mössbauer spectra (e.g., Refs. [76–80]), etc.

Experimental success [20,81–83] in measuring the reversal time of the magnetization **M** of an *individual* particle, and in verifying that behavior as a function of the damping parameter (effectively the coupling to the thermal bath) predicted by the Néel–Brown model (which in its most rudimentary form as considered by Néel is just an application of classical reaction rate theory) [15–18] has stimulated renewed interest in appropriately adapting the Kramers escape-rate theory [84] to classical macrospins. In particular, the Néel–Brown theory is in one of its aspects an adaptation of the Kramers escape-rate theory [84,85]. That theory was originally formulated for the escape rate out of a potential well for point Brownian particles moving in a one-dimensional extension x (so that they effectively have only one degree of freedom with the free motion governed by Newton's equations) and was extended to macrospin relaxation governed by a gyromagnetic-like equation which has two degrees of freedom, namely, the polar angles ϑ and φ of a macrospin. Here interaction of the macrospin with its thermal bath is accounted for by the Brownian motion, a factor which was ignored by Néel. Hence the verification of the Néel–Brown theory in the magnetic nanoparticle [20,81–83] context vividly illustrates the Kramers concept of a thermal relaxation process as that occurring over a potential barrier arising from the shuttling action of the Brownian motion. Yet another adaptation of the Kramers theory to nanomagnetism is to the relatively new subject of spintronics. This topic has been reviewed by Dunn *et al.* [86].

1.2 Motion of magnetic moments: classical treatment

1.2.1 *Larmor equation*

The magnetic moment $\boldsymbol{\mu}$ of a nanomagnet ultimately stems from the angular momentum of electrons in an atom, which for free atoms is a combination of both electron spin and orbital momentum. For solids the spin component dominates the magnetic moment while if a magnetic moment is placed in a magnetic field **H**, it experiences a torque **K** which may be expressed as the vector product

$$\mathbf{K} = \mu_0 \left[\boldsymbol{\mu} \times \mathbf{H} \right], \qquad (1.1)$$

where the permeability of free space $\mu_0 = 4\pi \cdot 10^{-7}\,\mathrm{JA^{-2}m^{-1}}$ in SI units.

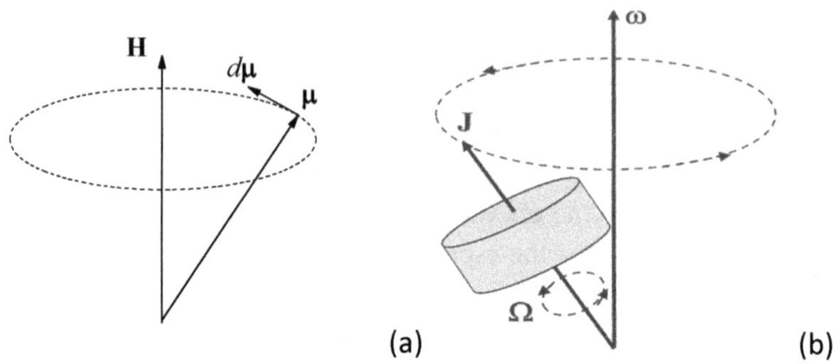

Fig. 1.1. a) Larmor precession of spin and magnetic moment vector; b) Precession of a top

In the classical physics, a magnetic moment may be pictured as a current loop. Moreover, for a static moment or a classical current loop, the torque **K** tends to align the magnetic moment with the magnetic field **H**, representing its lowest energy configuration. A typical example is a compass needle. However, if the magnetic moment arises rather from the orbital motion of an electron about a nucleus, **μ** is proportional to the angular momentum of the electron **J**

$$\boldsymbol{\mu} = -\gamma\mu_0^{-1}\mathbf{J}, \tag{1.2}$$

where γ is a constant called the gyromagnetic ratio. In the following, the gyromagnetic ratio is associated with the electron spin yielding $\gamma = 2.2 \cdot 10^5$ A^{-1}m s^{-1}. The torque so exerted causes a rate of change of angular momentum which is perpendicular to **J**, viz.,

$$\frac{d\mathbf{J}}{dt} = \mathbf{K} = \mu_0\left[\boldsymbol{\mu}\times\mathbf{H}\right]. \tag{1.3}$$

Hence **μ** *precesses* about the direction of **H** in the manner of a miniature top rather than *aligning* in that direction thus

$$\frac{d\boldsymbol{\mu}}{dt} = \gamma\left[\mathbf{H}\times\boldsymbol{\mu}\right]. \tag{1.4}$$

Equation (1.4) is the Larmor equation (named [2] after Sir Joseph Larmor) which describes precession of **μ** about **H** with an angular frequency

$$\boldsymbol{\omega} = \gamma\mathbf{H} \tag{1.5}$$

known as the *Larmor frequency.* To demonstrate this, we use a frame of reference rotating with respect to fixed axes with angular velocity represented by the vector $\boldsymbol{\omega}$. According to the Coriolis theorem of classical mechanics, the evolution rates of $\boldsymbol{\mu}$, as viewed from the *rotating* and fixed (*laboratory*) frames, respectively, are connected via

$$\begin{aligned}\dot{\boldsymbol{\mu}}_{\text{rot}} &= \dot{\boldsymbol{\mu}}_{\text{lab}} + [\boldsymbol{\mu} \times \boldsymbol{\omega}] \\ &= [(\gamma \mathbf{H} - \boldsymbol{\omega}) \times \boldsymbol{\mu}].\end{aligned} \tag{1.6}$$

Equation (1.6) will have the same form as Eq. (1.4), provided **H** is replaced by an effective field $\mathbf{H}_{\text{eff}} = \mathbf{H} - \boldsymbol{\omega}/\gamma$. For the choice $\boldsymbol{\omega} = \gamma\mathbf{H}$ that field vanishes, thus $\boldsymbol{\mu}$ is time-independent in the rotating frame. Therefore, viewed from the laboratory frame perspective $\boldsymbol{\mu}$ rotates about **H** with an angular velocity $\boldsymbol{\omega} = \gamma\mathbf{H}$. This enduring precession of $\boldsymbol{\mu}$ about an external magnetic field is called the *Larmor precession* (see Fig.1.1(a)) and is basically similar to the precession of a spinning top with angular velocity $\boldsymbol{\Omega}$ in a gravitational field (see Fig. 1.1(b)). Here the weight of the top causes a change in the angular momentum **J** that causes the top to precess with the angular velocity $\boldsymbol{\omega}$.

We can assume without loss of generality that the field direction defines the Z axis. Hence the components of $\boldsymbol{\mu}$ satisfy the simple precessional equations

$$\begin{aligned}\dot{\mu}_X &= -\mu_Y \omega_Z, \\ \dot{\mu}_Y &= \mu_X \omega_Z, \\ \dot{\mu}_Z &= 0,\end{aligned} \tag{1.7}$$

where $\omega_Z = \gamma H_Z$. Thus, the Z-component is unaltered, and the complete solution is

$$\begin{aligned}\mu_X(t) &= \mu_X(0)\cos\omega_Z t - \mu_Y(0)\sin\omega_Z t, \\ \mu_Y(t) &= \mu_X(0)\sin\omega_Z t + \mu_Y(0)\cos\omega_Z t, \\ \mu_Z(t) &= \mu_Z(0),\end{aligned} \tag{1.8}$$

so representing continuous rotation of the μ_X and μ_Y components.

1.2.2 *Magnetic properties of solids*

Hitherto we have merely considered an *isolated* magnetic moment. Now we shall discuss magnetic solids and their magnetic properties. The former may be classified by their response to externally applied magnetic fields as diamagnetic, paramagnetic, or ferromagnetic, and these responses differ greatly in strength.

Diamagnetism is a property common to all materials and *opposes* applied magnetic fields, however it is always very weak and may be explained as follows. The orbital motion of electrons creates tiny atomic current loops, which give rise to magnetic fields. If an external magnetic field is applied to a material, these minute current loops will tend to align so as to oppose the applied field. This phenomenon may be regarded as an atomic version of Lenz's law, viz., induced magnetic fields tend to oppose the change which created them. Materials in which this effect is the sole magnetic response are called *diamagnetic*. Moreover, materials with no elementary magnetic dipoles at all are diamagnetic. Clearly all materials are inherently diamagnetic, however, if the atoms have some net magnetic moment as in *paramagnetic* materials, or if there is *long-range ordering* of atomic magnetic moments as in *ferromagnetic* materials, these more powerful effects will always dominate. Indeed, diamagnetism constitutes the residual magnetic behavior when materials are neither paramagnetic nor ferromagnetic.

Paramagnetism, where it occurs, is significantly stronger than diamagnetism and causes magnetization in the direction of the applied field. Paramagnetic solids are those in which some of the atoms, ions or molecules constituting the solid possess a permanent magnetic dipole moment. These dipoles are isolated from one another in the solid, so that in effect, it is made up of noninteracting atomic magnets. In most materials these moments cancel, however in those classified as paramagnetic, the cancellation is incomplete. Without a magnetic field, the dipoles are arranged at random so that the solid exhibits no net magnetic moment. In a magnetic field, however, the elementary dipoles attempt to orient themselves parallel to the direction of magnetic induction in the solid, so enhancing the internal field within the solid and causing the observed paramagnetic effect. In general, the alignment of the dipoles is incomplete,

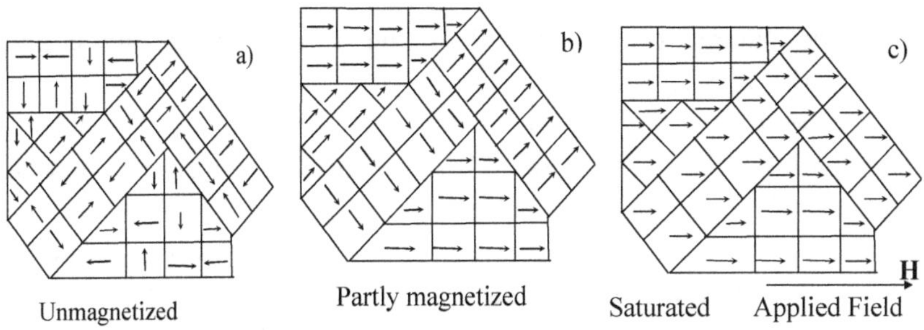

Fig. 1.2. (a) A ferromagnetic material, where the domains of the material are magnetized in the directions of easy magnetization. The arrows indicate the direction of the magnetization of each domain, where adjacent domains have opposite directions of magnetization, so that the overall magnetization of the material is negligible. If however we apply a uniform external magnetic field, in the direction shown, some of the domains, those with magnetic orientations perpendicular or opposite to the direction of the applied field, become unstable and may quickly rotate to another direction of easy magnetization in the same direction as the applied field **H**. This is illustrated in (b). If we were to increase the applied field to the point of saturation, we would have the situation (c).

due to both thermal effects and interaction with the surrounding atoms. Moreover, the orientation of the spins changes continuously.

Ferromagnetic effects are very large, causing magnetizations which are sometimes orders of magnitude greater than those expected from the applied field alone and are therefore much larger than either diamagnetic or paramagnetic effects. Iron, nickel, cobalt and some of the rare earths exhibit ferromagnetic behavior. On an atomic scale the *exchange interaction* in ferromagnetic bodies tends to align neighboring spins and arises due to quantum mechanical interaction at the atomic level. This interaction is remarkable in that it *locks* the magnetic moments of neighboring atoms into a rigid parallel order over a large number of atoms regardless of the thermal agitation which tries to randomize any atomic-level order. In addition, two other main interactions tend to produce small uniformly magnetized regions, which may be observed experimentally and are called *magnetic domains* (see Fig. 1.2). These comprise *dipole-dipole* interaction and *exchanged* (electrostatic) interactions. If the two interactions are compared, then at short ranges the electrostatic force is the most important, and any arrangement of parallel spins is of lowest energy. However, as the distance from a given dipole increases, the short-range electrostatic interaction is less than the dipole-dipole interaction. Consequently, at such distances, the system may *lower* its energy by reversing the spins. Therefore, the domains act as a balance between these two competing effects – short-range electrostatic and long-range dipolar. The domain sizes range from 0.1 mm to a few mm.

The theory that a ferromagnetic material consists (magnetically speaking) of elementary regions, each magnetized almost to saturation in some direction, was first proposed by Weiss [8] in 1907. He assumed that these regions coincided with the crystals of which the material was composed, an assumption which was subsequently refuted by Frenkel and Dorfman and by Heisenberg and Bloch [8], who realized that even a *single* crystal is comprised of these *minute permanent magnets* (magnetic domains). However, the exact nature of domains remained under debate until 1935 when Landau and Lifshitz [8] discovered them to be in the form of *elementary layers*.

Within the domain itself, the magnetic field is intense, however in a bulk sample the material will usually be unmagnetized because the *many domains will themselves be randomly oriented* with respect to one another. *Ferromagnetism* then manifests itself due to the observation that a small externally imposed magnetic field $\mathbf{H}'$ causes the many magnetic domains to align with each other in which case the material is said to be magnetized. Furthermore, the driving magnetic field will then be increased by a large factor which is usually expressed as a *relative permeability* for the material. Another commonly used magnetic quantity is the *magnetic susceptibility* specifying how much the relative permeability differs from unity. Ferromagnets will tend to remain partly magnetized following imposition of an external magnetic field. This tendency to "remember their magnetic history" is called *hysteresis*. The fraction of the saturation magnetization retained when the driving field is removed is called the *remanence* and is an important factor in the description of permanent magnets. Moreover, some materials can retain a state of magnetization almost indefinitely. These are called *magnetically hard* materials and are used to make permanent magnets. A ferromagnetic solid behaves as if it were surrounded by a magnetic field and will attract or repel other ferromagnets via the interactions of their respective magnetic fields. A ferromagnetic material also has *magnetocrystalline anisotropy*, meaning that the magnetic moments align more readily in certain crystallographic directions rather than in others. These are known as ‘easy’ and ‘hard’ directions respectively.

A *ferrimagnetic* material is one that has populations of atoms with *opposing* magnetic moments. However, in ferrimagnetic materials the opposing moments are unequal, and a spontaneous magnetization remains. This happens when the populations consist of different materials or ions (such as Fe^{2+} and Fe^{3+}). Ferrimagnetism is exhibited by ferrites and magnetic garnets. Indeed, the oldest known magnetic substance, magnetite (iron (II, III) oxide; Fe_3O_4), is a ferrimagnet; it was originally classified as a ferromagnet before Néel’s discovery of ferrimagnetism and antiferromagnetism in 1948 [87]. In materials exhibiting *antiferromagnetism*, the magnetic moments of atoms or molecules, which are usually related to the spins of electrons, align in a regular pattern with neighboring spins (on different sublattices) pointing in *opposite* directions which is just like ferromagnetism and ferrimagnetism, a manifestation of *ordered magnetism*.

In general, a particle of ferromagnetic material [88] below a certain critical size (typically 150 Å in radius) constitutes a *single-domain particle* meaning [8] that it is in a state of *uniform magnetization* for *any applied field*. We shall denote the magnetic dipole moment of such a particle by $\boldsymbol{\mu} = v\mathbf{M}$, where $\mathbf{M}$ is the magnetization and v is the volume of the particle. *Superparamagnetism* is yet another form of magnetism appearing in *small* ferromagnetic nanoparticles. Here the magnetic moment $\boldsymbol{\mu}$ is *not* that of a *single* atom but rather of a *single-domain particle* of volume v which may be of the order of 10^4–10^5 Bohr magnetons, so that *extremely large* moments are involved: hence the term *superparamagnetism*. Moreover single-domain particles will in general not be *isotropic* but will have *anisotropic* contributions to their total energy associated with the *external shape* of the particle, *imposed stress* or the *crystalline structure* itself. The superparamagnetism, or *thermal instability* of the magnetization, occurs if the thermal energy kT (k is Boltzmann’s constant, T is the temperature) is enough to change the orientation of $\boldsymbol{\mu}$ of the entire particle in spite of the anisotropy potential. Thus, the overall behavior is just like an ensemble of paramagnetic atoms. There is no hysteresis, merely saturation behavior.

The *static magnetic susceptibility* χ of diamagnetic and paramagnetic (superparamagnetic) substances, which is a dimensionless quantity, may be defined as

$$\mathbf{M} = \chi \mathbf{H}' \tag{1.9}$$

In anisotropic solids, **M** and **H′** are not necessarily parallel, therefore the magnetic susceptibility will then vary with direction in the crystal. Furthermore, the magnetic susceptibility of a diamagnetic substance is negative while that of a paramagnetic solid is positive. Finally, the magnetic susceptibility of ferromagnetic materials is generally very large.

1.2.3 *Magnetocrystalline anisotropy*

Most ferromagnetic solids are *magnetically anisotropic*. The *spin-orbit interaction* is the primary source of *magnetocrystalline anisotropy*. It basically stems from the orbital motion of the electrons which couples with the crystal electric field thereby giving rise to the first order contribution to magnetocrystalline anisotropy. The second order contribution arises due to the *mutual interaction* of the magnetic dipoles.

Some crystal systems have a *single axis* of high symmetry and the anisotropy of such crystals is called *uniaxial anisotropy*. The magnetocrystalline anisotropy energy is generally represented as an expansion in powers of the direction cosines of the magnetization (u_X, u_Y, u_Z) which are the components of a unit vector **u** directed along **M**. If the Z axis is taken to be the principal axis of symmetry of the crystal, the free energy density of the magnetic particle V is then

$$V = K(u_X^2 + u_Y^2) = K\sin^2\vartheta, \tag{1.10}$$

where K is an anisotropy constant. The parameter K has the units of energy density and depends on the composition. If $K > 0$, the directions of lowest energy are $u_Z = \pm 1$ ($\vartheta = 0$ and $\vartheta = \pi$), i.e., the polar axis (cf. Fig. 1.5). If the applied dc field $\mathbf{H}_0$ is assumed parallel to the polar axis, then the total free energy (per unit volume) is

$$V(\vartheta) = K\sin^2\vartheta - \mu_0 M_S H_0 \cos\vartheta, \tag{1.11}$$

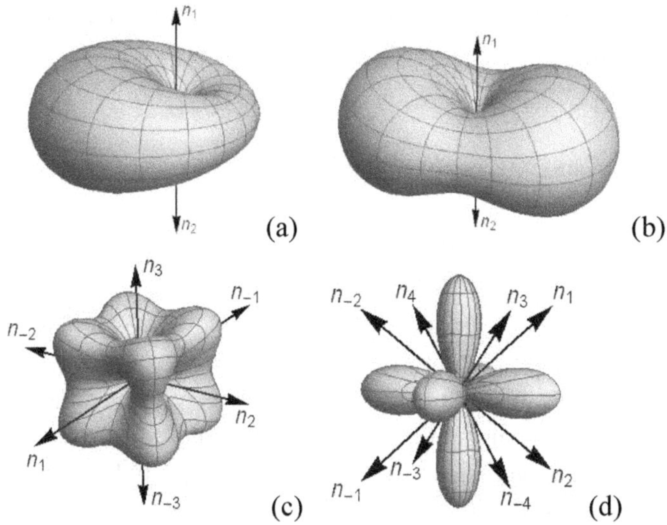

Fig. 1.3. (a) Uniaxial potential Eq. (1.12) with external field; (b) biaxial anisotropy potential Eq. (1.14); (c) cubic anisotropy potential Eq. (1.15) for $K_c > 0$; (d) cubic anisotropy potential for $K_c < 0$. The arrows denote the metastable magnetization orientations and n_i is the number of particles in the ith orientation.

where M_S is the *saturation magnetization*. However, by applying $\mathbf{H}_0$ at an angle ψ with respect to the easy axis, the symmetry is broken so that the free energy density will also depend on the azimuthal angle φ (see Figs. 1.3(a) and 1.5), viz.,

$$V(\vartheta,\varphi) = K\sin^2\vartheta - \mu_0 M_S H_0\left(\cos\psi\cos\vartheta + \sin\psi\sin\vartheta\cos\varphi\right). \quad (1.12)$$

The anisotropy potential of single-domain particles with *orthorhombic anisotropy* still has a *single easy* axis, however it also has a *hard* axis (direction of maximum energy) and an *intermediate* axis (direction associated with a saddle point of the energy). The free energy of such a particle V is (see Fig. 1.3(b))

$$V = K\left(u_X^2 + u_Y^2\right) + K_1 u_X^2 = K\sin^2\vartheta + K_1\sin^2\vartheta\cos^2\varphi, \qquad (1.13)$$

where K and K_1 are the anisotropy parameters. Here the easy axis is the Z direction, the intermediate axis is the Y direction and the hard axis is the X direction. Thus, *biaxial anisotropy* generates *azimuthally nonuniform* energy distributions with two saddle points. If the applied field $\mathbf{H}_0$ is assumed parallel to the Z axis, then the free energy is

$$V(\vartheta,\varphi) = K\sin^2\vartheta + K_1\sin^2\vartheta\cos^2\varphi - \mu_0 M_S H_0\cos\vartheta. \tag{1.14}$$

In a *cubic* crystal, the anisotropy potential is

$$\begin{aligned} V &= K_c\left(u_X^2u_Y^2 + u_X^2u_Z^2 + u_Y^2u_Z^2\right) = \frac{K_c}{2} - \frac{K_c}{2}\left(u_X^4 + u_Y^4 + u_Z^4\right) \\ &= \frac{K_c}{4}(\sin^4\vartheta\sin^2 2\varphi + \sin^2 2\vartheta), \end{aligned} \tag{1.15}$$

where K_c is the anisotropy constant, which may have either positive or negative values. A magnetic particle with *cubic anisotropy* has three or four easy axes, depending on the sign of K_c. If $K_c > 0$ the easy axes are the X, Y and Z axes (see 1.3(c)). If $K_c < 0$ there are four easy axes characterized by $X = \pm Y = \pm Z$ (see 1.3(d)).

Each anisotropy potential $V(\vartheta, \varphi)$ generates an effective magnetic field $\mathbf{H}$ proportional to the negative gradient of the free energy density V, viz.,

$$\mathbf{H} = -\frac{1}{\mu_0}\frac{\partial V}{\partial \mathbf{M}} \quad \text{or} \quad \frac{\partial V}{\partial \mathbf{M}} = \mathbf{i}\frac{\partial V}{\partial M_X} + \mathbf{j}\frac{\partial V}{\partial M_Y} + \mathbf{k}\frac{\partial V}{\partial M_Z}. \tag{1.16}$$

The field $\mathbf{H}$ which in a spherical coordinate system basis $\{\mathbf{e}_r, \mathbf{e}_\vartheta, \mathbf{e}_\varphi\}$ is

$$\mathbf{H} = -\frac{1}{\mu_0 M_S}\left(\frac{\partial V}{\partial \vartheta}\mathbf{e}_\vartheta + \frac{1}{\sin\vartheta}\frac{\partial V}{\partial \varphi}\mathbf{e}_\varphi\right) = -\frac{1}{\mu_0 M_S}\left(0, \frac{\partial V}{\partial \vartheta}, \frac{1}{\sin\vartheta}\frac{\partial V}{\partial \varphi}\right),$$

can be graphically visualised (see Fig 1.5). The effective field amplitude $H(\vartheta, \varphi) = |\mathbf{H}|$ and the direction of $\mathbf{H}$ are shown in Fig. 1.4 for various anisotropy potentials $V(\vartheta, \varphi)$.

1.2.4 *Landau–Lifshitz and Gilbert equations*

The first dynamical model for the precessional motion of the magnetization $\mathbf{M}$ of a single-domain ferromagnetic particle or macrospin was proposed by Landau and Lifshitz in 1935 [89]. They postulated that without damping the magnetization precesses about an effective magnetic

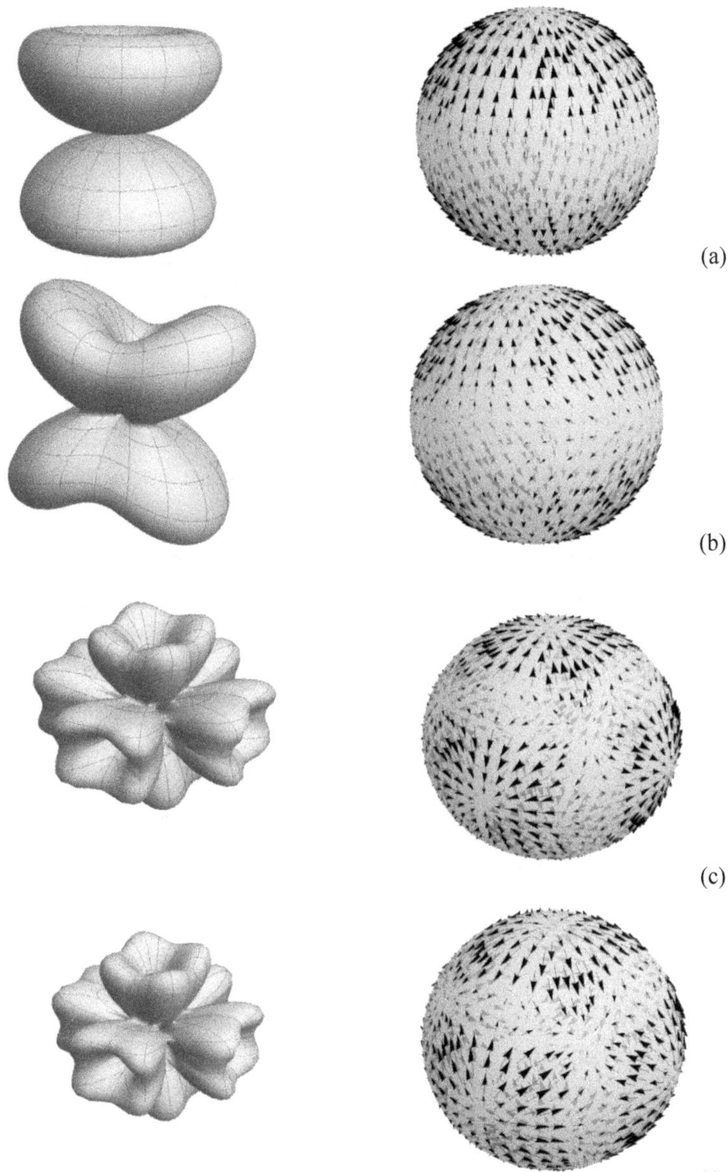

Fig. 1.4. The effective magnetic field **H** (amplitude and direction on the unit sphere) for (a) uniaxial anisotropy potential Eq. (1.10); (b) biaxial anisotropy potential Eq. (1.13); (c) cubic anisotropy potential Eq. (1.15) ($K_c > 0$); (d) cubic anisotropy potential Eq. (1.15) ($K_c < 0$).

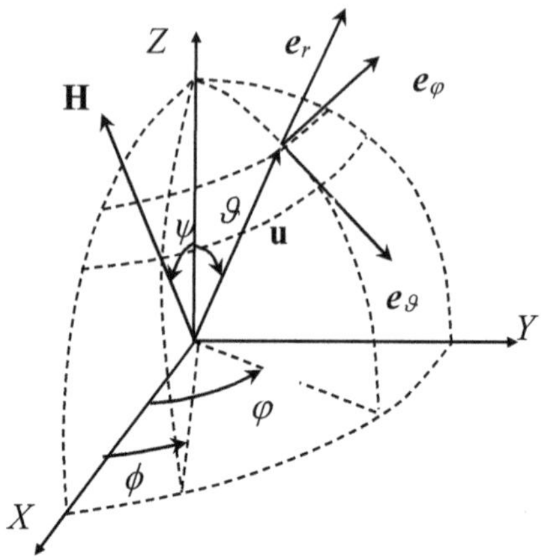

Fig. 1.5. Spherical polar coordinate system.

field **H** according to the gyromagnetic equation (1.4) with anisotropy phenomenologically accounted for via the effective magnetic field **H**. Assuming that the single-domain ferromagnetic particle is at its saturation magnetization M_S so that only the orientation of **M** can change we rewrite Eq. (1.4) in terms of a unit vector **u** directed along **M**, namely,

$$\mathbf{u} = \mathbf{M} / M_S = (\sin\vartheta\cos\varphi, \sin\vartheta\sin\varphi, \cos\vartheta) \tag{1.17}$$

(here ϑ and φ are the spherical coordinates as defined in Fig. 1.5). Hence, the kinematic equation becomes

$$\dot{\mathbf{u}} = \gamma[\mathbf{H} \times \mathbf{u}], \ \mathbf{H} = -\frac{1}{\mu_0 M_S}\frac{\partial V}{\partial \mathbf{u}}. \tag{1.18}$$

Here $\partial/\partial\mathbf{u}$ is the gradient on the surface of the unit sphere explicitly defined in spherical coordinates as

$$\frac{\partial}{\partial \mathbf{u}} = \frac{\partial}{\partial \vartheta}\mathbf{e}_\vartheta + \frac{1}{\sin\vartheta}\frac{\partial}{\partial\varphi}\mathbf{e}_\varphi. \tag{1.19}$$

Invariably the free energy density V comprises an anisotropy potential plus the Zeeman energy due to an external magnetic field. Taking as example a bistable potential (Fig. 1.3), a giant spin will orbit at constant energy (Stoner–Wohlfarth orbits) in one of the wells (see Fig. 1.6).

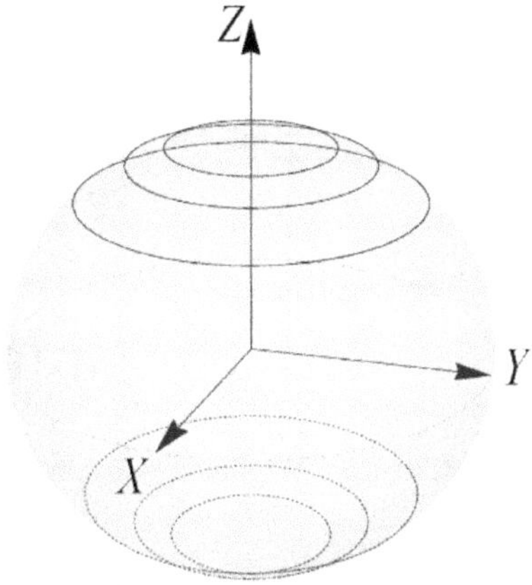

Fig. 1.6. Stoner–Wohlfarth orbits encircling an energy minimum along the polar positive and negative Z-axis.

So far the evolution of the magnetization as described by the gyromagnetic Eq. (1.18) is a *conservative* process, thus no energy is lost to the surrounding environment via the motion of the magnetization. Therefore, **M** must follow paths of *constant* energy, i.e., the Stoner–Wohlfarth orbits [14,90] (Fig. 1.6). Hence, it will continue to precess *ad infinitum* in a well of the potential provided that the energy of the applied field is less than the barrier height. However, by applying a strong enough field, the orbital precession will be *reversed*, that is flip over into the other potential well. Switching of the direction of precession in this way represents the process of erasure of magnetic recordings by Oersted fields. In practice, however, the nanoparticle loses energy to the microscopic degrees of freedom of its surroundings. Therefore, to model the effect of energy dissipation, Landau and Lifshitz in 1935 [89] introduced a damping torque opposing the precession so that the gyromagnetic equation becomes

$$\dot{\mathbf{u}} = \gamma[\mathbf{H}\times\mathbf{u}] - \lambda\big[\mathbf{u}\times[\mathbf{u}\times\mathbf{H}]\big], \tag{1.20}$$

where λ is a dimensionless damping parameter. Note that the additional term must be of a form such that the magnitude of the magnetization is preserved, i.e. $|\mathbf{M}| = M_S$, as may be deduced by scalar multiplication of both sides of Eq. (1.20) by **u** yielding

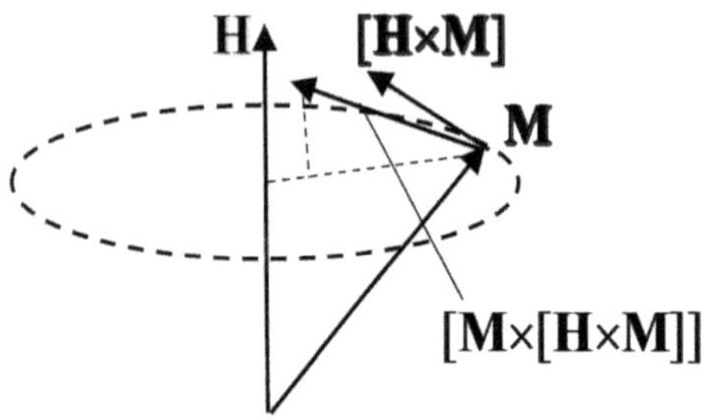

Fig. 1.7. Equation (1.20) tends to make **M** spiral towards the direction of the field **H**.

$$(\mathbf{u}\cdot\dot{\mathbf{u}})=\frac{1}{2}\frac{d}{dt}(\mathbf{u}\cdot\mathbf{u})=\frac{1}{2}\frac{d}{dt}|\mathbf{u}|^2=0. \tag{1.21}$$

Equation (1.20) obviously tends to make **M** spiral towards the direction of the field **H** (see Fig. 1.7).

A different approach was proposed by Gilbert in 1955 [91]. He observed that the conservative equation (1.4) can be derived from a *Lagrangian formulation*, where *the role of the generalized coordinates is played by the components of the magnetization vector* **M**. Therefore, the most natural way of treating phenomenological dissipation is to introduce an additional torque with components proportional to the time derivatives of the generalized coordinates

$$\dot{\mathbf{u}}=\gamma[\mathbf{H}\times\mathbf{u}]+\alpha[\mathbf{u}\times\dot{\mathbf{u}}], \tag{1.22}$$

where $\alpha > 0$ is the dimensionless damping constant, depending on the material (typical values are in the low damping range $0.001 < \alpha < 0.1$). Like the Landau–Lifshitz equation, the additional term introduced by Gilbert preserves the magnitude of the magnetization $|\mathbf{M}| = M_S$. The modified Eq. (1.22) is generally called the *Landau–Lifshitz–Gilbert* equation.

However, a marked difference between the Landau–Lifshitz Eq. (1.20) and the Landau–Lifshitz–Gilbert Eq. (1.22) exists although both are very similar from a mathematical point of view. Indeed, the Landau–Lifshitz form can be easily obtained from the Gilbert one. For example, by vector multiplication of both sides of Eq. (1.22) by **u**, we have

$$[\mathbf{u}\times\dot{\mathbf{u}}]=-\gamma[\mathbf{u}\times[\mathbf{u}\times\mathbf{H}]]+\alpha[\mathbf{u}\times[\mathbf{u}\times\dot{\mathbf{u}}]]. \qquad (1.23)$$

Then recalling the vector identity $[\mathbf{a}\times[\mathbf{b}\times\mathbf{c}]]=\mathbf{b}(\mathbf{a}\cdot\mathbf{c})-\mathbf{c}(\mathbf{a}\cdot\mathbf{b})$ and observing that $(\mathbf{u}\cdot\dot{\mathbf{u}})=0$, we have

$$[\mathbf{u}\times\dot{\mathbf{u}}]=-\gamma[\mathbf{u}\times[\mathbf{u}\times\mathbf{H}]]-\alpha\dot{\mathbf{u}}. \qquad (1.24)$$

Next by substitution of the latter equation into the right hand side of the Landau–Lifshitz–Gilbert equation (1.22), we have

$$\dot{\mathbf{u}}=-\gamma[\mathbf{u}\times\mathbf{H}]-\gamma\alpha[\mathbf{u}\times[\mathbf{u}\times\mathbf{H}]]-\alpha^2\dot{\mathbf{u}}. \qquad (1.25)$$

This equation can then be rearranged yielding

$$\dot{\mathbf{u}}=-\frac{\gamma}{1+\alpha^2}[\mathbf{u}\times\mathbf{H}]-\frac{\gamma\alpha}{1+\alpha^2}[\mathbf{u}\times[\mathbf{u}\times\mathbf{H}]], \qquad (1.26)$$

which is commonly referred to as the Gilbert equation rendered in the Landau–Lifshitz form. Clearly Eq. (1.26) and Eq. (1.20) are mathematically the same given the replacements

$$\gamma\rightarrow\frac{\gamma}{1+\alpha^2},\ \lambda=\frac{\gamma\alpha}{1+\alpha^2}. \qquad (1.27)$$

However careful comparison of the quantities γ and $\gamma/(1+\alpha^2)$ indicates that the Landau–Lifshitz (1.20) and Landau–Lifshitz–Gilbert equations (1.22) express different physics and so will become identical only in the limit of vanishing damping ($\alpha << 1$). For example, in the limit of infinite damping ($\lambda\rightarrow\infty$ in Eq. (1.20), $\alpha\rightarrow\infty$ in Eq. (1.22)), the Landau–Lifshitz and Landau–Lifshitz–Gilbert equations respectively yield $\dot{\mathbf{u}}\rightarrow\infty$ and $\dot{\mathbf{u}}\rightarrow 0$. Since the second result $\dot{\mathbf{u}}\rightarrow 0$ is in agreement with the intuitive notion that very large damping should result in a very slow motion while the first is not, one may conclude that the Landau–Lifshitz–Gilbert (1.22) equation is the more appropriate one to describe magnetization dynamics. Hence, we will use the latter equation in all that follows.

The Landau–Lifshitz–Gilbert equation shows that the energy of the system is no longer conserved instead being continuously dissipated by the drag torque. Thus, the Stoner–Wohlfarth orbits initially centered about the energy minima will now start to collapse until they become a

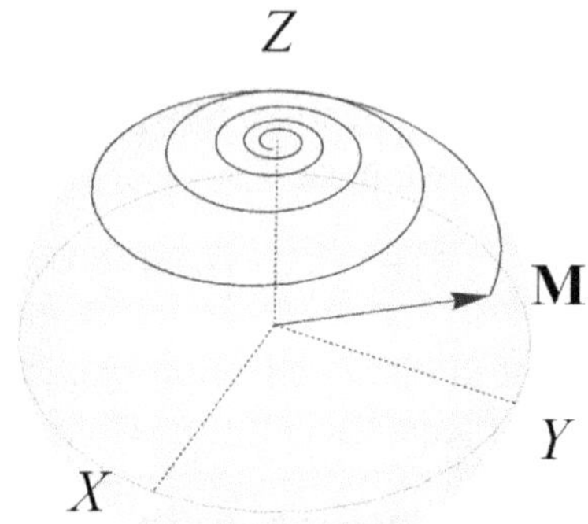

Fig. 1.8. Spiral motion described by the deterministic Landau–Lifshitz–Gilbert equation with $\gamma H_Z(1+\alpha^2)^{-1}=1$.

singularity (see Fig. 1.8). To demonstrate this, we solve Eq. (1.26) by using the spherical coordinate system basis $\{\mathbf{e}_r, \mathbf{e}_\vartheta, \mathbf{e}_\varphi\}$, where

$$\mathbf{u}=\begin{pmatrix}1\\0\\0\end{pmatrix},\ \dot{\mathbf{u}}=\begin{pmatrix}0\\\dot{\vartheta}\\\dot{\varphi}\sin\vartheta\end{pmatrix},\ \mathbf{H}=-\frac{1}{\mu_0 M_S}\begin{pmatrix}0\\\partial_\vartheta V\\\sin^{-1}\vartheta\,\partial_\varphi V\end{pmatrix}. \tag{1.28}$$

Thus Eq. (1.26) represents two simultaneous differential equations for ϑ and φ

$$\begin{aligned}\mu_0 M_S\dot{\vartheta}&=-\frac{\gamma}{1+\alpha^2}\frac{1}{\sin\vartheta}\frac{\partial V}{\partial\varphi}-\frac{\gamma\alpha}{1+\alpha^2}\frac{\partial V}{\partial\vartheta},\\ \mu_0 M_S\sin\vartheta\dot{\varphi}&=\frac{\gamma}{1+\alpha^2}\frac{\partial V}{\partial\vartheta}-\frac{\gamma\alpha}{1+\alpha^2}\frac{1}{\sin\vartheta}\frac{\partial V}{\partial\varphi}.\end{aligned} \tag{1.29}$$

The solution of Eq. (1.29), where the field direction **H** defines the Z axis, so that $V(\vartheta)=-\mu_0 M_S H_Z\cos\vartheta$, is

$$\tan\left(\tfrac{1}{2}\vartheta(t)\right)=e^{-\frac{t\gamma\alpha H_Z}{1+\alpha^2}}\tan\left(\tfrac{1}{2}\vartheta(0)\right),\ \varphi(t)=t\frac{\gamma H_Z}{1+\alpha^2}+\varphi(0). \tag{1.30}$$

Thus Eqs. (1.30) in combination describe a transient precession which ultimately ceases (spiral motion) (see Fig. 1.8).

Likewise, the solution of Eq. (1.29) for uniaxial anisotropy $V(\vartheta) = -K\cos^2\vartheta$ is

$$\begin{gathered}\tan\left(\vartheta(t)\right) = e^{-2h'Kt}\tan\left(\vartheta(0)\right),\\ \varphi(t) = \alpha^{-1}\left\{\operatorname{arc\,sinh}\left[e^{2h'Kt}\cot\vartheta(0)\right] - \operatorname{arc\,sinh}\left[\cot\vartheta(0)\right]\right\} + \varphi(0),\end{gathered} \tag{1.31}$$

where

$$h' = \frac{\alpha\gamma}{(1+\alpha^2)\mu_0 M_S}. \tag{1.32}$$

Thus, the pair of Eqs. (1.31) again describes spiral motion.

1.2.5 *Bloch equation*

The electrons are usually embedded in a dense medium, which may be either liquid or solid, so that the spins are coupled to all the other degrees of freedom, traditionally called 'the lattice' by reference to the crystalline lattice of solids. This coupling, called the *spin-lattice coupling*, encourages an evolution towards common thermal equilibrium of the spin system and the lattice.

Now suppose that a static field $\mathbf{H}_0$ is applied along the direction of the Z-axis. In thermal equilibrium with the lattice, the longitudinal magnetization M_Z has a nonvanishing value M_0, whereas the transverse components M_X and M_Y vanish. Now, starting from a nonequilibrium state, the evolution of M_Z towards M_0 modifies the energy of the spin system and corresponds to an exchange of energy with the lattice, whereas the decrease of the transverse magnetization components does not cause any such exchange of energy. Felix Bloch [2] assumed in a phenomenological approach that the evolution of the magnetization towards equilibrium was exponential, however embodying two distinct time constants: $T_\parallel$ for the longitudinal component and $T_\perp$ for the transverse components. The process of evolution of the spin system towards equilibrium with the lattice is then called *spin-lattice relaxation*, where the time constants $T_\parallel$ and $T_\perp$ are respectively the *longitudinal* and *transverse* relaxation times. Therefore, the relaxation equations are

$$\dot{M}_{X,Y} = -\frac{1}{T_\perp} M_{X,Y}, \quad \dot{M}_Z = -\frac{1}{T_\parallel}(M_Z - M_0). \tag{1.33}$$

In conventional experiments, one has in addition to the steady field $\mathbf{H}_0$, much smaller time-dependent fields which do not significantly alter the direction and magnitude of the total field, so that the spin-lattice evolution rate is the same as that in Eq. (1.33). By adding to the relaxation terms the evolution term resulting from the Larmor equation (1.4), we then have the *Bloch equations*

$$\dot{\mathbf{M}} = \gamma[\mathbf{H}_0 \times \mathbf{M}] - \frac{1}{T_\parallel}(M_Z - M_0)\mathbf{k} - \frac{1}{T_\perp}(M_X \mathbf{i} + M_Y \mathbf{j}), \tag{1.34}$$

where **i**, **j**, and **k** are unit vectors in the directions X, Y, and Z, respectively. In practice it is often convenient to use the frequency $\boldsymbol{\omega} = \gamma \mathbf{H}_0$ and write Eq. (1.34) as

$$\dot{\mathbf{M}} = [\boldsymbol{\omega} \times \mathbf{M}] - \frac{1}{T_\parallel}(M_Z - M_0)\mathbf{k} - \frac{1}{T_\perp}(M_X \mathbf{i} + M_Y \mathbf{j}). \tag{1.35}$$

As an illustration, we solve the Bloch equations for a constant field in the XZ plane, viz., $\mathbf{H}_0 = H_X \mathbf{i} + H_Z \mathbf{k}$. Calculation of the components of the cross product in Eq. (1.34) then renders the three component equations

$$\begin{gathered} \dot{M}_X = -\omega_Z M_Y - \frac{M_X}{T_\perp}, \quad \dot{M}_Y = \omega_Z M_X - \omega_X M_Z - \frac{M_Y}{T_\perp}, \\ \dot{M}_Z = \omega_X M_Y + \frac{M_0 - M_Z}{T_\parallel}, \end{gathered} \tag{1.36}$$

where $\omega_Z = \gamma H_Z$ and $\omega_X = \gamma H_X$. If $\omega_X = 0$, the solutions for $M_Z(t)$ and $M_\pm(t) = M_X(t) \pm iM_Y(t)$ are

$$M_\pm(t) = M_\pm(0)e^{(\pm i\omega_Z - 1/T_\perp)t}, \quad M_Z(t) = M_0 + (M_Z(0) - M_0)e^{-t/T_\parallel}. \tag{1.37}$$

Hence the general time-dependent solution for the *transverse* components, Eq. (1.37), represents sinusoidal terms as modified by a decay factor. The sinusoidal terms correspond to the *precessional* motion, and the damping factor originates in the *transverse relaxation* effect. The longitudinal component $M_Z(t)$ purely relaxes from its initial value to the equilibrium

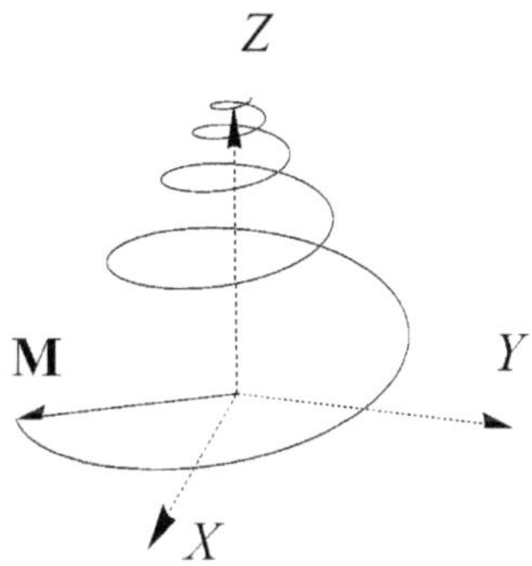

Fig. 1.9. Evolution of magnetization as given by Eqs.(1.37).

value M_0 while the transverse components $M_X(t)$ and $M_Y(t)$ decrease in magnitude (see Fig. 1.9). Thus, the magnitude $|\mathbf{M}|$ is not fixed. The steady state solution of Eqs. (1.36), i.e., for $\dot{M}_X = \dot{M}_Y = \dot{M}_Z = 0$, is given by

$$\begin{aligned} M_X &= M_0 \frac{\omega_X \omega_Z T_\perp^2}{1 + \omega_Z^2 T_\perp^2 + \omega_X^2 T_\perp T_\parallel}, \\ M_Y &= -M_0 \frac{\omega_X T_\perp}{1 + \omega_Z^2 T_\perp^2 + \omega_X^2 T_\perp T_\parallel}, \\ M_Z &= M_0 \frac{1 + \omega_Z^2 T_\perp^2}{1 + \omega_Z^2 T_\perp^2 + \omega_X^2 T_\perp T_\parallel}. \end{aligned} \tag{1.38}$$

Next by introducing new functions

$$f(t) = \frac{M_Z(t) - M_0}{M_Z(0) - M_0} \quad \text{and} \quad \varphi(t) = \frac{M_\pm(t)}{M_\pm(0)} \tag{1.39}$$

and then using the one-sided Fourier transform

$$\tilde{F}(i\omega) = \int_0^\infty F(t) e^{-i\omega t} dt, \tag{1.40}$$

we have

$$\tilde{f}(i\omega) = \frac{1}{i\omega + 1/T_\parallel} \quad \text{and} \quad \tilde{\varphi}(i\omega) = \frac{1}{i(\omega \mp \omega_Z) + 1/T_\perp}. \tag{1.41}$$

The real and imaginary parts of $\tilde{f}(i\omega)$ are shown in Fig. 1.10.

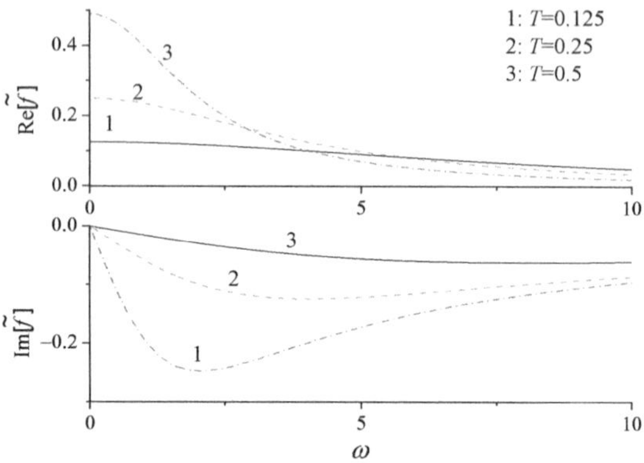

Fig. 1.10. The real and imaginary parts of $\tilde{f}(i\omega)$, Eq. (1.41), vs. ω, $T=T_{\|}$.

1.3 Stochastic motion of the magnetization

1.3.1 *Langevin equation for magnetic moment*

The Landau–Lifshitz–Gilbert equation ignores thermal fluctuations arising because the nanomagnet is *maintained at a finite temperature T*. If these are included the precessional motion would endure due to the energy provided by the heat bath. The theory of the thermal fluctuations of the magnetization, having been initiated by Néel [15,16], was further developed by Brown [17,18,92] who set it in the context of the general theory of stochastic processes. We reiterate that the thermal fluctuations and relaxation of the magnetization of nanoscale ferromagnets are vital in information storage [2,11], rock magnetism [19], biomedical applications (hyperthermia, etc.) [12,93,94], spintronics [95,96] as well as in magnetization reversal in isolated ferromagnetic nanoparticles [20].

To include thermal fluctuations Brown [17] in 1963 simply added a random isotropic noise magnetic field **h** to the Landau–Lifshitz–Gilbert equation (1.22) which in direct contrast to the *dissipative field* acts as a *source of energy* to the system

$$\dot{\mathbf{u}} = \gamma\left[(\mathbf{H}+\mathbf{h})\times\mathbf{u}\right]+\alpha\left[\mathbf{u}\times\dot{\mathbf{u}}\right]. \tag{1.42}$$

The random magnetic field is regarded as *spatially isotropic Gaussian white noise* so that

$$\overline{h_i(t_1)} = 0, \quad \overline{h_i(t_1)h_j(t_2)} = 2D\delta_{ji}\delta(t_1 - t_2), \tag{1.43}$$

where the noise-strength constant

$$D = \frac{\alpha kT}{v\gamma\mu_0 M_S} \tag{1.44}$$

is determined by imposing the Boltzmann equilibrium distribution of orientations [17,97], δ_{ji} is Kronecker's delta and $i, j = 1, 2, 3$ represent the cartesian axes of the laboratory coordinate system. The overbars denote statistical averages over a very large number of moments, which have all started with the same orientation (ϑ, φ) (here we again use spherical polar coordinates; see Fig. 1.5). The random field accounts for the thermal fluctuations of the magnetization of an *individual* single-domain particle. According to Eq. (1.42) the magnitude of the magnetization vector **M** does not fluctuate. However, since the random torque arising from the noise field *counteracts* the damping torque it can, if the temperature is high enough, *reverse* the direction of precession. The time taken for the reversal of the direction of precession over the anisotropy-Zeeman energy barrier is known as the *superparamagnetic (magnetization) relaxation time*. Equation (1.42) is known as the *magnetic Langevin equation*. This Langevin equation describes how magnetic recordings may degrade in a heat bath because thermal fluctuations cause unwanted magnetic reversal so that data stored in magnetic recordings is ultimately lost.

Equation (1.42) is often used to treat stochastic magnetization dynamics. We reiterate that Brown [17,92] justified using Eq. (1.42) because all the terms in it can be derived from a Lagrangian function and a Rayleigh dissipation function [91]. Moreover, Gilbert's form of Langevin equation (1.42) fits naturally into the Kramers escape-rate theory in all damping ranges if the *damping torque is regarded as the time average of a fluctuating torque, whose instantaneous value also contains a random term with statistical properties given by Eq. (1.43)*.

However, alternative forms of Langevin equations governing the magnetization $\mathbf{M}(t)$ have also been proposed in the literature. Two other frequently used Langevin equations for stochastic spin dynamics are the Landau–Lifshitz [89] and Kubo [98–100] forms, respectively,

$$\dot{\mathbf{u}} = \gamma[(\mathbf{H}+\mathbf{h})\times\mathbf{u}] - \gamma\alpha\left[\mathbf{u}\times\left[\mathbf{u}\times(\mathbf{H}+\mathbf{h})\right]\right] \tag{1.45}$$

and

$$\dot{\mathbf{u}} = \gamma[(\mathbf{H}+\mathbf{h})\times\mathbf{u}] - \gamma\alpha\left[\mathbf{u}\times\left[\mathbf{u}\times\mathbf{H}\right]\right]. \tag{1.46}$$

The difference between the two is that in the Kubo model, Eq. (1.46), the random field $\mathbf{h}(t)$ appears only in the gyromagnetic term.

The *precise form* of the Langevin equation ultimately determines the Fokker–Planck equation for the distribution function $W(\vartheta, \varphi, t)$ of the orientations of the magnetization vector **M** on the surface of a sphere of constant radius M_S. However, all the Langevin equations, Eqs. (1.42), (1.45) and (1.46) always yield very similar Fokker–Planck equations, the only difference lies in the definition of the free diffusion time. To demonstrate this, we give a detailed derivation of the Fokker–Planck equations corresponding to Eqs. (1.42), (1.45) and (1.46), and demonstrate that each of the three Langevin equations yield the *same* mathematical form for the corresponding Fokker–Planck equation. The only difference lies in the characteristic time constants. This difference is negligible at low damping, which is the case of most interest in superparamagnetism. At *high damping*, however, each of the models predicts completely different behavior. Therefore, attention must be paid to the particular damping region in which a model is valid.

1.3.2 *Fokker–Planck equation for magnetic moments in configuration space*

Here we directly derive the Fokker–Planck equation from the magnetic Langevin equation in the *Stratonovich interpretation* [101,102] since this definition represents the mathematical idealization of the physical process of magnetization relaxation of magnetic nanoparticles. Hence the conventional rules of calculus apply [102]. The particle is assumed to be uniformly magnetized (although such a "macrospin" approximation does not necessarily explain all aspects of the magnetization dynamics in nanomagnets, nevertheless many qualitative features needed to interpret experimental data are satisfactorily reproduced). Surface and memory effects are also ignored. The *instantaneous orientation* of the unit vector **u** may be described by a *representative point* on the surface of the unit sphere (1, ϑ, φ), where ϑ and φ are angular coordinates describing the orientation of **M**. Hence the thermal agitation causes continual changes in the orientation of the magnetization **M** so that as **M** changes its direction, the representative point moves over the surface of the sphere. Consequently, in an *ensemble* of such particles, the thermal agitation maintains a *distribution* of orientations. Following [97], consider now a statistical ensemble of identical magnetic particles and let $W(\vartheta, \varphi, t)\, d\Omega$ be the probability that **M** has orientation (ϑ, φ) within a solid angle $d\Omega = \sin\vartheta\, d\vartheta\, d\varphi$.

We first recall [101,102] that for Stratonovich stochastic differential equations with two random variables $\boldsymbol{\xi}(t) = \{\xi_1(t) = \vartheta(t), \xi_2(t) = \varphi(t)\}$, namely

$$\dot{\xi}_i(t) = F_i[\boldsymbol{\xi}(t),t] + g_{ij}[\boldsymbol{\xi}(t),t]h_j(t), \tag{1.47}$$

($i = 1, 2$; $j = x, y, z$), where the random field components are Gaussian random variables with white noise properties, Eq. (1.43), the associated Fokker–Planck equation for the probability density function $P(\vartheta, \varphi, t)$ is

$$\frac{\partial P}{\partial t} = -\frac{\partial}{\partial x_i}\left(D_i P\right) + \frac{\partial^2}{\partial x_i \partial x_j}\left(D_{ij} P\right), \tag{1.48}$$

where the drift D_i and the diffusion D_{ij} coefficients are

$$D_i = F_i + D g_{kj} \frac{\partial}{\partial x_k} g_{ij}, \quad D_{ij} = D g_{ik} g_{jk}, \tag{1.49}$$

and the constant D is given by Eq. (1.44). Everywhere, we shall use Einstein's convention, i.e., the summation is performed over indices appearing twice in the equations without writing down the summation sign.

To illustrate the general method of averaging (over an infinitesimal time) of Stratonovich stochastic differential equations and, in particular, to prove Eq. (1.49), we commence by interpreting the Langevin equations,

Eq. (1.47), for a set of 2 random variables $\boldsymbol{\xi}=(\xi_1,\xi_2)$, as the integral equation [97]

$$\xi_i(t+\Delta t)=x_i+\int_t^{t+\Delta t}\left\{F_i[\boldsymbol{\xi}(t'),t']+g_{ij}[\boldsymbol{\xi}(t'),t']h_j(t')\right\}dt', \qquad (1.50)$$

where $\xi_i(t')$ is a solution of Eq. (1.47) which at instant t has the sharp value $\xi_i(t)=x_i$, which are themselves random variables. We proceed by inserting the Taylor series expansions about the sharp values

$$F_i\left[\boldsymbol{\xi}(t'),t'\right]=F_i(\mathbf{x},t')+\left(\xi_k(t')-x_k\right)\frac{\partial}{\partial x_k}F_i(\mathbf{x},t')+\cdots,$$

$$g_{ij}\left[\boldsymbol{\xi}(t'),t'\right]=g_{ij}(\mathbf{x},t')+\left(\xi_k(t')-x_k\right)\frac{\partial}{\partial x_k}g_{ij}(\mathbf{x},t')+\cdots$$

into the integral equation, Eq. (1.50). We have

$$\begin{aligned}\xi_i(t+\Delta t)-x_i &= \int_t^{t+\Delta t}F_i(\mathbf{x},t')dt'+\int_t^{t+\Delta t}[\xi_k(t')-x_k]\frac{\partial}{\partial x_k}F_i(\mathbf{x},t')dt' \\ &+\int_t^{t+\Delta t}g_{ij}(\mathbf{x},t')h_j(t')dt'+\int_t^{t+\Delta t}h_j(t')[\xi_k(t')-x_k]\frac{\partial}{\partial x_k}g_{ij}(\mathbf{x},t')dt'+\cdots.\end{aligned} \qquad (1.51)$$

We iterate for the increment $\xi_k(t')-x_k$ under the integral signs in Eq. (1.51). Taking the average over an infinitesimal time and noting Eq. (1.43) and the δ-function property, Eq. (A.121) from Appendix A.1.10, we have

$$\begin{aligned}\frac{1}{\Delta t}\overline{\xi_i(t+\Delta t)-x_i} &= F_i(\mathbf{x},t+\Delta t\Theta_{iii}) \\ &+\frac{2D}{\Delta t}\delta_{jl}\int_t^{t+\Delta t}\int_t^{t'}\delta(t'-t'')g_{kl}(\mathbf{x},t'')\frac{\partial}{\partial x_k}g_{ij}(\mathbf{x},t')dt''dt'+\ldots \\ &=F_i(\mathbf{x},t+\Delta t\Theta_{iii})+Dg_{kj}(\mathbf{x},t+\Delta t\Theta_{ijk})\frac{\partial}{\partial x_k}g_{ij}(\mathbf{x},t+\Delta t\Theta_{ijk})+\ldots,\end{aligned} \qquad (1.52)$$

where Θ_{ijk} are constants $(0\le\Theta_{ijk}\le 1)$. Letting $\Delta t\to 0$ in Eq. (1.52), we have *the time evolution equation of the sharp values* x_i, namely

$$\frac{d}{dt}x_i=F_i(\mathbf{x},t)+Dg_{kj}(\mathbf{x},t)\frac{\partial}{\partial x_k}g_{ij}(\mathbf{x},t). \qquad (1.53)$$

The second term in the right-hand side of Eq. (1.53) is the *noise-induced drift* arising from the *multiplicative* noise term in Eq. (1.47). We remark that Eq. (1.53), also yields the drift coefficients $D_i = \dot{x}_i$ in the corresponding Fokker–Planck equation (1.48). The diffusion coefficients D_{ij} in that equation can be evaluated similarly, and are given by

$$D_{ij} = \lim_{\Delta t \to 0} \frac{\overline{\left[\xi_i(t+\Delta t) - x_i\right]\left[\xi_j(t+\Delta t) - x_j\right]}}{2\Delta t} = Dg_{ik}(\mathbf{x},t)g_{jk}(\mathbf{x},t). \quad (1.54)$$

Likewise, we can also determine the averaged equation for an *arbitrary differentiable* function $f(\boldsymbol{\xi})$. Indeed, because the rule for a change of variable in Stratonovich differential equations is just the same as in ordinary analysis [101,102], the equation of motion for an *arbitrary differentiable function* $f(\boldsymbol{\xi})$ may be obtained by simply cross-multiplying the ith Eq. (1.47) by $\partial f[\boldsymbol{\xi}(t')]/\partial \xi_i$ and then summing it. Thus, we obtain the *stochastic differential equation* for f:

$$\frac{d}{dt} f[\boldsymbol{\xi}(t')] = \left\{F_i[\boldsymbol{\xi}(t'),t'] + g_{ij}[\boldsymbol{\xi}(t'),t']h_j(t')\right\}\frac{\partial}{\partial \xi_i} f[\boldsymbol{\xi}(t')]. \quad (1.55)$$

As before, the stochastic differential equation, Eq. (1.55), must be regarded as the *stochastic integral equation*

$$f[\boldsymbol{\xi}(t+\Delta t)] - f(\mathbf{x}) = \int_t^{t+\Delta t} \left[F_i[\boldsymbol{\xi}(t'),t'] + g_{ij}[\boldsymbol{\xi}(t'),t']h_j(t')\right]\frac{\partial f[\boldsymbol{\xi}(t')]}{\partial \xi_i} dt'.$$

Thus, expanding the integrands in this equation in Taylor series, we obtain

$$\begin{aligned} f[\boldsymbol{\xi}(t+\Delta t)] - f(\mathbf{x}) &= \int_t^{t+\Delta t} F_i(\mathbf{x},t')\frac{\partial f}{\partial x_i} dt' \\ &+ \int_t^{t+\Delta t} [\xi_k(t') - x_k]\frac{\partial}{\partial x_k}\left[F_i(\mathbf{x},t')\frac{\partial f}{\partial x_i}\right]dt' + \int_t^{t+\Delta t} g_{ij}(\mathbf{x},t')\frac{\partial f}{\partial x_i} h_j(t')dt' \quad (1.56) \\ &+ \int_t^{t+\Delta t} [\xi_k(t') - x_k]\frac{\partial}{\partial x_k}\left[g_{ij}(\mathbf{x},t')\frac{\partial f}{\partial x_i}\right]h_j(t')dt' + \cdots. \end{aligned}$$

Substituting the increment $\xi_k(t') - x_k$, Eq. (1.51) into Eq. (1.56), we iterate yielding

$$f[\xi(t+\Delta t)] - f(\mathbf{x}) = \int_t^{t+\Delta t} F_i(\mathbf{x},t')\frac{\partial f}{\partial x_i}dt'$$

$$+\int_t^{t+\Delta t} \frac{\partial}{\partial x_k}\left[F_i(\mathbf{x},t')\frac{\partial f}{\partial x_i}\right]\int_t^{t'} h_k(\mathbf{x},t'')dt''dt' + \int_t^{t+\Delta t} g_{ij}(\mathbf{x},t')\frac{\partial f}{\partial x_i}h_j(t')dt'$$

$$+\int_t^{t+\Delta t} \frac{\partial}{\partial x_k}\left[F_i(\mathbf{x},t')\frac{\partial f}{\partial x_i}\right]\int_t^{t'} g_{kn}(\mathbf{x},t'')h_n(t'')dt''dt'$$

$$+\int_t^{t+\Delta t} \frac{\partial}{\partial x_k}\left[g_{ij}(\mathbf{x},t')\frac{\partial f}{\partial x_i}\right]h_j(t')\int_t^{t'} h_k(\mathbf{x},t'')dt''dt'$$

$$+\int_t^{t+\Delta t} \frac{\partial}{\partial x_k}\left[g_{ij}(\mathbf{x},t')\frac{\partial f}{\partial x_i}\right]h_j(t')\int_t^{t'} g_{kn}(\mathbf{x},t'')h_n(t'')dt''dt' + \cdots.$$

Then on averaging this equation, taking into account the properties of the white noise fields, Eq. (1.43), the Dirac δ-function property, Eq. (A.121) from Appendix A.1.10, and retaining only terms of order Δt, we have after obvious transformations

$$\frac{1}{\Delta t}\left(\overline{f[\xi(t+\Delta t)] - f(\mathbf{x})}\right) = F_i(\mathbf{x}, t+\Delta t\,\Theta_{iii}^{(1)})\frac{\partial f}{\partial x_i}$$
$$+Dg_{kj}(\mathbf{x}, t+\Delta t\,\Theta_{ijk}^{(2)})\frac{\partial}{\partial x_k}\left[g_{ij}(\mathbf{x}, t+\Delta t\,\Theta_{ijk}^{(2)})\frac{\partial f}{\partial x_i}\right] + \cdots, \tag{1.57}$$

where $\Theta_{ijk}^{(n)}$ are constants $(0 \le \Theta_{ijk}^{(n)} \le 1)$. Letting $\Delta t \to 0$ in Eq. (1.57), *we have the average evolution equation for f*(**x**):

$$\frac{df}{dt} = F_i\frac{\partial f}{\partial x_i} + Dg_{kj}\frac{\partial}{\partial x_k}\left(g_{ij}\frac{\partial}{\partial x_i}f\right). \tag{1.58}$$

The evolution equation, Eq. (1.58), for the sharp values, is one of the most important results of the Stratonovich calculus. We shall use it to construct differential-recurrence relations for the statistical moments (averaged spherical harmonics) in Section 1.3.5.

The functions F_i for the Landau–Lifshitz–Gilbert Eq. (1.42) are found as follows. Firstly recall the fact that this equation can be transformed into the Landau–Lifshitz form (see Eq. (1.26))

$$\dot{\mathbf{u}} = h'\mu_0 M_S \left\{ \alpha^{-1}\left[(\mathbf{H}+\mathbf{h})\times\mathbf{u}\right] + \left[\mathbf{u}\times\left[(\mathbf{H}+\mathbf{h})\times\mathbf{u}\right]\right]\right\}, \tag{1.59}$$

where the constant h' is given by Eq. (1.32). Next, we introduce a spherical polar coordinate system [104] as shown in Fig. 1.5 so that in the spherical polar coordinate basis $(\mathbf{e}_r, \mathbf{e}_\vartheta, \mathbf{e}_\varphi)$ the various vectors are given by Eq. (1.28) and the vector products in Eq. (1.59) are

$$\mathbf{h}\times\mathbf{u} = \begin{pmatrix} 0 \\ h_\varphi \\ -h_\vartheta \end{pmatrix}, \quad \mathbf{H}\times\mathbf{u} = \frac{1}{\mu_0 M_S}\begin{pmatrix} 0 \\ -\csc\vartheta\,\partial_\varphi V \\ \partial_\vartheta V \end{pmatrix}, \tag{1.60}$$

where the components $h_\vartheta(t)$ and $h_\varphi(t)$ of the random noise field $\mathbf{h}(t)$ in the spherical coordinate system are expressed in terms of its components $h_X(t)$, $h_Y(t)$, and $h_Z(t)$ in the cartesian one as [104]

$$\begin{aligned} h_\vartheta(t) &= h_X(t)\cos\vartheta(t)\cos\varphi(t) + h_Y(t)\cos\vartheta(t)\sin\varphi(t) - h_Z(t)\sin\vartheta, \\ h_\varphi(t) &= -h_X(t)\sin\varphi(t) + h_Y(t)\cos\varphi(t), \end{aligned} \tag{1.61}$$

while the components of the random field $h_X(t), h_Y(t), h_Z(t)$ each satisfy Eq. (1.43) Thus Eq. (1.59) is equivalent to two coupled nonlinear stochastic differential equations with *multiplicative* noise for the polar and azimuthal angles ϑ and φ:

$$\begin{aligned} \dot{\vartheta}(t) &= h'\mu_0 M_S\left[h_\vartheta(t) + \alpha^{-1}h_\varphi(t)\right] - h'\left(\frac{\partial V}{\partial\vartheta} + \alpha^{-1}\csc\vartheta\frac{\partial V}{\partial\varphi}\right), \\ \sin\vartheta(t)\dot{\varphi}(t) &= h'\mu_0 M_S\left[h_\varphi(t) - \alpha^{-1}h_\vartheta(t)\right] - h'\left(\csc\vartheta\frac{\partial V}{\partial\varphi} - \alpha^{-1}\frac{\partial V}{\partial\vartheta}\right). \end{aligned} \tag{1.62}$$

Equation (1.62) can be written as Eq. (1.47), where the functions $F_i(\vartheta,\varphi)$ are

$$\begin{aligned} F_1(\vartheta,\varphi) &= -h'\left(\frac{\partial V}{\partial\vartheta} + \frac{1}{\alpha\sin\vartheta}\frac{\partial V}{\partial\varphi}\right), \\ F_2(\vartheta,\varphi) &= -h'\left(\frac{1}{\sin^2\vartheta}\frac{\partial V}{\partial\varphi} - \frac{1}{\alpha\sin\vartheta}\frac{\partial V}{\partial\vartheta}\right), \end{aligned} \tag{1.63}$$

and the functions g_{ij} are given in Table 1.1.

Table 1.1. Coefficients g_{ij} for the Landau–Lifshitz–Gilbert model, Eq. (1.42)

g_{ij}	$i=1,\ \xi_1(t)=\vartheta(t)$	$i=2,\ \xi_2(t)=\varphi(t)$
$j=X$	$\mu_0 M_S h'\left(\cos\vartheta\cos\varphi-\alpha^{-1}\sin\varphi\right)$	$-\mu_0 M_S h'\dfrac{\alpha^{-1}\cos\vartheta\cos\varphi+\sin\varphi}{\sin\vartheta}$
$j=Y$	$\mu_0 M_S h'\left(\cos\vartheta\sin\varphi+\alpha^{-1}\cos\varphi\right)$	$-\mu_0 M_S h'\dfrac{\alpha^{-1}\cos\vartheta\sin\varphi-\cos\varphi}{\sin\vartheta}$
$j=Z$	$-\mu_0 M_S h'\sin\vartheta$	$\mu_0 M_S h'\alpha^{-1}$

Table 1.2. Coefficients g_{ij} for the Kubo model, Eq. (1.46)

g_{ij}	$i=1,\ \xi_1(t)=\vartheta(t)$	$i=2,\ \xi_2(t)=\varphi(t)$
$j=X$	$-\mu_0 M_S h'\left(\alpha+\alpha^{-1}\right)\sin\varphi$	$-\mu_0 M_S h'\left(\alpha+\alpha^{-1}\right)\cos\vartheta\cos\varphi\csc\vartheta$
$j=Y$	$\mu_0 M_S h'\left(\alpha+\alpha^{-1}\right)\cos\varphi$	$-\mu_0 M_S h'\left(\alpha+\alpha^{-1}\right)\cos\vartheta\sin\varphi\csc\vartheta$
$j=Z$	0	$\mu_0 M_S h'\left(\alpha+\alpha^{-1}\right)$

Table 1.3. Diffusion coefficients D_{11} and D_{22}

Equation	$vD_{11}/(kT)$	$vD_{22}/(kT)$
(1.42)	h'	$h'\csc^2\vartheta$
(1.45)	$h'(1+\alpha^2)^2$	$h'(1+\alpha^2)^2\csc^2\vartheta$
(1.46)	$h'(1+\alpha^2)$	$h'(1+\alpha^2)\csc^2\vartheta$

Table 1.4. Drift coefficients D_1 and D_2

Eq.	D_1	D_2
(1.42)	$h'\left[\dfrac{kT}{v}\cot\vartheta-\dfrac{\partial V}{\partial\vartheta}-\dfrac{1}{\alpha\sin\vartheta}\dfrac{\partial V}{\partial\varphi}\right]$	$-\dfrac{h'}{\sin\vartheta}\left(\dfrac{1}{\sin\vartheta}\dfrac{\partial V}{\partial\varphi}-\dfrac{1}{\alpha}\dfrac{\partial V}{\partial\vartheta}\right)$
(1.45)	$h'(1+\alpha^2)\left[(1+\alpha^2)\dfrac{kT}{v}\cot\vartheta-\dfrac{\partial V}{\partial\vartheta}-\dfrac{1}{\alpha\sin\vartheta}\dfrac{\partial V}{\partial\varphi}\right]$	$-\dfrac{h'(1+\alpha^2)}{\sin\vartheta}\left(\dfrac{1}{\sin\vartheta}\dfrac{\partial V}{\partial\varphi}-\dfrac{1}{\alpha}\dfrac{\partial V}{\partial\vartheta}\right)$
(1.46)	$h'(1+\alpha^2)\left[\dfrac{kT}{v}\cot\vartheta-\dfrac{\partial V}{\partial\vartheta}-\dfrac{1}{\alpha\sin\vartheta}\dfrac{\partial V}{\partial\varphi}\right]$	$-\dfrac{h'(1+\alpha^2)}{\sin\vartheta}\left(\dfrac{1}{\sin\vartheta}\dfrac{\partial V}{\partial\varphi}-\dfrac{1}{\alpha}\dfrac{\partial V}{\partial\vartheta}\right)$

The F_i for the Landau–Lifshitz model, Eq. (1.45), and the Kubo model, Eq. (1.46), are obtained from Eq. (1.63) via the replacement $h' \to h'(1+\alpha^2)$. The g_{ij} for Eq. (1.45) are obtained from Table 1.2 via the replacement $h' \to h'(1+\alpha^2)$, while the g_{ij} for Eq. (1.46) are given in Table 1.2. The diffusion coefficients D_{11} and D_{22} ($D_{12} = D_{21} = 0$) as evaluated from the definition Eq. (1.49) are given in Table 1.3, while D_1 and D_2 as also evaluated from Eq. (1.49) are given in Table 1.4. The general Stratonovich stochastic differential Eq. (1.47) with drift and the diffusion coefficients from Tables 1.3 and 1.4 then yields *the Fokker–Planck equation,* Eq. (1.48), for the probability density function $P(\vartheta, \varphi, t)$. This equation can be transformed to the corresponding equation for the *surface* probability density function $W(\vartheta, \varphi, t)$ given by (we recall that $P(\vartheta, \varphi, t)$ is related to $W(\vartheta, \varphi, t)$ via $P(\vartheta, \varphi, t) = \sin\vartheta\, W(\vartheta, \varphi, t)$)

$$\frac{\partial W}{\partial t} = \mathrm{L}_{FP} W, \tag{1.64}$$

where

$$\begin{aligned} \mathrm{L}_{FP} W = b\Delta W + \frac{a}{\sin\vartheta}\left\{\frac{\partial}{\partial\vartheta}\left[\left(\sin\vartheta\frac{\partial V}{\partial\vartheta} + \frac{1}{\alpha}\frac{\partial V}{\partial\varphi}\right)W\right]\right. \\ \left. + \frac{\partial}{\partial\varphi}\left[\left(\frac{1}{\sin\vartheta}\frac{\partial V}{\partial\varphi} - \frac{1}{\alpha}\frac{\partial V}{\partial\vartheta}\right)W\right]\right\} \end{aligned} \tag{1.65}$$

with the angular part of the Laplacian Δ defined as

$$\Delta = \frac{1}{\sin\vartheta}\frac{\partial}{\partial\vartheta}\left(\sin\vartheta\frac{\partial}{\partial\vartheta}\right) + \frac{1}{\sin^2\vartheta}\frac{\partial^2}{\partial\varphi^2} \tag{1.66}$$

(the terms in a/α and a are the *precessional* (gyromagnetic) ones (giving rise to ferromagnetic resonance, usually in the GHz range), and the *alignment* ones, respectively). As far as the constants a and b in this generic Fokker–Planck equation are concerned $a = h'$ and $b = kTh'/v$ for the Landau–Lifshitz–Gilbert Eq. (1.42), while $a = h'(1+\alpha^2)$ and $b = kTh'(1+\alpha^2)^2/v$ for the Landau–Lifshitz Eq. (1.45) and $a = h'(1+\alpha^2)$ and $b = kTh'(1+\alpha^2)/v$ for the Kubo Eq. (1.46).

The normalization condition for the surface probability density function $W(\vartheta, \varphi, t)$ is

$$\int_0^{2\pi}\int_0^{\pi} W(\vartheta,\varphi,t)\sin\vartheta d\vartheta d\varphi = 1. \tag{1.67}$$

The stationary (equilibrium) solution of the general Eq. (1.64), viz.

$$\mathrm{L}_{FP} W_0(\vartheta,\varphi) = 0 \tag{1.68}$$

is the Boltzmann distribution

$$W_0(\vartheta,\varphi) = Z^{-1} e^{-vV(\vartheta,\varphi)/(kT)},$$

where Z is the partition function given by

$$Z = \int_0^{2\pi}\int_0^{\pi} e^{-vV(\vartheta,\varphi)/(kT)} \sin\vartheta d\vartheta d\varphi.$$

The surface probability density function $W(\vartheta, \varphi, t)$ is related to the probability current $\mathbf{J}$ of representative points swarming over the surface of the unit sphere $(1, \vartheta, \varphi)$ via the continuity equation

$$\dot{W} + \mathrm{div}\mathbf{J} = 0, \tag{1.69}$$

where $\mathbf{J} = W\dot{\mathbf{u}}$. Equation (1.69) simply states that the swarming representative points are neither created nor destroyed, merely moving to new angular positions on the surface of the sphere [18]. The differential operator div**J** in the spherical polar coordinate basis $(\mathbf{e}_r, \mathbf{e}_\vartheta, \mathbf{e}_\varphi)$ is

$$\mathrm{div}\mathbf{J} = \frac{1}{\sin\vartheta}\left(\frac{\partial}{\partial\vartheta}\left(\sin\vartheta J_\vartheta\right) + \frac{\partial}{\partial\varphi} J_\varphi\right), \tag{1.70}$$

so from Eqs. (1.64) and (1.69) we have for the components of **J**

$$\begin{aligned} J_\vartheta &= -a\left[\left(\frac{\partial V}{\partial\vartheta} + \frac{1}{\alpha\sin\vartheta}\frac{\partial V}{\partial\varphi}\right)W + \frac{b}{a}\frac{\partial W}{\partial\vartheta}\right], \\ J_\varphi &= -a\left[\left(-\frac{1}{\alpha}\frac{\partial V}{\partial\vartheta} + \frac{1}{\sin\vartheta}\frac{\partial V}{\partial\varphi}\right)W + \frac{b}{a\sin\vartheta}\frac{\partial W}{\partial\varphi}\right]. \end{aligned} \tag{1.71}$$

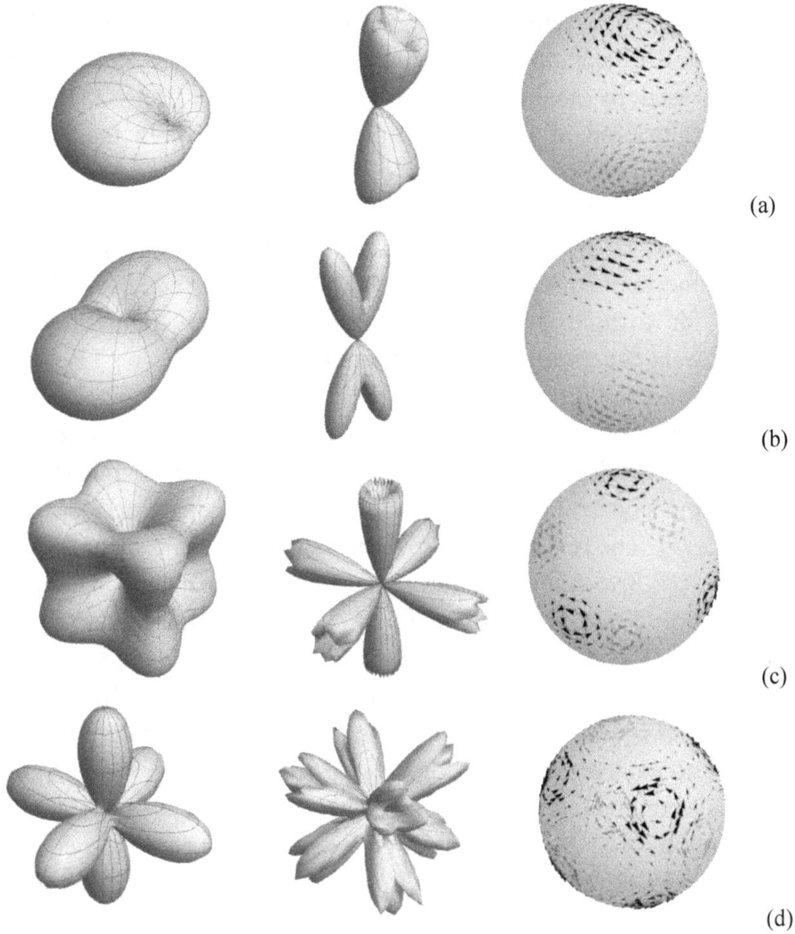

Fig. 1.11. The probability current **J** (amplitude and direction on the unit sphere) corresponding to the stationary distribution $W_0(\vartheta,\varphi) = Z^{-1}e^{-vV(\vartheta,\varphi)/(kT)}$ for (a) uniaxial potential Eq. (1.12) with external field; (b) biaxial potential Eq. (1.14); (c) cubic anisotropy potential Eq. (1.15) for $K_c > 0$; (d) cubic anisotropy potential for $K_c < 0$.

The absolute values of the probability current $|\mathbf{J}| = (J_\vartheta^2 + J_\varphi^2)^{1/2}$ corresponding to the stationary distribution $W_0(\vartheta,\varphi)$ for several anisotropy potentials are shown in Fig. 1.11.

A detailed discussion of the basic assumptions made in the derivation of the Fokker–Planck equation (1.64) is given elsewhere (e.g., [18,25,39,

92,105–109]). We remark that in developing his theory of the magnetization relaxation for superparamagnets (classical spins) Brown obviously used by analogy ideas originating in the Debye model of dielectric relaxation of polar dielectrics [97,110–112]. In the Fokker–Planck equation (1.64), the term in α^{-1} corresponds to the *precessional* (gyromagnetic) term in Eq. (1.42), giving rise to ferromagnetic resonance. However when $\alpha \to \infty$ (i.e., ignoring the gyromagnetic term) Eq. (1.64) has the same mathematical form as the *noninertial* rotational diffusion equation for a rigid body in an external potential (known as the Smoluchowski equation) in configuration space (ϑ, φ) (see, e.g., Ref. [97], Chapter 1 for details).

The reversal or switching time τ, namely, the longest relaxation time of the magnetization, may be defined as the inverse of the smallest nonvanishing eigenvalue λ_1 of the Fokker–Planck operator L_{FP} in Eq. (1.64) [13,97]. The reversal time may be calculated using three different approaches: (i) *Brownian dynamics simulations* via the magnetic Langevin equation, Eq. (1.42), (ii) via *numerical solutions* of the Fokker–Planck equation (1.64), and (iii) *approximate analytical solutions* of Eqs. (1.42) and (1.64) such as those for λ_1^{-1} yielded by escape-rate theory [84,85,186], the mean first-passage time method, etc. These complementary methods allow one to evaluate τ over wide ranges of temperature, damping, etc. Numerical methods and escape-rate theory are very useful for the determination of τ at low and high potential barriers, respectively. However, they have considerable limitations. For example, escape-rate theory cannot be used for low barriers, while numerical methods encounter substantial computational difficulties in the very low damping range.

In passing, we comment on the calculation of the escape rate from the Fokker–Planck Eq. (1.64). For the Landau–Lifshitz–Gilbert model, Eq. (1.42), a characteristic relaxation time of the magnetization relaxation (scaling factor) is the *free diffusion time* τ_N which is

$$\tau_N = \frac{1}{2b} = \frac{v}{2kTh'} = \frac{1+\alpha^2}{2D\gamma^2}. \tag{1.72}$$

Clearly, by comparing the parameters a and b in Eq. (1.64) for the Gilbert and Kubo models, the free diffusion time τ_N should be replaced by

$\tau_K = \tau_N(1+\alpha^2)^{-1}$. Consequently the Fokker–Planck equation for the Kubo model coincides with that of the Gilbert one, the *only difference being that the time* τ_N *is replaced by* τ_K. Similarly, for the Landau–Lifshitz model, the free diffusion time is given by $\tau_{LL} = \tau_N(1+\alpha^2)^{-2}$. Apparently, the three models for Brownian motion of classical spins all yield both the same mathematical form for the corresponding Fokker–Planck equations and the same infinite hierarchy of differential-recurrence equations for the statistical moments. Nevertheless, the one and only important difference lies in the free-diffusion time constants, a *difference that is negligible* for low damping, where $\tau_K \cong \tau_{LL} \cong \tau_N$ (the experimentally relevant damping range). However, only the Gilbert model, where the systematic and random terms in the stochastic equation, Eq. (1.42), viewed as the kinematic relation $\dot{\mathbf{u}} = \boldsymbol{\omega} \times \mathbf{u}$, are in the original Langevin form (i.e., *with the rate of change of the angular momentum systematically slowed down superimposed on which is a rapidly fluctuating white noise random torque*) *can be used in all damping ranges*. Clearly, neither the Kubo nor the Landau–Lifshitz models can hold for high damping, because they predict *unphysical* behavior of the observables (escape rates, reversal times, etc.). For example, for $\alpha \geq 1$, i.e., at intermediate and high damping (IHD) as defined by Kramers, both models yield IHD escape rates from a metastable state *higher* than the theoretical upper bound for the escape rate Γ^{TST} yielded by transition state theory (TST) (see Section 1.5.1). A simple explanation of this unphysical behavior of the IHD escape rates as yielded by both of these models relates to the qualitative behavior of the overbarrier current density of representative points J at the saddle point(s) for $\alpha \geq 1$. In each of these equations, $J^K, J^{LL} \sim \alpha$ *increases with increasing* α, i.e., *damping tends to enhance the overbarrier current density*. Thus, the *escape rate, which is proportional to the current density, also increases*. In contrast, in Gilbert's model, $J^G \sim \alpha / (1+\alpha^2)$ *decreases with increasing damping for* $\alpha \geq 1$, i.e., *damping tends to retard the current density* over the saddle point so that the *escape rate* $\sim J^G$ *also decreases*. Here $\Gamma^{IHD} / \Gamma^{TST} << 1$ for $\alpha >> 1$. However, in the low-damping range, $\alpha < 1$, the escape rates calculated from all three models, each have *the same* damping dependence as they must do. Thus, to avoid *unphysical behavior* of the escape rate for classical spins in the IHD range, $\alpha > 1$, the various (asymptotic) Kramers escape-rate formulas must only be used in conjunction with Gilbert's equation (which is valid for all α) and cannot be used with either the Landau–Lifshitz or Kubo equations. The reason being that both these equations *strictly apply only in the underdamped regime*, $\alpha < 1$, where energy-controlled diffusion dominates. However, some authors have in error used the *intrinsically underdamped* Landau–Lifshitz equation in conjunction with escape-rate formulas, *which are valid in the IHD range only*. Hence the ensuing escape-rate formulas lead to confusion both at *low* damping, where they *coincide* with the TST rate, and in the IHD damping range, where they predict *unphysical behavior*, namely, a rate in excess of the TST one.

Finally, in the more general treatment of the isotropic Brownian motion of a classical spin, the memoryless assumption, Eq. (1.43), for the random field correlation function $h_i(t_1)h_j(t_2)$ is discarded. Thus, the random field $\mathbf{h}(t)$ no longer has white noise properties, so that now Eq. (1.43) is replaced by

$$\overline{h_i(t)} = 0, \ \overline{h_i(t_1)h_j(t_2)} = 2kT\delta_{ij}\phi(t_1 - t_2), \tag{1.73}$$

and the (generalized) stochastic magnetic Langevin equation becomes [107,113]

$$\dot{\mathbf{u}}(t) = -\gamma\left[\mathbf{u}(t) \times (\mathbf{H}_{\text{ef}}(t) + \mathbf{h}(t))\right] + \left[\mathbf{u}(t) \times \int_0^t \phi(t - t')\dot{\mathbf{u}}(t')dt'\right]. \tag{1.74}$$

Here Eq. (1.74) takes into account memory effects and the random field correlation function $\phi(t - t')$ has the meaning of a memory function.

1.3.3 *Spin-transfer torque (STT) effects in the magnetization dynamics*

One of the most significant developments in thermally assisted magnetization reversal in nanomagnets since the seminal works of Néel [15,16] and Brown [17] has been the discovery of the STT effect [114,115]. The phenomenon exists because an electric current with spin polarization in a small (nanoscale) ferromagnet may transfer spin angular momentum between the current and the magnetization **M** giving rise to the macroscopic spin-torque on **M** so that the latter *may be altered by* *spin-polarized currents* [95,96,109,116]. The phenomenon underpins the novel subject of *spintronics* [96], i.e., *current-induced control over magnetic nanostructures*; applications include very high-speed current-induced magnetization switching by reversing the orientation of magnetic bits and using spin-polarized currents to manipulate steady state microwave oscillations [95,96,109,116]. These are maintained via the steady state magnetization precession due to STT so representing the conversion of dc input current into an ac output voltage. However, in STT devices, the thermal fluctuations cannot be ignored at finite temperatures because they lead to mainly noise-induced switching at currents far less than the critical switching current *without noise* as well as introducing randomness into the precessional orbits [117,118]. This phenomenon has been corroborated by many experiments (e.g., [119–124]) demonstrating that STT near room temperature alters thermally activated switching processes, these then exhibit a pronounced dependence on both material and geometrical parameters. Therefore, accurate solutions of generic STT models at *finite temperatures* are necessary both to critically assess such theories and to achieve further improvements in the design and interpretation of experiments, particularly due to the manifold practical applications in spintronics, random access memory technology, and so on. Recently various analytical and numerical approaches to the determination of the measurable parameters of STT devices such as the magnetization reversal time, etc. via the magnetic Langevin and/or Fokker–Planck equations including STT have been developed [86,109,125–138]. For the most part, they generalize methods originally developed for treating thermal fluctuations in nanoscale ferromagnets with zero STT such as stochastic (Brownian) dynamics simulations and the mean first-passage time method in the Kramers escape-rate theory for classical spins.

The archetypal model (Fig. 1.12) of a STT device is a nanostructure comprising two magnetic strata labeled the *free* and *fixed* layers, separated by a *nonmagnetic* conducting spacer. The fixed layer is much more strongly pinned along its orientation than the free one. On passing an electric current through the fixed layer it becomes *spin-polarized* which, as it encounters the free layer, induces a STT. Thus, the magnetization **M** of the free layer is altered. Both ferromagnetic layers are assumed to be uniformly magnetized. Although this approximation cannot explain all

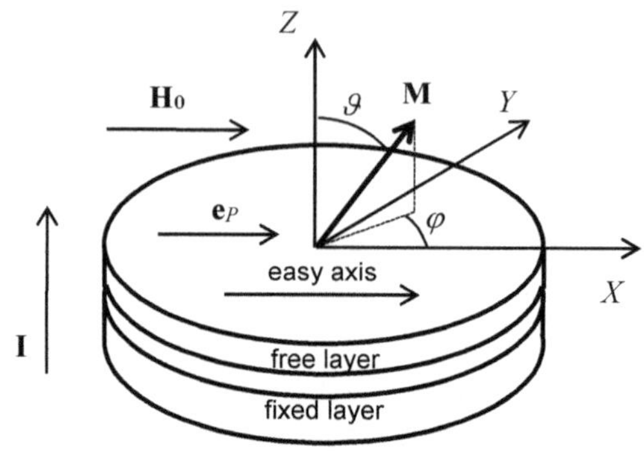

Fig. 1.12. Geometry of the problem: a STT device consists of two ferromagnetic strata labeled the *free* and *fixed* layers, respectively, and a normal conducting spacer all sandwiched on a pillar between two ohmic contacts [95,109]. Here **I** is the spin-polarized current, **M** is the magnetization of the free layer, and $\mathbf{H}_0$ is the dc bias magnetic field. The magnetization of the fixed layer is directed along the unit vector $\mathbf{e}_P$.

observations of the magnetization dynamics in spin-torque systems, nevertheless many qualitative features needed to interpret experimental data are satisfactorily reproduced. Consequently, the *current-induced magnetization dynamics* in the free layer may be described by the *Landau–Lifshitz–Gilbert–Slonczewski* equation including thermal fluctuations. This equation is simply the Landau–Lifshitz–Gilbert equation, Eq. (1.22), including the STT term augmented by a random magnetic field $\mathbf{h}(t)$ with the white noise properties, Eq. (1.43), thereby becoming a magnetic Langevin equation [95]:

$$\dot{\mathbf{M}} = \gamma\left[(\mathbf{H}+\mathbf{h})\times\mathbf{M}\right] + \frac{\alpha}{M_S}\left[\mathbf{M}\times\dot{\mathbf{M}}\right] + \frac{\gamma}{M_S}\left[\mathbf{M}\times\left[\mathbf{M}\times\mathbf{I}_S\right]\right], \tag{1.75}$$

while the effective magnetic field is $\mathbf{H} = -(\mu_0 M_S)^{-1}\partial V/\partial\mathbf{u}$ comprising the anisotropy and external applied fields. Furthermore V, constituting a conservative potential, is the free energy per unit volume of the free layer which is usually taken in the standard form of superimposed easy-plane and in-plane easy-axis anisotropies plus the Zeeman term, viz., [139]

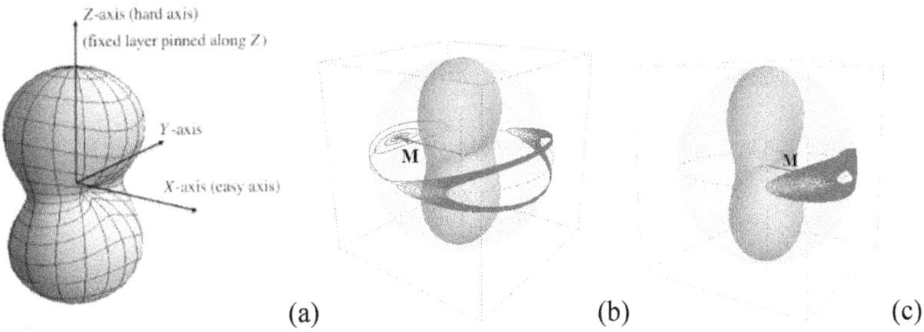

(a) (b) (c)

Fig. 1.13. (a) Biaxial anisotropy potential $V(\vartheta, \varphi)$ given by Eq. (1.76) with two saddle points and two nonequivalent wells separated by unequal barriers. (b) Current-induced trajectory of the magnetization escape. Solid line: numerical solution of the deterministic Eq. (1.75), i.e., omitting the random field **h**, for a spin-polarized current $J = 9.3$, damping $\alpha = 0.01$, and typical values of other model parameters ($D_{\|} = 0.034$, δ=20, $P = 0.3$, $\vartheta_p = \pi/2$ and $\vartheta_\xi = \pi/2$). The magnetization reversal over the saddle points separating one metastable state from another typically occurs after *many precessions* about the X axis. On overcoming the potential barrier, the magnetization decays to a new stable direction of precession so that in this reverse direction the current *accelerates* the decay by *increasing the effective damping*, as obvious from Eq. (1.75). (c) The same as in Fig. 1.13(b) for a *zero spin-polarized current* and the same values of the other model parameters. Here, only the damped precessions of the magnetization about the stable direction exist.

$$\begin{aligned} V(\vartheta,\varphi) &= \mu_0 M_S^2 D_{\|}\left[\delta u_Z^2 - u_X^2 - 2h(\gamma_1 u_X + \gamma_2 u_Y + \gamma_3 u_Z)\right] \\ &= \mu_0 M_S^2 D_{\|}\left[\delta \cos^2\vartheta - \sin^2\vartheta\cos^2\varphi \right. \\ &\left. -2h(\gamma_1 \cos\varphi \sin\vartheta + \gamma_2 \sin\varphi \sin\vartheta + \gamma_3 \cos\vartheta)\right] \end{aligned} \quad (1.76)$$

as shown in Fig. 1.13(a). Thus, the potential creates an energyscape with two minima and two saddle points effectively forcing the magnetization to align in a given direction in one or other of the energy minima in the equatorial or XY plane. The Z axis is obviously taken as the hard axis while the X axis is the easy one; ϑ and φ are angular coordinates describing the orientation of the moment **u** in the usual spherical polar coordinate system, while $\gamma_1 = \cos\varphi_\xi \sin\vartheta_\xi, \gamma_2 = \sin\varphi_\xi \sin\vartheta_\xi$, and $\gamma_3 = \cos\vartheta_\xi$ are the direction cosines of the applied field $\mathbf{H}_0$, $h = H_0/(2M_S D_{\|})$ is the external field parameter, and $\delta = D_\perp / D_{\|}$ is the *biaxiality parameter*, where $D_{\|}$ and $D_\perp$ account for both demagnetizing and magnetocrystalline anisotropy effects [109]. Notice that the STT *induced field* which we denote by $\mathbf{I}_S$ is

$$\mathbf{I}_S = \frac{1}{\mu_0 M_S}\frac{\partial \Phi}{\partial \mathbf{u}}. \tag{1.77}$$

The symbol Φ is the *nonconservative* potential due to STT and is

$$\Phi(\mathbf{u}) = J\frac{kT}{vc_P}\ln\left[1 + c_P(\mathbf{u}\cdot\mathbf{e}_P)\right], \tag{1.78}$$

where the unit vector $\mathbf{e}_P$ identifies the magnetization direction in the fixed layer, cf. Fig. 1.12, $J = b_P \hbar I/(|e|kT)$ is the dimensionless STT parameter, I is the *spin-polarized current* which is taken as positive if electrons flow from the free into the fixed layer, e is the electronic charge, $\hbar$ is Planck's reduced constant. Moreover, the coefficients b_P and c_P above are model dependent and are determined by the spin-polarization factor P ($0 < P < 1$)

$$\begin{aligned} b_P &= 4\left[3P^{-3/2}(1+P)^3 - 16\right]^{-1}, \\ c_P &= \left[3 - 16P^{3/2}(1+P)^{-3}\right]^{-1}, \end{aligned} \tag{1.79}$$

where $0 < b_P < 1/2$ and $1/3 < c_P < 1$ as P increases from 0 to 1. A typical value of the spin-induced current density for a 3 nm thick layer of cobalt (where $M_S \approx 1.4\cdot 10^6$ A m^{-1}) is of the order of 10^9 A cm^{-2} [109]. For the qualitative description of STT effects, the nonconservative potential Φ may be taken in the simplest form as

$$\Phi(\mathbf{u}) = J\frac{kT}{v}(\mathbf{e}_P\cdot\mathbf{u}). \tag{1.80}$$

The form of the nonconservative potential Φ (**u**), Eqs. (1.78) and (1.80), is shown in Fig.1.14.

Equation (1.75) contains precessional, damping and STT terms. A complete analysis taking all of them into account shows that the switching process is initiated via the excitation of a precessional motion about the initial state. Thereafter the cone angle of the trajectory *increases steadily* under the action of the STT, which opposes the damping torque (see Fig. 1.15). Nevertheless, whenever the magnetization reaches the *separatrix* (i.e. where a saddle point separates the initial and final states) switching occurs meaning that the magnetization relaxes towards the final state, and

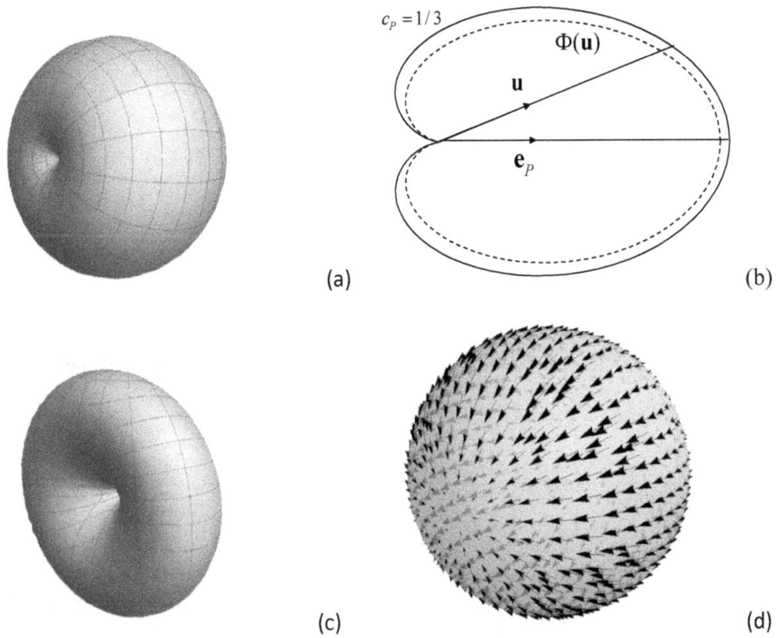

Fig. 1.14. (a) Form of the nonconservative potential $\Phi(\mathbf{u})$; (b) projection of the potential onto plane $Z = 0$: solid line Eq. (1.78), dashed line Eq. (1.80); (c) amplitude and (d) direction of STT induced field $\mathbf{I}_S$ Eqs. (1.77) and (1.78).

in contrast is now directed along a precessional trajectory with *decreasing* cone angle (see Fig. 1.13(b)). This process can only take place if the superposition of the external field and magnetocrystalline anisotropy gives rise to at least two stable states. For zero spin-polarized current and omitting noise torques, obviously only the damped precessions of the magnetization about the stable direction exist (see Fig. 1.13(c)).

Now the Langevin equation (1.75) can be rewritten as

$$\begin{aligned}\dot{\mathbf{u}} &= \frac{v}{2\tau_N kT\alpha}\left[\mathbf{u}\times\left(\frac{\partial}{\partial\mathbf{u}}(V-\alpha\Phi)-\mu_0 M_S\mathbf{h}\right)\right]\\ &+\frac{v}{2\tau_N kT}\left[\mathbf{u}\times\left[\mathbf{u}\times\left(\frac{\partial}{\partial\mathbf{u}}(V+\alpha^{-1}\Phi)-\mu_0 M_S\mathbf{h}\right)\right]\right],\end{aligned} \tag{1.81}$$

and Eq. (1.81) again leads to Brown's Fokker–Planck equation (now however incorporating STT) for the surface probability density $W(\vartheta, \varphi, t)$ of magnetization orientations $\mathbf{M}$ on the unit sphere, viz., [109,137,138]

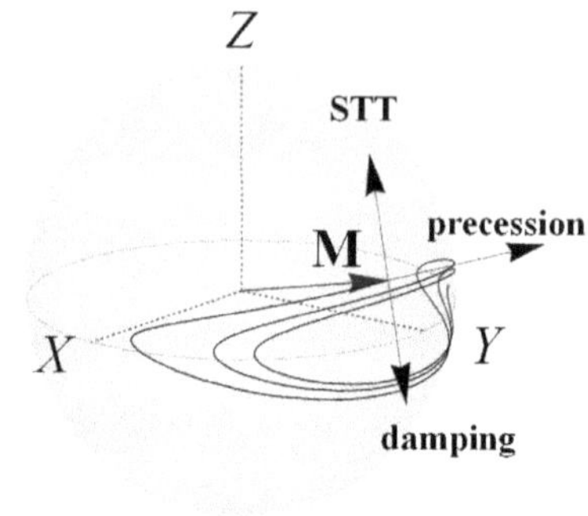

Fig. 1.15. The excitation of a precessional motion about the initial state.

$$\frac{\partial W}{\partial t}=\frac{1}{2\tau_{\mathrm{N}}\sin\vartheta}\frac{\partial}{\partial\vartheta}\left\{\sin\vartheta\left[\frac{\partial W}{\partial\vartheta}+\frac{vW}{kT}\left(\frac{\partial\left(V+\alpha^{-1}\Phi\right)}{\partial\vartheta}+\frac{1}{\sin\vartheta}\frac{\partial\left(\alpha^{-1}V-\Phi\right)}{\partial\varphi}\right)\right]\right\}$$
$$+\frac{1}{2\tau_{\mathrm{N}}\sin^2\vartheta}\frac{\partial}{\partial\varphi}\left[\frac{\partial W}{\partial\varphi}+\frac{vW}{kT}\left(\frac{\partial\left(V+\alpha^{-1}\Phi\right)}{\partial\varphi}-\sin\vartheta\frac{\partial\left(\alpha^{-1}V-\Phi\right)}{\partial\vartheta}\right)\right]. \quad (1.82)$$

For zero STT, the Fokker–Planck equation, Eq. (1.82), becomes Brown's Eq. (1.64). Now, invariably STT effects on the magnetization relaxation are significant only for low damping, $\alpha \ll 1$, because their magnitude is governed by the ratio J/α. Hence for $\alpha \geq 1$ the STT term in Eq. (1.75) does not influence the reversal process at all because it is then negligible compared to the damping and random field terms.

However, the VLD dynamics of the magnetization including STT may still be very different from those with zero STT. Due to STT the *Gilbert damping torque may be completely overcome*. Thus, reversal of the magnetization becomes possible even *in the absence* of thermal fluctuations. As far as the *joint* action of the STT and thermal fluctuations is concerned, the overall situation, albeit more complicated, is in some way reminiscent of that occurring in the resistively shunted junction model of a Josephson junction; electric analog of the motion of a Brownian particle in a tilted periodic potential (*v*. [97]). To see this, consider the bias current in the junction. This current represents a *nonconservative* electrical source giving rise to the motion in a tilted cosine periodic potential and so ensures that the stationary distribution is no longer the Boltzmann distribution. In like manner, in a ferromagnet subjected to spin-polarized current, the stationary distribution of magnetization orientations (which now depends both on the spin-polarized current and damping) is no longer a Boltzmann one.

1.3.4 *Method of statistical moments*

In magnetization relaxation, the relevant quantities are orientational averages involving the spherical harmonics $Y_{lm}(\vartheta,\varphi)$ which are defined as [104] (see Appendix A.1.7)

$$Y_{lm}(\vartheta,\varphi)=\sqrt{\frac{(2l+1)(l-m)!}{4\pi(l+m)!}}e^{im\varphi}P_l^m(\cos\vartheta),\ \ Y_{l-m}=(-1)^m Y_{lm}^*, \quad (1.83)$$

where $P_l^m(x)$ are the associated Legendre functions [104] defined by Eq. (A.69) in Appendix A.1.7) and the asterisk denotes the complex conjugate. The orthogonality and normalization conditions for the $Y_{lm}(\vartheta,\varphi)$ are also given in that Appendix. The spherical harmonics Y_{lm} are functions of the angles ϑ and φ. However, in the present application these angles are *themselves random variables* spatially distributed according to the surface probability density function W. Thus, by averaging Y_{lm} over W, we have the *averaged* spherical harmonics, namely $\langle Y_{lm}\rangle$, where $\langle\ \rangle$ designates the average, viz.,

$$\langle Y_{lm}\rangle(t)=\int_0^{2\pi}\int_0^{\pi}Y_{lm}(\vartheta,\varphi)W(\vartheta,\varphi,t)\sin\vartheta d\vartheta d\varphi. \quad (1.84)$$

The form of the Fokker–Planck Eq. (1.64) and the orthogonality of spherical harmonics suggests that we may seek a solution in the form of the Fourier–Laplace series [97]

$$W(\vartheta,\varphi,t)=\sum_{l,m}\langle Y_{lm}\rangle(t)Y_{lm}^*(\vartheta,\varphi), \quad (1.85)$$

where the Fourier coefficients $\langle Y_{lm}\rangle(t)$ are simply the *statistical moments* representing expectation values of the functions Y_{lm}. We now have from Eqs. (1.64) and (1.85) the moment system for the *averaged spherical harmonics* rendered as time-dependent differential-recurrence relations which have the form of infinite linear combinations, namely

$$\frac{d}{dt}\left\langle Y_{lm}\right\rangle(t) = \sum\nolimits_{l',m'} d_{l'm'lm}\left\langle Y_{l'm'}\right\rangle(t), \tag{1.86}$$

where the expansion coefficients

$$d_{l'm'lm} = \int_0^{2\pi}\int_0^{\pi} Y_{lm}(\vartheta,\varphi)\mathrm{L}_{\mathrm{FP}}Y^*_{l'm'}(\vartheta,\varphi)\sin\vartheta d\vartheta d\varphi \tag{1.87}$$

are the matrix elements of the Fokker–Planck operator $\mathrm{L_{FP}}$. In the *stationary* state, Eq. (1.86) becomes

$$\sum\nolimits_{l',m'} d_{l'm'lm}\left\langle Y_{l'm'}\right\rangle_0 = 0, \tag{1.88}$$

where the angular brackets $\langle\cdot\rangle_0$ mean averaging over the new equilibrium or if STT is included the out-of equilibrium time-independent stationary distribution $W_0\left(\vartheta,\varphi\right)$, which with or without STT can always be rendered as the Fourier series

$$W_0\left(\vartheta,\varphi\right) = \sum_{l,m}\left\langle Y^*_{lm}\right\rangle_0 Y_{lm}(\vartheta,\varphi). \tag{1.89}$$

Moreover, for *arbitrary* magnetocrystalline anisotropy, which can be expressed in terms of spherical harmonics as

$$\frac{v}{kT}V(\vartheta,\varphi) = \sum_{R=1}^{\infty}\sum_{S=-R}^{R} A_{RS}Y_{RS}(\vartheta,\varphi), \tag{1.90}$$

we have by assuming a solution in the form of the Fourier expansion Eq. (1.85) for the Fokker–Planck equation (1.64), an infinite hierarchy of differential-recurrence equations for the statistical moments $\left\langle Y_{lm}\right\rangle(t)$, viz., (details are in Ref. [97])

$$\tau_{\mathrm{N}}\frac{d}{dt}\left\langle Y_{lm}\right\rangle(t) = \sum_{s,r} e_{lml+rm+s}\left\langle Y_{l+rm+s}\right\rangle(t). \tag{1.91}$$

In Eq. (1.91), the coefficients $e_{lml'm\pm s} = \tau_{\mathrm{N}} d_{l'm\pm slm}$ are given by [97] (see Section 1.3.5)

$$
\begin{aligned}
e_{lml'm\pm s} &= -\frac{l(l+1)}{2}\delta_{ll'}\delta_{s0} + (-1)^m \frac{1}{4}\sqrt{\frac{(2l+1)(2l'+1)}{\pi}} \\
&\times \sum_{r=s}^{\infty} A_{r\pm s} \left\{ \frac{\left[l'(l'+1) - r(r+1) - l(l+1)\right]}{2\sqrt{2r+1}} C_{l0l'0}^{r0} C_{lml'-m\mp s}^{r\mp s} \right. \\
&- \frac{i}{\alpha}\sqrt{\frac{(2r+1)(r-s)!}{(r+s)!}} \sum_{\substack{L=s-\varepsilon_{rs} \\ \Delta L=2}}^{r-1} \sqrt{\frac{(L+s)!}{(L-s)!}} C_{l0l'0}^{L0} \\
&\left. \times \left[mC_{lml'-m\mp s}^{L\mp s} \pm s\sqrt{\frac{(l\mp m)(l\pm m+1)}{(L+s)(L-s+1)}} C_{lm\pm 1l'-m\mp s}^{L\mp s\pm 1} \right] \right\},
\end{aligned}
\tag{1.92}
$$

where $s \geq 0$ and $C_{lml'm'}^{rs}$ are the Clebsch–Gordan coefficients [104]. We remark that Eq. (1.92) determines the coefficients of the linear combination $e_{lml'm'}$ for *arbitrary* magnetocrystalline anisotropy and Zeeman energy densities. The Gilbert–Langevin equation, Eq. (1.42), can also be reduced to the moment system for $\langle Y_{lm} \rangle(t)$, Eq. (1.91), by an appropriate transformation of variables and by direct averaging (without recourse to the Fokker–Planck equation at all) of the stochastic equation thereby obtained [97,140]. Examples of calculations of the $e_{lml'm'}$ for various magnetocrystalline anisotropies, are available in Ref. [97].

We remark in passing that the Fokker–Planck equation (1.64) for the surface probability density function $W(\vartheta, \varphi, t)$ of orientations of the unit vector **u**, can also be solved by separation of the variables. This gives rise to a Sturm–Liouville problem so that $W(\vartheta, \varphi, t)$ can be written as

$$
W(\vartheta,\varphi,t) = W_0(\vartheta,\varphi) + \sum_{k=1}^{\infty} \Phi_k(\vartheta,\varphi) e^{-\lambda_k t}, \tag{1.93}
$$

where $\Phi_k(\vartheta,\varphi)$ and λ_k are the eigenfunctions and eigenvalues of the Fokker–Planck operator L_{FP} and $W_0(\vartheta,\varphi)$ is the stationary solution of that equation, i.e., $\mathrm{L}_{\mathrm{FP}} W_0 = 0,$ corresponding to the Boltzmann equilibrium distribution.

We have briefly summarized the derivation of the differential-recurrence relations for the statistical moments using the Fokker–Planck equation. Now we derive these equations explicitly, on this occasion not from the Fokker–Planck equation instead directly from the Gilbert form of the Langevin equation. To achieve this, we again use the Stratonovich definition [101,102] of the stochastic differential equations, Eq. (1.47), since this definition represents the *mathematical idealization of the physical process of magnetization relaxation of ferromagnetic nano-particles*. Hence, the conventional rules of calculus described, for example, in [102] apply. We have from Eq. (1.47) the stochastic differential equation of motion for an *arbitrary* function $f[\vartheta(t), \varphi(t)]$, viz.,

$$\begin{aligned}\frac{df}{dt} &= \frac{\partial f}{\partial \vartheta}\frac{d\vartheta}{dt} + \frac{\partial f}{\partial \varphi}\frac{d\varphi}{dt} \\ &= F_1 \frac{\partial f}{\partial \vartheta} + F_2 \frac{\partial f}{\partial \varphi} + \left(g_{1j}\frac{\partial f}{\partial \vartheta} + g_{2j}\frac{\partial f}{\partial \varphi} \right) h_j(t),\end{aligned} \tag{1.94}$$

where the functions F_i are given by Eqs. (1.63) and the g_{ij} are defined in Table 1.1. Therefore, by using the general theory of averaging of a stochastic differential equation of the Stratonovich type [97], we have *the average evolution equation,* Eq. (1.58). Here the *deterministic* drift term, namely, the first one on the right-hand side of Eq. (1.58), is given by

$$\begin{aligned}F_i \frac{\partial f}{\partial \xi_i} &= \lim_{\Delta t \to 0} \overline{\frac{1}{\Delta t}\int_t^{t+\Delta t} \left(F_1 \frac{\partial f}{\partial \vartheta} + F_2 \frac{\partial f}{\partial \varphi} \right) dt'} \\ &= -h'\left(\left\{ \nabla V - \alpha^{-1}[\mathbf{u} \times \nabla V] \right\} \cdot \nabla f \right) \\ &= \frac{v}{4kT\tau_{\mathrm{N}}}\left[V\Delta f + f\Delta V - \Delta(Vf) \right] + \frac{v}{2kT\alpha\tau_{\mathrm{N}}}\left([\mathbf{u} \times \nabla V] \cdot \nabla f \right),\end{aligned} \tag{1.95}$$

where Δ is the angular part of the Laplacian as defined by Eq. (1.66) and we have used an identity for the operator product $(\nabla V \cdot \nabla) f$, namely

$$2(\nabla V \cdot \nabla) f = \Delta(Vf) - V\Delta f - f\Delta V, \tag{1.96}$$

which simplifies the subsequent derivation of the differential-recurrence equations for the statistical moments. The overbar on the left-hand side of Eq. (1.95) means the statistical average during a *small time interval* Δt over an ensemble of magnetic dipoles which all have at time t the *same sharp* values ϑ and φ of the stochastic variables (processes) $\vartheta(t)$ and $\varphi(t)$ at the averaging instant t. The *noise-induced drift*, namely, the second term on the right-hand side of Eq. (1.58), is

$$\overline{\left(g_{1j}\frac{\partial f}{\partial\vartheta}+g_{2j}\frac{\partial f}{\partial\varphi}\right)h_j(t)}=\frac{\alpha kT}{v\gamma\mu_0 M_{\mathrm{S}}}\left[g_{1j}\frac{\partial}{\partial\vartheta}\left(g_{1j}\frac{\partial f}{\partial\vartheta}+g_{2j}\frac{\partial f}{\partial\varphi}\right)\right.$$
$$\left.+g_{2j}\frac{\partial}{\partial\varphi}\left(g_{1j}\frac{\partial f}{\partial\vartheta}+g_{2j}\frac{\partial f}{\partial\varphi}\right)\right]=\frac{1}{2\tau_{\mathrm{N}}}\Delta f. \tag{1.97}$$

In writing Eq. (1.97), we have utilized the white noise properties of the random Gaussian magnetic field **h**, Eq. (1.43). Finally, combining Eqs. (1.95) and (1.97), we have the desired evolution equation for *f*, viz.,

$$\tau_{\mathrm{N}}\dot{f}=\frac{1}{2}\Delta f+\frac{v}{4kT}\left[V\Delta f+f\Delta V-\Delta\left(Vf\right)\right]+\frac{v}{2\alpha kT}\left(\left[\mathbf{u}\times\nabla V\right]\cdot\nabla f\right). \tag{1.98}$$

In magnetic applications, the relevant quantities *f* correspond to averages involving the spherical harmonics Y_{lm} defined by Eq. (1.83). The general Eq. (1.98) as applied to the specific function Y_{lm} yields

$$\tau_{\mathrm{N}}\dot{Y}_{lm}=\frac{1}{2}\Delta Y_{lm}+\frac{v}{4kT}\left\{V\Delta Y_{lm}+Y_{lm}\Delta V-\Delta\left(VY_{lm}\right)\right.$$
$$\left.-\frac{1}{\alpha\sin\vartheta}\left[\frac{\partial V}{\partial\varphi}\frac{\partial Y_{lm}}{\partial\vartheta}-\frac{\partial V}{\partial\vartheta}\frac{\partial Y_{lm}}{\partial\varphi}\right]\right\}. \tag{1.99}$$

Mindful that our ultimate goal is to represent Eq. (1.99) as a differential-recurrence relation, we therefore first express the right-hand side of Eq. (1.99) in terms of the orbital angular momentum operators $\hat{\mathbf{L}}^2$, $\hat{L}_Z$, and $\hat{L}_\pm$ defined as [88,104] (see Appendix A.1.7)

$$\hat{\mathbf{L}}^2=-\Delta,\ \hat{L}_Z=-i\frac{\partial}{\partial\varphi},$$
$$\hat{L}_\pm=\pm e^{\pm i\varphi}\frac{\partial}{\partial\vartheta}+i\cot\vartheta e^{\pm i\varphi}\frac{\partial}{\partial\varphi}. \tag{1.100}$$

Equation (1.99) is now rendered in such a form that it may be radically simplified by using the theory of angular momentum [104], because the action of the angular momentum operators $\hat{L}_Z$, $\hat{L}_\pm$, $\hat{\mathbf{L}}^2$ on a spherical harmonic Y_{lm} is either to reproduce Y_{lm} itself or $Y_{lm\pm1}$, viz., [88]

$$\hat{\mathbf{L}}^2 Y_{lm}-l(l+1)Y_{lm},\hat{L}_Z Y_{lm}=mY_{lm},\hat{L}_\pm Y_{lm}=\sqrt{l(l+1)-m(m\pm1)}Y_{lm\pm1}. \tag{1.101}$$

Thus we have from Eq. (1.99) and utilizing Eq. (1.101)

$$\tau_{\mathrm{N}}\dot{Y}_{lm} = \frac{v}{4kT}\left[\hat{\mathbf{L}}^2\left(VY_{lm}\right) - V\hat{\mathbf{L}}^2 Y_{lm} - Y_{lm}\hat{\mathbf{L}}^2 V\right] - \frac{1}{2}\hat{\mathbf{L}}^2 Y_{lm}$$
$$+\frac{i}{4\alpha}\sqrt{\frac{3}{2\pi}}\left\{Y_{11}^{-1}\left[\left(\hat{L}_Z V_+\right)\left(\hat{L}_+ Y_{lm}\right) - \left(\hat{L}_+ V_+\right)\left(\hat{L}_Z Y_{lm}\right)\right]\right. \quad (1.102)$$
$$\left.+Y_{1-1}^{-1}\left[\left(\hat{L}_Z V_-\right)\left(\hat{L}_- Y_{lm}\right) - \left(\hat{L}_- V_-\right)\left(\hat{L}_Z Y_{lm}\right)\right]\right\},$$

where we have also used the following representation for the expansion of the (split) potential $vV/(kT) = V_+ + V_-$ which for convenience is written as the following series of spherical harmonics for each component

$$V_+ = \sum_{R=1}^{\infty}\sum_{S=0}^{R} A_{RS}Y_{RS},$$
$$V_- = \sum_{R=1}^{\infty}\sum_{S=-R}^{-1} A_{RS}Y_{RS}. \quad (1.103)$$

Furthermore, we also have

$$Y_{1\pm1}^{-1}Y_{l\pm m} = \sqrt{\frac{8\pi(2l+1)(l-m)!}{3(l+m)!}}\sum_{L=m-\varepsilon_{lm}}^{l-1}\sqrt{\frac{(2L+1)(L+m-1)!}{(L-m+1)!}}Y_{L\pm(m-1)}, \quad (1.104)$$

where $m > 0$, $\Delta L = 2$ and $\varepsilon_{lm} = 1$, if the indices l and m are of the same order of evenness; $\varepsilon_{lm} = 0$ otherwise. Moreover, Eq. (1.104) may also be regarded as the solution of the recurrence relation

$$Y_{1\pm1}^{-1}Y_{lm} = \sqrt{\frac{8\pi(2l-1)(2l+1)}{3(l\pm m)(l\pm m-1)}}Y_{l-1m\mp1} + \sqrt{\frac{(2l+1)(l\mp m)(l\mp m-1)}{(2l-3)(l\pm m)(l\pm m-1)}}Y_{1\pm1}^{-1}Y_{l-2m}$$

($\left|m\mp1\right| \le l-1$), which in turn follows by transposition from the well-known recurrence relation [104]

$$\sqrt{\frac{8\pi}{3}}Y_{1\pm1}Y_{lm} = \sqrt{\frac{(l\pm m+1)(l\pm m+2)}{(2l+1)(2l+3)}}Y_{l+1m\pm1} - \sqrt{\frac{(l\mp m-1)(l\mp m)}{(2l-1)(2l+1)}}Y_{l-1m\pm1}. \quad (1.105)$$

By using Eqs. (1.99), (1.101), and (1.104) in combination, we then have

$$\begin{aligned} Y_{1\pm1}^{-1}\hat{L}_Z Y_{l\pm m} &= \pm m\sqrt{\frac{8\pi(2l+1)(l-m)!}{3(l+m)!}}\sum_{L=m-\varepsilon_{lm}}^{l-1}\sqrt{\frac{(2L+1)(L+m-1)!}{(L-m+1)!}}Y_{L\pm(m-1)}, \\ Y_{1\pm1}^{-1}\hat{L}_\pm Y_{l\pm m} &= \sqrt{\frac{8\pi(2l+1)(l-m)!}{3(l+m)!}}\sum_{L=m+\varepsilon_{lm}}^{l-1}\sqrt{\frac{(2L+1)(L+m)!}{(L-m)!}}Y_{L\pm m} \end{aligned} \tag{1.106}$$

(here $m \geq 0$ and $\Delta L = 2$). Thus, with Eqs. (1.104)–(1.106), we can now transform the evolution Eq. (1.102) into a differential-recurrence equation for Y_{lm}, viz., [141]

$$\tau_N \frac{d}{dt} Y_{lm} = \sum\nolimits_{l',s} e_{lml'm+s} Y_{l'm+s}, \tag{1.107}$$

where the coefficients $e_{lml'm\pm s}$ are given by Eq. (1.92) for *arbitrary* magnetocrystalline anisotropy via the Clebsch–Gordan coefficients.

The basis set Y_{lm} in Eq. (1.107) are functions of the *sharp* values ϑ and φ at time t, which we reiterate are themselves random variables with probability density function W. Thus by averaging Eq. (1.107) over W, we have the moment system for the averaged spherical harmonics, namely

$$\tau_N \frac{d}{dt}\langle Y_{lm}\rangle(t) = \sum_{s,r} e_{lml+rm+s}\langle Y_{l+rm+s}\rangle(t), \tag{1.108}$$

where the angular brackets $\langle\ \rangle$ designate ensemble (spatial) averaging over W. We also have from Eq. (1.107) a similar set of differential-recurrence equations for the appropriate *equilibrium correlation functions* $c_{lm}(t)$, viz., [97]

$$\tau_N \dot{c}_{lm}(t) = \sum_{s,r} e_{lml+rm+s} c_{l+rm+s}(t), \tag{1.109}$$

where the $c_{lm}(t)$ are given by

$$\begin{aligned} c_{lm}(t) &= \langle F[\vartheta(0),\varphi(0)] Y_{lm}[\vartheta(t),\varphi(t)]\rangle_0 \\ &- \langle F[\vartheta(0),\varphi(0)]\rangle_0 \langle Y_{lm}[\vartheta(0),\varphi(0)]\rangle_0, \end{aligned} \tag{1.110}$$

$F(\vartheta,\varphi)$ is an arbitrary function of ϑ and φ, the angular brackets $\langle\ \rangle_0$ designate the spatial averaging over the equilibrium distribution W_0. Thus, the statistical moment systems, Eqs. (1.91) and (1.108), derived from the respective magnetic Fokker–Planck and Langevin equations are in complete agreement, and, moreover, prove the equivalence of the two approaches. We recall in passing that the differential-recurrence Eqs. (1.86) and (1.108) are valid for an *arbitrary* magnetocrystalline anisotropy potential.

In the presence of STT, the relevant magnetic Langevin equation is the Gilbert equation for the evolution of the magnetization **M** in the free layer (see Fig. 1.12) as modified by Slonczewski to include the STT, Eq. (1.81), where the precessional and alignment terms are now clearly delineated. The vector Langevin equation, Eq. (1.81), represents in spherical polar coordinates two coupled nonlinear stochastic differential equations for the polar and azimuthal angles ϑ and φ, viz.,

$$\frac{d\vartheta}{dt}=\frac{v}{2kT\tau_{\mathrm{N}}}\left\{\mu_0 M_{\mathrm{S}}\left[h_\vartheta(t)+\alpha^{-1}h_\varphi(t)\right]-\frac{\partial V+\alpha^{-1}\Phi}{\partial\vartheta}-\frac{\partial\alpha^{-1}V-\Phi}{\sin\vartheta\partial\varphi}\right\},$$

$$\frac{d\varphi}{dt}=\frac{v}{2kT\tau_{\mathrm{N}}}\left\{\mu_0 M_{\mathrm{S}}\frac{h_\varphi(t)-\alpha^{-1}h_\vartheta(t)}{\sin\vartheta}-\frac{\partial V+\alpha^{-1}\Phi}{\sin^2\vartheta\partial\varphi}+\frac{\partial\alpha^{-1}V-\Phi}{\sin\vartheta\partial\vartheta}\right\},$$

where the random field components $h_\vartheta(t)$ and $h_\varphi(t)$ are given by Eq. (1.61). For $\Phi=0$, these equations reduce to the Langevin equations for zero STT, Eq. (1.62). The desired Langevin equation for the evolution of the spherical harmonics is then given by

$$\frac{dY_{lm}}{dt}=\frac{\partial Y_{lm}}{\partial\vartheta}\frac{d\vartheta}{dt}+\frac{\partial Y_{lm}}{\partial\varphi}\frac{d\varphi}{dt}$$

$$=\frac{v}{2kT\tau_{\mathrm{N}}}\left\{\mu_0 M_{\mathrm{S}}\left(\left[h_\vartheta(t)+\alpha^{-1}h_\varphi(t)\right]\frac{\partial Y_{lm}}{\partial\vartheta}+\frac{h_\varphi(t)-\alpha^{-1}h_\vartheta(t)}{\sin\vartheta}\frac{\partial Y_{lm}}{\partial\varphi}\right)\right.$$

$$\left.-\left(\frac{\partial V+\alpha^{-1}\Phi}{\partial\vartheta}+\frac{\partial\alpha^{-1}V-\Phi}{\sin\vartheta\partial\varphi}\right)\frac{\partial Y_{lm}}{\partial\vartheta}-\left(\frac{\partial V+\alpha^{-1}\Phi}{\sin^2\vartheta\partial\varphi}-\frac{\partial\alpha^{-1}V-\Phi}{\sin\vartheta\partial\vartheta}\right)\frac{\partial Y_{lm}}{\partial\varphi}\right\}.$$

By averaging this Langevin equation over an infinitesimal time as before, we have an infinite hierarchy of differential-recurrence equations for the statistical moments $\langle Y_{lm}\rangle(t)$, Eq. (1.108), whcrc

$$e_{lml'm\pm s} = -\frac{l(l+1)}{2}\delta_{ll'}\delta_{s0} + (-1)^m \frac{1}{4}\sqrt{\frac{(2l+1)(2l'+1)}{\pi}} \sum_{r=s}^{\infty} \left\{ \left(A_{r\pm s} + \frac{B_{r\pm s}}{\alpha} \right) C^{r0}_{l0l'0} \right.$$

$$\times C^{r\mp s}_{lml'-m\mp s} \frac{l'(l'+1) - r(r+1) - l(l+1)}{2\sqrt{2r+1}} - \frac{i}{\alpha}\left(A_{r\pm s} - \alpha B_{r\pm s} \right) \sqrt{\frac{(2r+1)(r-s)!}{(r+s)!}}$$

$$\left. \times \sum_{\substack{L=s-\varepsilon_{rs} \\ \Delta L=2}}^{r-1} \sqrt{\frac{(L+s)!}{(L-s)!}} C^{L0}_{l0l'0} \left[m C^{L\mp s}_{lml'-m\mp s} \pm s \sqrt{\frac{(l\mp m)(l\pm m+1)}{(L+s)(L-s+1)}} C^{L\mp s\pm 1}_{lm\pm 1l'-m\mp s} \right] \right\},$$

(1.111)

and A_{rs} and B_{rs} are, respectively, the coefficients of the Fourier series expansions in terms of spherical harmonics of both the conservative free energy density $V(\vartheta, \varphi)$ and the nonconservative potential $\Phi(\vartheta, \varphi)$, viz.,

$$\frac{vV}{kT} = \sum_{r,s} A_{rs} Y_{rs},$$

$$\frac{v\Phi}{kT} = \sum_{r,s} B_{rs} Y_{rs}.$$

For $\Phi = 0$, i.e. $B_{rs} = 0$, Eq. (1.111) reduces to Eq. (1.92) derived for zero STT.

Just like zero STT, the same result may be obtained, albeit with more labor and in a less transparent manner, from the Fokker–Planck equation, Eq. (1.82). In the *stationary* state, the statistical moments $\langle Y_{lm} \rangle_0$ are independent of time. Here the angular brackets $\langle\ \rangle_0$ mean averaging over the out-of-equilibrium time-independent stationary distribution $W_0(\vartheta,\varphi)$, which can always be rendered as the Fourier series, Eq. (1.89). By analogy with the Boltzmann distribution, since we on intuitive grounds expect the stationary distribution $W_0(\vartheta,\varphi)$ to be formally similar to it, we may define the *effective potential* $V_{\mathrm{ef}}(\vartheta,\varphi)$ via

$$V_{\mathrm{ef}}(\vartheta,\varphi) = -v^{-1}kT \ln[W_0(\vartheta,\varphi)].$$

Having completed our preliminary discussion of STT, we show later (*v*. Section 1.4.6) how it may be incorporated into escape-rate theory and in Chapter 2 how it may be treated by solving the differential-recurrence equations (1.108) and (1.111) for the generic nanopillar model.

1.3.5 *Methods of solution of infinite sets of differential-recurrence equations*

The differential-recurrence Eq. (1.91) and (1.108) may always be written as the homogeneous matrix differential equation

$$\dot{\mathbf{C}}(t) + \mathbf{AC}(t) = 0, \tag{1.112}$$

where $\mathbf{C}(t)$ is an infinite column vector formed from the statistical moments $\langle Y_{lm} \rangle(t)$ and $\mathbf{A}$ is the infinite system matrix. The formal solution of the linear Eq. (1.112) is [142]

$$\mathbf{C}(t) = e^{-\mathbf{A}t}\mathbf{C}(0) = \mathbf{U}e^{-\mathbf{\Lambda}t}\mathbf{U}^{-1}\mathbf{C}(0), \tag{1.113}$$

where $\mathbf{\Lambda} = \mathbf{U}^{-1}\mathbf{AU}$ is a diagonal matrix with elements composed of all the eigenvalues λ_k of $\mathbf{A}$ and $\mathbf{U}$ is a right eigenvector matrix composed of all the eigenvectors of $\mathbf{A}$. Now having calculated $\mathbf{C}(t)$ from the matrix Eq. (1.113), we have all the statistical moments $\langle Y_{lm} \rangle(t)$. The numerical solution of Eq. (1.112) is determined by successively increasing the size of $\mathbf{A}$ until convergence is attained. Furthermore, the reversal time of the magnetization τ can be estimated via the inverse of the smallest nonvanishing eigenvalue λ_1 of the system matrix operator $\mathbf{A}$, viz.,

$$\tau = 1/\lambda_1. \tag{1.114}$$

Alternatively, the multi-term differential-recurrence Eqs. (1.91) and (1.108) can be solved by rearranging it as $(2n + 1)$-term matrix differential-recurrence relations admitting of an exact matrix continued fraction solution as follows. In general, one will have

$$\begin{aligned}\tau_\mathrm{N}\frac{d}{dt}\mathbf{C}_p(t) &= \mathbf{Q}_p^{(-n)}\mathbf{C}_{p-n}(t) + \ldots + \mathbf{Q}_p^{(-1)}\mathbf{C}_{p-1}(t)\\ &+\mathbf{Q}_p\mathbf{C}_p(t) + \mathbf{Q}_p^{(+1)}\mathbf{C}_{p+1}(t) + \ldots + \mathbf{Q}_p^{(+n)}\mathbf{C}_{p+n}(t),\end{aligned} \tag{1.115}$$

where the time-independent matrices $\mathbf{Q}_p^{(\pm i)}$ are composed of the elements $e_{lml'm'}$ and the column vectors $\mathbf{C}_p(t)$ are arranged in an appropriate way from $\langle Y_{lm} \rangle(t)$. In the simplest case $n = 1$, Eq. (1.115) becomes the three-term differential-recurrence equation

$$\tau_{\mathrm{N}} \frac{d}{dt}\mathbf{C}_n(t) = \mathbf{Q}_n^{-}\mathbf{C}_{n-1}(t) + \mathbf{Q}_n\mathbf{C}_n(t) + \mathbf{Q}_n^{+}\mathbf{C}_{n+1}(t). \qquad (1.116)$$

Taking the Laplace transform of Eq. (1.115), yields an *algebraic* $(2n+1)$-*term recurrence equation* in the s domain

$$\mathbf{Q}_p^{(-n)}\tilde{\mathbf{C}}_{p-n} + \ldots + \left(\mathbf{Q}_p - s\tau_{\mathrm{N}}\mathbf{I}\right)\tilde{\mathbf{C}}_p + \ldots + \mathbf{Q}_p^{(+n)}\tilde{\mathbf{C}}_{p+n} = -\tau_{\mathrm{N}}\mathbf{C}_p(0), \quad (1.117)$$

where $\mathbf{I}$ is the unit matrix and

$$\tilde{\mathbf{C}}_p = \tilde{\mathbf{C}}_p(s) = \int_0^\infty \mathbf{C}_p(t)e^{-st}dt.$$

For illustration, we consider the solution of a three-term recurrence equation using matrix iteration. The three-term algebraic recurrence relation (see Eq. (1.117) with $n = 1$) may be written as

$$\mathbf{Q}_p^{-}\tilde{\mathbf{C}}_{p-1} + \left(\mathbf{Q}_p - s\tau_{\mathrm{N}}\mathbf{I}\right)\tilde{\mathbf{C}}_p + \mathbf{Q}_p^{+}\tilde{\mathbf{C}}_{p+1} = -\tau_{\mathrm{N}}\mathbf{C}_p(0). \qquad (1.118)$$

Thus we may seek the solution of Eq. (1.118) (in the manner of a linear differential equation) as

$$\tilde{\mathbf{C}}_p = \mathbf{S}_p(s)\tilde{\mathbf{C}}_{p-1} + \mathbf{R}_p, \qquad (1.119)$$

where the matrix continued fraction $\mathbf{S}_p(s)$ as defined by the recurrence relation

$$\mathbf{S}_p(s) = \left[s\tau_{\mathrm{N}}\mathbf{I} - \mathbf{Q}_p - \mathbf{Q}_p^{+}\mathbf{S}_{p+1}(s)\right]^{-1}\mathbf{Q}_p^{-} \qquad (1.120)$$

represents the *complementary solution* of Eq. (1.118) [143]. The *particular solution* $\mathbf{R}_p(s)$ may be found by directly substituting Eq. (1.119) into Eq. (1.118). We have

$$\begin{aligned} &\mathbf{Q}_p^{-}\tilde{\mathbf{C}}_{p-1} + \left(\mathbf{Q}_p - s\tau_{\mathrm{N}}\mathbf{I}\right)\left[\mathbf{S}_p(s)\tilde{\mathbf{C}}_{p-1} + \mathbf{R}_p\right] \\ &+\mathbf{Q}_p^{+}\left\{\mathbf{S}_{p+1}(s)\left[\mathbf{S}_p(s)\tilde{\mathbf{C}}_{p-1} + \mathbf{R}_p\right] + \mathbf{R}_{p+1}\right\} = -\tau_{\mathrm{N}}\mathbf{C}_p(0) \end{aligned} \qquad (1.121)$$

or with Eq. (1.120)

$$\left[s\tau_{\mathrm{N}}\mathbf{I} - \mathbf{Q}_p - \mathbf{Q}_p^{+}\mathbf{S}_{p+1}(s)\right]\mathbf{R}_p - \mathbf{Q}_p^{+}\mathbf{R}_{p+1} = \tau_{\mathrm{N}}\mathbf{C}_p(0).$$

Thus, we have

$$\mathbf{R}_p = \mathbf{\Delta}_p(s)\left[\tau_\varepsilon \mathbf{C}_p(0) + \mathbf{Q}_p^+ \mathbf{R}_{p+1}\right], \tag{1.122}$$

where $\mathbf{\Delta}_p(s)$ is the matrix continued fraction defined as

$$\mathbf{\Delta}_p(s) = [s\tau_{\mathrm{N}}\mathbf{I} - \mathbf{Q}_p - \mathbf{Q}_p^+ \mathbf{\Delta}_{p+1}(s)\mathbf{Q}_{p+1}^-]^{-1}. \tag{1.123}$$

Equation (1.122) can then be solved by iteration yielding

$$\mathbf{R}_p = \tau_{\mathrm{N}} \sum_{n=0}^{\infty} \prod_{k=0}^{n} \mathbf{Q}_{p+k-1}^+ \, \mathbf{\Delta}_{p+k}(s) \mathbf{C}_{p+n}(0). \tag{1.124}$$

Next substituting Eq. (1.124) into Eq. (1.119) yields [144]

$$\tilde{\mathbf{C}}_p = \mathbf{S}_p(s)\tilde{\mathbf{C}}_{p-1} + \tau_{\mathrm{N}} \sum_{n=0}^{\infty} \prod_{k=0}^{n} \mathbf{Q}_{p+k-1}^+ \, \mathbf{\Delta}_{p+k}(s) \mathbf{C}_{p+n}(0). \tag{1.125}$$

Equation (1.125) represents the general solution of the three-term matrix differential-recurrence relation Eq. (1.118) rendered as a *sum of products of matrix continued fractions in the s domain*. The *exact solution* for the Laplace transform $\tilde{\mathbf{C}}_1$ is formally given by [144]

$$\tilde{\mathbf{C}}_1 = \tau_{\mathrm{N}} \left(\mathbf{C}_1(0)\mathbf{\Delta}_1(s) + \sum_{n=1}^{\infty} \prod_{k=1}^{n} \mathbf{Q}_k^+ \mathbf{\Delta}_{k+1}(s) \mathbf{C}_{n+1}(0) \right). \tag{1.126}$$

As a further example, we consider the solution of a five-term recurrence equation using direct matrix iteration instead of continued fractions. The matrix iterative solution method may be easily extended to $(2n + 1)$-term recurrence equations. The solution relies on the fact that $(2n + 1)$-term matrix algebraic recurrence relations can be also solved by using supermatrices containing submatrices as elements. Examples of such solutions will be given in Chapter 2 for given systems. For illustration we commence with the following five-term matrix recurrence equation

$$\mathbf{Q}_p^{--}\tilde{\mathbf{C}}_{p-2} + \mathbf{Q}_p^-\tilde{\mathbf{C}}_{p-1} + \left(\mathbf{Q}_p - s\tau_{\mathrm{N}}\mathbf{I}\right)\tilde{\mathbf{C}}_p + \mathbf{Q}_p^+\tilde{\mathbf{C}}_{p+1} + \mathbf{Q}_p^{++}\tilde{\mathbf{C}}_{p+2} = -\tau_{\mathrm{N}}\mathbf{C}_p(0), \tag{1.127}$$

which may be solved using matrix iteration as follows. First, we again seek the *complementary solution* ($\mathbf{C}_p(0) = 0$) of Eq. (1.127) as the sum

$$\tilde{\mathbf{C}}_p = \mathbf{S}_p \tilde{\mathbf{C}}_{p-1} + \mathbf{T}_p \tilde{\mathbf{C}}_{p-2}, \tag{1.128}$$

which for the particular cases $\tilde{\mathbf{C}}_{p+2}$ and $\tilde{\mathbf{C}}_{p+1}$ becomes

$$\tilde{\mathbf{C}}_{p+1} = \mathbf{S}_{p+1}\tilde{\mathbf{C}}_p + \mathbf{T}_{p+1}\tilde{\mathbf{C}}_{p-1} \text{ and } \tilde{\mathbf{C}}_{p+2} = \mathbf{S}_{p+2}\tilde{\mathbf{C}}_{p+1} + \mathbf{T}_{p+2}\tilde{\mathbf{C}}_p. \tag{1.129}$$

By substituting Eq. (1.129) into Eq. (1.127), we again have Eq. (1.128), save that $\mathbf{T}_p = \boldsymbol{\Delta}_p \mathbf{Q}_p^{--}, \mathbf{S}_p$, and $\boldsymbol{\Delta}_p$ are now rendered in explicit fashion as

$$\mathbf{S}_p = \boldsymbol{\Delta}_p \left(\mathbf{Q}_p^- + \left(\mathbf{Q}_p^+ + \mathbf{Q}_p^{++}\mathbf{S}_{p+2} \right) \boldsymbol{\Delta}_{p+1} \mathbf{Q}_{p+1}^{--} \right), \tag{1.130}$$

$$\boldsymbol{\Delta}_p = -\left[\left(\mathbf{Q}_p - s\tau_{\mathrm{N}}\mathbf{I} \right) + \mathbf{Q}_p^{++}\boldsymbol{\Delta}_{p+2}\mathbf{Q}_{p+2}^{--} + \left(\mathbf{Q}_p^+ + \mathbf{Q}_p^{++}\mathbf{S}_{p+2} \right)\mathbf{S}_{p+1} \right]^{-1}. \tag{1.131}$$

As a test by setting $\mathbf{Q}_p^{\pm\pm} = 0$, we regain the previously known result for the three-term recurrence equation, namely, Eq. (1.120). This procedure completes the *complementary solution*. To find the *particular solution* ($\mathbf{C}_p(0) \neq 0$), we represent $\tilde{\mathbf{C}}_p$ as

$$\tilde{\mathbf{C}}_p = \mathbf{S}_p \tilde{\mathbf{C}}_{p-1} + \boldsymbol{\Delta}_p \mathbf{Q}_p^{--} \tilde{\mathbf{C}}_{p-2} + \mathbf{R}_p. \tag{1.132}$$

Thus, substituting Eqs. (1.132) for $\tilde{\mathbf{C}}_p, \tilde{\mathbf{C}}_{p+1}$, and $\tilde{\mathbf{C}}_{p+2}$ into Eq. (1.127) we have that solution rendered as

$$\begin{aligned} \mathbf{R}_p = &-\left[\mathbf{Q}_p + \mathbf{Q}_p^{++}\boldsymbol{\Delta}_{p+2}\mathbf{Q}_{p+2}^{--} + \left(\mathbf{Q}_p^+ + \mathbf{Q}_p^{++}\mathbf{S}_{p+2} \right)\mathbf{S}_{p+1} \right]^{-1} \\ &\times \left[\tau_{\mathrm{N}}\mathbf{C}_p(0) + \left(\mathbf{Q}_p^+ + \mathbf{Q}_p^{++}\mathbf{S}_{p+2} \right)\mathbf{R}_{p+1} + \mathbf{Q}_p^{++}\mathbf{R}_{p+2} \right]. \end{aligned} \tag{1.133}$$

Again, as a check by setting $\mathbf{Q}_p^{\pm\pm} = 0$, we regain the previous result, Eq. (1.122), for the three-term recurrence equation.

One can also calculate via matrix continued fractions the smallest nonvanishing eigenvalue(s) λ_1 (yielding the reversal time of the magnetization) from the secular or characteristic equation [145]

$$\det\left(\lambda_1 \tau_{\mathrm{N}} \mathbf{I} - \mathbf{S} \right) = 0, \tag{1.134}$$

where the matrix $\mathbf{S}$ is defined as the continued fraction

$$\mathbf{S} = -\left[\mathbf{Q}_1 + \mathbf{Q}_1^+ \boldsymbol{\Delta}_2(0)\mathbf{Q}_2^- \right]\left[\mathbf{I} - \mathbf{Q}_1^+ \boldsymbol{\Delta}_2'(0)\mathbf{Q}_2^- \right]^{-1} \tag{1.135}$$

and the prime designates the derivative of $\mathbf{\Delta}_2(s)$ with respect to $\tau_N s$. Thus λ_1 is the smallest nonvanishing eigenvalue of the (formally) known matrix **S**.

We emphasize that in practical applications, such as magnetization reversal, these fractions are much better suited to numerical calculations than standard direct matrix inversion based on the representation of the infinite system of linear differential-recurrence relations for the averaged spherical harmonics as the matrix differential equation, Eq. (1.112). The simplification afforded by the matrix continued fraction method occurs because the *dimensions of the matrix continued fractions involved are all much smaller than that of the system matrix* **A** in Eq. (1.112). Thus, that method yields benchmark solutions enabling verification of results from complementary methods such as Brownian-dynamics simulations and escape-rate theory. Examples of applications of matrix continued fractions to magnetization relaxation in nanomagnets are given in, e.g., Refs. [31–35,37,38,47,69,146–164]. Some of these results will be summarized in Chapter 2. However, for very small $\alpha < 0.001$, *and/or very high potential barriers*, the continued-fraction method although *formally exact* may not be applicable at all because the matrices involved become ill-conditioned, meaning that numerical inversions are no longer possible. Thus, alternative approximate methods (e.g., escape-rate theory) must be sought which we now describe.

1.3.6 *Calculation of observables*

By numerically solving the foregoing matrix differential-recurrence relations, we can determine relevant observables, e.g., the characteristic times of the magnetization decay, the components of the complex magnetic susceptibility tensor $\hat{\chi}(\omega)$, the equilibrium correlation functions of the longitudinal and transverse components of the magnetization, and so on. The longitudinal and transverse components of the magnetic susceptibility tensor, $\chi_\parallel(\omega)$ and $\chi_\perp(\omega)$, of a superparamagnetic particle are defined using linear response theory as [97]

$$\frac{\chi_\gamma(\omega)}{\chi_\gamma} = 1 - i\omega \int_0^\infty e^{-i\omega t} C_\gamma(t)dt, \quad (\gamma = \parallel, \perp). \tag{1.136}$$

Here

$$\chi_\gamma = \frac{v}{kT}[\langle M_\gamma^2(0)\rangle_0 - \langle M_\gamma(0)\rangle_0^2] \tag{1.137}$$

are the components of the *static* magnetic susceptibility tensor, and $C_\gamma(t)$ is the *equilibrium correlation function* given by

$$C_\gamma(t) = \frac{\langle M_\gamma(0)M_\gamma(t)\rangle_0 - \langle M_\gamma(0)\rangle_0^2}{\langle M_\gamma^2(0)\rangle_0 - \langle M_\gamma(0)\rangle_0^2}, \quad (\gamma = \parallel, \perp) \tag{1.138}$$

where the angular brackets denote the equilibrium ensemble average defined as

$$\langle A\rangle_0 = \int_0^{2\pi}\int_0^{\pi} A(\vartheta,\varphi)W_0(\vartheta,\varphi)\sin\vartheta d\vartheta d\varphi, \tag{1.139}$$

$W_0(\vartheta,\varphi) = Z^{-1}e^{-vV(\vartheta,\varphi)/(kT)}$ is the equilibrium Boltzmann distribution function and Z is the partition function. The cartesian components of the magnetization M_X, M_Y, and M_Z, may be simply written in terms of spherical harmonics as

$$\begin{aligned} M_X &= M_S\sqrt{2\pi/3}\left[Y_{1-1}(\vartheta,\varphi) - Y_{11}(\vartheta,\varphi)\right], \\ M_Y &= iM_S\sqrt{2\pi/3}\left[Y_{1-1}(\vartheta,\varphi) + Y_{11}(\vartheta,\varphi)\right], \\ M_Z &= M_S\sqrt{4\pi/3}Y_{10}(\vartheta,\varphi) \end{aligned} \tag{1.140}$$

(these components are not independent since $M_X^2 + M_Y^2 + M_Z^2 = M_S^2$).

To characterize the time behavior of the equilibrium correlation function $C_\gamma(t)$, we may introduce the *integral relaxation time* $\tau_{\mathrm{int}}^\gamma$, viz.,

$$\tau_{\mathrm{int}}^\gamma = \int_0^\infty C_\gamma(t)dt, \tag{1.141}$$

i.e., the area under the normalized decay curve $C_\gamma(t)$, and the *effective relaxation time* $\tau_{\mathrm{ef}}^\gamma$ defined by

$$\tau_{\mathrm{ef}}^\gamma = -1/\dot{C}_\gamma(0), \tag{1.142}$$

which yields precise information on the initial decay of $C_\gamma(t)$. Also the relaxation times $\tau_{\text{int}}^{\gamma}$ and $\tau_{\text{ef}}^{\gamma}$ may equivalently be defined using the eigenvalues (λ_k^γ) of the Fokker–Planck operator because $C_\gamma(t)$ may formally be written as an infinite series of decaying exponentials, viz.,

$$C_\gamma(t) = \sum\nolimits_k c_k^\gamma e^{-\lambda_k^\gamma t}, \tag{1.143}$$

so that, from Eqs. (1.141)–(1.143),

$$\tau_{\text{int}}^{\gamma} = \sum\nolimits_k c_k^\gamma / \lambda_k^\gamma \text{ and } \tau_{\text{ef}}^{\gamma} = \left(\sum\nolimits_k \lambda_k^\gamma c_k^\gamma\right)^{-1}. \tag{1.144}$$

These times obviously contain contributions from *all* the eigenvalues λ_k^γ. Therefore, in general, to evaluate $C_\gamma(t)$, $\tau_{\text{int}}^{\gamma}$, and $\tau_{\text{ef}}^{\gamma}$ numerically, knowledge of all the λ_k^γ and c_k^γ is required. However, the effective relaxation times $\tau_{\text{ef}}^{\gamma}$ can also (and indeed much more easily) be evaluated in terms of equilibrium averages from the exact analytic equations [97]

$$\tau_{\text{ef}}^{\parallel} = 2\tau_{\text{N}} \frac{\langle M_\parallel^2 \rangle_0 - \langle M_\parallel \rangle_0^2}{M_{\text{S}}^2 - \langle M_\parallel^2 \rangle_0}, \quad \tau_{\text{ef}}^{\perp} = 2\tau_{\text{N}} \frac{\langle M_\perp^2 \rangle_0}{2M_{\text{S}}^2 - \langle M_\perp^2 \rangle_0}. \tag{1.145}$$

According to Eq. (1.136), the behavior of $\chi_\gamma(\omega)$ in the frequency domain is completely determined by the time behavior of $C_\gamma(t)$. Furthermore Eqs. (1.136) and (1.143) defining $C_\gamma(t)$ allow us to formally write the dynamic susceptibility $\chi_\gamma(\omega)$ as an infinite sum of Lorentzians

$$\frac{\chi_\gamma(\omega)}{\chi_\gamma} = \sum_k \frac{c_k^\gamma}{1 + i\omega / \lambda_k^\gamma}. \tag{1.146}$$

Now in both the low- ($\omega \to 0$) and high-frequency ($\omega \to \infty$) limits, we have from Eq. (1.146)

$$\frac{\chi_\gamma(\omega)}{\chi_\gamma} = 1 - i\omega\tau_{\text{int}}^{\gamma} + \cdots, \quad \frac{\chi_\gamma(\omega)}{\chi_\gamma} = -\frac{i}{\omega\tau_{\text{ef}}^{\gamma}} + \cdots. \tag{1.147}$$

Thus both the low- and high-frequency behavior of $\chi_\gamma(\omega)$ are completely determined by the integral $\tau_{\text{int}}^{\gamma}$ and effective $\tau_{\text{ef}}^{\gamma}$, relaxation times respectively, as in the dielectric case [97].

Considering the *longitudinal* relaxation, we may also use a two-mode approximation to understand the qualitative behavior of $\chi_{\parallel}(\omega)$. Here the *smallest* nonvanishing eigenvalue(s) $\lambda_1^{\parallel}$ of the Fokker–Planck operator may be directly associated with the *longest* relaxation (reversal) time of the magnetization τ, characterizing the *long-time* behavior of $C_{\parallel}(t) \sim e^{-t/\tau}$, $t >> \tau = 1/\lambda_1^{\parallel}$. The reversal time can be calculated numerically via Eq. (1.114) or Eq. (1.134). Thus $C_{\parallel}(t)$, which in general comprises an *infinite* number of decaying exponentials (see Eq. (1.143)), may be approximated by *just* two exponentials, viz.,

$$C_{\parallel}(t) \approx \Delta_1 e^{-t/\tau} + (1-\Delta_1)e^{-t/\tau_W}, \tag{1.148}$$

where Δ_1 and τ_W are expressed in terms of $\tau_{\text{int}}^{\parallel}, \tau_{\text{ef}}^{\parallel}$, and τ as

$$\Delta_1 = \frac{\tau_{\text{int}}^{\parallel} / \tau_{\text{ef}}^{\parallel} - 1}{\tau_{\text{int}}^{\parallel} / \tau - 2 + \tau / \tau_{\text{ef}}^{\parallel}}, \quad \tau_W = \frac{\tau_{\text{int}}^{\parallel} - \tau}{1 - \tau / \tau_{\text{ef}}^{\parallel}}. \tag{1.149}$$

Thus, the longitudinal dynamic susceptibility $\chi_{\parallel}(\omega)$, hitherto rendered in a relatively complicated form as an *infinite series* of Lorentzians, may be effectively approximated by a sum of *two* Lorentzians

$$\frac{\chi_{\parallel}(\omega)}{\chi_{\parallel}} \approx \frac{\Delta_1}{1+i\omega\tau} + \frac{1-\Delta_1}{1+i\omega\tau_W}. \tag{1.150}$$

The parameters Δ_1 and τ_W in Eq. (1.149) are determined in a manner ensuring the correct asymptotic behavior of $\chi_{\parallel}(\omega)$ in the extremes of very low and very high frequencies. However, if we desire only the *low-frequency* behavior of $\chi_{\parallel}(\omega)$, $\omega\tau \leq 1$, we may always use just a single-mode approximation, viz.,

$$\frac{\chi_{\parallel}(\omega)}{\chi_{\parallel}} \approx 1 - \frac{i\omega\tau_{\text{int}}^{\parallel}}{1+i\omega\tau}. \tag{1.151}$$

This equation predicts the correct asymptotic behavior of $\chi_{\parallel}(\omega)$ for low frequencies. Moreover, at low temperatures (i.e., for high potential barriers), the longest relaxation time τ can be explicitly evaluated by the Brown–Kramers escape-rate theory (see Section 1.5).

1.3.7 *Langevin and Fokker–Planck equations for magnetic moments in energy space*

The magnetic relaxation of macrospins with *nonseparable* and *nonadditive* Hamiltonians in the very-low-damping (VLD) regime can be studied in much the same way as that of point particles and rigid inertial rotators [86,165,166], where the Hamiltonians are *separable* and *additive*. Again, the reason is that the pronounced time separation between *fast* precessional and *slow* energy variables which exists in *lightly* damped closed phase-space trajectories (Stoner–Wohlfarth orbits) at energies near the barrier energy allows one to formulate a one-dimensional Fokker–Planck equation with the energy as independent variable. This equation governs the evolution of the energy distribution function and is essentially similar to that derived by Kramers [84] in studying the VLD noise-activated escape rate of Brownian particles from a potential well. Thus, for VLD, we have the magnetization reversal time in closed integral form the derivation of which may be described as follows.

By analogy with the earlier derivation of the energy-controlled diffusion equation for Newtonian point particles or rigid inertial rotators in the VLD regime given in [97], we shall parameterize the instantaneous magnetization direction by the *slow* energy variable $E = vV/(kT)$ and the *fast* precessional phase variable ϕ running uniformly along a *closed* Stoner–Wohlfarth orbit of energy E, implying that $d\phi = f_E\, dt$, where f_E is the frequency of precession in the potential well at a given energy E [86,165,166]. To transform the vector magnetic Langevin equation, Eq. (1.42), to scalar Langevin equations for the energy E and phase ϕ, notice that the energy-angle parametrization $\mathbf{M} = \mathbf{M}(E, \phi)$ is specified by the following relations [86,165,166], namely,

$$\frac{\partial \mathbf{M}}{\partial \phi} = \frac{\gamma}{f_E}[\mathbf{H}\times\mathbf{M}], \quad \frac{\partial \mathbf{M}}{\partial E} = -\frac{kT}{v\mu_0}\frac{[\mathbf{M}\times[\mathbf{H}\times\mathbf{M}]]}{\|[\mathbf{H}\times\mathbf{M}]\|^2}. \tag{1.152}$$

Clearly the *locally orthogonal* coordinates (E, ϕ) satisfy the orthogonality conditions $(\partial_{\mathbf{M}}E\cdot\partial_E\mathbf{M}) = 1$ and $(\partial_\phi\mathbf{M}\cdot\partial_E\mathbf{M}) = 0$. Equation (1.152) further implies that the equations of motion of $E\,(t)$ and $\phi\,(t)$ are [86]

$$\frac{dE}{dt}=\left(\frac{\partial E}{\partial \mathbf{M}}\cdot\dot{\mathbf{M}}\right)=-\frac{v\mu_0}{kT}\left(\mathbf{H}\cdot\dot{\mathbf{M}}\right),$$
$$\frac{d\phi}{dt}=\frac{f_E}{\gamma}\frac{\left([\mathbf{H}\times\mathbf{M}]\cdot\dot{\mathbf{M}}\right)}{\left|[\mathbf{H}\times\mathbf{M}]\right|^2}. \tag{1.153}$$

On substituting $\dot{\mathbf{M}}$ as given by the magnetic Langevin equation (1.42) into Eq. (1.153), we have

$$\frac{dE}{dt}=\frac{v\gamma\mu_0}{kT}\left[\left(\mathbf{H}\cdot[\mathbf{M}\times\mathbf{H}]\right)-\frac{\alpha}{M_S\gamma}\left(\mathbf{H}\cdot\left[\mathbf{M}\times\dot{\mathbf{M}}\right]\right)+\left(\mathbf{H}\cdot[\mathbf{M}\times\mathbf{h}]\right)\right],$$
$$\frac{d\phi}{dt}=f_E\left[1+\frac{\alpha}{M_S}\frac{\left([\mathbf{H}\times\mathbf{M}]\cdot\left[\mathbf{M}\times\dot{\mathbf{M}}\right]\right)}{\gamma\left|[\mathbf{H}\times\mathbf{M}]\right|^2}+\frac{\left([\mathbf{H}\times\mathbf{M}]\cdot[\mathbf{h}\times\mathbf{M}]\right)}{\left|[\mathbf{H}\times\mathbf{M}]\right|^2}\right]. \tag{1.154}$$

Equations (1.154) are just Langevin equations for the new *random variables* E and ϕ with *multiplicative noise*, which describe the precession of the magnetization subject to *weak frictional* forces and *weak internal* fluctuations since α and $\mathbf{h}(t)$ are small. Because the first term on the right hand side of Eq. (1.154) is zero due to the definition of the triple scalar product and in the VLD regime all terms of order α^2, α^3, etc. in Eqs. (1.154) may be neglected, on using Eq. (1.152), we then have the simplified equations [86,165,166]

$$\frac{dE}{dt}=-\frac{v\alpha\mu_0}{\gamma kTM_S}\left|\dot{\mathbf{M}}_{\mathrm{pr}}\right|^2+\frac{v\mu_0 f_E}{kT}\left(\frac{\partial\mathbf{M}}{\partial\phi}\cdot\mathbf{h}\right), \tag{1.155}$$

$$\frac{d\phi}{dt}=f_E-\frac{v\mu_0 f_E}{kT}\left(\frac{\partial\mathbf{M}}{\partial E}\cdot\mathbf{h}\right). \tag{1.156}$$

The Gilbert damping term drops out of Eq. (1.156) for the phase variable ϕ because the phase dynamics in the VLD regime are effectively governed by the uniform precession. Thus, in Eq. (1.155) and subsequently, $\dot{\mathbf{M}}_{\mathrm{pr}}$ must be interpreted in the *conservative* Larmor sense as $\dot{\mathbf{M}}_{\mathrm{pr}}=\gamma[\mathbf{H}\times\mathbf{M}]$. Hence the f_E can be calculated via the corresponding energy-dependent precession period $P_E=(f_E)^{-1}$. This is accomplished by expressing the time interval dt in terms of an element of orbit $d\mathbf{M}$ in the evolution equation for $\mathbf{M}$ in the absence of damping, Eq. (1.4). On taking the scalar products $\left(\dot{\mathbf{M}}\cdot[\mathbf{H}\times\mathbf{M}]\right)=\gamma\left|\mathbf{H}\times\mathbf{M}\right|^2$ and rearranging them, we have [86,165,166] for the elementary time interval

$$dt=\frac{1}{\gamma\left|\mathbf{H}\times\mathbf{M}\right|^2}\left([\mathbf{H}\times\mathbf{M}]\cdot d\mathbf{M}\right). \tag{1.157}$$

Therefore, we have the precession period P_E explicitly (by taking a closed line integral along a Stoner–Wohlfarth orbit of *constant* energy E), viz., [86,165,166] so that

$$\begin{aligned}P_E&=\oint_E\frac{\left([\mathbf{H}\times\mathbf{M}]\cdot d\mathbf{M}\right)}{\gamma\left|[\mathbf{H}\times\mathbf{M}]\right|^2}\\&=\frac{\mu_0 vM_S}{\gamma kT}\oint_E\frac{\sin\vartheta\dfrac{\partial E}{\partial\vartheta}d\varphi-\dfrac{1}{\sin\vartheta}\dfrac{\partial E}{\partial\varphi}d\vartheta}{\dfrac{1}{\sin^2\vartheta}\left(\dfrac{\partial E}{\partial\varphi}\right)^2+\left(\dfrac{\partial E}{\partial\vartheta}\right)^2}.\end{aligned} \tag{1.158}$$

For illustration, we take the simple uniaxial anisotropy potential $E(\vartheta)=\sigma\cos^2\vartheta$. Then by substituting $E(\vartheta)$ so defined into the general Eq. (1.158), we have

$$\begin{aligned}P_E&=\frac{v\mu_0 M_S}{2\gamma\sigma kT}\oint_E\frac{d\varphi}{\cos\vartheta}\\&=\frac{v\mu_0 M_S}{2\gamma\sigma kT\cos\vartheta_E}\int_0^{2\pi}d\varphi=\frac{\pi v\mu_0 M_S}{\gamma\sigma kT\cos\vartheta_E}.\end{aligned} \tag{1.159}$$

This result can be verified by direct solution of the (undamped) gyromagnetic equation of motion (1.18) which in our example of a uniaxial potential is

$$\frac{d\vartheta}{dt}=0,\ \frac{d\varphi}{dt}=\frac{2\gamma\sigma kT}{v\mu_0 M_S}\cos\vartheta. \tag{1.160}$$

The solution is

$$\vartheta(t)=\vartheta_E,\ \ \varphi(t)=\frac{2\gamma\sigma kT\cos\vartheta_E}{v\mu_0 M_S}t+\varphi(0). \tag{1.161}$$

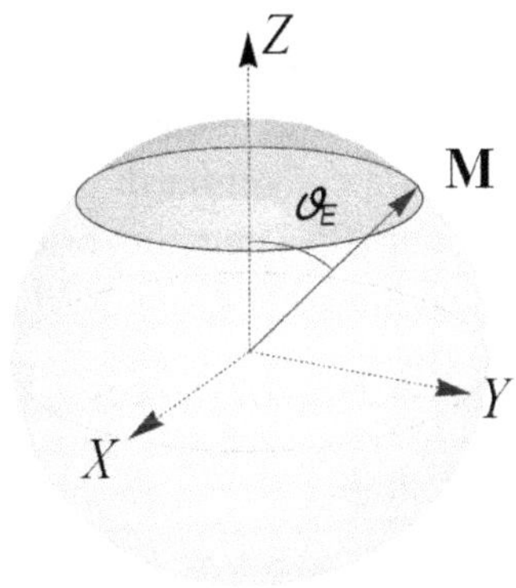

Fig. 1.16. Area of a spherical cap A_E enclosed by a Stoner–Wohlfarth orbit.

Thus, **M** precesses at a constant angle (colatitude) ϑ_E. The period of this precessional motion is then

$$P_E = \frac{2\pi}{\dot{\varphi}(t)} = \frac{\pi v \mu_0 M_S}{\gamma \sigma k T \cos\vartheta_E}. \tag{1.162}$$

Clearly, Eq. (1.162) is in full agreement with Eq. (1.159). We may also give a useful geometrical interpretation of the periodic time P_E as [86]

$$P_E = \frac{v}{\gamma k T M_S}\frac{dA_E}{dE} = \frac{v\mu_0}{\gamma k T M_S^2}\oint_E \left(\mathbf{M}\cdot\left[\frac{\partial \mathbf{M}}{\partial E}\times d\mathbf{M}\right]\right), \tag{1.163}$$

where A_E is the oriented area of the sphere enclosed by a Stoner–Wohlfarth orbit with energy E. On using $\partial\mathbf{M}/\partial E$ from Eq. (1.152) and noticing that

$$\left(\mathbf{M}\cdot\left[\frac{\partial \mathbf{M}}{\partial E}\times d\mathbf{M}\right]\right) = -\frac{kT}{v\mu_0\left|[\mathbf{H}\times\mathbf{M}]\right|^2}\left(\left[\mathbf{M}\times\left[\mathbf{M}\times[\mathbf{H}\times\mathbf{M}]\right]\right]\cdot d\mathbf{M}\right),$$

$$\left[\mathbf{M}\times\left[\mathbf{M}\times[\mathbf{H}\times\mathbf{M}]\right]\right] = M_S^2[\mathbf{M}\times\mathbf{H}],$$

we then have

$$\frac{\mu_0 v}{\gamma k T M_S^2}\oint_E \left(\mathbf{M}\cdot\left[\frac{\partial \mathbf{M}}{\partial E}\times d\mathbf{M}\right]\right) = \frac{1}{\gamma}\oint_E \frac{([\mathbf{M}\times\mathbf{H}]\cdot d\mathbf{M})}{|\mathbf{H}\times\mathbf{M}|^2}, \tag{1.164}$$

so verifying Eq. (1.163). Thus, we have the geometrical interpretation of the periodic time P_E as the time rate of change of the portion of the area of the sphere enclosed by a Stoner–Wohlfarth orbit of constant energy, e.g. consider the uniaxial potential. Then the oriented area of the sphere enclosed by the orbit of the precessing vector **M** is (see Fig. 1.16)

$$A_E = 2\pi\mu_0 M_S^2 (1 - \cos\vartheta_E), \tag{1.165}$$

while the orbital energy is simply

$$E = -\sigma\cos^2\vartheta_E + \text{const.} \tag{1.166}$$

Consequently, the periodic time is

$$P_E = \frac{v}{\gamma kTM_S}\frac{dA_E}{d\vartheta_E}\frac{d\vartheta_E}{dE} = \frac{\pi v\mu_0 M_S}{\gamma\sigma kT\cos\vartheta_E}, \tag{1.167}$$

agreeing with Eqs. (1.159) and (1.162).

We can also evaluate the mean power loss per period (cycle) of a Stoner–Wohlfarth orbit. We have

$$\overline{\dot{E}} = -\frac{v}{kT}\frac{\alpha\mu_0 f_E}{\gamma M_S}\int_0^{P_E}\left(\dot{\mathbf{M}}\cdot\dot{\mathbf{M}}\right)dt, \tag{1.168}$$

where the overbar represents the average over a period. However, the time is measured on a *closed* orbit and since $\overline{\dot{E}}$ is the mean power loss in a precessional cycle, we also have

$$\overline{\dot{E}} = -\frac{v\alpha\mu_0 f_E}{kT\gamma M_S}\oint_E\left(\dot{\mathbf{M}}\cdot d\mathbf{M}\right) = -f_E\overline{\delta E}, \tag{1.169}$$

where $\overline{\delta E}$ is the mean energy loss in such a cycle, i.e.,

$$\overline{\delta E} = \frac{\alpha v\mu_0}{\gamma kTM_S}\oint_E\left(\dot{\mathbf{M}}\cdot d\mathbf{M}\right). \tag{1.170}$$

The Fokker–Planck equation for the probability distribution function $P(E, \phi, t)$ corresponding to the Langevin equations (1.155) and (1.156) for energy and phase variables $\xi_1 = E$ and $\xi_2 = \phi$, which are formally

$$\dot{\xi}_i = F_i + \left(\mathbf{g}_i\cdot\mathbf{h}\right), \tag{1.171}$$

is (cf. Eq. (1.48))

$$\frac{\partial P}{\partial t} = -\sum_i \frac{\partial}{\partial \xi_i}\left[D_i P - \sum_j \frac{\partial}{\partial \xi_j}\left(D_{ij} P\right)\right], \tag{1.172}$$

with *drift* and *diffusion* coefficients D_i and D_{ij} given by

$$\begin{gathered} D_i = F_i + D\left(\mathbf{g}_1 \cdot \frac{\partial \mathbf{g}_i}{\partial E}\right) + D\left(\mathbf{g}_2 \cdot \frac{\partial \mathbf{g}_i}{\partial \phi}\right), \\ D_{ij} = D\left(\mathbf{g}_i \cdot \mathbf{g}_j\right). \end{gathered} \tag{1.173}$$

The diffusion constant itself D is given by Eq. (1.44) while F_i and $\mathbf{g}_i$ are defined by

$$\begin{gathered} F_1(E,\phi) = -\frac{v_0 \alpha \mu_0}{\gamma M_S kT}\left|\dot{\mathbf{M}}_{\mathrm{pr}}\right|^2, \quad F_2(E,\phi) = f_E, \\ \mathbf{g}_1(E,\phi) = \frac{v \mu_0 f_E}{kT}\frac{\partial \mathbf{M}}{\partial \phi}, \\ \mathbf{g}_2(E,\phi) = -\frac{v \mu_0 f_E}{kT}\frac{\partial \mathbf{M}}{\partial E}. \end{gathered} \tag{1.174}$$

Because [86,165,166]

$$\left(\mathbf{g}_2 \cdot \frac{\partial}{\partial \xi_i}\mathbf{g}_1\right) = -\left(\mathbf{g}_1 \cdot \frac{\partial}{\partial \xi_i}\mathbf{g}_2\right) \text{ and } \frac{\partial \mathbf{g}_2}{\partial \phi} = -\frac{\partial \mathbf{g}_1}{\partial E} + \mathbf{g}_1 \frac{\partial \ln f_E}{\partial E}, \tag{1.175}$$

we can write the drift and diffusion coefficients in Eq. (1.173) as

$$\begin{gathered} D_1 = h_1 + \frac{\partial}{\partial E}\left|\mathbf{g}_1\right|^2 - \frac{\partial \ln \Omega_E}{\partial E}\left|\mathbf{g}_1\right|^2, \\ D_2 = h_2 + \frac{\partial}{\partial \phi}\left|\mathbf{g}_2\right|^2, \\ D_{11} = D\left|\mathbf{g}_1\right|^2,\ D_{12} = D_{21} = 0,\ D_{22} = D\left|\mathbf{g}_2\right|^2. \end{gathered} \tag{1.176}$$

Finally substituting Eqs. (1.173)–(1.176) into Eq. (1.172) yields the Fokker–Planck equation for the evolution of P in terms of the two chosen state variables, namely slow E and fast ϕ, viz.,

$$\frac{\partial P}{\partial t}=\alpha\frac{\partial}{\partial E}\left[\frac{v\mu_0}{\gamma M_S kTf_E}\left|\dot{\mathbf{M}}_{\rm pr}\right|^2\left(1+\frac{\partial}{\partial E}\right)f_E P\right]$$
$$+\frac{\partial}{\partial\phi}\left[-f_E P+\frac{v\mu_0\alpha}{\gamma M_S kT}f_E^2\left|\frac{\partial\mathbf{M}}{\partial E}\right|^2\frac{\partial P}{\partial\phi}\right]. \qquad (1.177)$$

The time-dependent Fokker–Planck equation, Eq. (1.177), because two state variables E and ϕ are involved, is difficult to treat and cannot be solved by quadratures. However, in the VLD regime, the energy E diffuses very slowly over time, i.e., is *almost conserved*. In contrast, the phase ϕ varies very rapidly (i.e., the probability density function $P(E, \phi, t)$ nearly equilibrates in ϕ and slowly evolves in E). Thus, the dependence of $P(E, \phi, t)$ on the fast variable ϕ may be eliminated by exploiting the periodicity of $P(E, \phi, t)$ in ϕ along a precessional orbit. This is accomplished by averaging $P(E, \phi, t)$ over ϕ in other words along a *closed* trajectory of the energy yielding [86]

$$W(E,t)=\int_0^1 P(E,\phi,t)\,d\phi. \qquad (1.178)$$

Thus, on introducing the dimensionless action S_E defined as [86]

$$\begin{aligned}S_E&=\frac{v\mu_0}{f_E\gamma kTM_S}\int_0^1\left|\dot{\mathbf{M}}_{\rm pr}\right|^2 d\phi\\ &=\frac{v\mu_0}{\gamma kTM_S}\int_0^{1/f_E}\left|\dot{\mathbf{M}}_{\rm pr}\right|^2 dt \qquad (1.179)\\ &=\frac{v\mu_0}{M_S kT}\oint_E\left([\mathbf{H}\times\mathbf{M}]\cdot d\mathbf{M}\right)\end{aligned}$$

(since $d\phi=f_E\,dt$), we find that after averaging, Eq. (1.177) becomes the following energy-controlled diffusion for the probability density function $W(E, t)$ in energy space, namely [86,165,166]

$$\frac{\partial W}{\partial t}=\alpha\frac{\partial}{\partial E}\left[S_E\left(f_E W+\frac{\partial}{\partial E}(f_E W)\right)\right]. \qquad (1.180)$$

We conclude that the energy-controlled diffusion equation, Eq. (1.180), for spins closely resembles but is not identical to that for point Brownian particles in a potential given in [97]. The differences lie in the definitions of the energy E, damping coefficient α, frequency f_E, and action S_E. All the relevant equations are listed in [97]. Although the respective formulas appear similar, despite this the calculation of f_E and S_E for macrospins is very much more involved than that for Newtonian point particles because it must be undertaken in spherical polar coordinates and the *undamped motion is precession in space rather than the librational motion in a one-dimensional well.* Moreover, in the magnetization problem, the Stoner–Wohlfarth orbits, namely, phase-space orbits at *constant* energy inside the well, can have very complicated forms as *two* degrees of freedom are involved rather than just the *single* one of the original Kramers problem.

1.3.8 *Energy-controlled diffusion equation in the presence of STT*

Invariably STT effects on the magnetization relaxation in nanomagnets manifest themselves only at low damping, $\alpha \ll 1$, simply because their magnitude is governed by the ratio J/α. However, for $\alpha \geq 1$, the STT term in Eq. (1.75) influences but weakly the reversal process because it then becomes very small compared to the damping and random field terms. Just as for zero STT, the magnetization kinetics in the VLD regime may be described via the energy-controlled diffusion equation. Likewise, for nonzero STT, one may parameterize the instantaneous magnetization direction $\mathbf{M} = \mathbf{M}(E, \phi)$ by the *slow* dimensionless energy variable E and the *fast* phase variable ϕ [86] since in the VLD regime, the energy varies very slowly compared to ϕ. For the *slightly damped precession* in the presence of the random field $\mathbf{h}(t)$ with white noise properties given by Eq. (1.43), the magnetic Langevin equations for the *random variables* E and ϕ accounting for the STT can be written as [86,167] (cf. Section 1.3.3)

$$\frac{dE}{dt} = -\frac{v\mu_0}{kT}\left[\frac{\alpha}{\gamma M_S}\left|\dot{\mathbf{M}}_{\mathrm{pr}}\right|^2 - \frac{1}{M_S}\left(\mathbf{I}_S \cdot \left[\mathbf{M} \times \dot{\mathbf{M}}_{\mathrm{pr}}\right]\right) - f_E\left(\frac{\partial \mathbf{M}}{\partial \phi} \cdot \mathbf{h}\right)\right], \quad (1.181)$$

$$\frac{d\phi}{dt} = f_E - \frac{\gamma f_E M_S}{\left|\dot{\mathbf{M}}_{\mathrm{pr}}\right|^2}\left(\mathbf{I}_S \cdot \dot{\mathbf{M}}_{\mathrm{pr}}\right) - \frac{v\mu_0 f_E}{kT}\left(\frac{\partial \mathbf{M}}{\partial E} \cdot \mathbf{h}\right). \quad (1.182)$$

Here we have assumed that in the VLD regime and for small spin-polarized currents, all terms of order $\alpha^2, \alpha J, J^2$, etc. in Eqs. (1.181) and

(1.182), may be neglected. Therefore Eqs. (1.181) and (1.182) describe precession subject to both *weak* damping and STT torques and *weak* thermal fluctuations. On generalizing the results of Section 1.3.7 for zero STT, we have from the Langevin equations, Eqs. (1.181) and (1.182), the two-variable Fokker–Planck equation for the probability distribution function $P(E, \phi, t)$ namely [86,167]

$$\frac{\partial P}{\partial t}=\alpha\frac{\partial}{\partial E}\left[\frac{v\mu_0\left|\dot{\mathbf{M}}_{\mathrm{pr}}\right|^2}{\gamma M_S kTf_E}\left(1+\frac{\partial}{\partial E}\right)f_E P-\frac{v\mu_0}{\alpha M_S kT}\left(\mathbf{I}_S\cdot\left[\mathbf{M}\times\dot{\mathbf{M}}_{\mathrm{pr}}\right]\right)P\right]$$
$$-\frac{\partial}{\partial\phi}f_E\left[P-\frac{\gamma M_S}{\left|\dot{\mathbf{M}}_{\mathrm{pr}}\right|^2}\left(\mathbf{I}_S\cdot\dot{\mathbf{M}}_{\mathrm{pr}}\right)P-\frac{v\mu_0\alpha}{\gamma M_S kT}f_E\left|\frac{\partial\mathbf{M}}{\partial E}\right|^2\frac{\partial P}{\partial\phi}\right]. \quad (1.183)$$

Since in the VLD regime and on long time scales, $f_E >> 1$, ϕ is again a *fast* variable and E is *slow* and is almost conserved (so that $P(E, \phi, t)$ again nearly equilibrates in ϕ and slowly evolves in E), the ϕ dependence in Eq. (1.183) can again be eliminated by exploiting the periodicity of W in ϕ along a precessional orbit formally corresponding to averaging $P(E, \phi, t)$ over ϕ as defined by Eq. (1.178). Thus on introducing the magnetic action S_E given by Eq. (1.179) and the dimensionless work V_E defined as

$$\begin{aligned}V_E&=\frac{v\mu_0}{f_E M_S kT}\oint_0^1\left(\mathbf{I}_S\cdot\left[\mathbf{M}\times\dot{\mathbf{M}}_{\mathrm{pr}}\right]\right)d\phi\\&=\frac{v\mu_0}{M_S kT}\oint_0^{f_E^{-1}}\left(\mathbf{I}_S\cdot\left[\mathbf{M}\times\dot{\mathbf{M}}_{\mathrm{pr}}\right]\right)dt,\end{aligned} \quad (1.184)$$

we see that on averaging along a closed orbit the Fokker-Planck equation (1.183) becomes the energy-controlled diffusion equation for the distribution function $W(E,t)$ [86,167], viz.,

$$\frac{\partial W}{\partial t}=\alpha\frac{\partial}{\partial E}\left[S_E\left(1+\frac{\partial}{\partial E}\right)f_E W\right]-\frac{\partial}{\partial E}\left(V_E f_E W\right). \quad (1.185)$$

In the derivation of Eq. (1.185), it has been assumed throughout that α is independent of $\mathbf{M}$. However, the results may be also generalized to magnetization-dependent damping $\alpha(\mathbf{M})$ [86]. We remark that S_E and V_E

given by Eqs. (1.179) and (1.184) are related to the corresponding terms in the Langevin equation (1.181), namely,

$$-\frac{v\alpha\mu_0}{\gamma kTM_S}\left|\dot{\mathbf{M}}_{pr}\right|^2 \text{ and } \frac{v\mu_0}{kTM_S}\left(\mathbf{I}_S\cdot\left[\mathbf{M}\times\dot{\mathbf{M}}_{pr}\right]\right);$$

these terms account for the *energy dissipation due to the damping* and the *pumping of energy into the system due to the spin-polarized current* and determine the magnetization reversal process and Stoner–Wohlfarth orbits for the steady state precession of the magnetization [86].

1.4 Magnetization relaxation times

1.4.1 *Relaxation processes*

We now describe *magnetic after-effect behavior*; i.e., under what conditions an assembly of single-domain particles can achieve thermal equilibrium in a time that is *short* compared with the time of an experiment. One way of achieving equilibrium is by *physical rotation* of the particles; this can occur if they are suspended in a liquid carrier – this is a *ferrofluid.* Here the factor determining the rate of approach to equilibrium is the *viscosity of the medium in which the particles are suspended* – this mechanism may be treated using the Debye theory of orientational relaxation, as illustrated for example in [97] for electric dipoles. We remark that the conditions for the validity of the Debye theory are *much better satisfied for a ferrofluid than for electric dipoles* of a polar liquid, because the ferrofluid particles, *owing to their relatively large size*, are a much closer approximation to idealized Brownian rotators than are polar molecules. In addition, the dynamical behavior of the particles will manifest itself in the mid-radio frequency (medium wave) region rather than in the microwave range.

In a solid, physical rotation of the particles cannot take place. However, in 1949, Néel [15,16,168] conjectured (as seen) that, if a single-domain particle were *sufficiently small*, thermal fluctuations could cause its *direction of magnetization* to undergo a type of Brownian rotation, so that the stable magnetic behavior characteristic of a ferromagnet would be destroyed. Here, since the relaxing entity is the magnetic moment inside the particle, the inertia of the particle will of course play no role, because no *physical rotation* of the particle occurs, unlike in Debye relaxation of a polar molecule of a liquid. An example given by Brown [18] of a tape recording is of interest: we expect that if we put this recording on a shelf that it will stay in the same magnetic state; we would be surprised if it suddenly jumped from being a recording of Beethoven to a recording of Brahms. In principle, however, [18] the *apparent stability* of the recording is only one of *many* local minima of the free energy: thermal agitation can cause spontaneous jumps from one such state to another. The apparent stability [18] (*ferromagnetic behavior*) arises because our tape or magnet cannot get from one magnetic state to another without passing over an energy barrier which is very large in comparison with the thermal energy *kT*. Thus, the probability per unit time of a jump over such a barrier is so small that the mean time we would have to wait for it to occur *far exceeds our own lifetime*; we perceive stable ferromagnetic behaviour. However, if the barrier is neither very large nor very small in comparison with the noise strength *kT* (our case), then the specimen neither remains in a single stable state for a long time nor attains thermal equilibrium in a short time after a change in field: it instead undergoes a change of magnetization which is not completed instantaneously but lags behind the field. This is called *magnetic after-effect*, or *magnetic viscosity*, or *Néel relaxation*, and occurs only for *sufficiently fine* ferromagnetic particles. In order to illustrate the Néel mechanism [8], consider an assembly of aligned uniaxial particles subject to a field **H**, with potential energy given by Eq. (1.11). Thus, the particles are fully magnetized along the polar axis, which is the axis of symmetry. A sufficiently long time after the field is switched off, the remanence will vanish as

$$M_r(t) = M_r(0)e^{-t/\tau}, \qquad (1.186)$$

which is the longest-lived mode of the relaxation process (here τ is the reversal time of the magnetization). Néel [15,16,168] then suggested that, from transition state theory (TST) [169,170], τ may be evaluated as

$$\tau^{-1} = f_A e^{-\Delta E}, \qquad (1.187)$$

where the attempt frequency f_A is the frequency of the gyromagnetic precession [171] and $\Delta E = E(\vartheta_C) - E(\vartheta_A)$ is the energy barrier so that, by varying the volume or the temperature of the particles, τ can be made to vary from 10^{-9} s to *millions of years* (f_A^{-1} is often taken as small as 10^{-10}–10^{-11} s in practice). The presence of the exponential factor in Eq. (1.187) indicates that, in order to approach the zero remanence (corresponding to thermal equilibrium), a sufficient number of particles (magnetic moments) must be reversed by thermal activation over the energy barrier ΔE. The probability of such a process is proportional to $e^{-\Delta E}$. (Notice that an essentially similar approach was proposed by Debye in his famous book Polar Molecules [111].)

Néel's calculation of τ was later criticized by Brown [17] on two counts: (i) the system is not explicitly treated as a *gyromagnetic one*; and (ii) it relies on a *discrete orientation* approximation. Moreover, the damping dependence of the prefactor is ignored. Brown [17] (as seen) suggested that both of these difficulties could be resolved by constructing the Fokker–Planck equation for the probability distribution function of magnetic moment orientations on the unit sphere from the appropriate Langevin equation for the evolution of the magnetization **M**. He was then able to find, using the Kramers method, an approximate formula for τ in the high-barrier limit for the potential of Eq. (1.11) which agreed with Néel's formula except for the prefactor f_A. This is discussed in detail below.

It is apparent from Eq. (1.187) that the Néel relaxation time τ depends *exponentially on the particle volume*; hence there is a fairly well-defined particle radius above which the magnetization will appear stable. We consider the figures given by Bean and Livingston [8] for a spherical iron particle with uniaxial anisotropy $E \sin^2 \vartheta$. A particle of radius 115 Å will have a relaxation time of 10^{-1} s at 300 K, so that the moment will relax almost instantaneously. A particle of radius 150 Å in contrast will have a relaxation time of 10^9 s and so will be exceedingly stable (i.e., the moment will not reverse in this time). This situation corresponds to an energy barrier that is very large in comparison to kT where, for any reasonable measurement time [17], we may ignore thermal agitation and calculate the static magnetization by simply minimizing E with respect to the polar and azimuthal angles (ϑ, φ) for each value of an applied field $\mathbf{H}_0$. This is the well-known Stoner–Wohlfarth calculation [14]; it *leads to hysteresis because in certain field ranges two or more minima exist and transitions between them are neglected.* Here a typical potential would be [172,173]

$$E(\vartheta,\varphi)=\sigma\sin^2\vartheta-\xi(\cos\vartheta\cos\psi+\sin\psi\sin\vartheta\cos\varphi). \quad (1.188)$$

Here $\sigma = vK/(kT)$ and $\xi = v\mu_0 M_S H_0/(kT)$ are, respectively, the anisotropy and dc field parameters. The polar axis **k** is the easy axis of magnetization; the field $\mathbf{H}_0$ is applied in the *xz* plane at an angle ψ to the easy axis (see Fig. 1.5). Thus, in general, there will be only a narrow range of particle sizes for which the relaxation time will be of the order of experimental times, and for which measurable "magnetic viscosity" effects, manifesting themselves as an *observable* change of magnetization, *lagging behind field changes*, would be expected. Bean and Livingston have given a rough measure of the size of the particle for transition to stable behavior, taking $\tau = 10^2$ s, they find that the energy is $25kT$. The temperature at which this occurs for a given particle is called the *blocking temperature*. They obtain sizes of 40 Å for hcp cobalt, 125 Å for iron, 140 Å for fcc cobalt. Metallic cobalt occurs as two crystallographic structures: hcp (hexagonal close-packed) and fcc (face-centered cubic). We mention that, in an assembly consisting *solely* of single-domain particles, the remanence at a given temperature should be a measure of the amount of material with particle volume greater than the volume that is just stable at this temperature. Thus [8], by following the increase of remanence with decreasing temperature, we can find out how much material lies in various volume ranges, and so determine the particle size distribution.

Néel [15,16] was led to his solid-state mechanism of relaxation, that is *rotation of the magnetic moment inside the particle due to thermal agitation*, through the study of *paleomagnetism*. Much work had been done on measuring the intensity and direction of the remanent magnetization in rocks, with the view that such data would yield information on the strength and direction of the earth's magnetic field at the time the rock was formed, as is well illustrated by the elegant exposition of Blackett [174],

"The detailed study of the natural magnetization of rocks is likely to allow us to trace back to the beginning of geological time both the history of the earth's magnetic field and the motion of continental masses relative to each other and to the

> geographical pole. The ability of the magnetic rocks to "remember" an earlier magnetic field depends on their ability to exist in thermal equilibrium with the earth's field at a stage early in their formation but to be later "frozen" in a state of magnetization stable against later changes of the direction and strength of the field."

Thus [8] the establishment of thermal equilibrium of the rock magnetization may be accomplished by either the Debye or Néel mechanisms. The Debye-like mechanism occurs [8] in sedimentary rocks. There the particles align themselves with the direction of the earth's field by mechanical rotation, while the sediment is still wet and uncompressed. Later the sediment becomes relatively hard rock and this ferrofluid-like behavior is lost, so that Debye relaxation can no longer take place and the magnetization is thus stable against later changes in the earth's field, so preserving the direction of the earth's magnetic field from the epoch in which the sediment was laid down.

In igneous rocks, however, the material becomes magnetic by cooling through its Curie point [8], and there is no mechanical rotation of the particles. At high temperatures, however, the ratio of barrier height to thermal energy is such that Néel relaxation may occur. Clearly, this will be in the early stages of the formation of the rock (e.g., in mountain building periods). As the rocks cool, the magnetization will become stable: the particles will have cooled to a temperature below their blocking temperature and the magnetization will be thus stable against later changes in the earth's field, again preserving the direction of the earth's magnetic field from the epoch in which the rocks have formed. The relaxation times are now of the order of geological times. The rocks thus play the role of "magnetic fossils" (rocks from widely scattered parts of the world but of the same age showing the same magnetic patterns), giving evidence of a past reversal of the earth's magnetic field, records of a time when the northern magnetic pole resided where the southern magnetic pole is today. Reversals are believed to have occurred as many as 25 times during the last 5 million years, with the last such reversal having occurred about 730,000 years ago. In an apt phrase coined in a popular article by Dr. William Reville, "*the earth's magnetic chronicle is written in stone.*"

Another possible mechanism of magnetic relaxation over potential barriers, of much current interest, is *macroscopic quantum tunneling* of the magnetization ("macroscopic" because of the large number of spins involved), which was originally suggested by Bean and Livingston [8]. By this is meant the possibility of transitions of the magnetic moment at absolute zero from a state of complete alignment to a state of zero overall magnetization, owing to quantum tunneling [175–177] of the magnetic moment through the anisotropy potential barrier (see Ref. [20] for a recent review of single-particle measurements).

1.4.2 *Discrete orientation model*

Before we proceed to the more sophisticated treatment of Brown [17] based on the Langevin equation, we shall briefly describe the *discrete orientation model* for the calculation of the Néel relaxation time (cf. the solid-state Debye mechanism [111]). We shall suppose that the energy barriers are so large in comparison with kT that the magnetization is always along one of the directions (ϑ_i, φ_i) of easy magnetization (stable orientations along the local minima of the free energy). Nevertheless, the barriers are not so high as to preclude changes of orientation altogether. Thus, in orientation i there is a probability v_{ij} of a jump in orientation j in unit time. The frequencies v_{ij} depend on an anisotropy parameter K, the applied dc field H_0, and thermal energy kT. If we have a large number n of non-interacting particles, the number of particles n_i changes with time in accordance with the *master equation*

$$\dot{n}_i = \sum_{j\neq i}\left(v_{ji}n_j - v_{ij}n_i\right),\ n = \sum_i n_i. \tag{1.189}$$

The total number of particles is conserved ($\dot{n}=0$). Let us now suppose that we have only two orientations as for a uniaxial anisotropy given by Eq. (1.11). We let 1 refer to the positive orientation and 2 to the negative, so that Eq. (1.189) reduces to

$$\dot{n}_1 = -\dot{n}_2 = v_{21}n_2 - v_{12}n_1. \tag{1.190}$$

Hence, we have the evolution equation [110]

$$\frac{d}{dt}(n_2 - n_1) = -(v_{21} + v_{12})(n_2 - n_1) + (v_{12} - v_{21})n,$$

so that n_1 and n_2 and the relative magnetization $M / M_S = n_1 - n_2$ approach their final values when ν_{12} and ν_{21} are constant, according to the factor $e^{-t/\tau} = e^{-(\nu_{12}+\nu_{21})t}$ that is, with time constant or reversal time

$$\tau = (\nu_{12} + \nu_{21})^{-1}. \tag{1.191}$$

For two equivalent wells with one saddle point, where $\nu_{12} = \nu_{21} = \Gamma$, e.g., for uniaxial anisotropy with a transverse magnetic field (see Fig. 1.3(a)), we have $\tau = (2\Gamma)^{-1}$. However, for two equivalent wells with two saddle points, e.g., for biaxial anisotropy (see Fig. 1.3(b)), we have $\tau = (4\Gamma)^{-1}$. Here the factor 4 occurs because (i) there are *two* escape routes from the well over the saddle points, and (ii) *two* equivalent wells are involved in the relaxation process.

For more than two orientations, the case of greatest interest is when the free energy per unit volume arises from cubic anisotropy, namely

$$V = K_c(u_1^2 u_2^2 + u_2^2 u_3^2 + u_3^2 u_1^2), \tag{1.192}$$

where the u_i denote the direction cosines with respect to the cubic axes. For $K_c > 0$ (Fe-type crystals), V has minima at six orientations of type $\{100\}$ (i.e., $u_1 = 1, u_2 = u_3 = 0$, **M** along a cube edge of the lattice) as shown in Fig. 1.3(c). It has maxima at eight orientations of type $\{111\}$ (**M** along a body diagonal) and saddle points at twelve orientations of type $\{110\}$ (**M** along a face diagonal). For $K_c < 0$ (Ni-type crystals), the maxima and minima are interchanged. The values of V at the orientations $\{100\}$, $\{110\}$, and $\{111\}$ are 0, $K_c/4$, and $K_c/3$, respectively [108].

The case of *positive cubic anisotropy* $K_c>0$ is shown in Fig. 1.3(c). Let n_1, n_2, and n_3 denote the numbers of particles at the $\{100\}$, $\{010\}$, and $\{001\}$ orientations, respectively, and n_{-1}, n_{-2}, and n_{-3} denote the corresponding numbers in the opposite orientations [108] (see Fig. 1.3(c)). To get from orientation 1 to orientation 2, a particle must surmount an energy barrier with as lowest point, the saddle point at $\{110\}$. To get from orientation 1 to orientation −1, it must surmount two successive energy barriers. If the energy barriers are high, it will be unlikely to achieve this in just a *single* event, and thus we may take

$$\nu_{ij} = 0 \quad \text{for} \quad j = \pm i \quad \text{and} \quad \nu_{ij} = \Gamma \quad \text{for} \quad j \neq \pm i. \tag{1.193}$$

Equation (1.189) then becomes

$$\dot{n}_{\pm i} = \Gamma(n_j + n_k + n_{-j} + n_{-k} - 4n_{\pm i}), \ i \neq j \neq k, \ i,j,k \in \{1,2,3\}. \quad (1.194)$$

Setting $x_i = n_i + n_{-i}$ and $y_i = n_i - n_{-i}$, we obtain by addition and subtraction

$$\dot{x}_i = 2\Gamma(x_j + x_k - 2x_i), \ i \neq j \neq k, \ i,j,k \in \{1,2,3\}, \quad (1.195)$$

$$\dot{y}_i = -4\Gamma y_i, \ i \in \{1,2,3\}. \quad (1.196)$$

Hence each component of the magnetization, $M_i = M_S y_i / n$, decays with time constant $(4\Gamma)^{-1}$. By symmetry the equilibrium values attained at $t = \infty$ are $n_i = n/6$ so that $x_i = 1/3$ and $y_i = 0$, and the solutions of Eq. (1.196) are

$$y_i(t) = y_i(0)e^{-4\Gamma t}, \ i \in \{1,2,3\}. \quad (1.197)$$

To solve Eq. (1.195), we set $x_j + x_k = n - x_i$ so that

$$\dot{x}_i = 2\Gamma(n - 3x_i), \ i \in \{1,2,3\}. \quad (1.198)$$

The solutions are then

$$x_i(t) = \left(x_i(0) - \frac{n}{3}\right)e^{-6\Gamma t} + \frac{n}{3}, \ i \in \{1,2,3\}. \quad (1.199)$$

Thus the deviations of the $x_i(t)$ from their equilibrium values $n/3$ decay with the time constant $(6\Gamma)^{-1}$, while the time behavior of the $n_i(t)$ is now governed by *two* time constants, namely, $(4\Gamma)^{-1}$ and $(6\Gamma)^{-1}$.

The *negative cubic anisotropy* situation $K_c < 0$ is shown in Fig. 1.3(d). The stable magnetization orientations now correspond to the eight corners of a cube. Let n_1, n_2, n_3 and n_4 denote the numbers of particles at the {111}, $\{11\bar{1}\}$, $\{1\bar{1}1\}$, and $\{\bar{1}11]\}$ orientations respectively and n_{-1}, n_{-2}, n_{-3}, and n_{-4} denote the corresponding numbers in the opposite orientations [108]; see Fig. 1.3(d). Similar to the situation $K_c > 0$, we suppose that only one barrier at a time can be surmounted. If we let *i*.ADJ.*j* mean that the subscripts *i* and *j* correspond to adjacent minima and *i*.NA.*j* mean the opposite, then [108]

$$\nu_{ij} = 0, \quad i.\mathrm{NA}.j, \quad \nu_{ij} = \Gamma, \quad i.\mathrm{ADJ}.j. \quad (1.200)$$

Equation (1.189) then becomes [108]

$$\dot{n}_{\pm i} = \Gamma\left(n_j + n_k + n_l - 3n_i\right), \ \ i.\text{ADJ}.j,k,l \ \ i \neq j \neq k \neq l. \quad (1.201)$$

By setting $x_i = n_i + n_{-i}$ and $y_i = n_i - n_{-i}$, we obtain by addition and subtraction

$$\dot{x}_i = \Gamma(x_j + x_k + x_l - 3x_i), \ \ i \neq j \neq k \neq l, \quad (1.202)$$

$$\begin{aligned} \dot{y}_1 &= \Gamma(-3y_1 + y_2 + y_3 + y_4), \\ \dot{y}_2 &= \Gamma(y_1 - 3y_2 - y_3 - y_4), \\ \dot{y}_3 &= \Gamma(y_1 - y_2 - 3y_3 - y_4), \\ \dot{y}_4 &= \Gamma(y_1 - y_2 - y_3 - 3y_4). \end{aligned} \quad (1.203)$$

Setting $x_j + x_k + x_l = n - x_i$ in Eq. (1.202) yields

$$\dot{x}_i = \Gamma(n - 4x_i). \quad (1.204)$$

Hence the x_i approach their equilibrium value $n/4$ with time constant $(4\Gamma)^{-1}$ and Eq. (1.203) can be expressed in matrix form as

$$\frac{d}{dt}\begin{pmatrix} y_1 \\ y_2 \\ y_3 \\ y_4 \end{pmatrix} = \Gamma \begin{pmatrix} -3 & 1 & 1 & 1 \\ 1 & -3 & -1 & -1 \\ 1 & -1 & -3 & -1 \\ 1 & -1 & -1 & -3 \end{pmatrix} \begin{pmatrix} y_1 \\ y_2 \\ y_3 \\ y_4 \end{pmatrix}. \quad (1.205)$$

Assuming a solution of this linear differential equation of the form

$$y_i(t) = y_i(0)e^{-\lambda_i \Gamma t}, \ \ i \in \{1,2,3,4\}, \quad (1.206)$$

requires that λ_i is an eigenvalue of the system matrix in Eq. (1.205). The eigenvalues are the solutions of the deterministic equation [108]

$$\begin{vmatrix} -3+\lambda & 1 & 1 & 1 \\ 1 & -3+\lambda & -1 & -1 \\ 1 & -1 & -3+\lambda & -1 \\ 1 & -1 & -1 & -3+\lambda \end{vmatrix} = (\lambda - 2)^3(\lambda - 6) = 0. \quad (1.207)$$

The eigenvalues are $\lambda = 2$ and $\lambda = 6$. Thus, in general, the decay of the n_i is governed by *three-time* constants, namely, $(2\Gamma)^{-1}$, $(4\Gamma)^{-1}$, and $(6\Gamma)^{-1}$.

The X component of the magnetization M_X is proportional to $y_1 + y_2 + y_3 - y_4$. Next from Eqs. (1.203), we see that [108]

$$\frac{d}{dt}(y_1 + y_2 + y_3 - y_4) = -2\Gamma(y_1 + y_2 + y_3 - y_4), \tag{1.208}$$

so that the rate of change of $y_1 + y_2 + y_3 - y_4$ is given by -2Γ times the quantity itself; thus each component of the magnetization, M_X (or M_Y or M_Z) decays with time constant $(2\Gamma)^{-1}$.

Furthermore if ν_{ij}^0 is the frequency of oscillation of a particle in a potential well, the probability per second (relative frequency) for the flip of a particle from orientation *i* to orientation *j* is given by

$$\nu_{ij} = \nu_{ij}^0 e^{-(E_C - E_i)}, \quad (i = 1, j = 2 \text{ or } i = 2, j = 1), \tag{1.209}$$

where E_i is the free energy density in orientation *i*, and E_C is the free energy density at the top of the barrier between the orientations *i* and *j*; *v* is, as usual, the particle volume. The frequencies ν_{ij}^0, if they vary with temperature, are assumed to do so slowly in comparison with the exponential factor, and are often taken to be constant, although Néel [15,16] has calculated them explicitly (see Refs. [17,18,168]). We reiterate that, regardless of the precise form of the prefactor ν_{ij}^0, if the ratio v/T changes by a factor of less than three in a certain critical part of its range, the time constant, Eq. (1.191), changes from 10^{-1} to 10^9 s. Thus, to a good approximation [18], we may reaffirm that a *critical volume* v_C exists such that particles with $v < v_c$ are *superparamagnetic*. We reiterate that the discrete orientation model of overbarrier relaxation was originally proposed for dielectric relaxation in polar crystals by Debye [111] and extensively developed by Fröhlich [110].

1.4.3 *Mean first-passage time (MFPT)*

First-passage time is the time at which a random variable $\xi(t)$ first leaves a given domain [143]. Clearly the first-passage time is also a random variable, as it varies from realization to realization of the process, and we may speak of a distribution function of first-passage times so that one may compute a *mean first-passage time* (MFPT) [85,143,178–180]. The MPFT for a stochastic process is defined as the time when that process, starting from a given point, reaches a *predetermined* level for the first time.

For simplicity, we consider only the one-dimensional case of a random variable $\xi(t)$ starting at instant $t = 0$ with $\xi(0) = x'$ say; the first-passage time T_1 is then the time when $\xi(t)$ reaches a boundary for the first time. Suppose that the boundaries (x_1, x_2) are *absorbing*, then $\xi(T_1) = x_2$. Since the first-passage time T_1 varies from realization to realization of the process, T_1 is itself a random variable. The conditional probability distribution function underlying $\xi(t)$ is then $W(x, t|x', 0)$, so that $W\,dx$ is the probability for the random variable $\xi(t)$, starting at $t = 0$ from the fixed point $\xi(0) = x'$, to reach x at time t. Furthermore, if $\xi(t)$ reaches either x_1 or x_2 for the first time, we no longer count its realizations. Hence, the probability distribution function W must satisfy the Fokker–Planck equation

$$\frac{\partial W}{\partial t} = \mathrm{L}_{\mathrm{FP}}(x)W \tag{1.210}$$

with the boundary and initial conditions

$$W(x,t|x',0) = 0 \text{ for } x = x_1 \text{ and } x = x_2 \tag{1.211}$$

and

$$W(x,t|x',0) = \delta(x - x'). \tag{1.212}$$

The probability $Q(x, t)$ of realizations that have started at $x = x'$, and which have not yet reached either boundary in a given time t, is given by

$$Q(x',t) = \int_{x_1}^{x_2} W(x,t|x',0)dx. \tag{1.213}$$

The probability dQ of a realization reaching one of the boundaries in a time interval $[t, t + dt]$ is thus

$$dQ = \int_{x_1}^{x_2} \partial_t W(x,t|x',0)dxdt. \tag{1.214}$$

Thus, the distribution function $q(x',T_1)$ for the first-passage time T_1 is

$$q(x',T_1) = -\frac{dQ(x',T_1)}{dT_1} = -\frac{d}{dT_1}\int_{x_1}^{x_2} W(x,T_1|x',0)dx. \tag{1.215}$$

The moments of the first-passage time distribution are [143]

$$\langle T_1^n \rangle(x') = \int_0^\infty T_1^n q(x',T_1)dT = \int_{x_1}^{x_2} p_n(x,x')dx, \tag{1.216}$$

where $p_0(x,x') = \delta(x-x')$ since the particle was at $x = x'$ at time $t = 0$. Furthermore, by integrating by parts, we have

$$p_n(x,x') = n\int_0^\infty T_1^{n-1} W(x,T_1|x',0)dT \quad (n \geq 1). \tag{1.217}$$

Hence, using the Fokker–Planck operator L_{FP} in Eq. (1.217), we have the set of coupled differential equations

$$L_{FP}(x)p_n(x,x') = -np_{n-1}(x,x'). \tag{1.218}$$

The first equation of the set, Eq. (1.218), viz.,

$$L_{FP}(x)p_1(x,x') = -\delta(x-x')$$

describes the stationary probability density [143]. The formal solution of Eq. (1.218) is given by [85]

$$p_n(x,x') = -n![L_{FP}(x)]^{-n}\,\delta(x-x'). \tag{1.219}$$

Now because $W(x,t|x',0) = W(x,0|x',-t)$ also satisfies the *backward Kolmogorov equation*, namely [143]

$$\frac{\partial W}{\partial t} = L_{FP}^{\dagger}(x')W. \tag{1.220}$$

Therefore, we can replace the Fokker–Planck operator $L_{FP}(x)$ in Eqs. (1.218) and (1.219) by its *adjoint* operator $L_{FP}^{\dagger}(x')$ acting on x'. Hence to calculate the moments $\langle T_1^n \rangle(x')$, Eq. (1.216), we may also use the adjoint Eq. (1.220) yielding [143]

$$\mathrm{L}_{\mathrm{FP}}^{\dagger}(x')\langle T_1^n\rangle(x') = -n\langle T_1^{n-1}\rangle(x'), \quad n \geq 1. \tag{1.221}$$

Thus, the equation for the MFPT $\langle T_1 \rangle(x')$ is

$$\mathrm{L}_{\mathrm{FP}}^{\dagger}(x')\langle T_1 \rangle(x') = -1. \tag{1.222}$$

For further details see Hänggi *et al.* [85], where a historical overview is also given. However, like the calculation of the correlation time, the MFPT may be determined by quadratures for one-dimensional or quasi-one-dimensional potentials, otherwise the problem may only be solved in the high-barrier limit by regarding the MFPT as the inverse of the Kramers escape rate.

We shall now discuss the MFPT problem in the context of magnetization relaxation of fine ferromagnetic particles, as this is perhaps the most striking example of its application [179]. We recall that a sufficiently fine ferromagnetic particle with uniaxial anisotropy potential $E(\vartheta)$ which has circular symmetry (a type of continuous symmetry for a surface that can be rotated about an axis through any arbitrary angle and map onto itself) comprises a single magnetic domain that possesses two stable magnetic states (see Fig. 1.3(a)). The orientation of the magnetization of such a particle may undergo a reversal, i.e., cross the internal anisotropy potential barrier, due to thermal agitation; thus the relaxation is governed by a one-space-variable Fokker–Planck equation, Eq. (1.64), for the probability distribution function in the space of orientations of the magnetization only, viz.,

$$\frac{\partial W}{\partial t} = \frac{1}{2\tau_{\mathrm{N}} \sin\vartheta} \frac{\partial}{\partial \vartheta}\left[\sin\vartheta\left(\frac{\partial W}{\partial \vartheta} + W\frac{\partial E}{\partial \vartheta}\right)\right], \tag{1.223}$$

where the abbreviation $vV(\vartheta)/kT \rightarrow E(\vartheta)$ is used. The time to cross the potential barrier, i.e. the reversal time τ has usually been obtained from the inverse of the smallest nonvanishing eigenvalue λ_1 of the Fokker–Planck operator [17]. Thus, it has been assumed that the relaxation process across the barrier is described by the *single* eigenmode associated with λ_1, which, in turn, means that in the high-barrier limit Kramers' escape-rate theory [97] may be used to obtain an approximate expression for the reversal time (see Section 1.5). This formulation, however, has the disadvantage that it is impossible to give an exact analytical formula for the reversal time, which is valid for all barrier heights. Furthermore, particular difficulties arise for moderate barrier heights, where neither the low- nor high-barrier approximations may be safely used [179]. All the foregoing defects may, however, be overcome via the MFPT [85,102,143,178,179], because it allows one to give an *exact* finite integral representation of the reversal time as we shall now demonstrate (a very detailed discussion of MFPTs and their relation to λ_1^{-1} is given by Gardiner [102], Chapters 5 and 9). The corresponding problem of noninertial rotational Brownian motion and dielectric relaxation of nematic liquid crystals (see Ref. [97]) may be treated simply by replacing magnetic quantities by appropriate molecular analogs. However, the MFPT (for uniaxial magnetic nanoparticles only) has due to symmetry the interesting property that it is valid for all values of the damping parameter α, unlike that for the dielectric relaxation problem, which is valid in the overdamped (where inertial effects may be ignored) regime only.

1.4.4 *MFPT for circularly symmetric free energy*

Here we describe the solution for the MFPT for the particular case of *circularly symmetric potentials* $E(\vartheta)$ (in 3-dimensions circular symmetry about an axis is also called cylindrical symmetry) with two wells at $\vartheta_A = 0$ and $\vartheta_B = \pi$ separated by a potential barrier at ϑ_C Eq.(1.11). The MFPT, $\langle T_1 \rangle(\vartheta')$, for a magnetic moment starting at a definite orientation ϑ' to leave either of the regions ($\vartheta_A \le \vartheta \le \vartheta_C$) and ($\vartheta_C \le \vartheta \le \vartheta_B$) may be calculated by solving Eq. (1.222), which now becomes

$$\mathrm{L}_{\mathrm{FP}}^{\dagger}(\vartheta')\langle T_1 \rangle(\vartheta') = -1, \tag{1.224}$$

where the adjoint Fokker–Planck operator $\mathrm{L}_{\mathrm{FP}}^{\dagger}$ is given by

$$\mathrm{L}_{\mathrm{FP}}^{\dagger} = \frac{e^{E(\vartheta')}}{2\tau_{\mathrm{N}} \sin\vartheta'} \frac{\partial}{\partial\vartheta'}\left[\sin\vartheta' e^{-E(\vartheta')} \frac{\partial}{\partial\vartheta'}\right]. \tag{1.225}$$

In the first domain ($\vartheta_A \le \vartheta \le \vartheta_C$) Eq. (1.224) can be integrated subject to the boundary condition $\langle T_1 \rangle(\vartheta_C) = 0$ yielding

$$\langle T_1\rangle(\vartheta') = 2\tau_{\mathrm{N}} \int_{\vartheta'}^{\vartheta_C} \csc\vartheta e^{E(\vartheta)} \int_{\vartheta_A}^{\vartheta} e^{-E(\vartheta'')} \sin\vartheta'' d\vartheta'' d\vartheta. \tag{1.226}$$

For transitions from the second region ($\vartheta_C \le \vartheta \le \vartheta_B$) the lower limit ϑ_A in the inner integral in Eq. (1.226), must be replaced by ϑ_B. Thus setting $\vartheta' = \vartheta_A$ and $\vartheta' = \vartheta_B$ for the domains ($\vartheta_A \le \vartheta \le \vartheta_C$) and ($\vartheta_C \le \vartheta \le \vartheta_B$) respectively, we obtain from Eq. (1.226) exact formulas for the MFPTs $\tau_A = \langle T_1\rangle(\vartheta_A)$ and $\tau_B = \langle T_1\rangle(\vartheta_B)$ namely

$$\tau_i = 2\tau_{\mathrm{N}} \int_{\vartheta_i}^{\vartheta_C} \csc\vartheta e^{E(\vartheta)} \int_{\vartheta_i}^{\vartheta} e^{-E(\vartheta'')} \sin\vartheta'' d\vartheta'' d\vartheta, \quad i = A, B, \tag{1.227}$$

($\vartheta_C \le \vartheta \le \vartheta_B$) which are the times to reach the barrier region ϑ_C from the minima at ϑ_A and ϑ_B respectively.

Equation (1.227) can also be derived using the Fokker–Planck equation (1.223). We have $\partial J_\varphi / \partial\varphi = 0$ since $W = W(\vartheta)$. Hence, recalling that, in the quasistationary case $\dot{W} = 0$ and $\mathrm{div}\mathbf{J} = \sin^{-1}\vartheta\, \partial_\vartheta (J_\vartheta \sin\vartheta) = 0$, the total current over the barrier $J = 2\pi J_\vartheta \sin\vartheta$ is constant. Thus we obtain (see Eq. (1.71))

$$W\frac{\partial E}{\partial \vartheta} + \frac{\partial W}{\partial \vartheta} = e^{-E}\frac{\partial}{\partial \vartheta}\left(We^{E}\right) = -\frac{2\tau_{\mathrm{N}} J}{2\pi \sin\vartheta}. \tag{1.228}$$

Suppose now that W vanishes at the barrier angle $\vartheta = \vartheta_C$ (i.e., particles which arrive at this boundary are no longer counted), so that $W(\vartheta_C) = 0$, i.e., all the particles are absorbed there. Thus, we have

$$W(\vartheta) = \frac{\tau_{\mathrm{N}} J}{\pi} e^{-E(\vartheta)} \int_{\vartheta}^{\vartheta_C} \frac{e^{E(\vartheta')}}{\sin\vartheta'} d\vartheta'. \tag{1.229}$$

Furthermore, the number of particles N_i in the well i ($i = A, B$) is

$$\begin{aligned} N_i &= 2\pi \int_{\vartheta_i}^{\vartheta_C} W(\vartheta) \sin\vartheta d\vartheta \\ &= 2\tau_{\mathrm{N}} J \int_{\vartheta_i}^{\vartheta_C} e^{-E(\vartheta)} \sin\vartheta \int_{\vartheta}^{\vartheta_C} \frac{e^{E(\vartheta')}}{\sin\vartheta'} d\vartheta' d\vartheta. \end{aligned} \tag{1.230}$$

Thus, the mean first-passage time τ_i from the well i is, via the flux-over-population method [143,179],

$$\begin{aligned}\tau_i &= \frac{N_i}{J} \\ &= 2\tau_{\mathrm{N}} \int_{\vartheta_i}^{\vartheta_C} e^{-E(\vartheta)} \sin\vartheta \int_{\vartheta}^{\vartheta_C} \frac{e^{E(\vartheta')}}{\sin\vartheta'} d\vartheta' d\vartheta.\end{aligned} \tag{1.231}$$

On integrating by parts, we then obtain Eq. (1.227).

Now τ_A and τ_B are related to the corresponding *escape rates* Γ_A and Γ_B from the wells A and B via $\Gamma_A = 1/2\tau(\vartheta_A)$ and $\Gamma_B = 1/2\tau(\vartheta_B)$ (because there is a 50–50 chance that a spin on reaching the barrier may return) so that the overall reversal time τ is given by $\tau = (\Gamma_A + \Gamma_B)^{-1}$, we have [85]

$$\tau = \frac{2\tau_A \tau_B}{\tau_A + \tau_B}. \tag{1.232}$$

Hence the reversal time (longest relaxation time) of the magnetization in a uniaxial nanomagnet may be determined for all potential barrier heights provided the calculation is formulated as a MFPT problem rather than as finding λ_1^{-1} of the appropriate Fokker–Planck operator. By inspection, the MFPTs τ_A and τ_B given by Eq. (1.227) agree in all respects with those calculated via the flux-over-population method. The resulting finite integral representation of the solution leads to substantial mathematical and conceptual simplifications (as conjectured by Klein [180], who suggested that the calculation of the Kramers escape rate for the translational Brownian motion of a particle in a double-well potential is best posed as a MFPT problem as substantiated by the arguments given by Gardiner [102]). We were only able to obtain our exact solution for the MFPT because the form of the adjoint Fokker–Planck operator for circular symmetry, Eq. (1.225), allows one to integrate the differential equation, Eq. (1.224), for the MFPTs by quadratures. More examples of calculation of the MFPTs for stochastic systems are given in [97]. An example is the calculation of the VLD reversal time governed by the energy-controlled diffusion equation.

1.4.5 *MFPT in the VLD regime*

To evaluate the VLD reversal time, we consider an assembly of spins in a potential well and assume that spins are initially trapped near the bottom of the well (at point A). They are thermalized very rapidly in the well. In the true VLD case, $\alpha S_E << 1$, where the energy loss per cycle of a precessing spin is very much less than the thermal energy, the energy trajectories diffuse very slowly so that they differ but little from the purely streaming ones characterizing the undamped precessional motion in a well. Then due to thermal fluctuations, on a noisy trajectory very near the saddle energy a spin may have enough energy (gained from a fluctuation) to escape over the potential barrier at the saddle point C.

The energy-controlled diffusion equation (1.180) which is valid in the VLD regime represents the continuity equation

$$\frac{\partial W}{\partial t} + \frac{\partial J_p}{\partial E} = 0, \tag{1.233}$$

where the steady state probability current J_p is given by

$$J_p = -\alpha S_E \left(f_E W + \frac{\partial}{\partial E}(f_E W) \right). \tag{1.234}$$

Equation (1.233) is just another Fokker–Planck equation in a single-space variable E. The mean first-passage time may now be exactly calculated by quadratures. The time-independent solutions of Eq. (1.233) are of particular interest to us. Setting $\partial_t W = 0$, the two possible solutions are then $J_p = 0$ or $J_p = \text{constant}$. The stationary solution $J_p = 0$ yields the equilibrium distribution. By inspection of Eq. (1.233), the equilibrium distribution $W_0(E)$ satisfies the first order linear differential equation:

$$\frac{\partial W_0}{\partial E} + \left(1 + \frac{\partial \ln f_E}{\partial E}\right) W_0 = 0, \tag{1.235}$$

thus

$$W_0(E) = N \frac{e^{-E}}{f_E} \left(\int_{E_A}^{E_C} f_E^{-1} e^{-E} dE \right)^{-1}, \tag{1.236}$$

where N is the population in the well A.

The second (time-independent) solution $J_p = \text{const}$ represents the quasistationary solution. Equation (1.234) is a forced first order linear differential equation, which we rewrite as

$$\frac{\partial W}{\partial E}+\left(1+\frac{\partial \ln f_E}{\partial E}\right)W=-\frac{J_p}{\alpha f_E S_E}. \tag{1.237}$$

Considering the behavior of $W(E)$ at E_C and following Kramers [84], on assuming that $W(E_C) = 0$ (all spins that reach the barrier are absorbed), we have the particular solution by quadratures

$$W(E)=J_p\frac{e^{-E}}{\alpha f_E}\int_E^{E_C}\frac{e^{E'}dE'}{S_{E'}}. \tag{1.238}$$

To determine the population N in the well A, we merely integrate the quasi-stationary distribution $W(E)$ over the domain of the well energy so that

$$N=\int_{E_A}^{E_C}W(E)dE=\frac{J_p}{\alpha}\int_{E_A}^{E_C}\frac{e^{-E}}{f_E}\int_E^{E_C}\frac{e^{E'}dE'}{S_{E'}}dE, \tag{1.239}$$

which becomes after integrating by parts

$$N=\frac{J_p}{\alpha}\int_{E_A}^{E_C}\frac{e^{E}}{S_E}\int_{E_A}^{E}\frac{e^{-E'}dE'}{f_{E'}}dE. \tag{1.240}$$

We then have again via the flux-over-population method [85] the reversal time $\tau_{\text{VLD}} = N / J_p$ from a potential well A over the saddle point C [165,166]

$$\tau_{\text{VLD}}=\frac{1}{\alpha}\int_{E_A}^{E_C}\frac{e^{E}}{S_E}\int_{E_A}^{E}\frac{e^{-E'}dE'}{f_{E'}}dE. \tag{1.241}$$

This is just the time needed to reach a separatrix starting from the point A provided that all spins reaching C are absorbed, which is the boundary condition that W vanishes at $E = E_C$. Equation (1.241) can also be reduced to Eq. (4.48a) of [85] derived by Hänggi *et al.* (noting that $dE = f_E dS_E$). We remark that Eq. (1.241) can also be derived using the mean first-passage time [21,50,85,129] (see Section 1.4.4) via the differential equation for the MFPT $\tau(E)$, viz., [85]

$$\mathrm{L}_{\mathrm{FP}}^{\dagger}\tau(E) = -1 \tag{1.242}$$

with the boundary condition $\tau(E_C) = 0$, where $\mathrm{L}_{\mathrm{FP}}^{\dagger}$ is the adjoint Fokker–Planck operator. The MFPT (as seen) is the *average time* needed to reach the separatrix for the *first* time from a starting point E_0 inside the initial domain of attraction. In the VLD regime, this time $\tau(E_0)$ becomes essentially independent of E_0, i.e., $\tau(E_0) = \tau_{\mathrm{VLD}}$ for all starting configurations away from the neighborhood of the separatrix. We emphasize that for the calculation of τ_{VLD} from the exact Eq. (1.241) only a knowledge of the *deterministic dynamics* is required, because (to first order in α) S_E and f_E in Eq. (1.241) are always calculated via the deterministic Larmor equation, Eq. (1.4). We remark that the quadrature solution, Eq. (1.241), is valid for *all barrier heights* including *low* barriers, $\Delta E = E_C - E_A \leq 1$. However, in the high-barrier limit, $\Delta E \geq 1$, Eq. (1.241) can be considerably simplified (see Section 1.5.5).

1.4.6 *MFPT in the VLD regime in the presence of STT*

To evaluate the magnetization reversal time from Eq. (1.185), we consider an assembly of classical spins placed in a potential well with a minimum at point A. The energy-controlled diffusion equation, Eq. (1.185), represents the continuity equation (1.233), where the probability current J_p is given by

$$J_p = -\alpha S_E\left(1 - \frac{V_E}{\alpha S_E} + \frac{\partial}{\partial E}\right) f_E W. \tag{1.243}$$

Like zero STT (see Section 1.4.5), we consider the quasistationary solution of Eq. (1.233), namely, $J_p(E) = J_p = \text{const}$. Equation (1.243) is again a first order linear differential equation. Assuming that $W(E_C) = 0$, i.e., again all spins that reach the barrier are absorbed, we have the particular solution of Eq. (1.243) as

$$W(E) = \frac{J_p}{\alpha f_E}\int_E^{E_C} \frac{1}{S_{E'}} e^{E'-E-\int_E^{E'} \frac{V_{E''}}{\alpha S_{E''}} dE''} dE'. \tag{1.244}$$

Next to determine the population N in the well A, we integrate the quasi-stationary distribution $W(E)$ over the domain of the well energy so that

$$N = \int_{E_A}^{E_C} W(E)dE = \frac{J}{\alpha}\int_{E_A}^{E_C} \frac{e^{-E}}{f_E}\int_{E}^{E_C} \frac{1}{S_{E'}} e^{E' - \int_E^{E'} \frac{V_{E''}}{\alpha S_{E''}} dE''} dE'dE. \qquad (1.245)$$

We then have the VLD reversal time $\tau_{\text{VLD}} = N / J$ from a potential well A over the saddle point C via the flux-over-population method [85] exactly as before, namely

$$\tau_{\text{VLD}} = \frac{1}{\alpha}\int_{E_A}^{E_C} \frac{e^{-E}}{f_E}\int_{E}^{E_C} \frac{1}{S_{E'}} e^{E' - \int_E^{E'} \frac{V_{E''}}{\alpha S_{E''}} dE''} dE'dE \qquad (1.246)$$

or after integrating by parts [167]

$$\tau_{\text{VLD}} = \frac{1}{\alpha}\int_{E_A}^{E_C} \frac{1}{S_E} e^{E - \frac{1}{\alpha}\int_{E_A}^{E} \frac{V_{E''}}{S_{E''}} dE''} \int_{E_A}^{E} \frac{1}{f_{E'}} e^{-E' + \frac{1}{\alpha}\int_{E_A}^{E'} \frac{V_{E''}}{S_{E''}} dE''} dE'dE. \qquad (1.247)$$

Here τ_{VLD} is once again the time to reach a separatrix from the point A *provided that all spins there are absorbed*, which is the boundary condition that W vanishes at $E = E_C$. The inverse of τ_{VLD} also determines the escape rate from the well. We emphasize that for the calculation of τ_{VLD} from Eq. (1.247) again only a knowledge of the *deterministic dynamics* is required, i.e., S_E, V_E, and f_E in Eq. (1.247) are always calculated via the deterministic Larmor equation (1.4). The general MFPT equation (1.247) yields the reversal time via quadratures, which can, in principle, be evaluated for *any* anisotropy potential. Yet another merit of Eq. (1.247) is that it is valid in parameter ranges, where escape-rate equations (such as Eq. (1.209)) do not apply at all, e.g., for *low barriers*. Equation (1.247) is also valuable as a benchmark solution with which numerical calculations of the reversal time from the magnetic Langevin and/or Fokker–Planck equation in the VLD regime must agree.

The results of this section will be used to analyze STT effects on the magnetization reversal of nanoscale ferromagnets (see Chapter 2).

1.5 Escape-rate theory for classical spins

Many problems in physics, including the subfields chemical physics and biophysics, exist where one wishes to calculate the average time required for a particle, generated at some point and diffusing under the influence of a potential, to reach and thereafter be bound to a certain target. In chemical physics, for example, the escape over the potential barrier represents the breaking of a chemical bond [85]. The decay of metastable states of spins is (as seen) also essential for the treatment of the reversal of the magnetization of magnetic nanoparticles, molecular magnets and metal clusters, etc. We reiterate that all magnetic systems are characterized by two or more magnetic metastable states corresponding to different orientations of the magnetic moments. As seen the stability of such states subject to thermal fluctuations is an important issue in many practical contexts, e.g., assessing the limit above which magnetic recording devices may be miniaturized.

The historical origin of modern escape-rate theory stems from the 1880s when Arrhenius [85,169,181–183] proposed, on analyzing experimental data, that the rate coefficient in a chemical reaction obeys the law

$$\Gamma = v_0 e^{-E_a}, \tag{1.248}$$

where E_a denotes the threshold energy for activation and v_0 is a prefactor [181]. Ultimately as summarized in [85], this equation pictured chemical reactions as originated from an assembly of particles situated at the bottom of a potential well. Next rare members of this assembly will have enough energy to escape over the potential hill, owing to the shuttling action of thermal agitation, and will never return [85], thus constituting a model of a chemical reaction.

The Arrhenius formula corresponds to empirical observations and ignores dynamic considerations. Therefore, further development was necessary to understand the two parameters associated with this law, namely the pre-exponential factor v_0 and the activation energy E_a. TST more or less successfully addresses these two issues.

1.5.1 *Transition state theory for spins*

The TST escape rate Γ of particles that are initially trapped in a potential well at A, and that may subsequently, under the influence of thermal agitation, escape over a high barrier of height ΔE (>> 1) at C and never return to A (see Fig. 1.17), may be written using TST [85,182] as

$$\Gamma^{\mathrm{TST}} = \frac{\omega_A}{2\pi} e^{-\Delta E}. \tag{1.249}$$

Here the *attempt frequency*, ω_A, is the angular frequency of a particle executing oscillatory motion (i.e., libration) at the bottom of a well. The barrier arises from the potential function of some external force, which may be electrical, magnetic, gravitational, and so on. The formula represents an *attempt frequency times a Boltzmann factor*, which weights the escape from the well.

The TST approximation for thermally activated transitions in spin systems is based on the Landau–Lifshitz–Gilbert equation of motion and subsequent quadratic expansion of the energy surface at the minima and saddle points. While the flux out of the initial state vanishes at the saddle

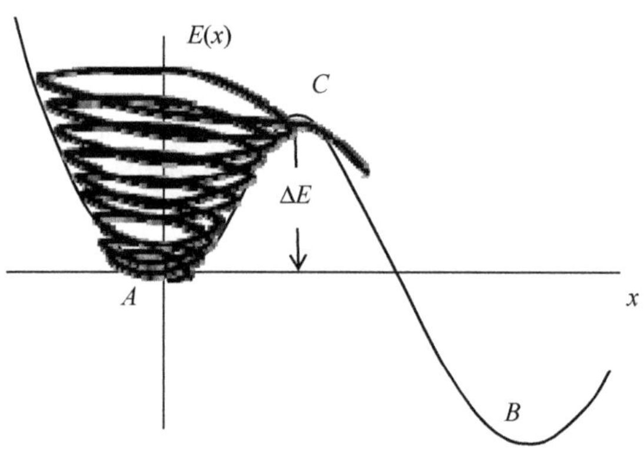

Fig. 1.17. Single-well potential function $E(x)$ as the simplest example illustrating noise assisted escape over a barrier. Particles are initially trapped in the well near the point A; they are then thermalized very rapidly in the well. The potential barrier at the point C is large enough to ensure that the rate of escape of particles Γ is very small. Due to thermal agitation, however, very few may attain enough energy to escape over the barrier into region B, whence they never return (a sink of probability).

points, the *integrated flux over the transition state is nonzero* and so yields an estimate of the rate.

We now describe the TST result for the particular case of potentials with circular symmetry given by Eq. (1.10). Here the solution significantly simplifies, because such a potential $V(\vartheta)$ is a function of one-space-variable only. In applying TST to classical spins (i.e., magnetic moments **M**) in circularly symmetric potentials the equation of motion of the magnetic moment **M** in the effective field $\mathbf{H} = -\partial V/\partial \mathbf{M}$ is as usual given by the Landau–Lifshitz–Gilbert equation (1.22). In the low-temperature limit (or high-barrier approximation), as far as TST is concerned, we merely have the flux-over-population:

$$\Gamma^{\mathrm{TST}} = \frac{I_{\mathrm{top}}}{Z_{\mathrm{well}}}. \tag{1.250}$$

where Z_{well} is the well partition function and I_{top} is the total current of (magnetization) representative points over the barrier. Furthermore, for the uniaxial anisotropy potential $V(\vartheta) = K\sin^2\vartheta$, the barrier top is situated at $\vartheta_C = \pi/2$ and the well bottom at $\vartheta_A = 0$. The well partition function and total current may then be estimated as (see Chapter 3, Section 3.2.6 for details)

$$\begin{aligned} Z_{\mathrm{well}} &\approx \frac{kT}{2vK} e^{-E(\vartheta_A)}, \\ I_{\mathrm{top}} &\approx \frac{\gamma kT}{2\pi v \mu_0 M_S} e^{-E(\vartheta_C)}. \end{aligned} \tag{1.251}$$

Hence, Eq. (1.250) yields the Néel formula [16] for escape from a *single* well, namely,

$$\Gamma^{\mathrm{TST}} = \frac{\omega_A}{2\pi} e^{-\Delta E}, \tag{1.252}$$

where $\Delta E = E(\vartheta_C) - E(\vartheta_A)$ and

$$\omega_A = \frac{2\gamma K}{\mu_0 M_S}$$

is the well (precession) frequency.

1.5.2 *Kramers' escape rate for Brownian particles*

A very unsatisfactory feature of the TST formula, Eq. (1.249), is that it predicts *escape in the absence of coupling to a heat bath*, contradicting the fluctuation–dissipation theorem. This defect was remedied, and reaction rate theory was firmly set in the framework of nonequilibrium statistical mechanics by the pioneering work of Kramers [84]. Thus, to take into account *nonequilibrium effects* in the barrier-crossing process which manifest themselves as a frictional dependence, i.e., a coupling to the heat bath, of the prefactor in the TST formula Eq. (1.249), he chose an assembly of Newtonian particles moving in a potential in a one-dimensional extension x (see Fig. 1.17). The fact that a typical particle of the assembly is embedded in a heat bath and interchanges energy with it is modelled by Brownian motion. In the single-particle distribution function, this represents all the remaining degrees of freedom of the system, consisting of the selected particle and the heat bath, which is in *perpetual thermal equilibrium* at temperature T. Kramers [84,181] assumed that the particles are initially trapped in a well near the minimum of the potential which is situated at the point $x = A$. They then receive energy from the surroundings, and the equilibrium Maxwell–Boltzmann distribution is rapidly attained in the well. Over a long period of time, however, rare particles may gain energy very slightly more than the barrier height ΔE. Kramers then assumed that these particles escape over the barrier $x = C$ (so that a *perturbation of the Maxwell–Boltzmann distribution in the well exists*) and reach a minimum at $x = B$, which is of lower energy than A, and once there, never return. We list the assumptions of Kramers:

(1) The particles are initially trapped in A (which is a *source* of probability).
(2) The barrier heights ΔE are very large compared with 1.
(3) A Maxwell–Boltzmann distribution is attained extremely rapidly in the well.
(4) Quantum effects are negligible.
(5) The escape of particles over the barrier is very slow (i.e., is a quasi-stationary process) so that the disturbance to the equilibrium Maxwell–Boltzmann distribution (assumption 3) is almost negligible at all times.

(6) Once a particle escapes over the barrier it practically never returns (i.e., B is a *sink* of probability).

(7) A typical particle of the reacting system may be modeled by the theory of Brownian motion, including the inertia of the particles.

We remark that assumption 5 relies heavily on assumption 2. If the barrier is too low, the particles escape too quickly to allow a Boltzmann distribution to be set up in the well. If in contrast the barrier is *high*, before many particles can escape, the Boltzmann distribution is already set up. As required by assumption 3, we therefore assume that ΔE is at least of the order 5.

The Kramers model, yielding explicit formulas for the escape rate for very low and intermediate-to-high dissipative coupling to the bath (thereby including nonequilibrium effects in the TST formula), is applicable in any physical system in which noise-activated escape from a potential well exists. It plays an essential role in fields as diverse as the dielectric relaxation of nematic liquid crystals [170], the magnetic relaxation of fine ferromagnetic particles [20], laser physics [184,185], and Josephson junctions [143]. Kramers' ultimate objective was to calculate the prefactor Λ in the escape rate, viz.,

$$\Gamma = \Lambda \Gamma^{\mathrm{TST}} \tag{1.253}$$

from a *microscopic model* of the motion of the particles incorporating fluctuations and dissipation (viz., an assembly of Brownian particles in a potential well). Thus Λ indicates that the prefactor is closely associated both with the *stochastic differential equation* underlying Brownian motion, i.e., the *Langevin equation* for the evolution of the random variables in the state space (position and momentum), and the *associated probability density diffusion equation* describing the evolution of the density of the realizations (phase points) of these random variables in phase (state) space. In this instance (Newtonian particles) the Fokker–Planck equation is called the *Fokker–Planck–Kramers equation* or the *Klein–Kramers equation*.

By supposing that the time derivative of the probability distribution function equals zero (quasistationary process), Kramers discovered two asymptotic formulas for the rate of escape from a well for a system with dynamics governed by the Langevin equation. The first is the IHD formula

$$\Gamma^{\text{IHD}} = \left(\sqrt{1 + \frac{\beta^2}{4\omega_C^2}} - \frac{\beta}{2\omega_C} \right) \Gamma^{\text{TST}}, \qquad (1.254)$$

where β denotes the damping coefficient per unit mass in the Langevin equation of the process and ω_C is the characteristic frequency of the inverted oscillator approximation to the potential near the barrier. In the IHD formula, the correction Λ to the TST result is essentially the positive eigenvalue (characterizing the unstable barrier-crossing mode) of the Langevin equation linearized about the saddle point of the potential. In the one-dimensional extension considered by Kramers, this is a one-dimensional maximum.

Equation (1.254) formally holds [181], when the energy loss per cycle of the motion of a particle librating in the well of the potential with energy equal to the barrier energy E_C, is much greater than kT. The energy loss per cycle of the motion of a barrier-crossing particle is $\beta S(E_C)$, where E_C is the energy contour through the saddle (barrier) point of the potential, and S is the action evaluated at $E = E_C$. This criterion effectively follows from the Kramers very-low-damping result (see below). The IHD asymptotic formula is derived by supposing that (i) the barrier is so high and the dissipative coupling to the bath so strong that a Maxwell–Boltzmann distribution always holds at the bottom of the well; and (ii) the Langevin equation may be linearized in the region very close to the potential barrier, meaning that all the coefficients in the corresponding Fokker–Planck equation for the joint probability density function are linear in the positions and velocities. If these simplifications can be made, then the evolution equation for the quasistationary probability density function, although it remains an equation in the two phase-space variables (x, p), may be integrated by introducing an independent variable which is a linear combination of the representative points x and p so that it becomes an ordinary differential equation in a single variable.

A particular case of the IHD formula, Eq. (1.254), is the very-high-damping (VHD) regime, $\beta/(2\omega_C) >> 1$, where the escape rate yielded by Eq. (1.254) becomes

$$\Gamma^{\text{VHD}} = \omega_C \beta^{-1} \Gamma^{\text{TST}}. \qquad (1.255)$$

Here the quasistationary solution for the probability distribution function may be obtained in integral form by quadratures, and its high-barrier limit (appropriate to the escape rate) may be determined by steepest descents (see Fig. 1.18). It is now unnecessary to linearize the Langevin equation about the saddle point (here a one-dimensional maximum), as the solution may be obtained more directly using the Smoluchowski equation [97].

For small friction β (such that $\beta S(E_C) << kT$, however, the IHD formula fails, predicting, like the TST formula, escape in the *absence of coupling to the bath*, because [181] the tacit assumption that the particles approaching the barrier from the depths of the well are in thermal equilibrium (so that the stationary Maxwell–Boltzmann solution applies) is violated (owing to the smallness of the dissipation of energy to the bath). Thus, *the spatial region of significant departure from the Maxwell–Boltzmann distribution in the well extends far beyond the region over which the potential may sensibly be approximated by an inverted parabola.*

Kramers showed how his second formula, valid in the VLD case, where the energy loss per cycle $\beta S(E_C)$ of a librating particle is very much less than kT, may be obtained by again reducing the evolution equation for the probability distribution function to a partial differential equation in a *single* spatial variable. This variable is the *energy* or, equivalently, the *action*. Here the energy trajectories diffuse so slowly that they do not differ significantly from the purely streaming ones characteristic of the undamped librational motion of a particle in a well with energy corresponding to the saddle energy E_C. Kramers solved the VLD problem by writing the evolution equation for the probability distribution function

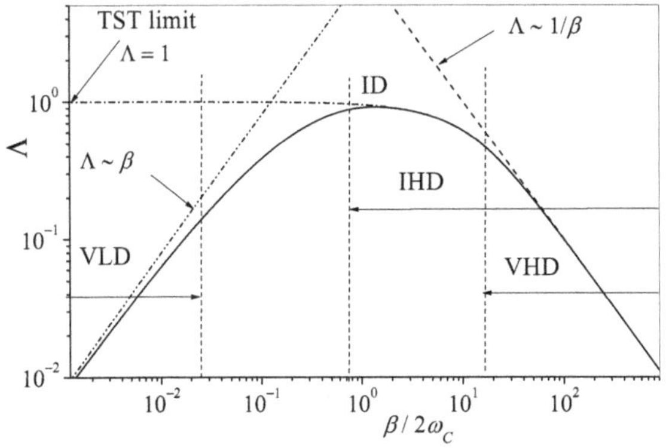

Fig. 1.18. Diagram of damping regions for the prefactor Λ in Eq.(1.253), namely, very low damping (VLD), intermediate (or moderate) damping, very high damping (VHD), and two crossovers between them. Kramers' turnover refers to the underdamped region between ID and VLD.

in angle-action (or angle-energy) variables (the angle is the *phase* or *instantaneous state* of the system along an energy trajectory) and taking a time average of the motion along a *closed* (streaming) energy trajectory in the well *infinitesimally close* to the saddle energy trajectory. Thus, by dint of thermal fluctuations, the resultant noisy trajectory may become the *separatrix* or the *open* trajectory on which the particle may exit the well. The average, being along a closed energy trajectory, is equivalent to an average over the *fast* phase variable; hence a diffusion equation, Eq. (1.180), in the *slow* energy (or action) variable emerges. Thus, once again, the time derivative of the probability distribution function (which is written as a function of the energy only, using the averaging procedure above) is *exponentially small* at the saddle point C. Hence, the quasistationary solution in the energy variable may be used yielding the Kramers' VLD formula:

$$\Gamma^{\mathrm{VLD}} = \frac{\beta S(E_C)}{kT}\Gamma^{\mathrm{TST}}. \tag{1.256}$$

This VLD formula holds when in the general Eq. (1.253) the prefactor $\Lambda \ll 1$, i.e., $\beta S(E_C) \ll kT$, unlike the TST result it vanishes when $\beta \to 0$, so that escape is now impossible without coupling to the bath. Thus, in all cases, analytical formulas for the escape rate exist because in the relevant damping regimes, the evolution equation for the probability distribution function may be reduced to an equation in a *single* coordinate.

In summary, the VLD formula demonstrates that escape is impossible without coupling to the bath otherwise the streaming energy trajectories are unaffected. Likewise, if the coupling to the bath is very large, the escape rate vanishes. Kramers made several estimates of the range of validity of both IHD and VLD formulas and the intermediate damping regime, where the TST Eq. (1.249) holds with a high degree of accuracy. He was, however, unable to give a formula in the underdamped region lying between IHD and VLD, as there $\beta S(E_C) \approx kT$ because no small perturbation parameter now exists. This long-standing problem, known as the Kramers turnover, was solved nearly 50 years later by Mel'nikov [186] and Mel'nikov and Meshkov [187], and elaborated upon by Grabert [188] and Pollak *et al.* [189]. All these investigators constructed an integral equation for the evolution of the energy distribution function which they solved using the Wiener–Hopf method [181,190,191], and so obtained an escape-rate formula valid for all values of the friction β [97], viz.,

$$\Gamma = A\left[\frac{\beta S(E_C)}{kT}\right]\Gamma^{\mathrm{IHD}}. \tag{1.257}$$

Here $A(\alpha S)$ is called the depopulation factor and is given by

$$A(\Delta) = e^{\frac{1}{2\pi}\int_{-\infty}^{\infty}\ln\left[1-e^{-\Delta()\lambda^2+1/4}\right]\frac{d\lambda}{\lambda^2+1/4}}. \tag{1.258}$$

1.5.3 *Langer's treatment of the IHD regime*

The original IHD treatment of Kramers pertained to a mechanical system of one degree of freedom specified by the coordinate x. However, the time evolution of the magnetic moment in a single-domain ferromagnetic particle is the two degree of freedom motion of the magnetization vector on the surface of a sphere governed by a *nonadditive, nonseparable* Hamiltonian, which in the simplest case is the magnetocrystalline anisotropy energy of the particle. Hence the Kramers theory should be extended to a multi-dimensional case. Such generalizations, having been instigated by Brinkman [192], were further developed by Landauer and Swanson [193]. However, the most complete treatment is due to Langer in 1969 [103], who considered the escape rate for multi-dimensional random processes in the IHD regime. Langer's method, since it can in effect be applied to a system of multiple degrees of freedom, is of much use in calculating relaxation times for fine particle magnetic systems in which other types of interaction, such as exchange and dipole-dipole coupling, also appear. We also emphasize that his treatment of the homogeneous nucleation problem contains within it the magnetic case of the Kramers' IHD calculation.

We shall consider the Fokker–Planck equation for a multi-dimensional random process governed by a state vector $\boldsymbol{\eta} = \{\eta_1, \eta_2, \ldots, \eta_{2N}\}$ which is [85,103]

$$\frac{\partial}{\partial t}W(\boldsymbol{\eta},t)=\sum_{i,n=1}^{2N}\frac{\partial}{\partial\eta_i}\left[M_{in}\left(\frac{\partial E}{\partial\eta_n}+\frac{\partial}{\partial\eta_n}\right)\right]W(\boldsymbol{\eta},t). \tag{1.259}$$

In Eq. (1.259), $E(\boldsymbol{\eta})$ is a Hamiltonian (energy) function having two minima at points A and B, separated by a saddle point C surrounded by two wells. One well, say that at B, is at a much lower energy than the other. The particles have to traverse the saddle point, which acts as a barrier at C. We again assume that the barrier height $\Delta E = E_C - E_A$ is very high (at least of the order of $5kT$), so that the *diffusion over the barrier is slow enough to ensure that an equilibrium distribution is established and maintained near A at all times*. The high barrier also ensures that the contribution to the flux over the saddle point emanates mainly from a small region around C. The $2N$ state variables $\boldsymbol{\eta}=\{\eta_1,\eta_2,\ldots,\eta_{2N}\}$ are parameters, which could equally well be the coordinates and momenta of a point in phase space, or angular coordinates describing the orientation of the magnetization vector of a magnetic nanoparticle. When the noise term in the Langevin equation is ignored, the system merely evolves according to the deterministic equation

$$\dot{\eta}_i=-\sum_n M_{in}\frac{\partial E}{\partial\eta_n}, \tag{1.260}$$

where M_{ij} are the matrix elements of the transport matrix $\mathbf{M}$, which, for simplicity, we shall assume to be constant. An example of such a system is the magnetization relaxation of a superparamagnetic particle.

We now define new matrices $\mathbf{D}$ and $\mathbf{A}$ by the equations

$$\mathbf{D}=\frac{1}{2}(\mathbf{M}+\mathbf{M}^{\mathrm{T}})\ \text{and}\ \mathbf{A}=\frac{1}{2}(\mathbf{M}-\mathbf{M}^{\mathrm{T}}), \tag{1.261}$$

where the symbol "T" means matrix transposition. Matrix $\mathbf{D}$ is called the *diffusion* matrix, which characterizes the thermal fluctuations due to the heat bath, while matrix $\mathbf{A}$ describes the motion in *the absence of the bath*, and if $\mathbf{D}$ is not identically zero, then the dissipation of energy satisfies [85]

$$\dot{E} = -\sum_{i,n} \frac{\partial E}{\partial \eta_i} D_{in} \frac{\partial E}{\partial \eta_n} \leq 0. \tag{1.262}$$

We consider, as before, a single well and suppose that, at finite temperatures, a Boltzmann distribution is set up and the density at equilibrium is

$$W_{\text{eq}}(\boldsymbol{\eta}) = \frac{1}{Z} e^{-E(\boldsymbol{\eta})}, \quad Z \equiv \int_{-\infty}^{\infty} \cdots \int_{-\infty}^{\infty} e^{-E} d\eta_1 \cdots d\eta_{2N}, \tag{1.263}$$

where Z is the partition function. The IHD escape rate for this multivariable problem may again be calculated by the flux-over-population method.

We make the following assumptions about the distribution $W(\boldsymbol{\eta})$:

(1) It obeys the stationary Fokker–Planck equation (i.e., $\partial_t W = 0$, which is (on linearization about the saddle point):

$$\sum_{i,n} \frac{\partial}{\partial \eta_i} M_{in} \left[\sum_k e_{nk} \left(\eta_k - \eta_k^S \right) + \frac{\partial}{\partial \eta_n} \right] W = 0, \tag{1.264}$$

where the e_{nk} are the coefficients in the Taylor expansion of the energy about the saddle point truncated at the second term, namely

$$E(\boldsymbol{\eta}) = E_C - \frac{1}{2} \sum_{i,n} e_{in} (\eta_i - \eta_i^C)(\eta_n - \eta_n^C), \tag{1.265}$$

$\boldsymbol{\eta} \approx \boldsymbol{\eta}^C$, and E_C is the value of the energy function at the saddle point (in the one degree of freedom extension considered by Kramers the saddle point is a one-dimensional maximum). Equation (1.265) in general constitutes the paraboloidal approximation to the potential in the vicinity of the saddle point. For example, in magnetic relaxation in a uniform field with uniaxial anisotropy, the energy surface near the saddle point will be a hyperbolic paraboloid [108]. Equation (1.264) is just the multi-dimensional Fokker–Planck equation linearized in the vicinity of the saddle point.

(2) Owing to the high barrier, like in the one-dimensional Kramers high-damping problem, a Boltzmann distribution is set up in the vicinity of the bottom of the well, i.e., at A, so:

$$W(\boldsymbol{\eta}) \approx W_{\mathrm{eq}}(\boldsymbol{\eta}), \quad \boldsymbol{\eta} \approx \boldsymbol{\eta}^{A}. \tag{1.266}$$

(3) Practically speaking, no particles have arrived at the far side of the saddle point, so we have the sink boundary condition

$$W(\boldsymbol{\eta}) = 0, \quad \boldsymbol{\eta} \text{ beyond } \boldsymbol{\eta}^{C}. \tag{1.267}$$

This is merely Kramers' supposition that only rare particles of the assembly can ever cross the barrier. Next, we make the substitution

$$W(\boldsymbol{\eta}) = g(\boldsymbol{\eta}) W_{\mathrm{eq}}(\boldsymbol{\eta}), \tag{1.268}$$

where the function g is known as the *crossover* function. Thus, we have from Eqs. (1.263) and (1.264), as an equation for g, namely

$$\sum_{i,n} M_{ni} \left[\sum_{k} e_{nk} \left(\eta_k - \eta_k^C \right) + \frac{\partial}{\partial \eta_n} \right] \frac{\partial}{\partial \eta_i} g(\boldsymbol{\eta}) = 0, \tag{1.269}$$

where $\boldsymbol{\eta} \approx \boldsymbol{\eta}^{C}$. We postulate that g may be written in terms of a single variable u, viz.,

$$g(u) = \frac{1}{2\pi} \int_{u}^{\infty} e^{-z^2/2} dz, \tag{1.270}$$

and we assume that u has the form of the linear combination

$$u = \sum_{i} U_i (\eta_i - \eta_i^C). \tag{1.271}$$

This is again just Kramers' method of forcing the multi-dimensional Fokker–Planck equation into an equation in a single variable u. We must now determine the coefficients U_i of the linear combination u of the η_i. This is accomplished as follows. We define the matrix $\tilde{\mathbf{M}} = -\mathbf{M}^{\mathrm{T}}$. Then we shall have the coefficients U_i of the linear combination as a solution of the *eigenvalue problem*

$$-\sum_{i,n} U_i \tilde{M}_{in} e_{nk} = \lambda_+ U_k. \tag{1.272}$$

The eigenvalue λ_+ is the *deterministic growth rate of a small deviation from the saddle point* and is the *positive* eigenvalue of the system matrix of the noiseless Langevin equations, linearized about the saddle point and thus characterizes the unstable barrier-crossing mode. Hence to calculate λ_+, all that is required is a knowledge of the energy landscape; Eq. (1.272) need not, in practice, be involved. Equation (1.272) is obtained by substituting the linear combination *u*, i.e., Eq. (1.271), into Eq. (1.269) for the crossover function, and requiring the resulting equation to be an ordinary differential equation in the single variable *u* with solution given by Eq. (1.270) (the details of this are given in [181]). Equation (1.272) may also be written in the matrix form

$$-\mathbf{U}^{\mathrm{T}}\tilde{\mathbf{M}}\mathbf{E}^{C} = \lambda_+ \mathbf{U}^{\mathrm{T}}. \tag{1.273}$$

(Hänggi *et al.* [85] describe this equation by stating that $\mathbf{U}^{\mathrm{T}}$ is a "left eigenvector" of the matrix $-\tilde{\mathbf{M}}\mathbf{E}^{C}$). The conventional eigenvalue equation of an arbitrary matrix **A** is $\mathbf{AX} = \lambda \mathbf{X}$. In the above terminology, **X** would be a "right eigenvector" of **A**. In Eq. (1.273), $\mathbf{E}^{C} \equiv (e_{ij})$ is the matrix of the second derivatives of the potential evaluated at the saddle point, which is used in the Taylor expansion of the energy near that point. The determinant of this matrix is called the Hessian. The normalization of U_i is fixed, so that

$$\lambda_+ = \sum_{i,n} U_i M_{in} U_n, \tag{1.274}$$

which is equivalent to

$$\sum_{i,n} U_i e_{in}^{-1} U_n = -1. \tag{1.275}$$

This condition ensures that the crossover function, Eq. (1.270), retains the form of an error function and so may describe diffusion over a barrier. Alternatively, one may say that the foregoing conditions require that the entry in the diffusion matrix in the (unstable) direction of flow is nonzero; that is, *we have current over the barrier and so particles escape the well.*

The Fokker–Planck equation, Eq. (1.259), is a continuity equation for the representative points, just as described earlier, so that

$$\partial_t W + \nabla \cdot \mathbf{J} = 0. \tag{1.276}$$

Thus, by inspection, we find that the current density becomes

$$j_i = -\sum_n M_{in}\left(\frac{\partial E}{\partial \eta_n} + \frac{\partial}{\partial \eta_n}\right)W \tag{1.277}$$

and we obtain, using Eqs. (1.263), (1.269), and (1.270) for the *stationary* current density, i.e., $\partial_t W = 0$,

$$j_i(\boldsymbol{\eta}) = \frac{1}{\sqrt{2\pi}}\sum_n M_{in}U_n W_{\text{eq}} e^{-u^2/2}. \tag{1.278}$$

We now utilize the condition stated above, namely that *the flux over the barrier simply emanates from a small region around the saddle point C*. We integrate the current density $j_i(\boldsymbol{\eta})$ over a plane containing the saddle point but not parallel to the flow of particles. The plane $u = 0$ will suffice here. Thus, the total current is

$$J = \sum_i \int_{u=0} j_i(\boldsymbol{\eta})dS_i. \tag{1.279}$$

Using Eq. (1.279) with the quadratic approximation of Eq. (1.265) for the energy near the saddle point, the integration for the total flux (current) now yields, after a long calculation [181],

$$J \approx \frac{1}{2\pi Z}\sum_{i,j} U_i M_{ij} U_j \left|\sum_{i,j} U_i e_{ij}^{-1} U_j \det\left(\frac{\mathbf{E}^C}{2\pi}\right)\right|^{-1/2} e^{-E_C}. \tag{1.280}$$

From Eqs. (1.274) and (1.275), we immediately obtain

$$J = \frac{\lambda_+}{2\pi Z}\left|\det\left[(2\pi)^{-1}\mathbf{E}^C\right]\right|^{-1/2} e^{-E_C}. \tag{1.281}$$

Now, we assume that the energy function near the bottom of the well A may again be written in the quadratic approximation as

$$E = E_A + \frac{1}{2}\sum_{i,j} a_{ij}\left(\eta_i - \eta_i^A\right)\left(\eta_j - \eta_j^A\right), \tag{1.282}$$

so that the number of particles in the well n_A is given by [181]

$$n_A = Z^{-1}\left\{\det[(2\pi)^{-1}\mathbf{E}^A]\right\}^{-1/2} e^{-E_A}, \tag{1.283}$$

where we have defined the matrix $\mathbf{E}^A = (a_{ij})$. Moreover the escape rate Γ, by the usual flux-over-population method, is defined to be $\Gamma = J / n_A$, and so from Eqs. (1.281) and (1.283), in terms of the *unique positive eigenvalue* λ_+ of the set of *noiseless* Langevin equations linearized about the saddle point, we have

$$\Gamma = \frac{\lambda_+}{2\pi}\sqrt{\det\left[\mathbf{E}^A\right]/\left|\det\left[\mathbf{E}^C\right]\right|}e^{-(E_C-E_A)}. \tag{1.284}$$

This is Langer's [103] expression for the escape rate for a multi-dimensional process in the IHD regime. The result again pertains to this limit because of our initial postulate that the potential near the saddle point may be approximated by an inverted paraboloid. Thus, Eq. (1.284) again fails for VLD because the region of deviation from the Boltzmann distribution prevailing in the depths of the well extends far beyond the narrow region at the top of the barrier in which the potential may sensibly be replaced by its quadratic approximation. In passing, we remark that rate theory at weak friction is generally known as "unimolecular rate theory" [85], the Kramers VLD regime treated earlier being an example of this. For a general discussion see Ref. [85].

1.5.4 *IHD escape rate for spins*

We recall that Kramers [84] obtained his IHD and VLD formulas for the escape rate for *point* particles. The analogous magnetic-spin problem, as formulated by Brown [17,18,92], differs fundamentally [86] from that of mechanical particles and rotators with separable and additive Hamiltonians. First, the magnetic system has two degrees of freedom, namely the polar ϑ and azimuthal φ angles; second, the undamped equation of motion of the magnetization of a single-domain ferromagnetic particle is the gyromagnetic equation; third, the Hamiltonian is not separable; fourth, inertial effects play no role. Nevertheless, the role of inertia in the mechanical system is essentially mimicked in the magnetic system for noncircularly symmetric potentials by the gyromagnetic term, which causes *coupling* or *entanglement* of the transverse (giving rise to ferromagnetic resonance) and longitudinal modes. Thus, to derive escape-rate formulas for superparamagnetic particles analogous to those for mechanical particles or inertial rotators, one must introduce in Brown's magnetic Fokker–Planck equation a noncircularly symmetric free energy density $E(\vartheta,\varphi)$, where *explicit* coupling between the two degrees of freedom now exists. Thus, both regimes of damping (IHD and VLD) can again occur, reflecting the fact that the dynamics of the transverse response affect the dynamics of the longitudinal one, and vice versa.

The application of Kramers' escape-rate theory to the magnetization relaxation of nanomagnets in the IHD regime has been given in detail by Smith and de Rozario [90], Brown [18], Klik and Gunther [106], and Geoghegan *et al.* [108]. Klik and Gunther [106] used Langer's method [103] (just described in Section 1.5.3) and realized that the various Kramers damping regimes also applied to magnetic relaxation of single-domain ferromagnetic particles because two canonical variables exist. In this section, we show in detail how Langer's method may be used to calculate the IHD magnetic relaxation time for a single-domain ferromagnetic particle for an arbitrary noncircularly symmetric potential of the magnetocrystalline anisotropy in that limit. Again, we deal with a normalized free energy density function, $E(\vartheta,\varphi)$, with minima at points A and B separated by a barrier (saddle point) at C. We remark, that the free energy is defined on a phase space which is a *closed* manifold (because the space (ϑ, φ) is the surface of the unit sphere) and that a local energy minimum is thus surrounded by two or more saddle points, depending on the symmetry of the problem [106]. The total probability flux away from the metastable minimum equals the sum of the fluxes through all the saddle points. In asymmetric cases, e.g., when an external field is applied, some of these fluxes become exponentially small and may safely be neglected. The total flux away from the metastable minimum is then dominated by *the energetically most favorable path.* The noiseless Landau–Lifshitz–Gilbert equation, Eq. (1.26), becomes in the coordinates p and φ $(p = \cos\vartheta)$ [18]

$$\dot{p} = \frac{1}{2\tau_N}\left(-(1-p^2)\frac{\partial E}{\partial p} + \frac{1}{\alpha}\frac{\partial E}{\partial \varphi}\right),$$
$$\dot{\varphi} = -\frac{1}{2\tau_N}\left(\frac{1}{\alpha}\frac{\partial E}{\partial p} + \frac{1}{1-p^2}\frac{\partial E}{\partial \varphi}\right). \quad (1.285)$$

We now linearize these equations about the saddle point and determine λ_+ from the transition matrix. Thus, expanding the normalized free energy density $E(p, \varphi)$ in a truncated Taylor series about the saddle point (p_C, φ_C), we obtain

$$E = E_C + \frac{1}{2}\left[E_{pp}^{(C)}(p-p_C)^2 + 2E_{p\varphi}^{(C)}(p-p_C)(\varphi-\varphi_C) + E_{\varphi\varphi}^{(C)}(\varphi-\varphi_C)^2\right], \quad (1.286)$$

where

$$E_{pp}^{(C)} = \frac{\partial^2 E}{\partial p^2}(p_C, \varphi_C),\ E_{p\varphi}^{(C)} = \frac{\partial^2 E}{\partial p \partial \varphi}(p_C, \varphi_C),\ E_{\varphi\varphi}^{(C)} = \frac{\partial^2 E}{\partial \varphi^2}(p_C, \varphi_C)$$

with the superscript (C) denoting evaluation at the saddle point. Thus, we have the linearized equations:

$$\begin{aligned} \frac{\partial E}{\partial p} &= (p-p_C)E_{pp}^{(C)} + (\varphi-\varphi_C)E_{p\varphi}^{(C)}, \\ \frac{\partial E}{\partial \varphi} &= (p-p_C)E_{p\varphi}^{(C)} + (\varphi-\varphi_C)E_{\varphi\varphi}^{(C)}. \end{aligned} \quad (1.287)$$

Suppose that the saddle point C of interest lies on the equator $p = 0$ and then make the transformation $\varphi - \varphi_C \to \varphi$. Thus, Eq. (1.285) becomes in the state space (φ, p)

$$\begin{pmatrix} \dot{\varphi} \\ \dot{p} \end{pmatrix} = \frac{1}{2\tau_N} \begin{pmatrix} -\alpha^{-1}E_{p\varphi}^{(C)} - E_{\varphi\varphi}^{(C)} & -\alpha^{-1}E_{pp}^{(C)} - E_{p\varphi}^{(C)} \\ -E_{p\varphi}^{(C)} + \alpha^{-1}E_{\varphi\varphi}^{(C)} & -E_{pp}^{(C)} + \alpha^{-1}E_{p\varphi}^{(C)} \end{pmatrix} \begin{pmatrix} \varphi \\ p \end{pmatrix}. \quad (1.288)$$

Equations (1.288) are the noiseless Langevin equations given by Klik and Gunther; see Ref. [106], Eq. (3.2). They have the form of the canonical Eqs. (1.260) and so Langer's general IHD equation, Eq. (1.284), may be used to calculate the IHD escape rate. The secular equation of Eq. (1.288) then yields the positive eigenvalue λ_+ as:

$$\begin{aligned} \lambda_+ = \frac{1}{4\tau_N}\Big\{ & -E_{pp}^{(C)} - E_{\varphi\varphi}^{(C)} \\ & + \sqrt{\left(E_{pp}^{(C)} + E_{\varphi\varphi}^{(C)}\right)^2 - 4\left(1+\alpha^{-2}\right)\left[E_{pp}^{(C)}E_{\varphi\varphi}^{(C)} - \left(E_{p\varphi}^{(C)}\right)^2\right]}\Big\}, \end{aligned} \quad (1.289)$$

where we observe that the α^{-2} term represents the explicit effect of the precessional term in the Gilbert equation on the longitudinal relaxation. This *mode coupling effect* is always present in a noncircularly symmetric potential, as the smallest nonvanishing eigenvalue of the Fokker–Planck operator will then always *intrinsically* depend on the damping. This is *quite unlike circular symmetry, where the damping only enters only via the free diffusion time*. If we now choose a *local* coordinate system (φ, p) at the saddle point, where the mixed derivative $E_{p\varphi} = 0$, then we obtain a more compact expression for λ_+, namely

$$\lambda_+ = \frac{1}{4\tau_N}\left\{ -E_{pp}^{(C)} - E_{\varphi\varphi}^{(C)} + \sqrt{[E_{pp}^{(C)} - E_{\varphi\varphi}^{(C)}]^2 - 4\alpha^{-2} E_{pp}^{(C)} E_{\varphi\varphi}^{(C)}} \right\}. \tag{1.290}$$

The well and saddle angular frequencies are, respectively, defined as

$$\begin{aligned} \omega_A &= \frac{\gamma kT}{v\mu_0 M_S}\sqrt{E_{pp}^{(A)} E_{\varphi\varphi}^{(A)} - (E_{p\varphi}^{(A)})^2}, \\ \omega_C &= \frac{\gamma kT}{v\mu_0 M_S}\sqrt{\left|E_{pp}^{(C)} E_{\varphi\varphi}^{(C)} - (E_{p\varphi}^{(C)})^2\right|}, \end{aligned} \tag{1.291}$$

where the superscripts (A) and (C) denote evaluation at the minimum of well A and at the saddle point C. Thus, with Eq. (1.284), we obtain the Klik and Gunther result [106], namely,

$$\Gamma_A^{\mathrm{IHD}} = \frac{\lambda_+ \omega_A}{2\pi\omega_C} e^{-(E_C - E_A)} = \frac{\lambda_+}{\omega_C}\Gamma_A^{\mathrm{TST}}. \tag{1.292}$$

This formula demonstrates the wide-ranging uses of Langer's method and shows clearly how, once the potential landscape is known, the IHD escape rate may be calculated. As seen, the range of validity of Eq. (1.292) is $\alpha S_{E_C} >> 1$, where S_{E_C} is the dimensionless magnetic action defined by Eq. (1.179). Obviously, in the limit $\alpha \to 0$, Eq. (1.292) predicts that

$$\Gamma_A^{\mathrm{IHD}}\Big|_{\alpha\to 0} \to \Gamma_A^{\mathrm{TST}}. \tag{1.293}$$

Initially, Eq. (1.292) was derived using an adaptation of Kramers' escape-rate theory, without any recourse to Langer's work, by Smith and de Rozario [90] and Brown [18]. In Brown's calculations, the Taylor series expansions of the free energy density E about the saddle point C and about the well minimum A are, respectively, $E = E_C + (c_1^{(C)}\varphi^2 + c_2^{(C)}p^2)/2$ and $E = E_A + (c_1^{(A)}\varphi^2 + c_2^{(A)}p^2)/2$, so that Brown's IHD result for the escape rate reads [18]

$$\Gamma_A^{\text{IHD}} = \frac{\Omega_C \omega_A}{2\pi\omega_C} e^{-(E_C - E_A)} = \frac{\Omega_C}{\omega_C}\Gamma_A^{\text{TST}}, \tag{1.294}$$

where

$$\Omega_C = \frac{1}{4\tau_{\text{N}}}\left[-c_1^{(C)} - c_2^{(C)} + \sqrt{(c_2^{(C)} - c_1^{(C)})^2 - 4\alpha^{-2}c_1^{(C)}c_2^{(C)}}\right] \tag{1.295}$$

is the damped saddle angular frequency. Obviously, the Brown IHD equation, Eq. (1.294), coincides with the Klik and Gunther result, Eq. (1.292).

However, Eq. (1.294) is simply a special case of Langer's extension [103] of the IHD Kramers escape rate to many degrees of freedom and non-separable Hamiltonians, in order to generalize Becker and Döring's calculations [194] of the rate of condensation of a supersaturated vapor. Using Langer's method [103], Γ_i^{IHD} can be estimated as (cf. Eq. (1.284))

$$\Gamma_i^{\text{IHD}} = \frac{\Omega_C}{2\pi}\frac{Z_C}{Z_i}, \tag{1.296}$$

where

$$Z_i \approx \int\int_{\text{well}} e^{-E(\vartheta,\varphi)} \sin\vartheta d\vartheta d\varphi \text{ and } Z_C \approx \int\int_{\text{saddle}} e^{-E(\vartheta,\varphi)} \sin\vartheta d\vartheta d\varphi$$

are the well and saddle partition functions, respectively. Like the Brown IHD equation, Eq. (1.294), the Langer escape rate, Eq. (1.296), is only valid in the IHD region, $\alpha \geq 1$. Thus, it *cannot be used* to estimate the reversal time for $\alpha < 1$. Indeed for vanishing damping, $\alpha \to 0$, the IHD escape rate Γ_i^{IHD} from Eq. (1.294) reduces to the TST escape rate Γ_i^{TST}, which is obviously independent of α. However, by direct analogy with point particles, this is not the *true* VLD regime or *energy-controlled diffusion*, where the energy loss per cycle of the almost-periodic noisy motion of the magnetization on the saddle point energy (escape) trajectory is much less than the thermal energy, as noted by Klik and Gunther [106].

1.5.5 VLD escape rate for spins by the Kramers method

For zero STT, the MFPT quadrature solution, Eq. (1.241), is valid for *all barrier heights* including *low* barriers, $\Delta E = E_C - E_A \leq 1$. In the *high-barrier* (or low-temperature) limit, $\Delta E >> 1$, Eq. (1.241) can be considerably simplified. Indeed, the main contribution to the inner integral of Eq. (1.241) emanates from the region near the bottom of the well because *the negative exponential dominates the integral in that region.* Furthermore, the precession frequency now satisfies $f_{E_A} \simeq f_A$, which is independent of E because of the paraboloid approximation for the potential near the bottom of the well. Thus

$$\int_{E}^{E_C} f_{E'}^{-1} e^{-E'} dE' \simeq f_A^{-1} \int_{E_A}^{\infty} e^{-E'} dE' = f_A^{-1} e^{-E_A}. \tag{1.297}$$

In contrast, the main contribution to the outer integral of Eq. (1.241) comes from the *positive exponential factor dominating the integrand near the saddle point of the potential.* Therefore, with Eq. (1.179) and using the approximation $S_{E'} \approx S_{E_C}$, where the action S_{E_C} is given by Eq. (1.304), we have

$$\int_{E_A}^{E_C} S_E^{-1} e^{E} dE \simeq S_{E_C}^{-1} \int_{-\infty}^{E_C} e^{E} dE = S_{E_C}^{-1} e^{E_C}. \tag{1.298}$$

Here we have entirely neglected the energy dependence of S_E. Moreover, the tacit assumption is again made that the separatrix lies *infinitesimally near* to the *closed* noiseless and undamped trajectory (Stoner–Wohlfarth orbit) with energy E_C. The critical energy is the energy required by a macrospin to just escape the well and the separatrix delineates the bounded precessional motion in the well from that outside it. Using Eqs. (1.297) and (1.298) in Eq. (1.241) yields the VLD reversal time from an individual well with a *single* escape path [165,166]

$$\tau_{\text{VLD}}^{\text{as}} \sim \frac{e^{E_C - E_A}}{\alpha f_A S_{E_C}} = \frac{\tau_{\text{TST}}}{\alpha S_{E_C}}, \tag{1.299}$$

where $\tau_{\text{TST}} \sim e^{E_C - E_A} / f_A$ is the TST reversal time. In the VLD regime, the spin system is only *very lightly* coupled to the bath. Thus, the energy loss per cycle of the almost-periodic noisy motion of the magnetization on the saddle-point energy (escape) trajectory is much less than the thermal energy, $\alpha S_{E_C} << 1$, so that $\tau_{\text{VLD}}^{\text{as}} >> \tau_{\text{TST}}$ for VLD. The asymptotic equation, Eq. (1.299), for the reversal time was obtained for the first time by Klik and Gunther [106] via the uniform asymptotic expansion of the MFPT method of Matkowsky *et al.* [195] and not via the energy-controlled diffusion equation.

To evaluate the reversal time both in the VLD and IHD regimes from Eqs. (1.292) and (1.299), we require *explicit* equations for the dimensionless magnetic action S_{E_C}, the well and saddle frequencies f_A and f_C along with the well and saddle energies E_A and E_C. This is accomplished by supposing [18] that the free energy per unit volume $E(\mathbf{M})$ of a single-domain particle has a multistable structure with a minimum at $\mathbf{n}_A^+$ and $\mathbf{n}_A^-$ separated by a potential barrier with a saddle point at $\mathbf{n}_C$. If $\mathbf{M}$ is close to the stationary points $\mathbf{n}_A^+$, $\mathbf{n}_A^-$, $\mathbf{n}_C$ and $(u_1^{(p)}, u_2^{(p)}, u_3^{(p)})$ (where $p = A^\pm, C$) denote the direction cosines of $\mathbf{M}$, then $E(\mathbf{M})$ can be approximated to second order in $u_1^{(p)}$ and $u_2^{(p)}$ via the Taylor series

$$E = E_p + \frac{1}{2}\left[c_1^{(p)}(u_1^{(p)})^2 + c_2^{(p)}(u_2^{(p)})^2\right] + \dots . \tag{1.300}$$

To determine the expansion coefficients $c_1^{(p)}$, $c_2^{(p)}$, and E_p, we first recall that the rotational transformation matrix $\mathbf{R}^{(p)}$ relating the reference polar coordinate system P and a new polar coordinate system P' with the origin at the stationary point $\mathbf{n}_p$, is given by [108]

$$\mathbf{R}^{(p)} = \begin{pmatrix} \cos\varphi_p \cos\vartheta_p & \sin\varphi_p \cos\vartheta_p & -\sin\vartheta_p \\ -\sin\varphi_p & \cos\varphi_p & 0 \\ \cos\varphi_p \sin\vartheta_p & \sin\varphi_p \sin\vartheta_p & \cos\vartheta_p \end{pmatrix},$$

so that the relationship between the direction cosines $u_n^{(p)}$ and $u_m'^{(p)}$ in the coordinate systems P and P' is given by

$$u_n^{(p)} = R_{1n}^{(p)} u_1'^{(p)} + R_{2n}^{(p)} u_2'^{(p)} + R_{3n}^{(p)} u_3'^{(p)}, \quad n = 1, 2, 3. \tag{1.301}$$

Because $u_3'^{(p)} \approx 1 - (u_1'^{(p)2} + u_2'^{(p)2})/2$, both E_p and the expansion coefficients $c_n^{(p)}$ can be evaluated from Eqs. (1.300) and (1.301) as

$$E_p = E_p(u_1^{(p)}, u_2^{(p)})\Big|_{u_1'^{(p)}, u_2'^{(p)}=0}, \quad c_n^{(p)} = \left.\frac{\partial^2 E}{\partial u_n'^{(p)2}}\right|_{u_1'^{(p)}, u_2'^{(p)}=0}. \tag{1.302}$$

The well precession and the saddle frequencies are then defined via the Taylor expansion Eq. (1.300) about the relevant stationary point as

$$f_A = \frac{\gamma kT}{2\pi v \mu_0 M_S}\sqrt{c_1^{(A)} c_2^{(A)}}, \quad f_C = \frac{\gamma kT}{2\pi v \mu_0 M_S}\sqrt{c_1^{(C)} c_2^{(C)}}. \tag{1.303}$$

In spherical polar coordinates $(\mathbf{e}_r, \mathbf{e}_\vartheta, \mathbf{e}_\varphi)$ (see Fig. 1.5), where $\mathbf{u} = \mathbf{e}_r$, $d\mathbf{u} = \mathbf{e}_\vartheta d\vartheta + \mathbf{e}_\varphi \sin\vartheta d\varphi$, and $\nabla E = \mathbf{e}_\vartheta \partial_\vartheta E + \mathbf{e}_\varphi \csc\vartheta \partial_\varphi E$, S_E from Eq. (1.179) simply reduces to

$$S_{E_C} = \oint_{E=E_C} \left[\frac{\partial E}{\partial \varphi}\frac{1}{\sin\vartheta} d\vartheta - \frac{\partial E}{\partial \vartheta}\sin\vartheta d\varphi\right]. \tag{1.304}$$

The line integral in Eq. (1.304) is taken along the critical energy trajectory (Stoner–Wohlfarth orbits) on which the magnetization (direction of precession) may reverse by passing through the saddle point(s).

In the presence of STT, the general MFPT equation, Eq. (1.247), is valid for *all barrier heights* including low barriers. However, in the high-barrier limit, Eq. (1.247) can be simplified as for zero STT also yielding Eq. (1.299) [167], where, however, the effective barrier ΔE is given by

$$\Delta E = E_C - E_A - \frac{1}{\alpha}\int_{E_A}^{E_C} \frac{V_E}{S_E} dE \tag{1.305}$$

and the work V_E is given by Eq. (1.184).

1.5.6 *Turnover escape rate for spins*

In [181,196], Coffey *at al.* have shown that the Mel'nikov–Meshkov formalism [186,187], bridging the VLD and IHD escape rates as a function of the dissipation parameter for mechanical particles can also be extended to include the magnetization relaxation of magnetic nanoparticles having noncircularly symmetric potentials of the magnetocrystalline anisotropy.

The equation bridging the VLD and IHD escape rates is given by [181,196]

$$\Gamma_A = A(\alpha S_{E_C})\Gamma_A^{\mathrm{IHD}}, \tag{1.306}$$

where $A(\Delta)$ is the depopulation factor given by Eq. (1.258), which interpolates between the VLD and ID regimes. Because

$$A(\Delta) \to 1 \text{ as } \Delta \to \infty \text{ and } A(\Delta)/\Delta \to 1 \text{ as } \Delta \to 0, \tag{1.307}$$

one may show [181] that Eq. (1.306) reduces, in the IHD and VLD regimes, to Eqs. (1.292) and (1.299), respectively. For a double-well potential, this prefactor in the intermediate-to-low-damping regime (crossover region), where neither IHD nor VLD treatments apply is

$$A(\alpha S_1, \alpha S_2) = \frac{A(\alpha S_1)A(\alpha S_2)}{A(\alpha S_1 + \alpha S_2)}, \tag{1.308}$$

where S_i is the magnetic action in the well i. For a bistable potential with *nonequivalent wells* an asymptotic equation bridging the VLD and TST magnetization reversal time from a single well for $\alpha \le 1$ and $\Delta E >> 1$ is given by [167,196]

$$\tau \approx \frac{\tau_{\mathrm{TST}}^{+}\tau_{\mathrm{TST}}^{-}A(\alpha S_{E_C}^{+} + \alpha S_{E_C}^{-})}{(\tau_{\mathrm{TST}}^{+} + \tau_{\mathrm{TST}}^{-})A(\alpha S_{E_C}^{+})A(\alpha S_{E_C}^{-})}, \tag{1.309}$$

where $\tau_{\mathrm{TST}}^{\pm} = e^{\Delta E_{\pm}} / f_A^{\pm}$, is the TST reversal time for either of the two wells. The asymptotic formula for the VLD reversal time *including STT* may be generalized to a wider damping range exactly as for nanomagnets with zero STT. Thus Eq. (1.309) is also valid for the magnetization reversal time including STT for $\alpha \le 1$ and $\Delta E_{\pm} >> 1$.

Notice that a second interpolation problem arises in the magnetic version of the Kramers escape rate, namely how to join circularly and noncircularly symmetric asymptotic expressions for the greatest relaxation time, in the limit of *small* departures from circular symmetry. This problem has been described in detail in Refs. [181,197] by considering the asymptotes generated by Eq. (1.188) for $\psi = 0$ and $\psi \neq 0$. We emphasize that in deriving all these formulas, it was assumed throughout that the potential is noncircularly symmetric. If the departures from circular symmetry become small, the noncircularly symmetric asymptotic formulas for the escape rate may be smoothly connected to the circularly symmetric ones by means of suitable interpolation integrals [181]. Finally, Eqs. (1.292), (1.299), and (1.306) may be used to experimentally verify the Kramers theory for magnetic particles. This has been accomplished using the sophisticated single-particle measurement techniques developed by Wernsdorfer [20].

1.6 Switching-field curves and surfaces

1.6.1 *Stoner–Wolhfarth astroids*

If one knows the reversal time of the magnetization **M** as a function of the *direction of an external magnetic field*, one can include thermal effects in the calculation of switching-field curves and/or surfaces. We recall that the first calculation of the magnetization reversal of single–domain ferromagnetic particles with uniaxial anisotropy subjected to an applied field was made by Stoner and Wohlfarth [14]. They made the hypothesis of *coherent rotation* of the magnetization and *zero* temperature, so that *thermally induced switching between the potential minima is completely ignored.* In the *simplest uniaxial anisotropy*, as considered by them, the magnetization reversal simply occurs at that particular value of the applied field (switching field) which destroys the bistable nature of the potential. The 2D parametric plot, of the parallel vs. the perpendicular component of the switching field, then yields the famous *critical (*or *limit of metastability) curves,* or *astroids* [70]. The calculation of the switching-field curves or surfaces at zero temperature has been generalized by Thiaville [71] to an *arbitrary* free-energy density potential. By fitting experimental switching-field curves and surfaces, one can then determine the free energy of a nanomagnet and the corresponding anisotropy constants (see Fig. 1.19).

The starting point of the Thiaville calculation is the *normalized* free-energy potential $\overline{V}(\mathbf{u})$ in the presence of a dc magnetic field $\mathbf{H}_0$ defined as

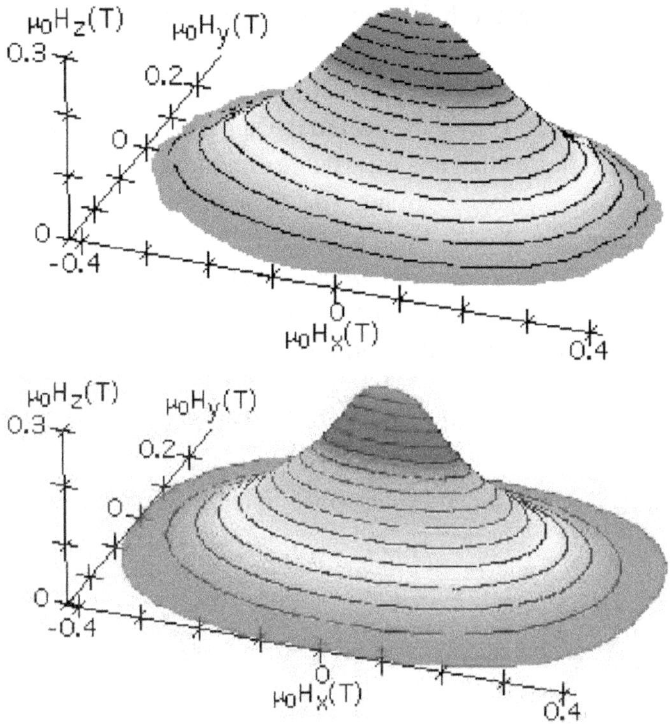

Fig. 1.19. 3D theoretical and experimental (measured at $T = 35$ mK with the microSQUID; upper figure) switching-field surfaces of a 3 nm cobalt cluster. These surfaces are symmetrical with respect to the $H_x - H_y$ plane and only the upper part ($\mu_0 H_z > 0$) is shown. Continuous lines on the surface are the contour lines on which $\mu_0 H_z$ is constant. The theoretical switching-field surface is calculated via Thiaville's method [71] for the free energy $V = -(K_4 / 4)(\sin^2 2\vartheta + \sin^4\vartheta \cos^2 2\varphi) - K_1 \cos^2\vartheta + K_2 \sin^2\vartheta \sin^2\varphi$ with the anisotropy constants $K_1 = 2.2{\cdot}10^5$ J/m³, $K_2 = 0.9{\cdot}10^5$J/m³, $K_4 = 0.1{\cdot}10^5$ J/m³. Reprinted from M. Jamet, W. Wernsdorfer, C. Thirion, D. Mailly, V. Dupuis, P. Mélinon, and A. Pérez, Phys. Rev. Lett. 86, 4676 (2001), with the permission of the American Physical Society.

$$\bar{V}(\mathbf{u}) = G(\mathbf{u}) - 2(\mathbf{u} \cdot \mathbf{h}), \tag{1.310}$$

where $\mathbf{h}$ is the normalized external field $\mathbf{H}_0 / H_K$ (H_K is the effective anisotropy field) and G is the normalized anisotropy without the external field $\mathbf{H}$. The unit vector $\mathbf{u} = \mathbf{M} / M_S$ is described by the polar angle ϑ relative to some reference axis OZ, and the azimuthal angle φ, while the orientation of the vector **h** is specified by the polar angle ψ and the azimuthal angle ϕ (see Fig. 1.5).

The switching field is characterized by the fact that both the first and second derivatives of $\overline{V}$ with respect to the angles ϑ and φ vanish so that this condition indicates that one metastable minimum and one saddle point in the potential $\overline{V}$ *merge*, giving rise to a point of inflexion. These conditions correspond to a *switching-field surface* in 3D space. This surface, as it *generalizes the critical curves* of the 2D problem of Stoner and Wohlfarth, is called the *limit of metastability surface*. At any point of that surface, $\overline{V}$ must satisfy the stationary conditions

$$\frac{\partial\overline{V}}{\partial\vartheta}=\frac{\partial G}{\partial\vartheta}-2(\mathbf{h}\cdot\mathbf{e}_\vartheta)=0,\ \frac{\partial\overline{V}}{\partial\varphi}=\frac{\partial G}{\partial\varphi}-2(\mathbf{h}\cdot\mathbf{e}_\varphi)\sin\vartheta=0,\quad(1.311)$$

so that the field vector **h** can be described in terms of a parameter λ, viz.,

$$\mathbf{h}=\lambda\mathbf{e}_r+\frac{1}{2}\frac{\partial G}{\partial\vartheta}\mathbf{e}_\vartheta+\frac{1}{2\sin\vartheta}\frac{\partial G}{\partial\varphi}\mathbf{e}_\varphi,\qquad(1.312)$$

where the unit vectors $\mathbf{e}_r$, $\mathbf{e}_\vartheta$, and $\mathbf{e}_\varphi$ (which form the orthonormal direct basis) are defined as [104]

$$\mathbf{e}_r=\begin{pmatrix}\sin\vartheta\cos\varphi\\ \sin\vartheta\sin\varphi\\ \cos\vartheta\end{pmatrix},\ \mathbf{e}_\vartheta=\begin{pmatrix}\cos\vartheta\cos\varphi\\ \cos\vartheta\sin\varphi\\ -\sin\vartheta\end{pmatrix},\ \mathbf{e}_\varphi=\begin{pmatrix}-\sin\varphi\\ \cos\varphi\\ 0\end{pmatrix}.\qquad(1.313)$$

The switching conditions are now determined by the equation

$$\frac{\partial^2\overline{V}}{\partial\vartheta^2}\frac{\partial^2\overline{V}}{\partial\varphi^2}-\left[\frac{\partial^2\overline{V}}{\partial\vartheta\partial\varphi}\right]^2=0.\qquad(1.314)$$

Because the second derivatives of $\overline{V}$ are given by

$$\begin{aligned}\frac{\partial^2\overline{V}}{\partial\vartheta^2}&=\frac{\partial^2 G}{\partial\vartheta^2}+2\lambda,\\ \frac{\partial^2\overline{V}}{\partial\varphi^2}&=\frac{\partial^2 G}{\partial\varphi^2}+\left(\cot\vartheta\frac{\partial G}{\partial\vartheta}+2\lambda\right)\sin^2\vartheta,\\ \frac{\partial^2\overline{V}}{\partial\vartheta\partial\varphi}&=\frac{\partial^2\overline{V}}{\partial\varphi\partial\vartheta}=\sin\vartheta\frac{\partial}{\partial\vartheta}\left(\frac{1}{\sin\vartheta}\frac{\partial G}{\partial\varphi}\right),\end{aligned}\qquad(1.315)$$

Eq. (1.314) simply reduces to a quadratic equation in the parameter λ, viz.,

$$4\lambda^2 + 2\lambda\left[\frac{1}{\sin^2\vartheta}\frac{\partial^2 G}{\partial\varphi^2} + \cot\vartheta\frac{\partial G}{\partial\vartheta} + \frac{\partial^2 G}{\partial\vartheta^2}\right] + \left[\frac{1}{\sin^2\vartheta}\frac{\partial^2 G}{\partial\varphi^2} + \cot\vartheta\frac{\partial G}{\partial\vartheta}\right]\frac{\partial^2 G}{\partial\vartheta^2} - \left[\frac{\partial}{\partial\vartheta}\left(\frac{1}{\sin\vartheta}\frac{\partial G}{\partial\vartheta}\right)\right]^2 = 0, \tag{1.316}$$

which has two distinct roots λ^+ and λ^- given by

$$\lambda^{\pm} = -\frac{1}{4}\left\{\left[\frac{1}{\sin^2\vartheta}\frac{\partial^2 G}{\partial\varphi^2} + \cot\vartheta\frac{\partial G}{\partial\vartheta} + \frac{\partial^2 G}{\partial\vartheta^2}\right] \mp \sqrt{\left[\frac{1}{\sin^2\vartheta}\frac{\partial^2 G}{\partial\varphi^2} + \cot\vartheta\frac{\partial G}{\partial\vartheta} - \frac{\partial^2 G}{\partial\vartheta^2}\right]^2 + 4\left[\frac{\partial}{\partial\vartheta}\left(\frac{1}{\sin\vartheta}\frac{\partial G}{\partial\varphi}\right)\right]^2}\right\}. \tag{1.317}$$

The semi-line, Eq. (1.312), defined by $\lambda > \lambda^+$ is the locus of the fields for which the magnetization is stable. However, for $\lambda = \lambda^+$ the metastable minimum in the potential $\overline{V}$ disappears so that the magnetization **M** can now escape from the potential well. Thus, the switching-field surface may be obtained from the vector $\mathbf{h}_S$ defined as [71]

$$\mathbf{h}_S = \lambda^+\mathbf{e}_r + \frac{1}{2}\frac{\partial G}{\partial\vartheta}\mathbf{e}_\vartheta + \frac{1}{2\sin\vartheta}\frac{\partial G}{\partial\varphi}\mathbf{e}_\varphi. \tag{1.318}$$

To illustrate the Thiaville geometrical method, we summarize in Fig. 1.20 the switching-field surfaces and curves for *uniaxial* anisotropy:

$$G_{\mathrm{un}}(\vartheta,\varphi) = \sin^2\vartheta, \tag{1.319}$$

biaxial anisotropy:

$$G_{\mathrm{b}}(\vartheta,\varphi) = \sin^2\vartheta + \delta\sin^2\vartheta\cos^2\varphi, \tag{1.320}$$

and *positive* and *negative cubic* anisotropies:

$$G_{\mathrm{c}\pm}(\vartheta,\varphi) = \pm(\sin^4\vartheta\sin^2 2\varphi + \sin^2 2\vartheta). \tag{1.321}$$

Furthermore, the transverse ($h_\perp$) and longitudinal ($h_\parallel$) components of the normalized 2D switching field for uniaxial, biaxial, and positive and negative cubic anisotropy are given by

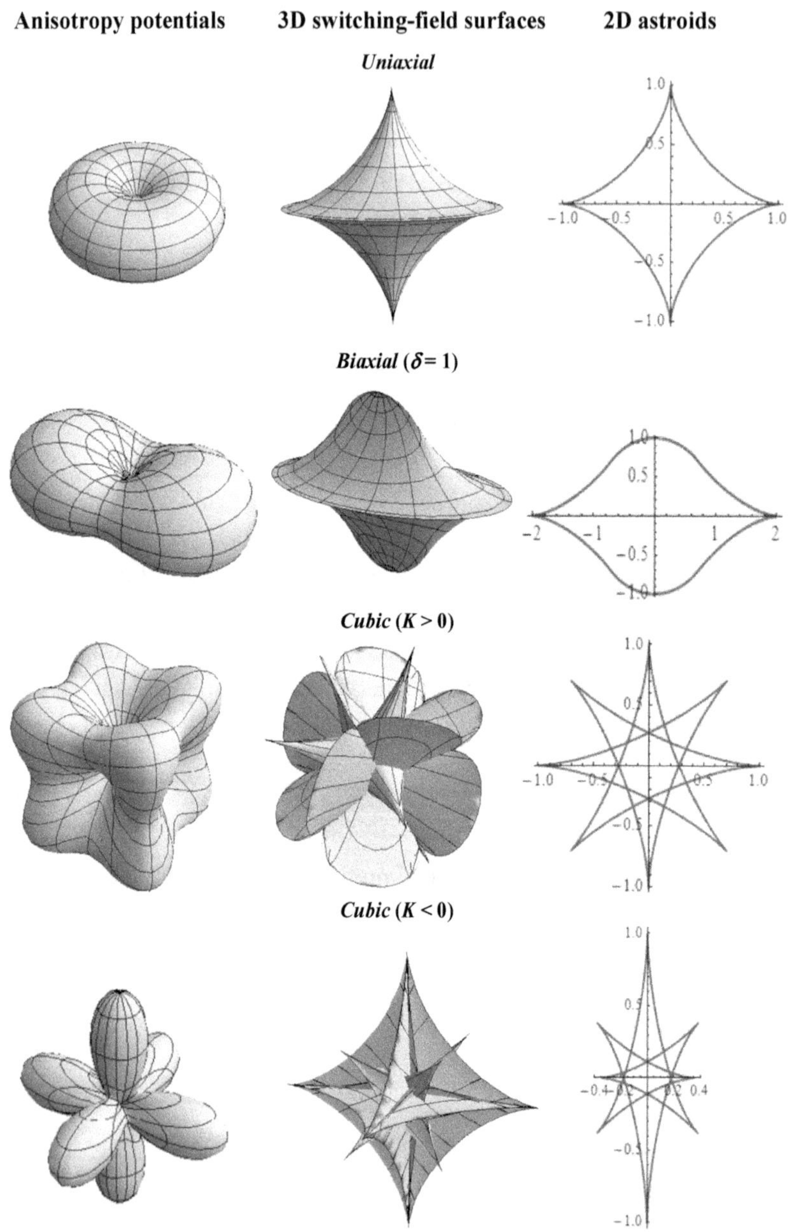

Fig. 1.20. 3D switching-field surfaces and 2D curves (astroids) for uniaxial, biaxial, and positive and negative cubic anisotropies.

$$h_{\perp}^{\text{un}} = \sin^3 \vartheta, \quad h_{\parallel}^{\text{un}} = -\cos^3 \vartheta, \tag{1.322}$$

$$\begin{aligned} h_{\perp}^{\text{b}} &= (1/2)\left[\delta + (1+\delta)\sin^2 \vartheta + \left|1-(1+\delta)\cos^2 \vartheta\right|\right]\sin \vartheta, \\ h_{\parallel}^{\text{b}} &= -(1/2)\left[1 + (1+\delta)\cos^2 \vartheta - \left|1-(1+\delta)\cos^2 \vartheta\right|\right]\cos \vartheta, \end{aligned} \tag{1.323}$$

$$h_{\perp}^{\text{c+}} = (2+3\cos 2\vartheta)\sin^3 \vartheta, \; h_{\parallel}^{\text{c+}} = (2-3\cos 2\vartheta)\cos^3 \vartheta, \tag{1.324}$$

$$\begin{aligned} h_{\perp}^{\text{c}-} &= \frac{1}{8\sqrt{2}}\sin^3 \vartheta(11 + 9\cos 2\vartheta - 3\left|1+3\cos 2\vartheta\right|), \\ h_{\parallel}^{\text{c}-} &= \frac{1}{32}(5 - 28\cos 2\vartheta - 9\cos 4\vartheta - 12\left|1+3\cos 2\vartheta\right|\sin^2 \vartheta)\cos \vartheta. \end{aligned} \tag{1.325}$$

The concept of limit-of-metastability surfaces and curves plays a fundamental role in micromagnetism, because it leads to an elegant graphical representation of the stability properties of nanoparticles [70,71]. Switching-field curves and surfaces have been measured experimentally [20,72,73,198] therefore yielding estimates of magnetic anisotropy in a single nanoparticle. The distinct anisotropy contributions (uniaxial, biaxial, cubic) can also be separated. Switching-field surface measurements [20,72,73,198] have also proved that the angular dependence of the switching field concurs with Brown's model.

1.6.2 *Effect of thermal agitation*

For zero temperature, magnetization reversal is only possible if the energy barrier is *fully suppressed.* At *finite* temperatures, the switching field becomes (intuitively) smaller, therefore, relaxation of the magnetization must now be accounted for. Moreover, *experimental* observation of the magnetization reversal depends both on the relaxation time of the cluster *and* on the measuring time τ_{m} of the experimental setup itself. Therefore, the magnetization reversal can be *experimentally* observed only if the relaxation time lies *in the time window of the experiment, or equivalently, if the relaxation rate is equal to the measuring frequency* $f_{\text{m}} = 1/(2\pi\tau_{\text{m}})$. Hence, for experimental observation of the magnetization reversal at finite temperatures, Néel's criterion for observation of the magnetization reversal must hold, namely,

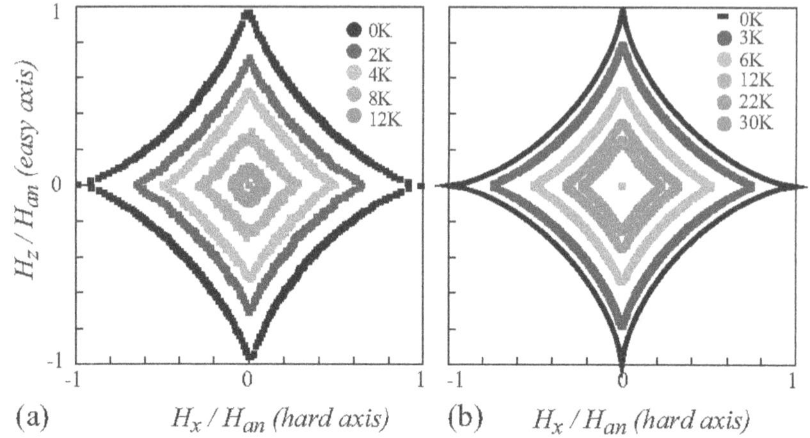

Fig. 1.21. Temperature dependence of the switching field of a 3 nm Co particle. (a) Experimental data measured with τ_m = 0.01 s and (b) Numerical calculations for a uniaxial anisotropy with α = 0.1. (Reproduced from Ref. [75] by permission of Elsevier).

$$\tau_{\mathrm{m}} = \tau(\psi, \phi, h_{\mathrm{S}}, \alpha, \text{anisotropy parameters}), \tag{1.326}$$

where τ is the reversal time of the magnetization and h_S is the normalized switching field. This equation can be solved numerically for h_S and τ_m for given values of ψ, ϕ, α and anisotropy parameters.

The temperature dependences of the switching-field curves and surfaces of magnetic nanoparticles have been measured experimentally [20,198] using the micro-SQUID technique. This technique allows one to measure the 3D switching-field surfaces as well as the temperature dependence of *individual* grains, simultaneously yielding their magnetic anisotropy as well as allowing one to probe the magnetization dynamics. Temperature effects on the switching-field curves can also be treated theoretically. For example, Vouille *et al.* [75] have calculated temperature–dependent switching curves by numerically solving the stochastic Landau–Lifshitz–Gilbert equation for uniaxial anisotropy. Moreover (as occurs in the experimental results) they have shown by numerical solution that Brown's diffusion model reproduces the principal features of the measured astroids [20], which continuously diminish with increasing temperature and ultimately vanish at the blocking temperature, thus corroborating the experimental results (see Fig. 1.21). They also

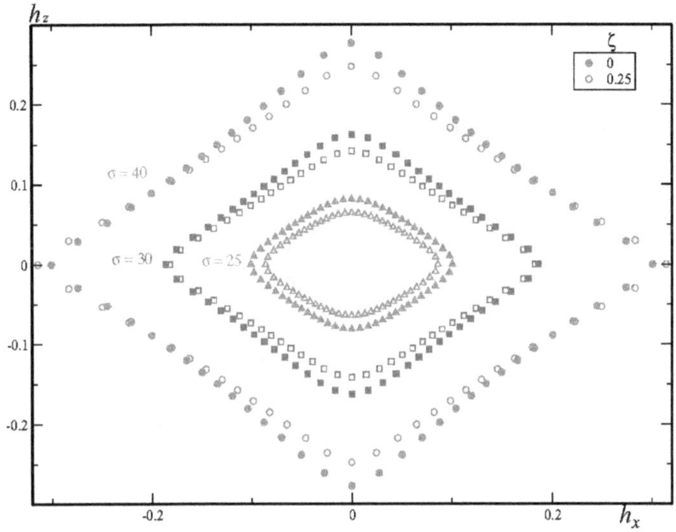

Fig. 1.22. Switching–field curves for mixed anisotropy for various values of the anisotropy parameter σ and $\zeta = 0$ (open symbols) and 0.25 (filled symbols). The magnetic field is applied in the *XZ* plane, and the easy axis coincides with the *Z* axis. (Reproduced from Ref. [199] by permission of the Institute of Physics.)

found that an Arrhenius law for the relaxation time τ corresponds closely to the calculations, permitting determination of the attempt frequency. The results also compare favorably with the escape-rate formulas for τ.

Déjardin *et al.* [199] have also estimated the temperature effects on the switching-field curves for a mixed anisotropy, viz.,

$$G_{\text{mix}}(\vartheta,\varphi) = \sin^2\vartheta - \zeta(\sin^4\vartheta\sin^2 2\varphi + \sin^2 2\vartheta). \qquad (1.327)$$

The mixed anisotropy allows one to approximately consider the surface effects on the magnetic properties of nanoparticles. At very low temperatures and with mixed parameter $\zeta = 0$ this anisotropy again yields the Stoner–Wohlfarth astroid [14]. The switching-field curves computed [199] using matrix continued fractions for different temperatures are shown in Fig. 1.22, where the measuring frequency has been taken as f_m = 100 Hz. Two observations can be made concerning the results shown in that figure. First, we observe a quasihomogeneous shrinking of the switching-field curve due to thermal fluctuations. In addition, the switching field decreases with temperature, owing to a decrease in the relaxation time. Moreover, Fig. 1.22 shows that, on increasing ζ from 0 to 0.25, a flattening of the switching-field curves exists; that is, the critical field *is reduced* in the longitudinal direction ($\psi = 0$) and *increased* in the transverse direction ($\psi = \pi/2$). Recalling that the mixed parameter ζ is related to the surface anisotropy, these results indicate how the latter affects the switching field. We remark that, when a magnetic field is applied in an *arbitrary* direction, as is necessary for the calculation of the switching field, the reversal time τ can always be calculated numerically (using matrix continued fractions) as λ_1^{-1}. For high barriers, where the computation of λ_1^{-1} by this method becomes extremely time-consuming, the analytic (asymptotic) escape-rate equations we have given may be used to estimate τ.

1.7 Ferrofluids

Ferrofluids are [200,201] *stable colloidal suspensions* of *single-domain* magnetic particles in a *liquid carrier*. Such fluids are composed of small (~150 Å) particles of ferromagnetic material, coated with a molecular layer of a surfactant, and suspended in an ordinary liquid. The coatings prevent the particles from adhering to each other, and thermal agitation keeps them suspended because of the ensuing Brownian motion.

The simplest model of a magnetic fluid is considered as a gas of noninteracting ferroparticles suspended in a liquid carrier. Consequently, its magnetic properties are analogous to those of a paramagnetic gas, i.e., the magnetization curve is described by the Langevin function, the static susceptibility, χ, by Curie's law $\chi \sim T^{-1}$, and its dispersion by the Debye formula for the complex magnetic susceptibility $\chi(\omega)$:

$$\frac{\chi(\omega)}{\chi} = \frac{1}{1 + i\omega\tau_c},$$

where τ_c is a characteristic relaxation time. In ferrofluids, both Debye relaxation with the Debye relaxation time τ_D (due to mechanical rotation of the fluid particles [97]) and Néel relaxation with a characteristic time τ

(due to rotation of the magnetic moment inside the particle) may occur. Theoretical studies have mainly assumed [200] that the Debye and Néel mechanisms may be treated *separately*, i.e., one has two extreme types of behavior: (a) the Debye relaxation mechanism, where the Néel relaxation mechanism is *blocked* or *frozen* in the particle, and (b) where the *mechanical rotation* of the particles is *frozen* and only the Néel mechanism is operative. The overall characteristic relaxation time τ_C is then supposed to obey the equation

$$\tau_c = \frac{\tau\tau_D}{\tau + \tau_D}. \tag{1.328}$$

A popular formula used in analysis, for example, is Brown's formula for the relaxation time τ of nanoparticles with uniaxial anisotropy (see Chapter 2) combined with the Debye time τ_D. The relaxation process has been described succinctly in Ref. [200] as follows:

> "There is a finite coupling between the orientation of the magnetic moment μ of a ferroparticle and the position of the particle itself (orientation of its crystal axes). Without this coupling the moment μ would be similar to a compass needle, where rotation of the instrument frame does not influence the behavior. Because of the above coupling, the reorientation of the vector μ may take place in two different ways: (1) rotation of μ within the particle with respect to its crystal axes, and (2) rotation of μ together with the particle with respect to the liquid matrix. Both processes – they proceed simultaneously – are of the rotary diffusion type. The efficiency of the internal (Néel) diffusion of the magnetic moment strongly depends on particle size, but that of the external one (the Brownian) depends strongly on the viscosity of the liquid carrier."

Along with these processes, two other factors are of great importance:

- *Polydispersedness of ferrofluids.* The actual distribution of the particle size in colloids results in a moderate (1–2 orders of magnitude) extension of the Debye relaxation spectrum, and an enormous (up to 13–15 orders!) extension of the Néel relaxation spectrum. Averaging over a particle-size distribution function may change the low-frequency Debye susceptibility $\chi(\omega)$ unrecognizably.
- *Blocking of the rotational degrees of freedom of the particles on solidification of a carrier liquid.* As the liquid matrix freezes, the

suspended particles lose their mechanical mobility. The Debye relaxation mechanism thus becomes ineffective.

We reiterate that one of the outstanding advantages of the ferrofluid system, as opposed to polar molecules, as a test of the Debye theory of relaxation, is that unlike electric dipoles, the ferrofluid particles closely approximate in size to actual Brownian particles. Thus, the conditions for the validity of the Debye theory are more closely satisfied by the ferrofluid particles.

A long-standing problem concerning ferrofluids is how in a single model one may include the solid-state mechanism of (longitudinal or Néel) relaxation by internal rotation of the magnetic dipole moment with respect to the crystalline axes inside the particle, the associated transverse modes (leading to ferromagnetic resonance), and the mechanical Brownian relaxation due to physical rotation of the ferrofluid particles in the carrier fluid. (We recall that the slowest mode of the longitudinal relaxation process describes the reversal of the magnetization over the potential barrier created by the internal anisotropy of the particle, which of course will be modified by an external applied field. We have seen that the time to cross the barrier is known as the Néel relaxation time and essentially follows the Arrhenius law.) Therefore, a major development in the theory of ferrofluids would be a model of particle reorientation which avoids the two extremes of a *frozen* Debye or a *frozen* Néel mechanism, instead taking account of both mechanisms *simultaneously*. Such a development has been inspired [202] by the discoveries of Fannin *et al.* [203,204], which suggest that both relaxation and ferromagnetic resonance behavior appear in magnetic fluids. The "egg" model of magnetic fluids [200], which is essentially a form of the itinerant oscillator model [205], represents an attempt to consider the combined behavior.

The question posed above was answered in part by Shliomis and Stepanov [200]. They showed that for uniaxial particles in weak applied magnetic fields and in the noninertial limit [97], the equations of motion of the ferrofluid particle incorporating both the internal and the Brownian relaxation processes *decouple* from each other. Thus, the reciprocal of the greatest relaxation time is the sum of the reciprocals of the Néel and Brownian relaxation times of both processes considered *independently*, i.e., those of a *frozen* Néel and a *frozen* Brownian mechanism. Here, the joint probability density function of the orientations of the magnetic moment and the particle in the fluid, i.e., the crystallographic axes, is merely the product of the individual probability density functions of the orientations of the axes and the particle; thus, the Fokker–Planck equation for the joint probability density function factorizes as do the statistical moments. The internal and Debye processes are therefore statistically independent. If the applied field is sufficiently strong, however, no such decoupling can take place. The Shliomis–Stepanov approach [200] to the ferrofluid relaxation problem, which is based on the Fokker–Planck equation, has come to be known in the literature on magnetism as the *egg model*. A different treatment has been given by Scherer and Matuttis [206] using a generalized Lagrangian formalism: however, in their discussion of the applications, they also limited themselves to a *frozen* Néel and a *frozen* Brownian mechanism, respectively.

Here, we reexamine the egg model (which is of course a 3D form of the itinerant oscillator model [97]) noting the ratio of the free Brownian diffusion (Debye) time to the free Néel diffusion time and discarding the assumption of a weak applied field. The results will then be used to demonstrate how the ferrofluid magnetic relaxation problem in the noninertial or high mechanical friction limit, is essentially similar to Néel relaxation in a uniform magnetic field applied at an oblique angle to the easy axis of magnetization (see Section 2.3.2). *Unlike in the solid-state mechanism however, the orientation of the field with respect to the easy axis is now a function of time, because of the physical rotation of the particle in the ferrofluid.* The fact that the behavior is essentially similar, in all other respects, to the solid-state oblique-field problem suggests that there is a strong intrinsic dependence of the greatest relaxation time on the damping (independent of that due to the free diffusion time), because of the coupling between longitudinal and transverse modes. Alternatively, unlike in circular symmetry, the set of eigenvalues which characterize the longitudinal relaxation now depends strongly on the damping. This *precession-dependent* or *mode-coupled* longitudinal relaxation is absent in the weak-field case [200]. Here, the equations of motion decouple into those describing a frozen Néel (pure Debye or Brownian) and a frozen Brownian (pure Néel) mechanism of relaxation, respectively. Thus, the Néel or longitudinal relaxation is governed by a circularly symmetric potential. Hence, there is no *intrinsic* dependence of the greatest relaxation time on the damping. Moreover, the longitudinal set of eigenvalues is independent of the damping, which enters only via the free (Néel) diffusion time. Thus, in the linear response to a weak applied field, the only effect of the fluid carrier is to further dampen, according to a Debye or Rocard (inertia-corrected Debye) mechanism [97], both the longitudinal and transverse responses of the solid-state mechanism. Furthermore, it will be demonstrated that now the transverse relaxation process in the ferrofluid, which may give rise to ferromagnetic resonance, may be accurately described by the solid-state transverse result for uniaxial anisotropy. This is so because the Brownian (Debye) relaxation time greatly exceeds the characteristic relaxation times of all the transverse modes. The latter conclusion is reinforced by the favorable agreement of the linear-response result with experimental observations [202] of the complex susceptibility of four ferrofluid samples.

To illustrate how precession-dependent relaxation effects may manifest themselves in a ferrofluid, it will be useful to briefly summarize the differences in relaxation behavior for circularly and noncircularly symmetric potentials of the magnetocrystalline anisotropy and applied field, when the Brownian relaxation mode is frozen. Thus, only the solid-state (Néel) mechanism is operative; that is, the magnetic moment of the nanoparticle may reorientate only with respect to the crystalline axes. Our starting point is the Landau–Lifshitz–Gilbert equation for the dynamics of the magnetization **M** of a single-domain ferromagnetic particle, namely

$$\dot{\mathbf{M}} = (1+\alpha^2)^{-1}\gamma\left\{[\mathbf{H}\times\mathbf{M}]-\alpha M_S^{-1}[\mathbf{M}\times[\mathbf{M}\times\mathbf{H}]]\right\} \qquad (1.329)$$

(all the definitions are as already given). Equation (1.329) is simply the Langevin equation of the solid-state orientation process. The field **H** may be written as $\mathbf{H} = \mathbf{H}_{ef} + \mathbf{h}(t)$, where

$$\mathbf{H}_{ef} = -\mu_0^{-1} v_m \partial_{\mathbf{m}} V \qquad (1.330)$$

is the conservative part of **H**, $\mathbf{m} = v_m\mathbf{M}$ is the magnetic moment of the nanoparticle and $\mathbf{h}(t)$ is the white-noise random field. By introducing $\mathbf{u} = \mathbf{m}/\mathrm{m}$, we find that Eq. (1.329) assumes the form of a kinematic relation involving the angular velocity $\boldsymbol{\omega}_e$ of the dipole vector **m**:

$$\dot{\mathbf{u}} = \boldsymbol{\omega}_e \times \mathbf{u} = \left(\boldsymbol{\omega}_L + \boldsymbol{\omega}_R\right) \times \mathbf{u}. \tag{1.331}$$

Here

$$\boldsymbol{\omega}_L = -\gamma(1+\alpha^2)^{-1}\left[\mathbf{H}_{ef} + \mathbf{h}(t)\right] \tag{1.332}$$

is the angular velocity of undamped (Larmor) precession of **m** in the field $\mathbf{H}_{ef}$, superimposed on which is the rapidly fluctuating field $\mathbf{h}(t)$, and

$$\boldsymbol{\omega}_R = \gamma\alpha(1+\alpha^2)^{-1}\left[\mathbf{u} \times \left(\mathbf{H}_{ef} + \mathbf{h}(t)\right)\right] \tag{1.333}$$

is the relaxational component of $\boldsymbol{\omega}_e$. Equations (1.332) and (1.333) differ from Eqs. (8) and (9) of Shliomis and Stepanov [200] because they contain the noise field and the factor $(1+\alpha^2)^{-1}$; since the Landau–Lifshitz–Gilbert equation is used rather than just the Landau–Lifshitz equation. The kinematic relation, Eq.(1.331), and the Langevin equations, Eqs. (1.332) and (1.333), are stochastic differential equations describing the motion of the dipole vector **m** relative to the crystallographic axes, i.e., internal or solid-state relaxation. Differential-recurrence relations for the statistical moments governing the dynamical behavior of **m** may be deduced from Eqs. (1.331)–(1.333), as described earlier. If, following Ref. [200] and allowing for the factor $(1+\alpha^2)^{-1}$, we now introduce the *magnetic viscosity*

$$\mu = \frac{\mu_0 M_S(1+\alpha^2)}{6\alpha\gamma}, \tag{1.334}$$

Eq. (1.333) then becomes

$$6\mu v_m \boldsymbol{\omega}_R = -v_m \mathbf{m} \times \frac{\partial V}{\partial \mathbf{m}} + \mu_0 \mathbf{m} \times \mathbf{h}(t). \tag{1.335}$$

Equation (1.335) will be the key equation in our discussion of precession-aided Néel relaxation.

The above discussion holds for arbitrary free energy density V. We now specialize to a particle with uniaxial anisotropy, so that

$$v_m V = -m\mu_0(\mathbf{u} \cdot \mathbf{H}_0) - K v_m (\mathbf{u} \cdot \mathbf{n})^2, \tag{1.336}$$

where $\mathbf{H}_0$ is the amplitude of the external uniform dc magnetic field (the polarizing field), $K > 0$ is the effective magnetic anisotropy constant, and $\mathbf{n}$ is a unit vector along the easy magnetization axis. The fact that $\mathbf{H}_0$ is not necessarily parallel to $\mathbf{n}$ means that the circular symmetry characteristic of uniaxial anisotropy will be broken. Thus, Eq.(1.335) becomes, say,

$$\begin{aligned} 6\mu v_m \boldsymbol{\omega}_{\mathrm{R}} - \mu_0 m[\mathbf{u}\times\mathbf{H}_0] - 2Kv_m(\mathbf{u}\cdot\mathbf{n})[\mathbf{u}\times\mathbf{n}] \\ = \mu_0 \mathbf{m}\times\mathbf{h}(t) + \boldsymbol{\lambda}_{\mathrm{R}}(t). \end{aligned} \tag{1.337}$$

Next, we write V explicitly as

$$v_m V = -m\mu_0 H_0[\cos\vartheta'\cos\vartheta + \sin\vartheta\sin\vartheta'\cos(\varphi-\varphi')] - v_m K\cos^2\vartheta, \tag{1.338}$$

where ϑ, φ are the polar angles of $\mathbf{u}$ with respect to the easy axis, which is the polar axis; ϑ', φ' are the polar angles of the external field direction $\mathbf{H}_0$, again with respect to the easy axis: in the solid-state problem, these are constants independent of time. As we have already seen, analytic expressions for the greatest relaxation time τ in the potential, Eq. (1.338), have been obtained in Ref. [207] using Mel'nikov's method as applied to nanomagnets [181]. If $\mathbf{H}_0$ is applied parallel to the easy axis $\mathbf{n}$, τ will be given by Brown's formula (see Chapter 2). If $\mathbf{H}_0 = 0$, τ is given by Brown's asymptotic formula for uniaxial anisotropy (see Chapter 2):

$$\frac{\tau}{\tau_{\mathrm{N}}} \sim \frac{\sqrt{\pi}e^{\sigma}}{2\sigma^{3/2}}, \tag{1.339}$$

where $\sigma = Kv_m/(kT)$. Thus, τ/τ_{N} is independent of the damping α, so that the mode-coupling effect completely disappears.

Now let the crystallographic axes rotate with angular velocity $\boldsymbol{\omega}_n$, corresponding to *physical rotation* of the ferrofluid particle as a result of the stochastic torques imposed by the liquid and the aligning action of $\mathbf{H}_0$ (we now have five degrees of freedom, viz., the dipole angles ϑ, φ as before, and the three Euler angles which, instead of being constant, are now functions of time because of the physical rotation of the easy axis). The relative angular velocity of the dipole and easy axis is then $\boldsymbol{\omega}_R - \boldsymbol{\omega}_n$, so that Eq.(1.337), describing the motion of $\mathbf{m}$, must be modified to

$$6\mu v_m(\boldsymbol{\omega}_{\mathrm{R}} - \boldsymbol{\omega}_n) - \mu_0 m[\mathbf{u}\times\mathbf{H}_0] - 2Kv_m(\mathbf{u}\cdot\mathbf{n})[\mathbf{u}\times\mathbf{n}] = \boldsymbol{\lambda}_{\mathrm{R}}(t). \tag{1.340}$$

The corresponding mechanical equation of motion of the particle is, by Newton's third law (the particle is treated as a rigid sphere; I is the moment of inertia of the sphere about a diameter),

$$I\dot{\boldsymbol{\omega}}_n + 6\mu v_m(\boldsymbol{\omega}_n - \boldsymbol{\omega}_{\mathrm{R}}) + \varsigma\boldsymbol{\omega}_n - 2Kv_m(\mathbf{u}\cdot\mathbf{n})[\mathbf{n}\times\mathbf{u}] = \boldsymbol{\lambda}_n(t) - \boldsymbol{\lambda}_{\mathrm{R}}(t), \quad (1.341)$$

where $\varsigma = 6\eta' v$ is the *mechanical* drag coefficient of the particle in the fluid, v is its hydrodynamic volume, η' is the viscosity of the fluid, and the random torque $\boldsymbol{\lambda}_n(t)$ arising from the fluid carrier has the usual white-noise properties

$$\langle \boldsymbol{\lambda}_n(t) \rangle = 0, \; \langle \lambda_n^{(i)}(t)\, \lambda_n^{(j)}(t') \rangle = 2kT\zeta\delta_{ij}\delta(t-t'), \quad (1.342)$$

where the indexes i, j = 1, 2, 3 refer to the orientation of the crystallographic axes relative to the laboratory cartesian axes. The $\boldsymbol{\lambda}_n(t)$ again has the white-noise properties [97], and we suppose that $\boldsymbol{\lambda}_n(t)$ and $\boldsymbol{\lambda}_{\mathrm{R}}(t)$ are uncorrelated. Thus, by addition of Eqs. (1.340) and (1.341), we have the mechanical equation

$$I\dot{\boldsymbol{\omega}}_n + \varsigma\boldsymbol{\omega}_n = \mu_0 m[\mathbf{u}\times\mathbf{H}_0] + \boldsymbol{\lambda}_n(t) \quad (1.343)$$

(see also [97] Eq. (11.1.1)). Equations (1.340) and (1.343) are, in general, coupled to each other inextricably by the external field term $\mathbf{u}\times\mathbf{H}_0$. However, if that term vanishes, they become:

$$6\mu v_m(\boldsymbol{\omega}_{\mathrm{R}} - \boldsymbol{\omega}_n) - 2Kv_m(\mathbf{u}\cdot\mathbf{n})[\mathbf{u}\times\mathbf{n}] = \boldsymbol{\lambda}_{\mathrm{R}}(t), \quad (1.344)$$

$$I\dot{\boldsymbol{\omega}}_n + \varsigma\boldsymbol{\omega}_n = \boldsymbol{\lambda}_n(t). \quad (1.345)$$

The equations of motion thus separate into that of $\mathbf{m}$ relative to the easy axis, Eq.(1.344), and that of the easy axis itself, Eq. (1.345). The orientational relaxation of a Brownian particle is governed by two characteristic times: the Debye relaxation time $\tau_{\mathrm{D}} = \varsigma/(2kT)$ and the frictional time $\tau_\eta = I/\varsigma$. Thus, the short-time dynamical behavior of Eq. (1.345) is governed by the inertial parameter [97]

$$a = \tau_\eta / \tau_{\mathrm{D}} = 2IkT\varsigma^{-2}, \quad (1.346)$$

which is of the order 10^{-5} for typical ferrofluid parameters [208]. Hence, one may entirely neglect inertial effects; this is quite unlike the situation in polar dielectrics, where these effects become progressively more important at high frequencies. If $a \to 0$, we have the noninertial response. This is treated by Shliomis and Stepanov [200], who were able to factorize the joint probability density function of the dipole and easy-axis orientations in the Fokker–Planck equation into the product of the two separate distributions. Thus, as far as the internal relaxation process is concerned, the single canonical variable treatment of Brown [17] still applies. Hence, no *intrinsic* coupling between the transverse and longitudinal modes exists, i.e., *the eigenvalues of the longitudinal relaxation process are independent of the magnetic dissipation parameter* α. The distribution function of the easy-axis orientations **n** is simply that of a free Brownian rotator excluding inertial effects.

The picture in terms of the decoupled Langevin equations, Eqs.(1.344) and (1.345) above, omitting the inertial term $I\dot{\boldsymbol{\omega}}_n$, in Eq. (1.345), is that the orientational correlation functions of the longitudinal and transverse components of the magnetization in the uniaxial potential,

$$V = -K\cos^2\vartheta, \tag{1.347}$$

are simply multiplied by the liquid state factor, e^{-t/τ_D}, of the Brownian (Debye) relaxation of the ferrofluid, stemming from Eq. (1.345). As far as ferromagnetic resonance (FMR) is concerned, we shall shortly demonstrate that this exponential factor is irrelevant. The longitudinal and transverse susceptibilities, characterizing the solid-state (Néel) process, are approximately described as (see Eq. (2.179))

$$\frac{\chi_\parallel(\omega)}{\chi_\parallel} = \frac{1}{1+i\omega\tau} \quad \text{and} \quad \frac{\chi_\perp(\omega)}{\chi_\perp} = \frac{(1+i\omega\tau_2)+\Delta}{(1+i\omega\tau_2)(1+i\omega\tau_\perp)+\Delta}, \tag{1.348}$$

where the Néel relaxation time τ is rendered reasonably accurately by Brown's uniaxial anisotropy asymptote, Eq.(1.339), if $\sigma \geq 2$. The transverse susceptibility, Eq.(1.348), which is derived using the effective-eigenvalue method [164], yields an accurate result for the transverse response provided that $\sigma \geq 5$ (see Section 2.4.5). The transverse effective eigenvalue solution, Eq. (1.348), fails for $\sigma < 5$, since at small-to-moderate barrier heights, a spread of the precession frequencies of the magnetization in the anisotropy field exists. Here the hierarchy must be solved exactly via matrix continued fractions (see Chapter 2). In Eq. (1.348), the transverse components of the magnetization are linear combinations of the autocorrelation functions (ACFs) of the spherical harmonics $Y_{11}(\vartheta,\varphi)$ and $Y_{1-1}(\vartheta,\varphi)$, while τ_2 is the effective relaxation time of the second-rank (quadrupole moment) relaxation mode, which is a linear combination of the ACFs of $Y_{21}(\vartheta,\varphi)$ and $Y_{2-1}(\vartheta,\varphi)$. The effective relaxation times $\tau_\perp$ and τ_2 decrease monotonically with σ, each having asymptotes $\tau_N > \sigma$, $\sigma >> 1$ (see Section 2.4).

From linear response theory, we have

$$\chi_\gamma(\omega) = f_\gamma(0) - i\omega\int_0^\infty f_\gamma(t)e^{-i\omega t}dt = f_\gamma(0) - i\omega\tilde{f}_\gamma(\omega), \quad (\gamma = \|, \perp), \quad (1.349)$$

where $f_\|(t)$ and $f_\perp(t)$ are the longitudinal and transverse after-effect functions; in the frequency domain (the tilde denoting the one-sided Fourier transform) they are, according to Eqs.(1.348),

$$\tilde{f}_\|(\omega) = \frac{\tau\chi_\|}{1+i\omega\tau} \quad \text{and} \quad \tilde{f}_\perp(\omega) = \frac{\chi_\perp\tau_\perp(1+i\omega\tau_2)}{(1+i\omega\tau_2)(1+i\omega\tau_\perp)+\Delta}. \quad (1.350)$$

The complex susceptibility of a ferrofluid in a weak applied field may be written directly from Eq. (1.350) and the Langevin equations, Eqs. (1.344) and (1.345), taking note of Eq. (1.346) and using the shift theorem for one-sided Fourier transforms [97]. Thus, we have

$$\chi_\|(\omega) = \frac{\chi_\|}{1+i\omega\mathrm{T}_\|}, \quad (1.351)$$

$$\chi_\perp(\omega) = \frac{\chi_\perp\left[i\omega\mathrm{T}_2 + 1 + \Delta\mathrm{T}_\perp\mathrm{T}_2/(\tau_\perp\tau_2)\right]}{(1+i\omega\mathrm{T}_2)(1+i\omega\mathrm{T}_\perp)+\Delta\mathrm{T}_\perp\mathrm{T}_2/(\tau_\perp\tau_2)}, \quad (1.352)$$

where the Néel relaxation time τ and the effective relaxation times $\tau_\perp$ and τ_2 are modified to

$$\mathrm{T}_\| = \frac{\tau\tau_\mathrm{D}}{\tau+\tau_\mathrm{D}}, \quad \mathrm{T}_\perp = \frac{\tau_\perp\tau_\mathrm{D}}{\tau_\perp+\tau_\mathrm{D}}, \quad \mathrm{T}_2 = \frac{\tau_2\tau_\mathrm{D}}{\tau_2+\tau_\mathrm{D}}. \quad (1.353)$$

In a weak measuring field, the particle anisotropy axes are oriented in a *random* fashion. Hence, the susceptibility averaged over particle orientations is given by

$$\chi(\omega) = (1/3)\chi_{\parallel}(\omega) + (2/3)\chi_{\perp}(\omega). \tag{1.354}$$

In ferrofluids, where the Néel mechanism is *blocked*, $\tau >> \tau_D$, and so we have $T_{\parallel} \cong \tau_D$ in Eq. (1.353). Furthermore, in Eq. (1.353) we may use the fact [164] that $\tau_{\perp}$ and τ_2 are monotonically decreasing functions of σ, and also that usually [208], the ratio of the free diffusion times is $\tau_N / \tau_D \sim 10^{-2}$, to ascertain which times may be neglected in Eq. (1.352). Thus, we deduce that $T_{\perp} \cong \tau_{\perp}$ and $T_2 \cong \tau_2$ in Eq. (1.353) for all σ. Hence, we may conclude, recalling the exact transverse relaxation solution for $\chi_{\perp}$ (Section 2.4) that the solid-state effective eigenvalue solution embodied in Eq. (1.348) can also accurately describe the ferromagnetic resonance in ferrofluids, with the exception of small σ (< 5). Then, the solid-state solution based on matrix continued fractions must be used. The conclusion agrees with that of Fannin [209] and Fannin *et al.* [210], who have analyzed experimental data on ferrofluids using Eq. (1.348). We also remark [202] that the very large σ (high-barrier) limit of Eq. (1.352) (the Landau–Lifshitz limit) agrees with the result of Scaife [211], who analyzed the problem using an entirely different method. In the high-magnetic-damping limit, $\alpha >> 1$, Eq. (1.352) reduces to a pure relaxation equation agreeing with Shliomis and Stepanov [200], namely their Eqs. (31) and (32).

We now very briefly describe possible inertial effects arising from the term $I\dot{\boldsymbol{\omega}}_n$ in Eq.(1.345). If they are included, the correlation functions pertaining to longitudinal and transverse motions will still be the product of the correlation functions of the rotational Brownian motion of a particle [200] and the solid-state correlation functions; however, the composite expressions will be much more complicated for arbitrary inertial parameter a, as in Eq. (1.346). The reason is that the orientational correlation functions for the rotational Brownian motion of a sphere may only be expressed as the inverse Laplace transform of an infinite continued fraction in the frequency domain [97,212].

Our treatment will apply not only to uniaxial anisotropy, but also to an arbitrary potential $V[(\mathbf{u}\cdot\mathbf{n}), t]$ of the magnetocrystalline anisotropy. Since the potential is a function of (**u·n**) only, the ACFs of the overall process (e.g., cubic anisotropy) will be the *product of the individual orientational ACFs of the freely rotating sphere and the solid-state mechanism.* However, unlike in uniaxial anisotropy, even though the correlation functions still factorize, substantial coupling between the transverse and longitudinal modes (which now have α dependent eigenvalues) will exist because the noncircular symmetry is now an *intrinsic* property of the particles. This phenomenon should be observable in measurements of the complex susceptibility of such particles.

We recall Eqs. (1.340) and (1.343). In the noninertial limit, Eq. (1.343) becomes

$$\varsigma \boldsymbol{\omega}_n = \mu_0 m[\mathbf{u}\times\mathbf{H}_0]+\boldsymbol{\lambda}_n(t) \tag{1.355}$$

and on eliminating $\boldsymbol{\omega}_n$ in Eq. (1.340) using Eq.(1.355), we have

$$\boldsymbol{\omega}_{\mathrm{R}} - \frac{\mu_0 m}{\zeta_m}[\mathbf{u}\times\mathbf{H}_0]-\frac{K}{3\mu}(\mathbf{u}\cdot\mathbf{n})[\mathbf{u}\times\mathbf{n}]=\frac{\boldsymbol{\lambda}_{\mathrm{R}}(t)}{6\mu v_m}+\frac{\boldsymbol{\lambda}_n(t)}{\varsigma}, \tag{1.356}$$

where $\zeta_m^{-1}=(6\mu v_m)^{-1}+\varsigma^{-1}$. Equation (1.356) is of the same form as Eq.(1.337) describing the motion of the dipole moment with a frozen ($\eta'\to\infty$) Brownian mechanism with a dc field $\mathbf{H}_0$ applied at an oblique angle to the easy axis. Thus, it appears that the effect of the fluid on the solid-state or internal mechanism of relaxation, is merely to alter the magnetic drag coefficient of the solid-state process. The corresponding change in the dimensionless damping coefficient α of that process is

$$\alpha' = \alpha\left(1+6\mu v_m/\varsigma\right)=\alpha\left(1+\tau_{\mathrm{N}}/\tau_{\mathrm{D}}\right). \tag{1.357}$$

Here the ratio $\tau_{\mathrm{N}}/\tau_{\mathrm{D}}$ represents the coupling between the magnetic and mechanical motions, arising from the nonseparable nature of the Langevin equations, Eqs.(1.355) and (1.356). Thus, the correction to the solid-state result imposed by the fluid is once again of the order 10^{-2}. Hence, despite the nonseparability of the equations of motion, the Néel relaxation time of the ferrofluid particle should still be accurately represented, in the IHD and VLD regimes, respectively, by the formulas already given in this chapter. Furthermore, Eq. (1.356) should be closely approximated by the solid-state relaxation equation

$$\boldsymbol{\omega}_{\mathrm{R}} - \mu_0 m(\mathbf{u}\times\mathbf{H}_0)\frac{1}{6\mu v_m} - \frac{K}{3\mu}(\mathbf{u}\cdot\mathbf{n})(\mathbf{u}\times\mathbf{n}) = \frac{\boldsymbol{\lambda}_{\mathrm{R}}(t)}{6\mu v_m}. \qquad (1.358)$$

Hence, like the solid-state problem, one would also expect the following effects to occur in a ferrofluid for a large dc bias field superimposed on which is a small ac field:

(a) A strong dependence of τ/τ_{N} on α (i.e., a frictional dependence of the smallest nonvanishing eigenvalue) unlike in the weak field case (circular symmetry), which is a signature of the coupling between the longitudinal and transverse modes.

(b) Suppression of the Néel and barrier-crossing modes in favor of the fast relaxation modes in the deep well of the bistable potential created by Eq.(1.338), if the bias field h_{c} exceeds a certain critical value [38,146]. This gives rise to a high-frequency Debye-like relaxation band.

(c) A very-high-frequency FMR peak arising from excitation of the transverse modes having frequencies close to the precession frequency.

In the longitudinal response, however, guided by the relaxation time Eq.(1.353), it will be necessary to have particles with $\tau < \tau_{\mathrm{D}}$, so that the characteristic α-dependence of the response is not masked by the Brownian process due to the damping imposed by the fluid.

We now return to Eq.(1.343), the mechanical equation of motion of the particles, which for small damping (if the small inertial term is retained) predicts a damped oscillation of fundamental frequency [208] $\omega_0 = \sqrt{mH_0 / I}$, appearing in the spectrum as a high-frequency resonant absorption. In ferrofluids, however, it has been estimated [208] that fields of order of magnitude 10^7 Oe are needed to observe oscillatory effects, which is higher than any field that can be obtained under terrestrial conditions. Hence, it appears that one may rule out this resonant mode of the motion.

It remains to determine the influence of $\mathbf{H}_0$ on the mechanical relaxation modes. An estimate of this may be made by recalling that Eq. (1.343) is basically the equation of motion of a rigid dipole in a strong

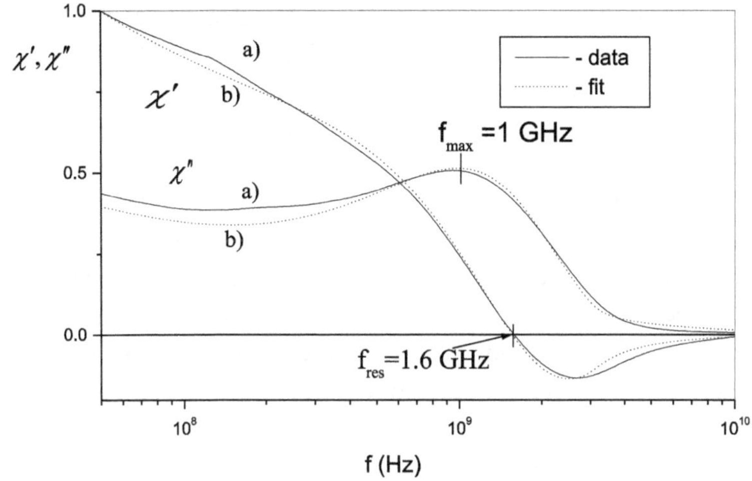

Fig. 1.23. Normalized plots of $\chi'(\omega)$ and $\chi''(\omega)$ (a) and corresponding fits (b) using Eqs. (1.351), (1.352), and(1.354) [202].

constant external field. Moreover, if inertial effects are neglected, it has been shown in Ref. [97], that the longitudinal and transverse effective relaxation times decrease monotonically with increasing field strength, having asymptotic behavior $\tau_{\parallel} \sim \tau_D / \xi$, $\tau_{\perp} \sim 2\tau_D / \xi$, when the parameter $\xi = v\mu_0 M_S H_0 / (kT) >> 1$. Thus, the principal effect of the external field in Eq.(1.355) is to reduce the Brownian relaxation time.

To support our theoretical discussions, we discuss room-temperature complex susceptibility data [202] for a ferrofluid consisting of magnetite particles suspended in isopar M (the surfactant is oleic acid). The particles have a median diameter of 9 nm and a bulk saturation magnetization of 0.4 T. Since no detailed experimental data for ferrofluid susceptibilities in a strong oblique polarizing field (of intensity 10 T and higher) superimposed on which is a weak ac field exists, we shall merely illustrate how the weak ac field susceptibilities (Eqs. (1.351), (1.352), and (1.354)), which incorporate the effect of the fluid carrier, compare favorably with experiment. The frequency dependence of the complex susceptibility has the same resonant form as that predicted by Eqs. (1.351), (1.352), and (1.354). Plot (a) of Fig. 1.23 shows the susceptibility components obtained over the wide frequency range of 50 MHz to 10 GHz, whilst plot (b) shows the fit obtained using Eqs. (1.351), (1.352), and (1.354). As magnetic fluids have a *distribution of both particle shape and size*, these parameters are accounted for by modifying the above equations to include a normal distribution of anisotropy constant, $\bar{K}$, and a Nakagami distribution of radii, r [209,210]. The value used for the damping parameter α, was 0.1, a figure within the range of values normally quoted [210].

Our approach to the combined mechanical and magnetic motions of a ferrofluid particle, and their mutual behavior, is based on rearrangement of the Langevin equation. It involves consideration of the various characteristic time scales, so that the physical effects of the fluid carrier on the magnetic relaxation may be explained without elaborate and detailed solution (which will always lead to supermatrix continued fractions) of the various equations describing the system. The relative orders of magnitude of the timescales involved determine which of the existing independent internal and Brownian mode solutions may be applied to the ferrofluid particle in any given situation. These considerations hold even in the strong-field case, where the variables cannot be separated in the Langevin equations. In particular, for zero or very weak external fields, it appears that the high-frequency behavior may be accurately modeled by the solid-state transverse response result, Eq. (1.348), as the Brownian relaxation time τ_D simply drops out of Eq. (1.352) due to the relative orders of magnitude of the various timescales. These are dictated by the ratio of the free diffusion times, and by the monotonic decrease of the effective relaxation times with barrier height. This appears to be the origin of the success of Eq. (1.348) in explaining the experimental results. We have made no assumptions other than that of the effective eigenvalue truncation of the set of differential-recurrence relations, since the Langevin equations will always decouple for zero or weak fields. Furthermore, for $\sigma < 5$ (where the effective-eigenvalue solution is not an accurate representation of the exact transverse susceptibility solution, see Section 2.4.5) the solution may always be found from the transverse set of differential-recurrence relations by using matrix continued fractions. The experimental data of the linear response, that is the weak ac field susceptibility, support our conjectures concerning the application of the solid-state transverse response result, Eq. (1.348), to magnetic fluids.

1.8 Quantum treatment of the motion of spins

Consider a classical point particle of mass m, of position $\mathbf{r}$ with respect to a given origin, and linear momentum $\mathbf{p} = m\mathbf{v}$, where $\mathbf{v}$ is the velocity of the particle, then the angular momentum of this particle is defined as the *moment of momentum*, viz., [104]

$$\mathbf{L} = \mathbf{r} \times \mathbf{p}. \tag{1.359}$$

In the classical mechanics, any general displacement of a rigid body (formally defined as a *system of particles* such that the *distances between the individual particles do not vary*) can be represented by a translation plus a rotation. Then the total angular momentum $\mathbf{J}$ divides into contributions from the translation of the center of mass and from the rotation about it. When the origin of the body-fixed system is taken as the center of mass of the rigid body, the angular momentum resulting from the motion *relative* to the center of mass is

$$\mathbf{J} = \sum_i m_i \mathbf{r}_i \times \mathbf{v}_i,$$

where $\mathbf{r}_i$ and $\mathbf{v}_i$ are, respectively, the radius vector and velocity of the ith particle relative to it. In the quantum mechanics, the *orbital* angular momentum operator may be defined in the coordinate representation as a dimensionless operator $\hat{\mathbf{L}} = -i[\mathbf{r} \times \nabla]$, where the operator ∇ (*nabla*) is the basic vector differential operator with cartesian components in the laboratory coordinate system

$$\nabla_X = \frac{\partial}{\partial X}, \; \nabla_Y = \frac{\partial}{\partial Y}, \; \nabla_Z = \frac{\partial}{\partial Z}.$$

The orbital angular momentum operator $\hat{\mathbf{L}}$ generates transformations of scalar spin-less wave functions under rotations of the coordinate systems [104]. The eigenfunctions of the operators $\hat{\mathbf{L}}^2$ and $\hat{L}_Z$ represent spherical harmonics $Y_{lm}(\vartheta, \varphi)$ depending on the polar angles ϑ and φ (see Appendix A.1.7). Besides having an angular momentum *associated with rotation in space*, many particles also possess an *intrinsic* angular momentum called *spin*, by direct analogy with a spinning top as suggested by Uhlenbeck and Goudsmit [5]. This is so for electrons, muons, quarks which are the constituents of the nucleus, etc. According to well-established convention, we write the spin angular momentum as $\hbar\hat{\mathbf{S}}$ and label $\hat{\mathbf{S}}$ the *dimensionless spin operator* (here $\hbar = h/(2\pi)$ and h is Planck's constant; see Appendix A.2). The spin operator $\hat{\mathbf{S}}$ is usually represented by a set of three (since the spin vector $\hat{\mathbf{S}}$ has three components) $(2S+1)\times(2S+1)$ *square matrices* $\hat{S}_X$, $\hat{S}_Y$, and $\hat{S}_Z$ with S being the spin number [104]. The simplest way of characterizing the components of a vectorial spin operator $\hat{\mathbf{S}}$ is merely to regard them as generators of rotations in a 'spin space' which is *intrinsic to each spin and bears no relation to the physical space.* This *Ansatz* is justified by the results flowing from it, as verified by innumerable experiments. Now atomic nuclei are complex aggregates of elementary particles so that they have a *total* angular momentum operator $\hat{\mathbf{J}} = \hat{\mathbf{L}} + \hat{\mathbf{S}}$ due to *coupling* of the orbital and spin momenta of their constituents. Since the *spatial* extension of atomic nuclei may be *neglected* then their total angular momentum appears to be linked to that of point particles. In the quantum mechanics, the *total* angular momentum operator $\hat{\mathbf{J}}$ is defined as an operator which generates transformations of wave functions and operators under infinitesimal rotations of the coordinate system. The eigenfunctions of the operators $\hat{\mathbf{J}}^2$ and $\hat{J}_Z$ represent, in general, the *tensor spherical harmonics* which describe the angular distribution and polarization of spin-S particles in a state with definite total angular momentum J, projection M, and orbital angular momentum L (see Ref. [104] for details). For a rigid symmetric top, the eigenfunctions of the operators $\hat{\mathbf{J}}^2$ and $\hat{J}_Z$ represent the Wigner D-functions $D^J_{MM'}(\alpha,\beta,\gamma)$ which depend on the Euler angles α, β, and γ (see Appendix A.1.8).

1.8.1 *Early theories*

We recall that the energy $\mathcal{E}$ of a magnetic moment $\boldsymbol{\mu}$ in a magnetic field $\mathbf{H}_0$ is just

$$\mathcal{E} = -\mu_0 (\boldsymbol{\mu} \cdot \mathbf{H}_0). \tag{1.360}$$

However, in quantum mechanics, the magnetic moment of a spin is no longer a vector; instead it is an *operator* $\hat{\boldsymbol{\mu}}$ proportional to the spin operator, viz.,

$$\hat{\boldsymbol{\mu}} = \gamma\hbar\mu_0^{-1}\hat{\mathbf{S}}. \tag{1.361}$$

Furthermore, as far as the classical energy, Eq. (1.360), is concerned, a corresponding quantum-mechanical Hamiltonian operator (called the Zeeman Hamiltonian) is given by

$$\hat{H} = -\gamma\hbar\left(\hat{\mathbf{S}}\cdot\mathbf{H}_0\right), \tag{1.362}$$

and if the magnetic field is applied along the direction *OZ*:

$$\hat{H} = -\gamma H_0 \hbar \hat{S}_Z. \tag{1.363}$$

The spin operator $\hat{S}_Z$ is itself quantized and takes on values *m* varying in steps of one unit from –*S* to +*S*, while each value of *m* corresponds to a quantum energy

$$\mathcal{E}_m = -\gamma\hbar H_0 m \tag{1.364}$$

(the spin number *S* may itself be integer or half-integer). We now evaluate the average value of the *Z* component of the magnetic dipole moment $\langle\mu_Z\rangle$ for a spin placed in an external dc field $\mathbf{H}_0$ applied along the Z-axis. Thus, the magnetic moment along that axis is

$$\langle\mu_Z\rangle = \mu_0^{-1}\sum_{m=-S}^{S}\gamma\hbar m P_m. \tag{1.365}$$

We suppose further that the ensemble of spins has come to equilibrium at temperature *T*. In an ensemble of identical spins, thermal equilibrium at temperature *T* corresponds to a Boltzmann distribution of populations of the various energy levels, viz.,

$$P_m = \frac{1}{Z_0}e^{\xi m/S},\quad Z_0 = \sum_{m=-S}^{S} e^{\xi m/S}, \tag{1.366}$$

where Z_0 is the partition function and $\xi = \beta\gamma\hbar H_0 S$ is a dimensionless dc field parameter. Thus, the average value of the *Z*-component of the magnetic moment $\langle\mu_Z\rangle$ depends on the population of these discrete levels at thermal equilibrium as follows

$$\langle\mu_Z\rangle = \frac{\mu}{SZ_0}\sum_{m=-S}^{S} m e^{\xi m/S} = \mu\frac{d}{d\xi}\ln Z_0 = \mu B_S(\xi), \tag{1.367}$$

where $\mu = \gamma\hbar S/\mu_0$ and

$$B_S(\xi) = \left(1+\frac{1}{2S}\right)\coth\left(\frac{2S+1}{2S}\xi\right) - \frac{1}{2S}\coth\left(\frac{\xi}{2S}\right) \tag{1.368}$$

is the *Brillouin function* [213]. This equation governing the equilibrium magnetization of a paramagnetic substance can be simplified if $\xi << 1$, holding at higher temperatures and smaller magnetic fields. Then

$$\langle \mu_Z \rangle = \frac{1}{3S}\beta\mu_0\mu^2(S+1)H_0. \tag{1.369}$$

The classical case, in which the magnetic dipole moments can rotate freely and the magnetic dipole $\boldsymbol{\mu}$ may lie in any direction with respect to that of the magnetic field $\mathbf{H}_0$, is equivalent to a *continuum of energy levels* with S tending to infinity. Here the Brillouin function simplifies to the *Langevin function* $L(\xi)$, named after Paul Langevin [171]

$$\begin{aligned}\lim_{S\to\infty} B_S(\xi) = L(\xi) &= \frac{1}{Z_{\text{cl}}}\int_0^\pi \cos\vartheta e^{\xi\cos\vartheta}\sin\vartheta d\vartheta \\ &= \coth\xi - \xi^{-1},\end{aligned} \tag{1.370}$$

where the dimensionless dc field parameter becomes $\xi = \beta\mu_0\mu H_0$ ($\gamma\hbar S \to \mu_0\mu$ as $S\to\infty$ and $\hbar \to 0$) and Z_{cl} is the classical partition function given by

$$Z_{\text{cl}} = \int_0^\pi e^{\xi\cos\vartheta}\sin\vartheta d\vartheta = 2\frac{\sinh\xi}{\xi}. \tag{1.371}$$

Then the mean dipole moment in the field direction is

$$\langle \mu_Z \rangle_{\text{cl}} = \mu L(\xi). \tag{1.372}$$

The behavior is then exactly analogous to that of an electric dipole in the Debye theory of the static electric susceptibility [214]. We shall reconsider this problem in Section 3.2.1 of Chapter 3 via the phase-space approach.

To undertake a spectroscopic study of such a system, one first induces a resonance by applying, in tandem with the large dc field $\mathbf{H}_0$, a small ac field $\mathbf{H}_1$ normal to $\mathbf{H}_0$ and of frequency ω. The ac field then induces transitions between the Zeeman levels. Thus, the populations of the various levels are modified by these transitions and vary according to

$$\frac{dP_m}{dt} = -U_{mm'}P_m + U_{m'm}P_{m'} \tag{1.373}$$

with $|m - m'| = 1$. On very general grounds the transition probabilities are symmetrical $U_{mm'} = U_{m'm}$. Hence via an appropriate weighted summation of Eq. (1.373) for various m, we have a simple equation for the evolution of the longitudinal component of the magnetic moment, viz.,

$$\frac{d}{dt}\langle \mu_Z \rangle = -w\langle \mu_Z \rangle, \tag{1.374}$$

where w is called the *saturation rate* proportional to the square of the ac field amplitude. Thus ignoring relaxation, ac-induced transitions tend to restore the Boltzmann distribution, Eq. (1.366), so that the populations evolve towards equalization and the longitudinal magnetization decays to zero. Moreover, considered as a function of frequency, it represents a bell-shaped curve centered on the Larmor frequency ω_0 of width comparable with that of the Zeeman levels (in frequency units). Saturation of the longitudinal magnetization corresponds to the increase of (Zeeman) energy gained from the ac field. While such ac-induced transitions tend to saturate the longitudinal magnetization, on the contrary, spin-lattice relaxation tries to restore it to its thermal equilibrium value μ_0 so that Eq. (1.374) must then be modified to

$$\frac{d}{dt}\langle \mu_Z \rangle = -w\langle \mu_Z \rangle - \frac{\langle \mu_Z \rangle - \mu_0}{T_1}, \tag{1.375}$$

where T_1 is a characteristic time associated with the approach to thermal equilibrium. Once w is large enough, $wT_1 \sim 1$, the power absorbed levels off irrespective of the increase in w (saturation effect). If w is known, one can then measure T_1 by observing the saturation effect [213].

1.8.2 *Density matrix formulation of spin relaxation*

In the classical mechanics, we consider the behavior of any given mechanical system of interest as it evolves in time from one *precisely defined* state to another. On the contrary in *statistical mechanics*, we have some knowledge of the system but not enough for a complete specification of the precise state. For this purpose, we *consider the average behavior of a collection of systems of the same structure as the one of actual interest but distributed over a range of different possible states*. We speak of such a collection as an *ensemble* of systems. Thus, we represent the *instantaneous state* of a dynamical system of N degrees of freedom by a phase point ($\mathbf{q}, \mathbf{p}$) in a $2N$ dimensional phase (state) space Γ of all the coordinates $\mathbf{q} = (q_1,\ldots,q_N)$ and momenta $\mathbf{p} = (p_1,\ldots,p_N)$ of the system. The instantaneous state of any system in an ensemble can then be regarded as being specified by the position of a *representative point* in the phase space, and the condition of the ensemble as a whole can be described by a "cloud" of density $f(\mathbf{q}, \mathbf{p}, t)$ of such representative points, one for each system in the ensemble. The behavior of the ensemble over time can then be associated with the "streaming" motion of the representative points as they describe trajectories in the phase space, in accordance with the laws of classical mechanics. The probability density function $f(\mathbf{q}, \mathbf{p}, t)$ defined in such a way that $f(\mathbf{q}, \mathbf{p}, t)\, d\mathbf{q}\, d\mathbf{p}$ is the probability at time t that the phase point will be inside a volume element $d\mathbf{q}\, d\mathbf{p}$ in phase space obeys the classical Liouville equation, namely, [163,183,215]

$$\frac{\partial f}{\partial t} = -\{H, f\} = -i\mathrm{L}f, \tag{1.376}$$

where $H = K + U$ is the Hamiltonian (total energy of the system), K and U are, respectively, kinetic and potential energies, $\{H,f\}$ denotes the *classical Poisson bracket*, viz.,

$$\{H, f\} = \sum_{i=1}^{N}\left(\frac{\partial f}{\partial q_i}\frac{\partial H}{\partial p_i} - \frac{\partial f}{\partial p_i}\frac{\partial H}{\partial q_i}\right)$$

and $\mathrm{L} = -i\{H,\ \}$ is the classical Liouville operator. From the principle of the *conservation of density in phase space*, it follows [215] that on considering the rate of change of density of phase points in the neighborhood of any selected *moving* phase point instead of in *the neighborhood* of a *fixed* point that the hydrodynamical derivative

$$\frac{df}{dt} = 0. \tag{1.377}$$

Hence the *density of phase points is always a constant along a phase-space trajectory*, viz.,

$$f\left(\mathbf{q}(0), \mathbf{p}(0), t = 0\right) = f\left(\mathbf{q}(t), \mathbf{p}(t), t\right), \tag{1.378}$$

or phase points purely *stream* in accordance with Hamilton's equations. Now, the evolution equation for a dynamical variable A $(\mathbf{q}, \mathbf{p})$ is

$$\frac{dA}{dt} = \sum_{i=1}^{N}\left(\frac{\partial A}{\partial q_i}\frac{\partial H}{\partial p_i} - \frac{\partial A}{\partial p_i}\frac{\partial H}{\partial q_i}\right) = \{H, A\} = i\mathrm{L}A. \tag{1.379}$$

The *expectation value* of A $(\mathbf{q}, \mathbf{p})$ at time t is then defined as

$$\langle A\rangle(t) = \int_{\Gamma} A(\mathbf{q}, \mathbf{p}) f(\mathbf{q}, \mathbf{p}, t) d\mathbf{q} d\mathbf{p} \tag{1.380}$$

with the normalization

$$\int_{\Gamma} f(\mathbf{q}, \mathbf{p}, t) d\mathbf{q} d\mathbf{p} = 1. \tag{1.381}$$

In the quantum mechanics, on the other hand, the probability density function as defined above is replaced by a *density matrix* $\hat{\rho}$ in *Hilbert space* to play a role somewhat similar to the density f in the classical statistical mechanics [215,216]. For convenience, we briefly recall the basic equations of the density matrix formalism [88,183,216,217].

The *state* of a quantum many-particle system characterized by a Hamiltonian operator $\hat{H}$ is represented by a *wave function* Ψ and the corresponding *state vector* $|\Psi\rangle$ in Hilbert space. The wave function Ψ satisfies the Schrödinger equation

$$\frac{\partial \Psi}{\partial t} = -\frac{i}{\hbar}\hat{H}\Psi, \tag{1.382}$$

also, in the quantum mechanics, a dynamical variable A must always correspond to a *Hermitian operator* $\hat{A}$ (because the eigenvalues associated with $\hat{A}$ must always be real) and the expectation value of $\hat{A}$ is formally defined as the scalar (inner) product [5,88]

$$\left\langle \hat{A} \right\rangle(t) = \left\langle \Psi(t) \right| \hat{A} \left| \Psi(t) \right\rangle, \tag{1.383}$$

where $\left\langle \Psi(t) \right| = \left| \Psi(t) \right\rangle^{\dagger}$ and the symbol † denotes the Hermitian conjugate [88]. Equation (1.383) implies that the wave function $\Psi(t)$ must be normalized to unity, i.e., the scalar product $\langle \Psi(t) | \Psi(t) \rangle = 1$. It further implies that the wave function may be expanded in the complete orthonormal basis of eigenstates $\{\phi_n\}$ of the Hamiltonian operator $\hat{H}$ as [5]

$$\Psi(t) = \sum_n c_n(t) \phi_n, \tag{1.384}$$

where the expansion coefficients $c_n(t)$ are given by the scalar product $c_n(t) = \left\langle \phi_n \middle| \Psi(t) \right\rangle$ with the normalization $\sum_n \left| c_n(t) \right|^2 = 1$, which follows directly from the normalization of $\Psi(t)$. Thus, one says [216] that the system is in the *pure* state. Hence, the *expectation value*, i.e. the scalar product, $\left\langle \hat{A} \right\rangle(t) = \left\langle \Psi(t) \right| \hat{A} \left| \Psi(t) \right\rangle$ of a dynamical variable A represented in the quantum mechanics by an operator $\hat{A}$ is given by [88,183,216]

$$\left\langle \hat{A} \right\rangle(t) = \sum_{n,n'} c_{n'}(t) c_n^*(t) A_{nn'} \equiv \sum_{n,n'} \rho_{n'n}(t) A_{nn'}, \tag{1.385}$$

where the expansion coefficients $A_{nn'} = \left\langle \phi_n \right| \hat{A} \left| \phi_{n'} \right\rangle$ and $\rho_{n'n}(t) = c_{n'}(t) c_n^*(t)$ are, respectively, the *matrix elements* of the operator $\hat{A}$ and the density operator $\hat{\rho}$ in the orthonormal basis of the eigenstates $\{\phi_n\}$ and the asterisks denote the complex conjugate. For a *pure* state, the density operator $\hat{\rho}$ can formally be written as the *outer product* [88,183,216]

$$\hat{\rho}(t) = \left| \Psi(t) \right\rangle \left\langle \Psi(t) \right| = \sum_{n,n'} c_{n'}(t) c_n^*(t) \left| \phi_{n'} \right\rangle \left\langle \phi_n \right|. \tag{1.386}$$

The density matrix operator defined by Eq. (1.386) is also Hermitian, i.e. $\hat{\rho}^{\dagger}(t) = \hat{\rho}(t)$, and *idempotent*, that is, it satisfies the condition $\hat{\rho}^2(t) = \hat{\rho}(t)$.

The density matrix operator for a quantum system in a *pure* state can also be applied with some modifications to a quantum system in a so-called *mixed* state, i.e., to a system, whose *actual* state $\psi(t) = \sum_n c_n(t) \phi_n$

is not *completely known* and only the probabilities $P_n = |c_n(t)|^2$ to be in any of the different states *n* can be evaluated [88,183]. An example is a system in thermal equilibrium, whose eigenfunctions constitute the orthonormal basis $\{\phi_n\}$ with probabilities P_n to be found in the different states *n* given by the Boltzmann distribution and all microscopic states compatible with these probabilities are assumed to be equally probable; here $\rho_{n'n} = P_n \delta_{n'n}$. The most obvious difference between pure and mixed states is encountered in their *diagonal representation*. In that representation the pure state density matrix will have only one nonzero element on its diagonal (equal to 1) whereas the mixed state density matrix must have at least two nonzero elements (whose sum is 1) [183]. Indeed, the overall situation has been well summarized by Gershenfeld [88] "Uncertainty is central to quantum mechanics through the unpredictability of the outcome of a measurement. It is also central to thermodynamics, in the probability distribution over the allowed states. These probabilities have very different characters though: the former is a fundamental limit while the latter reflects incomplete knowledge about a system with many degrees of freedom".

For our purposes, the most relevant properties of the density matrix operator $\hat{\rho}$, which are valid both for pure and mixed states are the following [88]: (i) $\hat{\rho}$ is a *positive definite operator*, i.e., $\langle \psi(t) | \hat{\rho} | \psi(t) \rangle \geq 0$ yielding the average probability to find the system in the state $|\psi(t)\rangle$; (ii) $\hat{\rho}$ is *Hermitian* and its diagonal elements ρ_{nn} are real, positive and represent the *average* probability that a system chosen at random from the ensemble would be found in the state specified by *n*. Moreover, $\hat{\rho}$ satisfies the normalization condition [88,215]

$$\mathrm{Tr}(\hat{\rho}) = 1, \tag{1.387}$$

where $\mathrm{Tr}(\hat{\rho})$ denotes its trace, while the quantum mechanical analog of the classical expression Eq. (1.380) for the mean value of a dynamical variable *A* is [5,88]

$$\langle \hat{A} \rangle = \mathrm{Tr}(\hat{\rho}\hat{A}). \tag{1.388}$$

Thus, the integrals over the whole phase space Γ in the classical Eqs. (1.380) and (1.381) are simply replaced by the traces in the quantum Eqs.

(1.387) and (1.388). Moreover, the time evolution of the density operator $\hat{\rho}$ is described by the quantum Liouville equation [88,183,215]

$$\frac{\partial \hat{\rho}}{\partial t} = -\frac{i}{\hbar}[\hat{H}, \hat{\rho}] = -i\mathrm{L}\hat{\rho}, \tag{1.389}$$

where $\hat{H}$ is the Hamiltonian operator and the square brackets denote the commutator, viz., $[\hat{H}, \hat{\rho}] = \hat{H}\hat{\rho} - \hat{\rho}\hat{H}$, while $\mathrm{L} = \hbar^{-1}[\hat{H},\ ]$ is the quantum Liouville operator. The formal solution of the operator Eq. (1.389) is [88,183,218]

$$\hat{\rho}(t) = e^{-\frac{i}{\hbar}\hat{H}t}\hat{\rho}(0)e^{\frac{i}{\hbar}\hat{H}t}. \tag{1.390}$$

Finally, the quantum evolution equation for an operator $\hat{A}$ is

$$\frac{d\hat{A}}{dt} = \frac{i}{\hbar}[\hat{H}, \hat{A}] = i\mathrm{L}\hat{A}. \tag{1.391}$$

Equations (1.389) and (1.391) describing the $\hat{\rho}$ and $\hat{A}$ evolutions in Hilbert space are quantum analogs of the classical Eqs. (1.376) and (1.379), respectively. We emphasize [215] that the quantum Liouville Eq. (1.389) for the evolution of the density matrix $\hat{\rho}$ has the same mathematical form as the classical Liouville equation (1.376) for the distribution function *f*.

Now the object of our interest is the dynamics of a spin characterized by the spin number *S* in contact with a *thermal bath or environment, with the latter being only of relevance to us insofar as it influences the spin dynamics*. Moreover, as in the classical case it is impossible to follow every variable of the composite spin-environment taken as a whole [219]. Hence, we desire a closed evolution equation for a *reduced density matrix* [88], where only *relevant* variables, i.e., those belonging to the *subspace comprising the spin*, appear explicitly. Then by tracing Eq. (1.389) over the bath variables, we shall ultimately have a *master or reduced density-matrix evolution equation* [183]. In other words, we can average (trace) out the environment of the spin so yielding a *statistical* description of the spin alone [220,221]. The ideas of course being entirely analogous to those used to derive an evolution equation for the single-particle distribution function in the classical statistical mechanics.

To treat the spin dynamics in a dissipative environment via the density matrix formalism, we loosely follow arguments advanced by Nitzan [183]. Thus, it is supposed from intuitive reasoning that the overall Hamiltonian $\hat{H}$ of the system (bath + spin) may be decomposed into three separate parts, viz.,

$$\hat{H} = \hat{H}_S + \hat{H}_B + \hat{H}_{SB}, \tag{1.392}$$

where the operators $\hat{H}_S, \hat{H}_{SB}$, and $\hat{H}_B$ are respectively the Hamiltonians of the spin alone, the spin-bath interactions, and the bath itself. Obviously the time evolution equation for the *overall* system-bath density matrix $\hat{\rho}_{SB}(t)$ of a spin system (*S*) in contact with a heat bath (*B*) may now be written taking account of the quantum Liouville Eq. (1.389) as

$$\frac{\partial \hat{\rho}_{SB}}{\partial t} = -\frac{i}{\hbar}\left[\hat{H}_S, \hat{\rho}_{SB}\right] - \frac{i}{\hbar}\left[\hat{H}_B, \hat{\rho}_{SB}\right] - \frac{i}{\hbar}\left[\hat{H}_{SB}, \hat{\rho}_{SB}\right]. \tag{1.393}$$

Obviously, our ultimate objective is to obtain an evolution equation for a *reduced* density matrix, describing the spin relaxation in contact with the thermal bath. Therefore, it is perfectly natural to first try to simplify Eq. (1.393) by considering the time evolution of the *overall* density matrix in the *interaction* representation. This representation is defined for an *arbitrary* operator $\hat{A}(t)$ by [183]

$$\hat{A}^I(t) = e^{i\hat{H}_0 t/\hbar}\hat{A}(t)e^{-i\hat{H}_0 t/\hbar}, \tag{1.394}$$

where $\hat{H}_0 = \hat{H}_S + \hat{H}_B$. Next taking the time derivative of $\hat{\rho}^I_{SB}(t)$ yields

$$\frac{\partial \hat{\rho}^I_{SB}}{\partial t} = \frac{i}{\hbar}\left[\hat{H}_0, \hat{\rho}^I_{SB}\right] + e^{iH_0 t/\hbar}\frac{\partial \hat{\rho}_{SB}}{\partial t}e^{-iH_0 t/\hbar}. \tag{1.395}$$

Therefore, substituting Eq. (1.393) into Eq. (1.395), we have the evolution equation for the overall system-bath density matrix $\hat{\rho}^I_{SB}(t)$ but on this occasion in the *interaction* representation [183]

$$\frac{\partial \hat{\rho}^I_{SB}}{\partial t} = -\frac{i}{\hbar}\left[\hat{H}^I_{SB}, \hat{\rho}^I_{SB}\right]. \tag{1.396}$$

Now recall that we desire the effect of the thermal environment on the dynamical behavior of a given spin and that the dynamics of the spin alone is given by $\hat{H}_S$. Therefore, a quite natural viewpoint is to consider the spin-bath interaction Hamiltonian $\hat{H}_{SB}$ itself as merely a *perturbation* of the *bare* dynamics of the spin. In order to formulate this idea, projection operators are used [88,163,183]. In problems involving a system interacting with its *equilibrium thermal environment,* a *projection operator* $\hat{P}$ is just a particular operator projecting the *overall* system-bath density operator $\hat{\rho}_{SB}$ *onto a product of the reduced density operator* of the system itself, namely $\hat{\rho}_S$, and the *equilibrium density operator* of the bath $\hat{\rho}_B^{eq}$, namely,

$$\hat{P}\hat{\rho}_{SB} = \hat{\rho}_B^{eq}\mathrm{Tr}_B\left(\hat{\rho}_{SB}\right) = \hat{\rho}_B^{eq}\hat{\rho}_S. \tag{1.397}$$

Here the operator $\hat{\rho}_S = \mathrm{Tr}_B\left(\hat{\rho}_{SB}\right)$ resulting from tracing over the bath variables designates a *reduced density operator* in contrast to the *overall* system-bath operator $\hat{\rho}_{SB}(t)$ and our task is then to find an evolution equation for $\hat{\rho}_S$. To achieve this objective since by definition $\mathrm{Tr}_B(\hat{\rho}_B^{eq}) = 1$, we need the operator identity $\hat{P}^2 = \hat{P}$, i.e., $\hat{P}$ is *idempotent* while the *complementary* projector is just $\hat{Q} = 1 - \hat{P}$. We also have the commutation relations

$$\hat{P}\left[\hat{H}_S, \hat{\rho}_{SB}\right] = \left[\hat{H}_S, \hat{P}\hat{\rho}_{SB}\right] = \hat{\rho}_B^{eq}\left[\hat{H}_S, \hat{\rho}_S\right], \tag{1.398}$$

$$\hat{P}\left[\hat{H}_B, \hat{\rho}_{SB}\right] = 0. \tag{1.399}$$

Equations (1.397)–(1.399) are also valid in the interaction representation implying that we can simply rewrite Eq. (1.396) for the evolution of the system-bath density matrix as the *two projected equations* [183]

$$\frac{\partial}{\partial t}\hat{P}\hat{\rho}_{SB}^I = -\frac{i}{\hbar}\hat{P}\left[\hat{H}_{SB}^I, (\hat{P}+\hat{Q})\hat{\rho}_{SB}^I\right], \tag{1.400}$$

$$\frac{\partial}{\partial t}\hat{Q}\hat{\rho}_{SB}^I = -\frac{i}{\hbar}\hat{Q}\left[\hat{H}_{SB}^I, (\hat{P}+\hat{Q})\hat{\rho}_{SB}^I\right]. \tag{1.401}$$

On using the projection Eq. (1.397), we further rewrite these equations as

$$\frac{\partial}{\partial t}\hat{\rho}_S^I = -\frac{i}{\hbar}\mathrm{Tr}_B\left(\left[\hat{H}_{SB}^I, \hat{\rho}_B^{eq}\hat{\rho}_S^I\right]\right) - \frac{i}{\hbar}\mathrm{Tr}_B\left(\left[\hat{H}_{SB}^I, \hat{Q}\hat{\rho}_{SB}^I\right]\right), \tag{1.402}$$

$$\frac{\partial}{\partial t}\hat{Q}\hat{\rho}_{SB}^I = -\frac{i}{\hbar}\hat{Q}\left[\hat{H}_{SB}^I, \hat{\rho}_B^{eq}\hat{\rho}_S^I\right] - \frac{i}{\hbar}\hat{Q}\left[\hat{H}_{SB}^I, \hat{Q}\hat{\rho}_{SB}^I\right]. \tag{1.403}$$

Formal integration of the complementary projection evolution equation Eq. (1.403) then yields

$$\hat{Q}\hat{\rho}^I_{SB}(t) = \hat{Q}\hat{\rho}^I_{SB}(0) \\ -\frac{i}{\hbar}\int_0^t \hat{Q}\left[\hat{H}^I_{SB}, \hat{\rho}^{eq}_B \hat{\rho}^I_S(t')\right]dt' - \frac{i}{\hbar}\int_0^t \hat{Q}\left[\hat{H}^I_{SB}, \hat{Q}\hat{\rho}^I_{SB}(t')\right]dt'. \tag{1.404}$$

This integral equation can be solved in the usual iterative fashion by successively inserting $\hat{Q}\hat{\rho}^I_{SB}(t)$ into the integrand in the third term on the right-hand side. Thus we have a perturbation expansion for our complementary projection operator $\hat{Q}\hat{\rho}^I_{SB}(t)$ in increasing powers of the system-bath Hamiltonian in the interaction representation $\hat{H}^I_{SB}$ which is accomplished by repeated iteration in the usual manner of perturbation theory. Hence neglecting all the higher orders of $\hat{H}^I_{SB}$, we have a drastically simplified version of Eq. (1.404), viz.,

$$\hat{Q}\hat{\rho}^I_{SB}(t) = -\frac{i}{\hbar}\int_0^t \hat{Q}\left[\hat{H}^I_{SB}, \hat{\rho}^{eq}_B \hat{\rho}^I_S(t')\right]dt'. \tag{1.405}$$

In writing Eq. (1.405) we have also disregarded the initial correlation term $\hat{Q}\hat{\rho}_I(0)$ in Eq. (1.404). The latter approximation is tantamount to assuming that $\hat{\rho}^I_{SB}(0)$ is in P space only, that is, *initially the system and the bath are uncorrelated* and also that the bath is *in thermal equilibrium*, or at the very least that the effect of initial correlations decays rapidly relative to the timescale on which the system is observed [183]. Inserting the (truncation) *Ansatz* Eq. (1.405) into the right-hand side of Eq. (1.402) then leads to the following *closed equation for the evolution of the reduced density operator in the interaction representation*

$$\frac{\partial \hat{\rho}^I_S(t)}{\partial t} = -\frac{i}{\hbar}\mathrm{Tr}_B\left(\left[\hat{H}^I_{SB}, \hat{\rho}^{eq}_B \hat{\rho}^I_S(t)\right]\right) \\ -\frac{1}{\hbar^2}\int_0^t \mathrm{Tr}_B\left(\left[\hat{H}^I_{SB}, \hat{Q}\left[\hat{H}^I_{SB}, \hat{\rho}^{eq}_B \hat{\rho}^I_S(t')\right]\right]\right)dt'. \tag{1.406}$$

Returning to the Schrödinger representation, Eq. (1.406) can be rewritten in that representation as the *reduced* equation

$$\frac{\partial\hat{\rho}_S}{\partial t} = -\frac{i}{\hbar}\left[\hat{H}_S + \hat{\bar{V}}, \hat{\rho}_S\right] + \mathrm{St}\left(\hat{\rho}_S\right), \tag{1.407}$$

where

$$\hat{\bar{V}} = \mathrm{Tr}_B\left(\hat{H}_{SB}\hat{\rho}_B^{eq}\right), \tag{1.408}$$

$$\mathrm{St}\left(\hat{\rho}_S\right) = -\frac{1}{\hbar^2}e^{-i\frac{\hat{H}_0 t}{\hbar}}\mathrm{Tr}_B\left(\left[\hat{H}_{SB}^I, \int_0^t (1-\hat{P})\left[\hat{H}_{SB}^I, \hat{\rho}_B^{eq}\hat{\rho}_S^I(t')\right]dt'\right]\right)e^{i\frac{\hat{H}_0 t}{\hbar}}. \tag{1.409}$$

By inspection of Eq. (1.407) it is apparent that the averaged operator $\hat{\bar{V}}$ now has a very simple interpretation, namely, as a mean potential that corrects the spin Hamiltonian $\hat{H}_S$ for the *average* effect of the bath. Such corrections are very important in chemical physics, for example, in determining solvent shifts of spectral lines. The shifts occur because the average solvent interaction may influence in a different way the energies of the ground and excited states of a solvent molecule. However, it is also clear that such *average* interactions can only affect the system eigenstates and energy levels themselves, and therefore cannot cause relaxation [183]. Thus relaxation phenomena are solely associated with the collision (relaxation) operator kernel $\mathrm{St}\left(\hat{\rho}_S\right)$ in Eq. (1.407). Hence, in addressing spin relaxation, we shall disregard the operator $\hat{\bar{V}}$ or else if unconvinced by this, one could consider a renormalized system Hamiltonian $\hat{H}_S$, including the energy shifts associated with the average effect of the bath.

Now we consider the term containing the projection operator $\hat{P}$ involved in the integrand of the collision term Eq. (1.409). Investigation of this term may be simplified if we assume that the interaction Hamiltonian $\hat{H}_{SB}^I$ is merely the *product* of system and bath operators, that is, $\hat{H}_{SB}^I = \hat{V}_S^I\hat{V}_B^I$. Considering the relevant term in Eq. (1.409), we then have [183] the simplified expression

$$\begin{aligned}
&\mathrm{Tr}_B\left(\left[\hat{H}_{SB}^I, \hat{P}\left[\hat{H}_{SB}^I, \hat{\rho}_B^{eq}\hat{\rho}_S^I(t')\right]\right]\right) \\
&= \mathrm{Tr}_B\left(\hat{V}_B^I\hat{\rho}_B^{eq}\right)\mathrm{Tr}_B\left(\hat{V}_B^I\hat{\rho}_B^{eq}\right)\left[\hat{V}_S^I, \left[\hat{V}_S^I, \hat{\rho}_S^I(t')\right]\right] \\
&= \left(\hat{\bar{V}}^B\right)^2\left[\hat{V}_S^I, \left[\hat{V}_S^I, \hat{\rho}^I(t')\right]\right],
\end{aligned} \tag{1.410}$$

where $\hat{\bar{V}}^{B}=\mathrm{Tr}_{B}\left(\hat{V}_{B}^{I}\hat{\rho}_{B}^{eq}\right)$. The mean value of the bath operator $\hat{\bar{V}}^{B}$ is zero for isotropic bath operators (however, it will give rise to corrections to bath correlation functions for *anisotropic* baths). Therefore, assuming for simplicity that the bath is *isotropic* and so neglecting this term in Eq. (1.409), we have the desired *reduced density matrix evolution equation* in Hilbert space describing the spin relaxation in contact with a thermal bath, viz.,

$$\frac{\partial\hat{\rho}_{S}}{\partial t}+\frac{i}{\hbar}\left[\hat{H}_{S},\hat{\rho}_{S}\right]=\mathrm{St}\left(\hat{\rho}_{S}\right), \tag{1.411}$$

where the collision term $\mathrm{St}\left(\hat{\rho}_{S}\right)$ (i.e., the quantum analog of Boltzmann's *Stosszahlansatz*) is given by (with an obvious change of variable in Eq. (1.409)) [183]

$$\mathrm{St}\left(\hat{\rho}_{S}\right)=-\frac{1}{\hbar^{2}}\int_{0}^{t}\mathrm{Tr}_{B}\left(\left[\hat{H}_{SB},e^{-i\hat{H}_{0}\tau/\hbar}\left[\hat{H}_{SB},\hat{\rho}_{B}^{eq}\hat{\rho}_{S}(t-\tau)\right]e^{i\hat{H}_{0}\tau/\hbar}\right]\right)d\tau. \tag{1.412}$$

The *reduced nature* of the evolution Eq. (1.411) now manifests itself in the *appearance of memory* meaning that the time evolution of the reduced density matrix $\hat{\rho}_{S}(t)$ at time t is determined not only by $\hat{\rho}_{S}(t)$ itself at time t but also by its past history, namely, $\hat{\rho}_{S}(t-\tau)$. The *nonlocal temporal* (or non-Markovian) behavior appears because the system evolves at time t in response to the *instantaneous state* of the bath at that time which in turn is determined by the history of the system-bath interaction. We saw that the closed Eq. (1.411) ultimately results from a *low order* perturbation expansion of the system-bath interaction [cf. Eq. (1.405)], so that its validity is expected to be limited to *weak system-bath coupling*. Furthermore, the neglect of *initial* spin-bath correlations, as expressed by dropping the term $\hat{Q}\hat{\rho}_{SB}^{I}(0)$ in Eq. (1.404) constitutes yet another approximation, or, more precisely, a *restriction on the choice of the initial nonequilibrium state*.

1.8.3 *Collision kernel in the Markov approximation*

The collision kernel $\mathrm{St}(\hat{\rho})$ in Eq. (1.412) can be further considerably simplified in the limit, where *the thermal bath dynamics are much faster than those of the spin* [183]. However, to implement this condition in the evolution equation (1.411), we require an *explicit* form for the spin-bath interaction Hamiltonian $\hat{H}_{SB}$. Various models for spin-bath interactions have been discussed in detail, e.g., in Refs. [183,222–228]. Here, we merely suppose for simplicity that the spin-bath interaction Hamiltonian operator $\hat{H}_{SB}$ in Eq. (1.412) has the rudimentary Zeeman form (see, e.g., [229–231]; in our notation)

$$\hat{H}_{SB} = -\gamma\hbar\left(\hat{\mathbf{S}}\cdot\hat{\mathbf{h}}\right) = -\gamma\hbar\sum_{\mu=-1}^{1}\hat{h}^{\mu}\hat{S}_{\mu}, \tag{1.413}$$

where the operator $\hat{\mathbf{h}} = (\hat{h}_X, \hat{h}_Y, \hat{h}_Z)$ represents the random noise field characterizing collisional damping due to the bath,

$$\hat{S}_0 = \hat{S}_Z,\ \hat{S}_{\pm 1} = \mp\frac{1}{\sqrt{2}}\left(\hat{S}_X \pm i\hat{S}_Y\right) \tag{1.414}$$

and

$$\hat{h}^0 = \hat{h}_Z,\ \hat{h}^{\pm 1} = \mp\frac{1}{\sqrt{2}}\left(\hat{h}_X \mp i\hat{h}_Y\right) \tag{1.415}$$

are, respectively, the *covariant and contravariant spherical components* of the spin operator $\hat{\mathbf{S}}$ and the field operator $\hat{\mathbf{h}}$. The properties of $\hat{\mathbf{S}}$ and $\hat{S}_{\mu}$ are explained in detail in Appendix A.2. The cartesian components of $\hat{\mathbf{S}}$ obey the following commutation relations

$$\left[\hat{S}_X, \hat{S}_Y\right] = i\hat{S}_Z,\ \left[\hat{S}_Y, \hat{S}_Z\right] = i\hat{S}_X,\ \left[\hat{S}_Z, \hat{S}_X\right] = i\hat{S}_Y, \tag{1.416}$$

which are merely those characterizing all angular momentum operators and hold for the orbital angular momentum as well as for spins larger than 1/2. We select the interaction operator $\hat{H}_{SB}$ in the Zeeman form of Eq. (1.413) merely because that particular choice obviously renders the evolution Eqs. (1.411) and (1.412) as *a quantum analog of the Brownian rotation of a magnetic dipole* [97] [cf. Eq. (1.42)]. Alternative forms of the Hamiltonian $\hat{H}_{SB}$ are discussed, for example, in Refs. [183,226–228]. Having so chosen the system-bath operator then on using the properties of the bath correlation functions, viz.,

$$\left\langle \hat{h}^{\mu}(t_1)\hat{h}^{-\nu}(t_2)\right\rangle = \delta_{\mu\nu}\mathrm{Tr}_B\left(\hat{h}^{\mu}(t_1)\hat{h}^{-\nu}(t_2)\rho_B^{eq}\right)$$

($\delta_{\mu\nu}$ is Kronecker's delta), the collision kernel Eq. (1.412) further simplifies to

$$\begin{aligned}\mathrm{St}\left(\hat{\rho}_S\right) = \sum_{\mu=-1}^{1}(-1)^{\mu}\int_0^t \Bigg\{ & C_{\mu,-\mu}(\tau)[e^{-\frac{i}{\hbar}\hat{H}_S\tau}\hat{S}_{-\mu}\hat{\rho}_S(t-\tau)e^{\frac{i}{\hbar}\hat{H}_S\tau},\hat{S}_{\mu}] \\ & + C_{\mu,-\mu}(-\tau)[\hat{S}_{-\mu},e^{-\frac{i}{\hbar}\hat{H}_S\tau}\hat{\rho}_S(t-\tau)\hat{S}_{\mu}e^{\frac{i}{\hbar}\hat{H}_S\tau}]\Bigg\}d\tau,\end{aligned} \tag{1.417}$$

where $C_{\mu,-\mu}(\pm\tau)$ are the correlation functions characterizing the properties of the equilibrium bath given by

$$\begin{aligned} C_{\mu,-\mu}(\tau) &= (-1)^{\mu}\gamma^2\left\langle \hat{h}^{\mu}(0)\hat{h}^{-\mu}(\tau)\right\rangle, \\ C_{\mu,-\mu}(-\tau) &= (-1)^{\mu}\gamma^2\left\langle \hat{h}^{\mu}(\tau)\hat{h}^{-\mu}(0)\right\rangle. \end{aligned} \tag{1.418}$$

Despite the foregoing choice of $\hat{H}_{SB}$ the elementary collision kernel Eq. (1.417) although now vastly simplified still describes, in general, the *non-Markovian* behavior of the spin as coupled to the bath, which is determined by the *history* of the spin-bath interaction. To avoid such memory effects, we may use the so-called Markovian limit [183], whereby the bath dynamics are supposed *much faster* than those of the spin (a similar *Ansatz* is essentially made in the classical theory of the Brownian motion [97]). Thus, the bath correlation functions $C_{\mu,-\mu}(\pm\tau)$ *decay to zero much faster than any other characteristic system time scale* of the system [183], therefore we can justly extend the upper limit of integration in Eq. (1.417) to infinity yielding the collision kernel in the more appealing form

$$\begin{aligned}\mathrm{St}\left(\hat{\rho}_S\right) = \sum_{\mu=-1}^{1}(-1)^{\mu}\int_0^{\infty} \{ & C_{\mu,-\mu}(\tau)[e^{-\frac{i}{\hbar}\hat{H}_S\tau}\hat{S}_{-\mu}e^{\frac{i}{\hbar}\hat{H}_S\tau}\hat{\rho}_S(t-\tau),\hat{S}_{\mu}] \\ & + C_{\mu,-\mu}(-\tau)[\hat{S}_{-\mu},\hat{\rho}_S(t-\tau)e^{-\frac{i}{\hbar}\hat{H}_S\tau}\hat{S}_{\mu}e^{\frac{i}{\hbar}\hat{H}_S\tau}]\}d\tau,\end{aligned} \tag{1.419}$$

which in turn is amenable to further simplification. However, tempting as it may seem, the rapidly decaying noise correlation functions $C_{\mu,-\mu}(\tau)$ and $C_{\mu,-\mu}(-\tau)$ in the collision kernel Eq. (1.419) cannot simply be replaced by Dirac-delta functions $\sim\delta(\tau)$ in order to avoid memory effects, because in the quantum case *the bath dynamics themselves may be significantly slower than the phase oscillations in the system* [183]. For example, in the free or bare system, i.e., without any coupling to the reservoir, so that $\mathrm{St}(\hat{\rho}_S)=0$, the solution of the evolution Eq. (1.411) is

$$\hat{\rho}_S(t)=e^{-\frac{i}{\hbar}\hat{H}_S t}\hat{\rho}_S(0)e^{\frac{i}{\hbar}\hat{H}_S t}. \tag{1.420}$$

Hence quantities involving *fast phase oscillations* of $\hat{\rho}_S(t)$ due to the phase factors $e^{\pm i\hat{H}_S t/\hbar}$ embodied in Eq.(1.420), or its remaining signature in the presence of system-bath coupling (i.e., when $\mathrm{St}(\hat{\rho}_S)\neq 0$), cannot simply be taken outside the integrals in the collision kernel Eq. (1.417). Therefore, to circumvent this problem, we again use the interaction representation of $\hat{\rho}_S(t)$, namely,

$$\hat{\rho}_S(t)=e^{-\frac{i}{\hbar}\hat{H}_S t}\hat{\rho}_S^I(t)e^{\frac{i}{\hbar}\hat{H}_S t}.$$

Thus, the collision operator $\mathrm{St}(\hat{\rho}_S)$ becomes in that representation

$$\begin{aligned}\mathrm{St}(\hat{\rho}_S)&=\sum_{\mu=-1}^{1}(-1)^{\mu}\\ &\times\int_0^{\infty}\{C_{\mu,-\mu}(\tau)[e^{-\frac{i}{\hbar}\hat{H}_S\tau}\hat{S}_{-\mu}e^{-\frac{i}{\hbar}\hat{H}_S(t-\tau)}\hat{\rho}_S^I(t-\tau)e^{\frac{i}{\hbar}\hat{H}_S(t-\tau)}e^{\frac{i}{\hbar}\hat{H}_S\tau},\hat{S}_{\mu}]\\ &+C_{\mu,-\mu}(-\tau)[\hat{S}_{-\mu},e^{-\frac{i}{\hbar}\hat{H}_S\tau}e^{-\frac{i}{\hbar}\hat{H}_S(t-\tau)}\hat{\rho}_S^I(t-\tau)e^{\frac{i}{\hbar}\hat{H}_S(t-\tau)}\hat{S}_{\mu}e^{\frac{i}{\hbar}\hat{H}_S\tau}]\}d\tau.\end{aligned} \tag{1.421}$$

To further simplify Eq. (1.421), we now assume that the relaxation of the noise correlation functions $C_{\mu,-\mu}(\tau)$ and $C_{\mu,-\mu}(-\tau)$ to zero as $\tau\to\infty$ is *fast* relative to the timescale on which the time-shifted density matrix $\hat{\rho}_S^I(t-\tau)$ in the interaction representation changes. We then make the approximation $\hat{\rho}_S^I(t-\tau)\approx\hat{\rho}_S^I(t)$ and subsequently return to $\hat{\rho}_S(t)$. Thus, we have our final simplified expression for the collision kernel

$$\begin{aligned}\mathrm{St}(\hat{\rho}_S)&=\sum_{\mu=-1}^{1}(-1)^{\mu}\int_0^{\infty}\{C_{\mu,-\mu}(\tau)[e^{-\frac{i}{\hbar}\hat{H}_S\tau}\hat{S}_{-\mu}e^{\frac{i}{\hbar}\hat{H}_S\tau}\hat{\rho}_S(t),\hat{S}_{\mu}]\\ &+C_{\mu,-\mu}(-\tau)[\hat{S}_{-\mu},\hat{\rho}_S(t)e^{-\frac{i}{\hbar}\hat{H}_S\tau}\hat{S}_{\mu}e^{\frac{i}{\hbar}\hat{H}_S\tau}]\}d\tau,\end{aligned} \tag{1.422}$$

which we see now involves only the reduced density matrix $\hat{\rho}_S(t)$ at time t and not the awkward shifted expression $\hat{\rho}_S(t-\tau)$.

The approximate Eq. (1.422) for the collision kernel corresponds to the traditional Redfield equation [224], which may be proved as follows. We first express $\mathrm{St}(\hat{\rho}_S)$ in the representation defined by the eigenstates $|\alpha\rangle$ of the spin Hamiltonian $\hat{H}_S$ and then rewrite the various commutation relations in Eq. (1.422) in terms of their respective matrix elements, viz.,

$$\begin{aligned}
\left[e^{-i\hat{H}_S\tau/\hbar}\hat{S}_{-\mu}e^{i\hat{H}_S\tau/\hbar}\hat{\rho}_S,\hat{S}_{\mu}\right]_{\alpha\alpha'} &= \sum_{\varepsilon,\varepsilon'}\left[\hat{S}_{-\mu}\right]_{\alpha\varepsilon}\left[\hat{S}_{\mu}\right]_{\varepsilon'\alpha'}\rho_{\varepsilon\varepsilon'}e^{-i(E_\alpha-E_\varepsilon)\tau/\hbar} \\
&\quad -\sum_{\varepsilon',\varepsilon}\left[\hat{S}_{\mu}\right]_{\alpha\varepsilon'}\left[\hat{S}_{-\mu}\right]_{\varepsilon'\varepsilon}\rho_{\varepsilon\alpha'}e^{-i(E_{\varepsilon'}-E_\varepsilon)\tau/\hbar}, \\
\left[\hat{S}_{-\mu},\hat{\rho}_S e^{-i\hat{H}_S\tau/\hbar}\hat{S}_{\mu}e^{i\hat{H}_S\tau/\hbar}\right]_{\alpha\alpha'} &= \sum_{\varepsilon,\varepsilon'}\left[\hat{S}_{-\mu}\right]_{\alpha\varepsilon}\left[\hat{S}_{\mu}\right]_{\varepsilon'\alpha'}\rho_{\varepsilon\varepsilon'}e^{i(E_{\alpha'}-E_{\varepsilon'})\tau/\hbar} \\
&\quad -\sum_{\varepsilon',\varepsilon}\left[\hat{S}_{\mu}\right]_{\varepsilon'\varepsilon}\left[\hat{S}_{-\mu}\right]_{\varepsilon\alpha'}\rho_{\alpha\varepsilon'}e^{i(E_\varepsilon-E_{\varepsilon'})\tau/\hbar}.
\end{aligned} \tag{1.423}$$

Hence, by introducing the matrix elements $R_{\alpha\alpha'\varepsilon\varepsilon'}$ of the relaxation operator $\mathrm{St}(\hat{\rho}_S)$ via

$$\left[\mathrm{St}(\hat{\rho}_S)\right]_{\alpha\alpha'} = \sum\nolimits_{\varepsilon\varepsilon'} R_{\alpha\alpha'\varepsilon\varepsilon'}\rho_{\varepsilon\varepsilon'},$$

it follows from Eq. (1.422) that we then have $R_{\alpha\alpha'\varepsilon\varepsilon'}$ in the eigenvector representation of the Hamiltonian $\hat{H}_S$, rendered in the Redfield form

$$\begin{aligned}
R_{\alpha\alpha'\varepsilon\varepsilon'} = \sum_{\mu=-1}^{1}(-1)^{\mu}\Big\{&\left(\tilde{C}_{\mu,-\mu}(i\omega_{\alpha\varepsilon})+\tilde{C}^{*}_{-\mu,\mu}(i\omega_{\alpha'\varepsilon'})\right)\left[\hat{S}_{-\mu}\right]_{\alpha\varepsilon}\left[\hat{S}_{\mu}\right]_{\varepsilon'\alpha'} \\
&-\delta_{\varepsilon'\alpha'}\sum_{\lambda}e^{\beta\hbar\omega_{\varepsilon\lambda}}\tilde{C}^{*}_{\mu,-\mu}(i\omega_{\varepsilon\lambda})\left[\hat{S}_{\mu}\right]_{\alpha\lambda}\left[\hat{S}_{-\mu}\right]_{\lambda\varepsilon} \\
&-\delta_{\alpha\varepsilon}\sum_{\lambda}e^{\beta\hbar\omega_{\varepsilon'\lambda}}\tilde{C}_{-\mu,\mu}(i\omega_{\varepsilon'\lambda})\left[\hat{S}_{\mu}\right]_{\varepsilon'\lambda}\left[\hat{S}_{-\mu}\right]_{\lambda\alpha'}\Big\}.
\end{aligned} \tag{1.424}$$

Here $\omega_{\alpha\beta}=(E_\alpha-E_\beta)/\hbar$ are the transition frequencies, E_α, E_ε are the eigenvalues of the energy, while the one-sided Fourier transforms $\tilde{C}_{\nu,\mu}(i\omega)$ of the various noise correlation functions (i.e., spectral densities) are

$$\tilde{C}_{\nu,\mu}(i\omega)=\int_0^\infty C_{\nu,\mu}(\tau)e^{-i\omega\tau}d\tau$$

with $C_{-\mu,\mu}(-\tau) = C^*_{\mu,-\mu}(\tau)$ and $\tilde{C}^*_{\mu,-\mu}(i\omega) = e^{-\beta\hbar\omega}\tilde{C}_{\mu,-\mu}(-i\omega)$. The latter relation between the spectral densities can be proved as follows (with $\hbar\omega + E_l = E_k$). We have by definition

$$\begin{aligned} e^{-\beta\hbar\omega}\tilde{C}_{\mu,-\mu}(-i\omega) &= \frac{\gamma^2 e^{-\beta\hbar\omega}}{Z_B}\int_0^\infty \sum_{k,l} e^{iE_l\tau/\hbar} h^{\mu}_{lk} e^{-iE_k\tau/\hbar} h^{-\mu}_{kl} e^{-\beta E_l} e^{i\omega\tau} d\tau \\ &= \frac{\gamma^2}{Z_B}\int_0^\infty \sum_{k,l} e^{iE_l\tau/\hbar} \left(h^{-\mu}_{lk}\right)^* e^{-iE_k\tau/\hbar} \left(h^{\mu}_{kl}\right)^* e^{-\beta E_k} e^{i\omega\tau} d\tau \\ &= \tilde{C}^*_{\mu,-\mu}(i\omega). \end{aligned} \tag{1.425}$$

The Redfield equation (1.424) describes the time evolution of the *reduced density matrix of a system coupled to an equilibrium bath.* The effect of the bath itself obviously enters via the matrix elements of the "relaxation operator" $R_{\alpha\alpha'\varepsilon\varepsilon'}$. Equation (1.424) governing this relaxation operator is written in the basis of eigenstates of the Hamiltonian $\hat{H}_S$ and has been obtained using three approximations: (i) neglect of initial correlations; (ii) the assumption of weak coupling, and (iii) the assumption of a distinct timescale separation between the (fast) bath variables and the (slow) system (spin) variables. All these approximations have been used to get the final Markovian form. We observe that the actual dynamics of the bath or environment *merely enter through the bath correlation functions* Eq. (1.418) *alone*. These functions are properties of the *equilibrium* bath only, regardless of whatever system it may be coupled to. We reiterate that although we have assumed that the bath dynamics are *fast on the timescale of the system dynamics*, the *details* of its dynamics do indeed matter, that is, we could not have simply assumed that the correlation functions themselves are merely of the form

$$C_{\mu,-\mu}(\tau) = 2D_{\mu}\delta(\tau)$$

as in the classical case in the white-noise approximation for random fields. The reason again is the fact that the *bath dynamics are usually slow relative to the phase oscillations* (which are obviously related to the inverse spacing between energy levels) in the system. Also, we must emphasize that solutions of Redfield-like equations for the density matrix even for simple systems (such as harmonic oscillators) do not necessarily satisfy the positivity property [183,232] (for some extreme, nevertheless physically acceptable initial conditions, this *breakdown of positivity* arises from the *omission of memory effects* in the early time evolution [183]). However, regarding relaxation of spin systems *close to thermal equilibrium*, the Redfield form is always positively defined; moreover, it also has a well-defined classical limit likewise invariably positively defined. Furthermore, for a wide range of initial distributions (for those considered here), the solutions of Redfield-like equations *invariably preserve positivity* hence under these conditions their use appears justifiable on physical grounds alone [183,233].

Ending this digression, we observe that the collision kernel also satisfies the equilibrium condition [224]

$$\sum_{\alpha}\left[\mathrm{St}\left(\hat{\rho}_{\mathrm{eq}}\right)\right]_{\alpha\alpha}=\sum_{\alpha,\varepsilon}R_{\alpha\alpha\varepsilon\varepsilon}\rho_{\varepsilon\varepsilon}^{\mathrm{eq}}=0, \tag{1.426}$$

where $\rho_{\varepsilon\varepsilon'}^{\mathrm{eq}}=\rho_{\varepsilon\varepsilon}^{\mathrm{eq}}\delta_{\varepsilon\varepsilon'}$. To prove this, we use Eq. (1.423) yielding

$$\begin{aligned}\sum_{\alpha}\left[\mathrm{St}\left(\hat{\rho}_{\mathrm{eq}}\right)\right]_{\alpha\alpha}&=\sum_{\alpha,\varepsilon}\sum_{\mu=-1}^{1}(-1)^{\mu}\int_{0}^{\infty}\Bigg\{C_{\mu,-\mu}(\tau)\Bigg(\left[\hat{S}_{-\mu}\right]_{\alpha\varepsilon}\left[\hat{S}_{\mu}\right]_{\varepsilon\alpha}\rho_{\varepsilon\varepsilon}^{\mathrm{eq}}e^{-i\frac{E_{\alpha}-E_{\varepsilon}}{\hbar}\tau}\\&\quad-\left[\hat{S}_{\mu}\right]_{\alpha\varepsilon}\left[\hat{S}_{-\mu}\right]_{\varepsilon\alpha}\rho_{\alpha\alpha}^{\mathrm{eq}}e^{-i\frac{E_{\varepsilon}-E_{\alpha}}{\hbar}\tau}\Bigg)+C_{\mu,-\mu}(-\tau)\\&\quad\times\Bigg(\left[\hat{S}_{-\mu}\right]_{\alpha\varepsilon}\left[\hat{S}_{\mu}\right]_{\varepsilon\alpha}\rho_{\varepsilon\varepsilon}^{\mathrm{eq}}e^{i\frac{E_{\alpha}-E_{\varepsilon}}{\hbar}\tau}-\left[\hat{S}_{\mu}\right]_{\alpha\varepsilon}\left[\hat{S}_{-\mu}\right]_{\varepsilon\alpha}\rho_{\alpha\alpha}^{\mathrm{eq}}e^{i\frac{E_{\varepsilon}-E_{\alpha}}{\hbar}\tau}\Bigg)\Bigg\}d\tau.\end{aligned}$$

Then by interchanging indexes via the replacement $\alpha\leftrightarrow\varepsilon$ in the second and fourth sums we have Eq. (1.426). Continuing with our analysis of $\mathrm{St}\left(\hat{\rho}_S\right)$ in the Redfield form, we can formally define diffusion matrix coefficients D_{μ} as

$$D_{\mu}=\tilde{C}_{\mu}^{sym}(\omega)\operatorname{sech}\frac{\beta\hbar\omega}{2}, \tag{1.427}$$

where the *symmetrized* spectral density, namely,

$$\tilde{C}_{\mu}^{sym}(\omega)=\frac{1}{2}\left[\tilde{C}_{\mu,-\mu}(-i\omega)+\tilde{C}_{\mu,-\mu}^{*}(i\omega)\right],$$

in turn, determines the spectrum of the *symmetrized* bath correlation functions [234] defined as

$$C_{\mu}^{sym}(\tau)=\frac{(-1)^{\mu}}{2}\gamma^{2}\left(\left\langle\hat{h}^{\mu}\hat{h}^{-\mu}(\tau)\right\rangle+\left\langle\hat{h}^{-\mu}(\tau)\hat{h}^{\mu}\right\rangle\right). \qquad (1.428)$$

Thus, with this definition, we have for the matrix elements $R_{\alpha\alpha'\varepsilon\varepsilon'}$ in the *eigenvector* representation

$$\begin{aligned} R_{\alpha\alpha'\varepsilon\varepsilon'} \approx \sum_{\mu=-1}^{1}(-1)^{\mu}D_{\mu}\Big\{ & e^{\beta\hbar\omega_{\varepsilon\alpha}/2}\left[\hat{S}_{-\mu}\right]_{\alpha\varepsilon}\left[\hat{S}_{\mu}\right]_{\varepsilon'\alpha'} \\ & + e^{\beta\hbar\omega_{\varepsilon'\alpha'}/2}\left[\hat{S}_{-\mu}\right]_{\alpha\varepsilon}\left[\hat{S}_{\mu}\right]_{\varepsilon'\alpha'} \qquad (1.429) \\ -\sum_{\lambda}\Big(\delta_{\varepsilon'\alpha'}e^{\beta\hbar\omega_{\varepsilon\lambda}/2}\left[\hat{S}_{-\mu}\right]_{\alpha\lambda}\left[\hat{S}_{\mu}\right]_{\lambda\varepsilon} & -\delta_{\alpha\varepsilon}e^{\beta\hbar\omega_{\varepsilon'\lambda}/2}\left[\hat{S}_{-\mu}\right]_{\varepsilon'\lambda}\left[\hat{S}_{\mu}\right]_{\lambda\alpha'}\Big)\Big\}. \end{aligned}$$

In the operator notation, Eq. (1.429) finally yields the collision kernel operator in the *symmetrized* Hubbard form [235]

$$\begin{aligned} \mathrm{St}\left(\hat{\rho}_{S}\right)=\sum_{\mu=-1}^{1}(-1)^{\mu}D_{\mu}\Big(& \left[\hat{S}_{\mu},\hat{\rho}_{S}e^{\beta\hat{H}_{S}/2}\hat{S}_{-\mu}e^{-\beta\hat{H}_{S}/2}\right] \\ & +\left[e^{-\beta\hat{H}_{S}/2}\hat{S}_{-\mu}e^{\beta\hat{H}_{S}/2}\hat{\rho}_{S},\hat{S}_{\mu}\right]\Big). \end{aligned} \qquad (1.430)$$

The vital difference between the eigenvector, Eq. (1.429), and the operator, Eq. (1.430), representations is that the operator picture is valid for an *arbitrary state* representation whereas Eq. (1.429) originates in contrast in the *energy state* representation [183].

The collision kernel $\mathrm{St}\left(\hat{\rho}_{S}\right)$ as rendered in the operator form Eq. (1.430) satisfies certain basic requirements constituting important consistency checks. These are as follows:

(i) the collision operator $\mathrm{St}\left(\hat{\rho}_{S}\right)$ is also Hermitian if $\hat{\rho}_{S}(t)$ and $\hat{H}_{S}$ are Hermitian;
(ii) the equilibrium density matrix $\hat{\rho}_{eq}=e^{-\beta\hat{H}_{S}}/\mathrm{Tr}\left(e^{-\beta\hat{H}_{S}}\right)$ renders $\mathrm{St}\left(\hat{\rho}_{eq}\right)=0$ thus ensuring that the spin system reaches a Boltzmann distribution at equilibrium while
(iii) the matrix exponents $e^{\pm\beta\hat{H}_{S}/2}$ in the collision kernel operator Eq. (1.430) ensure detailed balance;

(iv) $\mathrm{St}(\hat{\rho}_S)$ in Eq. (1.430) is invariant under rotations of the coordinate system;
(v) the Hubbard model [235] embodied in Eq. (1.430) can be generalized to time-dependent Hamiltonians (see the next subsection); finally,
(vi) the spin density matrix evolution Eq. (1.411) with $\mathrm{St}(\hat{\rho}_S)$ in the symmetrized form of Eq. (1.430) effectively constitutes the direct analog of the classical Fokker–Planck equation (1.64) for rotational diffusion of a classical magnetic dipole.

The relaxation operator written in the form of Eq. (1.430) is a reasonable approximation in the high-temperature limit [229,230]. Essentially, it follows from the equation of motion of the reduced density matrix in the rotating-wave approximation (long familiar in quantum optics, where counter rotating, rapidly oscillating terms, are averaged out [99]) provided the spin-bath interactions are taken in the *weak coupling limit* and for *Ohmic damping*. In which case, the correlation time τ_c characterizing the bath may indeed be regarded as very short so that the stochastic process originating in the bath is also Markovian ($\gamma H \tau_c \ll 1$, where H is the averaged amplitude of the random magnetic field). If the approximation rendered by Eq. (1.430) is invalid (e.g., throughout the very low-temperature region), alternative models for the spin-bath interactions should be used [183,226–228,234–238]. Nevertheless, we shall persevere with the model in the simplified form just described because despite its drawbacks our ultimate objective is merely to understand in *semiclassical fashion* how quantum effects alter the rotational Brownian motion and longitudinal relaxation of a classical giant spin at finite S. Moreover, that model as well as providing a qualitative description of the spin relaxation of a variety of systems can also be regarded as the *direct analog* of the Fokker–Planck equation formalism used by Brown [17,18] and others (e.g., [97–100,109,239–242]) to treat relaxation of classical macrospins.

We observe that the collision kernel Eq. (1.430) can be further simplified in the high-temperature limit, $\beta\hbar\omega_{\varepsilon\alpha} \ll 1$, by supposing that the coefficients D_μ of the diffusion matrix are *frequency-independent*, i.e.,

$$D_\mu = D_{-\mu} \approx \tilde{C}_\mu^{sym}(0). \tag{1.431}$$

In the time domain, this approximation merely corresponds to the representation of the bath correlation functions $C_\mu^{sym}(\tau)$ Eq. (1.428) as the sum of two delta functions, viz.,

$$C_\mu^{sym}(\tau) \approx D_\mu\left[\delta\left(\tau + i\beta\hbar/2\right) + \delta\left(\tau - i\beta\hbar/2\right)\right]. \tag{1.432}$$

In the classical limit $\hbar \to 0$, Eq. (1.432) obviously reduces to the familiar result for the classical bath correlation functions in the white-noise approximation for random fields, namely,

$$C_\mu^{cl}(\tau) = 2D_\mu \delta(\tau).$$

The derivation of Eq. (1.430) for $\mathrm{St}(\hat{\rho}_S)$ assumes that the spin Hamiltonian does not depend *explicitly* on the time [224]. However, Hubbard [235] has investigated the more general case of a *time-dependent* spin-Hamiltonian operator $\hat{H}_S = \hat{H}_S(t)$ and in so doing proposed the corresponding form for the collision kernel (in our notation)

$$\begin{aligned}\mathrm{St}(\hat{\rho}_S) &= \sum_{\mu=-1}^{1}\sum_r (-1)^\mu e^{i\omega_r^{-\mu}t} D_\mu(\omega_r^{-\mu}) \\ &\times\left\{ e^{\frac{\beta\hbar\omega_r^{-\mu}}{2}}\left[\hat{S}_\mu, \hat{\rho}_S \hat{U}^{-1}(t)\hat{S}_{-\mu}^r \hat{U}(t)\right]\right. \\ &\left. + e^{-\frac{\beta\hbar\omega_r^{-\mu}}{2}}\left[\hat{U}^{-1}(t)\hat{S}_{-\mu}^r \hat{U}(t)\hat{\rho}_S, \hat{S}_\mu\right]\right\},\end{aligned} \tag{1.433}$$

where the operators $\hat{S}_{\mu'}^r$ are the coefficients in the series expansion of the time-dependent spin operators $\hat{S}_{\mu'}(t) = \hat{U}(t)\hat{S}_{\mu'}\hat{U}^{-1}(t)$, namely,

$$\hat{S}_{\mu'}(t) = \sum_r \hat{S}_{\mu'}^r e^{i\omega_r^{\mu'}t}, \tag{1.434}$$

where $\omega_r^{\mu'}$ represents a parameter, while the operator $\hat{U}(t)$ is defined as

$$\hat{U}(t) = e^{\frac{i}{\hbar}\int_0^t \hat{H}_S(t')dt'}, \tag{1.435}$$

and D_μ is the spectral density of the correlation function of the bath as given by Eq. (1.427). Then by reconverting the result to operator form [see Eq. (1.434)], we now have a simplified form for the collision kernel, viz.,

$$\mathrm{St}\left(\hat{\rho}_S\right)=\sum_{\mu=-1}^{1}\sum_{r}(-1)^{\mu}D_{\mu}e^{i\omega_r^{-\mu}t}$$
$$\times\left(e^{\beta\hbar\omega_r^{-\mu}/2}\left[\hat{S}_{\mu},\hat{\rho}_S\hat{U}^{-1}(t)\hat{S}^{r}_{-\mu}U(t)\right]+e^{-\beta\hbar\omega_r^{-\mu}/2}\left[\hat{U}^{-1}(t)\hat{S}^{r}_{-\mu}\hat{U}(t)\hat{\rho}_S,\hat{S}_{\mu}\right]\right)$$
$$=\sum_{\mu=-1}^{1}(-1)^{\mu}D_{\mu}\left(\left[\hat{S}_{\mu},\hat{\rho}_S\hat{U}^{-1}(t)\hat{U}(t-i\beta\hbar/2)\hat{S}_{-\mu}\hat{U}^{-1}(t-i\beta\hbar/2)\hat{U}(t)\right]\right.$$
$$\left.+\left[\hat{U}^{-1}(t)\hat{U}(t+i\beta\hbar/2)\hat{S}_{-\mu}\hat{U}^{-1}(t+i\beta\hbar/2)\hat{U}(t)\hat{\rho}_S,\hat{S}_{\mu}\right]\right).\quad(1.436)$$

In Eq. (1.436), we encounter typical operator products of the form $\hat{U}^{-1}(t)\hat{U}(t\pm i\beta\hbar/2)$ given by

$$\hat{U}^{-1}(t)\hat{U}(t\pm i\beta\hbar/2)=e^{\frac{i}{\hbar}\int_{t}^{t\pm i\beta\hbar/2}\hat{H}_S(t')dt'}.\quad(1.437)$$

We then use for the value of the integral occurring in the exponent

$$\frac{i}{\hbar}\int_{t}^{t\pm i\beta\hbar/2}\hat{H}_S(t')dt'\approx\mp\frac{\beta}{2}\hat{H}_S(t).\quad(1.438)$$

Hence, we have made an approximation valid only in the high-temperature limit $\beta\hbar\omega<<1$ that is to say for *small* increments $\Delta t\sim\beta\hbar/2<<1$ in Eq. (1.438), we have supposed that the Hamiltonian $\hat{H}_S(t')$ does not alter significantly during the small time Δt. Thus, we can simply use the value of that operator at time t and so may place it outside the integral. By treating in like manner all other such time-dependent functions in Eq. (1.436), we have an equation similar to the previous Eq. (1.430). Thus, the Hubbard form of the collision kernel Eq. (1.436) with time-dependent Hamiltonian $\hat{H}_S(t)$ in the *high-temperature* limit simplifies to

$$\mathrm{St}\left(\hat{\rho}_S\right)=\sum_{\mu=-1}^{1}(-1)^{\mu}D_{\mu}\left(\left[\hat{S}_{\mu},\hat{\rho}_S e^{\beta\hat{H}_S(t)/2}\hat{S}_{-\mu}e^{-\beta\hat{H}_S(t)/2}\right]\right.$$
$$\left.+\left[e^{-\beta\hat{H}_S(t)/2}\hat{S}_{-\mu}e^{\beta\hat{H}_S(t)/2}\hat{\rho}_S,\hat{S}_{\mu}\right]\right.\quad(1.439)$$

Therefore as discussed above the form of the collision kernel given by Eq. (1.439) also corresponds to the *high-temperature limit* and *short correlation time* characteristic of the Markovian approximation.

1.8.4 *Method of statistical moments for quantum spins*

One of the most powerful techniques for the solution of the evolution equations governing the relaxation dynamics of classical and quantum spin systems is ultimately the *method of statistical moments* [97,104,143,217]. To summarize the principal equations of that method for quantum systems, we first recall that the polarization operators $\hat{T}_{LM}^{(S)}$ are widely used in describing spin states of a particle [104,217] (see Appendix A.2). Since the latter constitute an orthonormal basis in the space of $(2S+1)\times(2S+1)$ square matrices, the reduced density matrix $\hat{\rho}_S$ may be expanded in a finite sum of the $\hat{T}_{LM}^{(S)}$ as [104,217]

$$\hat{\rho}_S(t) = \sum_{L=0}^{2S}\sum_{M=-L}^{L} a_{LM}(t)\hat{T}_{LM}^{(S)}, \tag{1.440}$$

where the expansion coefficients $a_{L,M}(t)$ defined as the average

$$a_{LM}(t) = \left\langle \hat{T}_{LM}^{\dagger(S)} \right\rangle(t) = \mathrm{Tr}\left(\hat{\rho}_S(t)\hat{T}_{LM}^{\dagger(S)}\right) \tag{1.441}$$

are simply the *statistical moments* representing expectation values of the operators $\hat{T}_{LM}^{\dagger(S)}$ in a state described by the density matrix $\hat{\rho}_S$. Equation (1.441) follows immediately from the orthogonality properties of the operators $\hat{T}_{LM}^{(S)}$ [104].

However, the *formal* solution of the reduced spin density matrix evolution Eq. (1.411) with an *arbitrary* collision kernel $\mathrm{St}(\hat{\rho}_S)$ can also be written as the operator series Eq. (1.440). Therefore to find the corresponding statistical moments, we simply substitute the operator expansion Eq. (1.440) into the reduced density matrix evolution equation (1.411). We then have

$$\sum_{L=0}^{2S}\sum_{M=-L}^{L} \hat{T}_{LM}^{(S)}\frac{da_{LM}}{dt} = \sum_{L=0}^{2S}\sum_{M=-L}^{L}\left(-\frac{i}{\hbar}\left[\hat{H}_S, \hat{T}_{LM}^{(S)}\right] + \mathrm{St}\left(\hat{T}_{LM}^{(S)}\right)\right)a_{LM}, \tag{1.442}$$

where the collision operator is explicitly rendered as

$$\mathrm{St}\left(\hat{T}_{LM}^{(S)}\right) = \sum_{\mu=-1}^{1}(-1)^{\mu}D_{\mu}\left([\hat{S}_{\mu}, \hat{T}_{LM}^{(S)}e^{\frac{\beta}{2}\hat{H}_S}\hat{S}_{-\mu}e^{-\frac{\beta}{2}\hat{H}_S}] + [e^{-\frac{\beta}{2}\hat{H}_S}\hat{S}_{-\mu}e^{\frac{\beta}{2}\hat{H}_S}\hat{T}_{LM}^{(S)}, \hat{S}_{\mu}]\right).$$

However, the term in parenthesis on the right-hand side of Eq. (1.442) is in itself just another matrix operator and so it can also be formally expanded as a polarization operator series, yielding (the expansion coefficients which are as usual defined by the trace of that operator)

$$-\frac{i}{\hbar}\left[\hat{H}_S,\hat{T}_{LM}^{(S)}\right]+\mathrm{St}\left(\hat{T}_{LM}^{(S)}\right) \\ =\sum_{L'=0}^{2S}\sum_{M'=-L}^{L}\mathrm{Tr}\left\{\left(-\frac{i}{\hbar}\left[\hat{H}_S,\hat{T}_{LM}^{(S)}\right]+\mathrm{St}\left(\hat{T}_{LM}^{(S)}\right)\right)\hat{T}_{L'M'}^{\dagger(S)}\right\}\hat{T}_{L'M'}^{(S)}. \tag{1.443}$$

Consequently, by substituting the polarization operator expansion Eq. (1.443) into Eq. (1.442) and equating (by orthogonality) terms with the same $\hat{T}_{LM}^{(S)}$, we have the *formal* evolution equation for the statistical moments $a_{LM}(t)$, viz.,

$$\frac{d}{dt}a_{LM}=\sum_{L'=0}^{2S}\sum_{M'=-L}^{L}g_{L'M';LM}\,a_{L'M'}, \tag{1.444}$$

where the coefficients $g_{LM;L'M'}$ are formally defined as the average

$$g_{LM;L'M'}=\mathrm{Tr}\left\{-\frac{i}{\hbar}\left[\hat{H}_S,\hat{T}_{LM}^{(S)}\right]\hat{T}_{L'M'}^{\dagger(S)}+\mathrm{St}\left(\hat{T}_{LM}^{(S)}\right)\hat{T}_{L'M'}^{\dagger(S)}\right\}. \tag{1.445}$$

The coefficients $g_{L'M';LM}$ arising in Eq. (1.444) can be obtained from the coefficients $g_{LM;L'M'}$ defined in Eq. (1.445) via the replacements $L\leftrightarrow L'$ and $M\leftrightarrow M'$. Equations (1.444) and (1.445) are valid for an *arbitrary* collision kernel operator $\mathrm{St}(\hat{\rho}_S)$. However, for $\mathrm{St}(\hat{\rho}_S)$ given *by the symmetrized Hubbard form*, Eq. (1.430), the formal Eq. (1.445) reduces to the following simplified expression

$$g_{LM;L'M'}=\mathrm{Tr}\left\{\left(-\frac{i}{\hbar}\left[\hat{H}_S,\hat{T}_{LM}^{(S)}\right]+\sum_{\mu=-1}^{1}(-1)^{\mu}D_{\mu}\left(\left[\hat{S}_{\mu},\hat{T}_{LM}^{(S)}e^{\frac{\beta}{2}\hat{H}_S}\hat{S}_{-\mu}e^{-\frac{\beta}{2}\hat{H}_S}\right]\right.\right.\right. \\ \left.\left.\left.+\left[e^{-\frac{\beta}{2}\hat{H}_S}\hat{S}_{-\mu}e^{\frac{\beta}{2}\hat{H}_S}\hat{T}_{LM}^{(S)},\hat{S}_{\mu}\right]\right)\right)\hat{T}_{L'M'}^{\dagger(S)}\right\}. \tag{1.446}$$

Nowadays, the explicit calculation of the coefficients $g_{LM;L'M'}$ for a given collision operator is best accomplished using MATHEMATICA®. In general however we will always have a *finite* set of differential-recurrence equations for the averages, i.e. the *statistical moments*, $\left\langle \hat{T}_{LM}^{\dagger(S)} \right\rangle(t)$ as defined by Eq. (1.441), viz.,

$$\frac{d}{dt}\left\langle \hat{T}_{LM}^{\dagger(S)} \right\rangle(t) = \sum_{L'=0}^{2S} \sum_{M'=-L}^{L} g_{LM;L'M'} \left\langle \hat{T}_{L'M'}^{\dagger(S)} \right\rangle(t), \tag{1.447}$$

or because $\hat{T}_{LM}^{\dagger(S)} = (-1)^M \hat{T}_{L-M}^{(S)}$ we also have

$$\frac{d}{dt}\left\langle \hat{T}_{LM}^{(S)} \right\rangle(t) = \sum_{L',M'} (-1)^{M'-M} g_{L'-M';L-M} \left\langle \hat{T}_{L'M'}^{(S)} \right\rangle(t). \tag{1.448}$$

The differential-recurrence Eqs. (1.447) and (1.448) pose the *general* solution of the problem of the determination of the statistical moments in the *operator representation for arbitrary spin Hamiltonians*. Explicit forms of the coefficients $g_{LM;L'M'}$ resulting from the polarization operator expansion of the Liouville and collision terms in the density matrix evolution equation, e.g. Eq.(1.446), for specific cases of the Hamiltonian $\hat{H}_S$ and particular examples of solutions of the differential-recurrence equations Eqs. (1.447) and (1.448) will be given later. In general, the averaged Eqs. (1.447) and (1.448) are also very useful in obtaining differential-recurrence relations for phase-space observables.

The recurrence Eq. (1.447) may be formally solved using standard matrix inversion techniques. This is accomplished by observing that according to Eq. (1.447), the behavior of any selected average $a_{L'M'}(t) = \left\langle \hat{T}_{L'M'}^{\dagger(S)} \right\rangle(t)$ say, is coupled to that of all the others so forming a *finite* hierarchy of differential-recurrence equations because *unlike* the classical one yielding an *infinite* hierarchy, the index L now ranges only between 0 and $2S$. As in Section 1.3.5, the solution of such a multi-term recurrence relation may always be obtained [97] by rewriting it as a first-order linear matrix differential equation with constant coefficients. This is accomplished by first defining a supercolumn vector $\mathbf{C}(t)$ such that

$$\mathbf{C}(t) = \begin{pmatrix} \mathbf{c}_1(t) \\ \mathbf{c}_2(t) \\ \vdots \\ \mathbf{c}_{2S}(t) \end{pmatrix}, \quad \mathbf{c}_L(t) = \begin{pmatrix} a_{L-L}(t) \\ a_{L-L+1}(t) \\ \vdots \\ a_{LL}(t) \end{pmatrix}, \tag{1.449}$$

whereupon the evolution Eq. (1.447) of the average polarization operators just becomes the linear homogeneous matrix differential Eq. (1.112), where **A** is the $4S(S+1)\times 4S(S+1)$ transition supermatrix with matrix elements

$$(\mathbf{A})_{LL'} = \mathbf{G}^{S}_{LL'} \text{ and } \left[\mathbf{G}^{S}_{LL'}\right]_{MM'} = -g_{L'M';LM}. \tag{1.450}$$

The formal solution of Eq. (1.450) for the desired column vector $\mathbf{C}(t)$ is given by Eq. (1.113). Having calculated $\mathbf{C}(t)$ from the matrix Eq. (1.113), we then have all the statistical moments

$$a_{LM}(t) = \left\langle \hat{T}^{\dagger(S)}_{LM} \right\rangle(t) = (-1)^M \left\langle \hat{T}^{(S)}_{L-M} \right\rangle(t). \tag{1.451}$$

However, recalling that the spin operators themselves, viz., $\hat{S}_X, \hat{S}_Y$, and $\hat{S}_Z$, may always be expressed in terms of the $\hat{T}^{(S)}_{10}$ and $\hat{T}^{(S)}_{1\pm1}$ via Eq. (A.147) of Appendix A, we have the average components of these operators rendered in terms of the average polarization operators, viz.,

$$\begin{gathered} \left\langle \hat{S}_X \right\rangle(t) = a\left[\left\langle \hat{T}^{(S)}_{1-1} \right\rangle(t) - \left\langle \hat{T}^{(S)}_{11} \right\rangle(t)\right], \\ \left\langle \hat{S}_Y \right\rangle(t) = ia\left[\left\langle \hat{T}^{(S)}_{1-1} \right\rangle(t) + \left\langle \hat{T}^{(S)}_{11} \right\rangle(t)\right], \\ \left\langle \hat{S}_Z \right\rangle(t) = \sqrt{2}a\left\langle \hat{T}^{(S)}_{10} \right\rangle(t), \end{gathered} \tag{1.452}$$

where

$$a = \sqrt{S(S+1)(2S+1)/6}. \tag{1.453}$$

The differential-recurrence relation, Eq. (1.447), can also be solved by the matrix continued fraction method [97,143].

The density-matrix method as illustrated via the evolution equation for the polarization operators has hitherto constituted the standard approach to the treatment of spin relaxation and resonance phenomena. However, relaxation and resonance of spins interacting with a bath can also be treated in classical-like fashion via the relevant *quasiprobability* distribution function $W_S(\vartheta,\varphi,t)$ of spin orientations in a phase space (here configuration space) (ϑ, φ); ϑ and φ are the polar and azimuthal angles (see, e.g., [220,221,229–231,243–264]). The spin quasiprobability distribution function $W_S(\vartheta,\varphi,t)$ is considered in the following sections.

1.9 Phase-space formulation of magnetization relaxation

Despite the undoubted merits of the phase-space description of spin relaxation [229–231,243–245], it appears that the formulation has been relatively underexploited outside the realm of quantum optics [184,265]. Thus, applications to other spin systems are few including spin relaxation problems, etc. This may in part be due to the relatively complex mathematical manipulations, involved in comparison with the phase-space formulation for particles with separable and additive Hamiltonians. Therefore, it is now our purpose to provide a reasonably comprehensive and didactic account of the phase-space description of spin relaxation and other allied topics. We also show by using Wigner's quasiprobability distribution function formalism, how the existing methods of analysis of the Brownian motion of classical spins may be extended to include quantum effects. Thus, a semiclassical theory of spin relaxation will ensue.

The phase-space (generalized coherent state) treatment comprises the extension of Wigner's phase-space representation of the density operator [6,184,265,267–272] (originally developed to obtain quantum corrections to the classical equilibrium distributions for an assembly of point particles in the phase space of their positions and momenta) to describe spin systems, see, e.g., Refs. [229–231,243–245,273–275]. However, phase-space representations of quantum mechanical evolution equations also provide a formal means of treating quantum effects in dynamical systems transparently linking to the classical representations, so facilitating the calculation of quantum corrections to classical distribution functions in a rather intuitive way. Such representations generally based on the *coherent state representation* of the density matrix introduced by Glauber and Sudarshan (see [184]) and widely used in quantum optics [184,265] when applied to spin systems (e.g., [220,221, 229–231,243–263]) allow one to analyze quantum spin relaxation using [264] a master equation for a *quasiprobability distribution function* $W_S(\vartheta,\varphi,t)$ of spin orientations in a phase (here configuration) space (ϑ, φ) which is *classically meaningful.*

The consequent mapping of the quantum spin dynamics onto *c*-number quasiprobability density evolution equations clearly shows how these equations reduce to the Fokker–Planck equation in the classical limit [245,256,257]. The general phase-space distribution function formalism for spins was originally introduced by Stratonovich [273] for closed systems and further developed both for closed and open spin systems, [220,221,229–231,243–263]. It is entirely analogous to the Wigner distribution $W(q, p, t)$ in the phase space of positions and momenta (q, p) of a particle [6,184,265,267–271], which is merely a *certain (overlap) Fourier transform corresponding to a quasiprobability representation of the density matrix operator* $\hat{\rho}(t)$. However, particular differences arise for spins because of the angular momentum commutation relations. In general the distribution function $W_S(\vartheta,\varphi,t)$ of spin orientations in a configuration space (ϑ, φ), like the Wigner function $W(q, p, t)$ for the translational motion of a particle, enables one to calculate the expected value of a spin operator $\hat{A}$ in Hilbert space via the *corresponding c-number* (or Weyl) symbol $A(\vartheta, \varphi)$. Hence, quantum mechanical averages may be determined in a classical-like manner, because in effect the Weyl symbol represents a general method of transforming operators defined on Hilbert space into phase-space functions. The Wigner representation contains only such features as are common to both quantum and classical statistical mechanics and formally represents quantum mechanics as a statistical theory on classical phase space [267,270,271,275]. Therefore, it is especially suitable for the *development of semiclassical methods of solution*, for example, for analyzing two interacting systems, where one is treated quantum mechanically while the other is treated by using classical theory. Here the Wigner representation allows one to easily obtain quantum corrections to the classical results, in a manner which is naturally suited to the calculation of quantum corrections to the latter. The formalism is relatively easy to implement because as we shall see the form of the master equations governing the time evolution of phase-space distributions enables powerful computational techniques originally developed for the solution of Fokker–Planck equations for the rotational Brownian motion of classical magnetic dipoles (e.g., continued fractions, mean first-passage times, etc. [97,143]) to be seamlessly carried over into the quantum domain [256,257,260, 276,277].

1.9.1 *Quasiprobability distribution functions for particles*

To introduce the concept of a quasiprobability distribution function we recall that, in general, a classical dynamical system of one degree of freedom may be described by a phase-space probability distribution function $W(q, p, t)$ yielding the joint probability $W(q, p, t)dpdq$ that the system is in a volume element $dpdq$ centered around the phase-space point (q, p) specifying the canonical variables which are the coordinate q and the momentum p. However, in the quantum mechanical description of a dynamical system (because of the uncertainty principle) the phase-space coordinates (q, p) *cannot take on definite values simultaneously*. Therefore, the concept of such a function does not exist at all for a quantum system because the idea of a *sharp* phase point and a collection or ensemble of such sharp points has of itself no meaning. Nevertheless, it is still possible to utilize certain mathematical constructs called *quasiprobability distributions*, closely resembling the classical phase-space distribution functions. Such quasiprobability distributions have proven [6,184,265,267–271] to be very useful in a variety of physical applications as they provide fruitful insights into the connection between classical and quantum mechanics allowing one to express quantum mechanical averages in an intuitively appealing form which is very similar to that of classical averages. Thus, they are ideally suited to the study of the quantum-classical correspondence. Furthermore, they provide a useful tool for introducing quantum corrections to classical models of dissipation such as many body collisions, Brownian motion, escape-rate theory, etc. (see, e.g., Refs. [236,270,271,278–281]) as well as the connection between decoherence (viewed as loss of information from a quantum system to an environment) and the quantum to classical transition [282]. The first of these quasiprobability distributions was introduced by Wigner [6] in 1932 in order to study in semiclassical fashion quantum corrections to the Maxwell–Boltzmann distribution of classical statistical mechanics which *inter alia* elucidated the role played by tunneling effects at high temperatures in reaction rate theory [278–281]. In principle, the Wigner distribution function was meant to be a reformulation, using the concept of a quasiprobability distribution in *phase space*, of Schrödinger's wave mechanics, which describes quantum states in *configuration space*.

Before proceeding, we recall [275] the basic properties of the coordinate (position) and momentum operators $\hat{q}$ and $\hat{p} = -i\hbar\partial / \partial q$, used to describe a quantum mechanical system in Hilbert space with one degree of freedom [88,216,218] and with separable and additive Hamiltonian operator $\hat{H} = \hat{p}^2 / (2m) + \hat{V}(q)$:

(i) the operators $\hat{q}$ and $\hat{p}$ are *non-commutative*, i.e.,

$$[\hat{q}, \hat{p}] = \hat{q}\hat{p} - \hat{p}\hat{q} = i\hbar; \tag{1.454}$$

(ii) the *eigenvectors* $|q\rangle$ and $|p\rangle$ of the operators $\hat{q}$ and $\hat{p}$ obey the conditions:

$$\begin{gathered} \hat{q}|q\rangle = q|q\rangle, \; \hat{p}|p\rangle = p|p\rangle, \\ \langle q'|q''\rangle = \delta(q' - q''), \; \langle p'|p''\rangle = \delta(p' - p''), \\ \langle q|p\rangle = \frac{1}{\sqrt{2\pi\hbar}} e^{iqp/\hbar}, \; \langle p|q\rangle = \frac{1}{\sqrt{2\pi\hbar}} e^{-iqp/\hbar}, \\ \int |q\rangle\langle q| dq = \hat{I}, \; \int |p\rangle\langle p| dp = \hat{I}, \end{gathered} \tag{1.455}$$

where $\hat{I}$ is the unity operator and δ is the Dirac-delta function;

(iii) the wave functions in the *coordinate* and *momentum* representations are defined via a *state vector* $|\psi(t)\rangle$ in Hilbert space [88,275] as

$$\psi(q,t) = \langle q|\psi(t)\rangle, \; \psi(p,t) = \langle p|\psi(t)\rangle, \tag{1.456}$$

and, finally,

(iv) the quantum mechanical probability densities of the coordinates and momenta are given by

$$W(q,t) = |\langle q|\psi(t)\rangle|^2, \; W(p,t) = |\langle p|\psi(t)\rangle|^2. \tag{1.457}$$

For purposes of exposition, we consider following Puri [275] a one-dimensional dynamical system described *classically* by a phase-space distribution function $W(q, p, t)$. Then the *classical statistical average* of any function $A(q, p)$ is by definition (cf. Eq. (1.380))

$$\langle A\rangle_{cl}(t) = \int W(q, p, t) A(q, p) dq dp. \tag{1.458}$$

By analogy the quantum mechanical description of a system is contained in its Hilbert space density operator $\hat{\rho}(t)$, which determines the *quantum statistical average* of any function $A(\hat{q},\hat{p})$ of the coordinate and momentum operators $\hat{q}$ and $\hat{p}$ by the relation (cf. Eq. (1.388))

$$\langle A\rangle(t)=\mathrm{Tr}\left[\hat{\rho}(t)A(\hat{q},\hat{p})\right]. \tag{1.459}$$

However, the classical distribution function $W(q,p,t)$ may always be expressed in terms of the averages of a *complete* set of functions of q and p. This fact suggests that we may be able to construct a quantum analog of the classical distribution function merely by expressing the latter distribution in terms of the average of a suitably chosen complete set of functions and then *relabeling those classical averages as quantum mechanical ones*. To explore this possibility, we simply rewrite a typical classical distribution in integral form via the δ function as [275]

$$\begin{aligned}W(q,p,t)&=\int_{-\infty}^{\infty}\int_{-\infty}^{\infty}\delta(p-p')\delta(q-q')W(q',p',t)dq'dp'\\&=\frac{1}{4\pi^2}\int_{-\infty}^{\infty}\int_{-\infty}^{\infty}\int_{-\infty}^{\infty}\int_{-\infty}^{\infty}e^{ik(q-q')}e^{il(p-p')}W(q',p',t)dq'dp'dkdl,\end{aligned} \tag{1.460}$$

which is just the inverse of the classical average, viz.,

$$W(q,p,t)=\frac{1}{4\pi^2}\int_{-\infty}^{\infty}\int_{-\infty}^{\infty}e^{ikq}e^{ilp}\left\langle e^{-ikq}e^{-ilp}\right\rangle_{cl}dkdl. \tag{1.461}$$

Thus, we have expressed the classical distribution function $W(q,p,t)$ in terms of the *average* of a complete set of functions of e^{-ikq} and e^{-ilp} or in statistical terminology as the inverse Fourier transform of the characteristic function $\left\langle e^{-ikq}e^{-ilp}\right\rangle_{cl}$. Equation (1.461) indicates that to construct the quantum analog of the classical distribution $W(q,p,t)$, we must first replace the *classical dynamical variables* (q,p) by the Hilbert space *quantum operators* $(\hat{q},\hat{p})$ and then simply replace the classical average $\left\langle e^{-ikq}e^{-ilp}\right\rangle_{cl}$ or characteristic function in the integrand of Eq. (1.461) by the quantum average as defined by Eq. (1.459). Thus, we immediately have an intuitively appealing quantum analog of the classical phase-space distribution function, namely

$$W^{\mathrm{T}}(q,p,t)=\frac{1}{4\pi^2}\int_{-\infty}^{\infty}\int_{-\infty}^{\infty}e^{ikq}e^{ilp}\left\langle \mathrm{T}(e^{-ik\hat{q}},e^{-il\hat{p}})\right\rangle_{qm}dkdl \tag{1.462}$$

called a *quasiprobability distribution*. However, in writing Eq. (1.462) we must formally regard the classical product $e^{-ikq}e^{-ilp}$ in Eq. (1.461) as an *ordering operator* $\mathrm{T}(e^{-ik\hat{q}},e^{-il\hat{p}})$ because owing to the *non-commutativity* of the Hilbert space operators $\hat{q}$ and $\hat{p}$ several *different* quantum operator forms exist which must all be considered as *quantum analogs* of the *unique* classical product $e^{-ikq}e^{-ilp}$. The different operator forms must then be formally described by an ordering operator $\mathrm{T}(e^{-ik\hat{q}},e^{-il\hat{p}})$. Hence, *different choices* of the operator product $\mathrm{T}(e^{-ik\hat{q}},e^{-il\hat{p}})$ will ultimately lead to *different quantum phase-space distribution functions* $W^{\mathrm{T}}(q,p,t)$ [216,270,271].

The name *quasiprobability* is used to emphasize that such a distribution represents a merely mathematical construct and is not in itself a *true* phase-space distribution function as according to the uncertainty principle *no such joint distribution function can exist* for a quantum system [6,270,271]. Now to systematically investigate various operator orderings $\mathrm{T}(e^{-ik\hat{q}},e^{-il\hat{p}})$, it is convenient to express the quantum operators $\hat{q},\hat{p}$ in terms of the *creation* and *annihilation* operators $\hat{a}$ and $\hat{a}^\dagger$ defined as [245]

$$\hat{a}=\frac{\hat{q}+i\hat{p}}{\sqrt{2\hbar}},\ \hat{a}^\dagger=\frac{\hat{q}-i\hat{p}}{\sqrt{2\hbar}}, \tag{1.463}$$

so that the position and momentum operators become via elementary manipulation $\hat{q}=\sqrt{\hbar/2}(\hat{a}+\hat{a}^\dagger)$, $\hat{p}=i\sqrt{\hbar/2}(\hat{a}^\dagger-\hat{a})$ and then consider the new ordering operator $\mathrm{T}(e^{i\xi\hat{a}},e^{i\xi^*\hat{a}^\dagger})$ which may be rewritten explicitly in compact form as [184,275]

$$\mathrm{T}(e^{i\xi\hat{a}},e^{i\xi^*\hat{a}^\dagger})=e^{s|\xi|^2/2}e^{i\xi\hat{a}+i\xi^*\hat{a}^\dagger}. \tag{1.464}$$

Here the parameter s is a complex number. The particular ordering for $s=0$ is usually called the *Weyl ordering* or the *symmetric ordering* and corresponds to the *Wigner function*. The quasiprobability distribution function for $s=1$ is known as the *P-function* $W^P(q,p,t)$ while that for $s=-1$ is known as the *Q-function* $W^Q(q,p,t)$ [245,248,275]. The *Q*- and *P*-representations, and the Wigner function representation arise from three different aims. In the Q-representation, it is desired to create a quasiprobability density for the quantum system by using the *diagonal matrix elements of the density operator* as a basis for this purpose. On the other hand, the P-representation arises from the desire to represent the density operator as an *ensemble of coherent states* [184,218]. In its earliest manifestation a coherent (minimum uncertainty) state is a quantum state of the quantum harmonic oscillator which is a state having dynamics most closely resembling the oscillatory behavior of a classical harmonic oscillator. The concept was originally used by Schrödinger [283] in 1926 to show that certain solutions of the Schrödinger equation satisfy the correspondence principle of Bohr [283]. Such states arise in many physical systems and in general are expressed as eigenvalues of the lowering operator [284]. Finally, the Wigner function yields a *joint* quasidistribution for *canonically conjugate* variables p and q, which in many respects resembles a classical probability distribution. All these phase-space representations have their own particular advantages and disadvantages [285] with the particular choice depending on the problem to be solved. At first, we restrict ourselves to the Wigner function for particles. For spins, the Q- and P-representations as well as the Wigner function representation will be considered in detail in Section 1.9.6 *et seq.*

1.9.2 *The Wigner distribution function for particles*

Wigner originally [6] appears to have simply written down his Fourier integral expression for the quantity now known as the Wigner function which he called the probability function of the simultaneous values for the coordinates q and for the momenta p in phase space and justified it by observing that when integrated with respect to p it yields the correct quantum mechanical probability for the coordinate (position) while when integrated with respect to q it yields the correct quantum mechanical probability for the momentum. However, by directly proceeding from a statistical view point, Moyal [286] has shown how the Wigner function follows naturally from inversion of a characteristic function $M(\mu, \nu)$ for canonically conjugate variables such as position and momentum in a state described by a state vector $|\psi(t)\rangle$ in the Hilbert space. Moyal [286], in view of his strong background in statistics, starts by defining a *characteristic function operator* [cf. Eq. (1.461) *et seq*.], namely,

$$\mathbf{M}_{\hat{q},\hat{p}}(\mu,\nu) = e^{i\mu\hat{q}+i\nu\hat{p}}. \tag{1.465}$$

The *characteristic function* $M(\mu,\nu)$ in a state specified by the state vector $|\psi(t)\rangle$ in Hilbert space is then given according to the usual definition [184] as the scalar product:

$$\begin{aligned} M(\mu,\nu) &= \left\langle e^{i\mu\hat{q}+i\nu\hat{p}} \right\rangle \\ &= \mathrm{Tr}\left(\hat{\rho}e^{i\mu\hat{q}+i\nu\hat{p}}\right) = \left\langle \psi(t)\middle| e^{i\mu\hat{q}+i\nu\hat{p}} \middle|\psi(t)\right\rangle \\ &= \int_{-\infty}^{\infty} \psi^*(q,t)e^{i\mu\hat{q}+i\nu\hat{p}}\psi(q,t)dq, \end{aligned} \tag{1.466}$$

where $\psi(q,t) = \langle q|\psi(t)\rangle$ is the wave function in the coordinate representation and $\hat{q}$ and $\hat{p} = -i\hbar\partial/\partial q$ are the canonically conjugate coordinate and momentum operators. Because the noncommuting quantum operators $\hat{p}$ and $\hat{q}$ satisfy the Baker–Campbell–Hausdorff identity [88,268], viz.,

$$e^{\hat{A}+\hat{B}} = e^{\hat{A}}e^{\hat{B}}e^{-[\hat{A},\hat{B}]/2}, \tag{1.467}$$

the characteristic function operator $\mathbf{M}_{\hat{q},\hat{p}}(\mu,\nu)$ above then becomes

$$\mathbf{M}_{\hat{q},\hat{p}}(\mu,\nu) = e^{i\hbar\mu\nu/2}e^{i\mu\hat{q}}e^{i\nu\hat{p}} = e^{-i\hbar\mu\nu/2}e^{i\nu\hat{p}}e^{i\mu\hat{q}} \tag{1.468}$$

since the commutator of the operators $\hat{q}$ and $\hat{p}$ is given by Eq. (1.454). Next, since we may eliminate the operator $\hat{p}$ via the shift theorem [218]

$$\begin{aligned} e^{i\nu\hat{p}}\psi(q) &= \psi(q) + \hbar\nu\frac{\partial}{\partial q}\psi(q) + \ldots \\ &= \psi(q+\nu\hbar), \end{aligned} \tag{1.469}$$

we have therefore the following simplified form of the characteristic function, Eq. (1.466) now involving q and ν only

$$M(\mu,\nu) = \int_{-\infty}^{\infty} \psi^*(q,t)e^{i\mu(q+\nu\hbar/2)}\psi(q+\nu\hbar,t)dq, \tag{1.470}$$

which with the obvious replacement $q \to q - v\hbar/2$ yields the characteristic function as the *overlap* integral

$$M(\mu,v) = \int_{-\infty}^{\infty} \psi^*(q-u,t)e^{i\mu q}\psi(q+u,t)dq, \tag{1.471}$$

where $u = v\hbar/2$. The phase-space quasidistribution $W(q, p, t)$ is thus formally by Fourier inversion (according to the usual statistical definition)

$$\begin{aligned}W(q,p,t) &= \frac{1}{4\pi^2}\int_{-\infty}^{\infty}\int_{-\infty}^{\infty} M(\mu,v)e^{-i\mu q - ivp}d\mu dv \\ &= \frac{1}{2\hbar\pi^2}\int_{-\infty}^{\infty}\int_{-\infty}^{\infty}\int_{-\infty}^{\infty} e^{-i\mu(q-x)}\psi^*(x-u,t)e^{-2iup/\hbar}\psi(x+u,t)dudxd\mu \\ &= \frac{1}{\pi\hbar}\int_{-\infty}^{\infty} \psi^*(q-u,t)e^{-2iup/\hbar}\psi(q+u,t)du,\end{aligned} \tag{1.472}$$

where we have used a property of the delta function, viz.,

$$\frac{1}{2\pi}\int_{-\infty}^{\infty} e^{-i\mu(q-x)}d\mu = \delta(q-x). \tag{1.473}$$

The function $W(q, p, t)$ is real but not *everywhere positive* so that it is indeed a quasiprobability distribution. Equation (1.472) is the Wigner function $W(q, p, t)$. It represents the Fourier transform of the overlap expression, namely, a type of spatial autocorrelation function of the wave function ψ in coordinate space [216] and it was also introduced independently in 1948 by Ville [287,288] in the entirely different context of time/frequency analysis where it roughly indicates how the spectral density for a *nonstationary* process varies with time. Here the overlap function is in a physical sense the product of the wavefunction at the advanced point $(q + u)$ and its conjugate at the retarded point $(q - u)$, viz., [216]

$$O(q,u,t) = \psi^*(q-u,t)\psi(q+u,t),$$

where $2u$ is the separation distance. The Wigner function then represents [216] the Fourier components of the overlap function which in turn measure the distribution of plane waves at each value of q having different de Broglie wavelengths $\lambda = h/p$ in the overlap function.

Equation (1.472) holds if ψ evolves according to the Schrödinger equation

$$i\hbar\frac{\partial\psi}{\partial t} = -\frac{\hbar^2}{2m}\frac{\partial^2\psi}{\partial q^2} + V(q)\psi \tag{1.474}$$

for a particle of mass m moving in a one-dimensional extension q and subject to a potential $V(q)$. Integration of $W(q, p, t)$ with respect to the momentum p yields

$$\int_{-\infty}^{\infty} W(q,p,t)dp = |\psi(q,t)|^2, \tag{1.475}$$

i.e., the correct quantum mechanical probability for the coordinate q. Likewise, integration of W with respect to q, yields the correct quantum mechanical probability for the momentum p, viz.,

$$\int_{-\infty}^{\infty} W(q,p,t)dq = |\Phi(p,t)|^2, \tag{1.476}$$

where $\Phi(p, t)$ is the wave function of the momentum given by the Fourier transform of the wave function over coordinate space, viz.,

$$\Phi(p,t) = \int_{-\infty}^{\infty} \psi(q,t)e^{-ipq}dq.$$

It follows from Eq. (1.475) that if the wave function $\psi(q, t)$ is normalized to unity then

$$\int_{-\infty}^{\infty}\int_{-\infty}^{\infty} W(q,p,t)dqdp = 1. \tag{1.477}$$

By recalling [216] the definition of the density matrix $\hat{\rho}$ for a *pure* state as the outer product $\hat{\rho}(q_1,q_2,t) = \langle q_1|\psi(t)\rangle\langle\psi(t)|q_2\rangle = \psi(q_1,t)\psi^*(q_2,t)$ and introducing the replacements $y/2 = (q_1 - q_2)/2$ and $q = (q_1 + q_2)/2$, we then have the Wigner function, Eq. (1.472) rendered as the overlap Fourier transform of the density matrix, viz., [216]

$$W(q,p,t) = \frac{1}{2\pi\hbar}\int_{-\infty}^{\infty}\hat{\rho}(q+y/2,q-y/2,t)e^{-ipy/\hbar}dy$$
$$= \frac{1}{2\pi\hbar}\int_{-\infty}^{\infty}\hat{\rho}(q-y/2,q+y/2,t)e^{ipy/\hbar}dy. \tag{1.478}$$

Moreover, the inverse map of Eq. (1.478) is the transform yielding the mapping from the *phase space back to operators in Hilbert space* [270,271], viz.,

$$\hat{\rho}(t) = \frac{\hbar}{2\pi}\int_{-\infty}^{\infty}\int_{-\infty}^{\infty}\int_{-\infty}^{\infty}\int_{-\infty}^{\infty}W(q,p,t)e^{ix(\hat{q}-q)+iy(\hat{p}-p)}dqdpdxdy. \tag{1.479}$$

In contrast to a single-particle distribution for a pure state Eq. (1.472), the definition of the Wigner function $W(q, p, t)$ via the density operator $\hat{\rho}(t)$, Eq. (1.478), is valid both for *pure* and *mixed* states. The only difference lies in the definition of $\hat{\rho}(t)$, e.g., for a pure state $|\psi(t)\rangle$, the density operator is given by the outer product $\hat{\rho}(t) = |\psi(t)\rangle\langle\psi(t)|$, while for a mixed state, the density operator is defined as the series [183,216]

$$\hat{\rho}(t) = \sum_n P_n |\psi_n(t)\rangle\langle\psi_n(t)|, \tag{1.480}$$

where P_n is the probability of each *individual* quantum state $|\psi_n(t)\rangle$. The corresponding Wigner function of a *mixed* state $W(q, p, t)$ is then [183]

$$W(q,p,t) = \sum_n P_n W_n(q,p,t), \tag{1.481}$$

where $W_n(q,p,t)$ is the Wigner function for the individual state $|\psi_n(t)\rangle$. In particular, the definition, Eq. (1.481), applies to a canonical ensemble of particles at temperature T, where $P_n = e^{-\beta E_n}/Z$ is the probability to find the particle in an energy state E_n and Z is the partition function.

The Wigner function $W(q, p, t)$ exhibits most of the properties of a classical phase-space distribution, including the fact that the expectation value $\langle\hat{A}\rangle(t) = \mathrm{Tr}(\hat{\rho}\hat{A})$ of a quantum operator $\hat{A}$ may be calculated in *classical fashion via the corresponding Weyl symbol* $A(q, p)$ as [270,271]

$$\langle\hat{A}\rangle(t) = \int_{-\infty}^{\infty}\int_{-\infty}^{\infty}W(q,p,t)A(q,p)dq\,dp, \tag{1.482}$$

where that symbol is explicitly given by

$$A(q,p)=\int_{-\infty}^{\infty} e^{-ipy/\hbar}\langle q+y/2|\hat{A}|q-y/2\rangle dy, \tag{1.483}$$

which in other words is merely the Weyl symbol of the Hilbert space operator $\hat{A}$ *expressed in the position q basis*. The Weyl symbol can alternatively be expressed in terms of the matrix elements of the operator $\hat{A}$ in the *momentum p basis* as

$$A(q,p)=\int_{-\infty}^{\infty} e^{-iqy/\hbar}\langle p+y/2|\hat{A}|p-y/2\rangle dy. \tag{1.484}$$

The Weyl symbol also implies the lemma that the trace of the *product of two operators* $\hat{A}$ and $\hat{B}$ is given by the integral over the phase space (q, p) of the *product of their individual Weyl symbols* $A(q, p)$ and $B(q, p)$, viz., [270,271]

$$\mathrm{Tr}\langle\hat{B}\hat{A}\rangle=\frac{1}{2\pi\hbar}\int_{-\infty}^{\infty}\int_{-\infty}^{\infty} B(q,p)A(q,p)dq\,dp. \tag{1.485}$$

Equations (1.478) and (1.485) (for the specific case $\hat{B}\equiv\hat{\rho}$) immediately yield Eq. (1.482) for the average of a quantum operator.

We may now determine from the overlap Fourier transform Eq. (1.472) given by Wigner the partial differential equation governing the time evolution of the Wigner function. Taking the time derivative of the Wigner function as defined by Eq. (1.472), we have

$$\begin{aligned}\frac{\partial W}{\partial t}=\frac{1}{\pi\hbar}\int_{-\infty}^{\infty}\Bigg[&\psi^*(q+u,t)\frac{\partial\psi(q-u,t)}{\partial t}\\ &+\frac{\partial\psi^*(q+u,t)}{\partial t}\psi(q-u,t)\Bigg]e^{2ipu/\hbar}du.\end{aligned} \tag{1.486}$$

Since the wave function ψ in coordinate space evolves according to the Schrödinger equation, Eq. (1.474), substitution of that equation into Eq. (1.486) gives

$$
\begin{aligned}
\frac{\partial W}{\partial t} = &-\frac{i}{2m\pi}\int_{-\infty}^{\infty}\left\{\psi(q-u)\frac{\partial^2\psi^*(q+u)}{\partial u^2}-\psi^*(q+u)\frac{\partial^2\psi(q-u)}{\partial u^2}\right. \\
&\left.-\frac{2m}{\hbar^2}\psi(q-u)\psi^*(q+u)\left[V(q+u)-V(q-u)\right]\right\}e^{\frac{2ipu}{\hbar}}du \\
= &-\frac{p}{m}\frac{\partial W}{\partial q}+\frac{i}{\pi\hbar^2}\int_{-\infty}^{\infty}\psi(q-u)\psi^*(q+u)\left[V(q+u)-V(q-u)\right]e^{\frac{2ipu}{\hbar}}du.
\end{aligned}
\tag{1.487}
$$

Here we have utilized the derivative property

$$
\frac{\partial^2\psi(q\pm u)}{\partial q^2}=\frac{\partial^2\psi(q\pm u)}{\partial u^2}
$$

and have performed one integration by parts with respect to u in the terms which do not involve the potential functions $V(q \pm u)$. Equation (1.487) is by inspection identical to the classical Liouville equation in the force-free case $V(q) = 0$ and also concurs with that given by Hillery *et al.* [269]. We now expand $V(q + u) - V(q - u)$ in Eq. (1.487) in a Taylor series about the point q yielding

$$
V(q+u)-V(q-u)=2\sum_{n=1}^{\infty}\frac{u^{2n-1}}{(2n-1)!}\frac{\partial^{2n-1}V}{\partial q^{2n-1}}. \tag{1.488}
$$

Substituting Eq. (1.488) into Eq. (1.487) we have

$$
\begin{aligned}
\frac{\partial W}{\partial t}+\frac{p}{m}\frac{\partial W}{\partial q} &= \frac{2i}{\pi\hbar}\int_{-\infty}^{\infty}\sum_{n=1}^{\infty}\frac{u^{2n-1}}{(2n-1)!}\frac{\partial^{2n-1}V}{\partial q^{2n-1}}\psi^*(q+u)\psi(q-u)e^{\frac{2ip}{\hbar}u}du \\
&= -\frac{\partial V}{\partial q}\frac{\partial W}{\partial p}-\sum_{r=1}^{\infty}\frac{(i\hbar/2)^{2r}}{(2r+1)!}\frac{\partial^{2r+1}V}{\partial q^{2r+1}}\frac{\partial^{2r+1}W}{\partial p^{2r+1}}.
\end{aligned}
\tag{1.489}
$$

Here we have used the obvious property

$$
u^{2n-1}e^{2ipu/\hbar}=\left(\frac{\hbar}{2i}\right)^{2n-1}\frac{\partial^{2n-1}}{\partial p^{2n-1}}e^{2ipu/\hbar}. \tag{1.490}
$$

Thus, we have the time evolution equation for the Wigner distribution function $W(q, p, t)$, viz.,

$$\frac{\partial W}{\partial t}+\hat{M}_W W=0, \tag{1.491}$$

where the operator $\hat{M}_W$ is defined as

$$\hat{M}_W W=\frac{p}{m}\frac{\partial W}{\partial q}-\frac{\partial V}{\partial q}\frac{\partial W}{\partial p}-\sum_{r=1}^{\infty}\frac{(i\hbar/2)^{2r}}{(2r+1)!}\frac{\partial^{2r+1}V}{\partial q^{2r+1}}\frac{\partial^{2r+1}W}{\partial p^{2r+1}}. \tag{1.492}$$

Equation (1.491) often known as the *Wigner–Moyal equation* [216] is simply a quantum analog of the classical Liouville equation. Equations (1.491) and (1.492) also hold if the system is in a mixed state represented by a density matrix $\hat{\rho}$.

Wigner [6] originally calculated quantum correction terms to the classical stationary distribution functions for a system with n degrees of freedom. However, for illustrative purposes, we shall only consider a system with $n=1$. As an example, we determine the stationary (equilibrium) solution of Eq. (1.491) for an assembly of noninteracting particles each of mass m moving in a one-dimensional extension q in a potential $V(q)$ at temperature T. Each particle is characterized by the energy

$$E(q,p)=\frac{p^2}{2m}+V(q). \tag{1.493}$$

Following Wigner [6], we develop the stationary distribution function $W_{\text{eq}}(q,p)$ in a power series in $\hbar$

$$W_{\text{eq}}(q,p)=W_0(q,p)+\hbar^2 W_2(q,p)+\hbar^4 W_4(q,p)+..., \tag{1.494}$$

where $W_0(q,p)=e^{-\beta E(q,p)}$ is the (unnormalized) Maxwell–Boltzmann distribution. By substituting Eq. (1.494) into Eq. (1.491), the functions $W_2(q,p)$ and $W_4(q,p)$ are easily evaluated. Thus, we have [272]

$$W_{\text{eq}}(q,p)=e^{-\beta E(q,p)}\left[1+\Lambda\left(\frac{\beta p^2}{m}V^{(2)}-3V^{(2)}+\beta\left(V^{(1)}\right)^2\right)+...\right], \tag{1.495}$$

where $\Lambda=\hbar^2\beta^2/(24m)$ is a characteristic quantum parameter and $V^{(m)}\equiv d^mV/dq^m$. The above equations are written explicitly to $o(\hbar^2)$. In like manner, higher order quantum correction terms to the Wigner stationary distribution $W_{\mathrm{eq}}(q,p)$ may be calculated. Thus, $W_{\mathrm{eq}}(q,p)$ can be given, in principle to any desired degree *r* of $\hbar^{2r}$. In general, quantum effects give rise to *non-Gaussian behavior* of the equilibrium phase-space distribution function $W_{\mathrm{eq}}(q,p)$ also unlike the classical distribution it is usually no longer *separable* and *additive* in the position and momentum.

Obviously, the calculation of the equilibrium Wigner distribution $W_{\mathrm{eq}}(q,p)$ is a tedious task for an arbitrary potential $V(q)$. However, in some cases, $W_{\mathrm{eq}}(q,p)$ can be found in closed form (various methods for the calculation of Wigner functions are described, e.g., in Refs. [265,268–271]). A famous example is the quantum harmonic oscillator, where the potential is

$$V(q) = \frac{1}{2} m\omega_0^2 q^2 \tag{1.496}$$

(ω_0 is the angular frequency of the oscillator). Here perturbation theory may be entirely avoided because the evolution equation for the Wigner function Eq. (1.491) for the potential (1.496) now coincides with the corresponding classical Liouville equation and the unnormalized equilibrium Wigner function $W_{\mathrm{eq}}(q,p)$ can be written in the exact Gaussian form [268,269,272,289]

$$W_{\mathrm{eq}}(q,p) = \frac{\operatorname{sech}\left(\beta\hbar\omega_0/2\right)}{2\pi\hbar} e^{-\frac{q^2}{2\langle q^2\rangle_{\mathrm{eq}}}-\frac{p^2}{2\langle p^2\rangle_{\mathrm{eq}}}}, \tag{1.497}$$

where

$$\begin{aligned} \left\langle q^2 \right\rangle_{\mathrm{eq}} &= \frac{\hbar}{2m\omega_0}\coth\frac{\beta\hbar\omega_0}{2}, \\ \left\langle p^2 \right\rangle_{\mathrm{eq}} &= \frac{m\hbar\omega_0}{2}\coth\frac{\beta\hbar\omega_0}{2}. \end{aligned} \tag{1.498}$$

However the Wigner function $W_{\mathrm{eq}}(q,p)$ from Eq. (1.497) is in fact a *superposition* of the Wigner functions $W_n(q,p)$ for the *pure* states of a harmonic oscillator, viz., [268,269,272]

$$W_{\mathrm{eq}}(q,p) = \sum_{n=0}^{\infty} W_n(q,p) e^{-\beta\hbar\omega_0(n+1/2)}. \tag{1.499}$$

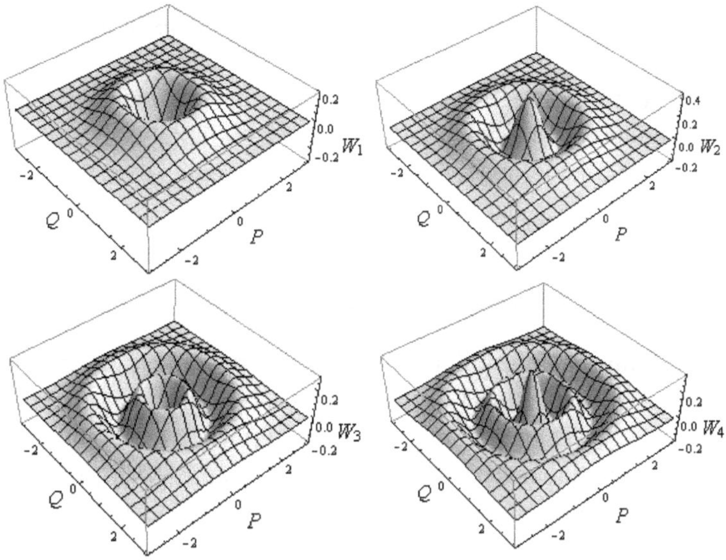

Fig. 1.24. 3D plots of the Wigner functions W_n of a harmonic oscillator vs. the normalized coordinate $Q = q\sqrt{m\omega_0/\hbar}$ and momentum $P = p/\sqrt{m\hbar\omega_0}$ for the pure states $n = 1, 2, 3$, and 4.

Here $W_n(q,p)$ and the corresponding eigenfunctions of the harmonic oscillator $\psi_n(q)$ are given by the well-known analytic equations [265]

$$\begin{aligned} W_n(q,p) &= \frac{1}{\pi\hbar}\int_{-\infty}^{\infty} \psi_n(q+u)\psi_n^*(q-u)e^{-2ipu/\hbar}du \\ &= \frac{(-1)^n}{\pi\hbar}e^{-\frac{p^2+m^2\omega_0^2q^2}{m\hbar\omega_0}} L_n\left[\frac{2(p^2+m^2\omega_0^2q^2)}{m\hbar\omega_0}\right], \end{aligned} \tag{1.500}$$

$$\psi_n(q) = \frac{1}{\sqrt{2^n n!}}\left(\frac{m\omega_0}{\pi\hbar}\right)^{1/4} e^{-\frac{m\omega_0 q^2}{2\hbar}} H_n\left(q\sqrt{\frac{m\omega_0}{\hbar}}\right), \tag{1.501}$$

where $H_n(z)$ and $L_n(z)$ are the Hermite and Laguerre polynomials, respectively [290] (see Appendix A.1.4). For illustration, the Wigner functions W_n of a harmonic oscillator for the pure states $n = 1, 2, 3$ and 4, Eq. (1.500), are shown in Fig. 1.24. In contrast to the behavior of the equilibrium Wigner function $W_{\text{eq}}(q,p)$, Eq. (1.497), the *individual* Wigner functions $W_n(q,p)$ of the pure states can take on *negative* values so that it is then obviously impossible to interpret W_n as a true probability distribution.

Equilibrium Wigner functions have also been calculated for other simple quantum systems such as a particle in an infinite square well [286], the Morse oscillator [291–294], the anharmonic quartic oscillator [295,296], the inverted harmonic oscillator [297], etc.

1.9.3 *Application to transition state theory*

As seen the simplest description of thermally activated escape of a particle with the total energy $E(q, p)$ given by Eq. (1.493) over a potential barrier is provided by TST. In the simplest form of TST, two assumptions are made [85,97]. First thermal equilibrium prevails in the well (for example, through the action of Maxwell's demon [215] which keeps replenishing the lost particles at the source) so that the metastable state is represented by a canonical equilibrium distribution (unlike in the Kramers [84] treatment of the escape rate, where *nonequilibrium effects* due to the loss of particles from the well are accounted for using the theory of Brownian motion, automatically leading to frictional dependence of the transmission coefficient). Secondly, a particle is supposed never to return to the well once it has crossed the potential barrier. The first assumption means that friction, i.e., dissipation to the bath does not affect the escape rate at all. Thus, the system in effect is a closed classical one. Nevertheless, according to Mel'nikov [186,298] the results of classical TST should also be applicable in a *wide range of dissipation* for which thermal noise is strong enough to thermalize the escaping particles yet not so strong as to affect particle motion across the top of the potential barrier, i.e., a Maxwell–Boltzmann distribution still holds there. In the Kramers model, this is the so-called *intermediate damping* case (cf. Fig. 1.13.2 of [97]). In the treatment of Kramers, however, which explicitly involves an *open* classical system with fluctuation-dissipation due to the bath described by the Brownian motion *Stosszahlansatz*, he shows that for sufficiently weak friction the escape rate is suppressed because of the consequent *depletion of the well population* while for strong friction the escape rate is also suppressed due to the consequent *slowing down of the particle motion at the barrier top*. The suggestion that quantum mechanical tunneling might

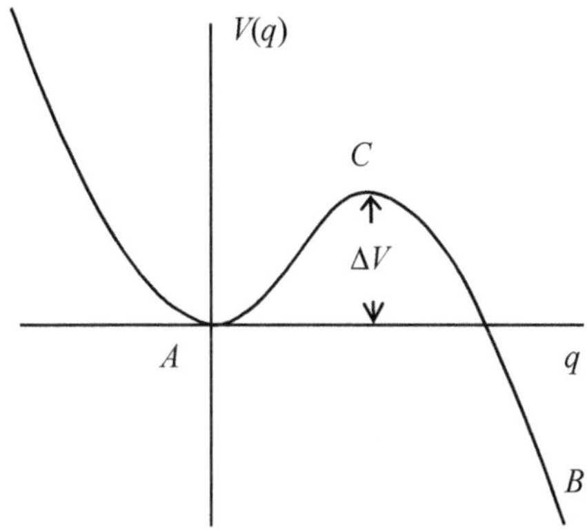

Fig. 1.25. Single well potential function as the simplest example of escape over a barrier. Particles are initially trapped in the well near the point A by a high potential barrier at the point C. They very rapidly thermalize in the well. However due to thermal agitation very few may attain enough energy to escape over the barrier into region B whence they never return. The height of the barrier $\Delta V = V(q_C) - V(q_A)$ is supposed to be large compared with the thermal energy kT.

play a significant role in some chemical reactions was first made in 1927 by Hund [280] almost at the inception of quantum mechanics, while the first guess at a quantum transition state theory again appears to have been made by Wigner. He proposed a quantum generalization of the classical TST [278–280], where the reaction rate Γ is given via the flux-over-population method [85], viz.,

$$\Gamma = I_C / Z_A, \tag{1.502}$$

where

$$Z_A = \iint_{\text{well}} W_{\text{eq}}(q,p)dpdq \text{ and } I_C = \iint_{\text{barrier}} J_C(q,p)dpdq. \tag{1.503}$$

are, respectively, the well partition function and the total current of the particles at the top of the barrier point C (see Fig. 1.25). Here the current density *exactly* at the barrier point J_C is given by

$$J_C(q,p) = \theta(p)\delta(q - q_C)\dot{q}W_{\text{eq}}(q,p), \tag{1.504}$$

where $\theta(x)$ and $\delta(x)$ are the unit-step and Dirac-delta functions, respectively, and $\dot{q} = \partial E / \partial p = p/m$ is the particle velocity. Now we assume that near the summit C and near the bottom A of the well the potential $V(q)$ may be approximated by inverted harmonic oscillator and harmonic oscillator potentials, respectively, namely,

$$V(q) \approx \begin{cases} V(q_C) - \frac{1}{2} m\omega_C^2 (q - q_C)^2, \\ V(q_A) + \frac{1}{2} m\omega_A^2 (q - q_A)^2, \end{cases} \tag{1.505}$$

where $\omega_C = \sqrt{|V''(q_C)|/m}$ and $\omega_A = \sqrt{V''(q_A)/m}$. Thus, near the bottom of the well, $W_{\text{eq}}(q,p)$ is approximated by that of a harmonic oscillator with $\omega_0 = \omega_A$ [see Eq. (1.497) above] meaning that the well partition function Z_A can now be explicitly evaluated via Gaussian integrals as [272]

$$\begin{aligned} Z_A &\approx \frac{e^{-\beta V(q_A)}}{2\pi\hbar \cosh(\beta\hbar\omega_A/2)} \int_{-\infty}^{\infty}\int_{-\infty}^{\infty} e^{-\frac{2}{\hbar\omega_A}\left(\frac{m}{2}\omega_A^2 q^2 + \frac{1}{2m}p^2\right)\tanh\left(\frac{\beta\hbar\omega_A}{2}\right)} dq dp \\ &= \frac{e^{-\beta V(q_A)}}{2\sinh(\hbar\omega_A\beta/2)}. \end{aligned} \tag{1.506}$$

Here the limits of integration with respect to q may be formally extended to ± infinity without significant error since the particles are almost all concentrated at the bottom A. Near the top of the barrier, the Wigner function $W_{\text{eq}}(q,p)$ is approximated by that of an *inverted* harmonic oscillator with $\omega_0 = i\omega_C$ in Eq. (1.497) so that at C it is given by [272]

$$W_{eq}^C(q_C, p) \approx \frac{e^{-\beta V(q_C) - \frac{\tan(\beta\hbar\omega_C/2)}{m\hbar\omega_C} p^2}}{2\pi\hbar \cos(\beta\hbar\omega_C/2)}. \tag{1.507}$$

Thus, we have from Eqs. (1.503) and (1.507) for the current at the barrier

$$\begin{aligned} I_C &\approx \frac{e^{-\beta V(q_C)}}{2\pi\hbar m \cos(\beta\hbar\omega_C/2)} \int_0^{\infty} p e^{-\frac{\tan(\beta\hbar\omega_C/2)}{m\hbar\omega_C} p^2} dp \\ &= \frac{\omega_C e^{-\beta V(q_C)}}{4\pi \sin(\beta\hbar\omega_C/2)}. \end{aligned} \tag{1.508}$$

Finally, substituting Eqs. (1.506) and (1.508) into Eq. (1.502), we obtain [272]

$$\Gamma \approx \frac{\omega_A}{2\pi} \Xi e^{-\beta \Delta V}, \tag{1.509}$$

where $\Delta V = V(q_C) - V(q_A)$ is the barrier height and

$$\begin{aligned} \Xi &= \frac{\omega_C}{\omega_A} \frac{\sinh(\hbar\beta\omega_A / 2)}{\sin(\hbar\beta\omega_C / 2)} \\ &= 1 + \frac{\beta^2\hbar^2}{24}\left(\omega_C^2 + \omega_A^2\right) + \ldots \end{aligned} \tag{1.510}$$

is the quantum correction to the classical TST result. The lowest order quantum correction to this pre-exponential factor was that first obtained by Wigner [278] (see also [279]). He emphasized that the quantum factor Ξ represents an *effective lowering* of the potential barrier so *enhancing* the escape rate. According to Wigner [278], Eq. (1.509) constitutes the quantum correction to classical TST at high temperatures. In the context of quantum dissipation, one may infer that Eq. (1.509) also represents the extension of the intermediate damping Kramers escape rate (for which classical TST provides a reasonably accurate approximation) to include quantum effects. An important feature of Eq. (1.509) not apparent in the first order in $\hbar^2$ approximation is that the prefactor Ξ *diverges* at a *crossover temperature* T_c given by $T_c = \hbar\omega_C / (2\pi k)$. The divergence occurs because the parabolic (or inverted oscillator) approximation for the potential is only valid near the top of the barrier. However, at very low temperatures $T << T_c$, where the particle is near the bottom of the well, the parabolic approximation to the barrier shape is no longer sufficient [186,298]. In contrast for $T > T_c$, transitions near the barrier top dominate so that the parabolic approximation is accurate [186,298]. Moreover, the simple approximation appearing on the right-hand side of Eq. (1.510) should hold with a reasonable degree of accuracy. This approximation also appears to agree with the experimental results of Bouchaud *et al.* [299].

We have just demonstrated how the quantum escape rate in the absence of dissipation, namely Eq. (1.509), may be obtained by Wigner's perturbation method. However, that result may be obtained in much more succinct fashion without using perturbation theory at all by recalling that the rate constant may be written as [280,300]

$$\Gamma = \frac{1}{Z_A} \int_{-\infty}^{\infty} w(E) e^{-\beta E} dE, \tag{1.511}$$

where the quantity

$$w(E) = \frac{1}{1 + e^{-2\pi(E - V_C)/(\hbar\omega_C)}} \tag{1.512}$$

is the *quantum transmission coefficient* (ignoring dissipation) of a parabolic barrier [218]. We again approximate the potential near the top of the barrier by that of an inverted harmonic oscillator, which holds good at the barrier as well as at a small distance below it. Hence, one can regard the integral in Eq. (1.511) as having infinite limits. Thus, one finds on evaluating that integral with these limits that the escape rate is given by

$$\Gamma = \frac{\hbar\omega_C}{2Z_A \sin(\beta\hbar\omega_C / 2)} e^{-\beta V_C}. \tag{1.513}$$

Then using Eq. (1.506), we immediately have the quantum TST escape-rate Eq. (1.509) from Eq. (1.513).

Quantum TST as formulated for particles with *separable* and *additive* Hamiltonians in the manner just described has also been applied [301] in the context of magnetization reversal to the escape rate of the giant spin model of single-domain ferromagnetic particles. This model describes a ferromagnetic particle with uniaxial anisotropy with external fields applied parallel and perpendicular to the anisotropy axis. Tunneling in such a model will be caused by the transverse field [301]. However, the Hamiltonian of the model (which is not separable and additive as the canonical variables are now the polar angles ϑ and φ specifying the orientation of the magnetization vector) may be mapped [301] onto that of a mechanical particle moving in a double-well potential. Hence, the quantum TST rate described above which is a close approximation to the exact escape rate in the intermediate damping region may also be used to study *thermally assisted tunneling* of the magnetization of a single-domain ferromagnetic particle.

1.9.4 *Application to quantum Brownian motion*

We have seen that the classical theory of the Brownian motion in a potential is ubiquitous in many areas of physics and chemistry, particularly those dealing with the nature of metastable states and the rates at which these states decay. Typical examples are current-voltage characteristics of Josephson junctions, the rate of condensation of a supersaturated vapor, dielectric and Kerr effect relaxation in liquids and nematic liquid crystals, dynamic light scattering, chemical reaction rate theory in condensed phases, superparamagnetic relaxation, polymer dynamics, nuclear fission and fusion and so on [85,97,103,302]. All these phenomena in one way or another depend on the nucleation and growth of some characteristic disturbance within a metastable system, e.g., condensation of a saturated vapor is initiated by the formation of a sufficiently large droplet of the liquid. If this droplet is big enough it will be more likely to grow than to dissipate and will bring about condensation of the entire sample [103]. In many cases, particularly at low temperatures, a theory of dissipation based on the classical Brownian motion may be inadequate because it ignores quantum effects. Quantum noise arising from quantum fluctuations is also important in nanoscale and biological systems. We mention [302] the noise-assisted tunneling and transfer of electrons and quasiparticles. The characteristics of such quantum noise vary strongly with temperature and at high temperatures a crossover to Johnson–Nyquist noise which is essentially governed by the classical Brownian motion takes place. Yet another aspect of the subject, which has come to the fore in recent years, is the evolution of *macroscopic* quantum variables such as the decay of a zero-voltage state in a biased Josephson junction, flux quantum transitions in a SQUID [302] and the possible reversal by quantum tunneling of the magnetization of a magnetic nanoparticle. Recall that it was conjectured by Bean and Livingston [8] that the magnetization may reverse by quantum tunneling through the internal magnetocrystalline anisotropy barrier of the particle instead of by the conventional mechanism of thermally agitated jumping over the barrier, namely, Néel relaxation [16].

All these considerations necessitate the development of a theory of quantum Brownian motion, which directly addresses the issue of the quantum-classical correspondence [303] in terms of a quantum analog of the classical Fokker–Planck equation. Such an evolution equation would allow dynamical parameters such as escape rates, correlation times, susceptibilities, etc. to be calculated in terms of the eigensolutions of that equation in a manner analogous to those of the Fokker–Planck equation. Moreover, it would be possible to compare asymptotic solutions for parameters such as escape rates yielded by reaction rate theory with the corresponding quantities calculated from the quantum master equation. The description of quantum mechanics in terms of phase-space distributions as advanced by Wigner is also an ideal starting point for the formulation of semiclassical quantum master equations of *open* quantum systems [285]. For quantum Brownian motion, the evolution equation for $W(q, p, t)$ (henceforth called the *Wigner–Fokker–Planck* equation) for the translational Brownian motion of a particle in a potential $V(q)$, can be derived by proceeding to the high-temperature limit. This procedure is equivalent to treating the system as *a quantum mechanical particle embedded in a classical bath and is effected by regarding the Brownian particle as bilinearly coupled to a bath of harmonic oscillators in thermal equilibrium at temperature T* [189,304]. The most convenient way of describing the influence of the bath is via the *spectral density* characterizing the coupling to the bath of oscillators. The oscillators then represent the normal modes of the bath. We remark that the effect of friction is regarded as being entirely equivalent to the bilinear coupling to these normal modes. Quantization of the bath of oscillators [304] then yields the following semiclassical master equation for the Wigner distribution function $W(q, p, t)$, which has intuitive appeal [305–314]

$$\frac{\partial W}{\partial t} + \hat{M}_W W = \mathrm{St}(W). \tag{1.514}$$

In this equation the operator $\hat{M}_W$ is given by Eq. (1.492) and the collision kernel operator $\mathrm{St}(W)$ is

$$\mathrm{St}(W) = \frac{\partial}{\partial p}\left(D_p pW + D_{pp}\frac{\partial W}{\partial p} + D_{qp}\frac{\partial W}{\partial q}\right) + \frac{\partial}{\partial q}\left(D_{qq}\frac{\partial W}{\partial q}\right). \tag{1.515}$$

Here $D_p, D_{pp}, D_{qp}, D_{qq}$ are coordinate, momentum, and time-dependent parameters (diffusion coefficients). The left-hand side of Eq. (1.514) again comprising the Wigner–Moyal operator is of course the quantum analog of the Liouville equation, while St(W) characterizes the interaction of the Brownian particle with the thermal bath at temperature T; the collision kernel St (W) being the analog of the collision kernel (*Stosszahlansatz*) in the classical kinetic theory. Conditions for the validity of the master Eq. (1.514) are discussed elsewhere (e.g., Refs. [289] and [315]). In the classical limit, $\hbar \to 0$, Eq. (1.514) reduces to the familiar Klein–Kramers (Fokker–Planck) equation for the translational Brownian motion of a particle in a potential $V(q)$ namely,

$$\frac{\partial W}{\partial t}+\frac{p}{m}\frac{\partial W}{\partial q}-\frac{\partial V}{\partial q}\frac{\partial W}{\partial p}=\frac{\zeta}{m}\frac{\partial}{\partial p}\left(pW+\frac{m}{\beta}\frac{\partial W}{\partial p}\right) \tag{1.516}$$

and the various coefficients $D_p, D_{pp}, D_{qp}, D_{qq}$ merely become

$$D_p=\frac{\zeta}{m},\ D_{pp}=\frac{\zeta}{\beta},\ D_{qp}=D_{qq}=0, \tag{1.517}$$

where ζ is the drag coefficient of a Brownian particle.

Now being in possession of the functional form of the master Eq. (1.514) for a Brownian particle, the next crucial step is to determine $D_p, D_{pp}, D_{qp}, D_{qq}$. With this end in mind we shall select the extension to the semiclassical case of the simple heuristic idea originally used by Einstein, Smoluchowski, Langevin, and Kramers to determine diffusion coefficients in the classical theory of the Brownian motion [97,143]. Thus to obtain the explicit form of $D_p, D_{pp}, D_{qp}, D_{qq}$ in Eq. (1.515), we first recall Wigner's results for the unnormalized equilibrium distribution $W_{\text{eq}}(q,p)$ developed in a power series in $\hbar^2$ Eq. (1.495). This equilibrium distribution, being a stationary solution of the Wigner–Moyal Eq. (1.491), must be the equilibrium solution of the generic master Eq. (1.514), i.e., it must also satisfy $\text{St}(W_{\text{eq}})=0$. Hence if $W_{\text{eq}}(q,p)$ from Eq. (1.495) is to satisfy $\text{St}(W_{\text{eq}})=0$, , the coefficients D_p, D_{qp}, and D_{qq} should remain as in Eq. (1.517) and only D_{pp} must be altered to read [272,316,317] (for simplicity, we retain only the leading quantum correction terms)

$$D_{pp}=\frac{\zeta}{\beta}+\frac{\zeta\beta\hbar^2}{12m}\frac{\partial^2 V}{\partial q^2}+\ldots. \tag{1.518}$$

Thus, the explicit form of the Wigner–Fokker–Planck equation up to $o(\hbar^2)$ is [272,316,317]

$$\begin{aligned}&\frac{\partial W}{\partial t}+\frac{p}{m}\frac{\partial W}{\partial q}-\frac{\partial V}{\partial q}\frac{\partial W}{\partial p}+\frac{\hbar^2}{24}\frac{\partial^3 V}{\partial q^3}\frac{\partial^3 W}{\partial p^3}+\dots\\&=\frac{\zeta}{m}\frac{\partial}{\partial p}\left[pW+\left(\frac{m}{\beta}+\frac{\beta\hbar^2}{12}\frac{\partial^2 V}{\partial q^2}+\dots\right)\frac{\partial W}{\partial p}\right].\end{aligned} \tag{1.519}$$

The phase-space master equation (1.519) has been derived in [272,316,317] assuming that $D_p, D_{pp}, D_{qp}, D_{qq}$ in Eq. (1.515) are time-independent. We observe that in the high-temperature limit, $\beta \to 0$, Eq. (1.519) obviously reduces to the Caldeira–Leggett master equation [304], where the collision kernel St (W) has the same form as in the classical Fokker–Planck equation (1.516). The evolution equation for the density matrix $\hat{\rho}$ in Hilbert space corresponding to Eq. (1.519) is again to order $\hbar^2$ [318]

$$\begin{aligned}&\frac{\partial\hat{\rho}}{\partial t}+\frac{i}{\hbar}\left[\hat{H},\hat{\rho}\right]\\&=-\frac{\zeta}{m}\left(\frac{i}{2\hbar}[\hat{q},\hat{p}\hat{\rho}+\hat{\rho}\hat{p}]+\frac{m}{\beta\hbar^2}\left[\hat{q},[\hat{q},\hat{\rho}]\right]+\frac{\beta}{6}\left[\hat{q},\left[\partial_q\hat{V},\hat{\rho}\right]\right]\right).\end{aligned} \tag{1.520}$$

We also observe that the imposition of the Wigner phase-space distribution $W_{\mathrm{eq}}(q,p)$ as the equilibrium solution of Eq. (1.514) so yielding a diffusion coefficient D_{pp}, which depends on the derivatives of the potential, appears to be the quantum analog of the *Ansatz* of a Maxwell–Boltzmann stationary distribution for the classical Klein–Kramers equation, Eq. (1.516). Furthermore, the condition $\mathrm{St}(W_{\mathrm{eq}})=0$ is equivalent to the property of the collision kernel St (W) in the classical kinetic theory, whereby the reduced or single particle phase-space distribution function $W(q, p, t)$ obeys the kinetic equation

$$\frac{\partial W}{\partial t}+\frac{p}{m}\frac{\partial W}{\partial q}-\frac{\partial V}{\partial q}\frac{\partial W}{\partial p}=\mathrm{St}(W). \tag{1.521}$$

Here the equilibrium Maxwell–Boltzmann distribution function $W_{\mathrm{eq}}(q,p)\sim e^{-\beta E(q,p)}$ always satisfies the condition $\mathrm{St}(W_{\mathrm{eq}})=0$. In particular, this is so for the Fokker–Planck equation Eq. (1.516). In the quantum case, this idea has been used before, e.g., by Gross and Lebowitz [319] in formulating quantum kinetic models of impulsive collisions. According to [319], for a system with a Hamiltonian $\hat{H}$, the equation governing the time behavior of the density matrix $\hat{\rho}$ is given by the (reduced) Eq. (1.411), where the collision kernel operator $\mathrm{St}(\hat{\rho})$ satisfies the condition $\mathrm{St}(\hat{\rho}_{\mathrm{eq}})=0$, where $\hat{\rho}_{eq}$ is the equilibrium density matrix. The condition $\mathrm{St}(\hat{\rho}_{\mathrm{eq}})=0$ has also been used by Redfield [224] [see Eq. (1.426)] in calculating the matrix elements of the relaxation operator $\mathrm{St}(\hat{\rho})$. Now we have just evaluated $D_p, D_{pp}, D_{qp}, D_{qq}$ for frequency-independent damping, meaning that in Eq. (1.514) they are *independent* of the time [308,311]. In the high-temperature limit, this approximation may be used in a wide range of the model parameters both in the limits of weak and strong damping (a detailed discussion of the validity of this approximation is given by Grabert [320]). However, for the parameter range, where such an approximation is invalid, e.g., throughout the very low-temperature region, alternative methods should be used.

The evolution Eq. (1.519) obtained for an *arbitrary* potential $V(q)$ radically simplifies for the harmonic potential, Eq. (1.496), i.e., the quantum Brownian oscillator model, viz., [272]

$$\frac{\partial W}{\partial t}+\frac{p}{m}\frac{\partial W}{\partial q}-m\omega_0^2 q\frac{\partial W}{\partial p}=\frac{\zeta}{m}\frac{\partial}{\partial p}\left(pW+D_{pp}\frac{\partial W}{\partial p}\right), \qquad (1.522)$$

where our heuristic generalization of the Einstein procedure yields D_{pp} in the closed form

$$D_{pp}=\left\langle p^2\right\rangle_{\mathrm{eq}}=\frac{m\hbar\omega_0}{2}\coth\frac{\beta\hbar\omega_0}{2}. \qquad (1.523)$$

The master Eq. (1.522) coincides in all respects with that of Agarwal [314], who first developed a detailed theory of the Brownian motion of a quantum oscillator for the *weak-coupling* case [see his equation (2.11) with $\lambda = 0$]. Furthermore, Eq. (1.522) is the same as the Fokker–Planck equation (here the Klein–Kramers equation) for a classical Brownian oscillator [272] except that the diffusion coefficient D_{pp} is altered to include the quantum effects. It is known, however, that according to the theory of quantum dissipation as applied to a quantum Brownian oscillator, both the stationary distribution as well as the corresponding averages $\langle q^2 \rangle$ and $\langle p^2 \rangle$ *depend on damping* (appropriate equations are given in Chapter 6 of [321]). In the approximation of Ohmic damping with Drude's regularization, these equations read [321]

$$\langle q^2 \rangle = \frac{1}{m\beta} \sum_{n=-\infty}^{\infty} \frac{1}{\omega_0^2 + v_n^2 + \dfrac{\omega_D \zeta / m}{1 + \omega_D / v_n}},$$
$$\langle p^2 \rangle = \frac{m}{\beta} \sum_{n=-\infty}^{\infty} \frac{\omega_0^2 + \dfrac{\omega_D \zeta / m}{1 + \omega_D / v_n}}{\omega_0^2 + v_n^2 + \dfrac{\omega_D \zeta / m}{1 + \omega_D / v_n}}, \tag{1.524}$$

where $v_n = 2\pi |n| / (\hbar\beta)$ and ω_D is a cutoff frequency (a Drude regularization is necessary as in pure Ohmic damping $\langle p^2 \rangle$ diverges [321]). However, both of these equations reduce to Eq. (1.498) either for vanishing damping ($\zeta / m \to 0$) or in the high-temperature limit ($\beta\hbar\omega_0 \to 0$). Moreover, the difference between the damping-dependent and damping-independent equations is negligible for $\zeta / (m\omega_0) < 0.1$ (which obviously fulfills the condition for the existence of damped oscillations and/or narrow spectral lines). Furthermore, for $\beta\hbar\zeta / m \le 1$, Eq. (1.522) (i.e., the Agarwal model) may be used as an approximate description of the kinetics of a quantum oscillator.

We conclude that the merit of the phase-space formalism for the quantum Brownian motion in a potential is that it originates in the master Eq. (1.514). This equation is a partial differential equation in phase space akin to the Fokker–Planck equation and so operators are not involved at all. Moreover, the main advantage of the phase-space approach now becomes quite apparent, namely it provides a master equation that may be solved using the methods [143,239] already associated with the classical theory of the Brownian motion in a potential, allowing one to study the quantum–classical correspondence for dissipative systems (see, e.g., [277,316–318,322–328]). Many other examples of the Wigner function representation of the density matrix in various applications in physics and chemistry may be found in the books [184,216,265,267,270,271] and references cited therein.

We now turn our attention to phase-space representations for spins demonstrating how the Wigner–Moyal formulation (representation) of quantum mechanics as a statistical theory on classical phase space can be successfully applied to spin systems.

1.9.5 *Quasiprobability distribution functions for spins*

By way of background to our discussion, we recall that in providing a phase-space description of spin systems, Stratonovich in 1956 [273] introduced the Wigner quasiprobability distribution function $W_S(\vartheta,\varphi,t)$ of the spin orientations in the configuration space of the polar and the azimuthal angles (ϑ, φ). This idea formed part of an elegant general discussion of *c*-number quasiprobability distributions for quantum systems in a representation space based on the *symmetry properties* of the underlying group. Examples are the Heisenberg–Weyl group for particles and the SU(2) group for rotations. The *c*-number representation for spins is especially important in treating spin relaxation phenomena. There *the spin orientation distribution is defined as the linear invertible bijective map onto the representation space comprised of the trace of the product of the system density matrix and the irreducible tensor operators with matrix elements in the spherical basis representation given via the Clebsch-Gordan coefficients*. Alternative quasiprobability distribution functions for spins have also been proposed using the spin coherent-state representation of the density matrix [245–248,251–254,274,275] introduced by Glauber and Sudarshan and commonly used in quantum optics (see, e.g., [184,265,275]). Moreover, Várilly and Gracia-Bondía [274] have shown that the spin coherent-state representation approach is equivalent to the Stratonovich formalism. In view of the importance and generality of Stratonovich's study of representation distributions for quantum systems, we shall summarize the general principles underlying such representations as given by him. Then we shall apply them to derive the Wigner and other representation distributions for systems with symmetries described by the SU(2) rotation group.

Stratonovich [273] defines in abstract fashion the "representation distribution" in the representation space M by the following requirements:

1. The space, in which the "representation distribution" is defined, has a clear *classical* meaning, for example, phase space (q, p) or the configuration space of orientations (ϑ, φ).
2. The representation distribution can be expressed *linearly* in terms of the density matrix $\hat{\rho}$. This requirement is directly related to the linearity of the whole apparatus of quantum theory [88], i.e., it relates to the statistical interpretation of the theory. The density matrix $\hat{\rho}$ just like any other quantum mechanical operator $\hat{A}$, has associated with it a c-number or Weyl function in the representation space M, viz.,

$$\rho(M,t) = \mathrm{Tr}\{\hat{\rho}(t)\hat{w}(M)\}, \tag{1.525}$$

   and, in general, for $\hat{A}$ we have

$$A(M) = \mathrm{Tr}\{\hat{A}\hat{w}(M)\}, \tag{1.526}$$

   comprising the *direct* mapping of the quantum operator $\hat{A}$ in Hilbert space onto the representation space via the kernel $\hat{w}$, which is just an operator depending on the point M as a parameter. For example, the point M in the representation space could be a point on the unit sphere (ϑ, φ).
3. The representation distribution must be *real*, i.e., in general, to a Hermitian operator $\hat{A}$ there must correspond a *real c-number* (Weyl symbol) $A(M)$. This requirement amounts to the condition that the bijective operator $\hat{w}(M)$ must itself be *Hermitian* for all points M.
4. Statistical averaging of the c-number $A(M)$ over the representation distribution *must* give the same results as the rule for averaging of Hilbert space operators, that is,

$$\left\langle \hat{A} \right\rangle(t) = \mathrm{Tr}\left(\hat{\rho}(t)\hat{A}\right) = \int A(M)\rho(M,t)dM. \tag{1.527}$$

Moreover, we have as a representation of a Hilbert space operator $\hat{A}$ the *inverse* map

$$\hat{A} = \int A(M)\hat{w}(M)dM \tag{1.528}$$

as can be easily verified by forming $\mathrm{Tr}\left(\hat{\rho}\hat{A}\right)$ via Eq. (1.528) and then using Eq. (1.525) thereby yielding Eq. (1.527).

These transformation rules immediately tell one how to get the operator relation from the *c*-number representation and vice-versa. Thus, condition 4 is equivalent to the requirement that both the *direct* Eq. (1.526) and *inverse* maps Eq. (1.528) are accomplished via the *same kernel* $\hat{w}(M)$, i.e., the mapping given by Eq. (1.526) is *bijective* (one-to-one-onto).

Next by regarding any given operators say $\hat{A}, \hat{B}, \ldots$ as elements of a complex Euclidean space with a scalar product given by the trace, viz.,

$$\left(\hat{A}, \hat{B}\right) = \mathrm{Tr}\left(\hat{A}\hat{B}^{\dagger}\right),$$

we can then introduce an orthonormal basis of operators $\{\hat{A}_1, \hat{A}_2, \ldots\}$ with orthogonality relation

$$\mathrm{Tr}\left(\hat{A}_i\hat{A}_j^{\dagger}\right) = \delta_{ij}. \tag{1.529}$$

Thus, the bijective kernel operator $\hat{w}(M)$ may itself be represented by its expansion in the orthonormal basis of operators as

$$\hat{w}(M) = \sum_i \hat{A}_i^{\dagger} A_i(M), \tag{1.530}$$

where by definition the *c*-number expansion coefficients $A_i(M)$ are given by the direct map [cf. Eq. (1.526)]

$$A_i\left(M\right) = \mathrm{Tr}\left\{\hat{A}_i\hat{w}(M)\right\}. \tag{1.531}$$

The requirement embodied in Eq. (1.529) is entirely equivalent to the *c*-number orthogonality relation

$$\int A_i\left(M\right) A_j^*\left(M\right) dM = \delta_{ij}. \tag{1.532}$$

According to Eq. (1.527) the normalization condition for the density matrix $\hat{\rho}$, viz. $\mathrm{Tr}(\hat{\rho}) = 1$, must become

$$\int \rho(M,t)\mathrm{Tr}\{\hat{w}(M)\}\, dM = 1. \tag{1.533}$$

Thus, as the normalized distribution we must obviously chose a c-number function $W(M, t)$ such that

$$W(M,t) = \mathrm{Tr}\{\hat{w}(M)\}\,\rho(M,t) \qquad (1.534)$$

and so the averaging rule Eq. (1.527) becomes in terms of W

$$\langle \hat{A} \rangle(t) = \int W(M,t)A(M)\left[\mathrm{Tr}\{\hat{w}(M)\}\right]^{-1} dM. \qquad (1.535)$$

Stratonovich's abstract treatment presented above may be used to derive specific distributions. The Wigner distribution $W(q, p, t)$ defined by Eq. (1.478) is an important explicit example. Here Eq. (1.478) can also be rewritten in terms of a bijective map $\hat{w}$ using the density matrix $\hat{\rho}(q_1, q_2, t)$ in the (q_1, q_2) notation [216] as

$$W(q,p,t) = \int_{-\infty}^{\infty}\int_{-\infty}^{\infty} \hat{\rho}(q_1,q_2,t)\,\hat{w}(q,p,q_1,q_2)\,dq_1 dq_2, \qquad (1.536)$$

where the kernel operator $\hat{w}(q,p,q_1,q_2)$ is now given by [273]

$$\hat{w}(q,p,q_1,q_2) = \frac{1}{\pi\hbar} e^{i(q_1-q_2)p/\hbar}\,\delta\left(q - \frac{q_1+q_2}{2}\right). \qquad (1.537)$$

The transformation Eq. (1.536) may be verified as follows. We introduce new variables $q_1 = u + y$, $q_2 = u - y$, then by the Jacobian of the transformation, the properties of the δ function and the replacement $y \to y/2$ we have Eq. (1.478). Likewise, Eq. (1.528) with Eq. (1.537) yields the transformation back to Hilbert space again by the properties of the δ function. Thus, the Wigner distribution $W(q, p, t)$ for the translational motion is derived merely by applying the principles of homogeneity and equivalence of directions embodied in the symmetries of the Heisenberg–Weyl group combined with the notion of a classical phase space. This group-theoretic argument should be compared with the intuitive method of Wigner who appears to have arrived at his quasidistribution by *ad hoc* reasoning insofar as its marginal distributions yield the correct marginal quantum mechanical probabilities for either the positions or the momenta.

1.9.6 *Spin phase-space distribution functions*

Stratonovich [273] originally introduced the spin phase-space distribution function for zero dissipation, i.e., for closed systems. This function was further developed both for closed and open spin systems (e.g., [229–231, 243,244,246–254,274,275]) and is entirely analogous to the translational Wigner distribution $W(q,p,t)$ in phase space (q,p), which is the quasiprobability representation of the density operator except that *certain vital differences arise because of the angular momentum commutation relations*. The basic ideas may be summarized as follows [275]. First, we recall that the *classical distribution function* $W(\vartheta,\varphi,t)$ of magnetic moment orientations on the surface of the unit sphere can be expanded in the usual *infinite* Fourier–Laplace series of the spherical harmonics $Y_{LM}(\vartheta,\varphi)$ as [97] [cf. Eq. (1.84)]

$$W(\vartheta,\varphi,t)=\sum_{L=0}^{\infty}\sum_{M=-L}^{L}Y_{LM}(\vartheta,\varphi)\left\langle Y_{LM}^{*}\right\rangle(t), \tag{1.538}$$

where the expansion coefficients (statistical moments) $\left\langle Y_{LM}^{*}\right\rangle(t)$ are

$$\left\langle Y_{LM}^{*}\right\rangle(t)=\int_0^{\pi}\int_0^{2\pi}Y_{LM}^{*}(\vartheta,\varphi)W(\vartheta,\varphi,t)\sin\vartheta d\vartheta d\varphi \tag{1.539}$$

(the asterisk denotes the complex conjugate) because the $Y_{LM}(\vartheta,\varphi)$ are orthogonal and constitute a complete set in configuration space (ϑ,φ) (various definitions and properties of the $Y_{LM}(\vartheta,\varphi)$ are discussed in detail in Appendix A.1.7). However, the $Y_{LM}(\vartheta,\varphi)$, which are alternatively just the components of some irreducible tensor of rank L, *could also be formally defined in terms of the commutation relations* [104]

$$\left[\hat{L}_{\mu},Y_{LM}(\vartheta,\varphi)\right]=\sqrt{L(L+1)}C_{LM1\mu}^{LM+\mu}Y_{LM+\mu}(\vartheta,\varphi) \tag{1.540}$$

yielding the three separate relations $(\mu=0,\pm1)$, viz., [104]

$$\begin{gathered}\hat{L}_0Y_{LM}(\vartheta,\varphi)=MY_{LM}(\vartheta,\varphi),\\ \hat{L}_{\pm1}Y_{LM}(\vartheta,\varphi)=\mp\sqrt{\frac{L(L+1)-M(M\pm1)}{2}}Y_{LM\pm1}(\vartheta,\varphi),\end{gathered} \tag{1.541}$$

where $C_{LM1\mu}^{LM+\mu}$ is the Clebsch–Gordan coefficient [104], $\hat{L}_\mu$ are the components of the orbital angular momentum operator $\hat{\mathbf{L}}$ in the spherical basis defined by Eq. (A.77) from Appendix A.1.7. However, the angular momentum operators $\hat{L}_\mu$ satisfy the *same* commutation relations, Eq. (A.156) from Appendix A.2, as the spherical components $\hat{S}_\mu$ of the spin operator $\hat{\mathbf{S}}$ given by Eq. (A.137) from Appendix A.2. Therefore, the quasiprobability distribution function for spins $W_S(\vartheta,\varphi,t)$ may be obtained in the manner of Eqs. (1.458) *et seq.* (pertaining to particles) merely by replacing the *classical* average $\left\langle Y_{LM}^{*}\right\rangle(t)$ in the classical distribution Eq. (1.538) by the *quantum mechanical expectation values* of appropriate operators, which *must transform under rotation* of the coordinate system in exactly the same way as the spherical harmonics $Y_{LM}(\vartheta,\varphi)$. Therefore, we must seek *specific operators* in *Hilbert space* corresponding to the $Y_{LM}(\vartheta,\varphi)$ such that *they have commutation relations with the spin spherical component operators* $\hat{S}_\mu$ *which are the same as those*, Eq. (1.540), *between the spherical harmonics* $Y_{LM}(\vartheta,\varphi)$ *and the spherical component operators* $\hat{L}_\mu$ *of the angular momentum operator* $\hat{\mathbf{L}}$ [5,104]. The particular operators with these commutation properties are the polarization operators $\hat{T}_{LM}^{(S)}$ [104]; see Eq. (A.147) *et seq.* in Appendix A.2. This intrinsic correspondence becomes obvious on comparing Eq. (1.541) defining the spherical harmonics with the commutation Eqs. (A.156)–(A.158) from Appendix A.2 of the polarization and spin operators yielding [104]

$$[\hat{S}_0,\hat{T}_{LM}^{(S)}]=M\hat{T}_{LM}^{(S)},$$
$$[\hat{S}_{\pm1},\hat{T}_{LM}^{(S)}]=\mp\sqrt{\frac{L(L+1)-M(M\pm1)}{2}}\hat{T}_{LM\pm1}^{(S)}. \tag{1.542}$$

We may go into more detail in the following manner. We have already seen that the density matrix $\hat{\rho}_S$ of the spin characterized by the spin number S may be expanded as a finite series of the polarization operators $\hat{T}_{LM}^{(S)}$ [104,217], namely,

$$\hat{\rho}_S(t)=\sum_{L=0}^{2S}\sum_{M=-L}^{L}\hat{T}_{LM}^{(S)}\left\langle \hat{T}_{LM}^{\dagger(S)}\right\rangle(t). \tag{1.543}$$

Therefore, to construct the quasiprobability distribution function $W_S(\vartheta,\varphi,t)$ of the polar and the azimuthal angles (ϑ, φ) corresponding to the density matrix $\hat{\rho}_S$ as defined by Eq. (1.543), we first, in the light of the previous paragraph, *formally* postulate $W_S(\vartheta,\varphi,t)$ as a *finite* series of spherical harmonics $Y_{LM}(\vartheta,\varphi)$ in the representation space, viz.,

$$\frac{2S+1}{4\pi}W_S(\vartheta,\varphi,t)=\sum_{L=0}^{2S}\sum_{M=-L}^{L}Y_{LM}(\vartheta,\varphi)\left\langle Y_{LM}^*\right\rangle(t), \tag{1.544}$$

where the expansion coefficients $\left\langle Y_{LM}^*\right\rangle(t)$ are

$$\left\langle Y_{LM}^*\right\rangle(t)=\frac{2S+1}{4\pi}\int_0^{\pi}\int_0^{2\pi}Y_{LM}^*(\vartheta,\varphi)W_S(\vartheta,\varphi,t)\sin\vartheta d\vartheta d\varphi \tag{1.545}$$

due to the orthogonality property of the $Y_{LM}(\vartheta,\varphi)$ given by Eq. (A.77) (see Appendix A.1.7). In general Eq. (1.544) merely represents a *quantum analog* of the expansion Eq. (1.538) of the *classical* distribution function $W(\vartheta, \varphi, t)$ of spin orientations in configuration space in spherical harmonics. Thus, following the above argument, it appears that both the quasiprobability (Wigner) distribution function on the sphere $W_S(\vartheta,\varphi,t)$ and the corresponding kernel operator $\hat{w}(\vartheta,\varphi)$ of the Wigner–Stratonovich bijective map onto phase space defined by Eq. (1.525), viz.,

$$W_S(\vartheta,\varphi,t)=\mathrm{Tr}\left\{\hat{\rho}_S(t)\hat{w}(\vartheta,\varphi)\right\}, \tag{1.546}$$

can be obtained by merely replacing the averages of spherical harmonics $\left\langle Y_{LM}^*\right\rangle(t)$ over the classical-like representation space distribution defined by Eq. (1.544) with the quantum mechanical expectation values of the polarization operators $\left\langle \hat{T}_{LM}^{\dagger(S)}\right\rangle(t)$. *However, the replacement must be achieved in such a way as to preserve commutation relations according to the prescription indicated* by Eq. (1.545), i.e., we must have the fundamental relation underpinning the whole theory, viz.,

$$\left\langle Y_{LM}^*(\vartheta,\varphi)\right\rangle(t)=\frac{2S+1}{4\pi}\Omega_{LM}^S\left\langle \hat{T}_{LM}^{\dagger(S)}\right\rangle(t), \tag{1.547}$$

where Ω_{LM}^S is a constant yet to be determined. Substitution of Eq. (1.547) into Eq. (1.544) immediately yields the desired Eq. (1.544) in representation space in the (now) *finite* series form

$$W_S(\vartheta,\varphi,t) = \mathrm{Tr}\{\hat{\rho}_S \hat{w}(\vartheta,\varphi)\} = \sum_{L=0}^{2S} \sum_{M=-L}^{L} \Omega_{LM}^{S} Y_{LM}(\vartheta,\varphi)\left\langle \hat{T}_{LM}^{\dagger(S)} \right\rangle(t). \quad (1.548)$$

Equation (1.548) implies that as before *different* forms of the quasiprobability distribution function $W_S(\vartheta,\varphi,t)$ exist corresponding to *different* choices of the constant Ω_{LM}^{S} in Eq. (1.547) entirely analogous to Eq. (1.464) for the translational case. Thus, we have quasiprobability distributions for spins merely *by identifying the average* $\left\langle Y_{LM}^{*} \right\rangle(t)$ *of the spherical harmonic* $Y_{LM}(\vartheta,\varphi)$ *over* (ϑ, φ) *as an average* $\left\langle \hat{T}_{LM}^{\dagger(S)} \right\rangle(t)$ *of the polarization operator* $\hat{T}_{LM}^{(S)}$ *via the spin density matrix* $\hat{\rho}_S$ [275]. To find explicit equations for the prefactor Ω_{LM}^{S} with the ultimate aim of introducing three different quasiprobability distribution functions analogous to those already described for particles [cf. Eq. (1.464)], we first consider the SU(2) *coherent or minimum uncertainty states*, which may be put into context as follows. Many spin relaxation problems can be dealt with as the interaction of an assembly of spins with electromagnetic fields or as those arising from internal anisotropy potentials so that a *particular* set of spin states has to be selected. However, the choice of a particular representation is motivated rather by convenience than by necessity. Thus, in this section, we define the *spin coherent states* and discuss their properties following the didactic Ref. [246]. The particular advantage of the *coherent state representation* of the density matrix is that in spin systems it allows one to analyze in *semi-classical fashion* spin relaxation phenomena using a quasiprobability distribution function $W_S(\vartheta,\varphi,t)$ of spin orientations in a phase (here configuration) space of the polar and the azimuthal angles (ϑ, φ).

The spin coherent state representation may be explained as follows: first we recall that in quantum mechanics, the spin operator $\hat{\mathbf{S}}$ is usually represented by a set of three square $(2S+1)\times(2S+1)$ matrices $\hat{S}_X, \hat{S}_Y$, and $\hat{S}_Z$ with S being the spin number, while the spin basis functions $\chi_{Sm} = |S,m\rangle$, describing the particular states with *definite* spin S and spin projection m onto the Z-axis, are *eigenfunctions* of the spin operators $\hat{\mathbf{S}}^2$ and $\hat{S}_Z$ [104] (the details of the properties of the spin operators $\hat{\mathbf{S}}, \hat{S}_Z$, etc. and the spin functions χ_{Sm} are described in Appendix A.2). Then the spin coherent states $|\psi_\kappa\rangle$ for a single particle of spin S are defined as [246] (in our notation)

$$|\psi_\kappa\rangle = N^{-1/2} e^{\kappa \hat{S}_{-1}} \chi_{SS}. \tag{1.549}$$

Moreover, the spin coherent states are analogous to the coherent states of a harmonic oscillator (cf. the title of Erwin Schrödinger's fundamental paper "Der stetige Übergang von der Mikro- zur Makromechanik" [283]). In such spin states, the *average of the spin-angular momentum follows classical-like dynamics even though the three individual components of the spin angular momentum do not commute*. Thus, these states are eminently suited to the semi-classical calculations considered here. In Eq. (1.549) κ is taken as complex valued, N is a normalization factor, $\hat{S}_{-1} = 2^{-1/2}(\hat{S}_X - i\hat{S}_Y)$ is the spin spherical component operator, and χ_{SS} is the *spin ground state* such that $\hat{S}_Z \chi_{SS} = S\chi_{SS}$. Expanding the exponential operator $e^{\kappa \hat{S}_{-1}}$ in Eq. (1.549) consequently yields the expansion of the state $|\psi_\kappa\rangle$ in the orthonormal basis of the spin eigenstates χ_{Sm}, viz., [246]

$$|\psi_\kappa\rangle = \frac{1}{\sqrt{N}} \sum_{p=0}^{\infty} \left(\kappa \hat{S}_{-1}\right)^p \chi_{SS} = \frac{1}{\sqrt{N}} \sum_{p=0}^{2S} \kappa^p \sqrt{\frac{p!(2S)!}{2^p(2S-p)!}} \chi_{SS-p}, \tag{1.550}$$

mindful that the operator $\hat{S}_{-1}$ *acting on the ground state* χ_{SS} creates *spin deviations* of the form [246]

$$\left(\hat{S}_{-1}\right)^p \chi_{SS} = \sqrt{\frac{p!(2S)!}{2^p(2S-p)!}} \chi_{SS-p}; \; 0 \le p \le 2S. \tag{1.551}$$

Furthermore, the normalization factor N in Eq. (1.550) is determined by the equality (the scalar product)

$$\langle \psi_\kappa | \psi_\kappa \rangle = N^{-1} \sum_{p=0}^{2S} \frac{(2S)! |\kappa|^{2p}}{2^p \, p!(2S-p)!} = N^{-1}\left(1 + |\kappa|^2/2\right)^{2S} = 1, \tag{1.552}$$

so that the normalized state $|\psi_\kappa\rangle$ is then given by

$$|\psi_\kappa\rangle = \left(1 + |\kappa|^2/2\right)^{-S} \sum_{p=0}^{2S} \kappa^p \sqrt{\frac{p!(2S)!}{2^p(2S-p)!}} \chi_{SS-p}. \tag{1.553}$$

However, *the spin coherent states are not themselves orthogonal* so that the scalar product of two *distinct* states $|\psi_\kappa\rangle$ and $|\psi_\lambda\rangle$ is no longer zero; instead it is

$$\langle\psi_\lambda|\psi_\kappa\rangle = \left(1+\frac{1}{2}|\lambda|^2\right)^{-S}\left(1+\frac{1}{2}|\kappa|^2\right)^{-S}\sum_{p=0}^{2S}\frac{(2S)!\lambda^{*p}\kappa^p}{2^p p!(2S-p)!}$$
$$= \frac{\left(2+\lambda^*\kappa\right)^{2S}}{\left(2+|\lambda|^2\right)^S\left(2+|\kappa|^2\right)^S} \tag{1.554}$$

and so we have

$$|\langle\psi_\lambda|\psi_\kappa\rangle|^2 = \left(1-\frac{2|\lambda-\kappa|^2}{(2+|\lambda|^2)(2+|\kappa|^2)}\right)^{2S}. \tag{1.555}$$

If we write in the definition Eq. (1.553)

$$p = S-m,\ \kappa = \sqrt{2}e^{i\varphi}\tan(\vartheta/2),\quad 0\le\vartheta<\pi, 0\le\varphi<2\pi, \tag{1.556}$$

then *the normalized states* $|\psi_\kappa\rangle = |S,\vartheta,\varphi\rangle$ can be explicitly written as

$$|S,\vartheta,\varphi\rangle = \cos^{2S}\frac{\vartheta}{2}e^{\sqrt{2}e^{i\varphi}\tan(\vartheta/2)\hat{S}_{-1}}\chi_{SS}$$
$$= \cos^{2S}\frac{\vartheta}{2}\sum_{m=-S}^{S}\sqrt{\frac{(2S)!}{(S+m)!(S-m)!}}\left(e^{i\varphi}\tan\frac{\vartheta}{2}\right)^{S-m}\chi_{Sm}. \tag{1.557}$$

Furthermore, the *spin coherent state vector* $|S, \vartheta, \varphi\rangle$ as defined by Eq. (1.557) can also be written in the *equivalent* form [104]

$$|S,\vartheta,\varphi\rangle = \tilde{\chi}_{SS}(\vartheta,\varphi)e^{iS\varphi},$$

where the $\tilde{\chi}_{S\lambda}(\vartheta,\varphi)$ constitute the *helicity basis functions* defined as

$$\tilde{\chi}_{S\lambda}(\vartheta,\varphi) = \sum_{m=-S}^{S} D^S_{m\lambda}(\varphi,\vartheta,0)\chi_{Sm},$$

and $D^S_{m\lambda}(\varphi,\vartheta,\psi)$ are the Wigner D functions [104] (see Appendix A.1.8). Both forms of the state vector $|S, \vartheta, \varphi\rangle$ are equivalent due to the trigonometric identity [104]

$$D^S_{m\pm S}(\varphi,\vartheta,0)e^{\pm iS\varphi} = \sqrt{\frac{(2S)!}{(S+m)!(S-m)!}}\cos^{2S}\frac{\vartheta}{2}\left[\pm e^{\pm i\varphi}\tan\frac{\vartheta}{2}\right]^{S\mp m}.$$

The coherent states $|S, \vartheta, \varphi\rangle$ so defined form *a complete set* with the completeness relation given by the *outer product*

$$\frac{2S+1}{4\pi}\int_0^{\pi}\int_0^{2\pi}|S,\vartheta,\varphi\rangle\langle S,\vartheta,\varphi|\sin\vartheta d\vartheta d\varphi$$

$$=\frac{2S+1}{2}\sum_{m=-S}^{S}\frac{(2S)!\chi_{Sm}\chi_{Sm}^{\dagger}}{(S+m)!(S-m)!}\int_0^{\pi}\cos^{4S}\frac{\vartheta}{2}\tan^{2(S-m)}\frac{\vartheta}{2}\sin\vartheta d\vartheta \quad (1.558)$$

$$=\sum_{m=-S}^{S}\chi_{Sm}\chi_{Sm}^{\dagger}=\hat{I},$$

where $\hat{I}$ is the $(2S+1)\times(2S+1)$ identity matrix, while the *overlap* of two coherent states $|S, \vartheta, \varphi\rangle$ and $|S, \vartheta', \varphi'\rangle$ say is [246]

$$\langle S,\vartheta',\varphi'|S,\vartheta,\varphi\rangle$$
$$=\left[\cos\tfrac{1}{2}\vartheta\cos\tfrac{1}{2}\vartheta'+e^{i(\varphi-\varphi')}\sin\tfrac{1}{2}\vartheta\sin\tfrac{1}{2}\vartheta'\right]^{2S}, \quad (1.559)$$

and so in vector form

$$|\langle S,\vartheta',\varphi'|S,\vartheta,\varphi\rangle|=2^{-S}\left[1+(\mathbf{u}\cdot\mathbf{u}')\right]^{S}, \quad (1.560)$$

where $\mathbf{u}$ and $\mathbf{u}'$ are unit vectors in the directions specified by the spherical polar coordinates $(\vartheta, \varphi)$and (ϑ', φ'), respectively.

In addition, the matrix elements $\langle S,\vartheta,\varphi|\hat{T}_{LM}^{(S)}|S,\vartheta,\varphi\rangle$ of the polarization operator $\hat{T}_{LM}^{(S)}$ in the spin coherent state representation may be expressed in terms of the spherical harmonics $Y_{LM}(\vartheta,\varphi)$ via the trigonometric expansion of the latter, viz., Eq. (A.84) from Appendix A.1.7 and Eq. (A.143) from Appendix A.2, thereby ultimately yielding the matrix elements of the $\hat{T}_{LM}^{(S)}$ in the following intuitively appealing compact closed form

$$\langle S,\vartheta,\varphi|\hat{T}_{LM}^{(S)}|S,\vartheta,\varphi\rangle$$

$$=(2S)!\cos^{4S}\frac{\vartheta}{2}\sum_{m,m'=-S}^{S}\frac{e^{-i(m-m')\varphi}(\tan\vartheta/2)^{2S-m-m'}(\chi_{Sm'}^{\dagger}\hat{T}_{LM}^{(S)}\chi_{Sm})}{\sqrt{(S+m')!(S-m')!(S+m)!(S-m)!}} \quad (1.561)$$

$$=\sqrt{\frac{4\pi}{2S+1}}C_{SSL0}^{SS}Y_{LM}(\vartheta,\varphi),$$

where C^{SS}_{SSL0} is the Clebsch–Gordan coefficient [104] explicitly defined as

$$C^{SS}_{SSL0} = \frac{(2S)!\sqrt{2S+1}}{\sqrt{(2S-L)!(2S+L+1)!}}. \tag{1.562}$$

Thus, by using Eqs. (1.561) and (1.562), the definitions of the operators of the cartesian components of the spin operators $\hat{S}_X, \hat{S}_Y, \hat{S}_Z$, Eq. (A.147), in terms of the polarization operators $\hat{T}^{(S)}_{10}$ and $\hat{T}^{(S)}_{1\pm1}$ from Appendix A.2, and the definitions of the spherical harmonics $Y_{10}(\vartheta,\varphi)$ and $Y_{1\pm1}(\vartheta,\varphi)$ of rank 1 from Appendix A.1.7, we have the matrix elements of $\hat{S}_X$, $\hat{S}_Y$, and $\hat{S}_Z$ rendered in the *spin coherent state* representation as the closed forms

$$\begin{aligned} \langle S,\vartheta,\varphi|\hat{S}_Z|S,\vartheta,\varphi\rangle &= S\cos\vartheta, \\ \langle S,\vartheta,\varphi|\hat{S}_X|S,\vartheta,\varphi\rangle &= S\sin\vartheta\cos\varphi, \\ \langle S,\vartheta,\varphi|\hat{S}_Y|S,\vartheta,\varphi\rangle &= S\sin\vartheta\sin\varphi. \end{aligned} \tag{1.563}$$

In this manner, we then have the matrix elements of the spin operator $\hat{\mathbf{S}}$ itself rendered in the simple intuitively appealing classical-like form

$$\langle S,\vartheta,\varphi|\hat{\mathbf{S}}|S,\vartheta,\varphi\rangle = S\mathbf{u}. \tag{1.564}$$

In other words, *the spin operator* $\hat{\mathbf{S}}$ *in the coherent state representation is simply the direct quantum analog of the classical magnetic dipole vector* $\boldsymbol{\mu} = \mu\,\mathbf{u}$.

We shall now use the spin coherent state representation of the spin density matrix just discussed to derive transformation kernels for the spin *Q*-, *P*-, and Wigner quasiprobability distributions. Proceeding we first consider the so-called *Q-function representation.* The *Q*-function representation distribution corresponding to a spin density matrix $\hat{\rho}_S$ in Hilbert space may be defined in a manner analogous to that for particles [cf. Eq. (1.464) with $s=-1$] i.e. via the *diagonal matrix elements* of the density matrix in the spin coherent state representation [275]. Thus, we may then form the matrix elements $\langle S,\vartheta,\varphi|\hat{\rho}_S|S,\vartheta,\varphi\rangle$ of $\hat{\rho}_S$ as defined by the polarization operator expansion Eq. (1.440) in the spin coherent state representation $|S,\vartheta,\varphi\rangle$ as the linear combination [mindful that in Eq. (1.440) the statistical moments are given by $a_{L,M}(t) = \left\langle \hat{T}^{\dagger(S)}_{LM}\right\rangle(t)$]

$$\langle S,\vartheta,\varphi|\hat{\rho}_S|S,\vartheta,\varphi\rangle = \sum_{L=0}^{2S}\sum_{M=-L}^{L}\langle S,\vartheta,\varphi|\hat{T}_{LM}^{(S)}|S,\vartheta,\varphi\rangle\left\langle \hat{T}_{LM}^{\dagger(S)}\right\rangle(t), \quad (1.565)$$

where the matrix elements $\langle S,\vartheta,\varphi|\hat{T}_{LM}^{(S)}|S,\vartheta,\varphi\rangle$ are explicitly given by the closed form Eq. (1.561). Then via the replacement prescription in terms of the averages of spherical harmonics for averages of polarization operators involving the density matrix $\hat{\rho}_S$ embodied in the fundamental Eq. (1.547) and using the identity (1.561) for the matrix elements of the $\hat{T}_{LM}^{(S)}$ in terms of Y_{LM}, we have from Eq. (1.565) the matrix elements $\langle S,\vartheta,\varphi|\hat{\rho}_S|S,\vartheta,\varphi\rangle$ of $\hat{\rho}_S$ rendered as the *finite* sum of spherical harmonics

$$\begin{aligned}&\langle S,\vartheta,\varphi|\hat{\rho}_S|S,\vartheta,\varphi\rangle\\ &= \sqrt{\frac{4\pi}{2S+1}}\sum_{L=0}^{2S}\sum_{M=-L}^{L}\left(\Omega_{LM}^{S}\right)^{-1}C_{SSL0}^{SS}Y_{LM}(\vartheta,\varphi)\left\langle Y_{LM}^{*}\right\rangle(t).\end{aligned} \quad (1.566)$$

If we let the prefactor be

$$\Omega_{LM}^{S} = \sqrt{\frac{4\pi}{2S+1}}C_{SSL0}^{SS}, \quad (1.567)$$

then in accordance with the classical-like representation distribution, Eq. (1.544), *we have the final form of the matrix elements of the density operator in the spin coherent state representation*

$$\langle S,\vartheta,\varphi|\hat{\rho}_S|S,\vartheta,\varphi\rangle = \sum_{L=0}^{2S}\sum_{M=-L}^{L}Y_{LM}(\vartheta,\varphi)\left\langle Y_{LM}^{*}\right\rangle(t), \quad (1.568)$$

that is, we have with Eq. (1.544) the phase-space distribution $W_S^Q(\vartheta,\varphi,t)$ defined in terms of the matrix elements of the density operator as

$$W_S^Q(\vartheta,\varphi,t) = \frac{4\pi}{2S+1}\langle S,\vartheta,\varphi|\hat{\rho}_S|S,\vartheta,\varphi\rangle.$$

Thus, this quasiprobability distribution function $W_S^Q(\vartheta,\varphi,t)$ comprises the *diagonal matrix elements* of the density operator $\hat{\rho}_S$ in the *spin coherent state representation* and is finally given explicitly by (returning to averages of the polarization operators)

$$W_S^Q(\vartheta,\varphi,t) = \sqrt{\frac{4\pi}{2S+1}}\sum_{L=0}^{2S}\sum_{M=-L}^{L}C_{SSL0}^{SS}Y_{LM}(\vartheta,\varphi)\left\langle \hat{T}_{LM}^{\dagger(S)}\right\rangle(t). \quad (1.569)$$

The normalization of $W_S^Q(\vartheta,\varphi,t)$ occurring in Eq. (1.569) can be found noticing the orthogonality property of the spherical harmonics, Eq. (A.78) from Appendix A.1.7, and Eq. (1.562) for $L=0$. We have

$$\int_0^\pi\int_0^{2\pi} W_S^Q(\vartheta,\varphi,t)\sin\vartheta d\vartheta d\varphi = \frac{4\pi}{2S+1}. \tag{1.570}$$

Next, on this occasion, we consider the *P-function* $W_S^P(\vartheta,\varphi,t)$, which is formally defined in terms of the density matrix $\hat{\rho}_S$ in the *spin coherent state representation* via the inverse Wigner–Stratonovich map as [275]

$$\hat{\rho}_S(t) = \frac{2S+1}{4\pi}\int_0^\pi\int_0^{2\pi} W_S^P(\vartheta,\varphi,t)\left|S,\vartheta,\varphi\right\rangle\left\langle S,\vartheta,\varphi\right|\sin\vartheta d\vartheta d\varphi. \tag{1.571}$$

However, this definition of $\hat{\rho}_S(t)$ as an inverse map implies that the average polarization operator $\left\langle \hat{T}_{LM}^{\dagger(S)}\right\rangle(t) = \mathrm{Tr}[\hat{\rho}_S(t)\hat{T}_{LM}^{\dagger(S)}]$ is now explicitly given by

$$\begin{aligned}\left\langle \hat{T}_{LM}^{\dagger(S)}\right\rangle(t) &= \frac{2S+1}{4\pi}\int_0^\pi\int_0^{2\pi} W_S^P(\vartheta,\varphi,t)\mathrm{Tr}\left(\left|S,\vartheta,\varphi\right\rangle\left\langle S,\vartheta,\varphi\right|\hat{T}_{LM}^{\dagger(S)}\right)\sin\vartheta d\vartheta d\varphi \\ &= \frac{2S+1}{4\pi}\int_0^\pi\int_0^{2\pi} W_S^P(\vartheta,\varphi,t)\left\langle S,\vartheta,\varphi\right|\hat{T}_{LM}^{\dagger(S)}\left|S,\vartheta,\varphi\right\rangle\sin\vartheta d\vartheta d\varphi \\ &= \sqrt{\frac{4\pi}{2S+1}}C_{SSL0}^{SS}\left\langle Y_{LM}^*\right\rangle(t)\end{aligned} \tag{1.572}$$

because in the integrand

$$\mathrm{Tr}\left(\left|S,\vartheta,\varphi\right\rangle\left\langle S,\vartheta,\varphi\right|\hat{T}_{LM}^{\dagger(S)}\right) = \left\langle S,\vartheta,\varphi\right|\hat{T}_{LM}^{\dagger(S)}\left|S,\vartheta,\varphi\right\rangle.$$

This relation is in turn based on Eqs. (A.143), (A.153), and (A.159) from Appendix A.2 and the identity contained in Eq.(1.561). Comparing Eqs. (1.572) and (1.547) then yields for the P-function simply

$$\Omega_{LM}^S = \sqrt{\frac{2S+1}{4\pi}}\left(C_{SSL0}^{SS}\right)^{-1}. \tag{1.573}$$

For purposes of symmetry and to satisfy the normalization condition Eq. (1.570), we define the P-function $W_S^P(\vartheta,\varphi,t)$ as (cf. Eq. (1.569))

$$W_S^P(\vartheta,\varphi,t)=\sqrt{\frac{4\pi}{2S+1}}\sum_{L=0}^{2S}\sum_{M=-L}^{L}\left(C_{SSL0}^{SS}\right)^{-1}Y_{LM}(\vartheta,\varphi)\left\langle\hat{T}_{LM}^{\dagger(S)}\right\rangle(t), \quad (1.574)$$

which differs from the Q function representation Eq. (1.569) only through the reciprocal of the Clebsch–Gordan coefficients involved in the summand. Finally, the *Wigner function for spins* is defined as the function, which is its *own conjugate*, i.e., the function obtained by setting [275]

$$\Omega_{LM}^S=1. \quad (1.575)$$

Again for purposes of symmetry and to satisfy the normalization relation Eq. (1.570), we define the Wigner function $W_S^W(\vartheta,\varphi,t)$ as

$$W_S^W(\vartheta,\varphi,t)=\sqrt{\frac{4\pi}{2S+1}}\sum_{L=0}^{2S}\sum_{M=-L}^{L}Y_{LM}(\vartheta,\varphi)\left\langle\hat{T}_{LM}^{\dagger(S)}\right\rangle(t). \quad (1.576)$$

By introducing $W_S^{(s)}(\vartheta,\varphi,t)=\mathrm{Tr}[\hat{\rho}_S(t)\hat{w}_s(\vartheta,\varphi)]$, where a new parameter s takes on the values $s=-1, 0, 1$ for Q-, Wigner-, and P-functions, respectively, we can rewrite all of the representation space mappings Eqs. (1.569), (1.574), and (1.576) in compact form as the single mapping [253]

$$W_S^{(s)}(\vartheta,\varphi,t)=\sqrt{\frac{4\pi}{2S+1}}\sum_{L=0}^{2S}\sum_{M=-L}^{L}\left(C_{SSL0}^{SS}\right)^{-s}Y_{LM}(\vartheta,\varphi)\left\langle\hat{T}_{LM}^{\dagger(S)}\right\rangle(t), \quad (1.577)$$

where $\hat{w}_s(\vartheta,\varphi)$ above is the Wigner–Stratonovich kernel of the bijective transformation given by

$$\hat{w}_s(\vartheta,\varphi)=\sqrt{\frac{4\pi}{2S+1}}\sum_{L=0}^{2S}\sum_{M=-L}^{L}\left(C_{SSL0}^{SS}\right)^{-s}Y_{LM}^*(\vartheta,\varphi)\hat{T}_{LM}^{(S)} \quad (1.578)$$

such that [cf. Eqs. (1.529)-(1.534)]

$$\hat{w}_s(\vartheta,\varphi)=\hat{w}_s^\dagger(\vartheta,\varphi),\ \mathrm{Tr}(\hat{w}_s)=1,\ \frac{2S+1}{4\pi}\int_0^\pi\int_0^{2\pi}\hat{w}_s(\vartheta,\varphi)\sin\vartheta d\vartheta d\varphi=\hat{I}.$$

Due to the orthogonality properties of the spherical harmonics Eq. (A.78) from Appendix A.1.7, all the phase-space distributions embodied in Eq. (1.577) can also be written in yet more compact form as the finite series

$$\frac{2S+1}{4\pi}W_S^{(s)}(\vartheta,\varphi,t)=\sum_{L=0}^{2S}\sum_{M=-L}^{L}Y_{LM}(\vartheta,\varphi)\left\langle Y_{LM}^{*}\right\rangle^{(s)}(t), \quad (1.579)$$

where

$$\left\langle Y_{LM}^{*}\right\rangle^{(s)}(t)=\sqrt{\frac{4\pi}{2S+1}}\left(C_{SSL0}^{SS}\right)^{-s}\left\langle \hat{T}_{LM}^{\dagger(S)}\right\rangle(t). \quad (1.580)$$

The quantum distribution function $(2S+1)W_S^{(s)}(\vartheta,\varphi,t)/4\pi$ given by the general finite series of Eq. (1.579) clearly has a similar form to the conventional Fourier expansion of the classical orientational distribution $W(\vartheta, \varphi, t)$, Eq. (1.538), and indeed reduces to it in the classical limit, $S\to\infty$. Furthermore, the quantum distribution Eq. (1.579) in representation space is a general result *valid* for an *arbitrary* spin system described by a density matrix $\hat{\rho}_S(t)$. Therefore, the phase-space representation allows one (as already discussed for particles) to describe spin systems via a *quasiprobability density function* $W_S^{(s)}(\vartheta,\varphi,t)$ of spin orientations in the phase (here configuration) space (ϑ, φ). The virtue of such a mapping of the density matrix onto $W_S^{(s)}(\vartheta,\varphi,t)$ as extensively used in quantum optics (see, e.g., Ref. [265]) is that one can now predict how $W_S^{(s)}(\vartheta,\varphi,t)$ *evolves as a function of spin number S*. In addition for large spins, $(2S+1)W_S^{(s)}(\vartheta,\varphi,t)/4\pi$ as defined by Eq. (1.579) transparently reduces to the classical distribution $W(\vartheta, \varphi, t)$, Eq. (1.538), thereby naturally linking the quantum and classical regimes.

Knowing the Wigner–Stratonovich kernel $\hat{w}_s(\vartheta,\varphi)$ of the bijective map, we can also calculate the Weyl symbol $A^{(s)}(\vartheta,\varphi)$ of any desired spin operator $\hat{A}$ as (see the definition Eq. (1.526))

$$A^{(s)}(\vartheta,\varphi)=\mathrm{Tr}\left(\hat{A}\hat{w}_s(\vartheta,\varphi)\right). \quad (1.581)$$

Conversely, as we shall immediately justify via Eqs. (1.584)–(1.586) below, the operator $\hat{A}$ can be reconstructed from its Weyl symbol $A^{(s)}(\vartheta,\varphi)$ via the inverse Wigner–Stratonovich map

$$\hat{A}=\frac{2S+1}{4\pi}\int_0^{\pi}\int_0^{2\pi}A^{(-s)}(\vartheta,\varphi)\hat{w}_s(\vartheta,\varphi)\sin\vartheta d\vartheta d\varphi. \quad (1.582)$$

In particular, via the inverse map Eq. (1.582) the density matrix $\hat{\rho}_S(t)$ (like any other quantum operator) may then be neatly expressed as

$$\hat{\rho}_S(t) = \frac{2S+1}{4\pi}\int_0^{\pi}\int_0^{2\pi} W_S^{(-s)}(\vartheta,\varphi,t)\hat{w}_s(\vartheta,\varphi)\sin\vartheta d\vartheta d\varphi. \qquad (1.583)$$

Thus knowing the phase-space distribution $W_S^{(s)}(\vartheta,\varphi,t)$, $\hat{\rho}_S(t)$ in Hilbert space can then be directly reconstructed via the particular inverse Wigner–Stratonovich map Eq. (1.583). This procedure may be justified, as we shall now see, because the Weyl symbols of any two spin operators $\hat{A}$ and $\hat{B}$ will provide the overlap relation [cf. Eq. (1.485) for particles]

$$\frac{2S+1}{4\pi}\int_0^{\pi}\int_0^{2\pi} A^{(s)}(\vartheta,\varphi)B^{(-s)}(\vartheta,\varphi)\sin\vartheta d\vartheta d\varphi = \mathrm{Tr}\left(\hat{A}\hat{B}\right), \qquad (1.584)$$

leading directly to Eqs. (1.582) and (1.583). To justify Eq. (1.584), we substitute the Weyl symbols of the operators $\hat{A}$ and $\hat{B}$ which are defined by the transformation Eq. (1.581) into Eq. (1.584), then we use the operator series expression for the kernel Eq. (1.578) of the bijective map and the orthogonality property of spherical harmonics Eq. (A.78) from Appendix A.1.7. Thus, we have from Eqs. (1.581) and (1.584) the lemma:

$$\begin{aligned}
&\frac{2S+1}{4\pi}\int_0^{\pi}\int_0^{2\pi} A^{(s)}(\vartheta,\varphi)B^{(-s)}(\vartheta,\varphi)\sin\vartheta d\vartheta d\varphi \\
&= \sum_{L,L'=0}^{2S}\sum_{M,M'=-L}^{L}\left(C_{SSL'0}^{SS}/C_{SSL0}^{SS}\right)^s \mathrm{Tr}\left(\hat{A}\hat{T}_{LM}^{(S)}\right)\mathrm{Tr}\left(\hat{B}\hat{T}_{L'M'}^{\dagger(S)}\right) \\
&\qquad\times\int_0^{\pi}\int_0^{2\pi} Y_{LM}^*(\vartheta,\varphi)Y_{L'M'}(\vartheta,\varphi)\sin\vartheta d\vartheta d\varphi \\
&= \sum_{L=0}^{2S}\sum_{M=-L}^{L}\mathrm{Tr}\left(\hat{A}\hat{T}_{LM}^{(S)}\right)\mathrm{Tr}\left(\hat{B}\hat{T}_{LM}^{\dagger(S)}\right) = \sum_{L=0}^{2S}\sum_{M=-L}^{L}\sum_{m,n,i,j=-S}^{S} A_{mi}B_{nj}\left[\hat{T}_{LM}^{(S)}\right]_{im}\left[\hat{T}_{LM}^{\dagger(S)}\right]_{jn} \\
&= \sum_{m,n,i,j=-S}^{S} A_{mi}B_{nj}\sum_{L=0}^{2S}\sum_{M=-L}^{L}(-1)^M\frac{2L+1}{2S+1}C_{SnL-M}^{Sj}C_{SmLM}^{Si} \\
&= \sum_{m,n=-S}^{S} A_{mn}B_{nm} = \mathrm{Tr}\left(\hat{A}\hat{B}\right).
\end{aligned} \qquad (1.585)$$

Here we have used Eq. (A.143) from Appendix A.2 for the matrix elements of the polarization operators $[\hat{T}_{LM}^{(S)}]_{im}$ and $[\hat{T}_{LM}^{\dagger(S)}]_{jn}$ as well as the following property of Clebsch–Gordan coefficients [104]

$$\sum_{L=0}^{2S}\sum_{M=-L}^{L}(-1)^{M}\frac{2L+1}{2S+1}C_{SnL-M}^{Sj}C_{SmLM}^{Si}=\sum_{L=0}^{2S}\sum_{M=-L}^{L}C_{S-jSn}^{LM}C_{S-mSi}^{LM}=\delta_{in}\delta_{mj}.$$

Thus, according to Eq. (1.584), with obvious replacements the *average* value $\langle\hat{A}\rangle=\mathrm{Tr}(\hat{\rho}_S\hat{A})$ of a spin operator $\hat{A}$ is rendered in the simple phase-space integral form

$$\langle\hat{A}\rangle=\frac{2S+1}{4\pi}\int_0^{\pi}\int_0^{2\pi}A^{(-s)}(\vartheta,\varphi)W_S^{(s)}(\vartheta,\varphi,t)\sin\vartheta d\vartheta d\varphi. \qquad (1.586)$$

Here the observable $\langle\hat{A}\rangle$ calculated from Eq. (1.586) is *independent* of the particular value of the parameter s chosen, i.e., the Q-, Wigner-, and P-functions will all yield identical results for $\langle\hat{A}\rangle$ as they must do.

In summary, the *one to one correspondence between the quantum state in the Hilbert space and a real representation space function* first envisaged for the closed system in the spin context by Stratonovich [273], *formally represents the quantum mechanics of a spin as a statistical theory in the classically-meaningful representation space of polar angles* (ϑ, φ). This conclusion is arrived at essentially in the manner of Wigner [6] whom we recall formally represented the quantum mechanics of a particle with separable and additive Hamiltonian $\hat{H}=\hat{p}^2/2m+V(\hat{q})$ as a statistical theory in phase space with the canonical variables (p, q). Clearly, the average value of a quantum spin operator may be calculated from Eq. (1.586) like in classical mechanics. Thus, the Stratonovich representation for spins [273] just as the Wigner representation for particles is particularly well suited to the development of semiclassical methods of solution allowing one to obtain *quantum corrections for finite spin number S in a manner closely analogous to the classical case*, $S\to\infty$ (see, e.g., [97]). We emphasize that besides spin relaxation of assemblies of noninteracting spins in contact with the thermal bath, the phase-space formalism can also be applied to related problems such as spin waves, interacting spins with Heisenberg coupling, etc. (see e.g., [246,275,328]).

1.9.7 *Weyl symbols of some spin operators*

As we have seen any spin operator $\hat{A}$ may be associated via the general Eq. (1.586) for $\langle\hat{A}\rangle$ with its Weyl symbol (*c*-number function) $A^{(s)}(\vartheta,\varphi)$ in the (classically-meaningful) representation space. Here we explicitly evaluate the *Weyl symbols* of the spin operators $\hat{S}_X$, $\hat{S}_Y$, $\hat{S}_Z$, and $\hat{\mathbf{S}}$ and also those of some other simple model spin Hamiltonians $\hat{H}_S$. Using the definition of the spin operators in terms of the polarization operators given by Eq. (A.143) from Appendix A.2, the Weyl symbols $S_X^{(s)}(\vartheta,\varphi)$, $S_Y^{(s)}(\vartheta,\varphi)$, $S_Z^{(s)}(\vartheta,\varphi)$, and $\mathbf{S}^{(s)}(\vartheta,\varphi)$ of the corresponding spin operators $\hat{S}_X$, $\hat{S}_Y$, $\hat{S}_Z$, and $\hat{\mathbf{S}}$ can be calculated from the general mapping for $A^{(s)}(\vartheta,\varphi)$ that is Eq. (1.581) for $s=0,\pm 1$ via the orthogonality property of the polarization operators defined by Eq. (A.160) from Appendix A.2. Thus, we have their Weyl symbols rendered as the following simple maps from Hilbert space onto phase space [253]

$$\hat{S}_X \to S_X^{(s)}(\vartheta,\varphi) = \sqrt{\frac{2\pi S(S+1)}{3}} \sum_{L=0}^{2S} \sum_{M=-L}^{L} \left(C_{SSL0}^{SS}\right)^{-s} Y_{LM}^{*}(\vartheta,\varphi) \times \mathrm{Tr}\left\{(\hat{T}_{1-1}^{(S)} - \hat{T}_{11}^{(S)})\hat{T}_{LM}^{(S)}\right\} = S\left(\frac{S+1}{S}\right)^{(s+1)/2} \sin\vartheta\cos\varphi, \tag{1.587}$$

$$\hat{S}_Y \to S_Y^{(s)}(\vartheta,\varphi) = i\sqrt{\frac{2\pi S(S+1)}{3}} \sum_{L=0}^{2S} \sum_{M=-L}^{L} \left(C_{SSL0}^{SS}\right)^{-s} Y_{LM}^{*}(\vartheta,\varphi) \times \mathrm{Tr}\left\{(\hat{T}_{1-1}^{(S)} + \hat{T}_{11}^{(S)})\hat{T}_{LM}^{(S)}\right\} = S\left(\frac{S+1}{S}\right)^{(s+1)/2} \sin\vartheta\sin\varphi, \tag{1.588}$$

$$\hat{S}_Z \to S_Z^{(s)}(\vartheta,\varphi) = \sqrt{\frac{4\pi S(S+1)}{3}} \sum_{L=0}^{2S} \sum_{M=-L}^{L} \left(C_{SSL0}^{SS}\right)^{-s} Y_{LM}^{*}(\vartheta,\varphi) \mathrm{Tr}\left\{\hat{T}_{10}^{(S)}\hat{T}_{LM}^{(S)}\right\}$$

$$= \sqrt{\frac{4\pi}{3}} S\left(\frac{S+1}{S}\right)^{(s+1)/2} Y_{10}(\vartheta,\varphi) = S\left(\frac{S+1}{S}\right)^{(s+1)/2} \cos\vartheta. \tag{1.589}$$

Clearly, these phase-space mappings of the spin component operators bear a very close resemblance to the corresponding classical quantities. Moreover, we have the overall compact form of the phase-space mapping of the spin operator $\hat{\mathbf{S}}$, viz.,

$$\hat{\mathbf{S}} \to \mathbf{S}^{(s)}(\vartheta,\varphi) = S\left(\frac{S+1}{S}\right)^{(s+1)/2} \begin{pmatrix} \sin\vartheta\cos\varphi \\ \sin\vartheta\sin\varphi \\ \cos\vartheta \end{pmatrix} = S\left(\frac{S+1}{S}\right)^{(s+1)/2} \mathbf{u}. \quad (1.590)$$

Since the magnetic moment operator $\hat{\boldsymbol{\mu}}$ is simply defined in terms of the spin operator $\hat{\mathbf{S}}$ as $\hat{\boldsymbol{\mu}} = \mu_0^{-1}\gamma\hbar\hat{\mathbf{S}}$, it then follows that the *Weyl symbol* $\boldsymbol{\mu}^{(s)} = \mu^{(s)}\mathbf{u}$ *of* $\hat{\boldsymbol{\mu}}$ *has essentially the form of the magnetic moment vector* $\boldsymbol{\mu}$ *for a classical spin* $\left[\mu^{(s)} = \mu_0^{-1}\gamma\hbar S(1+S^{-1})^{(s+1)/2}\right]$.

As further examples of the mapping procedure from Hilbert space onto phase space, we evaluate the Weyl symbols for uniaxial, biaxial, cubic, and mixed anisotropy Hamiltonians defined, respectively, as

$$\beta\hat{H}_S^{\text{un}} = -\frac{\sigma}{S^2}\hat{S}_Z^2, \quad (1.591)$$

$$\beta\hat{H}_S^{\text{bi}} = -\frac{\sigma}{S^2}\hat{S}_Z^2 + \frac{\delta}{S^2}\left(\hat{S}_X^2 - \hat{S}_Y^2\right), \quad (1.592)$$

$$\beta\hat{H}_S^{\text{cub}} = -\frac{\sigma_{\text{c}}}{2S^4}\left(\hat{S}_X^4 + \hat{S}_Y^4 + \hat{S}_Z^4\right), \quad (1.593)$$

$$\beta\hat{H}_S^{\text{mix}} = -\frac{\sigma_1}{S^2}\hat{S}_Z^2 - \frac{\sigma_2}{S^4}\hat{S}_Z^4 + \frac{\chi}{S^4}\left(\hat{S}_{+1}^4 + \hat{S}_{-1}^4\right), \quad (1.594)$$

where σ, δ, $\sigma_{\text{c}}, \sigma_1, \sigma_2$, and χ are dimensionless anisotropy parameters. We cite the following practical examples: the uniaxial Hamiltonian Eq. (1.591) is commonly used, e.g., to describe the magnetic properties of the dodecanuclear manganese molecular cluster Mn12 with $S = 10$, $\sigma_1 TS^{-2} = 0.6 \div 0.7\,\text{K}$ [329,330], while the biaxial Hamiltonian $\hat{H}_S^{\text{bi}}$, Eq. (1.592), is used to describe the magnetic properties of an octanuclear iron(III) molecular cluster Fe8 [20,331] with $S = 10$, $\sigma TS^{-2} = 0.275\,\text{K}$ and $\delta TS^{-2} = 0.046\,\text{K}$. Finally the cubic anisotropy Hamiltonian Eq. (1.593) contributes to the mixed anisotropy Hamiltonian Eq. (1.594), which is used to describe the magnetic properties of the dodecanuclear manganese molecular cluster Mn12 with $\chi TS^{-4} = \pm 3\cdot 10^{-5}\,K$, $\sigma_1 TS^{-2} = 0.56\,K$, $\sigma_2 TS^{-4} = 1.1\cdot 10^{-3}\,K$, and $S = 10$ [331].

Here for simplicity, we only evaluate the Weyl symbols for the Hamiltonians Eqs. (1.591)–(1.594) for the Q-function corresponding to

$s=-1$ (for $s=0$ and $s=+1$ the calculations can be accomplished in like manner). Thus, the Weyl symbols $H_S^{\text{un}}(\vartheta,\varphi)$, $H_S^{\text{bi}}(\vartheta,\varphi)$, $H_S^{\text{cub}}(\vartheta,\varphi)$, and $H_S^{\text{mix}}(\vartheta,\varphi)$ corresponding to the respective Hamiltonians Eqs. (1.591)–(1.594) can now be calculated from the general finite series representation of the kernel $\hat{w}_s$ [cf. Eq. (1.578)]

$$\begin{aligned} H_S^{(-1)}(\vartheta,\varphi) &= \mathrm{Tr}\left\{\hat{H}_S \hat{w}_{-1}(\vartheta,\varphi)\right\} \\ &= \sqrt{\frac{4\pi}{2S+1}}\mathrm{Tr}\left\{\hat{H}_S \sum_{L=0}^{2S}\sum_{M=-L}^{L} C_{SSL0}^{SS} Y_{LM}^*(\vartheta,\varphi)\hat{T}_{LM}^{(S)}\right\} \end{aligned} \tag{1.595}$$

[cf. the mapping Eq. (1.581) for the phase-space representation of an arbitrary spin operator $\hat{A}$]. Hence, we obtain from the Hamiltonian operators given by Eqs. (1.591)–(1.594) and the general mapping Eq. (1.595) after some algebra involving both products of the polarization operators and their orthogonality relations as described in Appendix A.2 the following *explicit* maps from Hilbert space onto phase space

$$\beta\hat{H}_S^{\text{un}} \underset{s=-1}{\rightarrow} \beta H_S^{\text{un}}(\vartheta,\varphi) = -\sigma\frac{S-1/2}{S}\cos^2\vartheta - \frac{\sigma}{2S}, \tag{1.596}$$

$$\begin{aligned} &\beta\hat{H}_S^{\text{bi}} \underset{s=-1}{\rightarrow} \beta H_S^{\text{bi}}(\vartheta,\varphi) \\ &= -\frac{2S-1}{2S}\left(\sigma\cos^2\vartheta - \delta\cos 2\varphi\sin^2\vartheta\right) - \frac{\sigma}{2S}, \end{aligned} \tag{1.597}$$

$$\begin{aligned} &\beta\hat{H}_S^{\text{cub}} \underset{s=-1}{\rightarrow} \beta H_S^{\text{cub}}(\vartheta,\varphi) = -\sigma_c\frac{2S^3+3S-1}{4S^3} \\ &+\sigma_c\frac{(S-1/2)(S-1)(S-3/2)}{4S^3}\left(\sin^2 2\vartheta + \sin^4\vartheta\sin^2 2\varphi\right), \end{aligned} \tag{1.598}$$

$$\begin{aligned} &\beta\hat{H}_S^{\text{mix}} \underset{s=-1}{\rightarrow} \beta H_S^{\text{mix}}(\vartheta,\varphi) \\ &= -\frac{1}{4S}\left(2\sigma_1 + \sigma_2\frac{3S-1}{S^2}\right) - \frac{S-1/2}{S}\left(\sigma_1 + \sigma_2\frac{3S-2}{S^2}\right)\cos^2\vartheta \\ &-\frac{(S-1/2)(S-1)(S-3/2)}{2S^3}\left[2\sigma_2\cos^4\vartheta - \chi\sin^4\vartheta\cos 4\varphi\right]. \end{aligned} \tag{1.599}$$

Again, the Weyl symbols of these quantum Hamiltonians resemble the classical free energies of the corresponding magnetocrystalline anisotropies (see, e.g., Ref. [239]). All these Weyl symbols will be used later on in Chapter 3.

1.9.8 *Master equation and statistical moment equations for spin relaxation in phase space*

By transforming the reduced density operator evolution which is Eq. (1.411) into phase space via the Wigner–Stratonovich map Eqs. (1.577) and (1.583), the phase-space evolution (master) equation for $W_S^{(s)}(\vartheta,\varphi,t)$ may be *formally* written as

$$\frac{\partial W_S^{(s)}}{\partial t} = \mathrm{L}_S W_S^{(s)}, \tag{1.600}$$

where L_S is the phase-space differential operator *corresponding* to the operator in Hilbert space $-i[\hat{H}_S,\hat{\rho}_S]/\hbar+\mathrm{St}(\hat{\rho}_S)$ in the density matrix evolution Eq. (1.411). Although L_S will have, in general, a very complicated form even for circular symmetry (see Chapter 3), nevertheless the phase-space master Eq. (1.600) still has some obvious practical advantages over the reduced density matrix evolution Eq. (1.411) because it is now possible to treat the spin relaxation for arbitrary spin number S in the manner of classical spins [e.g., Eqs. (1.587)–(1.590)] (see Appendix A.3 for a specific nontrivial example). Indeed, in the classical limit, the phase-space evolution equation (1.600) just becomes the Fokker–Planck equation for the distribution function $W_{\mathrm{cl}}(\vartheta,\varphi,t)$ of the orientations of classical spins thereby naturally linking the quantum and classical regimes. The analogy between the quantum and classical formulations for spins arising from the Wigner–Stratonovich map obviously enables the previously developed powerful methods of solution of classical Fokker–Planck equations for the rotational Brownian motion of classical magnetic dipoles (e.g., continued fractions, mean first-passage times, etc. [97]) to be carried over to the quantum domain [220,221,256-264].

As seen, the formal solution of the evolution Eq. (1.411) for the reduced density matrix can be written, using the polarization operators $\hat{T}_{LM}^{(S)}$ and the statistical moments $\left\langle \hat{T}_{LM}^{\dagger(S)} \right\rangle(t)$, as the finite linear combination given by Eq. (1.440) [104,217]. The $\left\langle \hat{T}_{LM}^{\dagger(S)} \right\rangle(t)$ for a given Hamiltonian can then be evaluated (usually after lengthy operator algebra) from the differential-recurrence Eq. (1.447). However, the statistical moment method can also be applied in analogous fashion to the purely *formal* phase-space master equation (1.600) because the phase-space distribution $W_S^{(s)}(\vartheta,\varphi,t)$ may be written for arbitrary S in terms of a *finite* linear combination of the spherical harmonics, Eq. (1.579). Then by using the identity Eq. (1.580) relating the average spherical harmonics $\left\langle Y_{LM}^* \right\rangle^{(s)}(t)$ and the average polarization operators $\left\langle \hat{T}_{LM}^{\dagger(S)} \right\rangle(t)$, the differential-recurrence equations for $\left\langle Y_{LM}^* \right\rangle^{(s)}(t)$ can be obtained by algebraic transformation from the differential-recurrence Eq. (1.447) for the evolution of the average polarization operators so that the latter just becomes

$$\frac{d}{dt}\left\langle Y_{LM}^* \right\rangle^{(s)}(t) = \sum_{L',M'} p_{L'M';LM} \left\langle Y_{L'M'}^* \right\rangle^{(s)}(t). \tag{1.601}$$

Here the expansion coefficients

$$p_{L'M';LM} = \int_0^{\pi}\int_0^{2\pi} Y_{LM}(\vartheta,\varphi) \mathrm{L}_S Y_{L'M'}^*(\vartheta,\varphi) \sin\vartheta d\vartheta d\varphi \tag{1.602}$$

constitute the matrix elements of the phase-space operator L_S given by

$$p_{L'M';LM} = \left(\frac{C_{SSL0}^{SS}}{C_{SSL'0}^{SS}} \right)^{-s} g_{L'M';LM}, \tag{1.603}$$

where the coefficients $g_{L'M';LM}$ which depend upon the precise form of the Hamiltonian operator $\hat{H}_S$ are defined by the averages indicated by Eq. (1.445). For classical spins, $S \to \infty$, the explicit equation for the matrix elements $p_{L'M';LM}$ for an *arbitrary* free energy has been derived in Ref. [331]. Clearly, Eq. (1.601) is just a phase-space correspondent of Eq. (1.447) governing the evolution of the average polarization operators $\left\langle \hat{T}_{LM}^{\dagger(S)} \right\rangle(t)$. Now Eq. (1.601) written as a linear matrix differential equation can be solved either by direct matrix diagonalization, involving the calculation of the eigenvalues and corresponding eigenvectors of the system matrix, or by the computationally efficient (matrix) continued fraction method [97,143]. However, due to the identity $Y_{lm}^* = (-1)^m Y_{l-m}$

[104], the conjugate Eq. (1.601) can also be rewritten as an evolution equation for the statistical moments $\langle Y_{LM}\rangle^{(s)}(t)$, viz.,

$$\frac{d}{dt}\langle Y_{LM}\rangle^{(s)}(t)=\sum_{L',M'}(-1)^{M'-M}\,p_{L'-M';L-M}\langle Y_{L'M'}\rangle^{(s)}(t). \qquad (1.604)$$

Now the phase-space evolution Eq. (1.601) for the average spherical harmonics $\langle Y_{LM}^{*}\rangle^{(s)}(t)$ and its Hilbert space correspondent Eq. (1.447) for the average polarization operators $\langle \hat{T}_{LM}^{\dagger(S)}\rangle(t)$ are equivalent. Thus, having determined $\langle Y_{LM}^{*}\rangle^{(s)}(t)$ via the phase-space Eqs. (1.579) and (1.600), we also have the reduced density matrix $\hat{\rho}_S$ from the polarization operator expansion Eqs. (1.447) and (1.448) *without formally solving* its evolution Eq. (1.411). Vice-versa, having calculated $\langle \hat{T}_{LM}^{\dagger(S)}\rangle(t)$ from the density matrix evolution Eqs. (1.411) and (1.447), we also have the phase-space distribution $W_S^{(s)}(\vartheta,\varphi,t)$ from Eqs. (1.579) and (1.580) *without solving* the phase-space evolution Eq. (1.600).

According to the finite series phase-space representation Eq. (1.579), all the statistical moments $\langle Y_{LM}\rangle^{(s)}(t)$ are required (in general) to evaluate the phase-space distribution $W_S^{(s)}(\vartheta,\varphi,t)$ for given spin number S. However, for the calculation of *particular* observables only a *few* moments may in practice be necessary. For example, in evaluating the *average* spin operators $\langle \hat{S}_X\rangle(t)$, $\langle \hat{S}_Y\rangle(t)$, and $\langle \hat{S}_Z\rangle(t)$ only the spherical harmonic averages $\langle Y_{10}\rangle^{(s)}(t)$ and $\langle Y_{1\pm1}\rangle^{(s)}(t)$ are required according to the Weyl symbols Eqs. (1.587)–(1.589), namely, [cf. Eq. (1.452) for these averages in terms of polarization operators]

$$\begin{aligned}\langle \hat{S}_X\rangle(t)&=\sqrt{\frac{2\pi}{3}}S\left(\frac{S+1}{S}\right)^{(s+1)/2}\left[\langle Y_{1-1}\rangle^{(s)}(t)-\langle Y_{11}\rangle^{(s)}(t)\right],\\ \langle \hat{S}_Y\rangle(t)&=i\sqrt{\frac{2\pi}{3}}S\left(\frac{S+1}{S}\right)^{(s+1)/2}\left[\langle Y_{1-1}\rangle^{(s)}(t)+\langle Y_{11}\rangle^{(s)}(t)\right],\\ \langle \hat{S}_Z\rangle(t)&=\sqrt{\frac{4\pi}{3}}S\left(\frac{S+1}{S}\right)^{(s+1)/2}\langle Y_{10}\rangle^{(s)}(t).\end{aligned} \qquad (1.605)$$

The above results are both formal and general. Specific applications of Eqs. (1.600)–(1.605) are given in Chapter 3, where we shall demonstrate how the phase-space distributions for *particular* spin systems can be determined. Notice that the representation space analysis for spins is intrinsically much more complicated than that for particles because both spin and polarization operators are involved. Furthermore, the Wigner correspondents must always be evaluated *from first principles for a given Hamiltonian* which in general is *neither separable nor additive*.

1.10 Conclusion and Summary

In this chapter, mainly of an introductory nature, we have summarized the principal elements of the theory (both classical and quantum) of stochastic spin dynamics as applied to the magnetization of fine particles. Thus, starting with the Larmor equation of the familiar deterministic theory we have justified, using the classical theory of the Brownian motion, the magnetic Langevin and Fokker–Planck equations and have given the general solution for the magnetization response in terms of the statistical moments $\langle Y_{LM} \rangle (t)$ for an *arbitrary* anisotropy-Zeeman energy. Moreover, the moment method may be easily extended to include spin torque-transfer effects and the formal general solution for the observables may be determined via matrix continued fractions. We have also described how the various Kramers escape-rate calculations for Brownian particles may be adapted to classical spins thereby yielding closed form asymptotic solutions for the overbarrier relaxation time which *inter alia* serve as a useful check on the accuracy of the semi-analytic numerical solutions obtained via continued fractions. We have also laid emphasis on the energy-controlled diffusion regime leading to the very low damping escape rate which is the most relevant one for magnetic relaxation both with and without STT. Thus, we have hopefully prepared the reader for the various important classical cases which we shall treat in depth in Chapter 2.

Proceeding to quantum effects we began by introducing the density matrix formulation for the calculation of expectation values of quantum operators in Hilbert space and how it arises naturally from the quantum mechanical Liouville equation. We then indicated how the reduced density matrix evolution equation describing the spin relaxation in Hilbert space under the influence of a thermal bath, may be justified by using reduction methods as explained by Nitzan [183]. Moreover, we have treated (by adapting methods originally due to Hubbard [235]) instances where the spin Hamiltonian operator does not depend explicitly on the time. Secondly, we have considered time-dependent instances, corresponding to the high temperatures and short correlation times embodied in the Markov approximation, leading to the direct quantum analog of the classical Fokker–Planck equation for the rotational diffusion of a magnetic dipole. Then (with the declared objective of extending the method of statistical moments to quantum systems so as to study the magnetization evolution as a function of spin size) we expanded the solution of the reduced spin density matrix evolution equation as a finite sum of polarization operators $\hat{T}_{LM}^{(S)}$. This expansion after some manipulation, renders the solution for the moments $\left\langle \hat{T}_{LM}^{(S)} \right\rangle(t)$ in a manner exactly analogous to the moments $\left\langle Y_{LM} \right\rangle(t)$ of the classical case so that $\left\langle \hat{T}_{LM}^{(S)} \right\rangle(t)$ may then be calculated by matrix inversion again as in the classical case. Here we have recalled that the $\hat{T}_{LM}^{(S)}$ are merely the quantum analogs in spin (Hilbert) space of the spherical harmonics Y_{LM} in configuration (phase) space and obey similar commutation relations, when the latter are regarded as just the components of some irreducible tensor of rank L. A fact which is absolutely necessary for the phase-space representation (ϑ, φ) of the Hilbert space reduced density matrix evolution equation.

We next describe the basic concepts of phase-space quasiprobability density evolution equations for particles of mass m with *separable* and *additive* Hamiltonians $\hat{H} = \hat{p}^2/(2m) + \hat{V}$ moving in a one-dimensional extension, via the overlap Fourier transform of the density matrix $\hat{\rho}$. This transform to position q and momentum p variables maps the density operator $\hat{\rho}$ in Hilbert space onto the quasiprobability representation W in phase space (q, p) and is accompanied by a well-defined inverse transform, mapping phase space or c-number representations back to Hilbert space (cf. Eqs. (1.478)–(1.479)). Moreover, the inverse transform implies that the trace of the product of two quantum operators is the integral over the phase space of the product of their Weyl symbols (see Eq. (1.485)). *Clearly this formulation allows one to calculate quantum mechanical expectation values for particles in terms of their c-number representations, in the same way as the corresponding classical averages*, via the Weyl symbol $A(q, p)$ of the operator $\hat{A}$ (see Eq. (1.482)). Having so discussed the main features of the phase-space representation for one-dimensional systems with symmetry described by the Heisenberg–Weyl group we derive the Wigner–Moyal equation for particles which is simply a quantum analog of the classical Liouville equation. Then we show how the former equation in the time-independent case may be used to calculate quantum corrections to the classical equilibrium distribution in the perturbative manner of Wigner [6]. Subsequently following some examples of the calculation of Wigner functions for simple systems, the Wigner function formulation is extended to quantum Brownian motion with collision kernel given by Eq. (1.515). The time-dependent treatment is realized by postulating that the equilibrium Wigner distribution is the stationary solution of the time-dependent Wigner–Moyal equation, so yielding the various drift and diffusion parameters in the collision term St (W) of that equation in the manner long familiar in the classical theory of Brownian motion.

Turning now to quasiprobability distribution functions for spins we have comprehensively based our treatment on Stratonovich's [273] general discussion of c-number quasiprobability distributions for quantum systems in a representation space based on the symmetry properties of the underlying group. For spins this is the SU(2) group for rotations indicating that the Hilbert space operators which transform under rotation in the same way as the $Y_{LM}(\vartheta,\varphi)$, are the polarization operators $\hat{T}_{LM}^{(S)}$. Thus we have the desired quasiprobability distribution function for spins in the form of Eq. (1.548). In other words, we have obtained quasiprobability distribution function for spins by identifying the average of the spherical harmonic $Y_{LM}(\vartheta,\varphi)$ over the representation space (ϑ, φ) as an average of the polarization operator $\hat{T}_{LM}^{(S)}$ via the spin density matrix $\hat{\rho}_S$. Therefore, we have the representation space distribution $W_S(\vartheta,\varphi,t)$ for spins in classical-like finite series form in S via the Wigner–Stratonovich transformation Eq. (1.547), namely, Eq. (1.548). Now since many spin relaxation problems involve interaction of an assembly of spins with electromagnetic or anisotropy fields and since we are interested in *semiclassical* behavior we select the spin coherent (minimum uncertainty) states which being a complete set ultimately render the matrix elements $\langle S,\vartheta,\varphi|\hat{\rho}_S|S,\vartheta,\varphi\rangle$ of the density matrix $\hat{\rho}_S$ as a finite series of spherical harmonics in the spin coherent state representation given (for the

Q-function) by Eq. (1.569) and indeed for all three forms (Q-, P-, Wigner) of the possible representation distributions by the general form Eqs. (1.579)–(1.580). This general form $W_S^{(s)}(\vartheta,\varphi,t)$ ($s = 0, \pm 1$) directly relates to the Fourier expansion of the classical orientational distribution $W(\vartheta, \varphi, t)$ and, moreover, reduces to it in the classical limit, where the spin number $S \to \infty$. Having determined the representation distributions, we then showed how they lead to the average value of a spin operator $\hat{A}$ in terms of its Weyl symbol $A^{(s)}(\vartheta,\varphi)$ (cf. Eq. (1.586)) because $\hat{A}$ can be reconstructed from $A^{(s)}(\vartheta,\varphi)$ via the inverse Wigner–Stratonovich map, Eq. (1.582). Thus, we have demonstrated that the one-to-one correspondence between the quantum state in the Hilbert space and a real representation space function first envisaged for the closed system in the spin context by Stratonovich [273], formally represents the quantum mechanics of a spin as a statistical theory in the representation space of polar angles (ϑ, φ). We have illustrated the calculation of Weyl symbols via the general map Eq. (1.581), namely,

$$A^{(s)}(\vartheta,\varphi)=\mathrm{Tr}\left(\hat{A}\hat{w}_s(\vartheta,\varphi)\right)$$

(where $\hat{w}_s$ is the kernel of the Wigner–Stratonovich transformation) for several Hamiltonians and their spin operators commonly used in fine particle magnetism. Finally, we have formally demonstrated how the time-dependent phase-space master or evolution equation for the average spherical harmonics and its Hilbert space correspondents for the average polarization operators are equivalent. This ends our description of the general foundations of spin relaxation which is treated in detail for various classical spin problems in Chapter 2 and for the quantum ones described in Chapter 3.

Chapter 2

Magnetization relaxation and resonance in nanomagnets

2.1 Introduction

We shall now consider the rotational diffusion and resonant behavior of classical spins in an external potential of the Zeeman-magnetocrystalline anisotropy energy, from the point of view of the magnetization dynamics of nanomagnets. We have seen that these are characterized by thermal instability of the magnetization due to thermal agitation (see Chapter 1, Sections 1.3), resulting in *superparamagnetism* because each nanomagnet behaves as a *paramagnetic atom of enormous magnetic moment* ($\sim 10^4$–10^5 Bohr magnetons). Thus, the magnetization is visualized as a (magnetization) stochastic process due to thermal fluctuations which may then be treated using the theory of the rotational Brownian motion. Hence from a mathematical point of view analysis of the ensuing stochastic magnetization dynamics will rely on two complementary approaches based on the magnetic Langevin and Fokker–Planck equations, respectively. The magnetic Langevin equation is just the Gilbert equation augmented by stochastic terms owing to the Brownian motion while the associated magnetic Fokker–Planck equation governs the spatial distribution of magnetization orientations on the unit sphere and is derived from the Langevin equation using the *Stratonovich interpretation rule*. Moreover, spin-torque effects may be easily incorporated in either approach (see Chapter 1, Sections 1.3.3). We shall also show using the foregoing methods how the stochastic magnetization dynamics manifests itself in the dynamic susceptibilities (linear and nonlinear), in stochastic resonance, in dynamic hysteresis, in magnetization reorientation times for different anisotropies, etc. We shall also describe various numerical methods for the computation of these and other characteristics of nanomagnets as well as various closed form approximations for the numerical results and their range of applicability. Finally, from a theoretical viewpoint the new knowledge obtained by investigating such nonequilibrium processes in nanomagnets will yield valuable insights leading to further development of nonequilibrium statistical physics. From a practical viewpoint the results will suggest more methods for manipulating spin degrees of freedom and for designing novel spintronic devices.

Here we wish to analyze in detail the dynamical behavior of the magnetization, i.e., the giant classical spin response for a given anisotropy and Zeeman energy density. This analysis involves the determination of the greatest relaxation (reversal) time of the magnetization characterized by the inverse of the smallest nonvanishing eigenvalue associated with the slowest (overbarrier) mode as well as all the other eigenvalues and the corresponding eigenfunctions governing phenomena such as magnetization decay, stochastic resonance, dynamic magnetic hysteresis, etc. Thus, it is essential to have an easily implemented and, moreover, foolproof method of determining the differential-recurrence relations for the statistical moments which ultimately govern these phenomena. The former are generated either by separating the variables in the Fokker–Planck equation for the surface transition probability distribution of magnetization orientations on the unit sphere or by averaging the magnetic Langevin equation (written as the equation of motion of a spherical harmonic) over an infinitesimal time subject to a sharp initial distribution of orientations over its realizations in phase (representation) space, afterwards performing a second (spatial) average over a distribution of the sharp values. The averaging must be done for the particular free energy density potential encountered in a given problem. However, for all but the simplest (i.e., circularly symmetric) potentials, this task poses a complicated algebraic problem originating in the enormous number of terms generated by the product of the spherical harmonics of the same argument but of different rank. Nevertheless, the algebraic problem has been largely circumvented (*v.* Chapter 1, Section 1.3.4). This is accomplished by posing the problem for an *arbitrary* anisotropy-Zeeman potential which may be expanded in spherical harmonics. Then the general problem may be treated either by averaging the Langevin equation, written in the Stratonovich interpretation, as the stochastic equation of motion of a spherical harmonic, viz. $Y_{lm}(\vartheta,\varphi)$, given a sharp set of initial conditions, over its realizations in phase space. The alternative method is to separate the variables in the accompanying Fokker–Planck equation by expanding the transition probability as a Fourier series in the $Y_{lm}(\vartheta,\varphi)$ (the principal steps are shown schematically in Table 2.1).

Table 2.1. (a) Principal steps of the Fokker–Planck equation method

Fokker–Planck equation for the surface probability distribution of orientations $\dot{W}(\vartheta,\varphi,t)=\mathrm{L}_{\mathrm{FP}}W(\vartheta,\varphi,t)$
Subsequent substitution in the Fokker–Planck equation of the expansion of the distribution of spin orientations as a Fourier series of the $Y_{lm}(\vartheta,\varphi)$, viz., $W(\vartheta,\varphi,t)=\sum_{l,m}c_{lm}(t)Y^{*}_{lm}(\vartheta,\varphi)$ then yields the ordinary differential-recurrence relation for the statistical moments $c_{lm}(t)=\int_0^{2\pi}\int_0^{\pi}Y_{lm}(\vartheta,\varphi)W(\vartheta,\varphi,t)d\vartheta d\varphi\equiv\langle Y_{lm}\rangle(t)$, viz., $\dot{c}_{lm}(t)=\sum_{l',m'}q_{lml'm'}c_{l'm'}(t)$

(b) Principal steps of the Langevin equation method

The Langevin equation for polar angles represents the stochastic differential equations: $\dot{\vartheta}(t)=f_1(\vartheta,\varphi)$ and $\dot{\varphi}(t)=f_2(\vartheta,\varphi)$
Hence the Langevin equation for spherical harmonics becomes: $\dot{Y}_{lm}(\vartheta(t),\varphi(t))=\dot{\vartheta}(t)\partial_{\vartheta}Y_{lm}+\dot{\varphi}(t)\partial_{\varphi}Y_{lm}=f\left[\vartheta(t),\varphi(t)\right]$
Averaging of stochastic variables over their realizations in an infinitesimal time given, a sharp set ϑ, φ of initial values yields differential-recurrence equations for the evolution of $Y_{lm}(\vartheta,\varphi)$: $\dot{Y}_{lm}(\vartheta,\varphi)=\sum_{l',m'}q_{lml'm'}Y_{l'm'}(\vartheta,\varphi)$
Second averaging over a spatial distribution of the initial values ϑ and φ yields the differential-recurrence relation for the statistical moments $c_{lm}=\langle Y_{lm}\rangle(t)$: $\frac{d}{dt}c_{lm}(t)=\sum_{l',m'}q_{lml'm'}c_{l'm'}(t)$

In either approach, the right-hand side of the Langevin equation for $\dot{Y}_{lm}(\vartheta,\varphi)$ or the Fokker–Planck equation $\dot{W} = \mathrm{L}_{\mathrm{FP}}W$ expanded as a Fourier series of the $Y_{lm}(\vartheta,\varphi)$ (where the Fourier coefficients are time dependent) must initially be written in a form such that only a *product* of the spherical harmonics is involved. This step is accomplished via the theory of angular momentum as outlined explicitly in Section 1.2.5. The advantage of rendering all the terms as such products is that one then may use the Clebsch–Gordan representation [104] of a product of spherical harmonics as a sum (linear combination) of spherical harmonics of different ranks, where the coefficients of the sum are given by the Clebsch–Gordan coefficients. Then the recurrence relations of the spherical harmonics yield the desired multi-term differential-recurrence-equations for the statistical moments (averaged spherical harmonics) in terms of the Clebsch–Gordan coefficients pertaining to the particular free energy density chosen. Since the Clebsch–Gordan coefficients can be easily handled nowadays, e.g., via MATHEMATICA®, the tedious and error-prone manual process of finding the recurrence relations for a given potential may be entirely automated. These scalar recurrence relations can then in turn be converted into a matrix three-term differential-recurrence relation with the formal solution rendered via matrix continued fractions in the frequency domain (cf. Eq. (1.126) of Chapter 1) so yielding the desired observables.

The reader may still quibble about the amount of mathematical manipulation needed to obtain even a formal solution of the evolution equation for an arbitrary potential, let alone the details of particular cases such as cubic anisotropy, etc. Nevertheless, it is necessary to treat them in the automated manner just described, otherwise one has no reliable way of finding accurate numerical solutions. The ultimate purpose being to determine the range of accuracy of, e.g., the various closed form asymptotic results for the greatest relaxation time provided by the Kramers escape-rate theory, etc. This is so because the escape rate yields in the appropriate damping range the asymptotic behavior of the smallest nonvanishing eigenvalue of the characteristic equation of the dynamical system as determined from the governing differential-recurrence relations. Moreover, numerical solutions of the recurrence relations for the statistical moments set benchmarks with which the results of all other methods must concur, e.g., Langevin dynamics simulations, escape-rate theory, etc.

The chapter is arranged as follows. We first evaluate the magnetization reversal time τ at very low damping (VLD) using the mean first-passage time (Section 2.2). We then compute τ for all values of the damping using escape-rate theory (Section 2.3). As particular examples, we evaluate τ for uniaxial, cubic, biaxial, and mixed (i.e., uniaxial + cubic) anisotropy potentials. Here the effect of an external dc magnetic field $\mathbf{H}_0$ on the reversal time of biaxial and uniaxial nanomagnets is also investigated. Furthermore, we consider STT effects in the magnetization reversal via the generic model introduced in Chapter 1, Section 1.3.3. Next (Section 2.4), exact (in terms of matrix continued fractions) and simple approximate analytical solutions of various problems appearing in the magnetization relaxation of uniaxial nanomagnets, subjected to a strong uniform magnetic field $\mathbf{H}_0$ are given. These include the longitudinal and transverse responses, their characteristic times, dynamic susceptibilities, and magnetic stochastic resonance. Then, we treat the magnetization relaxation in nanomagnets with noncircularly symmetric magnetic anisotropy (Section 2.5). Two examples are considered in detail, namely uniaxial nanomagnets subjected to an external dc magnetic field $\mathbf{H}_0$ of *arbitrary* orientation, and nanomagnets with cubic anisotropy (both positive and negative). In all cases, both matrix continued-fraction and simple analytic solutions for the relaxation times and dynamic susceptibilities are given. Then in Section 2.6, we evaluate the nonlinear ac stationary response of the magnetization of nanoparticles in superimposed ac and dc bias magnetic fields of arbitrary strength and orientation. The dynamic magnetic hysteresis in uniaxial and biaxial nanomagnets is treated in Section 2.7. STT effects in the magnetization reversal and the linear and nonlinear ac stationary response are treated in Section 2.8. The magnetization reversal and ac stationary responses of antiferromagnetic nanoparticles are considered in Section 2.9. *Finite barrier corrections* for the FMR frequency of nanomagnets with various magnetocrystalline anisotropies are treated in Section 2.10. In Section 2.11, the relaxation time of the magnetization and other related observables for *two interacting* magnetic moments (macrospins) coupled via exchange and dipole-dipole interactions are calculated. All results are presented in a form suitable for comparison with experimental data.

2.2 Reversal time of the magnetization at VLD: the MFPT approach

We now seek formulas for the magnetization reversal time of nanoparticles with a noncircularly symmetric free energy density, $E(\vartheta, \varphi)$. Consequently, *the dynamics of the transverse response affect the dynamics of the longitudinal response and vice versa*. We first show how escape rates may be calculated in the VLD regime, $\alpha \ll 1$, where the *dynamics are almost entirely determined by the pure gyromagnetic equation*. This damping range is in practice the most important because both experimental and theoretical estimates yield small values of α (of the order of 0.001–0.1) [36,82,83,332,333]. The calculation of the reversal time in the entire damping range will be given in Section 2.3. For clarity, we first calculate the escape rate for circularly symmetric potentials which has the important property *that it is valid for all damping parameters* α as it depends only on one canonical variable.

2.2.1 *Reversal time for circularly symmetric potentials*

For circularly symmetric potentials of the magnetocrystalline anisotropy and Zeeman energy no *intrinsic* dynamical coupling between the longitudinal and the transverse modes of motion exists because the longitudinal modes are governed by a *single* state variable only, namely, the colatitude ϑ, i.e., the polar angle of **M**. The second state variable, namely, the azimuthal angle φ of **M**, gives rise only to a steady precession of **M**. The ensuing decoupling between the transverse and longitudinal modes existing for circular symmetry leads to an exact single-variable Fokker–Planck equation in ϑ, Eq. (1.223). Hence the Kramers escape-rate theory for point particles with one degree of freedom [84,85] may be easily adapted to such spins; the result being an expression for the escape rate for spins in circularly symmetric potentials valid for all values of the damping parameter α. We remark, however, that the *exact* Fokker–Planck equation,

Eq. (1.223), in the single variable ϑ, which has the same mathematical form as the Debye equation for the noninertial rotational diffusion of a sphere [97], *arises not from the strong damping of the angular momentum as in the approximate Smoluchowski equation pertaining to the latter case, rather it follows from the circular symmetry of the potential, i.e., from geometry,*

$$E(\vartheta) = \sigma(\sin^2 \vartheta - 2h\cos\vartheta), \tag{2.1}$$

where $\sigma = vK/(kT)$ and $h = \mu_0 M_S H_0 / (2K)$ are, respectively, the dimensionless anisotropy and external dc bias field parameters. The biased uniaxial potential Eq. (2.1) has two minima at $\vartheta_A = 0$ and $\vartheta_B = \pi$ separated by a potential barrier at $\vartheta_C = \arccos(-h)$. The circularly symmetric potential, Eq. (2.1), was first used by Néel [15,16], who gave an expression for the reversal time using TST and the discrete orientation model approximation of Debye and Fröhlich [97]. This potential was further studied and, moreover, the results flowing from it set in the context of the general theory of stochastic processes by Brown [17,18] who obtained approximate expressions for the reversal time τ in the limit of large and small σ using the Kramers escape-rate method and perturbation theory, respectively. Aharoni [334], Garanin *et al.* [23], Coffey *et al.* [146,148,335], and Klik and Yao [336] and others (see Refs. [337–339]) have reconsidered this problem and have developed various techniques (variational methods, mean first-passage time, etc.; see Chapter 1, Section 1.4.4) to determine the reversal times.

Using the MFPT approach (see Chapter1, Sections 1.4.3–1.4.5), the reversal time in the potential Eq. (2.1) valid for all α is given by Eqs. (1.227) and (1.232). Therefore, for the biased uniaxial potential Eq. (2.1), we have from Eq. (1.227) two distinct MFPTs from the wells A and B [179,340,341], viz.

$$\tau_A = 2\tau_N \int_{-h}^{1} \frac{e^{-\sigma(z^2+2hz)}}{1-z^2} \int_{z}^{1} e^{\sigma(z'^2+2hz')} dz' dz, \tag{2.2}$$

$$\tau_B = 2\tau_N \int_{-1}^{-h} \frac{e^{-\sigma(z^2+2hz)}}{1-z^2} \int_{-1}^{z} e^{\sigma(z'^2+2hz')} dz' dz. \tag{2.3}$$

For high potential barriers $\sigma(1-h^2) >> 1$, Eqs. (2.2) and (2.3) then yield the well-known asymptotic formula [17,92] for the reversal time due to Brown (details in [179]), viz.,

$$\tau_{\text{as}} = \frac{2\tau_A \tau_B}{\tau_A + \tau_B} \sim \frac{\tau_{\text{N}} \sqrt{\pi}}{\sigma^{3/2}(1-h^2)[(1+h)e^{-\sigma(1+h)^2} + (1-h)e^{-\sigma(1-h)^2}]}. \tag{2.4}$$

In the limit $h \to 0$, Eq. (2.4) reduces to the asymptotic formula deduced by Brown [17] for uniaxial nanomagnets, viz.,

$$\tau_{\text{as}} \sim \tau_{\text{N}} \frac{\sqrt{\pi}}{2\sigma^{3/2}} e^{\sigma}. \tag{2.5}$$

The conditions for the applicability of Eqs. (2.4) are $0 \le h < 1$ and $\sigma(1-h^2) >> 1$. Equation (2.4) is also valid for all damping α.

The manifold methods available for circularly symmetric potentials cannot be used if the free energy density is a function of both ϑ and φ so breaking the circular symmetry. Nevertheless, in the VLD regime, it is still possible to write an energy-controlled diffusion equation in a *single* (energy) variable so that it may be solved by quadratures yielding the reversal time via the MFPT (see Section 1.4). Thus, the VLD reversal time valid *for arbitrary free energy E* may be determined from Eq. (1.241). Furthermore, STT effects can be considered via Eq. (1.247).

2.2.2 *VLD reversal time for uniaxial anisotropy with an oblique field*

By way of practical illustration of Eq. (1.241), we consider a *uniaxial* anisotropy potential augmented by the Zeeman term due to an external magnetic field $\mathbf{H}_0$. With the dc magnetic field $\mathbf{H}_0$ applied at an angle ψ to the easy axis, the symmetry is entirely broken so that the free energy E also depends on the azimuthal angle φ (see Fig. 2.1), viz.

$$E(\vartheta,\varphi) = \sigma \sin^2 \vartheta - 2h\sigma(\cos\psi \cos\vartheta + \sin\psi \sin\vartheta \cos\varphi). \tag{2.6}$$

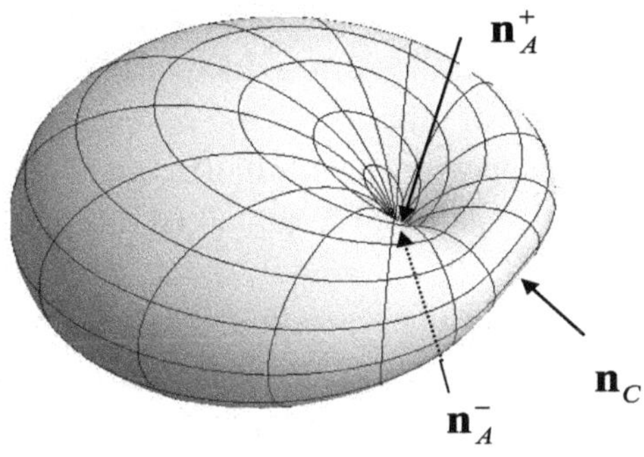

Fig. 2.1. 3D plot of the oblique field potential $E(\vartheta, \varphi)/\sigma$, Eq. (2.6), for the field parameter $h = 0.5$ and $\psi = \pi/2$.

A detailed treatment of this oblique-field problem has been given by Coffey *et al.* [108,342,343], Kennedy [344], Kalmykov *et al.* [32,38,207], and Fukushima *et al.* [345,346] These numerical calculations of the reversal time also agree with computer simulations [347–349] and with experiments [82,83], emphasizing the vital importance of an accurate determination of its damping dependence wherever it is possible as is so in the VLD case. The free energy density E, Eq. (2.6), has a bistable structure with minima at $\mathbf{n}_A^+ = \mathbf{n}_1$ and $\mathbf{n}_A^- = \mathbf{n}_2$ separated by a potential barrier with a saddle point at $\mathbf{n}_C$. The saddle point lies generally in the equatorial region, while $\mathbf{n}_A^+$ and $\mathbf{n}_A^-$ lie in the north and south polar regions, respectively (see Fig. 2.1). In general, the potential Eq. (2.6) retains its *asymmetric bistable* form for $0 < h < h_c$ and $\psi \neq \pi/2$, i.e., the function $E(\vartheta, 0)$ has two minima and one saddle point. Here h_c is some critical (nucleation) value of h at which the potential Eq. (2.6) loses its bistable character [97]. The critical value h_c follows from the condition for a point of inflexion of $E(\vartheta, 0)$ namely $\partial_\vartheta E = \partial_{\vartheta\vartheta}^2 E = 0$, yielding two simultaneous trigonometric equations:

$$\sin 2\vartheta = -2h_c \sin(\vartheta - \psi), \quad \cos 2\vartheta = -h_c \cos(\vartheta - \psi). \tag{2.7}$$

The relevant solution of Eq. (2.7) in the range $(0, \pi)$ is $\tan\vartheta = -(\tan\psi)^{1/3}$ so that

$$h_c = (\cos^{2/3}\psi + \sin^{2/3}\psi)^{-3/2}. \tag{2.8}$$

The two minima have, in general, different energies, so that the energy barriers $\Delta E_{1,2} = E(\vartheta_C, 0) - E(\vartheta_{1,2}, 0)$ are unequal. For $h = h_c$, the second minimum becomes a point of inflexion; if $h > h_c$, the two-well structure of the potential entirely disappears. For $\psi \neq 0$, it is not at all easy to evaluate the barrier heights as a function of h and ψ (see below). However, for particular values of ψ if $\mathbf{H}_0$ is parallel to the easy axis, e.g., $\psi = 0$ corresponding to circular symmetry, we find that $\Delta E_{1,2} = \sigma(1 \pm h)^2$. For $\psi = \pi/2$ the barriers ΔE_1 and ΔE_2 coincide, so that $\Delta E_{1,2} = \sigma(1 - h)^2$ and the potential assumes a *symmetric* bistable form. The stationary points of the potential energy $E(\vartheta, \varphi)$ occur for $\varphi = 0$ and $\varphi = \pi$. The stationary point for $\varphi = \pi$ corresponds to a maximum of $E(\vartheta, \varphi)$ in Eq. (2.6), and so is of no interest to us. The stationary points at $\varphi = 0$, however, correspond to a *saddle point* C at ϑ_C, and two minima at ϑ_1 and ϑ_2 for $h < h_c$, where h_c is some critical (nucleation) value of h at which the potential Eq. (2.6) loses its bistable character (to be determined below). The saddle point C is generally in the equatorial region. The two equilibrium directions of the magnetization and their associated polar angles ϑ_1 and ϑ_2 lie in the x–z plane ($\varphi = 0$) and are determined by the conditions for minima of $E(\vartheta, 0)$ namely $\partial_\vartheta E = 0$, $\partial^2_{\vartheta\vartheta} E > 0$. The position of the saddle point C is determined from the conditions $\partial_\vartheta E(\vartheta, 0) = 0$, $\partial^2_{\vartheta\vartheta} E(\vartheta, 0) < 0$.

Now, to calculate the VLD reversal time from the general VLD solution Eq. (1.241), we need only the *deterministic* equations of motion of the magnetization. For the uniaxial anisotropy potential, Eq. (2.6), the gyromagnetic equation, Eq. (1.4), governing the *deterministic* dynamics can be rewritten in terms of the components (u_X, u_Y, u_Z) of the unit vector $\mathbf{u}$ along the direction of magnetization $\mathbf{M}$ as

$$\tau_0 \frac{d}{dt} \begin{pmatrix} u_X \\ u_Y \\ u_Z \end{pmatrix} = \begin{pmatrix} -(u_Z + h\cos\psi) u_Y \\ (u_Z + h\cos\psi) u_X - h\sin\psi u_Z \\ h\sin\psi u_Y \end{pmatrix}, \tag{2.9}$$

where

$$\tau_0 = \frac{\tau_N}{\sigma(\alpha + \alpha^{-1})} = \frac{\mu_0 M_S}{2\gamma K} \tag{2.10}$$

is a precession time constant. For $\gamma = 2.2\cdot10^5$ m $A^{-1}s^{-1}$, $\mu_0 = 4\pi\cdot10^{-7}$ J m^{-1} A^{-2}, $M_S \approx 1.4\cdot10^6$ A m^{-1} and $K \approx 2\cdot10^5$ J m^{-3} (cobalt), we estimate $\tau_0 \approx 2\cdot10^{-11}$ s. The components u_X, u_Y, u_Z are subject to the constraint $|\mathbf{u}| = 1$. Furthermore, the trajectories of the precessional dynamics must also obey the additional constraint (energy conservation)

$$\varepsilon = -u_Z^2 - 2h\cos\psi u_Z - 2h\sin\psi u_X = \text{const}, \tag{2.11}$$

where $\varepsilon = E/\sigma$ is the normalized free energy.

To find the two *nonequivalent* minima $\varepsilon_A^{\pm}$ and the saddle point ε_C ($|h| < h_c \le 1$) of the normalized free energy (2.11) the equation $\partial_\vartheta E(\vartheta, 0) = 0$ may be rewritten as a quartic equation in $\cos\vartheta$:

$$(\cos\vartheta + h\cos\psi)^2(1-\cos^2\vartheta) = (h\cos\vartheta\sin\psi)^2. \tag{2.12}$$

The roots of this quartic equation, corresponding to two minima and a saddle point, viz., $\cos\vartheta_A^{\pm}$ and $\cos\vartheta_C$ have a complicated algebraic form (see Ref. [108] for details). However, they may be written as convergent Taylor series to any desired order of h ($h < h_c(\psi) \le 1$) [207], viz.,

$$\begin{aligned}
\cos\vartheta_A^{\pm} &= \pm1 \mp \frac{h^2}{2}\sin^2\psi\left[1 \mp 2h\cos\psi + \frac{h^2}{8}\left(13 + 11\cos 2\psi\right) \mp \cdots\right], \\
\cos\vartheta_C &= -h\cos\psi + \frac{h^2}{2}\sin 2\psi\left[1 - h\sin\psi + \frac{h^2}{4}\left(3 - \cos 2\psi\right) + \cdots\right],
\end{aligned} \tag{2.13}$$

also allowing one to calculate $\varepsilon_A^{\pm}$, ε_C, and the well precession frequencies $f_A^{\pm}$ via converging Taylor series, viz.,

$$\begin{aligned}
\varepsilon_A^{\pm} &= -1 \mp 2h\cos\psi - h^2\sin^2\psi\left(1 \mp h\cos\psi + h^2\cos\psi^2 - \ldots\right), \\
\varepsilon_C &= -2h\sin\psi + h^2\cos^2\psi\left(1 + h\sin\psi - 3h^2\sin^2\psi + \ldots\right), \\
f_A^{\pm} &= \frac{1}{2\pi\tau_0}\left\{1 \pm h\cos\psi - (h^2/2)\sin^2\psi\right. \\
&\quad \left.\times\left[1 \mp 3h\cos\psi + (h^2/8)(21 + 19\cos 2\psi) + \ldots\right]\right\},
\end{aligned} \tag{2.14}$$

First, we must again define two *individual* MFPTs, namely, τ_+ from the deeper well (region I: $\varepsilon_A^{+} \le \varepsilon \le \varepsilon_C$) and τ_- from the shallow well (region II: $\varepsilon_A^{-} \le \varepsilon \le \varepsilon_C$). These times are unequal in general, i.e., $\tau_+ \ne \tau_-$.

Thus, Eq. (1.241) from Chapter 1 as specialized to uniaxial anisotropy with the double-well potential Eq. (2.6) becomes for each of the two wells

$$\tau_{\pm} = \frac{\sigma^2}{\alpha} \int_{\varepsilon_A^{\pm}}^{\varepsilon_C} \frac{e^{\sigma\varepsilon}}{S_{\varepsilon}^{\pm}} \int_{\varepsilon_A^{\pm}}^{\varepsilon} \frac{e^{-\sigma\varepsilon'} d\varepsilon'}{f_{\varepsilon'}^{\pm}} d\varepsilon, \tag{2.15}$$

where

$$S_{\varepsilon}^{\pm} = 2\tau_0\sigma \int_0^{1/f_\varepsilon} \left[\dot{u}_{X\pm}^2(t) + \dot{u}_{Y\pm}^2(t) + \dot{u}_{Z\pm}^2(t)\right] dt. \tag{2.16}$$

Hence recalling that *unlike the isolated well two* escape rates Γ_+ and Γ_- from each well now exist, where τ_+ and τ_- are related to the corresponding escape rates from the individual wells via [18,239] $\Gamma_+ = (2\tau_+)^{-1}$ and $\Gamma_- = (2\tau_-)^{-1}$, the overall reversal time of the magnetization $\tau_{\text{VLD}} = (\Gamma_+ + \Gamma_-)^{-1}$, which for high barriers represents the inverse of the smallest nonvanishing eigenvalue λ_1 of the Fokker–Planck operator, is then given by [239]

$$\tau_{\text{VLD}} = \frac{2\tau_+\tau_-}{\tau_+ + \tau_-}. \tag{2.17}$$

Thus τ_{VLD} can be determined from Eqs. (2.15) and (2.17) by calculating the energy-dependent frequency $f_{\varepsilon}^{\pm}$ and action $S_{\varepsilon}^{\pm}$. To accomplish this for the deeper well, we first introduce a parameter p_ε defined as

$$\begin{aligned} p_\varepsilon^2 &= (u_Z + h\cos\psi)^2 + 2h\sin\psi\, u_X \\ &= (h\cos\psi)^2 - \varepsilon = \text{const.} \end{aligned} \tag{2.18}$$

Next, we introduce a new function $u(t)$ related to $u_X(t)$, $u_Y(t)$, and $u_Z(t)$ via

$$\begin{aligned} u_X &= \frac{p_\varepsilon^2}{2h\sin\psi}\left(1-u^2\right), \\ u_Y &= \sqrt{1-\left(p_\varepsilon u - h\cos\psi\right)^2 - \frac{p_\varepsilon^4}{4h^2\sin^2\psi}\left(1-u^2\right)^2}, \\ u_Z &= -h\cos\psi + p_\varepsilon u. \end{aligned} \tag{2.19}$$

The function $u(t)$ satisfies an equation of generic form

$$\frac{du}{dt} = \frac{p_\varepsilon}{2\tau_0}\sqrt{\Phi(u)}, \tag{2.20}$$

where $\Phi(u)$ is the fourth-order polynomial given by

$$\begin{aligned}\Phi(u) &= 4h^2 p_\varepsilon^{-4} \sin^2\psi\left[1-\left(p_\varepsilon u - h\cos\psi\right)^2\right]-\left(1-u^2\right)^2 \\ &= (e_1-u)(u-e_2)(u-e_3)(u-e_4)\end{aligned} \tag{2.21}$$

and in the factorization e_1, e_2, e_3 and e_4 are the roots of $\Phi(u)$. We do not give here the rather complicated explicit equations for the roots of $\Phi(u)$ because using MATHEMATICA® they can be easily calculated both analytically and numerically. Since in the deeper well, u varies in the interval $e_1 \le u \le e_2$ and [350]

$$\int_u^{e_1} \frac{dx}{\sqrt{(e_1-x)(x-e_2)(x-e_3)(x-e_4)}} = \frac{2F\left(\varphi|m_\varepsilon\right)}{\sqrt{(e_1-e_3)(e_2-e_4)}}, \tag{2.22}$$

where $F(\varphi|m)$ is the incomplete elliptic integral [290] (see Appendix A.1.9),

$$\sin\varphi = \sqrt{\frac{(e_2-e_4)(e_1-u)}{(e_1-e_2)(u-e_4)}}, \quad m_\varepsilon = \frac{(e_1-e_2)(e_3-e_4)}{(e_1-e_3)(e_2-e_4)}, \tag{2.23}$$

we then have in terms of the elliptic function $\mathrm{sn}(u|m)$ [290]:

$$\frac{(e_2-e_4)[e_1-u(t)]}{(e_1-e_2)[u(t)-e_4]} = \mathrm{sn}^2\left(\omega_\varepsilon t + w|m_\varepsilon\right)$$

so that the solution of Eq. (2.20) is

$$u(t) = \frac{e_1(e_2-e_4)+e_4(e_1-e_2)\mathrm{sn}^2\left(\omega_\varepsilon t + w|m_\varepsilon\right)}{e_2-e_4+(e_1-e_2)\mathrm{sn}^2\left(\omega_\varepsilon t + w|m_\varepsilon\right)}. \tag{2.24}$$

Here Jacobi's doubly periodic elliptic function $\mathrm{sn}(u|m)$ [290] has the real period $4K(m)$, where $K(m)$ is the complete elliptic integral of the first kind [290] (see Appendix A.1.9), w is an initial phase, and ω_ε is the frequency parameter defined as

$$\omega_\varepsilon = \frac{p_\varepsilon}{4\tau_0}\sqrt{(e_1 - e_3)(e_2 - e_4)}. \tag{2.25}$$

Hence, the energy-dependent precession frequency f_ε^+ of the magnetization is

$$f_\varepsilon^+ = \frac{p_\varepsilon}{16\tau_0 K(m_\varepsilon)}\sqrt{(e_1 - e_3)(e_2 - e_4)}, \tag{2.26}$$

while the action S_ε^+ from the general definition, Eq. (2.16), is given by

$$\begin{aligned} S_\varepsilon^+ = \frac{\tau_0\sigma}{2}\int_0^{1/f_\varepsilon^+} \Big\{&4h^2\sin^2\psi - p^4 + 4p^3 h\cos\psi u(t) \\ &+2p^2\left[2 - 2h^2\cos^2\psi - p^2\right]u^2(t) + 4p^3 h\cos\psi u^3(t) - p^4u^4(t)\Big\}dt \end{aligned} \tag{2.27}$$

[noting Eqs. (2.9) and (2.19)]. Clearly the calculation of S_ε^+ from Eq. (2.27) always reduces to evaluating integrals of the form

$$f_\varepsilon^+ \int_0^{1/f_\varepsilon^+} \left[\frac{b_\varepsilon + \mathrm{sn}^2(\omega_\varepsilon t + w|m_\varepsilon)}{a_\varepsilon + \mathrm{sn}^2(\omega_\varepsilon t + w|m_\varepsilon)}\right]^N dt = \sum_{n=0}^{N} \frac{N!(b_\varepsilon - a_\varepsilon)^n}{n!(N-n)!} I(n) \tag{2.28}$$

for $N = 1, 2, 3$, and 4, where the various integrals $I(n)$ are defined by

$$I(n) = \frac{1}{4K(m_\varepsilon)}\int_0^{4K(m_\varepsilon)} \frac{dt}{[a_\varepsilon + \mathrm{sn}^2(\omega_\varepsilon t + w|m_\varepsilon)]^n}. \tag{2.29}$$

Furthermore, for $n = 1, 2, 3$, and 4, the $I(n)$ can be expressed in terms of complete elliptic integrals by Eqs. (A.115)–(A.119) in Appendix A.1.9 (using the table of integrals from Ref. [350]).

To calculate the VLD reversal time τ_{VLD}, we need f_ε^+ and Δ_ε^+, Eqs. (2.26) and (2.27), corresponding to the deeper well, because S_ε^- and f_ε^- for the shallow well can be formally obtained by replacing *in all the preceding equations h* by $-h$, i.e., $S_\varepsilon^-(h) = S_\varepsilon^+(-h)$ and $f_\varepsilon^-(h) = f_\varepsilon^+(-h)$. Hence we have $\tau_+ = \tau_+(h)$ and $\tau_- = \tau_+(-h)$ so that τ_{VLD}, Eq. (2.17), is given by

$$\tau_{\mathrm{VLD}} = \frac{2\tau_+(h)\tau_+(-h)}{\tau_+(h) + \tau_+(-h)}. \tag{2.30}$$

The action S_C (see Eq. (2.16)) at the saddle point energy ε_C is also given by Eq. (1.304) of Chapter 1, where the contour integral is taken along the critical energy trajectory on which the magnetization may possibly reverse by passing through the saddle point(s) of the energy. It can again be evaluated via converging Taylor series to any desired order of h [207]

$$S_C^{\pm} = \sigma\sqrt{h\sin\psi}\left[16-(104/3)h\sin\psi + h^2\left(1-21\cos 2\psi\right)+...\right] \\ \pm 2\pi\sigma h^2 \sin 2\psi\left(4-3h\sin\psi+...\right). \tag{2.31}$$

The above results considerably simplify when the external field $\mathbf{H}_0$ is applied *along* the easy axis, $\psi = 0$, so that the problem becomes circularly symmetric. Therefore the frequency $f_\varepsilon^+\big|_{\delta\to 0}$ from Eq. (2.26) reduces to the closed algebraic form

$$f_\varepsilon^+\big|_{\delta\to 0} = \frac{1}{2\pi\tau_0}\sqrt{h^2-\varepsilon}. \tag{2.32}$$

Furthermore, we have from Eqs. (2.27) and (2.32) the dimensionless action as

$$S_\varepsilon^{\pm}\big|_{\delta\to 0} = 4\pi\sigma\sqrt{h^2-\varepsilon}\left(\varepsilon - 2h^2 + 1 \pm 2h\sqrt{h^2-\varepsilon}\right). \tag{2.33}$$

We then have from Eq. (2.15)

$$\tau_{\pm} = \frac{\tau_0\sigma}{2\alpha}\int_{-1\mp 2h}^{h^2}\int_{-1\mp 2h}^{\varepsilon}\frac{e^{-\sigma\varepsilon'}d\varepsilon'}{\sqrt{h^2-\varepsilon'}}\frac{e^{\sigma\varepsilon}}{(\varepsilon+1-2h^2)\sqrt{h^2-\varepsilon}\pm 2h(h^2-\varepsilon)}d\varepsilon. \tag{2.34}$$

Notice that the MFPT for the circularly symmetric potential given in Eq. (1.12) also reduces to Eq. (2.34) as it must. For $h = 0$, Eq. (2.34) also yields

$$\tau_{\text{VLD}} = \frac{\tau_0\sigma}{2\alpha}\int_{-1}^{0}\frac{e^{\sigma\varepsilon}}{(1+\varepsilon)\sqrt{-\varepsilon}}\int_{-1}^{\varepsilon}\frac{e^{-\sigma\varepsilon'}}{\sqrt{-\varepsilon'}}d\varepsilon' d\varepsilon \tag{2.35}$$

and with the transformation $\sigma\varepsilon \to -z$, we have the exact τ_{VLD} for uniaxial nanomagnets in the absence of an external field, viz.,

$$\tau_{\text{VLD}} = \tau_0\frac{\sigma\sqrt{\pi}}{2\alpha}\int_0^{\sigma}\frac{e^{-z}[\text{erfi}(\sqrt{\sigma})-\text{erfi}(\sqrt{z})]}{(\sigma-z)\sqrt{z}}dz, \tag{2.36}$$

where erfi (z) is the error function of imaginary argument defined in Appendix A.1.3 [290].

Referring again to the general noncircularly symmetric case we have in the high-barrier limit, $\sigma(\varepsilon_C-\varepsilon_A^-)>>1$, from Eqs. (1.309), (2.15) and (2.17) the closed form asymptote

$$\tau_{\mathrm{VLD}}^{\mathrm{as}}=\frac{2\alpha^{-1}e^{\sigma(\varepsilon_C-\varepsilon_A^-)}}{f_A^-S_C^-+f_A^+S_C^+e^{-\sigma(\varepsilon_A^--\varepsilon_A^+)}}. \tag{2.37}$$

Here the energy-independent precession frequencies $f_A^{\pm}$ pertaining to the bottoms of the wells and the dimensionless actions $S_C^{\pm}$ pertaining to the saddle region are given in series form below [note that by definition $(\varepsilon_C-\varepsilon_A^-)\geq 0$ and $(\varepsilon_A^--\varepsilon_A^+)\geq 0$]. However as $\psi\to 0$, despite the agreement of the *exact* integral solutions, the *asymptotic* escape-rate Eq. (2.37) derived for a *noncircularly symmetric* potential cannot be used to determine the reversal time for high barriers ($\sigma>>1$) for a *circularly symmetric* potential. Instead one must proceed via an asymptotic expansion based on the (exact) noncircularly symmetric Eq. (2.34) thereby leading directly to Brown's asymptotic formula, Eq. (2.4).

2.2.3 *VLD reversal time for biaxial anisotropy*

By way of yet another practical illustration of the MFPT Eq. (1.241), we consider a *biaxial* anisotropy potential augmented by the Zeeman term due to an external dc magnetic field $\mathbf{H}_0$ applied along the easy axis of magnetization, viz., [90,151,152,351,352]

$$E(\vartheta,\varphi)=-\sigma\left(\cos^2\vartheta-\delta\sin^2\vartheta\cos^2\varphi+2h\cos\vartheta\right). \tag{2.38}$$

In general, the noncircularly symmetric potential, Eq. (2.38), has two *nonequivalent wells* and two *equivalent saddle points* (see Fig. 2.2). The biaxial anisotropy may yield an appreciable contribution to the free energy density of magnetic nanoparticles [20]. In particular, Eq. (2.38) describes the magnetic anisotropy energy of a spheroidal single-domain particle, with the axis of symmetry inclined at a certain angle to the easy anisotropy axis of the particle as well that of elongated particles, where easy- and

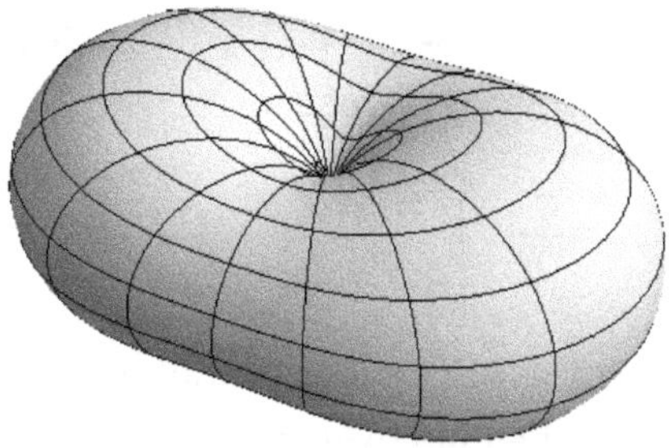

Fig. 2.2. Biaxial anisotropy potential $E(\vartheta,\varphi)/\sigma$, Eq. (2.38), for $\delta = \Delta/\sigma = 0.5$ and $h = 0.2$.

hard-axis anisotropy terms coexist [351]. Furthermore, the biaxial potential is commonly used in spintronic applications [109,126,139] to represent the free energy density of a nanopillar in the standard form of superimposed easy-plane and in-plane easy axis anisotropies (see Section 1.3.3 of Chapter 1).

For the biaxial anisotropy potential, Eq. (2.38), the gyromagnetic equation, Eq. (1.4), can be rewritten in terms of the cartesian components of $\mathbf{u} = (u_X, u_Y, u_Z)$ as [cf. Eqs. (2.9)]

$$\tau_0 \frac{d}{dt}\begin{pmatrix} u_X \\ u_Y \\ u_Z \end{pmatrix} = \begin{pmatrix} -(u_Z + h)u_Y \\ \left[(1+\delta)u_Z + h\right]u_X \\ -\delta u_X u_Y \end{pmatrix}. \tag{2.39}$$

The components u_X, u_Y, u_Z are again not independent being subject to the normalization condition $|\mathbf{u}| = 1$ and energy conservation

$$\varepsilon = -u_Z^2 - 2hu_Z + \delta u_X^2 = \text{const}, \tag{2.40}$$

where $\varepsilon = E/\sigma$ is the normalized free energy, the possible value of which is limited by the conditions $-1-2h \le \varepsilon \le \delta + h^2/(1+\delta)$.

If $|h|<1$, the potential Eq. (2.38) has two *nonequivalent* wells with minima $\varepsilon_A^{\pm} = \sigma(-1 \mp 2h)$ at $u_Z = \pm 1$ (i.e., $\vartheta = 0, \pi$) and two equivalent saddle points at $\varepsilon_C = h^2$ (see Fig. 2.2). Thus, we must again define two *individual* MFPTs, namely, τ_+ from the deeper well around the energy minimum at $u_Z = 1$ (region I: $-1-2h \le \varepsilon \le h^2$) and τ_- from the shallow well around the energy minimum at $u_Z = -1$ (region II: $-1+2h \le \varepsilon \le h^2$). These times are again unequal in general, i.e., $\tau_+ \neq \tau_-$. Thus, Eq. (1.241) as specialized to biaxial anisotropy becomes

$$\tau_\pm = \frac{\sigma^2}{\alpha} \int_{-1\mp 2h}^{h^2} \frac{e^{\sigma\varepsilon}}{S_\varepsilon^\pm} \int_{-1\mp 2h}^{\varepsilon} \frac{e^{-\sigma\varepsilon'} d\varepsilon'}{f_{\varepsilon'}} d\varepsilon, \tag{2.41}$$

where $S_\varepsilon^\pm$ is defined by Eq. (2.16). The overall reversal time which must be used for comparison with numerical results is then given by Eq. (2.17).

The solutions of the equation $\partial_\vartheta E|_{\varphi=\pi/2} = 0$ are $\vartheta_A^+ = 0$, $\vartheta_A^- = \pi$ and $\cos\vartheta_C = -h$. Thus, the well precession frequencies $f_A^\pm$, minima $\varepsilon_A^\pm$ and saddle point ε_C are

$$f_A^\pm = \frac{1}{2\pi\tau_0}\sqrt{(1\pm h+\delta)(1\pm h)},\ \varepsilon_A^\pm = \sigma(-1\mp 2h),\ \varepsilon_C = \sigma h^2. \tag{2.42}$$

Moreover, as in uniaxial anisotropy with an oblique field, we first have to calculate the energy-dependent frequency $f_\varepsilon^\pm$ and dimensionless action $S_\varepsilon^\pm$. We note that Eq. (2.40) and the normalization $|\mathbf{u}| = 1$ lead to the equality

$$\left(u_Z + \frac{h}{\delta+1}\right)^2 + \frac{\delta}{\delta+1}u_Y^2 = \frac{\delta - \varepsilon + h^2/(1+\delta)}{\delta+1} = p_\varepsilon^2. \tag{2.43}$$

Equations (2.39) and (2.43), again prompt us to introduce a new function $u(t)$ related to $u_X(t), u_Y(t)$, and $u_Z(t)$ via

$$\begin{pmatrix} u_X(t) \\ u_Y(t) \\ u_Z(t) \end{pmatrix} = \begin{pmatrix} p_\varepsilon u(t) - h(\delta+1)^{-1} \\ p_\varepsilon\sqrt{(1+\delta^{-1})[1-u^2(t)]} \\ p_\varepsilon\sqrt{\delta^{-1}[u(t)-e_+][u(t)-e_-]} \end{pmatrix}, \tag{2.44}$$

where

$$e_\pm = -\frac{h\delta}{p_\varepsilon(\delta+1)} \pm \frac{1}{p_\varepsilon}\sqrt{h^2-\varepsilon}. \tag{2.45}$$

In the deeper well, u varies in the interval $e_+ \le u \le 1$, while in the shallow well it varies in the interval $-1 \le u \le e_-$. By substituting Eqs. (2.44) into Eq. (2.39), we have the differential equation

$$\tau_0 \frac{du}{dt} = -p_\varepsilon \sqrt{(\delta+1)(1-u^2)(u-e_+)(u-e_-)}. \tag{2.46}$$

Again, by introducing a new function $x(t)$ defined via

$$u(t) = \frac{a_\varepsilon - x^2(t)}{a_\varepsilon + x^2(t)}, \tag{2.47}$$

where $a_\varepsilon = (1+e_+)/(1-e_+)$, we can once more rearrange Eq. (2.46) in standard form [350] as

$$\frac{dx}{dt} = \omega_\varepsilon \sqrt{(1-x^2)(1-m_\varepsilon x^2)}, \tag{2.48}$$

where

$$m_\varepsilon = \frac{(1+e_-)(1-e_+)}{(1-e_-)(1+e_+)}, \quad \omega_\varepsilon = \frac{p_\varepsilon}{2\tau_0}\sqrt{(\delta+1)(1+e_+)(1-e_-)}. \tag{2.49}$$

Clearly $0 \le m_\varepsilon \le 1$ for $-1+2h < \varepsilon < h^2$ and $-1 \le m_\varepsilon \le 0$ for $-1-2h < \varepsilon < -1+2h$. The solution of Eq. (2.48) is given in terms of Jacobi's elliptic function $\mathrm{sn}\,(u\,|\,m)$ as [290]

$$x(t) = \mathrm{sn}\left(\omega_\varepsilon t + w \middle| m_\varepsilon\right), \tag{2.50}$$

where w is a constant of integration (initial phase). Consequently, the function $u(t)$ is given by

$$u(t) = \frac{a_\varepsilon - \mathrm{sn}^2\left(\omega_\varepsilon t + w \middle| m_\varepsilon\right)}{a_\varepsilon + \mathrm{sn}^2\left(\omega_\varepsilon t + w \middle| m_\varepsilon\right)}. \tag{2.51}$$

The energy-dependent precession frequency f_ε of the magnetization is

$$f_\varepsilon = \frac{\omega_\varepsilon}{4K(m_\varepsilon)} = \frac{p_\varepsilon}{8\tau_0 K(m_\varepsilon)}\sqrt{(\delta+1)(1+e_+)(1-e_-)}, \tag{2.52}$$

while S_ε^+ is given by [noting Eqs. (2.39) and (2.43)]

$$\begin{aligned} S_\varepsilon^+ &= 2\tau_0\sigma \int_0^{1/f_\varepsilon} \left[\dot{u}_{X+}^2(t) + \dot{u}_{Y+}^2(t) + \dot{u}_{Z+}^2(t)\right] dt \\ &= \frac{2\sigma p_\varepsilon^2}{\tau_0} \int_0^{1/f_\varepsilon} \left\{(1+\delta)\varepsilon - h^2 + 2hp_\varepsilon(1+\delta)u(t) + (1+\delta-h^2)u^2(t)\right\} dt. \end{aligned} \tag{2.53}$$

Next with Eqs. (A.116)–(A.117) from Appendix A.1.9, we have

$$S_\varepsilon^+ = \frac{16\sigma p_\varepsilon K(m_\varepsilon)\sqrt{1+\delta}}{\sqrt{(1+e_+)(1-e_-)}}\left\{\varepsilon - \frac{h^2}{1+\delta} - 2hp_\varepsilon\left(1-2\frac{\Pi(-a_\varepsilon^{-1}|m_\varepsilon)}{K(m_\varepsilon)}\right)\right.$$
$$\left. + \frac{1+\delta-h^2}{1+\delta}\left(\frac{1-a_\varepsilon}{1+a_\varepsilon} + 2\frac{a_\varepsilon E(m_\varepsilon)-(1-a_\varepsilon^2 m_\varepsilon)\Pi(-a_\varepsilon^{-1}|m_\varepsilon)}{(1+a_\varepsilon)(1+a_\varepsilon m_\varepsilon)K(m_\varepsilon)}\right)\right\}. \tag{2.54}$$

For the shallow well, S_ε^+ can again be formally obtained by replacing h by $-h$ in all equations, i.e., $S_\varepsilon^-(h) = S_\varepsilon^+(-h)$. Thus, we have $\tau_+(h)$ and $\tau_- = \tau_+(-h)$ so that τ_{VLD} is given by Eq. (2.30). At the saddle point ε_C, $S_C^\pm$ can be evaluated in closed form from Eq. (2.54) as

$$S_C^\pm = 8\delta\sigma\left[1-h^2(1+\delta)^{-1}\right]$$
$$\times\left\{\sqrt{\frac{1-h^2}{\delta}} + \frac{h}{\sqrt{1+\delta}}\arctan\left[\frac{h}{\sqrt{(1-h^2)(1+\delta^{-1})}}\right] \pm \frac{h\pi}{2}\right\}. \tag{2.55}$$

Having calculated f_ε and $S_\varepsilon^\pm$ the time τ_{VLD} can be determined from Eq. (2.41). Furthermore, in the high-barrier limit, Eq. (2.41) can again be considerably simplified yielding for τ_{VLD} the closed form asymptotic Eq. (2.37). Likewise, the *energy-independent* precession frequencies $f_A^\pm$ pertaining to the bottom of the wells and the dimensionless actions $S_{\varepsilon_C}^\pm$ are rendered by the closed form Eqs. (2.42) and (2.55).

The above results can be further simplified in two important cases, namely, $\delta \to 0$ (uniaxial anisotropy) and $h \to 0$ (zero external field). For *zero external field* $h = 0$, the biaxial free energy is a double-well potential with two *equivalent* wells, where only the region $-1 \le \varepsilon \le 0$ is appropriate to escape because the energy of a separatrix trajectory is now $\varepsilon_C = 0$, $S_\varepsilon^- = S_\varepsilon^+ = S_\varepsilon$. Therefore, $\tau_+ = \tau_- = \tau$, so that the overall reversal time is then $\tau_{\text{VLD}} = \tau$. We have for $h \to 0$

$$m_\varepsilon = a_\varepsilon^{-2} = \left(\frac{\sqrt{\delta-\varepsilon}-\sqrt{-(\delta+1)\varepsilon}}{\sqrt{\delta-\varepsilon}+\sqrt{-(\delta+1)\varepsilon}}\right)^2,$$

$$\omega_\varepsilon = \frac{\sqrt{\delta(1+\varepsilon)}}{2\tau_0},\; p_\varepsilon = \sqrt{\frac{\delta-\varepsilon}{\delta+1}},\; e_\pm = \pm\sqrt{-\frac{\varepsilon(\delta+1)}{\delta-\varepsilon}}.$$

We also have from Eqs. (2.52) and (2.54) for $h = 0$

$$f_\varepsilon\big|_{h\to 0} = \frac{\sqrt{\delta(1+\varepsilon)}}{8\tau_0 K(a_\varepsilon^{-2})}, \tag{2.56}$$

$$S_\varepsilon^\pm\big|_{h\to 0} = S_\varepsilon = \frac{16\sigma\sqrt{\delta-\varepsilon}}{1+e_+}\left\{\left(\varepsilon - \frac{a_\varepsilon - 1}{a_\varepsilon + 1}\right)K(a_\varepsilon^{-2}) + \frac{2a_\varepsilon^2 E(a_\varepsilon^{-2})}{\left(1+a_\varepsilon\right)^2}\right\}. \tag{2.57}$$

Using known identities from the theory of elliptic functions, viz., [353]

$$K(a_\varepsilon^{-2}) = \frac{1}{2}\left(1+\sqrt{m'}\right)K(m) \text{ and } E(a_\varepsilon^{-2}) = \frac{1-\sqrt{m'}}{m}\left[\sqrt{m'}K(m) + E\left(m\right)\right],$$

where the parameters a_ε, m', and m are related via

$$\sqrt{m'} = \frac{a_\varepsilon - 1}{a_\varepsilon + 1} = \sqrt{-\frac{(\delta+1)\varepsilon}{\delta-\varepsilon}} \text{ and } m = 1 - m' = \frac{\delta+\delta\varepsilon}{\delta-\varepsilon},$$

we obtain from Eqs. (2.56) and (2.57)

$$f_\varepsilon^{-1} = \frac{4\tau_0}{\sqrt{\delta-\varepsilon}}K\left(\frac{\delta+\delta\varepsilon}{\delta-\varepsilon}\right),$$

$$S_\varepsilon = 8\sigma\sqrt{\delta-\varepsilon}\left[E\left(\frac{\delta+\delta\varepsilon}{\delta-\varepsilon}\right) + \varepsilon K\left(\frac{\delta+\delta\varepsilon}{\delta-\varepsilon}\right)\right].$$

Having calculated f_ε and S_ε, τ_{VLD} can be written in integral form as

$$\tau_{\mathrm{VLD}} = \frac{\tau_0\sigma}{2\alpha}\int_{-1}^{0}\frac{e^{\sigma\varepsilon}\int_{-1}^{\varepsilon}K\left(\frac{\delta+\delta\varepsilon'}{\delta-\varepsilon'}\right)\frac{e^{-\sigma\varepsilon'}}{\sqrt{\delta-\varepsilon'}}d\varepsilon'}{\sqrt{\delta-\varepsilon}\left[E\left(\frac{\delta+\delta\varepsilon}{\delta-\varepsilon}\right) + \varepsilon K\left(\frac{\delta+\delta\varepsilon}{\delta-\varepsilon}\right)\right]}d\varepsilon. \tag{2.58}$$

Here $K(m)$ and $E(m)$ are the complete elliptic integrals of the first and second kind, respectively [290].

Referring again to the general noncircularly symmetric case, we have in the high-barrier limit $\sigma >> 1$ for $h = 0$ that τ_{VLD} from Eq. (2.58) is closely approximated by [151,152]

$$\tau_{\mathrm{VLD}}^{\mathrm{as}} \sim \frac{\tau_0 \pi e^{\sigma}}{4\alpha\sigma\sqrt{\delta(1+\delta)}}. \tag{2.59}$$

Equation (2.59) follows from Eqs. (2.41), (2.42) and (2.55) for $h = 0$. For uniaxial anisotropy ($\delta \to 0$), where $u(t) \to 1$, $e_{\pm} \to \pm 1$, $\omega_{\varepsilon} \to \tau_0^{-1}\sqrt{h^2 - \varepsilon}$, $u_Z(t) \to \sqrt{h^2 - \varepsilon} - h$, and $K(m_{\varepsilon}) \to \pi/2$, $f_{\varepsilon}\big|_{\delta\to 0}$ from Eq. (2.52) and $S_{\varepsilon}^{\pm}\big|_{\delta\to 0}$ from Eq. (2.53) are given by Eqs. (2.32) and (2.33), respectively. Equation (2.41) in the limit $\delta \to 0$ reduces to the uniaxial Eq. (2.34).

2.2.4 *Comparison of analytical, asymptotic, and numerical results*

We now compare τ_{VLD} for uniaxial and biaxial anisotropy as yielded by the *exact integral sol*utions Eqs. (2.15), (2.17), and (2.41) with the general asymptotic VLD escape rate $\tau_{\mathrm{VLD}}^{\mathrm{as}}$ predicted by Eq. (2.37) and its particular cases embodied in Eqs. (2.13), (2.14), (2.42), (2.31), and (2.55). We also compare these results with the inverse of the smallest nonvanishing eigenvalue λ_1 of the Fokker–Planck operator calculated numerically by the matrix continued fraction method [97,32,38,151,152] (see Section 2.5). For uniaxial and biaxial anisotropies, the results are shown in Figs. 2.3 and 2.4 as functions of the inverse-temperature parameter σ for various field parameters h and typical values of other model parameters. Clearly, for high barriers, $\sigma > 5$, the *temperature dependence* has the customary Arrhenius behavior $\tau_{\mathrm{VLD}} \sim e^{\sigma(\varepsilon_C - \varepsilon_A^-)}$, i.e., *exponentially increasing* with *decreasing temperature*; the slope of $\tau_{\mathrm{VLD}}(T^{-1})$ being markedly dependent on *h*, since the barrier height of the shallow well is strongly influenced by that parameter and significantly decreases with increasing *h*. In contrast for *low* barriers, $\sigma < 3$, $\tau_{\mathrm{VLD}}(T^{-1})$ may deviate considerably from Arrhenius-like behavior. Nevertheless, both τ_{VLD} and λ_1^{-1} lie very close to each other for virtually all σ. Also in the high-barrier limit, $\tau_{\mathrm{VLD}}^{\mathrm{as}}$ from Eq. (2.37) and its particular cases Eqs. (2.14), (2.31), (2.42), and (2.55) provide an accurate approximation to both λ_1^{-1} and τ_{VLD}. For low barriers, $\sigma < 3$, $\tau_{\mathrm{VLD}}^{\mathrm{as}}$ deviates considerably from each of the foregoing results as expected, hence it cannot then be used to calculate τ_{VLD}.

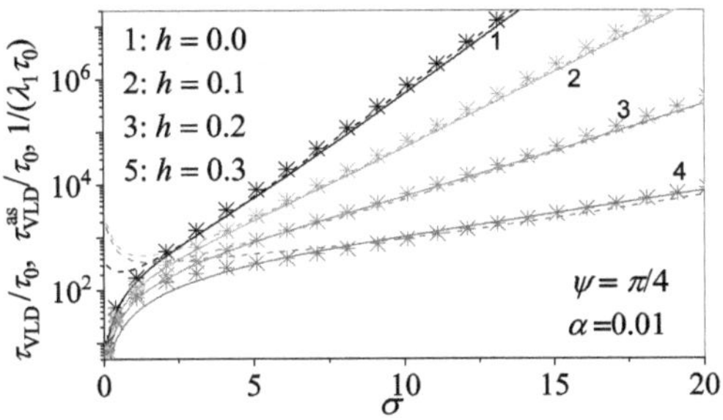

Fig. 2.3. $\tau_{\mathrm{VLD}}/\tau_0$, $\tau_{\mathrm{VLD}}^{\mathrm{as}}/\tau_0$, and $(\tau_0\lambda_1)^{-1}$ vs. the barrier height (inverse temperature) parameter $\sigma = vK/(kT)$ for *uniaxial anisotropy*, Eq. (2.6) for various values of the field parameter h for $\psi = \pi/4$ and $\alpha = 0.01$. Solid line: τ_{VLD} from the solution embodied in Eqs. (2.15), (2.17), (2.26)–(2.37) and (A.115)–(A.119) from Appendix A.1.9. Asterisks: numerical solution for the inverse of the smallest nonvanishing eigenvalue $(\tau_0\lambda_1)^{-1}$ of the Fokker-Planck operator [32,38,97]. Dashed line: the VLD asymptotic Eqs. (2.4) (for $h = 0$ only) and the more general cases Eqs. (2.14), (2.31), and (2.37).

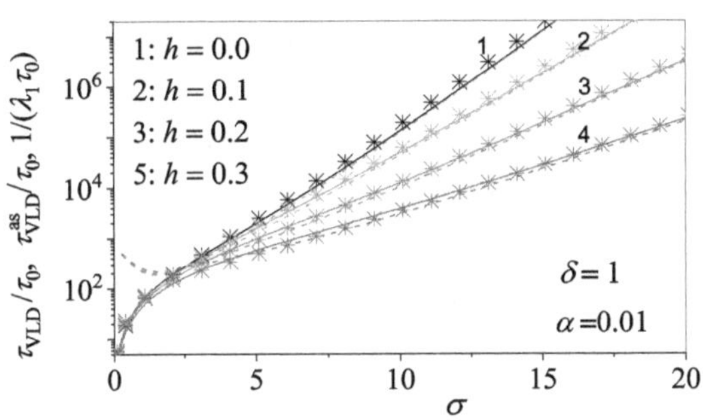

Fig. 2.4. $\tau_{\mathrm{VLD}}/\tau_0$, $\tau_{\mathrm{VLD}}^{\mathrm{as}}/\tau_0$, and $(\tau_0\lambda_1)^{-1}$ vs. the barrier height (inverse temperature) parameter $\sigma = vK/(kT)$ for *biaxial anisotropy*, Eq. (2.38), with $\delta = 1$ and $\alpha = 0.01$. Solid line: τ_{VLD} from the exact solution embodied in Eqs. (2.17), (2.41), (2.52), and (2.54). Asterisks: numerical solution for $(\tau_0\lambda_1)^{-1}$ of the Fokker-Planck operator [97,151,152]. Dashed line: the VLD asymptotic Eqs. (2.37), (2.42), and (2.55).

Now we consider yet another problem known as *uniaxial crossovers*. These are encountered in those parameter ranges, where the departures from circular symmetry are small, and exist in the calculation of the reversal time via escape-rate theory [181]. We illustrate by plotting in Fig. 2.5 τ_{VLD} for uniaxial anisotropy superimposed on the Zeeman term, Eq. (2.6), as a function of the oblique angle ψ for $\sigma = 15$ and $h = 0.2$. Thus we

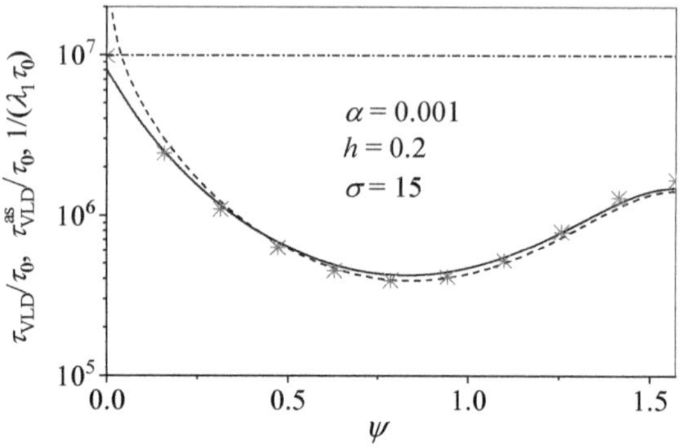

Fig. 2.5. τ_{VLD} / τ_0, τ_{VLD}^{as} / τ_0, and $(\tau_0\lambda_1)^{-1}$ vs. the oblique angle ψ for *uniaxial anisotropy*, i.e., Eq. (2.6) for $\sigma = 15$ and $\alpha = 0.01$. Solid line: exact τ_{VLD} from Eqs. (2.15), (2.17), (2.26)-(2.37) and (A.115)-(A.119) from Appendix A.1.9. Asterisks: numerical solution for $(\tau_0\lambda_1)^{-1}$ of the Fokker-Planck operator [32,38,97]. Dashed line: the VLD asymptotic Eqs. (2.13), (2.14), (2.31), and (2.37). Dashed-dotted line: Brown's asymptotic Eq. (2.4).

see that in the interval $0 \le \psi \le \pi/2$ [note that $\tau(\psi) = \tau(\pi - \psi)$], τ_{VLD} strongly depends on ψ having a minimum at $\psi \approx \pi/4$ and deviating considerably from the uniaxial asymptote, Eq. (2.4). However in the limit $\psi \to 0$, τ_{VLD}^{as} from the asymptotic Eq. (2.37) *diverges* because now the dimensionless action $S_{\varepsilon_C}^{\pm} \to 0$ (uniaxial crossover) thus it cannot be used to calculate the reversal time. In contrast, the *exact* MFPT Eqs. (2.15) and (2.17) yield values very close both to the numerical results and Brown's asymptote from Eq. (2.4).

However, the general Eq. (1.241) yielding the VLD reversal time in integral form via quadratures can, in principle, be evaluated for *any* anisotropy potential. Therefore, the outstanding merit of Eq. (1.241) is that it is valid in parameter ranges, where asymptotic escape-rate equations [like Eqs. (2.4), (2.37), and (2.59)] do not apply at all, i.e., both for *low barriers* and *uniaxial crossovers* [181,196,239,354,355]. Furthermore, for certain anisotropy potentials such as uniaxial and biaxial anisotropies, etc., the τ_{VLD} can be written as integrals of known functions [see, e.g., Eqs. (2.34), (2.36), and (2.58)]. Equation (1.241) is also valuable as a benchmark solution with which numerical calculations of the reversal time from the magnetic Langevin and/or Fokker–Planck equation in the VLD regime must always agree.

2.2.5 *STT effect on the magnetization reversal time at VLD*

As discussed in Section 1.2.3, due to the STT effect, an electric current with spin polarization in a small (nanoscale) ferromagnet may *transfer* spin angular momentum between the current and the magnetization **M.** This gives rise to a macroscopic spin-torque on **M** so that the latter *may be altered by spin-polarized currents.* Consequently, the dynamics of **M** including STT may be very different from those without STT. *Due to STT the Gilbert damping torque may be completely overcome* so that reversal of the magnetization becomes possible even *in the absence* of thermal fluctuations. As far as the joint action of STT and thermal fluctuations are concerned, the stationary distribution of magnetization orientations in a ferromagnet subjected to spin-polarized current now depends both on the spin-polarized current and damping (see Section 1.2.3). Consequently, the external current and damping parameters now govern the reversal time and the effect of the STT may be as much as several orders of magnitude even for relatively weak alternating currents. For *nonzero* STT, the reversal time in nanomagnets has been evaluated in Refs. [129,130]. As far as the verification of the formulas for the reversal time with nonzero STT is concerned, a comparison with the numerical solutions of the Fokker–Planck equation has been given in Ref. [167].

Here we shall use the results of Refs. [129,130,165–167,356] treating simultaneously STT and dc bias field effects in nanomagnets. We calculate the reversal time for the free energy density written in the standard form of superimposed easy-plane and in-plane easy-axis anisotropies with the dc bias field $\mathbf{H}_0$ imposed along the easy X-axis, Eq. (1.76), namely [109,139] (see Fig. 2.6)

$$\begin{aligned}\sigma^{-1}E(\vartheta,\varphi) &= \varepsilon(\vartheta,\varphi)\\ &= \delta\cos^2\vartheta - \sin^2\vartheta\cos^2\varphi - 2h\sin\vartheta\cos\varphi.\end{aligned} \tag{2.60}$$

We shall suppose for simplicity that the unit vector $\mathbf{e}_P$, *defining the orientation of the magnetization of the fixed layer* (see Fig. 1.12), is applied along the easy X-axis, so that the approximate Eq. (1.80) for the nonconservative potential Φ becomes

$$\Phi = Jv^{-1}kTu_X = Jv^{-1}kT\sin\vartheta\cos\varphi. \tag{2.61}$$

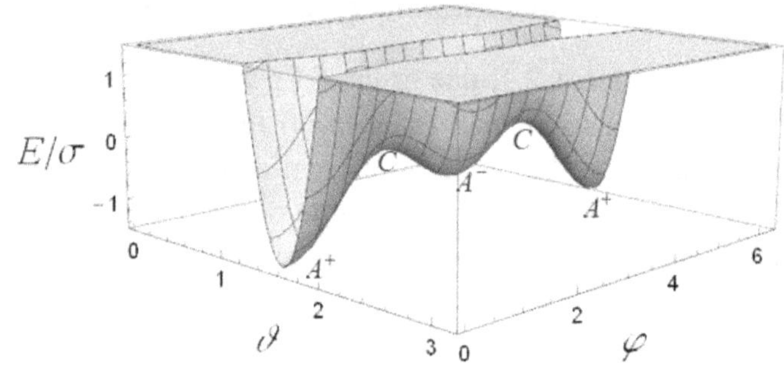

Fig. 2.6. Free energy potential of the free layer presented in the standard form of superimposed easy-plane and in-plane easy-axis anisotropies, Eq. (2.60), at $\delta = 20$ and $h = 0.2$.

If $|h| < 1$ the normalized free energy density $\varepsilon(\vartheta, \varphi)$ from Eq. (2.60) has two *nonequivalent* wells with minima $\varepsilon_A = -1 \pm 2h$ at $u_X = \mp 1$ and two equivalent saddle points $\varepsilon_C = h^2$ at $u_X = -h$ (see Fig. 2.6). To evaluate τ_{VLD} we must once again define two individual reversal times by using Eq. (1,259), namely, τ_+ from the deeper well around the energy minimum at $u_X = +1$ $(-1-2h \le \varepsilon \le h^2)$ and τ_- from the shallow well around the energy minimum at $u_X = -1$ $(-1+2h \le \varepsilon \le h^2)$. These times are unequal in general, i.e., $\tau_+ \neq \tau_-$, so that Eq. (1.247), as specialized to the potential Eq. (2.60), becomes

$$\tau_\pm = \frac{\sigma^2}{\alpha} \int_{-1\mp 2h}^{h^2} \frac{1}{S_\varepsilon^\pm} e^{\sigma\varepsilon - \frac{\sigma}{\alpha}\int_{-1\mp 2h}^{\varepsilon} \frac{V_{\varepsilon'}^\pm}{S_{\varepsilon'}^\pm} d\varepsilon'} \int_{-1\mp 2h}^{\varepsilon} \frac{1}{f_{\varepsilon'}^\pm} e^{-\sigma\varepsilon' + \frac{\sigma}{\alpha}\int_{-1\mp 2h}^{\varepsilon'} \frac{V_{\varepsilon''}^\pm}{S_{\varepsilon''}^\pm} d\varepsilon''} d\varepsilon' d\varepsilon. \quad (2.62)$$

We require as before explicit equations for $f_\varepsilon^\pm$, the action $S_\varepsilon^\pm$ and the work $V_\varepsilon^\pm$ done by the spin torque. For the biaxial anisotropy potential, Eq. (2.60), the gyromagnetic equation, Eq. (1.4), governing the *deterministic* dynamics of the magnetization can be written as a set of three coupled differential equations [cf. Eq. (2.39)]:

$$\tau_0 \frac{d}{dt} \begin{pmatrix} u_X \\ u_Y \\ u_Z \end{pmatrix} = \begin{pmatrix} \delta u_Z u_Y \\ -[(1+\delta)u_X + h]u_Z \\ (u_X + h)u_Y \end{pmatrix}. \quad (2.63)$$

However, the difference between Eq. (2.63) and Eq. (2.39) lies only in the orientation of the axes. Thus, the solutions of Eq. (2.63) are

$$\begin{pmatrix} u_X(t) \\ u_Y(t) \\ u_Z(t) \end{pmatrix} = \begin{pmatrix} p_\varepsilon u(t) - h(\delta+1)^{-1} \\ p_\varepsilon \sqrt{(1+\delta^{-1})[1-u^2(t)]} \\ p_\varepsilon \sqrt{\delta^{-1}[e_+ - u(t)][e_- - u(t)]} \end{pmatrix}, \tag{2.64}$$

where $u(t)$ is given by Eq. (2.51). Again, the only difference is in the definition of p_ε^2 which becomes

$$\begin{aligned} p_\varepsilon^2 &= \frac{\delta-\varepsilon}{\delta+1} + \frac{h^2}{(\delta+1)^2} \\ &= \left(u_X(t) + \frac{h}{\delta+1}\right)^2 + \frac{\delta}{\delta+1} u_Y^2(t) = \text{const.} \end{aligned} \tag{2.65}$$

Noticing Eqs. (2.63), we have the action S_ε^+ from Eq. (1.179) and the work V_ε^+ from Eq. (1.184) in the deeper well as

$$\begin{aligned} S_\varepsilon^+ &= 2\tau_0 \sigma \int_0^{1/f_\varepsilon^+} \left[\dot{u}_X^2(t) + \dot{u}_Y^2(t) + \dot{u}_Z^2(t)\right] dt = \frac{2\sigma p_\varepsilon^2}{\tau_0} \\ &\times \left[\frac{\varepsilon(\delta+1) - h^2}{f_\varepsilon^+} + 2hp_\varepsilon(\delta+1) \int_0^{1/f_\varepsilon^+} u(t)dt + (1+\delta-h^2) \int_0^{1/f_\varepsilon^+} u^2(t)dt \right], \end{aligned} \tag{2.66}$$

$$\begin{aligned} V_\varepsilon^+ &= J \int_0^{1/f_\varepsilon^\pm} \left[\dot{u}_Z(t) u_Y(t) - \dot{u}_Y(t) u_Z(t)\right] dt \\ &= \frac{J}{\tau_0} \left[\frac{hp_\varepsilon^2}{f_\varepsilon^+} + p_\varepsilon \left(\varepsilon + 1 - \frac{2h^2}{\delta+1}\right) \int_0^{1/f_\varepsilon^+} u(t)dt + hp_\varepsilon^2 \int_0^{1/f_\varepsilon^+} u^2(t)dt \right], \end{aligned} \tag{2.67}$$

where the precession frequency f_ε^+ at energy ε is defined as

$$f_\varepsilon^+ = \frac{\omega_\varepsilon}{4K(m_\varepsilon)} = \frac{p_\varepsilon}{8\tau_0 K(m_\varepsilon)} \sqrt{(1+\delta)(1+e_+)(1-e_-)} \tag{2.68}$$

and $K(m)$ is the complete elliptic integral of the first kind (Jacobi's elliptic function $\operatorname{sn}(u|m)$ has the period $4K(m)$ [290]). Since all integrals involving the elliptic function $\mathrm{sn}(u|m)$ are simply of the generic form given by Eq. (2.29) [350], we have

$$\frac{1}{4K(m)}\int_0^{4K(m)}\frac{a-\mathrm{sn}^2(u\,|\,m)}{a+\mathrm{sn}^2(u\,|\,m)}du=2aI(1)-1,$$
$$\frac{1}{4K(m)}\int_0^{4K(m)}\left[\frac{a-\mathrm{sn}^2(u\,|\,m)}{a+\mathrm{sn}^2(u\,|\,m)}\right]^2du=1-4aI(1)+4a^2I(2), \tag{2.69}$$

and so on, where $I(1)$ and $I(2)$ are given in terms of the complete elliptic integrals of the first, $K(m)$, second, $E(m)$, and third kind, $\Pi(n\,|\,m)$ [290] by Eqs. (A.116) and (A.117) from Appendix A.1.9. Hence, all the integrals in Eqs. (2.66) and (2.67) can be evaluated analytically in terms of these known functions as

$$S_\varepsilon^+=\frac{2\sigma p_\varepsilon^2(1+\delta)}{\tau_0 f_\varepsilon^+}\left\{\frac{1+\delta-h^2}{(1+\delta)(1+a_\varepsilon)}\left[1-a_\varepsilon+\frac{2a_\varepsilon E(m_\varepsilon)}{(1+a_\varepsilon m_\varepsilon)K(m_\varepsilon)}\right]+\varepsilon\right.$$
$$\left.+2hp_\varepsilon-\frac{h^2}{1+\delta}+\left[4hp_\varepsilon-2\frac{(1+\delta-h^2)(1-a_\varepsilon^2m_\varepsilon)}{(1+\delta)(1+a_\varepsilon)(1+a_\varepsilon m_\varepsilon)}\right]\frac{\Pi\left(-a_\varepsilon^{-1}|m_\varepsilon\right)}{K\left(m_\varepsilon\right)}\right\}, \tag{2.70}$$

$$V_\varepsilon^+=\frac{J}{\tau_0 f_\varepsilon^+}\left\{h+\frac{h^3}{(\delta+1)^2}-\frac{h(\varepsilon+1)}{\delta+1}+p_\varepsilon\left(\varepsilon+1-\frac{2h^2}{\delta+1}\right)\left[2\frac{\Pi(-a_\varepsilon^{-1}\,|\,m_\varepsilon)}{K(m_\varepsilon)}-1\right]\right.$$
$$\left.+\frac{hp_\varepsilon^2}{1+a^{-1}{}_\varepsilon}\left[a_\varepsilon^{-1}-1+2\frac{a_\varepsilon^{-1}E(m_\varepsilon)-(a_\varepsilon^{-2}-m_\varepsilon)\Pi(-a_\varepsilon^{-1}\,|\,m_\varepsilon)}{(a_\varepsilon^{-1}+m_\varepsilon)K(m_\varepsilon)}\right]\right\}. \tag{2.71}$$

For the shallow well, S_ε^-, V_ε^-, and f_A^- can be obtained by merely replacing h by $-h$ and J by $-J$ in all the equations for S_ε^+, V_ε^+, and f_A^+. Thus, we have $\tau_-=\tau_+(-h,-J)$. Recalling that τ_- and τ_+ are as usual related to the corresponding escape rates from the individual wells via $\Gamma_+=(2\tau_+)^{-1}$ and $\Gamma_-=(2\tau_-)^{-1}$ [85], we see that the VLD reversal time is given by $\tau_{\mathrm{VLD}}=(\Gamma_++\Gamma_-)^{-1}$, namely

$$\tau_{\mathrm{VLD}}=\frac{2\tau_+\tau_-}{\tau_++\tau_-}. \tag{2.72}$$

In the high-barrier limit, we also have from Eqs. (1.305), (2.68), (2.70), and (2.71) the simple asymptotic equation

$$\tau_{\pm} \sim \frac{e^{\Delta E^{\pm}}}{\alpha f_A^{\pm} S_{\varepsilon_C}^{\pm}}, \tag{2.73}$$

where

$$\Delta E^{\pm} = \sigma(1 \pm h)^2 - \frac{\sigma}{\alpha} \int_{\varepsilon_A^{\pm}}^{\varepsilon_C} \frac{V_{\varepsilon}^{\pm}}{S_{\varepsilon}^{\pm}} d\varepsilon, \tag{2.74}$$

$$f_A^{\pm} = \frac{1}{2\pi\tau_0} \sqrt{(1 \pm h + \delta)(1 \pm h)}, \tag{2.75}$$

$$S_{\varepsilon_C}^{\pm} = 8\delta\sigma\left(1 - h^2(1+\delta)^{-1}\right) \times \left\{ \sqrt{\frac{1-h^2}{\delta}} + \frac{h}{\sqrt{1+\delta}} \arctan\left[h \Big/ \sqrt{(1-h^2)(1+\delta^{-1})} \right] \pm \frac{h\pi}{2} \right\}. \tag{2.76}$$

At zero dc bias field, i.e., $h \to 0$, where

$$f_{\varepsilon}^{\pm}\Big|_{h\to 0} = \frac{\sqrt{\delta - \varepsilon}}{4\tau_0 K(m_{\varepsilon})}, \quad m_{\varepsilon} = \frac{\delta(1+\varepsilon)}{\delta - \varepsilon},$$

$$S_{\varepsilon}^{\pm}\Big|_{h\to 0} = 8\sigma\sqrt{\delta - \varepsilon}\left[\varepsilon K(m_{\varepsilon}) + E(m_{\varepsilon})\right], \quad V_{\varepsilon}^{\pm}\Big|_{h\to 0} = \pm 2\pi J \frac{1+\varepsilon}{\sqrt{1+\delta}},$$

we see that Eq. (2.62) coincides with the corresponding equations of Taniguchi *et al.* [129,130].

Moreover for zero STT, i.e., $J = 0$, and nonzero dc bias field, the high barrier asymptotic Eq. (2.73) yields the known results

$$\tau_{\pm}^{\text{as}} \sim \frac{e^{\sigma(1\pm h)^2}}{\alpha f_A^{\pm} S_{\varepsilon_C}^{\pm}}, \quad \sigma(1-h)^2 >> 1. \tag{2.77}$$

Furthermore, for zero STT and zero dc bias field ($h = 0$), the free energy Eq. (2.38) represents a double-well potential with two *equivalent* wells. Therefore, $\tau_+ = \tau_- = \tau$, thus the overall reversal time is then $\tau_{\text{VLD}} = \tau$, which may be written in the closed integral form

$$\tau_{\text{VLD}} = \frac{\tau_0 \sigma}{2\alpha} \int_{-1}^{0} \frac{e^{\sigma\varepsilon}}{\sqrt{\delta - \varepsilon}\left[E(m_{\varepsilon}) + \varepsilon K(m_{\varepsilon})\right]} \int_{-1}^{\varepsilon} \frac{K(m_{\varepsilon'}) e^{-\sigma\varepsilon'}}{\sqrt{\delta - \varepsilon'}} d\varepsilon' d\varepsilon. \tag{2.78}$$

For $\delta = 0$ (uniaxial anisotropy), we have from Eq. (2.58)

$$\begin{aligned}\tau_{\mathrm{VLD}} &= \frac{\tau_0 \sigma}{2\alpha} \int_{-1}^{0} \frac{e^{\sigma\varepsilon}}{\sqrt{-\varepsilon}\left(1+\varepsilon\right)} \int_{-1}^{\varepsilon} \frac{e^{-\sigma\varepsilon'} d\varepsilon'}{\sqrt{-\varepsilon'}} d\varepsilon \\ &= \tau_0 \frac{\sqrt{\pi\sigma}}{\alpha} \int_{-1}^{0} \frac{\left[\operatorname{erfi}(\sqrt{\sigma}) - \operatorname{erfi}(z\sqrt{\sigma})\right] e^{-\sigma z^2}}{1-z^2} dz.\end{aligned} \tag{2.79}$$

Equation (2.79) agrees with the MFPT Eq. (2.36). For high barriers, $\sigma >> 1$, τ_{VLD} from both Eqs. (2.78) and (2.79) are closely approximated by the known formula [239]

$$\tau_{\mathrm{VLD}}^{\mathrm{as}} \sim \frac{\tau_0 \pi e^{\sigma}}{4\alpha\sigma\sqrt{\delta(1+\delta)}} \quad \left(\sigma\delta > 0.5\right) \tag{2.80}$$

and Brown's asymptotic equation [17]

$$\tau_{\mathrm{VLD}}^{\mathrm{as}} = \frac{\tau_{\mathrm{TST}}}{4\alpha\sqrt{\pi\sigma}} \left(1 + \frac{1}{\sigma} + \ldots\right), \tag{2.81}$$

where $\tau_{\mathrm{TST}} = \pi\mu_0 M_S e^{\sigma} / (K\gamma)$.

In Fig. 2.7, we again compare τ_{VLD} yielded by the exact integral solutions, Eqs. (2.62) and (2.72), with both the asymptotic VLD escape rate $\tau_{\mathrm{VLD}}^{\mathrm{as}}$ Eqs. (2.73)–(2.76) and the inverse of the smallest nonvanishing eigenvalue λ_1 calculated numerically by matrix continued fractions. All the calculations have been done for damping parameters corresponding to the true VLD regime, $\alpha S_{E_C} << 1$ for all barrier values. The spin-polarization factor selected is $P = 0.3$ yielding $b_P = 0.166$. Furthermore, for $D_{\|} = 0.034$, $\gamma = 2.2\cdot 10^5$ m A^{-1}s^{-1}, $M_S \approx 1.4\cdot 10^6$ A m^{-1} (cobalt), we have $\tau_0 = 4.8\cdot 10^{-11}$ s. The magnetization reversal time is shown in Fig. 2.7 as a function of the anisotropy (inverse-temperature) parameter σ for various spin-polarized current parameters J, damping parameters α and dc bias field parameters h and typical values of the other model parameters. Apparently, τ_{VLD} and λ_1^{-1} again lie very close to each other for virtually all σ. Also, in the high-barrier limit (large σ), $\tau_{\mathrm{VLD}}^{\mathrm{as}}$ from the asymptotic Eqs. (2.73)–(2.76) provides an accurate approximation to both λ_1^{-1} and τ_{VLD}. However, for $\sigma < 5$, $\tau_{\mathrm{VLD}}^{\mathrm{as}}$ deviates considerably from each of these so that it cannot now be used to calculate the reversal time. Moreover, the

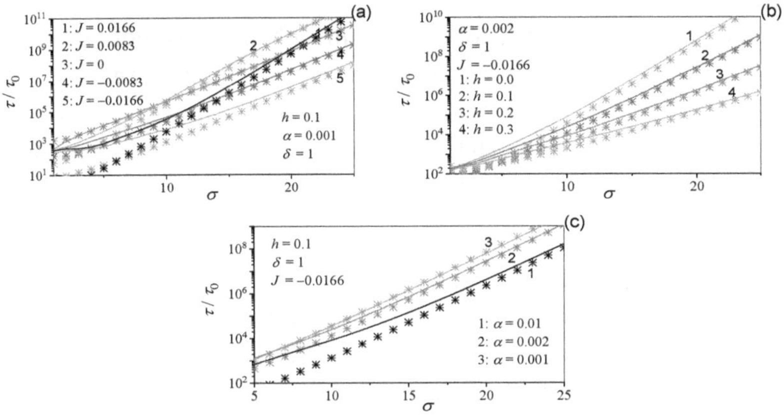

Fig. 2.7. Normalized times $\tau/\tau_0, \tau_{as}/\tau_0$, and $(\tau_0\lambda_1)^{-1}$ vs. the anisotropy (inverse temperature) parameter σ for various spin-polarized currents J (a), dc field parameters h (b), and damping coefficients α (c). Solid lines: numerical solution for the inverse of the smallest nonvanishing eigenvalue $(\tau_0\lambda_1)^{-1}$ of the Fokker–Planck operator [167]. Dashed lines: τ/τ_0 from the MFPT Eqs. (2.62) and (2.72). Asterisks: the VLD asymptotic Eqs. (2.73)–(2.76).

STT effects are clearly governed by the ratio J/α, indeed by altering J/α the ensuing variation of τ may be as much as several orders of magnitude (Fig. 2.7a). In particular τ may greatly exceed or, on the other hand, be much less than the value without STT, viz., $J = 0$. Moreover, the increase or decrease in τ is entirely governed by the *direction* of the current, i.e., by the *sign* of J as indeed expected. The temperature dependence of τ can be understood via the *effective potential barriers* $\Delta E^{\pm}$ in Eq. (2.73). Evidently, for large anisotropy parameters, $\sigma > 5$, the *temperature dependence* of τ has the customary Arrhenius behavior $\tau \sim e^{\Delta E^{\pm}}$, i.e., exponential increase with decreasing temperature. Here τ is markedly dependent on both the ratio J/α and the dc bias field parameter h because the barrier height of the shallow well is strongly influenced by both J/α and h.

The exact VLD Eq. (2.62) is valid in those parameter ranges where escape-rate equations such as Eqs. (2.73)–(2.76) do not apply at all, e.g., for *low barriers* and is also valuable as a benchmark solution with which numerical calculations of the reversal time from the magnetic Langevin and/or Fokker–Planck equation in the VLD regime must agree. We saw that the range of validity of the results obtained is defined by the inequality, $\alpha S_{E_C} << 1$. However, the various damping regimes (VLD, IHD, etc.) have been defined *solely* in the context of *asymptotic* calculations of the magnetization reversal time τ and unlike numerical solutions for λ_1^{-1} must all be treated separately [97] (see Sections 1.4 and 1.5 of Chapter 1).

2.3 Magnetization reversal time in nanomagnets with noncircularly symmetric potentials: Kramers' escape-rate theory approach

First, we recall results for the reversal time of the magnetization τ of single-domain nanoparticles from the Kramers escape-rate theory. As shown in Chapter 1, Section 1.5, in order to estimate the reversal time of the magnetic moment in a nanoscale ferromagnet over the internal anisotropy potential barrier, i.e., the inverse of the smallest nonvanishing eigenvalue λ_1 of the Fokker–Planck operator in Eq. (1.65), Brown [17,18] adapted to magnetic relaxation the ingenious method proposed by Kramers [84,85] for thermally activated escape of Brownian particles out of a potential well (Chapter 1, Section 1.13). We recall that Kramers' basic idea is to calculate the prefactor Λ in an Arrhenius-like equation for the escape rate Γ over the potential barrier ΔE, namely,

$$\Gamma = \Lambda f_A e^{-\Delta E},$$

where the attempt frequency, f_A, is the frequency of a Brownian particle of mass m performing oscillatory motion at the bottom of a well. Referring to magnetic relaxation, in his first calculations of approximate expressions for the reversal time $\tau \sim \Gamma^{-1}$, Brown [17] confined himself to *circularly symmetric potentials* of the magnetocrystalline anisotropy and Zeeman energy. Hence, there exists no dynamical coupling between the longitudinal and the transverse modes of motion, so that the longitudinal modes are governed by a *single* state variable, namely the colatitude ϑ, i.e., the polar angle of the magnetization vector **M**. The second state variable, namely the azimuthal angle φ of **M**, merely gives rise to a steady precession of that vector. Noting the decoupling between the transverse and longitudinal modes existing for circular symmetry, which results in an *exact* single-variable Fokker–Planck equation in ϑ, Eq. (1.223), Brown demonstrated that the Kramers escape-rate theory for Newtonian particles may be adapted to yield the escape rate of spins in *circularly symmetric* potentials, which is valid for all values of α. We recall, however, that the *exact* Fokker–Planck equation, Eq. (1.223), in the single state variable ϑ does not arise from strong damping of the momentum (which, in the mechanical Brownian particle problem, gives rise to the *approximate* Fokker–Planck equation in configuration space, known as the Smoluchowski equation); rather, it follows from the *circular symmetry* of the potential, i.e., from geometry.

We recall that Kramers [84,85] obtained so-called IHD and VLD formulas for the escape rate for point particles, assuming throughout that the energy barrier is much greater than the thermal energy, so that the concept of an escape rate applies. He mentioned in his paper, however, that he could not find a general (asymptotic) method of attack in order to obtain a formula which would be valid for any damping regime. As shown in Chapter 1, this problem, known as the *Kramers turnover*, was solved by Mel'nikov and Meshkov [186,187]. They obtained the escape rate in the whole damping range by expressing the energy distribution function in the underdamped region (extending from zero to intermediate damping) at a given action as an integral equation, which may be rewritten as a Wiener–Hopf equation. The solution of this equation, *simply* multiplied by the IHD escape rate, yields an integral formula for the relaxation time which is valid for all values of the damping, and constitutes a solution of the Kramers turnover problem for mechanical particles. By extending the Mel'nikov–Meshkov approach, Grabert [188] and Pollak *et al.* [189] later presented a complete solution of the Kramers turnover problem, showing that their turnover formula can be obtained without *ad hoc* interpolation between the low- and high-damping regimes. We also recall that the analogous magnetic-spin problem, as formulated by Brown [17,18], differs fundamentally from that of mechanical particles with separable and additive Hamiltonians. First, the magnetic system has two degrees of freedom, namely the polar ϑ and azimuthal φ angles; second, the undamped equation of motion of the magnetization of a single-domain ferromagnetic particle is the gyromagnetic equation; third, the

Hamiltonian is not separable; fourth, inertial effects play no role. Nevertheless, the role of inertia in the mechanical system is essentially mimicked in the magnetic system for noncircularly symmetric potentials by the gyromagnetic term, which causes *coupling* or *entanglement* of the transverse and longitudinal modes. Thus, in order to derive escape-rate formulas for magnetic nanoparticles equivalent to those for mechanical particles, one must consider in Brown's Fokker–Planck equation, Eq. (1.65), a *noncircularly symmetric free energy density*, where both regimes of damping (IHD and VLD) can occur. We saw in Chapter 1, Section 1.5, that IHD formulas for noncircularly symmetric potentials were first derived by Smith and de Rozario [90] and Brown [18]. Subsequently, in 1990, Klik and Gunther [106] realized that the various Kramers damping regimes also applied to the magnetization relaxation of single-domain ferromagnetic particles and derived the corresponding VLD formula for the reversal time. Furthermore, they emphasized that the magnetic IHD calculations [18,90] are, in effect, merely a special case of Langer's general treatment of the decay of metastable states of systems with many degrees of freedom [103] (see Chapter 1, Section 1.5.3). Finally, Coffey *et al.* [196] have shown that the Mel'nikov–Meshkov method [186,187] for bridging the VLD and IHD Kramers escape rates as a function of the dissipation parameter for mechanical particles, can be extended to single-domain ferromagnetic nanoparticles (classical spins).

In this section, we estimate reversal times of the magnetization for classical nanomagnets with various anisotropy potentials. In order to assess the (damping) range of applicability of the analytic formulas so obtained, we must compare them with the results of numerical solutions of the Fokker-Planck equation, Eq. (1.65).

2.3.1 *Basic escape-rate equations for nanomagnets*

As shown in Chapter 1, Section 1.5, in evaluating the IHD escape rate of a nanomagnet, it is supposed that its normalized free energy density $E(\mathbf{M})$ has minima at $\mathbf{n}_A$ and $\mathbf{n}_B$ separated by a potential barrier with a saddle point at $\mathbf{n}_C$. If the magnetization $\mathbf{M}$ is close to a stationary point $\mathbf{n}_p$ and $(u_1^{(p)}, u_2^{(p)}, u_3^{(p)})$ are the direction cosines of $\mathbf{M}$, then E can be expanded in a Taylor series to second order in $u_1^{(p)}$ and $u_2^{(p)}$ as

$$E \approx E_p + \frac{1}{2}\left[c_1^{(p)}\left(u_1^{(p)}\right)^2 + c_2^{(p)}\left(u_2^{(p)}\right)^2\right]. \tag{2.82}$$

The calculation of the parameters E_p and $c_{1,2}^{(p)}$ is described in Chapter 1, Section 1.5.5. Then the Fokker–Planck equation may be solved in the saddle point region [18] yielding the IHD escape rate Γ_A^{IHD} from the well A over a saddle point C as (see Chapter 1, Section 1.5.4)

$$\Gamma_A^{\mathrm{IHD}} = \frac{\omega_A \lambda_+(\alpha)}{2\pi\omega_C} e^{-\Delta E_A}, \tag{2.83}$$

where

$$\omega_A = \frac{\gamma kT}{v\mu_0 M_S}\sqrt{c_1^{(A)} c_2^{(A)}} \text{ and } \omega_C = \frac{\gamma kT}{v\mu_0 M_S}\sqrt{-c_1^{(C)} c_2^{(C)}} \tag{2.84}$$

are, respectively, the well and saddle angular frequencies (cf. Eq. (1.303) of Chapter 1), $\Delta E_A = E_C - E_A$ is the dimensionless barrier height,

$$\lambda_+ = \frac{1}{4\tau_0(\alpha + \alpha^{-1})}\left[-c_1^{(C)} - c_2^{(C)} + \sqrt{(c_2^{(C)} - c_1^{(C)})^2 - 4\alpha^{-2} c_1^{(C)} c_2^{(C)}}\right] \tag{2.85}$$

is the unique positive eigenvalue of the set of noiseless magnetic Langevin equations linearized about the saddle point, and

$$\tau_0 = \frac{\tau_N}{\alpha + \alpha^{-1}} = \frac{v\mu_0 M_S}{2kT\gamma} \tag{2.86}$$

is the damping-independent normalizing time (do not confuse with τ_0 from Eq. (2.10)). For vanishing damping, $\alpha \to 0$, the IHD escape rate Γ_A^{IHD} from Eq. (2.83) reduces to the TST escape rate Γ_A^{TST},

$$\Gamma_A^{\mathrm{TST}} = f_A e^{-\Delta E_A}, \tag{2.87}$$

$(f_A = \omega_A / 2\pi)$, which is independent of α. However, this is not the *true* VLD regime, where the energy loss per cycle of the almost-periodic noisy motion of the magnetization on the saddle-point energy (escape) trajectory is much less than the thermal energy. Instead, it comprises the *intermediate damping regime* corresponding to the TST result. Recognizing this, Klik and Gunther [106,107] derived the *correct* VLD magnetization escape rate, viz., (see Chapter 1, Section 1.5.5)

Table 2.2. Range of validity of asymptotic formulas for the escape rate Γ_A.

Escape rates	Γ_A^{TST}	Γ_A^{IHD}	Γ_A^{VLD}	Γ_A
Equations	(2.87)	(2.83)	(2.88)	(2.89)
Ranges of validity	$\alpha S_C \sim 1$	$\alpha S_C > 1$	$\alpha S_C << 1$	all α
References	[15]	[18]	[106,107]	[181, 196]

$$\Gamma_A^{\mathrm{VLD}} = \alpha S_C f_A e^{-\Delta E_A}, \tag{2.88}$$

where the action S_C at the saddle-point energy E_C is given by Eq. (1.304) of Chapter 1. The conditions of applicability of these IHD and VLD solutions for nanomagnets are precisely defined by $\alpha S_C \to 1$ and $\alpha S_C << 1$, respectively. In the turnover region, $0.01 < \alpha S_C < 1$, by extending the Mel'nikov–Meshkov formalism, Coffey *et al.* [181, 196] have obtained for the escape rate Γ_A from a *single* well

$$\Gamma_A = A(\alpha S_C)\Gamma_A^{\mathrm{IHD}} = A(\alpha S_C)\frac{\omega_A \lambda_+(\alpha)}{2\pi\omega_C} e^{-\Delta E_A}, \tag{2.89}$$

where the magnetization depopulation factor A is defined as

$$A(\Delta) = e^{\frac{1}{\pi}\int_0^\infty \frac{\ln\{1-\exp[-\Delta(\lambda^2+1/4)]\}}{\lambda^2+1/4} d\lambda}. \tag{2.90}$$

The turnover Eq. (2.89) may be deemed universal, insofar as it describes the magnetization escape rate Γ_A for all damping α. The behavior of $A(\alpha S_C)$ from Eq. (2.90) as a function of α, namely, $A(\alpha S_C) \to 1$ as $\alpha \to \infty$ and $A(\alpha S_C) \to \alpha S_C$ as $\alpha \to 0$, ensures that both IHD and VLD regimes of the escape rate, Eqs. (2.83) and (2.88), respectively, are correctly reproduced. In practical applications, the precise conditions of validity of the escape-rate formulas must always be taken into account. First, they only hold in the high-barrier limit and for the Taylor approximation, Eq. (2.82), to the free energy density in the vicinity of the stationary points. For convenience, the damping ranges in which the various escape-rate formulas are valid are summarized in Table 2.2.

We emphasize that the (overall) *reversal time* of the magnetization τ and the *individual inverse escape times* Γ_A^{-1} may differ considerably from each other. In general, both depend on the energyscape as well as on the particular damping regime. However, the relation between τ and Γ_A can be found using the discrete orientation model (see Section 1.4.2). For example, for a double-well potential with *two nonequivalent wells* and *one saddle point*, τ is given by (see Section 1.5.6)

$$\tau = \tau_{\text{IHD}} \frac{A(\alpha S_1 + \alpha S_2)}{A(\alpha S_1)A(\alpha S_2)}, \tag{2.91}$$

where

$$\tau_{\text{IHD}} = \left(\Gamma_1^{\text{IHD}} + \Gamma_2^{\text{IHD}}\right)^{-1}, \tag{2.92}$$

Γ_1^{IHD} is the escape rate from well 1 to well 2, Γ_2^{IHD} is the escape rate from well 2 to well 1, and the escape rates Γ_i^{IHD} are defined by Eq. (2.83).

All the results we have presented in this section for the reversal time pertain to the memoryless (white-noise) limit of the stochastic magnetization dynamics. If long-time memory is included, all the basic equations, e.g., Gilbert's equation, escape-rate equations, etc., must be generalized in an appropriate manner (see, for example, Refs. [107,113], where a generalized Gilbert equation taking into account memory effects is given). For a generalized (non-Markovian) Langevin description of the stochastic spin dynamics, the escape rates may differ from the predictions of the Kramers theory (see, for example, Refs. [107,113]). Nevertheless, as conjectured in Ref. [107], if memory effects are included, the high-barrier asymptotes for the reversal time should hold with an *effective* damping constant.

2.3.2 *Uniaxial nanomagnet in an oblique dc magnetic field*

A 3D plot of the free energy potential $E(\vartheta, \varphi)/\sigma$, Eq. (2.6), of an uniaxial nanomagnet in an oblique uniform dc magnetic field $\mathbf{H}_0$ is shown in Fig. 2.1. A detailed treatment of the VLD reversal time of such a nanomagnet has been given in Section 2.2.2. The turnover formula for τ (valid in all damping ranges) for the oblique field potential, Eq. (2.6), with two nonequivalent wells is given by Eq. (2.91). For the free energy potential $E(\vartheta, \varphi)$, Eq. (2.6), the equations for $c_1^{(p)}, c_2^{(p)}$ and E_p become [108]

$$\begin{aligned} c_1^{(p)} &= 2\sigma\left[\cos 2\vartheta_p + h\cos(\vartheta_p - \psi)\right], \\ c_2^{(p)} &= 2\sigma\left[\cos^2\vartheta_p + h\cos(\vartheta_p - \psi)\right], \\ E_p &= \sigma\left[\sin^2\vartheta_p - 2h\cos(\vartheta_p - \psi)\right], \end{aligned} \tag{2.93}$$

where ϑ_p are the solutions of the equation $\sin 2\vartheta = 2h\sin(\psi - \vartheta)$. Now to derive an explicit analytic formula for τ, we require the parameters $\Delta E_{1,2}$, $\omega_{1,2}$, ω_C, and $c_{1,2}^{(C)}$ in the escape-rate Eqs. (2.91) and (2.83). This may be accomplished as follows. First, the equation $\partial_\vartheta E(\vartheta, 0) = 0$ may be rewritten as the quartic Eq. (2.12) for $\cos\vartheta$. The roots of Eq. (2.12), corresponding to the minima and saddle point, viz., $\cos\vartheta_A^{\pm} = \cos\vartheta_{1,2}$ and $\cos\vartheta_C$ are given by Eq. (2.13). The corresponding Taylor series for these can be obtained from Eqs. (2.93) and (2.13). We have [207]

$$\begin{aligned} \Delta E_{1,2} &= \sigma\Big[1 \pm 2h(\cos\psi \mp \sin\psi) + h^2 \\ &\quad + \frac{1}{2}h^3\sin 2\psi(\cos\psi \mp \sin\psi) + \frac{1}{2}h^4\sin^2 2\psi \cdots\Big], \\ \omega_{1,2} &= \frac{\sigma\gamma kT}{v\mu_0 M_S}\left\{2 \pm 2h\cos\psi - h^2\sin^2\psi\left[1 \mp 3h\cos\psi + \frac{h^2}{8}(21 + 19\cos 2\psi) + \cdots\right]\right\}, \\ c_1^{(C)} &= 2\sigma h\sin\psi\left\{1 + h^2\cos^2\psi\left[\frac{1}{2} + h\sin\psi + \frac{3h^2}{16}(5 - 3\cos 2\psi) + \cdots\right]\right\}, \\ c_2^{(C)} &= 2\sigma\left[-1 + h\sin\psi + h^2\cos^2\psi + \frac{5h^3}{4}\sin 2\psi\cos\psi + h^4\sin^2 2\psi + \cdots\right], \\ \omega_C &= \frac{\sigma\gamma kT}{v\mu_0 M_S}\sqrt{h\sin\psi}\Big[2 - h\sin\psi - \frac{h^2}{8}(3 + \cos 2\psi) - \frac{h^3}{16}(19 + 17\cos 2\psi)\sin\psi \\ &\quad - \frac{h^4}{512}(351 + 28\cos 2\psi - 283\cos 4\psi) + \cdots\Big]. \end{aligned} \tag{2.94}$$

The actions $S_{1,2} = S_C^{\pm}$ are given by Eq. (2.31). Thus, using $\Delta E_{1,2}$, $\omega_{1,2}$, ω_C, $c_{1,2}^{(C)}$, and $S_{1,2} = S_C^{\pm}$, one can estimate τ from the turnover equation, Eq. (2.91). In the VLD regime ($\alpha \ll 1$), Eq. (2.91) yields the VLD asymptote τ_{VLD} for the reversal time of the magnetization, viz.,

$$\tau_{\mathrm{VLD}} = \alpha^{-1}\tau_{\mathrm{TST}}(S_1^{-1} + S_2^{-1}), \tag{2.95}$$

where the TST reversal time τ_{TST} is given by

$$\tau_{\mathrm{TST}} = \frac{2\pi}{\omega_1 e^{-\Delta E_1} + \omega_2 e^{-\Delta E_2}}. \tag{2.96}$$

The limiting value $h\sin\psi = 0$ corresponds to uniaxial anisotropy. Here the normalized reversal time τ/τ_N is independent of α and is in consequence given by Brown's formula (2.4). For $\psi = \pi/2$ (transverse field), when the potential, Eq. (2.6), has *two equivalent wells*, Eq. (2.91) can be vastly simplified [108,354,355]. We have

$$\tau = \frac{2\tau_0(\alpha + \alpha^{-1})\pi\sqrt{h}e^{\sigma(1-h)^2}A(2\alpha S_i)}{\sigma\sqrt{1+h}[1 - 2h + \sqrt{1 + 4h(1-h)\alpha^{-2}}]A^2(\alpha S_i)}, \tag{2.97}$$

where

$$S_i = \sigma\sqrt{h}\left(16 - \frac{104}{3}h + 22h^2 - 3h^3 + \frac{7}{24}h^4 + \frac{1}{16}h^5 + \cdots\right). \tag{2.98}$$

We emphasize that Eqs. (2.91) and (2.97) are invalid for very small $h\sin\psi$, viz., $h\sin\psi < 0.03$ [181,359]. Then the dependence of E on the azimuthal angle φ is weak (i.e., the potential, Eq. (2.6), is almost circularly symmetric), and *all* escape paths from the potential wells are thus *roughly equivalent*. To estimate τ one can now use the following relation derived in Ref. [359] via perturbation theory (details are in Ref. [181]):

$$\tau \cong \tau_{\mathrm{B}}\left\{1 + h^2\sigma^2\left[1 + 2\left(2\sigma\alpha^2 e\right)^{1/(2\sigma\alpha^2)}\gamma\left(1 + \frac{1}{2\sigma\alpha^2}, \frac{1}{2\sigma\alpha^2}\right)\right]\right\}^{-1}, \tag{2.99}$$

where $\tau_{\mathrm{B}} = \tau_{\mathrm{as}}$ is the reversal time for a uniaxial (circularly symmetric) potential given by Eq. (2.5) and $\gamma(a,z)$ is the incomplete gamma function defined by Eq. (A.4) in Appendix A.1.1. It can be shown [359] that the expression in square brackets in Eq. (2.99) is about 1 for $\alpha >> 1$ and about $\approx \alpha^{-1}(\pi/\sigma)^{1/2}$ for $\alpha << 1$. The range of validity of Eq. (2.99) is determined by the inequalities $h^2\sigma^2 << 1$, $\alpha > 4h^2\sigma^{3/2}$, and $\sigma \approx 4$.

The turnover equation (2.91) allows one to estimate the reversal time over wide ranges of the model parameters (see Fig. 2.8). In this figure, τ from the turnover, Eq. (2.91), VLD, Eq. (2.95), and TST, Eq. (2.96), is

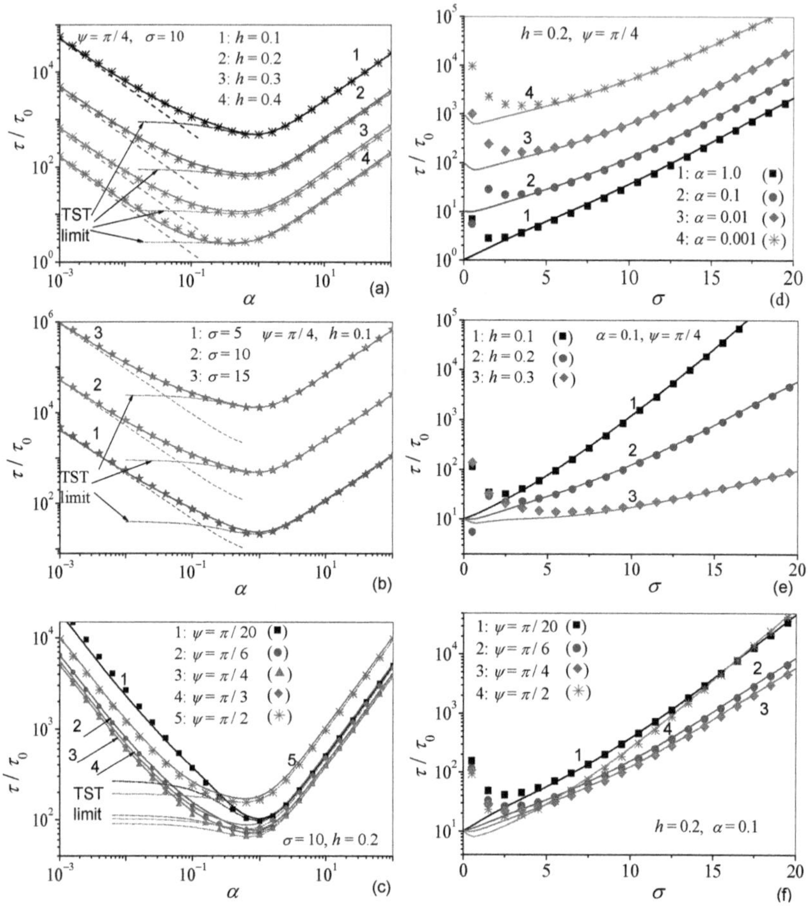

Fig. 2.8. (Left) τ/τ_0 vs. damping parameter α (a) for $\sigma = 10$, $\psi = \pi/4$, and various dc field parameters h, (b) for $h = 0.1$, $\psi = \pi/4$, and various inverse-temperature parameters σ, and (c) for $h = 0.2$, $\sigma = 10$, and various ψ. Solid lines: matrix continued fraction solution [32]; dashed lines: the VLD asymptotes; dashed-dotted lines: the IHD reversal time; symbols: Eq. (2.91). (Right) τ/τ_0 vs. σ (d) for $h = 0.2$, $\psi = \pi/4$, and various α, (e) for $\alpha = 0.1$, $\psi = \pi/4$, and various h, and (f) for $\alpha = 0.1$, $h = 0.2$, and various angles ψ. Solid lines: matrix continued fraction solution [32] (see Section 2.5.1); symbols: Eq. (2.91).

compared with numerical calculations yielded by matrix continued fractions [32] (see Section 2.5.1). The TST (Néel) time τ_{TST} from Eq. (2.96) is independent of damping α, and so may only be used to estimate τ for uniaxial nanomagnets in a *very narrow* range of α. However, in the VLD and VHD regimes, the deviation between τ from Eq. (2.91) and τ_{TST} from Eq. (2.96) can again be of several orders of magnitude; see Fig. 2.8. In the high-barrier limit, the turnover Eq. (2.91) provides an accurate approximation to τ for all α. Equation (2.91) can be used to deduce experimental values of α and to reproduce the angular variation of the switching field of *individual* nanoparticles. It can also be used for comparison with switching curves measured on a single (Co) nanoparticle [20,72,73]. Hence it should be possible [82,83] to determine α by fitting the theory to the experimental switching-field curves (see Section 1.6.2).

The above results for the reversal time were obtained for an *individual* uniaxial particle thus constituting a *bare* relaxation time. However, to calculate the relaxation time of an *assembly* of uniaxial nanoparticles, where the easy axes are *randomly* oriented in space, averaging over the oblique angle ψ must also be undertaken [172,173]. Observing that $\tau(\psi) = \tau(\pi - \psi)$, the averaging of $\tau(\psi)$ over particle easy axis orientations is accomplished numerically using Gaussian quadratures [290], namely

$$\bar{\tau} = \int_0^{\pi/2} \tau(\psi)\sin\psi \, d\psi = \frac{\pi}{4}\sum_{i=1}^{n} w_i \tau(\psi_i)\sin\psi_i, \tag{2.100}$$

where

$$w_i = \frac{2(1-x_i^2)}{\left[(n+1)P_{n+1}(x_i)\right]^2}, \psi_i = \frac{\pi}{4}(x_i + 1), \tag{2.101}$$

and x_i is the i-th root of the Legendre polynomial $P_n(x)$. In Fig. 2.9, the relaxation time $\bar{\tau}$ of the assembly is compared with the bare relaxation time $\tau = \tau(\psi)$ for the oblique angles $\psi = 0$, $\pi/4$, and $\pi/2$. Notice that the calculation of $\bar{\tau}$ from the escape-rate equations, Eq. (2.91), can be drastically simplified by assuming that the ψ dependent parameters $\Delta E_{1,2}$, $\omega_{1,2}$, and $S_{1,2}$ appearing under the averaging sign in Eq. (2.100) may be simply replaced by their averages. We then have from Eqs. (2.91) the *average* relaxation time:

$$\bar{\tau} \sim \frac{2\pi A(\alpha\overline{S_1} + \alpha\overline{S_2})}{(\overline{\omega_1} e^{-\overline{\Delta E_1}} + \overline{\omega_2} e^{-\overline{\Delta E_2}})A(\alpha\overline{S_1})A(\alpha\overline{S_2})}. \tag{2.102}$$

Now $\overline{\Delta E_{1,2}}$, $\overline{\omega_{1,2}}$, and $\overline{S_{1,2}}$ as appearing in Eq. (2.102) can be calculated analytically by averaging Eq. (2.31) over ψ, namely

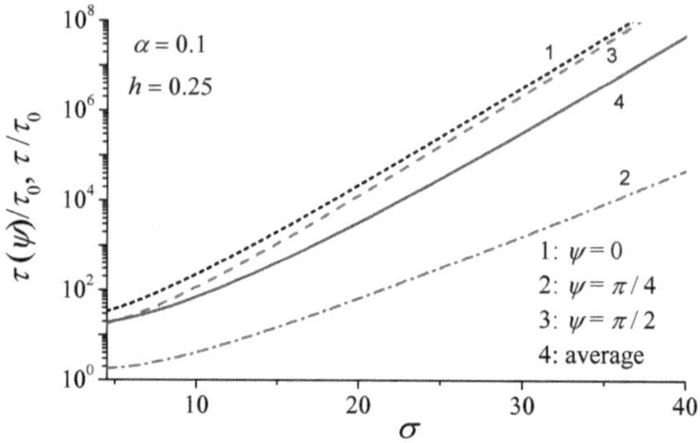

Fig. 2.9. Bare magnetization relaxation time $\tau(\psi)/\tau_0$ for various angles ψ and average $\bar{\tau}/\tau_0$ [Eq. (2.100): solid line] vs the inverse-temperature parameter $\sigma = vK/(kT)$ for the field parameter $h = 0.25$ and damping $\alpha = 0.1$.

$$\overline{\Delta E_{1,2}} = \sigma\left[1 + h\left(\pm 1 - \frac{\pi}{2}\right) + h^2 + \ldots\right],$$

$$\overline{\omega_{1,2}} = \tau_0^{-1}\left(1 \pm \frac{h}{2} - \frac{h^2}{3} + \ldots\right), \tag{2.103}$$

$$\overline{S_{1,2}} = \sigma\frac{8}{3}\sqrt{\frac{h}{2\pi}}\left[\Gamma^2(1/4) - \frac{39}{32}\Gamma^2(-1/4)h \pm (2\pi h)^{3/2} + \ldots\right],$$

where $\Gamma(z)$ is the gamma function. Consequently, by using Eqs. (2.102)–(2.122), we have the average reversal time $\bar{\tau}$ of the assembly in analytic fashion. Figure 2.10 illustrates that the approximate Eq. (2.102) provides an accurate approximation to $\bar{\tau}$ as calculated via matrix continued fractions. Clearly from Fig. 2.10, $\bar{\tau}$ strongly depends on α (which cannot be predicted by TST) and the external field parameter h. Thus $\bar{\tau}$ *is the key parameter characterizing the thermal instability of the magnetization*. Consequently, estimates of that time are required in numerous physical applications, e.g., to accurately evaluate linear and nonlinear dynamic susceptibilities, the specific power loss in dynamic magnetic hysteresis, the signal-to-noise ratio in stochastic resonance, etc. These can all be evaluated using Eq. (2.102). In deriving our results, applying for *random* orientation of the easy axes, we suppose that all particles are identical. To account for the *polydispersity* of the particles of a real sample, one must

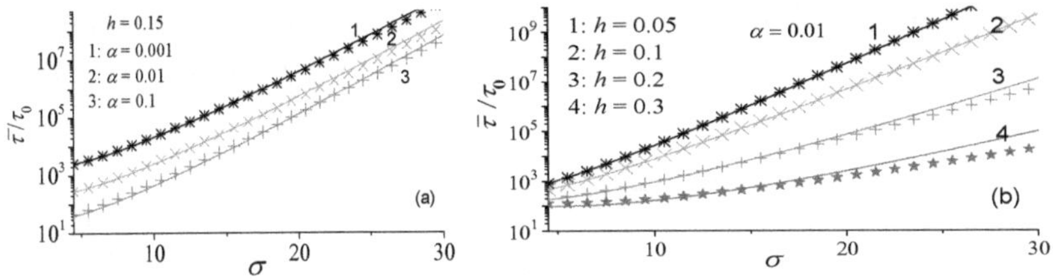

Fig. 2.10. $\bar{\tau}/\tau_0$ vs the inverse-temperature parameter $\sigma = vK/(kT)$ (a) for $h = 0.15$ and various values of the damping parameter α and (b) for $\alpha = 0.01$ and various values of the field parameter h. Solid lines: Eq. (2.100), where $\tau(\psi)$ is calculated via matrix continued fractions [32]. Symbols: Eqs. (2.102).

also impose in fitting all the quantities of interest (susceptibility, etc.) an additional averaging procedure over the appropriate distribution function of *particle volumes*.

2.3.3 *Biaxial nanomagnet in a uniform dc magnetic field applied along the easy axis*

For zero dc bias field, τ for orthorhombic crystals has been calculated by Smith and de Rozario [90] in the IHD regime, and by Kalmykov *et al.* [151,152,354,355] for all α. The appropriate formula for the IHD reversal time has been given by Braun [351] for the basically similar problem of the magnetization dynamics in elongated biaxial particles subjected to a strong dc magnetic field. Furthermore, Ouari and Kalmykov [152] have evaluated τ in a dc bias field for all α. For $0 < h < 1$ the anisotropy potential, Eq. (2.38), has two *nonequivalent wells* and two *equivalent saddle points* (see Fig. 2.2). The reversal time is formally given by the turnover formula, Eq. (2.91). For biaxial nanomagnets, the discrete-state orientation model implies (see Section 1.4.2) that in the IHD regime, the mean magnetization decays with the relaxation time τ_{IHD} which is given by

$$\tau_{\text{IHD}} = \left[2\left(\Gamma_1^{\text{IHD}} + \Gamma_2^{\text{IHD}}\right)\right]^{-1}, \tag{2.104}$$

where Γ_i^{IHD} is the escape rate from the well i for the IHD regime over one saddle point given by Eq. (2.83) and the factor 2 occurs because there are *two* magnetization escape routes from the well over the saddle points. For biaxial anisotropy, Eq. (2.38), the parameters $c_1^{(p)}$, $c_2^{(p)}$, and E_p appearing in Eq. (2.83) are given by

$$c_1^{(p)} = 2\sigma\left(\cos 2\vartheta_p + h\cos\vartheta_p\right),$$
$$c_2^{(p)} = 2\sigma\left(\delta + \cos^2\vartheta_p + h\cos\vartheta_p\right), \tag{2.105}$$
$$E_p = \sigma\left(\sin^2\vartheta_p - 2h\cos\vartheta_p\right).$$

The angles ϑ_p are the solutions of the equation $\partial_\vartheta E\big|_{\varphi=\pi/2} = 0$. These are $\vartheta_1 = 0$, $\vartheta_2 = \pi$, and $\cos\vartheta_C = -h$. Thus, we have

$$\Delta E_{1,2} = \sigma(1 \pm h)^2,$$
$$c_1^{(i)} = 2\sigma(1 \pm h),$$
$$c_2^{(i)} = 2\sigma(1 + \delta \pm h), \tag{2.106}$$
$$c_1^{(C)} = -2\sigma(1 - h^2),\ c_2^{(C)} = 2\sigma\delta.$$

The escape rate Γ_2^{IHD} is now explicitly given by

$$\Gamma_2^{\mathrm{IHD}}(h) = \frac{\sigma e^{-(1-h)^2\sigma}}{2\pi\tau_0(\alpha + \alpha^{-1})}\sqrt{\frac{1-h+\delta}{\delta(1+h)}}$$
$$\times\left[1 - h^2 - \delta + \sqrt{(1-h^2+\delta)^2 + 4\delta(1-h^2)\alpha^{-2}}\right] \tag{2.107}$$

and $\Gamma_1^{\mathrm{IHD}}(h) = \Gamma_2^{\mathrm{IHD}}(-h)$. The dimensionless actions $S_{1,2}$ are given by the contour integral, Eq. (1.304) of Chapter 1, taken along the separatrixes which are determined by the equation $E(\vartheta, \varphi) = E_C = \sigma(1+h^2)$, where E_C is the value of E at the saddle points. This equation is

$$\cos^2\vartheta(1 + \delta\cos^2\varphi) \pm 2h\cos\vartheta + h^2 - \delta\cos^2\varphi = 0.$$

Appropriate solutions of the above equation yield two separatrixes

$$\cos\vartheta_{1,2}(\varphi)\big|_{E=E_C} = \frac{\mp h + \cos\varphi\sqrt{\delta(1-h^2) + \delta^2\cos^2\varphi}}{1 + \delta\cos^2\varphi}.$$

The actions $S_{1,2}$ can be evaluated analytically from Eq. (1.304) over the path $E = E_C$ between two saddle points as [152] (cf. Eq. (2.76))

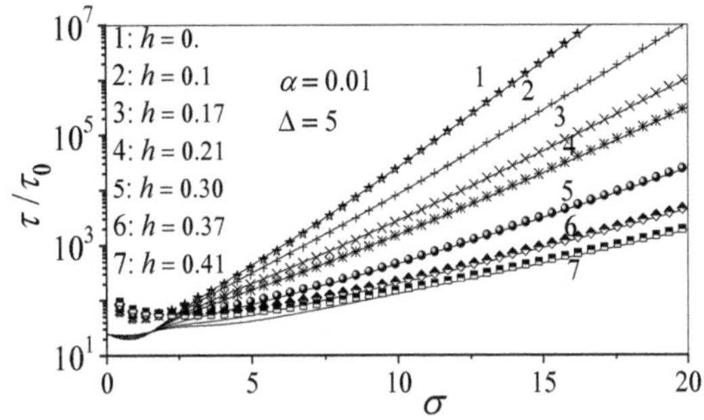

Fig. 2.11. Reversal time τ/τ_0 vs. the barrier-height parameter σ for various values of the field parameter h and $\alpha = 0.01$ and $\Delta = \sigma\delta = 5$. Solid lines 1–7: matrix continued fraction solution [152]. Symbols: turnover, Eq. (2.91).

$$S_{1,2} = 2\sigma\left(1-h^2+\delta\right)$$
$$\times \int_{-\pi/2}^{\pi/2} \left\{ \frac{1-h^2+\delta(1+h^2)\cos^2\varphi}{\sqrt{1-h^2+\delta\cos^2\varphi}} \pm 2h\sqrt{\delta}\cos\varphi \right\} \frac{\cos\varphi d\varphi}{\left(1+\delta\cos^2\varphi\right)^2}$$
$$= \frac{4\delta\sigma\left(1-h^2+\delta\right)}{(1+\delta)^{3/2}} \tag{2.108}$$
$$\times \left\{ \left[\left(1+\frac{1}{\delta}\right)(1-h^2)\right]^{1/2} + h\arctan\left[h\left[(1-h^2)\left(1+\frac{1}{\delta}\right)\right]^{-1/2}\right] \pm \frac{h\pi}{2} \right\}.$$

Equation (2.91) also yields, with $h = 0$, [354,355]

$$\tau = \frac{\tau_0(\alpha+\alpha^{-1})\pi e^{\sigma} A(8\alpha\sigma\sqrt{\delta})}{\sigma\sqrt{1+1/\delta}\left[1-\delta+\sqrt{(1+\delta)^2+4\delta/\alpha^2}\right]A^2(4\alpha\sigma\sqrt{\delta})}. \tag{2.109}$$

Equations (2.91), (2.107)–(2.108) enable us to evaluate τ from Eq. (2.91) *for all* α. The time τ as predicted by the turnover equation, Eq. (2.91), and $1/\lambda_1$ as calculated by the matrix continued fraction method for a biaxial potential [152], are shown in Figs. 2.11 and 2.12. Once again, in the high-barrier limit, $\sigma >> 1$, the turnover equation (2.91) provides a good approximation to τ for all α. We emphasize that the asymptotic Eq. (2.91) is invalid for $\delta = 0$ corresponding to uniaxial anisotropy; here in contrast τ is correctly given by Brown's equation, Eq. (2.4).

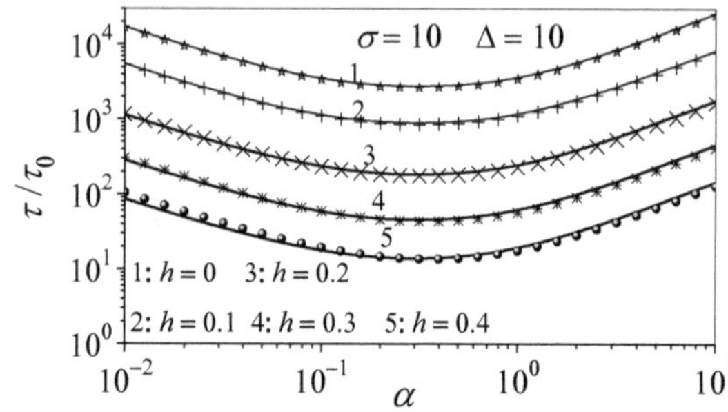

Fig. 2.12. Reversal time τ/τ_0 vs. the damping parameter α for various values of h for $\Delta = \sigma\delta = 10$ and $\sigma = 10$. Solid lines 1–5: matrix continued fraction solution [152]. Symbols: turnover, Eq. (2.91).

2.3.4 *Reversal time for cubic anisotropy*

The several low-order eigenvalues for the cubic anisotropy potential, Eq. (1.15), were first calculated numerically in the IHD regime by Aharoni [360] and Eisenstein and Aharoni [361,362]. The IHD formulas for the escape rates for cubic crystals were derived by Smith and de Rozario [90] and by Brown [18]. Klik and Gunther [106,107] have also derived corresponding formulas for the VLD escape rate. The magnetization reversal time was also estimated in Refs. [33–35,354,355]. Furthermore, the effect of a dc bias field on the reversal time was studied in Refs. [31] and [363].

For a positive anisotropy constant, $K_c > 0$ (Fe-type crystals), the potential in Eq. (1.15) has six minima (wells), eight maxima and twelve saddle points (see Fig. 1.3(c)). Then the turnover formula, Eq. (2.89), yields

$$\tau = \frac{\tau_{\mathrm{IHD}}}{A(\alpha S_i)}. \tag{2.110}$$

As shown (Section 1.4.2 of Chapter 1), for the discrete orientation model, the mean magnetization decays with time constant $\tau_{\mathrm{IHD}} = (4\Gamma_i^{\mathrm{IHD}})^{-1}$. Here Γ_i^{IHD} is the IHD escape rate from the well i, Eq. (2.83). For the cubic anisotropy potential Eq. (1.15) the coefficients $c_1^{(p)}$, $c_2^{(p)}$, and E_p are now

$$c_1^{(p)} = 8\sigma\cos 4\vartheta_p,\; c_2^{(p)} = 2\sigma(3+\cos 4\vartheta_p),\; E_p = \sigma\sin^2 2\vartheta_p. \tag{2.111}$$

Here $\sigma = vK_c/(4kT)$ is the dimensionless anisotropy (inverse-temperature) parameter, the ϑ_p are the solutions of the trigonometric equation $\partial_\vartheta E|_{\varphi=0} = 0$, viz., $\vartheta_1 = 0$, $\vartheta_2 = \pi/2$, $\vartheta_3 = \pi$ (wells) and $\vartheta_C = \pi/4$, $\vartheta_C = 3\pi/4$ (saddle points). Thus, we obtain

$$\begin{gathered} c_1^{(i)} = c_2^{(i)} = 8\sigma, c_1^{(C)} = 4\sigma,\ c_2^{(C)} = -8\sigma, \\ \Delta E_i = E_C - E_i = \sigma, \end{gathered} \tag{2.112}$$

and

$$\tau_{\mathrm{IHD}} = \frac{1}{4\Gamma_i^{\mathrm{IHD}}} = \frac{\tau_0(\alpha+\alpha^{-1})\pi e^{\sigma}}{2\sqrt{2}\sigma(\sqrt{9+8/\alpha^2}+1)}. \tag{2.113}$$

To evaluate the action S_C given by Eq. (1.304) in the turnover equation, Eq. (2.110), we again need an explicit solution for the critical trajectory (separatrix). For the well with a minimum at $\vartheta = 0$, this critical trajectory, passing between the two saddle points at $\varphi_C = 0$, $\vartheta_C = \arccos 2^{-1/2}$ and $\varphi_C = \pi/2$, $\vartheta_C = \arccos 2^{-1/2}$ is determined from the trigonometric equation $\sin^4\vartheta \sin^2 2\varphi + \sin^2 2\vartheta = 1$, where the physically meaningful solution is

$$\cos\vartheta(\varphi)\big|_{E=E_C} = \sqrt{\frac{1+\sin 2\varphi}{2+\sin 2\varphi}}. \tag{2.114}$$

The dimensionless action at the saddle point energy S_i is defined by Eq. (2.35). Hence, we have from Eqs. (1.304) and (2.114) [181]

$$\begin{aligned} S_i &= 12\sigma \int_0^{\pi/2} \sin 2\varphi (1+\sin 2\varphi)^{1/2} (2+\sin 2\varphi)^{-5/2} d\varphi \\ &= \frac{8\sqrt{2}\sigma}{9}. \end{aligned} \tag{2.115}$$

Thus, using Eqs. (2.110), (2.113), and (2.115), we have the turnover formula [354,355]

$$\tau = \frac{\tau_0(\alpha+\alpha^{-1})\pi e^{\sigma}}{2\sqrt{2}\sigma(\sqrt{9+8/\alpha^2}+1)A(8\sqrt{2}\sigma\alpha/9)}. \tag{2.116}$$

In the VLD regime ($\alpha \ll 1$), Eq. (2.116) yields the VLD asymptote, viz.,

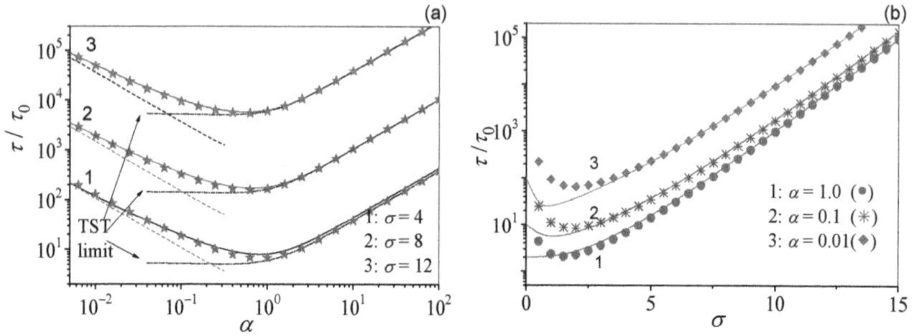

Fig. 2.13. Cubic anisotropy, $\sigma > 0$ (Fe-type crystals): (a) τ/τ_0 vs. α for various σ. Solid lines: matrix continued fraction solution [33,34]; dashed lines: the VLD equation, Eq. (2.117); dashed-dotted lines: the IHD equation, Eq. (2.113); stars: the turnover equation, Eq. (2.116); (b) τ/τ_0 vs. σ for various α. Solid lines: matrix continued fraction solution [33,34] (see Section 2.5.2); diamonds: the VLD equation, Eq. (2.117); circles: the IHD equation, Eq. (2.113); asterisks: the turnover equation, Eq. (2.116).

$$\tau_{\text{VLD}} = \frac{\tau_0 9\pi e^{\sigma}}{64\alpha\sqrt{2}\sigma^2}. \tag{2.117}$$

By formally setting $\alpha = 0$ in Eq. (2.113), we obtain the TST reversal time τ_{TST} predicted by the Néel theory, namely,

$$\tau_{\text{TST}} = \frac{\tau_0 \pi e^{\sigma}}{8\sigma}. \tag{2.118}$$

The analytic equation (2.116) so obtained allows one to easily estimate the reversal time τ for all damping α. In Fig. 2.13, τ as evaluated from Eq. (2.116), (2.117), and (2.118) is compared with numerical calculations using the matrix continued fraction technique [33,34]. The TST (Néel) time τ_{TST} from Eq. (2.118) is independent of damping, and again may be used to estimate τ for nanomagnets with positive cubic anisotropy *only in the very narrow range of damping,* $\alpha \sim 1$. In the VLD and VHD regimes, deviations between τ from Eq. (2.116) and τ_{TST} from Eq. (2.118) can be of *several orders of magnitude* (see Fig. 2.13).

For negative anisotropy constant, $\sigma < 0$ (Ni-type crystals), the maxima and minima are interchanged, and the barrier height is now given by $v|K_c|/(12kT)$. By using the discrete orientation model (see Section 1.4.2 of Chapter 1), we see that the mean magnetization decays with time constant $1/(2\Gamma_i^{\text{IHD}})$.

Therefore, since

$$c_1^{(p)} = -|\sigma|(8\cos 2\vartheta_p - 3\sin^2 2\vartheta_p),$$
$$c_2^{(p)} = -2|\sigma|(\cos 2\vartheta_p + 3\cos 4\vartheta_p),$$
$$E_p = -|\sigma|(\sin^4 \vartheta_p + \sin^2 2\vartheta_p),$$

where $p = C, 1, 2$, $\vartheta_1 = \arctan(2^{1/2})$, $\vartheta_2 = \pi - \arctan(2^{1/2})$ (wells) and $\vartheta_C = \pi/2$ (saddle point) are the solutions of the trigonometric equation $\partial_\vartheta E|_{\varphi=\pi/4} = 0$, so that

$$\Delta E_i = E_C - E_i = |\sigma|/3,$$
$$c_1^{(i)} = c_2^{(i)} = 16|\sigma|/3,\ c_1^{(C)} = 8|\sigma|,\ c_2^{(C)} = -4|\sigma|,$$

we then have [18]

$$\tau_{\mathrm{IHD}} = \frac{1}{2\Gamma_i^{\mathrm{IHD}}} = \frac{3\tau_0(\alpha+\alpha^{-1})\pi e^{|\sigma|/3}}{2\sqrt{2}|\sigma|(\sqrt{9+8/\alpha^2}-1)}. \tag{2.119}$$

The explicit turnover formula for τ is then [354,355]

$$\tau = \frac{\tau_{\mathrm{IHD}}}{A(\alpha S_i)} = \frac{3\tau_0(\alpha+\alpha^{-1})\pi e^{|\sigma|/3}}{2\sqrt{2}|\sigma|(\sqrt{9+8/\alpha^2}-1)A(\alpha|\sigma|8\sqrt{2}/9)}, \tag{2.120}$$

where $S_i = 8\sqrt{2}|\sigma|/9$ just as $K_c > 0$ while the VLD asymptote ($\alpha << 1$) is

$$\tau_{\mathrm{VLD}} = \tau_0 \frac{27\pi e^{|\sigma|/3}}{64\sqrt{2}\sigma^2\alpha}. \tag{2.121}$$

By setting $\alpha = 0$ in Eq. (2.119), we obtain

$$\tau_{\mathrm{TST}} = \tau_0 \frac{3\pi e^{|\sigma|/3}}{8|\sigma|}. \tag{2.122}$$

In Fig. 2.14, τ rendered by Eqs. (2.120), (2.121), and (2.122) is again compared with that from numerical calculations using matrix continued fractions [33,34,35] (see Section 2.5.2). The TST relaxation time τ_{TST} from Eq. (2.122) is independent of damping, and thus may be used to

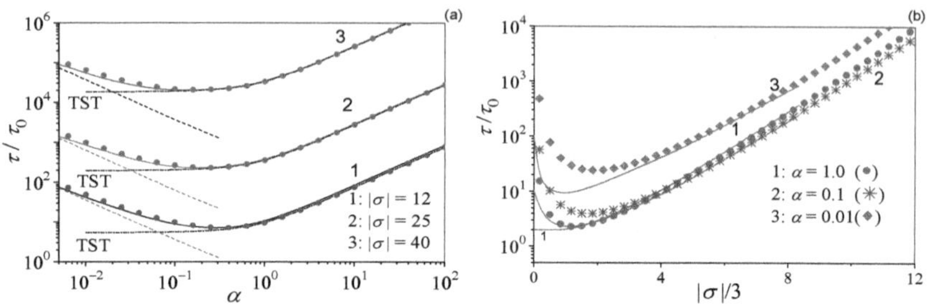

Fig. 2.14. Cubic anisotropy, $\sigma < 0$ (Ni-type crystals): (a) τ/τ_0 vs. α for various σ. Solid lines: matrix continued fraction solution [33–35]; dashed lines: the VLD equation, Eq. (2.121); dashed-dotted lines: the IHD equation, Eq. (2.119); filled circles: the turnover equation Eq. (2.120). (b) τ/τ_0 vs. $|\sigma|/3$ for various damping α. Solid lines: matrix continued fraction solution [33–35] (see Section 2.5.2); symbols: Eq. (2.120).

estimate τ for nanomagnets with negative cubic anisotropy, however again only in the very narrow range of damping, $0.2 < \alpha < 1$. In the VLD and VHD regimes, deviations between τ from Eq. (2.120) and τ_{TST} from Eq. (2.122) can again be of *several orders of magnitude* (see Fig. 2.14).

2.3.5 *Mixed anisotropy: breakdown of the paraboloid approximation*

The salient feature of the IHD equation, Eq. (2.83), is the elliptic and hyperbolic paraboloid approximation, i.e., Eq. (1.300), for the normalized free energy density $E(\mathbf{M})$ near the relevant stationary points. However, certain situations exist where either the well or the damped saddle frequencies or both are zero, so that the IHD Eq. (2.83) now predicts zero escape rate. This obviously incorrect result may occur (i) when $E(\mathbf{M})$ is approximately circularly symmetric, leading to the *uniaxial crossover* phenomenon, where the saddle points become *simple maxima*, or (ii) if the paraboloidal approximation, Eq. (1.300), fails. The breakdown of the paraboloidal approximation (ii) is encountered, for example, for $E(\mathbf{M})$ with either flat saddles or flat well bottoms, or both. An example is *mixed* uniaxial and cubic anisotropy

$$E = \sigma\left[\sin^2\vartheta + \frac{\zeta}{4}(\sin^4\vartheta\sin^2 2\varphi + \sin^2 2\vartheta)\right], \qquad (2.123)$$

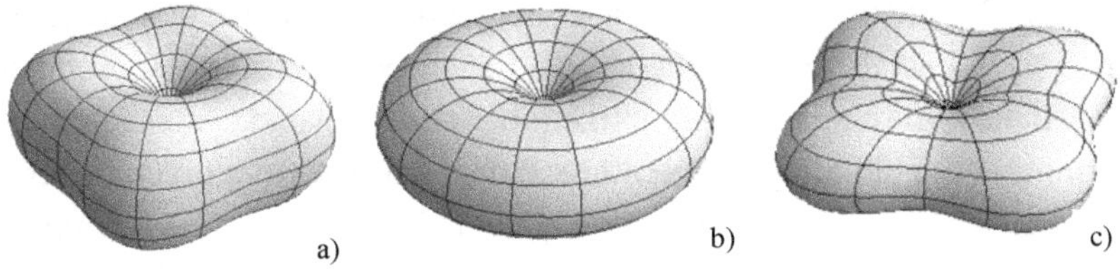

a) b) c)

Fig. 2.15. 3D plots of the mixed anisotropy potential for $\zeta = 1$ (a: flat saddles), $\zeta = 0$ (b: uniaxial anisotropy), and $\zeta = -1$ (c: flat wells).

where $\sigma = vK_u/(kT)$, K_u denotes the uniaxial anisotropy constant, and $\zeta = K_c/K_u$ is the cubic-to-uniaxial anisotropy ratio, which may be either positive or negative. For $|\zeta|/4 >> 1$ Eq. (2.123) represents *cubic* anisotropy. For $|\zeta| \leq 1$ Eq. (2.123) represents in contrast a double-well potential with two equivalent wells and four equivalent saddle points; these saddle points disappear at $\zeta = 0$, corresponding to a uniaxial anisotropy (see Fig. 2.15). For $\zeta = -1$ the well frequency $\omega_i = 0$, while for $\zeta = 1$ the damped saddle frequency $\lambda_+ = 0$. These values of ζ are of interest, since the existing escape-rate formulas described in Section 2.3.1 cannot be used to estimate the reversal time without modification. The mixed anisotropy equation, Eq. (2.123), appears in various applications. For example, in the "effective macrospin" model [364], where a many-spin cluster is mapped onto a macrospin representing the net moment of the cluster, with the corresponding energy comprising mixed uniaxial and cubic anisotropies. Here the effective anisotropy energy landscape depends on the size and shape of the cluster, the crystalline structure of the medium, and other physical parameters such as the exchange coupling and local anisotropy constants. The model provides a compromise between the *macrospin* (based on the Stoner–Wohlfarth concept of *coherent rotation of all the atomic spins*) and the *many-spin approaches*. The effective constants of the model, e.g., the parameter ζ etc., must, however, be computed from microscopic considerations to account for the crystallographic structure, the shape of the particle, the number of spins, etc., so that the model can be directly compared with experimental data or numerical simulations. Furthermore, the model may also be applied [199] to the switching-field curves of isolated Co nanoclusters characterized by mixed anisotropy (see Section 1.6) [72,73]. The mixed anisotropy energy,

Eq. (2.123), also occurs in paleomagnetism and rock magnetism [365,366]. Hence, thermal relaxation in such potentials is important for both thermo-remanent magnetization and related measurements determining the blocking temperature(s) characteristic of a given material. Moreover, the anisotropy energy, Eq. (2.123), permits many remanent states to coexist for $|\zeta| > 1$ thereby leading to *multiple blocking temperatures and a transition towards a single blocking temperature*. The magnetization relaxation rate problem for particles with mixed anisotropy, Eq. (2.123), was implicitly identified by Smith and de Rozario [90], Brown [18], and Dormann *et al.* [78], while Newell [365, 366] evaluated the IHD escape rate from Eq. (2.83), noting the absurd prediction of a vanishing escape rate for $|\zeta| = 1$. Now, insofar as the rate calculation in the vicinity of non-parabolic stationary points is concerned, a method for point Brownian particles with separable and additive Hamiltonians has been suggested by Talkner and Ryter [367] and Hänggi *et al.* [85]. This has been applied to a Brownian single-axis inertial rotator in a periodic potential with *non-parabolic* barriers [368], as well as to a nanoparticle with mixed anisotropy [369], yielding a turnover formula for the magnetization reversal time τ.

Following the exposition of Ref. [369], we shall, for purposes of illustration, confine our treatment to anisotropy ratios $|\zeta| \le 1$. The calculations can be extended to $|\zeta| > 1$ without major difficulties. For $0 < |\zeta| \le 1$, i.e., for a double-well potential with two equivalent wells and four equivalent saddle points, τ is given in terms of the depopulation factor $A(S)$ and the escape rate Γ_1^{IHD} from a single well as

$$\tau = \frac{A(2\alpha S)}{8\Gamma_1^{\mathrm{IHD}} A^2(\alpha S)}, \tag{2.124}$$

since $S_1 = S_2 = S$, $\Gamma_1^{\mathrm{IHD}} = \Gamma_2^{\mathrm{IHD}}$ and the IHD Γ_1^{IHD} refers to *one* saddle point only. The factor 8 occurs because (i) there are *four* magnetization escape routes from the well over the saddle points, and (ii) *two* equivalent wells are involved in the relaxation process.

To evaluate the action S in Eq. (2.124) from Eq. (1.304), we first need an explicit solution for the separatrix. For the distinct cases of *positive* and *negative* cubic anisotropy, i.e., $0 < \zeta \leq 1$ and $-1 \leq \zeta \leq 0$, the respective separatrixes are determined by the trigonometric equations

$$\begin{aligned} \sin^2 \vartheta + \zeta(\sin^4 \vartheta \sin^2 2\varphi + \sin^2 2\vartheta)/4 &= 1, \\ \sin^2 \vartheta - |\zeta|(\sin^4 \vartheta \sin^2 2\varphi + \sin^2 2\vartheta)/4 &= 1 - |\zeta|/4. \end{aligned} \quad (2.125)$$

The physically meaningful solutions of these equations are

$$\cos\vartheta(\varphi)\big|_{E=E_C} = \sqrt{\frac{\zeta(3+\cos 4\varphi) - 4 + 4\sqrt{(\zeta-1)^2 + \zeta \sin^2 2\varphi}}{\zeta(7+\cos 4\varphi)}},$$

$$\cos\vartheta(\varphi)\big|_{E=E_C} = \sqrt{\frac{|\zeta|(3+\cos 4\varphi) + 4 - 2\sqrt{4 + 8|\zeta| + |\zeta|(|\zeta| - 4)\sin^2 2\varphi}}{|\zeta|(7+\cos 4\varphi)}}.$$

(2.126)

By substituting Eqs. (2.125) and (2.126) into Eq. (1.304) of Chapter 1, we have the actions S analytically as Taylor series expansions up to any desired order of ζ in two distinct regions, viz.,

$$S = \sigma\sqrt{\zeta}\left(1 - \frac{\zeta}{6} + \frac{\zeta^2}{8} + \frac{17\zeta^3}{240} - \frac{\zeta^4}{128} - \cdots\right) \quad (2.127)$$

for *positive* cubic anisotropy and

$$S = \sigma\sqrt{|\zeta|}\left(1 + \frac{7}{6}|\zeta| - \frac{15}{32}\zeta^2 + \frac{363}{640}|\zeta|^3 - \frac{1569}{2048}\zeta^4 - \cdots\right) \quad (2.128)$$

for *negative* cubic anisotropy. Furthermore, S can also be computed numerically from Eq. (1.304).

Thus, the only remaining quantity in Eq. (2.124) is Γ_i^{IHD}, which cannot be evaluated by naively applying the IHD equation, Eq. (2.83), and requires separate analysis for $0 < \zeta \leq 1$ and $-1 \leq \zeta \leq 0$. In the first instance, the *hyperbolic* paraboloidal approximation at the saddle point breaks down, so that the *Kramers method of determining the crossover function between the wells needs to be modified.* In the second instance, the *elliptic* paraboloidal approximation at the bottom of the wells breaks down, so that the *calculation of well populations via steepest descents* needs to be modified. We treat both cases separately.

Despite the breakdown of the hyperbolic paraboloidal approximation near the saddle point, the IHD magnetization escape rate can still be expressed as a flux-over-population. To see this, we first recall that, in general, in IHD the picture is that inside the well the distribution function of **M** (t) is almost the equilibrium Boltzmann one. However, very near the saddle, the distribution *deviates* from the Boltzmann one because of the quasistationary reversal of **M** (t) over the saddle point. The saddle point (separatrix) region where nonequilibrium prevails is very small, and the saddle point itself is a stationary point; therefore, near that point the Fokker–Planck equation may be written in terms of the direction cosines of **M** (t) (see also Section V of Ref. [108]) as

$$\dot{W} \approx \frac{1}{2\tau_{\mathrm{N}}} \sum_{i=1,2} \left[\frac{\partial}{\partial u_i} \left(W \frac{\partial E}{\partial u_i} + \frac{\partial W}{\partial u_i} \right) - \frac{1}{\alpha} \left(\frac{\partial E}{\partial u_1} \frac{\partial W}{\partial u_2} - \frac{\partial W}{\partial u_1} \frac{\partial E}{\partial u_2} \right) \right]. \quad (2.129)$$

Here, by approximating E near a saddle point to *fourth* order in the direction cosines (u_1, u_2), we have E as

$$E(u_1, u_2) \approx \sigma - \sigma(1-\zeta)u_1^2 + \zeta\sigma u_2^2 - \zeta\sigma\left(u_1^4 + u_1^2 u_2^2 + u_2^4\right). \quad (2.130)$$

However the barrier-crossing process is *exponentially slow*, therefore we may now assume a *quasistationary* solution of Eq. (2.129) in the separatrix region, of the form

$$W(u_1, u_2, t) = w(u_1, u_2) e^{-\Gamma_i^{\mathrm{IHD}} t}, \quad (2.131)$$

leading to the partial differential equation

$$2\tau_{\mathrm{N}} \Gamma_i^{\mathrm{IHD}} w \approx \frac{1}{\alpha} \left(\frac{\partial E}{\partial u_1} \frac{\partial w}{\partial u_2} - \frac{\partial w}{\partial u_1} \frac{\partial E}{\partial u_2} \right) - \sum_{i=1,2} \frac{\partial}{\partial u_i} \left(w \frac{\partial E}{\partial u_i} + \frac{\partial w}{\partial u_i} \right). \quad (2.132)$$

Thus, by integrating this equation with respect to the direction cosines u_1 and u_2, limiting the integration to a *single well* and then using Green's theorem in the (u_1, u_2) plane, we may *formally* obtain Γ_i^{IHD} as the closed-line integral along the saddle-point contour

$$\Gamma_i^{\mathrm{IHD}} \approx \frac{1}{2\tau_{\mathrm{N}} Z_i} \oint_{\mathrm{well}\, i} e^{-E} \left\{ \left(\frac{g}{\alpha} \frac{\partial E}{\partial u_2} + \frac{\partial g}{\partial u_1} \right) du_2 + \left(\frac{g}{\alpha} \frac{\partial E}{\partial u_1} - \frac{\partial g}{\partial u_2} \right) du_1 \right\}. \quad (2.133)$$

Here $g = e^E w$ is again called the crossover function; this concept (as seen) was originally used by Kramers to obtain the IHD solution of the Klein–Kramers equation for point particles, by converting that equation near the saddle into an ordinary differential equation. The partition function Z_i represents the well population, where the elliptic paraboloidal approximation for the energy near the bottom of the well still holds, so that Z_i may be evaluated (as usual) by steepest descents. For E given by Eq. (2.123), we obtain

$$Z_i = \int_0^{2\pi}\int_0^{\pi/2} e^{-E(\vartheta,\varphi)} \sin\vartheta d\vartheta d\varphi \approx \frac{\pi}{\sigma(1+\zeta)}. \tag{2.134}$$

To evaluate the escape rate, we require an expression for the relevant Kramers crossover function $g = e^E w$ and its first derivatives at the well boundary, along with a suitable parameterization of the well boundary itself. Since the distribution function w must always be finite, and because a large barrier is assumed, the left-hand side of Eq. (2.132) almost vanishes by quasistationarity. Hence, in terms of the Kramers crossover function g, we have

$$\left[\frac{\partial E}{\partial u_1} + \alpha \frac{\partial E}{\partial u_2}\right]\frac{\partial g}{\partial u_2} - \left[\frac{\partial E}{\partial u_2} - \alpha \frac{\partial E}{\partial u_1}\right]\frac{\partial g}{\partial u_1} \approx \alpha \left[\frac{\partial^2 g}{\partial u_1^2} + \frac{\partial^2 g}{\partial u_2^2}\right]. \tag{2.135}$$

Following Hänggi *et al.* [85], we now seek g as [369]

$$g(u_1,u_2) = C^{-1} \int_{-\infty}^{\rho(u_1,u_2)} e^{-A_1 z^2 - A_2 z^4} dz,\ C = \int_{-\infty}^{\infty} e^{-A_1 z^2 - A_2 z^4} dz, \tag{2.136}$$

where A_1 and A_2 are unknown coefficients accounting for both the *shape* of the saddle region *and the energy loss at the saddle*, and where the upper limit $\rho(u_1, u_2)$ is a function to be determined. We note that the Kramers IHD calculation, as adapted to magnetization reversal by Brown, corresponds to setting $A_2 = 0$ in Eq. (2.136), and dropping altogether the fourth-order terms in the Taylor series expansion of the free energy density in Eq. (2.130). Here in contrast, we must have $A_2 \neq 0$ *in order to account for the nonparaboloidal shape of the saddle*, and all terms of the fourth-order Taylor expansion of the free energy density in Eq. (2.130) are therefore retained in Eqs. (2.135) *et seq*. In succinct terms, because of the *nonparaboloidal* shape of the saddle region, the crossover function *deviates* from the error function originally used by Kramers [84] for *parabolic* barriers.

One must now transform the partial differential equation Eq. (2.135) in the two variables (u_1,u_2), into an ordinary differential equation in the single variable ρ. We may accomplish this, following Kramers, by implicitly seeking the upper limit ρ as a linear combination of u_1 and u_2 in the saddle region. The details are as follows. By substituting the crossover function, Eq. (2.136), into Eq. (2.135), we obtain a nonlinear partial differential equation for $\rho(u_1, u_2)$, viz.,

$$\left[\frac{\partial E}{\partial u_2}-\alpha\frac{\partial E}{\partial u_1}\right]\frac{\partial \rho}{\partial u_1}-\left[\frac{\partial E}{\partial u_1}+\alpha\frac{\partial E}{\partial u_2}\right]\frac{\partial \rho}{\partial u_2}$$
$$\approx \alpha\left(2A_1\rho+4A_2\rho^3\right)\left[\left(\frac{\partial \rho}{\partial u_1}\right)^2+\left(\frac{\partial \rho}{\partial u_2}\right)^2\right]-\alpha\left[\frac{\partial^2 \rho}{\partial u_1^2}+\frac{\partial^2 \rho}{\partial u_2^2}\right]. \tag{2.137}$$

Next, we seek the conditions under which Eq. (2.137) can be reduced to an ordinary differential equation for ρ. By introducing, again following Kramers [84,108], a new variable $r = qu_1 + u_2$ in Eq. (2.137), we have

$$\left(\frac{q-\alpha}{1+q^2}\frac{\partial E}{\partial u_2}-\frac{1+\alpha q}{1+q^2}\frac{\partial E}{\partial u_1}\right)\frac{d\rho}{dr}=-\alpha\frac{d^2\rho}{dr^2}+\alpha\left(2A_1\rho+4A_2\rho^3\right)\left(\frac{d\rho}{dr}\right)^2,$$

where we reiterate that all the terms of Eq. (2.130) should be included when computing the derivatives $\partial E/\partial u_i$ unlike the Kramers–Langer–Brown IHD calculation, where the fourth-order terms are *dropped.* Since ρ is implicitly sought as a *linear combination* of u_1 and u_2, it follows that ρ and r are proportional to each other, so that the second-order derivative $d^2\rho/dr^2$ is zero. Hence, the above equation becomes

$$\alpha\left(1+q^2\right)\left(2A_1\rho+4A_2\rho^3\right)\frac{d\rho}{dr}=\zeta\sigma\left(q-\alpha\right)$$
$$\times\left(u_2-2u_1^2u_2-4u_2^3\right)+\sigma\left(1+\alpha q\right)\left[(1-\zeta)u_1+2\zeta u_1u_2^2+4\zeta u_1^3\right], \tag{2.138}$$

where we have used Eq. (2.130). Next, expressing u_1 as $(r - u_2)/q$ in the left-hand side of Eq. (2.138) and expanding in powers of $r - u_2$ we may choose q so that the terms *linear* in u_2 *cancel*, and then set $u_2 = 0$ in the remaining terms. The quadratic equation determining q is $q(q - \alpha)\zeta = (1 + \alpha q)(1 - \zeta)$. Since this equation has two real roots (one positive and one negative), one must choose the root that permits normalization of the crossover function g, i.e., q as rendered by

$$q = \frac{1}{2\zeta}\left[\alpha + \sqrt{\alpha^2 + 4\zeta(1-\zeta)}\right], \tag{2.139}$$

which represents the *unstable barrier-crossing mode*. Thus, Eq. (2.138) becomes the *proper* first-order nonlinear ordinary differential equation

$$\alpha\left(1+q^2\right)\left(A_1\rho+2A_2\rho^3\right)\frac{d\rho}{dr} = (q-\alpha)\zeta\sigma r - 2(1+\alpha q)\zeta\sigma q^{-3}r^3.$$

Integrating this equation with the boundary condition $\rho(0) = 0$ yields

$$\begin{aligned}(q-\alpha)\zeta\sigma r^2 - (1+\alpha q)\zeta\sigma q^{-3}r^4 \\ = \alpha(1+q^2)(A_1\rho^2 + A_2\rho^4).\end{aligned} \tag{2.140}$$

The physically meaningful solution of this biquadratic equation is $\rho = r$, provided that the coefficients A_1 and A_2 are given by

$$A_1 = -\frac{\zeta\sigma(1-\alpha^{-1}q)}{1+q^2}, \quad A_2 = -\frac{\zeta\sigma(q+\alpha^{-1})}{q^3(1+q^2)}. \tag{2.141}$$

These expressions for ρ, A_1 and A_2 determine the crossover function g.

According to Brown [18] and Geoghegan *et al.* [108], the well boundary is parameterized by $u_1 = 0$. Therefore, setting $du_1 = 0$ in the contour integral, Eq. (2.133), and retaining the parabolic approximation *only* in the factor $e^{-E(0,u_2)}$, Eq. (2.133) finally becomes [369]

$$\begin{aligned}\Gamma_i^{\mathrm{IHD}} &\approx \frac{\sigma(1+\zeta)(1+\alpha q)}{2\pi\tau_0(1+\alpha^2)C}\int_{-\infty}^{\infty} e^{-(\zeta\sigma+A_1)u_2^2 - A_2u_2^4}du_2 \\ &= \sqrt{1+\frac{\zeta\sigma}{A_1}}\,\frac{\sigma(1+\zeta)(1+\alpha q)K_{1/4}\left[(A_1+\zeta\sigma)^2/(8A_2)\right]}{\tau_0 2\pi\left(1+\alpha^2\right)K_{1/4}\left[A_1^2/(8A_2)\right]}e^{\frac{\zeta\sigma(\zeta\sigma+2A_1)}{8A_2}-\sigma},\end{aligned} \tag{2.142}$$

where $0 \le \zeta \le 1$ and $K_{1/4}(z)$ is a modified Bessel function of the second kind [290] (see Appendix A.1.6). Here we have used

$$\int_0^\infty e^{-az^2-bz^4}\,dz = \frac{1}{4}\sqrt{a/b}\,e^{a^2/(8b)}K_{1/4}\left(a^2/(8b)\right). \tag{2.143}$$

For $\zeta = 1$, when $A_1 \to 0$, because (see Appendix A.1.6)

$$e^x K_{1/4}(x) \approx \Gamma(1/4)(x/2)^{-1/4}/2 + \cdots, \tag{2.144}$$

where $\Gamma(z)$ is the gamma function [290], Eq. (2.142) becomes [369]

$$\Gamma_i^{\text{IHD}} = \frac{\alpha\sigma^{5/4}}{\tau_0\pi\Gamma(1/4)}K_{1/4}(\alpha^4\sigma/8)e^{-\sigma(1-\alpha^4/8)}, \tag{2.145}$$

which for $\alpha^4\sigma/8 >> 1$ reduces to

$$\Gamma_i^{\text{IHD}} = \frac{2\sigma^{3/4}e^{-\sigma}}{\tau_0\alpha\pi^{1/2}\Gamma(1/4)}. \tag{2.146}$$

This completes the solution of the flat-saddle problem.

We now consider *negative anisotropy ratio*, $-1 \le \zeta \le 0$, where the free-energy potential, Eq. (2.123), near the bottom of a well may not be approximated by an elliptic paraboloid, which will affect the well partition function. Nevertheless, Γ_i^{IHD} can still be estimated from Langer's IHD equation, Eq. (1.296). First, we recall that near the saddle for $-1 \le \zeta \le 0$ the hyperbolic paraboloid approximation still holds, so that the saddle partition function Z_C and the damped saddle frequency λ_+ can be evaluated as usual. Hence, we have

$$Z_C = \frac{2\pi e^{-\sigma(1-|\zeta|/4)}}{\sqrt{|\zeta|(2+|\zeta|)}}, \lambda_+ = \frac{\sigma\left[2-|\zeta|+\sqrt{(3|\zeta|+2)^2+8\alpha^{-2}|\zeta|(2+|\zeta|)}\right]}{4\tau_0(\alpha+\alpha^{-1})}. \tag{2.147}$$

The calculation of the well partition function Z_i, where the elliptic paraboloid approximation fails, can be accomplished using Eq. (2.130). Hence, with Eq. (2.143), we have an accurate approximation for Z_i, viz.,

$$Z_i \approx \frac{(1-|\zeta|)}{4|\zeta|}e^{\sigma(1-|\zeta|)^2/4|\zeta|}K_{1/4}^2\left[\sigma(1-|\zeta|)^2/(8|\zeta|)\right], \tag{2.148}$$

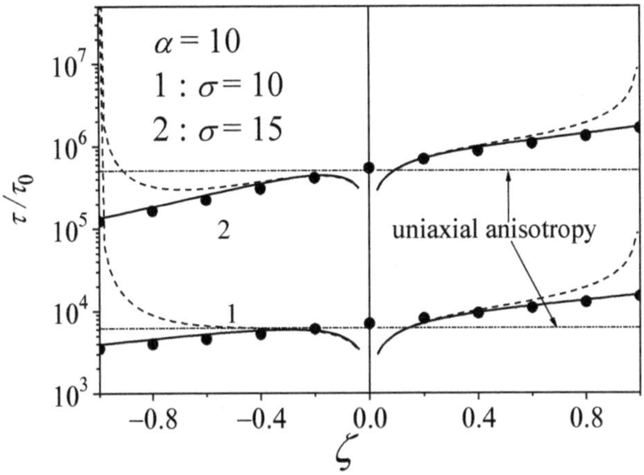

Fig. 2.16. Reversal time τ/τ_0 vs. ζ for $\alpha = 10$ and $\sigma = 10, 15$. Filled circles: matrix continued fraction solution [369]. Solid lines: the turnover Eq. (2.124). Dashed lines: the IHD Eqs. (2.149) and (2.150). Dashed-dotted line: Brown's uniaxial asymptote, Eq. (2.5).

which approximates the well partition function Z_i with an error of the order of 5% for $|\zeta| < 1$. Thus, by substituting Eqs. (2.147) and (2.148) into Langer's equation, Eq. (1.296), we finally obtain

$$\tau_0 \Gamma_i^{\mathrm{IHD}} = e^{-\sigma \frac{1+2|\zeta|}{4|\zeta|}} \sqrt{\frac{|\zeta|}{2(2+|\zeta|)}}$$
$$\times \frac{\left[2-|\zeta|+\sqrt{(2+3|\zeta|)^2+8\alpha^{-2}|\zeta|(2+|\zeta|)}\right]}{(\alpha+\alpha^{-1})(1-|\zeta|)K_{1/4}^2\left[(1-|\zeta|)^2\sigma/(8|\zeta|)\right]}. \tag{2.149}$$

From Eq. (2.144), Eq. (2.149) for $\zeta = -1$ becomes

$$\tau_0 \Gamma_i^{\mathrm{IHD}} = \frac{\sqrt{\sigma/6}(\alpha+\sqrt{25\alpha^2+24})}{(1+\alpha^2)\Gamma^2(1/4)} e^{-3\sigma/4}. \tag{2.150}$$

This equation, combined with the turnover, Eq. (2.124), completes the *flat minimum* magnetization escape rate.

The results of the numerical [369] and asymptotic (from Eqs. (2.124) (turnover), (2.142), (2.145), (2.149), and (2.150) (all pertaining to IHD) calculations of the normalized reversal time τ/τ_0 as functions of the anisotropy ratio parameter ζ and α are shown in Figs. 2.16 and 2.17.

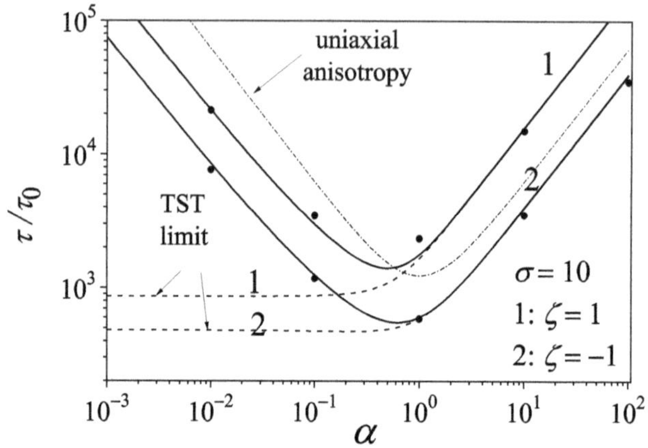

Fig. 2.17. Reversal time τ/τ_0 vs. α for $\zeta = -1$ and $\zeta = 1$. Filled circles: matrix continued fraction solution [369]. Solid lines: the turnover Eq. (2.124). Dashed lines: the IHD Eqs. (2.150) and (2.145). Dashed-dotted line: Brown's uniaxial asymptote, Eq. (2.5).

Figure 2.17 shows that the turnover equation, Eq. (2.124), describes the behavior of the reversal time in the *entire* dissipation range for both $\zeta = -1$ and $\zeta = 1$. Moreover, Eq. (2.124) is valid in the range $0.2 \le |\zeta| \le 1$. Here, the usual IHD formula, Eq. (2.83), based on the paraboloidal approximation, does not describe the relaxation rate at all (as is apparent from Fig. 2.16). In fact, for mixed anisotropy, the IHD Eq. (2.83) yields

$$\tau_0 \Gamma_i^{\mathrm{IHD}} = \sigma\left(1-|\zeta|\right)e^{-\sigma(1-|\zeta|/4)} \times \frac{\left[2-|\zeta|+\sqrt{(2+3|\zeta|)^2+8\alpha^{-2}|\zeta|(2+|\zeta|)}\right]}{4\pi(\alpha+\alpha^{-1})\sqrt{2|\zeta|\left(2+|\zeta|\right)}} \tag{2.151}$$

for $-1 \le \zeta \le 0$ and

$$\Gamma_i^{\mathrm{IHD}} = \frac{\sigma\left(1+\zeta\right)\left[1-2\zeta+\sqrt{1+4\alpha^{-2}\zeta(1-\zeta)}\right]e^{-\sigma}}{4\pi\tau_0(\alpha+\alpha^{-1})\sqrt{\zeta(1-\zeta)}} \tag{2.152}$$

for $0 \le \zeta \le 1$. By inspection, Eq. (2.151) predicts *zero* escape rate for $|\zeta| = 1$, while yielding an *infinite* escape rate as $\zeta \to 0$. Figures 2.16 and 2.17 also indicate that Eqs. (2.124), (2.142), (2.145), (2.149), and (2.150) correctly reproduce the behavior of the relaxation time for $0.2 \le |\zeta| \le 1$ in contrast to the IHD Eq. (2.83). Notice that the correct escape-rate equation for the uniaxial case, $\zeta = 0$, is in fact provided by Brown's *uniaxial* anisotropy formula, Eq. (2.5), for the reversal time.

The turnover formula, Eq. (2.124), is valid for *all* α. However, as $\zeta \to 0$, the circular symmetry is regained, and the azimuthal dependence of the distribution function disappears. The saddle region now becomes *very wide*, so that the method of Garanin *et al.* [359] should be used; as the Mel'nikov–Meshkov method (on which Eq. (2.124) is based) fails since the action S is now zero, once again yielding zero escape rate. By comparing the uniaxial asymptotes in Eq. (2.5) with Eq. (2.124), as shown in Fig. 2.16, we see that τ for *pure* uniaxial anisotropy can differ by as much as an order of magnitude from τ for *mixed* anisotropy as rendered by Eq. (2.124). This behavior may be attributed to the difference in the prefactors between the circularly and noncircularly symmetric results. In particular, for $\zeta = 1$ and $\alpha \ll 1$ Eq. (2.145) predicts the *same* barrier height as a uniaxial nanomagnet. However, the dependence of the prefactor on α and σ renders a different reversal time.

The methods just described may be extended to other potentials for which the second-order Taylor expansion coefficients of the potential from Eq. (1.300) vanish at the relevant stationary points.

2.3.6 *STT effects in the thermally assisted magnetization reversal*

The salient feature of the results of Section 2.2.5 is that they apply for VLD only so that the inequality $\alpha S_{E_C} \ll 1$ holds meaning that the energy loss per cycle of the almost-periodic precessional motion at the critical energy is much less than the thermal energy. However, for typical values of material parameters, S_{E_C} may be very high ($>10^3$), thus this inequality is satisfied only for $\alpha < 10^{-3}$. In addition, both experimental and theoretical estimates suggest *higher* values of α of the order of 0.001–0.1 (see, e.g., Refs. [82,83,109,332,333]), implying that the VLD asymptotic results are no longer valid as they will now differ substantially from the true value of the reversal time. The foregoing considerations suggest that the asymptotic calculations for STT should be extended to include both the VLD and intermediate damping regimes. Coffey *et al.* [181], Déjardin *et al.* [196], and Kalmykov *et al.* [151,152,354] have generalized the Kramers theory to classical spins (nanomagnets) with zero STT. However, that theory can also be used to include STT effects. In other words the zero STT results cf. Refs. [181,196] may be extended to cover the *damping dependence of STT effects* (in the magnetization reversal of nanoscale ferromagnets via escape-rate theory) in the most important damping ranges, namely VLD and ID where $\alpha \leq 1$. In Section 1.4 we have shown that *at zero STT* the magnetization reversal time for damping up to intermediate values, $\alpha \leq 1$, can be evaluated via the turnover formula for the escape rate interpolating between VLD and IHD escape rates, namely, (see Eqs. (1.306) and (1.308))

$$\tau \approx \frac{A(\alpha S_{E_C}^{+} + \alpha S_{E_C}^{-})}{\Gamma^{\mathrm{TST}} A(\alpha S_{E_C}^{+}) A(\alpha S_{E_C}^{-})}. \tag{2.153}$$

Here the overall TST escape rate Γ^{TST} is determined via the *individual* escape rates $\Gamma_{\pm}^{\mathrm{TST}}$ from each of the two wells as

$$\begin{aligned} \Gamma^{\mathrm{TST}} &= \Gamma_{+}^{\mathrm{TST}} + \Gamma_{-}^{\mathrm{TST}} \\ &= 2\left(f_A^{+} e^{-\Delta E^{+}} + f_A^{-} e^{-\Delta E^{-}}\right). \end{aligned} \tag{2.154}$$

Here the factor 2 occurs again because *two* magnetization escape routes from each well over the *two* saddle points exist and the corresponding well precession frequencies are given by Eq. (2.42) from Section 2.2.3). For the bistable potential, the spin-polarized current dependent effective barrier heights $\Delta E^{\pm}$ are given by (cf. Eq. (2.74))

$$\Delta E^{\pm} = \sigma(1 \pm h)^2 \mp \alpha^{-1} J F^{\pm}(\delta, h), \tag{2.155}$$

where

$$F^{\pm}(\delta, h) = \frac{\sigma}{J} \int_{\varepsilon_A^{\pm}}^{\varepsilon_C} \frac{V_{\varepsilon}^{\pm}}{S_{\varepsilon}^{\pm}} d\varepsilon \tag{2.156}$$

with

$$\varepsilon = E/\sigma, \ \ \varepsilon_A^{\pm} = E_A^{\pm}/\sigma = -1 \mp 2h, \ \varepsilon_C = E_C/\sigma = h^2.$$

The action S_ε^+ and the work V_ε^+ parameters are defined by Eqs. (2.70) and (2.71). Furthermore in the parameter ranges $0<h<1$ and $\delta>1$, the integral in Eq. (2.156) can be accurately evaluated analytically using an interpolation function for $V_\varepsilon^\pm / S_\varepsilon^\pm$ between its two limiting values $V_{\varepsilon_A}^\pm / S_{\varepsilon_A}^\pm$ and $V_{\varepsilon_C}^\pm / S_{\varepsilon_C}^\pm$ at $\varepsilon_A^\pm = -1 \mp h$ and $\varepsilon_C = h^2$, respectively, namely

$$\frac{V_\varepsilon^\pm}{S_\varepsilon^\pm} \approx \frac{V_{\varepsilon_A}^\pm}{S_{\varepsilon_A}^\pm} + \left(\frac{V_{\varepsilon_C}^\pm}{S_{\varepsilon_C}^\pm} - \frac{V_{\varepsilon_A}^\pm}{S_{\varepsilon_A}^\pm}\right)\left(\frac{\varepsilon - \varepsilon_A}{\varepsilon_C - \varepsilon_A}\right)^{\beta^{-1}-1}, \tag{2.157}$$

where $\beta \approx 0.38$ is an interpolation parameter yielding the best fit of $V_\varepsilon^\pm / S_\varepsilon^\pm$ in the interval $\varepsilon_A \le \varepsilon \le \varepsilon_C$. These limiting values may then be calculated explicitly yielding after tedious algebra:

$$\frac{\sigma}{J}\frac{V_{\varepsilon_A}^\pm}{S_{\varepsilon_A}^\pm} = \frac{1}{2+\delta \pm 2h} \tag{2.158}$$

and

$$\frac{\sigma}{J}\frac{V_{\varepsilon_C}^\pm}{S_{\varepsilon_C}^\pm} = \pm\frac{1}{2\delta h}\left\{1 - \frac{(1-h^2)(1+\delta)}{(1-h^2+\delta)}\right.$$
$$\left.\times\left[1 \pm \frac{2h}{\sqrt{(1-h^2)(1+\delta^{-1})}}\arctan\sqrt{\frac{\sqrt{\delta^{-1}(1-h^2)+1} \pm h}{\sqrt{\delta^{-1}(1-h^2)+1} \mp h}}\right]^{-1}\right\}. \tag{2.159}$$

Hence, we have from Eq. (2.156) a simple analytic formula for the functions $F^\pm(\delta, h)$, namely,

$$F^\pm(\delta,h) \approx \frac{\sigma}{J}\left\{\left[(1\pm h)^2 - \beta\right]\frac{V_{\varepsilon_A}^\pm}{S_{\varepsilon_A}^\pm} + \beta\frac{V_{\varepsilon_C}^\pm}{S_{\varepsilon_C}^\pm}\right\}$$
$$= \frac{(1\pm h)^2 - \beta}{2+\delta\pm 2h} \pm \frac{\beta}{2\delta h}\left[1 - \frac{(1-h^2)(1+\delta)}{(1-h^2+\delta)}\right. \tag{2.160}$$
$$\left.\times\left(1 \pm \frac{2h}{\sqrt{(1-h^2)(1+\delta^{-1})}}\arctan\sqrt{\frac{\sqrt{\delta^{-1}(1-h^2)+1} \pm h}{\sqrt{\delta^{-1}(1-h^2)+1} \mp h}}\right)^{-1}\right].$$

Thus, the decay rate τ^{-1} from Eq. (2.153) becomes

$$\tau^{-1} \approx \frac{A(\alpha S_{E_C}^{+})A(\alpha S_{E_C}^{-})}{\pi\tau_0 A(\alpha S_{E_C}^{+} + \alpha S_{E_C}^{-})}\left[\sqrt{(1+h+\delta)(1+h)}e^{-\sigma(1+h)^2+\frac{J}{\alpha}F^{+}(\delta,h)} + \sqrt{(1-h+\delta)(1-h)}e^{-\sigma(1-h)^2-\frac{J}{\alpha}F^{-}(\delta,h)}\right]. \tag{2.161}$$

For zero STT, $J = 0$, Eq. (2.161) is just the known result of the Néel–Brown theory [151,152,239,354] pertaining to classical magnetic moments with superimposed easy-plane and in-plane easy-axis anisotropies plus the Zeeman term due to the applied magnetic field (see Section 2.3.2). However in contrast to zero STT (for normalized spin currents $J \neq 0$) τ depends on α not only via the depopulation factors $A(\alpha S_{E_C}^{\pm})$ but also in *exponential* fashion via the *spin-polarized current dependent effective barrier heights* $\Delta E^{\pm}$. This behavior occurs because parts of the arguments of the exponentials in Eq. (2.161), namely, Eq. (2.160), are markedly dependent on the ratio J/α and the dc bias field parameter. We observe that the turnover Eq. (2.161) also yields an asymptotic estimate for the inverse of the smallest nonvanishing eigenvalue of the Fokker–Planck operator L_{FP}. In addition, one may as usual determine two *individual* reversal times, namely, τ_+ from the deeper well around the energy minimum at $\varphi_A = 0$ and τ_- from the shallow well around the energy minimum at $\varphi_A = \pi$ as

$$\tau_{\pm} \approx \frac{2\pi\tau_0}{A(\alpha S_{E_C}^{\pm})\sqrt{(1\pm h+\delta)(1\pm h)}}e^{\sigma(1\pm h)^2 \pm \frac{J}{\alpha}F^{\pm}(\delta,h)}. \tag{2.162}$$

The individual times are in general unequal, i.e., $\tau_+ \neq \tau_-$. In deriving Eqs. (2.161) and (2.162), all terms of order α^2, αJ, J^2, etc. are once again neglected. This hypothesis is true only for the *underdamped regime*, $\alpha \ll 1$, and *weak* spin-polarized currents, $J \ll 1$. However, despite these restrictions Eqs. (2.161) and (2.162) still yield accurate estimates for τ for much higher values of α and J.

The reversal time τ can also be calculated *numerically* via the statistical moment method (v. Section 1.3.4) developed in Ref. [137,138] for nonzero STT. Thus, one can evaluate τ numerically via λ_1 of the Fokker–Planck operator L_{FP} by using matrix continued fractions as described in Ref. [145]. The ranges of applicability of both escape-rate theory and the

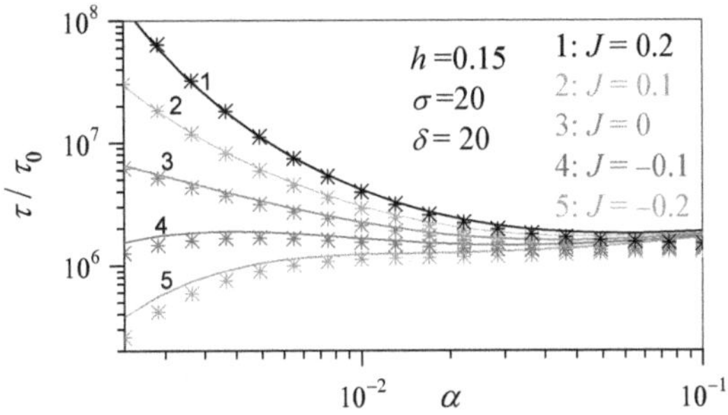

Fig. 2.18. Reversal time τ/τ_0 vs the damping parameter α for various values of the spin-polarized current parameter J. Solid lines: numerical calculations of the inverse of the smallest nonvanishing eigenvalue $(\tau_0\lambda_1)^{-1}$ of the Fokker–Planck operator. Asterisks: the turnover formula, Eq. (2.161).

(exact) matrix continued fraction method are in some sense complementary because the former cannot be used for low potential barriers, $\Delta E < 3$, while the latter encounters substantial computational difficulties in the VLD range $\alpha < 10^{-4}$ for *very high* potential barriers $\Delta E > 25$. Thus, in the foregoing sense, both methods have their uses for the determination of τ for low and very high potential barriers, respectively. Nevertheless, in certain (wide) ranges of model parameters either method yields accurate results for the reversal time (here these ranges are $5 < \sigma < 30$, $\sigma\delta > 3$ and $\alpha > 10^{-4}$). Then the *numerically exact benchmark* solution provided by matrix continued fractions allows one to test the accuracy of the analytical escape-rate equations given above.

The anisotropy and spin-polarization parameters are everywhere taken as in [109], namely, $D_{\|} = 0.034$, δ=20, and P =0.3. Figures 2.18 and 2.19, compare τ as yielded by the asymptotic escape-rate Eq. (2.161) with λ_1^{-1} of the Fokker–Planck operator L_{FP} as calculated numerically via matrix continued fractions [137]. Apparently, τ as rendered by the turnover equation (2.161) and λ_1^{-1} both lie very close to each other in the high-barrier limit, where the asymptotic Eq. (2.161) should provide an accurate approximation to λ_1^{-1}. In Fig. 2.18, τ is plotted as a function of α for various J. As far as STT effects are concerned they are governed (as seen) by the ratio J/α thus *by altering* J/α *the ensuing variation of* τ *may exceed several orders of magnitude* (Fig. 2.18). Invariably for $J \ll 1$, i.e., the

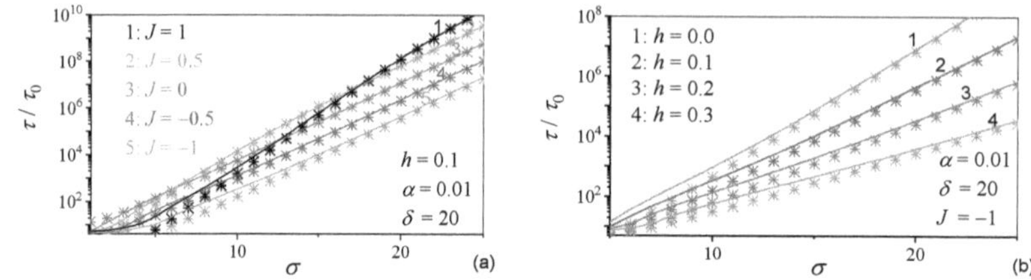

Fig. 2.19. Reversal time τ/τ_0 vs. the anisotropy (inverse-temperature) parameter σ for various spin-polarized currents J (a) and various dc bias field parameters h (b). Solid lines: numerical solution for the inverse of the smallest nonvanishing eigenvalue $(\tau_0\lambda_1)^{-1}$ of the Fokker–Planck operator. Asterisks: Eq. (2.161).

condition of applicability of Eq. (2.161), STT effects on the magnetization relaxation are pronounced only for very low damping, $\alpha << 1$. For $\alpha \geq 1$, i.e., high damping, STT barely influences the reversal process at all because the STT term in Eq. (1.81) is then *small compared to the damping and random field terms*. Otherwise, τ may greatly exceed or, in contrast, be very much less than that for zero STT, i.e., $J = 0$ (see Fig. 2.18), e.g., as J decreases from positive values, τ at first exponentially increases attaining a maximum at some critical value of the spin-polarized current. Next τ smoothly switches over to exponential decrease as $|J|$ is further increased through negative values of J [137]. The temperature, external dc bias field, and damping dependence of τ may be understood via the *effective potential barriers* $\Delta E^{\pm}$ in Eq. (2.161). For example, for σ>5 the *temperature dependence* has the customary Arrhenius behavior $\tau \sim e^{\Delta E^-}$, where ΔE^- is markedly dependent on J/α (see Fig. 2.19), while the slope of $\tau(T^{-1})$ significantly decreases as the dc field parameter h increases. This behavior may be attributed to lowering of the barrier height ΔE^- owing to the action of the external field (see Fig. 2.19b). Furthermore, although the range of applicability of Eqs. (2.161) and (2.162) is ostensibly confined to weak spin-polarized currents, $J << 1$, they may still yield accurate estimates for the reversal time for much higher values of J far exceeding this condition (see Fig. 2.19).

However, the turnover Eqs. (2.161) and (2.162), bridging the Kramers VLD and ID escape rates as a function of the damping for point Newtonian particles [186,187] as extended by Coffey *et al.* [181,196] to nanoscale ferromagnets still apply. Therefore they allow one (via the further extension to include STT embodied in Eq. (2.161)) to accurately evaluate STT effects in the reversal time of a nanomagnet driven by spin-polarized current in the highly relevant ID to VLD damping range, a fact which may be very significant in interpreting many STT experiments. This (underdamped) range is characterized by $\alpha \le 1$ and the asymptotic escape rates so obtained concur with independent numerical results [137]. Two particular merits of the escape-rate equations for the reversal time are that (i) they are relatively simple (i.e., expressed via elementary functions) and (ii) that they can be used in those parameter ranges, where numerical methods (such as matrix continued fractions [137]) may not succeed at all, e.g., for *very high barriers,* $\Delta E > 25$.

2.4 Magnetization relaxation in uniaxial nanomagnets

We commence with the free energy density of a uniaxial nanomagnet (see Fig. 1.3 of Chapter 1). One of the most important types of circularly symmetric potential relevant to superparamagnetism occurs if a uniaxial particle is subjected to a dc bias field $\mathbf{H}_0$. In general, the field is applied at *some angle to* the easy axis of the magnetization thereby automatically posing a noncircularly symmetric problem. However, to preserve circular symmetry and the consequent mathematical simplifications which occur because only one space variable ϑ is involved, it is often assumed [17,18,23,43,44,67,78,80,334,336] that the field is also collinear with the easy (Z) axis, so that the normalized potential is simply

$$\begin{aligned} E(\vartheta) &= \sigma \sin^2 \vartheta - \xi \cos \vartheta \\ &= -\xi \sqrt{\frac{4\pi}{3}} Y_{10} - 4\sigma \sqrt{\frac{\pi}{45}} Y_{20} - \frac{\sigma}{3}, \end{aligned} \tag{2.163}$$

where $\sigma = vK/(kT)$ and $\xi = v\mu_0 M_S H_0 / (kT)$ are, respectively, the dimensionless anisotropy and external field parameters. The longitudinal magnetization relaxation in the potential given by Eq. (2.163) may then be treated as merely a particular case of the Eqs. (1.92), (1.107) and (2.163). Using the explicit representation of the Clebsch–Gordan coefficients, we have the five-term differential recurrence equation for the sharp values of the spherical harmonics [97,104]

$$
\begin{aligned}
\tau_{\mathrm{N}} \frac{d}{dt} Y_{nm} =& -\left[\frac{n(n+1)}{2} + \frac{i\xi m}{2\alpha} - \sigma \frac{n(n+1)-3m^2}{(2n-1)(2n+3)}\right] Y_{nm} \\
& + \frac{\sigma(n+1)}{2n-1}\sqrt{\frac{(n^2-m^2)[(n-1)^2-m^2]}{(2n-3)(2n+1)}} Y_{n-2m} \\
& + \left(\xi \frac{n+1}{2} + \frac{im\sigma}{\alpha}\right)\sqrt{\frac{n^2-m^2}{4n^2-1}} Y_{n-1m} \\
& - \left(\xi \frac{n}{2} - \frac{im\sigma}{\alpha}\right)\sqrt{\frac{(n+1)^2-m^2}{(2n+3)(2n+1)}} Y_{n+1m} \\
& - \frac{n\sigma}{2n+3}\sqrt{\frac{[(n+2)^2-m^2][(n+1)^2-m^2]}{(2n+5)(2n+1)}} Y_{n+2m},
\end{aligned}
\tag{2.164}
$$

where $\tau_{\mathrm{N}} = v\mu_0 M_S(1+\alpha^2)/(2kT\gamma\alpha)$ is the characteristic free diffusion time. Noticing Eq. (A.67) from Appendix A.1.7, Eq. (2.164) yields for $m = 0$ on spatial averaging the infinite hierarchy of differential-recurrence relations for the longitudinal equilibrium correlation functions $f_n^{\|}(t)$, viz.,

$$
\tau_{\mathrm{N}} \frac{d}{dt} f_n^{\|} = d_n f_n^{\|} + \xi a_n \left(f_{n-1}^{\|} - f_{n+1}^{\|}\right) + c_n f_{n-2}^{\|} + e_n f_{n+2}^{\|}, \tag{2.165}
$$

where

$$
f_n^{\|}(t) = \left\langle \cos\vartheta(0) P_n\left(\cos\vartheta(t)\right)\right\rangle_0 - \left\langle \cos\vartheta(0)\right\rangle_0 \left\langle P_n\left(\cos\vartheta(0)\right)\right\rangle_0 \tag{2.166}
$$

and

$$
\begin{aligned}
a_n &= \frac{n(n+1)}{2(2n+1)}, \\
d_n &= -\frac{n(n+1)}{2} + \frac{\sigma n(n+1)}{(2n-1)(2n+3)}, \\
c_n &= \frac{\sigma n(n^2-1)}{4n^2-1}, \\
e_n &= -\frac{\sigma n(n+1)(n+2)}{(2n+1)(2n+3)}.
\end{aligned}
\tag{2.167}
$$

Equation (2.165) clearly indicates that now the gyromagnetic terms do *not influence the longitudinal relaxation at all*. In like manner for $m = 1$, we have the set of differential-recurrence relations for the transverse correlation functions $f_n^{\perp}(t)$, viz., [148]

$$\tau_{\mathrm{N}} \frac{d}{dt} f_n^{\perp} = d_n f_n^{\perp} + a_n f_{n-1}^{\perp} - a_n' f_{n+1}^{\perp} + c_n f_{n-2}^{\perp} + e_n f_{n+2}^{\perp}, \tag{2.168}$$

where

$$f_n^{\perp}(t) = \langle \cos\varphi(0) P_1^1[\cos\vartheta(0)] e^{i\varphi(t)} P_n^1[\cos\vartheta(t)] \rangle_0 \tag{2.169}$$

and

$$\begin{aligned}
a_n &= \xi \frac{n+1}{2n+1}\left(\frac{n+1}{2} - \frac{i\sigma}{\alpha\xi}\right), \\
a_n' &= \frac{\xi n}{2n+1}\left(\frac{n}{2} + \frac{i\sigma}{\alpha\xi}\right), \\
c_n &= \frac{\sigma n(n+1)^2}{4n^2-1}, \\
d_n &= -\frac{n(n+1)}{2} - \frac{i\xi}{2\alpha} + \sigma \frac{n(n+1)-3}{(2n-1)(2n+3)}, \\
e_n &= -\frac{\sigma n^2(n+1)}{(2n+1)(2n+3)}.
\end{aligned} \tag{2.170}$$

Here the gyromagnetic term now influences the transverse relaxation and may also cause ferromagnetic resonance. We shall presently demonstrate how Eqs. (2.165) and (2.168) governing the relaxation of the magnetization in uniaxial magnetic nanoparticles, may be exactly solved using matrix continued fractions. We shall also illustrate their application to actual problems pertaining to their dynamics (e.g., reversal times of the magnetization, ferromagnetic resonance, stochastic resonance, etc.).

Uniaxial anisotropy $E(\vartheta) = \sigma \sin^2 \vartheta$ represents a special case; $\xi = 0$, where the energy depends only on the polar angle ϑ, irrespective of the angle φ. Numerous aspects of the magnetization relaxation in uniaxial nanomagnets have been treated using this potential; for these see Refs. [17,18,21,22,24–26,29,30,36,40,53,56,60–63,67,80,92,105,108,334,370, 371].

This uniaxial potential has been used to analyze nonlinear magnetic susceptibilities [29,30,39,145,153–156], stochastic resonance [56,60–68], dynamic hysteresis [43,44,47,159,160], Mössbauer spectra [78,80], and other related parameters of fine particle systems. Here we briefly summarize the principal findings. The longitudinal integral $\tau_{\text{int}}^{\|}$, effective $\tau_{\text{ef}}^{\|}$, and longest $\tau = \lambda_1^{-1}$, relaxation times are given by [97]:

$$\frac{\tau_{\text{int}}^{\|}}{\tau_{\text{N}}} = \frac{3e^{\sigma}}{\sigma^2 M(3/2,5/2,\sigma)} \int_0^1 \frac{\cosh[\sigma(1-z^2)]-1}{1-z^2} dz, \tag{2.171}$$

$$\frac{\tau_{\text{ef}}^{\|}}{\tau_{\text{N}}} = \frac{2\langle\cos^2\vartheta\rangle_0}{1-\langle\cos^2\vartheta\rangle_0} = \frac{2S+1}{1-S}, \tag{2.172}$$

$$\tau \approx \tau^{\text{MFPT}} = \tau_{\text{N}} \sqrt{\frac{\pi}{\sigma}} \int_{-1}^{0} \frac{\operatorname{erfi}(\sqrt{\sigma}) - \operatorname{erfi}(z\sqrt{\sigma})}{1-z^2} e^{-\sigma z^2} dz, \tag{2.173}$$

where

$$S = \langle P_2 \rangle_0 = \frac{3e^{\sigma}}{4\sigma M(1/2,3/2,\sigma)} - \frac{3}{4\sigma} - \frac{1}{2}. \tag{2.174}$$

Here $M(a,b,z)$ are Kummer's functions [290] (see Appendix A.1.2). Therefore, in the high-barrier (low-temperature) limit, $\sigma >> 1$, $\tau_{\text{int}}^{\|}, \tau_{\text{ef}}^{\|}$, and τ can all be evaluated using well known properties of these functions yielding the asymptotic formulas ([97], Section 7.7.3)

$$\tau_{\text{int}}^{\|} \approx \tau_{\text{N}} \frac{\sqrt{\pi}}{2} e^{\sigma} \sigma^{-3/2} \left(1 + \frac{1}{\sigma} + \frac{3}{2\sigma^2} + \cdots\right), \tag{2.175}$$

$$\tau_{\text{ef}}^{\|} = \tau_{\text{N}} \left(1 + \frac{2}{5}\sigma + \cdots\right), \tag{2.176}$$

$$\tau \approx \tau_{\text{N}} \frac{\sqrt{\pi}}{2} e^{\sigma} \sigma^{-3/2} \left(1 + \frac{1}{\sigma} + \frac{7}{4\sigma^2} + \cdots\right). \tag{2.177}$$

In addition, for practical calculations of both $\tau_{\text{int}}^{\|}$ and τ one may use the simple empirical equation valid for all σ [371]

$$\frac{1}{\tau} \approx \frac{1}{\tau_{\text{int}}^{\|}} \approx \frac{\sigma}{\tau_{\text{N}}(e^{\sigma}-1)} \left(2^{-\sigma} + \frac{2\sigma^{3/2}}{\sqrt{\pi}(1+\sigma)}\right). \tag{2.178}$$

The quantity of most interest is the longitudinal complex susceptibility $\chi_{\parallel}(\omega)=\chi_{\parallel}'(\omega)-i\chi_{\parallel}''(\omega)$, which can be calculated either via exact continued fractions [97] or else from the two-mode approximation Eq. (1.150) (see Section 1.3.6), where $\tau_{\text{int}}^{\parallel}$, $\tau_{\text{ef}}^{\parallel}$, and τ are given by Eqs. (2.171), (2.172) and (2.173), respectively. The transverse dynamic susceptibility $\chi_{\perp}(\omega)$ can be also calculated via continued fractions [148] or via an approximate analytic formula which has been obtained by Raikher and Shliomis [105] using the effective-eigenvalue method [97]

$$\frac{\chi_{\perp}(\omega)}{\chi_{\perp}}=\frac{1+\Delta+i\omega\tau_2}{(1+i\omega\tau_2)(1+i\omega\tau_{\perp})+\Delta}. \tag{2.179}$$

Here $\Delta=\alpha^{-2}\sigma(\tau_2/\tau_{\text{N}})(1-\tau_{\perp}/\tau_{\text{N}})$, $\chi_{\perp}$ is the transverse static susceptibility, and $\tau_{\perp}$ and τ_2 are the effective relaxation times, which in turn can be expressed in terms of the order parameter $S=\langle P_2\rangle_0$, [97] as

$$\begin{aligned}\frac{\tau_{\perp}}{\tau_{\text{N}}}&=\frac{1-S}{1+S/2},\\ \frac{\tau_2}{\tau_{\text{N}}}&=\frac{3S}{2\sigma+S(\sigma-6)}.\end{aligned} \tag{2.180}$$

For $\sigma>>1$, when $\tau_{\perp}\sim\tau_{\text{N}}\sigma^{-1}$, $\tau_2\sim\tau_{\text{N}}\sigma^{-1}$, $\Delta\sim\sigma^2\alpha^{-2}$, Eq. (2.179) reduces to the Landau–Lifshitz equation [89]

$$\frac{\chi_{\perp}(\omega)}{\chi_{\perp}}=\frac{(1+\alpha^2)\omega_{\text{pr}}^2+i\alpha\omega\omega_{\text{pr}}}{(1+\alpha^2)\omega_{\text{pr}}^2-\omega^2+2i\alpha\omega\omega_{\text{pr}}}, \tag{2.181}$$

where $\omega_{\text{pr}}=\sigma(\alpha\tau_{\text{N}})^{-1}\approx\gamma H_{\text{an}}$ is the precession frequency and $H_{\text{an}}=2K/\mu_0 M_{\text{S}}$ is the magnitude of the anisotropy field.

The second type of circularly symmetric potential relevant in superparamagnetism occurs if a dc bias field $\mathbf{H}_0$ is superimposed on the uniaxial anisotropy potential, Eq. (2.163) (see Section 2.3.2). The longitudinal and transverse complex susceptibilities have been calculated in Refs. [146,148]. Various aspects of the uniaxial potential with an external dc field have also been treated, e.g., in Refs. [23,336–339]. For the uniaxial potential E given by Eq. (2.163), the longitudinal magnetization relaxation is described by the *single-variable* Fokker–

Planck equation, Eq. (1.223), so that the integral relaxation time $\tau_{\text{int}}^{\|}$ may be expressed in closed-integral form as (see Appendix A.4, Eq. (A.243))

$$\tau_{\text{int}}^{\|} = \frac{2\tau_{\text{N}}/Z}{\langle\cos^2\vartheta\rangle_0 - \langle\cos\vartheta\rangle_0^2} \int_{-1}^{1} \left[\int_{-1}^{z} (z' - \langle\cos\vartheta\rangle_0) e^{\sigma z'^2 + \xi z'} dz' \right]^2 \frac{e^{\sigma z^2 + \xi z}}{1 - z^2} dz, \quad (2.182)$$

where the partition function Z is given by by Eq. (A.50) in Appendix A.1.5. Furthermore, the effective relaxation time $\tau_{\text{ef}}^{\|}$ defined by Eq. (1.145) is expressed in terms of equilibrium averages or, in turn, in terms of known functions by

$$\begin{aligned} \tau_{\text{ef}}^{\|} &= 2\tau_{\text{N}} \frac{\langle\cos^2\vartheta\rangle_0 - \langle\cos\vartheta\rangle_0^2}{1 - \langle\cos^2\vartheta\rangle_0} \\ &= \tau_{\text{N}} \frac{2\langle P_2\rangle_0 + 1 - 3\langle P_1\rangle_0^2}{1 - \langle P_2\rangle_0}, \end{aligned} \quad (2.183)$$

where the equilibrium averages $\langle P_1\rangle_0$ and $\langle P_2\rangle_0$ are given by Eqs. (A.48)–(A.50) in Appendix A.1.5.

For either $\sigma = 0$ (isotropic particles) or $\xi = 0$ (zero dc bias field), Eqs. (2.183) and (2.182) for $\tau_{\text{ef}}^{\|}$ and $\tau_{\text{int}}^{\|}$ can be considerably simplified. For $\xi = 0$, $\tau_{\text{int}}^{\|}$ and $\tau_{\text{ef}}^{\|}$ are given by Eqs. (2.171) and (2.172), respectively, while for $\sigma = 0$, both $\tau_{\text{int}}^{\|}$ and $\tau_{\text{ef}}^{\|}$ may be rendered as the closed forms, viz., [338]

$$\begin{aligned} \tau_{\text{int}}^{\|} &= \frac{2\tau_{\text{D}}}{\langle\cos^2\vartheta\rangle_0 - \langle\cos\vartheta\rangle_0^2} \int_{-1}^{1} \left[\int_{-1}^{z} (z' - \langle\cos\vartheta\rangle_0) e^{\xi z'} dz' \right]^2 \frac{e^{-\xi z}}{1 - z^2} dz \\ &= \frac{\tau_{\text{D}} \operatorname{csch}\xi}{\xi(1 + \xi^{-2} - \coth^2\xi)} \int_{-1}^{1} \left[z - \coth\xi + e^{-\xi(1+z)}(1 + \coth\xi) \right]^2 \frac{e^{\xi z} dz}{1 - z^2}, \end{aligned} \quad (2.184)$$

$$\begin{aligned} \frac{\tau_{\text{ef}}^{\|}}{\tau_{\text{D}}} &= -\frac{f_1(0)}{\tau_{\text{D}} \dot{f}_1(0)} = \left[1 + \frac{\xi f_2(0)}{3 f_1(0)} \right]^{-1} \\ &= \frac{\xi}{L(\xi)} - \xi L(\xi) - 2. \end{aligned} \quad (2.185)$$

Moreover, for small σ and ξ, we have the Taylor series expansions [338]:

$$\frac{\tau_{\text{int}}^{\|}}{\tau_{\text{N}}} = 1 + \frac{2}{5}\sigma + \frac{18}{175}\sigma^2 + \frac{164}{7875}\sigma^3 - \left(\frac{1}{9} + \frac{442}{4725}\sigma\right)\xi^2 + \cdots,$$
$$\frac{\tau_{\text{ef}}^{\|}}{\tau_{\text{N}}} = 1 + \frac{2}{5}\sigma + \frac{16}{175}\sigma^2 + \frac{32}{2625}\sigma^3 - \left(\frac{2}{15} + \frac{206}{1575}\sigma\right)\xi^2 + \cdots, \quad (2.186)$$
$$\lambda_1 \tau_{\text{N}} = 1 - \frac{2}{5}\sigma + \frac{48}{875}\sigma^2 - \frac{32}{21875}\sigma^3 + \left(\frac{1}{10} + \frac{1}{875}\sigma\right)\xi^2 + \cdots.$$

The above equations are obtained from Eqs. (2.183), (2.182), and (2.194).

2.4.1 *Longitudinal relaxation and dc field effect*

In Section 2.4, we wrote the differential-recurrence equation, Eq. (2.165), for the longitudinal equilibrium correlation functions $f_n^{\|}(t)$, Eq. (2.171), governing the longitudinal relaxation of a uniaxial nanomagnet subjected to a dc bias field applied along the easy axis of the magnetization. For $\xi \neq 0$, Eq. (2.165) does not decouple into separate sets for even- and odd-index $f_n^{\|}(t)$. The first-rank function $f_1^{\|}(t)$ governs the longitudinal response. The set of equations Eq. (2.165) may be solved to yield the relaxation behavior of $f_n^{\|}(t)$ by either of two methods. The first one is to simply arrange the set in the matrix form (see Section 1.3.5)

$$\dot{\mathbf{X}}(t) = \mathbf{A}\mathbf{X}(t), \quad (2.187)$$

where the system matrix **A** and the column vector **X** (t) are determined by Eq. (2.165) and are given by

$$\mathbf{A} = \frac{1}{\tau_{\text{N}}}\begin{pmatrix} d_1 & -\xi a_1 & e_1 & 0 & 0 & 0\ldots \\ \xi a_2 & d_2 & -\xi a_2 & e_2 & 0 & 0\ldots \\ c_3 & \xi a_3 & d_3 & -\xi a_3 & e_3 & 0\ldots \\ 0 & c_4 & \xi a_4 & d_4 & -\xi a_4 & e_4\ldots \\ \vdots & \vdots & \vdots & \vdots & \vdots & \vdots\;\vdots \end{pmatrix}, \quad \mathbf{X}(t) = \begin{pmatrix} f_1^{\|}(t) \\ f_2^{\|}(t) \\ \vdots \\ f_l^{\|}(t) \\ \vdots \end{pmatrix}. \quad (2.188)$$

The second method consists in arranging the five-term scalar differential-recurrence equation, Eq. (2.165), as a three-term matrix differential-recurrence equation which can be solved exactly in terms of matrix continued fractions (see Section 1.3.5).

2.4.1.1 *Matrix solution*

This method proceeds as follows. In Eq. (2.188), the dimension l of the state vector $\mathbf{X}(t)$ is taken large enough (equal to L, say) to achieve the desired degree of accuracy. We have explicitly the linear combination

$$\mathbf{X}(t) = b_1 e^{-\lambda_1 t}\mathbf{g}_1 + b_2 e^{-\lambda_2 t}\mathbf{g}_2 + \cdots + b_L e^{-\lambda_L t}\mathbf{g}_L, \tag{2.189}$$

where λ_i are the eigenvalues of the system matrix $\mathbf{A}$ and the coefficients b_i are to be determined from the initial condition vector $\mathbf{X}(0)$. All the relaxation modes $f_n^{\parallel}$ may then be found from Eq. (2.189). In particular, Eq. (2.189) yields the correlation function $f_1^{\parallel}$ as a finite sum of decaying exponentials

$$f_1^{\parallel}(t) \approx \sum_{k=1}^{L} A_k e^{-\lambda_k t}. \tag{2.190}$$

The longitudinal dynamic susceptibility $\chi_{\parallel}(\omega)$ of the particle is then given by a finite sum of Lorentzians, namely

$$\frac{\chi_{\parallel}(\omega)}{\chi_{\parallel}} = 1 - i\omega \int_0^{\infty} C_{\parallel}(t) e^{-i\omega t} dt \approx \sum_{k}^{L} \frac{c_k}{1 + i\omega / \lambda_k}, \tag{2.191}$$

where

$$C_{\parallel}(t) = \frac{f_1^{\parallel}(t)}{f_1^{\parallel}(0)}, \quad c_k = \frac{A_k}{\sum_k A_k}, \tag{2.192}$$

and $\chi_{\parallel}$ is the static susceptibility given by

$$\chi_{\parallel} = \frac{v\mu_0 M_S^2}{kT} f_1^{\parallel}(0) = \frac{v\mu_0 M_S^2}{kT} (\langle \cos^2 \vartheta \rangle_0 - \langle \cos \vartheta \rangle_0^2). \tag{2.193}$$

The longest relaxation time of the magnetization $\tau \approx \lambda_1^{-1}$ is determined via the smallest nonvanishing eigenvalue λ_1 corresponding to the smallest root of the characteristic equation:

$$\det(\lambda \mathbf{I} - \mathbf{A}) = 0. \tag{2.194}$$

The correlation time $\tau_{\text{int}}^{\parallel}$ and the effective relaxation times $\tau_{\text{ef}}^{\parallel}$ may be calculated in terms of A_k and λ_k as

$$\tau_{\text{int}}^{\|} = \frac{\sum_{k=1}^{L} A_k / \lambda_k}{\sum_{k=1}^{L} A_k} \quad \text{and} \quad \tau_{\text{ef}}^{\|} = \frac{\sum_{k=1}^{L} A_k}{\sum_{k=1}^{L} A_k \lambda_k}. \tag{2.195}$$

Equivalently, the linear matrix differential Eq. (2.187) can be formally solved using the Laplace transform. Thus, we obtain

$$\tilde{\mathbf{X}}(s) = \begin{pmatrix} \tilde{f}_1^{\|}(s) \\ \tilde{f}_2^{\|}(s) \\ \tilde{f}_3^{\|}(s) \\ \vdots \end{pmatrix} = (s\mathbf{I} - \mathbf{A})^{-1} \mathbf{X}(0) = (s\mathbf{I} - \mathbf{A})^{-1} \begin{pmatrix} f_1^{\|}(0) \\ f_2^{\|}(0) \\ f_3^{\|}(0) \\ \vdots \end{pmatrix}. \tag{2.196}$$

Because of well-known identities [290] for the Legendre polynomials, Eqs. (A.36) and (A.37) from Appendix A.1.5, the initial conditions $f_l^{\|}(0) = \langle P_1 P_l \rangle_0 - \langle P_1 \rangle_0 \langle P_l \rangle_0$ may be determined from the equilibrium averages $\langle P_l \rangle_0$ as

$$f_l^{\|}(0) = \frac{l+1}{2l+1} \langle P_{l+1} \rangle_0 + \frac{l}{2l+1} \langle P_{l-1} \rangle_0 - \langle P_l \rangle_0 \langle P_1 \rangle_0 . \tag{2.197}$$

Thus the $\langle P_l \rangle_0$ can be evaluated *numerically* by simply calculating the inverse of the system matrix $\mathbf{A}^{-1}$. We demonstrate as follows. First, note that they must satisfy the five-term recurrence relation

$$d_l \langle P_l \rangle_0 + c_l \langle P_{l-2} \rangle_0 + e_l \langle P_{l+2} \rangle_0 + \xi a_l \left[\langle P_{l-1} \rangle_0 - \langle P_{l+1} \rangle_0 \right] = 0 \tag{2.198}$$

which is simply Eq. (2.165) with $\dot{f}_l^{\|} = 0$. The recurrence relation, Eq. (2.198), is just the *inhomogeneous* set $\mathbf{AF}^0 = \mathbf{B}$ with

$$\mathbf{F}^0 = \begin{pmatrix} \langle P_1 \rangle_0 \\ \langle P_2 \rangle_0 \\ \vdots \\ \langle P_l \rangle_0 \\ \vdots \end{pmatrix} \quad \text{and } \mathbf{B} = \begin{pmatrix} \xi / 3 \\ 2\sigma / 5 \\ 0 \\ 0 \\ \vdots \end{pmatrix}. \tag{2.199}$$

Thus, having calculated all the equilibrium averages $\langle P_l \rangle_0$ from the formal expression $\mathbf{F}^0 = \mathbf{A}^{-1}\mathbf{B}$, we have the initial value vector $\mathbf{X}(0)$ and, consequently, $\tilde{\mathbf{X}}(s)$ in Eq. (2.196). Having calculated $\tilde{f}_1(s)$, we then have the dynamic susceptibility $\chi_{\|}(\omega)$, Eq. (2.191), and the relaxation times $\tau_{\text{int}}^{\|}$ and $\tau_{\text{ef}}^{\|}$, Eq. (2.195).

2.4.1.2 *Matrix continued fraction solution*

The advantage of posing the problem in this way is that *formally exact* expressions may be written for the Laplace transform of the after-effect function, the correlation time, and the complex susceptibility. Equation (2.165) can be arranged as a matrix three-term differential-recurrence relation, Eq. (1.116), namely,

$$\tau_{\text{N}}\dot{\mathbf{C}}_n(t) = \mathbf{Q}_n^- \mathbf{C}_{n-1}(t) + \mathbf{Q}_n \mathbf{C}_n(t) + \mathbf{Q}_n^+ \mathbf{C}_{n+1}(t), \tag{2.200}$$

where the 2 × 2 matrices $\mathbf{Q}_n^{\pm}$ and $\mathbf{Q}_n$ are given by

$$\mathbf{Q}_n^- = \begin{pmatrix} c_{2n-1} & \xi a_{2n-1} \\ 0 & c_{2n} \end{pmatrix}, \mathbf{Q}_n = \begin{pmatrix} d_{2n-1} & -\xi a_{2n-1} \\ \xi a_{2n} & d_{2n} \end{pmatrix}, \mathbf{Q}_n^+ = \begin{pmatrix} e_{2n-1} & 0 \\ -\xi a_{2n} & e_{2n} \end{pmatrix}, \tag{2.201}$$

and

$$\mathbf{C}_n(t) = \begin{pmatrix} f_{2n-1}^{\|}(t) \\ f_{2n}^{\|}(t) \end{pmatrix}. \tag{2.202}$$

The *exact* solution for the Laplace transform $\tilde{\mathbf{C}}_1(s)$ of Eq. (2.200), which is simply a particular case of the archetypal matrix three-term differential-recurrence relation, is given by Eq. (1.126) (see Section 1.3.5), namely,

$$\tilde{\mathbf{C}}_1(s) = \tau_{\text{N}}\left(\mathbf{C}_1(0)\boldsymbol{\Delta}_1(s) + \sum_{n=1}^{\infty}\prod_{k=1}^{n} \mathbf{Q}_k^+ \boldsymbol{\Delta}_{k+1}(s) \right), \tag{2.203}$$

where the matrix continued fraction $\boldsymbol{\Delta}_n(s)$ is defined by the recurrence relation, Eq. (1.123) in Chapter 1 and the initial-condition column vectors $\mathbf{C}_n(0)$ may be determined using Eqs. (2.197) and (2.198). We have

$$\mathbf{C}_n(0) = \left[\begin{pmatrix} 0 & \frac{2n-1}{4n-1} \\ 0 & 0 \end{pmatrix} + \begin{pmatrix} -\langle P_1 \rangle_0 & \frac{2n}{4n-1} \\ \frac{2n}{4n+1} & -\langle P_1 \rangle_0 \end{pmatrix} + \begin{pmatrix} 0 & 0 \\ \frac{2n+1}{4n+1} & 0 \end{pmatrix} \mathbf{S}_{n+1}(0) \right] \cdots \mathbf{S}_1(0) \begin{pmatrix} 0 \\ 1 \end{pmatrix}.$$

where as usual $\mathbf{S}_n(0) = \mathbf{\Delta}_n(0)\mathbf{Q}_n^-$. Having calculated $\tilde{f}_1^{\|}(s)$, we can then evaluate both the integral relaxation (correlation) time

$$\tau_{\text{int}}^{\|} = \lim_{s\to 0} \int_0^\infty C_{\|}(t) e^{-st} dt = \tilde{C}_{\|}(0) = \frac{\tilde{f}_1^{\|}(0)}{f_1^{\|}(0)} \tag{2.204}$$

and the dynamic susceptibility $\chi_{\|}(\omega)$, Eq. (2.191). Furthermore, the longest relaxation time of the magnetization $\tau \approx \lambda_1^{-1}$ can be evaluated from Eq. (1.134) of Chapter 1.

2.4.2 *Characteristic times and magnetic susceptibility*

In Fig. 2.20(a), we plot the smallest nonvanishing eigenvalue λ_1 calculated from the characteristic equation, Eq. (2.194), as a function of the anisotropy parameter σ for various dc bias field parameters h, showing that our results concur with the corresponding Fig. 1 of Aharoni [334]. A 45 × 45 matrix **A** was enough to obtain λ_1 accurate to three significant digits for the range of parameters chosen. In Fig. 2.20(b), we also plot $1/\tau_{\text{int}}^{\|}$ computed from Eq. (2.204) for different values of h. The most astonishing result is that $1/\tau_{\text{int}}^{\|}$ is markedly greater than λ_1 above a certain critical value h due to the depletion effect (a discussion of this effect follows in Section 2.4.3). This behavior is further emphasized in Table 2.3. More insight into the disparity between λ_1^{-1} and $\tau_{\text{int}}^{\|}$ may be gained from the general equation, Eq. (2.195). Clearly, the calculation of $\tau_{\text{int}}^{\|}$ from Eq. (2.195) requires knowledge of a sufficiently large set of

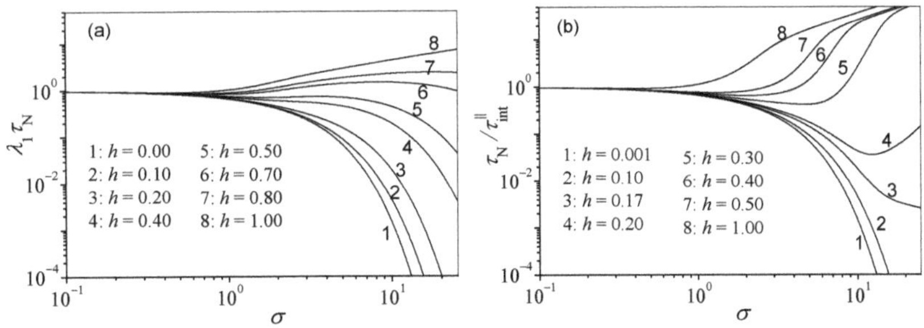

Fig. 2.20. The smallest nonvanishing eigenvalue λ_1 (a) and the inverse of the correlation time $\tau_{\text{int}}^{\|}$ (b) as functions of σ for various values of the field parameter h.

Table 2.3. Product of λ_1 and $\tau_{\text{int}}^{\|}$ for various barrier heights σ and fields h.

h	$\tau_{\text{int}}^{\|}\lambda_1$						
	$\sigma = 0.2$	$\sigma = 0.5$	$\sigma = 1$	$\sigma = 2$	$\sigma = 5$	$\sigma = 10$	$\sigma = 20$
0.01	1.	0.999	0.998	0.994	0.988	0.996	0.999
0.1	1.	0.999	0.997	0.99	0.974	0.964	0.725
0.2	1.	0.999	0.995	0.979	0.896	0.455	0.00134
0.4	1.	0.997	0.986	0.926	0.331	0.00887	0.000315
0.5	0.999	0.996	0.979	0.878	0.153	0.0171	0.00183
0.7	0.999	0.992	0.959	0.746	0.113	0.0482	0.0176
1.0	0.998	0.984	0.913	0.544	0.176	0.119	0.084

Table 2.4. Amplitudes A_k and the corresponding eigenvalues λ_k for various field parameters h; $\sigma = 10$.

k	$h = 0.01$		$h = 0.2$		$h = 0.4$	
	A_k	$\lambda_k\tau_N$	A_k	$\lambda_k\tau_N$	A_k	$\lambda_k\tau_N$
1	0.858	0.00147	0.0019	0.0192	1.11×10^{-6}	0.192
2	9.05×10^{-8}	8.06	2.01×10^{-6}	7.32	6.68×10^{-8}	7.12
3	0.00283	12.3	6.32×10^{-5}	12.7	7.86×10^{-7}	14.4
4	4.9×10^{-5}	14.8	0.0014	17.6	4.10×10^{-4}	22.6
5	8.07×10^{-4}	19.2	6.68×10^{-4}	20.3	1.06×10^{-3}	23.6
6	2.18×10^{-6}	25.2	1.01×10^{-4}	26.6	2.90×10^{-5}	31.2
7	3.61×10^{-5}	32.1	2.98×10^{-5}	33.4	2.27×10^{-5}	37.7
8	8.78×10^{-8}	40.0	5.84×10^{-6}	41.2	6.14×10^{-6}	44.8
10	1.92×10^{-9}	58.9	1.5×10^{-7}	60.0	2.06×10^{-7}	63.5

eigenvalues $\{\lambda_k\}$ and their corresponding amplitudes $\{A_k\}$, as shown in Table 2.4. The reason for the disparity between $\tau_{\text{int}}^{\|}$ and λ_1^{-1} now becomes obvious: at short to intermediate times, the high-frequency (intrawell) decay modes cannot be neglected, as they contribute significantly to the correlation time. Indeed, it is apparent from Table 2.4, that for $h\sigma \geq 0.1$, $\tau_{\text{int}}^{\|}$ cannot be approximated by λ_1^{-1} this is in marked contrast to $h = 0$,where the A_k are negligible for all $k > 1$ (see Table 2.4). However, for all σ, λ_1 is effectively available in closed finite integral form, because

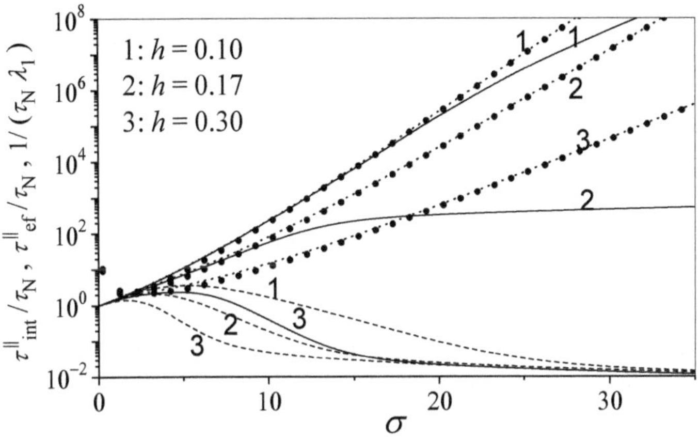

Fig. 2.21. $\tau^{\|}_{int}$ (Eq. (2.182); solid lines), $\tau^{\|}_{ef}$ (Eq. (2.183); dashed lines), and λ_1^{-1} (Eq. (2.194); dotted lines) vs. σ for various values of h. Filled circles: asymptote Eq. (2.4).

it can be approximated by the inverse of the mean first-passage time $\lambda_1 \approx \tau_{MFPT}^{-1}$, (see Chapter 1, Section 1.4.4). Moreover, for $\sigma(1-h)^2 > 3$, λ_1 can be approximated by Brown's asymptotic formula Eq. (2.4) [17,92]. For $\sigma(1-h)^2 < 2$, this formula is inadequate (see Fig. 2.21). However, in this range of σ, λ_1^{-1} can be approximated by the effective relaxation time $\tau^{\|}_{ef}$, Eq. (2.183).

We now compare each of the time constants λ_1^{-1}, $\tau^{\|}_{int}$, and $\tau^{\|}_{ef}$ in Fig. 2.21 for different h. It appears that the most interesting effect is the behavior of the integral relaxation time $\tau^{\|}_{int}$ as a function of the barrier-height parameter σ for sufficiently large bias parameter h. This may be summarized as follows when h exceeds a certain critical value $h_c \approx 1/6$, $\tau^{\|}_{int}$ no longer has an activation character for large σ (solid curve 2 in Fig. 2.21). At this critical value of h, the relaxation switches from being dominated by the slowest overbarrier mode to being dominated by the fast intrawell modes. Thus, $\tau^{\|}_{int}$ decreases as the height of the potential barrier increases. This effect was discovered numerically by Coffey *et al.* [146] and later explained quantitatively by Garanin [337].

Having determined λ_1^{-1}, $\tau^{\|}_{int}$, and $\tau^{\|}_{ef}$, one can also calculate the dynamic susceptibility $\chi_{\|}(\omega)$ from the *two-mode approximation,* Eq. (1.150) (see Section 1.3.6), where the asymptotic estimates for $\chi_{\|}$, Δ_1, and τ_W may be evaluated from the equations [337]

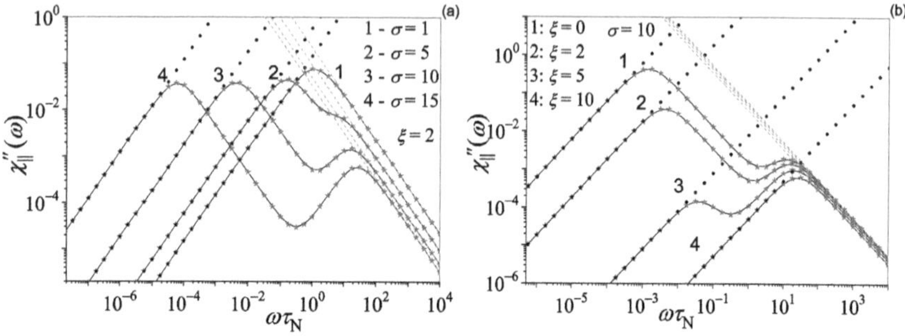

Fig. 2.22. $\chi''_{\parallel}(\omega)$ vs. $\omega\tau_N$ as evaluated from the matrix continued fraction solution (solid lines) and from the two-mode approximation Eq. (1.150) (symbols) for various σ (a) and ξ (b) with numerical values of $\tau_{int}^{\parallel}$, $\tau_{ef}^{\parallel}$, and λ_1^{-1} from Table 2.4. Dotted and dashed lines: the low- and high-frequency asymptotes, Eq. (1.147).

Table 2.5. Numerical values of $\tau_{int}^{\parallel}$, Eq. (2.182), $\tau_{ef}^{\parallel}$, Eq. (2.183), λ_1^{-1}, Eq. (2.194), and τ_W, Eq. (1.149).

$\sigma = 10$	$\xi = 0$	$\xi = 2$	$\xi = 5$	$\xi = 10$
$(\tau_N\lambda_1)^{-1}$	693.92	232.54	26.826	2.2650
$\tau_{int}^{\parallel}/\tau_N$	691.02	224.15	3.3982	0.03865
$\tau_{ef}^{\parallel}/\tau_N$	16.644	1.6899	0.05654	0.03839
τ_W/τ_N	0.07143	0.06146	0.04949	0.03838
$\xi = 2$	$\sigma = 1$	$\sigma = 5$	$\sigma = 10$	$\sigma = 15$
$(\tau_N\lambda_1)^{-1}$	0.97627	6.3866	232.54	16484
$\tau_{int}^{\parallel}/\tau_N$	0.89117	5.7200	224.15	16221
$\tau_{ef}^{\parallel}/\tau_N$	0.75275	1.0975	1.6899	2.3663
τ_W/τ_N	0.28660	0.13831	0.06146	0.03774

$$\frac{\tau_N}{\tau_W} \sim 2\sigma + \xi - \frac{5+h}{1+h}, \qquad \chi_{\parallel}(1-\Delta_1) \sim \frac{1}{(2\sigma+\xi)^2}. \tag{2.205}$$

The asymptotic estimate for λ_1 is given by Eq. (2.4). Numerical calculations of the normalized complex susceptibility $(v\mu_0 M_S^2/(kT) = 1)$

from Eqs. (1.150), and those from the exact matrix continued fraction solution, are compared in Fig. 2.22. The parameters used are given in Table 2.5. Clearly, the agreement between the numerical calculations and the two-mode approximation is very good. Similar (or even better) agreement exists for all ξ and σ. The low- and high-frequency asymptotes are also shown in Fig. 2.22. The figures demonstrate that the low- and high-frequency behaviors of the magnetic loss $\chi''_{\parallel}(\omega)$ are completely determined by $\tau^{\parallel}_{\text{int}}$ and $\tau^{\parallel}_{\text{ef}}$, respectively.

2.4.3 *Depletion effect in a biased bistable potential*

If we have a random variable, the quantities of most interest are its expectation value and its variance, and, as far as the dynamics are concerned, the autocorrelation function (ACF). The ACF measures the correlation between the value of a random variable at time $t_1 = 0$, and its value at time $t_2 = t$. It may be shown (as seen) from linear response theory that the decay of the magnetization of a nanomagnet following a small change ΔH in the applied field directed along the easy axis is $M_{\parallel}(t) = M_0 C_{\parallel}(t)$, where $M_0 = v\mu_0 \Delta H M_S^2 / (kT)$ and

$$C_{\parallel}(t) = \langle \cos\vartheta(0)\cos\vartheta(t)\rangle_0 - \langle \cos\vartheta(0)\rangle_0^2 \tag{2.206}$$

is the ACF of the longitudinal component of the magnetization (see Section 1.3.6 of Chapter 1). The zeros on the angular braces represent the averages without the small perturbation ΔH. The pertinent property of the ACF is the integral relaxation time $\tau^{\parallel}_{\text{int}}$ defined as the normalized area under the curve of $C_{\parallel}(t)$, Eq. (2.204), which is identical with the correlation time of linear response theory. The integral relaxation time includes the contributions of all the modes of the decay of the magnetization, and so is a *global* characteristic of the relaxation process. Therefore to evaluate τ_{int}, one would, in principle, have to determine all the eigenvalues λ_k ($k = 1, 2,\ldots$) of the Fokker–Planck operator, which may be accomplished by solving the recurrence relations generated by the Fokker–Planck or averaged Langevin equations. We have shown (see Section 1.4.4) how the longest relaxation time $\tau = \lambda_1^{-1}$ associated with the longest-lived mode (λ_1 is the smallest nonvanishing eigenvalue) may be evaluated using the MFPT method. Moreover, the integral relaxation time τ_{int} is virtually identical with the escape time τ if the configuration of the system is such that the contributions of all the other modes, save the longest lived one, are negligible. In other cases, however, τ_{int} may differ substantially from τ, and so cannot be used to estimate the escape time [179,337]. To demonstrate this effect, we consider the integral relaxation time in a biased bistable potential, Eq. (2.163), shown in Fig. 2.1. Now the population of the shallower of the two potential wells may be substantially decreased by the uniform bias force. This has a profound effect on τ_{int} because, at a certain critical value of the bias force (however much less than that required to destroy the bistable character of the potential), a switchover of τ_{int} from Arrhenius to non-Arrhenius behavior will take place. The effect was originally discovered for magnetic relaxation in 1995 by Coffey *et al.* [146] for the biased uniaxial anisotropy potential Eq. (2.163) by numerical solution of the Fokker–Planck equation (1.223). The solution of that equation for τ_{int} indicated that, at fields well below the critical field at which the bistable character of the potential is destroyed, the integral relaxation time ceases to be dominated by λ_1^{-1}. Thus, the escape time τ no longer dominates τ_{int} for bias fields in excess of a certain critical field. Hence, the *contribution of relaxation modes to* τ_{int}, *other than the barrier-crossing mode, becomes significant*. The physical interpretation of this behavior was later given by Garanin [337]. It appears to be a universal feature of the relaxation in biased bistable potentials [31,335] (see also [179] for a review).

Following the elegant exposition of Garanin [337], we may suppose [335] that $C_{\|}(t)$ is the *sum* of the *fast* relaxation processes in the deep well plus the *slow* overbarrier process from the shallower well into the deep well, so that $C_{\|}(t)$ may be approximated as (see Section 1.3.6)

$$C_{\|}(t) \approx \Delta_{\mathrm{W}} e^{-t/\tau_{\mathrm{W}}} + \Delta_{\mathrm{B}} e^{-\lambda_1 t}, \tag{2.207}$$

where Δ_{W} is the population of particles in the deep well and Δ_{B} arises from the (rare) population of particles crossing over the barrier. The relaxation time τ_{W} is associated with the fast relaxation processes in the well. The inverse of the well time τ_{W}^{-1} is approximately equal to $2\alpha\omega_{\mathrm{pr}} \approx 2\sigma\tau_{\mathrm{N}}^{-1}(1+h)$ [337]. Equation (2.204) may be now used to determine the integral relaxation time. We have, from Eqs. (2.204) and (2.207)

$$\tau_{\text{int}}^{\parallel} \approx \frac{1}{C_{\parallel}(0)} \int_0^{\infty} \left(\Delta_{\text{W}} e^{-t/\tau_{\text{W}}} + \Delta_{\text{B}} e^{-\lambda_1 t} \right) dt = \frac{\tau_{\text{W}} \Delta_{\text{W}} + \Delta_{\text{B}} / \lambda_1}{\Delta_{\text{W}} + \Delta_{\text{B}}}. \quad (2.208)$$

However, in the high-barrier limit, escape over a barrier is an *exponentially slow process*, therefore since only rare particles escape, it follows that $\Delta_{\text{B}} << \Delta_{\text{W}}$; thus, we can write

$$\tau_{\text{int}}^{\parallel} = \tau_{\text{W}} + (\Delta_{\text{B}} / \Delta_{\text{W}}) \lambda_1^{-1}. \quad (2.209)$$

Since λ_1^{-1} is exponentially large in the high-barrier limit, one would expect that $\tau_{\text{int}}^{\parallel}$ would be dominated by the second term, unless $\Delta_{\text{B}} / \Delta_{\text{W}}$ has *negative exponential* type behavior so that $\tau_{\text{int}}^{\parallel}$ would then be *dominated* by the fast relaxation time τ_{W}. We shall show that this is indeed the case if the reduced field h exceeds a critical value [335,337], and we shall further show that the behavior is *independent* of the precise nature of the prefactor.

To facilitate this, we will derive an expression for the total population $\Delta_{\text{B}} + \Delta_{\text{W}}$ based on our analysis of the ensemble averages using the partition function. We have

$$C_{\parallel}(0) = \langle z^2 \rangle_0 - \langle z \rangle_0^2 = \Delta_{\text{W}} + \Delta_{\text{B}} \quad (2.210)$$

with $z = \cos\vartheta$. We can now evaluate $\langle z \rangle_0$ and $\langle z^2 \rangle_0$ via the partition function

$$Z = \int_{-1}^{1} e^{\sigma z^2 + \xi z} dz. \quad (2.211)$$

We have [337]

$$\langle z \rangle_0 = \frac{1}{Z} \frac{\partial Z}{\partial \xi} = \frac{Z'}{Z} \text{ and } \langle z^2 \rangle_0 = \frac{1}{Z} \frac{\partial^2 Z}{\partial \xi^2} = \frac{Z''}{Z}. \quad (2.212)$$

Substituting these expressions in Eq. (2.210), we have the total population

$$\Delta_{\text{W}} + \Delta_{\text{B}} = Z'' / Z - (Z' / Z)^2. \quad (2.213)$$

Thus, the vital quantity in calculating averages (and, hence, the integral relaxation time), is Z as defined by Eq. (2.211). We shall now indicate, using the asymptotic results, how approximate expressions for Z may be calculated for the biased uniaxial potential, Eq. (2.163), expressions which of course hold for potential barrier heights well in excess of kT. We have, from Eq. (2.211) evaluated at the ith minimum, $i=1,2$,

$$
\begin{aligned}
Z(\vartheta_i) &\sim \int_{\text{near } \vartheta_i} e^{-E(\vartheta)} \sin\vartheta d\vartheta \\
&\approx e^{-E(\vartheta_i)} \int_{\vartheta_i}^{\infty} (\vartheta-\vartheta_i) e^{-(\vartheta-\vartheta_i)^2 \partial^2_{\vartheta\vartheta} E(\vartheta_i)/2} d\vartheta = e^{-E(\vartheta_i)} \big/ \partial^2_{\vartheta\vartheta} E(\vartheta_i),
\end{aligned}
\tag{2.214}
$$

where $E(\vartheta)$ and $\sin\vartheta$ are approximated near the well bottom ϑ_i by the leading nonvanishing terms in their Taylor series (this is permissible because almost all the magnetic dipoles are situated near the minimum at ϑ_i so that $\vartheta-\vartheta_i << 1$), namely $E(\vartheta) \approx E(\vartheta_i) + E''_{\vartheta\vartheta}(\vartheta_i)(\vartheta-\vartheta_i)^2/2$ and $\sin\vartheta \approx \vartheta-\vartheta_i$. We now take the second derivative of the potential and make the appropriate substitutions for the numerator and denominator. Thus, we can express Z in terms of the dimensionless anisotropy and external field parameters σ and ξ, respectively. Following Ref. [337], since $\partial^2_{\vartheta\vartheta} E = 2\sigma\cos 2\vartheta + \xi\cos\vartheta$, then evaluating Eq. (2.214) at $\vartheta = 0$ will yield an expression for the portion of the overall partition function (population) corresponding to the deeper well, which we call $Z_+ = e^{\sigma+\xi}/(2\sigma+\xi)$. On the other hand if we evaluate Eq. (2.214) at $\vartheta = \pi$, i.e., where the particles are located in the shallower well, we will have the portion of the overall partition function corresponding to the shallower well, which we call $Z_- = e^{\sigma-\xi}/(2\sigma-\xi)$. The overall partition function, however, is effectively the sum of these two individual contributions, i.e., $Z \approx Z_+ + Z_-$. The asymptotic method which we have employed is useful since it is difficult to evaluate the integrals exactly when anisotropy is included, because they are not expressible as elementary functions.

The *total* population $\Delta_B + \Delta_W$ can now be evaluated using our high-barrier approximation for Z,

$$
\Delta_W + \Delta_B = \frac{Z''_+ + Z''_-}{Z_+ + Z_-} - \left(\frac{Z'_+ + Z'_-}{Z_+ + Z_-}\right)^2. \tag{2.215}
$$

The essence of our calculation is *also to evaluate* $\Delta_B + \Delta_W$ from our *individual* expressions for Z_+ and Z_-. It is important that we are mindful of the fact that very few particles cross over the barrier once the relaxation process has been initiated. Therefore, the variance may also be approximately estimated as being *the sum of the variances in each well if it were isolated.* The individual terms in this expression would not, however, contain any exponentials, as they will always cancel in a formula such as y'/y or y''/y, if y is given by a single exponential term, as is so, for each isolated well. Therefore, the *difference* between the variance of the *overall* process calculated by using the *exact* $Z = Z_+ + Z_-$ and the variance calculated by using Z_+ and Z_- *separately* gives Δ_B. Thus, by determining which terms correspond to Δ_W, and which correspond to Δ_B, we have τ_{int}, as expressed by Eq. (2.208). For example, consider $Z_\pm = A_\pm e^{\sigma \pm \xi}$, where $A_+ = (2\sigma + \xi)^{-1}$ and $A_- = (2\sigma - \xi)^{-1}$ represent the prefactors of this particular system. It is obvious that Z'/Z contains exponential factors while the isolated well moments $Z'_\pm / Z_\pm$ do not. Moreover, we need not be concerned with the precise form of $A_\pm$ as they will simply be algebraic prefactors for any given system.

We now reach our final result by considering the second term on the right of Eq. (2.215). If we ignore the irrelevant algebraic prefactors, then it is obvious that $Z_-/Z_+ \cong e^{-2\xi}$ and $Z_+/Z_- \cong e^{2\xi}$, so that

$$\frac{Z'_+ + Z'_-}{Z_+ + Z_-} = \frac{Z'_+/Z_+}{1 + Z_-/Z_+} + \frac{Z'_-/Z_-}{1 + Z_+/Z_-} \cong \frac{Z'_+}{Z_+} + e^{-2\xi}\frac{Z'_-}{Z_-}. \tag{2.216}$$

Clearly, if we apply the same method to $(Z''_+ + Z''_-)/(Z_+ + Z_-)$, we obtain

$$\frac{Z''_+ + Z''_-}{Z_+ + Z_-} \cong \frac{Z''_+}{Z_+} + e^{-2\xi}\frac{Z''_-}{Z_-}. \tag{2.217}$$

By substituting these two equations into Eq. (2.215), we then have

$$\Delta_W + \Delta_B \cong \frac{Z''_+}{Z_+} - \left(\frac{Z'_+}{Z_+}\right)^2 + \frac{Z''_-}{Z_-}e^{-2\xi} - 2\frac{Z'_+}{Z_+}\frac{Z'_-}{Z_-}e^{-2\xi}. \tag{2.218}$$

In writing this equation, we have declared an $e^{-4\xi}$ term small and irrelevant. Equation (2.218) now yields, by inspection, the terms corresponding to Δ_W and Δ_B. Clearly, the first two terms on the right-hand side correspond to the variance in the deeper well, and must then correspond to Δ_W. The third and fourth terms of Eq. (2.218) must correspond to Δ_B because, from our initial reasoning, they can only have their origin in the partition function Z of the *entire* system and so must be due to the combined effect of the two wells. We can thus infer that $\Delta_B / \Delta_W \approx e^{-2\xi} = e^{-4\sigma h}$, while $\lambda_1^{-1} \sim e^{\sigma(1-2h+h^2)}$ (see Eq. (2.4)), so that the overall contribution of the second term is $\lambda_1^{-1} \Delta_B / \Delta_W \sim e^{\sigma(1-6h+h^2)}$, which verifies the analytical result first given by Garanin [337] in 1996 and confirmed by earlier (1994) [146] numerical calculations of the integral relaxation time. Thus, the relaxation time can change its sign from *positive exponential* to *negative exponential* behavior if $h_c \approx 0.17$. Hence, the overall relaxation time switches from being dominated by growing exponential behavior ($h < 0.17$), to being dominated by the (algebraic) first term pertaining to relaxation in the deep well. Such behavior, with h_c dependent on the *precise* form of the potential, appears to be a universal feature of bistable potentials subjected to a bias force; see Refs. [179,335].

The depletion effect we have just described may be attributed to the bistable nature of the external potential.

2.4.4 *Magnetic stochastic resonance in uniaxial nanomagnets*

Stochastic resonance (SR) is a well-known, remarkable effect allowing effective control of the quality of periodic signals passing through noisy systems [372]. As a manifestation of cross-coupling between stochastic and deterministic motions, it is ubiquitous in physics (e.g., optics, mechanics of solids, superconductivity, fine particle magnetism, surface science), communication engineering (optimal detection and tracing of signals) as well as in various branches of chemistry and biology; recently SR has also been detected in behavioral and related problems. Comprehensive reviews are available in Refs. [55,372]. Now the simplest theoretical model of SR [55,97] is a heavily damped (so that inertial effects can be ignored) Brownian particle in a bistable potential which is excited by an ac driving force of frequency Ω *close to the rate of Kramers transitions between the wells nevertheless with amplitude insufficient to induce the transitions by itself* (see Fig.2.23). Consequently, switching may occur only through the *combined* effect of the deterministic ac force

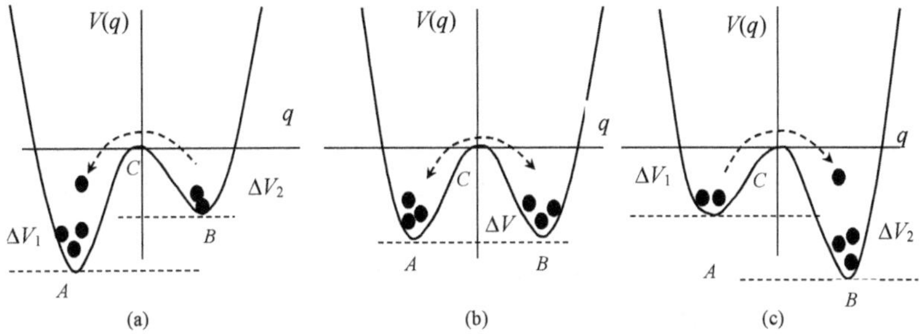

Fig. 2.23. Double-well potential as used in stochastic resonance [97]. The minima are located at A and B. In the absence of a periodic forcing function (b), the barrier heights ΔV_1 and ΔV_2 are both simply ΔV, so that the potential is symmetric. The periodic forcing function causes the double-well potential to tilt back and forth, thereby raising and lowering the potential barriers of the right and left wells, respectively, in an antisymmetric cyclic fashion; (a) and (c).

and the noise. The spectral density $\Phi(\omega)$ of the motion at the signal frequency $\omega = \Omega$ is then evaluated, and the resulting signal-to-noise ratio (SNR) is analyzed as a function of the noise intensity D showing that the curve SNR(D) has a bell-like shape, i.e., it passes through a maximum thus exhibiting *stochastic resonance.* The maximum in the SNR is interpreted as originating in the remarkable ability of noise to enhance the intensity of the interwell hoppings in the system.

Superparamagnetism of fine particles is yet another important manifestation of SR. Here the magnetic anisotropy is the source of multistability for the magnetic moment of the particle while the thermal fluctuations supplied by the bath which is at finite temperature are the source of the noise so giving rise to *magnetic stochastic resonance*. The superparamagnetic behavior of ferromagnetic nanoparticles which has until recently being mostly of academic interest, has now become of practical importance in technology because of the ever-decreasing size of the particles used in magnetic recording. Once the size of the particle is reduced to nanoscale dimensions, superparamagnetism automatically imposes a physical limit on the miniaturization of the bit-recording elements. Hitherto in micromagnetism, the size of the particles used ensured that thermal effects occurring at room temperature could safely be ignored. However, due to the drastic size reduction, a superparamagnetic nanoparticle is now regarded as an acceptable candidate for c and read/write heads [373]. The magnetic SR may be succinctly defined as the *enhancement* of the signal-to-noise ratio (or the spectral power amplification coefficient) due to noise [374]. Here a magnetically anisotropic ferromagnetic nanoparticle at finite temperature subjected to a weak external ac field exhibits the essential features of mechanical SR and is described by Brown's Langevin-Fokker–Planck model.

Now we quantitatively discuss *stochastic resonance* (SR) in uniaxial superparamagnetic particles which was first predicted theoretically [60–63,67] and shortly afterwards observed experimentally [57–59]. The magnetic energy of a uniaxial single-domain ferromagnetic particle is given by Eq. (2.163). Without a dc bias field, the magnetic moment has two equivalent stable orientations at $\vartheta = 0$ and $\vartheta = \pi$, so that it is a prime example of a bistable system subjected to noise. The rate of transitions between the potential wells is controlled by the usual parameter $\sigma = vK/(kT)$, relating the height Kv of the magnetic anisotropy barrier to the thermal energy kT. Assuming that the barrier height is fixed, one may then regard the inverse anisotropy parameter σ^{-1} as the *dimensionless temperature*, i.e., the noise intensity. A dc bias field $\mathbf{H}_0$ applied parallel to the anisotropy axis of the particle breaks the bidirectional symmetry of the potential. However, the two-minima profile of the potential $E(\cos\vartheta)$ survives as long as the bias field parameter $h = \mu_0 M_S H_0 / (2K) \le 1$.

Proceeding, we first recall the basic results of SR theory for magnetic systems in weak ac magnetic fields [60–63]. According to linear response theory [55,62,63], the Fourier components of the magnetization M_ω and the applied external ac magnetic field H_ω are related via the dynamic susceptibility $\chi(\omega)$ as

$$M_\omega = \chi(\omega) H_\omega, \tag{2.219}$$

where $\chi(\omega)$ may now be regarded as a transfer function. Thus, the spectral density $\Phi_M(\omega)$ of the forced magnetization oscillations in an applied ac field $H(t) = H\cos\Omega t$, where Ω is near the frequency of the Kramers transitions is [62,63]

$$\Phi_M(\omega) = \frac{1}{4}\mu_0 H^2 \left|\chi(\Omega)\right|^2 \left[\delta(\omega+\Omega) + \delta(\omega-\Omega)\right]. \tag{2.220}$$

Also because of the parity condition we have $\chi^*(\omega) = \chi(-\omega)$ and, from the definition of the spectral density at the excitation frequency Ω [375]

$$\Phi_M^{(\mathrm{sign})}(\Omega) = 2 \lim_{\Delta\omega \to 0} \int_{\Omega-\Delta\omega}^{\Omega+\Delta\omega} \Phi_M(\omega) d\omega, \tag{2.221}$$

we have the spectral density of the signal

$$\Phi_M^{(\mathrm{sign})}(\Omega) = \frac{1}{2}\mu_0 H^2 |\chi(\Omega)|^2. \tag{2.222}$$

Now, according to linear response theory [163], the *noise-induced* part of the magnetization spectral density is [62,63]

$$\Phi_M^{(\mathrm{noise})}(\Omega) = \frac{kT\chi''(\Omega)}{v\pi\Omega}. \tag{2.223}$$

Here the particle volume v appears explicitly because we are using the magnetization, not the magnetic moment. Since the total spectral density $\Phi_M(\Omega)$ is equal to $\Phi_M^{(\mathrm{signal})}(\Omega) + \Phi_M^{(\mathrm{noise})}(\Omega)$, Eqs. (2.222) and (2.223) yield the SNR as [62,63]

$$\mathrm{SNR} = \frac{\Phi_M^{(\mathrm{sign})}}{\Phi_M^{(\mathrm{noise})}} = \frac{\pi v \mu_0 \Omega H^2 |\chi(\Omega)|^2}{2kT\chi''(\Omega)}. \tag{2.224}$$

This result Eq. (2.224) demonstrates that the calculation of the SNR is, in general, reduced to merely determining the dynamic susceptibility, which is a fundamental dynamical characteristic of any relaxing system.

Since the longitudinal dynamic susceptibility $\chi(\Omega)$ of uniaxial particles has already been obtained for $H_0 = 0$ and $H_0 \neq 0$, one may formally use these solutions for $\chi(\Omega)$ in Eq. (2.224), because we treat the susceptibility of a *single* particle and not that of an assembly. However, deeper insight can be achieved via the eigenvalue representation for $\chi(\Omega)$ at the signal frequency Ω, as yielded by Eq. (2.191) and rewritten here as

$$\chi(\Omega) = \chi_{\parallel} \sum_{k=1}^{\infty} \frac{c_k}{1 + i\Omega/\lambda_k}, \tag{2.225}$$

where c_k are the normalized weight coefficients. To emphasize the temperature dependence of the SNR, we rewrite it as

$$\mathrm{SNR} = \frac{\pi\sigma}{2\tau_\mathrm{N}}\left(\frac{\mu_0 M_\mathrm{S} H}{K}\right)^2 R(\Omega), \tag{2.226}$$

where the dimensionless function $R(\Omega)$ is [97]

$$\begin{aligned} R(\Omega) &= \tau_\mathrm{N}\sigma(\langle\cos^2\vartheta\rangle_0 - \langle\cos\vartheta\rangle_0^2)\frac{\Omega|\chi(\Omega)|^2}{\chi_\parallel \chi''(\Omega)} \\ &= \frac{\tau_\mathrm{N}\sigma(\langle\cos^2\vartheta\rangle_0 - \langle\cos\vartheta\rangle_0^2)}{\sum_{k=1}^{\infty}\frac{c_k\lambda_k}{\lambda_k^2+\Omega^2}}\left\{\left[\sum_{k=1}^{\infty}\frac{c_k\lambda_k^2}{\lambda_k^2+\Omega^2}\right]^2 + \left[\sum_{k=1}^{\infty}\frac{\Omega c_k\lambda_k}{\lambda_k^2+\Omega^2}\right]^2\right\}. \end{aligned} \tag{2.227}$$

Equation (2.226) shows that the relevant quantity is $R(\Omega)$. In general, $R(\Omega)$, besides its obvious dependence on the noise intensity (temperature) and on the constant (bias) field strength H_0, depends on the frequency of the exciting field Ω. In the limits of low and high signal frequencies, $\Omega \to 0$ and $\Omega \to \infty$, Eq. (2.227) yields

$$R(0) = \frac{\tau_\mathrm{N}}{\tau_\mathrm{int}^{\parallel}}B, \quad R(\infty) = \frac{\tau_\mathrm{N}}{\tau_\mathrm{ef}^{\parallel}}B, \tag{2.228}$$

where $\tau_\mathrm{int}^{\parallel}$ and $\tau_\mathrm{ef}^{\parallel}$ are the integral and effective relaxation times as determined from Eqs. (2.182) and (2.183) and $B = \sigma(\langle\cos^2\vartheta\rangle_0 - \langle\cos\vartheta\rangle_0^2)$ is the normalized static susceptibility. For $\Omega \to 0$, Eq. (2.228) may also be derived by observing that the low-frequency part of the dynamic susceptibility $\chi(\Omega)$ is given by Eq. (1.147). Thus, in this limit, the SNR from Eq. (2.224) is independent of the driving frequency Ω and simply reduces to

$$\mathrm{SNR} = \frac{\pi v \mu_0 H^2 \chi_\parallel}{2kT\tau_\mathrm{int}^{\parallel}}. \tag{2.229}$$

The function R vs. the temperature parameter $\sigma^{-1} = kT/(Kv)$ is shown in Fig. 2.24. For zero bias $H_0 = 0$, stochastic resonance is most pronounced in the *adiabatic limit* $\Omega \to 0$ (see curve 1 of Fig. 2.24(a)). This case is relatively straightforward, since by setting $\Omega = 0$, Eq. (2.227) merely reduces to the simple Eq. (2.228). Here, the principal maximum of R is attained at $\sigma \sim 1$, moreover $R \to 0$ as $\sigma \to \infty$. (Here $\tau_\mathrm{int}^{\parallel}$ is very close to

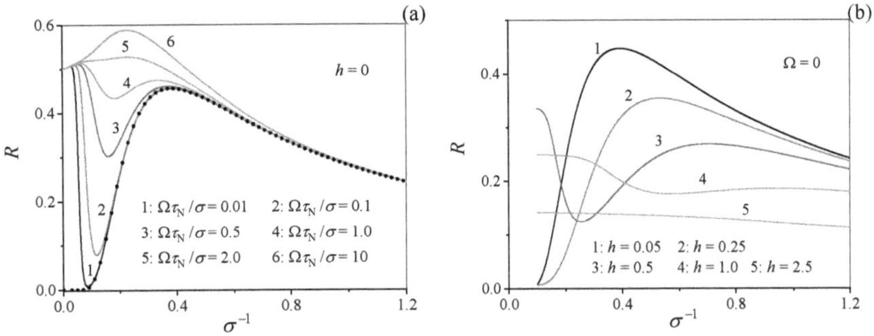

Fig. 2.24. (a) Signal-to-noise ratio as a function of dimensionless temperature for zero dc bias field ($h = 0$) and various frequency parameters $\Omega\tau_N/\sigma$. Solid curves: $\Omega\tau_N/\sigma = 0.01$ (1), 0.1 (2), 0.5 (3), 1 (4), 2 (5), 10 (6). Circles: the adiabatic limit, $\Omega = 0$. (b) Signal-to-noise ratio at nonzero values of the applied dc bias field in the adiabatic limit, $\Omega = 0$, for various dc bias field parameters $h = 0.05$ (1), 0.25 (2), 0.5 (3), 1 (4), and 2.5 (5).

λ_1^{-1} for all σ, so that either parameter may be used.) Indeed (even for any *finite* Ω exceeding the adiabatic limit) since λ_1^{-1} is *exponentially large* in σ for $\sigma >> 1$ (i.e., increasing exponentially as the system is cooled, while all the other λ_k^{-1} simply have an algebraic dependence on σ), the ratio Ω/λ_1 invariably tends to infinity *for low temperatures.* Thus, the overbarrier transition effectively vanishes. Nevertheless, the magnetic moment, although now confined to a particular potential well, is not *completely immobilized*: it may still *rotate inside the well.* This motion is very sensitive to the exact shape of the potential near the bottom of the well, and so is determined by the entire (infinite) eigenvalue spectrum. Thus $R \to 1/2$ as $\sigma \to \infty$. This finite limit is unambiguously confirmed by numerical calculations in Fig. 2.24. For bias fields $H_0 \neq 0$ (see Fig. 2.24(b)), *the maximum of the SNR shifts to higher temperatures as the dc field strength increases,* because such fields substantially alter the temperature dependence of the static susceptibility. For $H_0 = 0$ (in which case $\langle\cos\vartheta\rangle_0 = 0$) in response to an increase in temperature, the function $B/\sigma = \langle\cos^2\vartheta\rangle_0 - \langle\cos\vartheta\rangle_0^2$ *decreases* from 1 to 1/3. However, for $H_0 \neq 0$, the effect of *saturation* of the longitudinal magnetization is crucial, as it causes B/σ to tend to zero at zero temperature. Thus, by raising the temperature, B/σ *increases* from zero to the limiting value 1/3 at $H_0 = 0$. Consequently, the static susceptibility B acquires an intrinsic

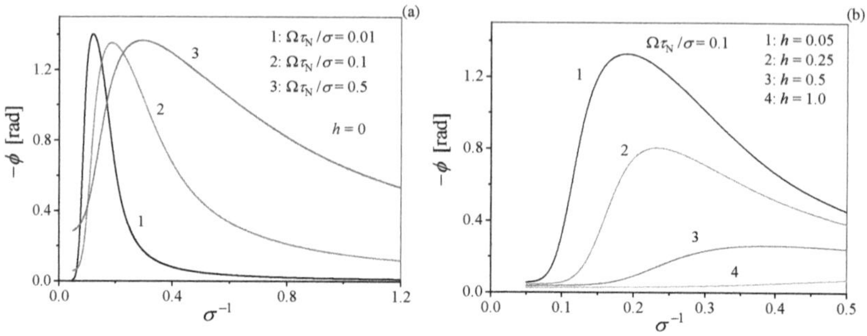

Fig. 2.25. (a) Phase shift as a function of dimensionless temperature σ^{-1} for zero dc bias field ($h = 0$) and various frequency parameters $\Omega\tau_N/\sigma = 0.01$ (1), 0.1 (2), 0.5 (3). (b) Phase shift vs. σ^{-1} for the following dc bias field strength parameters $h = 0.05$ (1), 0.25 (2), 0.5 (3), 1 (4) and the frequency parameter $\Omega\tau_N/\sigma = 0.1$

temperature-dependent maximum. This specific temperature-dependent feature of the static susceptibility ultimately counteracts the shift in the SNR maximum.

Likewise, one can study the phase shift $\phi(\sigma, \Omega)$ of the magnetic SR [62,63] which may be defined in the usual way as

$$\phi(\sigma,\Omega) = -\arctan(\chi''/\chi'). \tag{2.230}$$

Here the phase has its conventional meaning, namely the lag between the stimulus and the output signal [62,63]. Calculations of $\phi(\sigma, \Omega)$ for a superparamagnetic particle are presented in Fig. 2.25. In a manner similar to the effect on the SNR curves, the bias field has a *destructive effect on the noise dependence of the phase*. This tendency is shown in Fig. 2.25. Note that the nucleation occurring at $h = 1$ is reflected in this graph by an almost-complete flattening and eventual disappearance of the curve, in other words, it ultimately vanishes.

Here SR in uniaxial superparamagnetic particles in weak ac fields and in the simplest configuration (i.e., where the direction of the ac and dc magnetic fields coincides with the easy axis of the magnetization) has been studied. However, the study may be extended to other interesting cases such as nonlinear SR in strong ac fields [64,65], noncircularly symmetric potentials [29,30], and so on. Estimates of the SR effect in nanomagnets are given in [57–59,66,160]. We also consider SR in a cubic potential in Section 2.5.2.

2.4.5 *Transverse response: ferromagnetic resonance*

We now evaluate the *transverse susceptibility* $\chi_{\perp}(\omega)$ of a single-domain ferromagnetic particle subjected to a uniform magnetic field. Like the longitudinal response, we give the *exact* solution [148] for $\chi_{\perp}(\omega)$ and then compare the results with various approximate solutions. The appropriate infinite hierarchy of differential-recurrence equations for the transverse equilibrium correlation functions $f_n^{\perp}(t)$, Eq. (2.169), is given by Eq. (2.168). Then as with the longitudinal response, the five-term scalar differential-recurrence relation, Eq. (2.168), can be transformed to the matrix three-term differential-recurrence Eq. (2.200), where the matrices $\mathbf{Q}_n^{\pm}$ and $\mathbf{Q}_n$ are given by

$$\mathbf{Q}_n^{-}=\begin{pmatrix} c_{2n-1} & a_{2n-1} \\ 0 & c_{2n} \end{pmatrix},\ \mathbf{Q}_n=\begin{pmatrix} d_{2n-1} & -a'_{2n-1} \\ a_{2n} & d_{2n} \end{pmatrix},\ \mathbf{Q}_n^{+}=\begin{pmatrix} e_{2n-1} & 0 \\ -a'_{2n} & e_{2n} \end{pmatrix}, \quad (2.231)$$

and

$$\mathbf{C}_n(t)=\begin{pmatrix} f_{2n-1}^{\perp}(t) \\ f_{2n}^{\perp}(t) \end{pmatrix}. \quad (2.232)$$

The *exact* solution of Eq. (2.200) for the Laplace transform $\tilde{\mathbf{C}}_1(s)$ is given by Eq. (2.203), where the initial-condition vectors $\mathbf{C}_n(0)$ are defined as

$$\mathbf{C}_n(0)=\begin{pmatrix} f_{2n-1}^{\perp}(0) \\ \\ f_{2n}^{\perp}(0) \end{pmatrix}=\begin{pmatrix} \dfrac{n(2n-1)}{4n-1}\left(\langle P_{2n-2}\rangle_0-\langle P_{2n}\rangle_0\right) \\ \dfrac{n(2n+1)}{4n+1}\left(\langle P_{2n-1}\rangle_0-\langle P_{2n+1}\rangle_0\right) \end{pmatrix}. \quad (2.233)$$

The $\mathbf{C}_n(0)$ may also be evaluated via matrix continued fractions. Having calculated $\tilde{\mathbf{C}}_1(s)$, we have the transverse susceptibility $\chi_{\perp}(\omega)$, which is formally expressed using the Laplace transform of $f_n^{\perp}(t)$ as

$$\frac{\chi_{\perp}(\omega)}{\chi_{\perp}}=1-i\omega\frac{\tilde{f}_1^{\perp}(i\omega)+\tilde{f}_1^{\perp *}(-i\omega)}{2\,\mathrm{Re}[\tilde{f}_1^{\perp}(0)]}, \quad (2.234)$$

where $\chi_{\perp}=v\mu_0M_S^2(1-S)/(3kT)$ is the static transverse susceptibility, $S=\langle P_2\rangle_0$ is given by Eq. (A.49) in Appendix A.1.5, and the asterisk denotes the complex conjugate.

The matrix continued fraction solution so obtained allows one to assess the range of validity of the various approximate solutions. Now most analytical results have been obtained by the effective-eigenvalue method. For example, for $\xi=0$ and $\sigma\neq 0$, Raikher and Shliomis [105] derived Eq. (2.179) for $\chi_{\perp}(\omega)$. Later, Raikher and Stepanov [376] gave a unified description of both homogeneous and inhomogeneous mechanisms for intrinsic FMR absorption-line broadening in uniaxial nanomagnets. Furthermore, by using that method in the limit $\sigma=0$ and $\xi\neq 0$, Garanin *et al.* [23] obtained an equation for the circular magnetic susceptibility, which can be rearranged as an expression for the transverse complex susceptibility, viz.,

$$\frac{\chi_{\perp}(\omega)}{\chi_{\perp}}=\frac{|\lambda|^2+i\omega\tau_{\mathrm{N}}\lambda'}{|\lambda|^2-\omega^2\tau_{\mathrm{N}}^2+2i\omega\tau_{\mathrm{N}}\lambda'}, \tag{2.235}$$

where λ is the dimensionless complex effective eigenvalue given by

$$\lambda=\lambda'+i\lambda''=\frac{\xi^2+1-\xi\coth\xi}{2(\xi\coth\xi-1)}-\frac{i\xi}{2\alpha}. \tag{2.236}$$

In high applied fields ($\xi>>1$), Eq. (2.235) again reduces to the Landau–Lifshitz equation, Eq. (2.181), with $\omega_{\mathrm{pr}}=\xi(2\alpha\tau_{\mathrm{N}})^{-1}$.

Equation (2.235) was derived for $\xi\neq 0$, $\sigma=0$ only. Since λ is the effective eigenvalue for the equilibrium ACF $f_1^{\perp}(t)$, Eqs. (2.235) and (2.236) can be generalized for the case $\sigma\neq 0$ and $\sigma<<\xi$ as follows. According to Eq. (2.168), the first equation of the infinite hierarchy is

$$\tau_{\mathrm{N}}\dot{f}_1^{\perp}(t)+\left[1-\frac{i\xi}{2\alpha}+\frac{\sigma}{5}\right]f_1^{\perp}(t)+\left(\frac{\xi}{6}-\frac{i\sigma}{3\alpha}\right)f_2^{\perp}(t)+\frac{2\sigma}{15}f_3^{\perp}(t)=0. \tag{2.237}$$

The effective eigenvalue implies that the solution of Eq. (2.237) may be approximated by a *single* exponential, namely, $f_1^{\perp}(t)=f_1^{\perp}(0)e^{-\lambda t/\tau_N}$. Here the (complex) effective eigenvalue λ is given by

$$\lambda=-\frac{\tau_{\mathrm{N}}\dot{f}_1^{\perp}(0)}{f_1^{\perp}(0)}=1+\frac{\sigma}{5}-\frac{i\xi}{2\alpha}+\left(\frac{\xi}{6}-\frac{i\sigma}{3\alpha}\right)\frac{f_2^{\perp}(0)}{f_1^{\perp}(0)}+\frac{2\sigma}{15}\frac{f_3^{\perp}(0)}{f_1^{\perp}(0)}. \tag{2.238}$$

Equation (2.238) now simplifies to

$$\lambda = \lambda' + i\lambda'' = \frac{2 + \langle P_2 \rangle_0}{2(1 - \langle P_2 \rangle_0)} - i\frac{3\langle P_1 \rangle_0}{2\alpha(1 - \langle P_2 \rangle_0)}, \tag{2.239}$$

where $\langle P_1 \rangle_0$ and $\langle P_2 \rangle_0$ are given by Eqs. (A.48)–(A.50) in Appendix A.1.5. For $\sigma = 0$, Eq. (2.239) reduces to Eq. (2.236) since

$$\langle P_1 \rangle_0 = \coth\xi - \frac{1}{\xi} = L(\xi), \langle P_2 \rangle_0 = 1 - \frac{3L(\xi)}{\xi}. \tag{2.240}$$

Applying this method to another particular case, namely $\xi \neq 0$ and $\xi << \sigma$, requires *two* effective eigenvalues, corresponding to the first two equations of the infinite hierarchy, Eq. (2.168) since resonance now arises due to the *coupling* between $f_1^{\perp}(t)$ and $f_2^{\perp}(t)$. This behavior is termed "entanglement" of the dipole and quadrupole branches of the response by Raikher and Shliomis [105]. Thus, we obtain the two coupled ordinary differential equations:

$$\begin{aligned} \tau_{\mathrm{N}} \frac{d}{dt} f_1^{\perp}(t) + \lambda_1^{\mathrm{ef}} f_1^{\perp}(t) + \left(\frac{\xi}{6} - \frac{i\sigma}{3\alpha}\right) f_2^{\perp}(t) = 0, \\ \tau_{\mathrm{N}} \frac{d}{dt} f_2^{\perp}(t) + \lambda_2^{\mathrm{ef}} f_2^{\perp}(t) - 3\left[\left(1 - \lambda_1^{\mathrm{ef}}\right)\left(\frac{\xi}{\sigma} - \frac{i}{\alpha}\right) + \frac{\xi}{2}\right] f_1^{\perp}(t) = 0, \end{aligned} \tag{2.241}$$

where the effective eigenvalues are given by

$$\begin{aligned} \lambda_1^{\mathrm{ef}} &= 1 - \frac{i\xi}{2\alpha} + \frac{\sigma}{5} + \frac{2\sigma}{15} \frac{f_3^{\perp}(0)}{f_1^{\perp}(0)} \\ &= \frac{\xi^2}{4\sigma} - \frac{1}{2} - \frac{i\xi}{2\alpha} + 3\frac{2\sigma - \xi\langle P_1 \rangle_0}{4\sigma(1 - \langle P_2 \rangle_0)}, \end{aligned} \tag{2.242}$$

$$\begin{aligned} \lambda_2^{\mathrm{ef}} &= 3 - \frac{\sigma}{7} - \frac{i\xi}{2\alpha} + \frac{12\sigma}{35} \frac{f_4^{\perp}(0)}{f_2^{\perp}(0)} \\ &= \frac{\xi^2}{2\sigma} - \frac{i\xi}{2\alpha} - 1 - \sigma + 3\frac{2\sigma\langle P_1 \rangle_0 - \xi\langle P_2 \rangle_0}{\xi(\langle P_2 \rangle_0 - 1) + 3\langle P_1 \rangle_0}. \end{aligned} \tag{2.243}$$

Here, $\langle P_1 \rangle_0$ and $\langle P_2 \rangle_0$ are given by Eqs. (A.48) and (A.49) from Appendix A.1.5 and in the derivation of Eq. (2.241) we have used the equation

$$f_3^{\perp}(t) = \left(\lambda_1^{\mathrm{ef}} - 1 - \frac{\sigma}{5} + \frac{i\xi}{2\alpha}\right)\frac{15}{2\sigma} f_1^{\perp}(t) \tag{2.244}$$

(in accordance with Eq. (2.242), the equality embodied in Eq. (2.244) is assumed to hold at any time t). In the limit ξ, $\sigma \to 0$, the effective eigenvalues have behavior

$$\lambda_1^{\mathrm{ef}} \approx 1 - \frac{i\xi}{2\alpha} + \frac{\sigma}{5} + \frac{8\sigma^2}{175} + \frac{4\sigma\xi^2}{175} + \cdots, \lambda_2^{\mathrm{ef}} \approx 3 - \frac{i\xi}{2\alpha} - \frac{\sigma}{7} + \frac{16\sigma^2}{147} + \frac{8\sigma\xi^2}{441} + \cdots.$$

Furthermore, using the one-sided Fourier transform, we can solve Eqs. (2.240) and (2.241) for $\tilde{f}_1^{\perp}(i\omega)$, yielding

$$\frac{\tilde{f}_1^{\perp}(i\omega)}{\tau_{\mathrm{N}} f_1^{\perp}(0)} = \frac{(\lambda_2^{\mathrm{ef}} + i\omega\tau_N) - \delta}{(\lambda_1^{\mathrm{ef}} + i\omega\tau_N)(\lambda_2^{\mathrm{ef}} + i\omega\tau_N) + \Delta}, \tag{2.245}$$

where

$$\Delta = \left(\frac{\xi}{2} - \frac{i\sigma}{\alpha}\right)\left[\left(1 - \lambda_1^{\mathrm{ef}}\right)\left(\frac{\xi}{\sigma} - \frac{i}{\alpha}\right) + \frac{\xi}{2}\right],$$

$$\delta = \left(\frac{\xi}{6} - \frac{i\sigma}{3\alpha}\right)\frac{f_2^{\perp}(0)}{f_1^{\perp}(0)} = \left(\frac{\xi}{4\sigma} - \frac{i}{2\alpha}\right)\left(\frac{3\langle P_1\rangle_0}{1 - \langle P_2\rangle_0} - \xi\right).$$

By substituting Eq. (2.245) into Eq. (2.234), we then have $\chi_{\perp}(\omega)$, which for the particular case $\xi = 0$ yields Eq. (2.179).

Furthermore, for large damping $\alpha \geq 1$ (i.e., ignoring precessional motion altogether) the transverse spectrum may be accurately described by the Debye equation [25,105,108]

$$\frac{\chi_{\perp}(\omega)}{\chi_{\perp}} = \frac{1}{1 + i\omega\tau_{\mathrm{N}} / \lambda'}, \tag{2.246}$$

where $\lambda' = \left(1 + \langle P_2\rangle_0 / 2\right) / \left(1 - \langle P_2\rangle_0\right)$ is the effective eigenvalue.

Typical spectra of the real and imaginary parts of $\chi_{\perp}(\omega)$ are plotted in Figs. 2.26 and 2.27. $\chi_{\perp}(\omega)$ exhibits pure relaxational behavior for small anisotropy and field parameters (ξ, $\sigma \approx 0$) or large damping ($\alpha > 1$). For small damping, the spectra, as expected, have a pronounced resonant peak and strongly depend on α (Fig. 2.26). The comparison of the

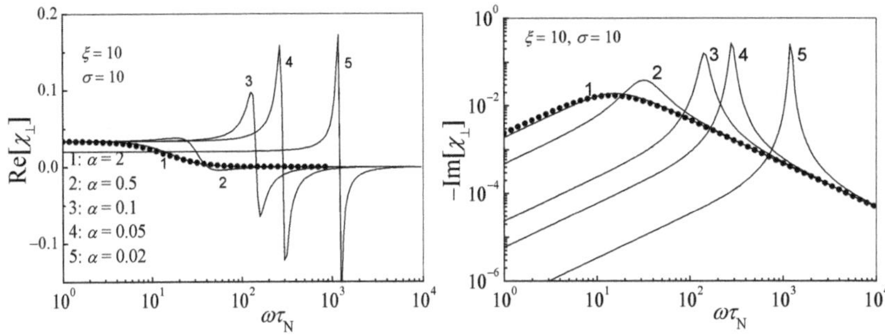

Fig. 2.26. $\mathrm{Re}[\chi_\perp(\omega)]$ and $-\mathrm{Im}[\chi_\perp(\omega)]$ (solid lines) vs. $\omega\tau_N$ for $\xi = 10$ and $\sigma = 10$ and various values of α. Filled circles: the effective-eigenvalue solution, Eq. (2.246).

effective-eigenvalue solutions with the exact results allows us to estimate the accuracy of the former. In Fig. 2.26, the matrix continued fraction solution is compared with the overdamped solution, Eq. (2.246), while in Fig. 2.27, the exact calculations are compared with the effective-eigenvalue solution, Eqs. (2.234) and (2.245). It is obvious that both solutions concur only for $\sigma \approx 0$ and for $\sigma >> 1$. However, for σ lying between 1 and 5, the effective-eigenvalue solution fails to describe the transverse response at all. The explanation appears to be as follows: at small to moderate barrier heights, *a spread of the precession frequencies of the magnetization in the anisotropy field essentially exists*. To a certain extent, this effect is analogous to inhomogeneous broadening, as it considerably exceeds the true damping [23,376]. Therefore, it is impossible to describe *asymmetric* absorption and dispersion curves by the *usual single resonance spectrum*, which is predicted by the effective eigenvalue. However, for $\sigma >> 1$, so that the anisotropy potential may be approximated by a harmonic one, the system may be accurately described as just a single resonator with characteristic frequency ω_{pr} given by the Landau–Lifshitz equation, Eq. (2.181). In contrast, for very weak anisotropy $\sigma \approx 0$ and/or $\sigma << \xi$, the effective-eigenvalue solution, as in longitudinal relaxation, concurs with the exact solution *for all values* of the field parameter ξ (see Fig. 2.27), since we now have *natural resonance* with angular frequency $\omega_{\mathrm{pr}} = \xi(2\alpha\tau_N)^{-1}$. Thus, *when the external magnetic field considerably exceeds the anisotropy field, or that field is close to zero*, the effective-eigenvalue solution may accurately describe the transverse relaxation.

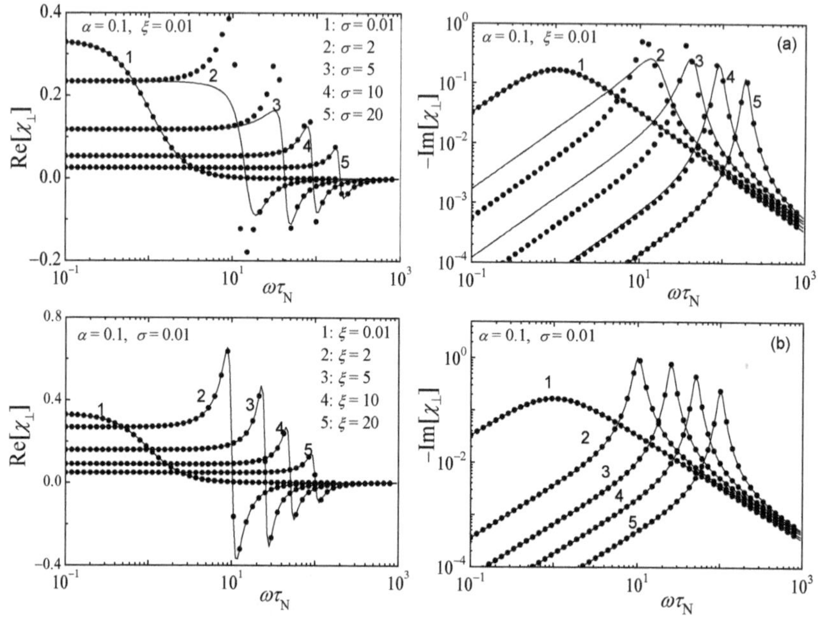

Fig. 2.27. (a) Exact (solid lines) and effective-eigenvalue (filled circles, Eqs. (2.234) and (2.245)) solutions for $\mathrm{Re}[\chi_\perp(\omega)]$ and $-\mathrm{Im}[\chi_\perp(\omega)]$ vs. $\omega\tau_\mathrm{N}$ for $\alpha = 0.1$ and $\xi = 0.01$ and various values of σ. (b) The same as Fig. 2.27(a) for $\alpha = 0.1$ and $\sigma = 0.01$ and various ξ.

However, if the influence of a constant magnetic field is negligible, this method requires a careful investigation of its range of applicability before proceeding with it. Otherwise, it is possible to arrive at erroneous results, as clearly demonstrated in Fig. 2.27.

Thus, the transverse response of an ensemble of noninteracting uniaxial nanomagnets can be evaluated using matrix continued fractions, as for the longitudinal response. Furthermore, we have demonstrated that the effective eigenvalue, yielding the simple analytical Eq. (2.245), describes the main features of the transverse complex susceptibility; the sole exception is the intermediate-barrier range $\sigma \sim 1$ to 5 and $\xi \leq 1$, where a spread of the precession frequencies in the anisotropy potential field essentially exists.

Here, we have treated magnetization relaxation and resonance phenomena in nanomagnets with *circularly symmetric* anisotropy potentials. However, this conventional assumption of circular symmetry is barely realizable under experimental conditions, because the easy axes of the particles are randomly oriented in space. Furthermore, the *intrinsic* interplay between precession and thermoactivation cannot be included at all, because in circular symmetry we reiterate that *no intrinsic dynamical coupling* between the longitudinal and transverse (or precessional) modes of motion exists. Next, we shall generalize our treatment by abandoning the assumption of circular symmetry.

2.5 Magnetization relaxation in nanomagnets with noncircularly symmetric anisotropy

2.5.1 *Uniaxial nanomagnet in an oblique field*

In Section 2.4, we studied the effect of an external uniform magnetic field $\mathbf{H}_0$ on the magnetization relaxation of magnetic nanoparticles having simple uniaxial anisotropy, if $\mathbf{H}_0$ and the easy axis $\mathbf{n}$ of the uniaxial nanomagnet are *collinear*, i.e., $\mathbf{n} \parallel \mathbf{H}_0$. However, in general, $\mathbf{H}_0$ has an *arbitrary* orientation with respect to the easy axis so that we must study the magnetization dynamics when the assumption that $\mathbf{n} \parallel \mathbf{H}_0$ is abandoned. Here the free energy potential for a uniaxial nanomagnet is given by the noncircularly symmetric form

$$E = -\sigma\cos^2\vartheta - \xi\cos\Theta, \qquad (2.247)$$

where Θ is the angle between the vectors $\mathbf{H}_0$ and $\mathbf{M}$, so that

$$\cos\Theta = \gamma_1\sin\vartheta\cos\varphi + \gamma_2\sin\vartheta\sin\varphi + \gamma_3\cos\vartheta, \qquad (2.248)$$

and $\gamma_1, \gamma_2, \gamma_3$ are the direction cosines of the vector $\mathbf{H}_0$. We recall that in such a general formulation, the reversal time of the magnetization has already been estimated in Section 2.3.2 using the Kramers–Brown escape-rate theory. A detailed treatment of the oblique-field problem for uniaxial superparamagnetic nanoparticles has been given by Coffey *et al.* [108, 342,343], Kennedy [344], Kalmykov *et al.* [32,38,207] and Fukushima *et al.* [345,346]. In this section, we follow the exposition in Refs. [32,38].

The normalized free energy of a uniaxial nanomagnet in a dc magnetic field at an angle ψ to the easy axis is given by Eq. (2.247) (see Fig. 2.1).

Here we can obviously express the normalized potential, Eq. (2.247), as a sum of spherical harmonics $Y_{lm}(\vartheta, \varphi)$, viz.,

$$E(\vartheta,\varphi) = A_{1-1}Y_{1-1}(\vartheta,\varphi) + A_{10}Y_{10}(\vartheta,\varphi) + A_{11}Y_{11}(\vartheta,\varphi) + A_{20}Y_{20}(\vartheta,\varphi) + \text{const} \tag{2.249}$$

with expansion coefficients A_{rs} given by

$$A_{10} = -2\xi\gamma_3\sqrt{\frac{\pi}{3}}, \quad A_{1\pm1} = \pm\xi(\gamma_1 \mp i\gamma_2)\sqrt{\frac{2\pi}{3}}, \quad A_{20} = -4\sigma\sqrt{\frac{\pi}{45}}. \tag{2.250}$$

Hence via Eqs. (1.92) and (2.250) and the explicit representation for the Clebsch–Gordan coefficients, we can calculate the nonzero elements $e_{lml'm'}$. Therefore, the differential-recurrence relation, Eq. (1.108), for the statistical moments $c_{nm}(t) = \langle Y_{nm}\rangle(t) - \langle Y_{nm}\rangle_0$ (expectation values of the spherical harmonics $Y_{lm}(\vartheta, \varphi)$), becomes [32,108,342]

$$\begin{aligned}\tau_N \frac{d}{dt}c_{nm}(t) &= v_{nm}c_{n-2m}(t) + z_{nm}c_{n+2m}(t) + w^-_{nm}c_{n-1m-1}(t) \\ &+ w_{nm}c_{n-1m}(t) + w^+_{nm}c_{n-1m+1}(t) + x^-_{nm}c_{nm-1}(t) + x_{nm}c_{nm}(t) \\ &+ x^+_{nm}c_{nm+1}(t) + y^-_{nm}c_{n+1m-1} + y_{nm}c_{n+1m}(t) + y^+_{nm}c_{n+1m+1}(t),\end{aligned} \tag{2.251}$$

with

$$x_{nm} = \sigma\frac{n(n+1) - 3m^2}{(2n-1)(2n+3)} - \frac{n(n+1)}{2} + i\frac{m\xi\gamma_3}{2\alpha},$$

$$x^+_{nm} = -\left(x^-_{n-m}\right)^* = i\frac{\xi(\gamma_1 - i\gamma_2)}{4\alpha}\sqrt{(n+m+1)(n-m)},$$

$$y_{nm} = -\left(\xi\gamma_3\frac{n}{2} - i\frac{\sigma m}{\alpha}\right)\sqrt{\frac{(n+1)^2 - m^2}{(2n+1)(2n+3)}},$$

$$y^+_{nm} = -\left(y^-_{n-m}\right)^* = \frac{n\xi(\gamma_1 - i\gamma_2)}{4}\sqrt{\frac{(n+m+1)(n+m+2)}{(2n+1)(2n+3)}},$$

$$w_{nm} = \left(\xi\gamma_3\frac{n+1}{2} + i\frac{\sigma m}{\alpha}\right)\sqrt{\frac{n^2 - m^2}{(2n+1)(2n-1)}},$$

$$w^+_{nm} = -\left(w^-_{n-m}\right)^* = \frac{\xi(n+1)(\gamma_1 - i\gamma_2)}{4}\sqrt{\frac{(n-m)(n-m-1)}{(2n+1)(2n-1)}},$$

$$z_{nm} = \frac{-\sigma n}{2n+3}\sqrt{\frac{[(n+2)^2 - m^2][(n+1)^2 - m^2]}{(2n+1)(2n+5)}},$$

$$v_{nm} = \frac{\sigma(n+1)}{2n-1}\sqrt{\frac{[n^2 - m^2][(n-1)^2 - m^2]}{(2n+1)(2n-3)}}.$$

In writing Eq. (2.251), we have used the fact that the equilibrium averages $\langle Y_{lm}\rangle_0$ satisfy the time-independent relation

$$\sum_{l',m'} e_{lml'm'} \langle Y_{l'm'} \rangle_0 = 0. \tag{2.252}$$

Here the angular brackets $\langle\ \rangle_0$ denote the equilibrium average, defined as

$$\langle A \rangle_0 = \int_0^{2\pi}\int_0^{\pi} A(\vartheta,\varphi) W_0(\vartheta,\varphi) \sin\vartheta d\vartheta d\varphi, \tag{2.253}$$

where $W_0(\vartheta,\varphi) = Z^{-1}e^{\sigma\cos^2\vartheta + \xi\cos\Theta}$ is the equilibrium distribution function and Z is the partition function.

By solving Eq. (2.251) for the first-rank relaxation functions $c_{10}(t)$ and $c_{1\pm1}(t)$, we have all relevant quantities, namely the integral, effective and reversal times of the magnetization, the dynamic susceptibility, dynamic hysteresis, etc. Indeed, according to linear response theory, when a small uniform external field $\mathbf{H}_1$ parallel to $\mathbf{H}_0$ ($v\mu_0(\mathbf{M}\cdot\mathbf{H}_1) << kT$) has been switched off at time $t = 0$, the decay of the longitudinal component of the magnetization $\langle M_\parallel \rangle(t)$ of a uniaxial nanoparticle is given by

$$\langle M_\parallel \rangle(t) - \langle M_\parallel \rangle_0 = \chi_\parallel H_1 C_\parallel(t), \tag{2.254}$$

where $C_\parallel(t)$ is the normalized relaxation function defined as

$$C_\parallel(t) = \frac{\gamma_3 c_{10}(t) - \sqrt{2}\,\mathrm{Re}[(\gamma_1 - i\gamma_2)c_{11}(t)]}{\gamma_3 c_{10}(0) - \sqrt{2}\,\mathrm{Re}[(\gamma_1 - i\gamma_2)c_{11}(0)]}, \tag{2.255}$$

and $\chi_\parallel$ is the static magnetic susceptibility given by

$$\chi_\parallel = \sqrt{\frac{4\pi}{3}}\left(\gamma_3 c_{10}(0) - \sqrt{2}\,\mathrm{Re}\left[(\gamma_1 - i\gamma_2)c_{11}(0)\right]\right)\frac{M_S}{H_1}.$$

The complex susceptibility $\chi_{\parallel}(\omega) = \chi'_{\parallel}(\omega) - i\chi''_{\parallel}(\omega)$ of the particle is given by (see Section 1.3.6 of Chapter 1)

$$\chi_{\parallel}(\omega) = \chi_{\parallel}\left[1 - i\omega\tilde{C}_{\parallel}(i\omega)\right], \tag{2.256}$$

where

$$\tilde{C}_{\parallel}(i\omega) = \frac{\sqrt{2}\gamma_3\tilde{c}_{10}(i\omega) - (\gamma_1 - i\gamma_2)\tilde{c}_{11}(i\omega) + (\gamma_1 + i\gamma_2)\tilde{c}_{1-1}(i\omega)}{\sqrt{2}\gamma_3 c_{10}(0) - (\gamma_1 - i\gamma_2)c_{11}(0) + (\gamma_1 + i\gamma_2)c_{1-1}(0)}. \tag{2.257}$$

Other relevant quantities are as usual the integral and effective relaxation times $\tau_{\text{int}}^{\parallel}$ and $\tau_{\text{ef}}^{\parallel}$, viz.,

$$\tau_{\text{int}}^{\parallel} = \tilde{C}_{\parallel}(0), \quad \tau_{\text{ef}}^{\parallel} = 2\tau_{\text{N}}\frac{\langle\cos^2\Theta\rangle_0 - \langle\cos\Theta\rangle_0^2}{1 - \langle\cos^2\Theta\rangle_0}. \tag{2.258}$$

Thus, we must evaluate the one-sided Fourier transforms of $c_{10}(t)$ and $c_{1\pm1}(t)$. This may be done, as before, using matrix continued fractions: the eleven-term recurrence equation, Eq. (2.251), can be written as the matrix three-term differential-recurrence equation, Eq. (2.200), where the column vector $\mathbf{C}_n(t)$ and supermatrices $\mathbf{Q}_n^{\pm}$, $\mathbf{Q}_n$ may be set up as follows

$$\mathbf{C}_0(t) = \mathbf{0},\ \mathbf{C}_1(t) = \begin{pmatrix} c_{2-2}(t) \\ c_{2-1}(t) \\ c_{20}(t) \\ c_{21}(t) \\ c_{22}(t) \\ c_{1-1}(t) \\ c_{10}(t) \\ c_{11}(t) \end{pmatrix},\ \mathbf{C}_n(t) = \begin{pmatrix} c_{2n-2n}(t) \\ c_{2n-2n+1}(t) \\ \vdots \\ c_{2n2n}(t) \\ c_{2n-1-2n+1}(t) \\ c_{2n-1-2n+2}(t) \\ \vdots \\ c_{2n-12n-1}(t) \end{pmatrix}, \tag{2.259}$$

$$\begin{gathered} \mathbf{Q}_n = \begin{pmatrix} \mathbf{X}_{2n} & \mathbf{W}_{2n} \\ \mathbf{Y}_{2n-1} & \mathbf{X}_{2n-1} \end{pmatrix},\ \mathbf{Q}_n^+ = \begin{pmatrix} \mathbf{Z}_{2n} & \mathbf{Y}_{2n} \\ \mathbf{0} & \mathbf{Z}_{2n-1} \end{pmatrix}, \\ \mathbf{Q}_n^- = \begin{pmatrix} \mathbf{V}_{2n} & \mathbf{0} \\ \mathbf{W}_{2n-1} & \mathbf{V}_{2n-1} \end{pmatrix} \text{ with } \mathbf{Q}_1^- = \begin{pmatrix} \mathbf{V}_2 \\ \mathbf{W}_1 \end{pmatrix}. \end{gathered} \tag{2.260}$$

The supermatrices $\mathbf{Q}_n, \mathbf{Q}_n^{\pm}$ comprise five distinct submatrices $\mathbf{V}_l, \mathbf{W}_l$, $\mathbf{X}_l, \mathbf{Y}_l$, and $\mathbf{Z}_l$ which have dimensions $(2l+1) \times (2l-3)$, $(2l+1) \times (2l-1)$,

(2*l*+1) × (2*l*+1), (2*l*+1) × (2*l*+3) and (2*l*+1) × (2*l*+5), respectively. The elements of these submatrices are explicitly given by

$$\begin{aligned}
(\mathbf{X}_l)_{nm} &= \delta_{n-2m} x^-_{l,-l+m+1} + \delta_{nm} x_{l,-l+m-1} + \delta_{n+2m} x^+_{l,-l+m-3}, \\
(\mathbf{Y}_l)_{nm} &= \delta_{n-1m} y^-_{l,-l+m} + \delta_{n+1m} y_{l,-l+m-2} + \delta_{n+3m} y^+_{l,-l+m-4}, \\
(\mathbf{W}_l)_{nm} &= \delta_{n-3m} w^-_{l,-l+m+2} + \delta_{n-1m} w_{l,-l+m} + \delta_{n+1m} w^+_{l,-l+m-2}, \\
(\mathbf{V}_l)_{nm} &= \delta_{n-2m} v_{l,-l+m+1}, \quad (\mathbf{Z}_l)_{nm} = \delta_{n+2m} z_{l,-l+m-3}.
\end{aligned} \tag{2.261}$$

Recall that the matrix continued fraction solution of Eq. (2.200) for the Laplace transform $\tilde{\mathbf{C}}_1(s)$ is formally given by Eq. (2.203), where the initial condition vectors $\mathbf{C}_n(0)$ can also be evaluated from the appropriate continued fractions. Here the initial values $c_{nm}(0)$ contained in the column vector $\mathbf{C}_n(0)$ (see Eq. (2.259)) are, in the linear approximation,

$$\begin{aligned}
c_{nm}(0) &= \frac{\int_0^{2\pi}\int_0^{\pi} Y_{nm}(\vartheta,\varphi) e^{\sigma\cos^2\vartheta + (\xi+\xi_1)\cos\Theta} \sin\vartheta \, d\vartheta \, d\varphi}{\int_0^{2\pi}\int_0^{\pi} e^{\sigma\cos^2\vartheta + (\xi+\xi_1)\cos\Theta} \sin\vartheta \, d\vartheta \, d\varphi} - \langle Y_{nm}\rangle_0 \\
&\approx \xi_1 \Bigg\{ \frac{\gamma_1 + i\gamma_2}{2} \left[\sqrt{\frac{(n-m+2)(n-m+1)}{(2n+3)(2n+1)}} \langle Y_{n+1m-1}\rangle_0 - \sqrt{\frac{(n+m)(n+m-1)}{(2n-1)(2n+1)}} \langle Y_{n-1m-1}\rangle_0 \right] \\
&+ \frac{\gamma_1 - i\gamma_2}{2} \left[\sqrt{\frac{(n-m)(n-m-1)}{(2n-1)(2n+1)}} \langle Y_{n-1m+1}\rangle_0 - \sqrt{\frac{(n+m+2)(n+m+1)}{(2n+3)(2n+1)}} \langle Y_{n+1m+1}\rangle_0 \right] \\
&+ \gamma_3 \left[\sqrt{\frac{(n-m+1)(n+m+1)}{(2n+3)(2n+1)}} \langle Y_{n+1m}\rangle_0 + \sqrt{\frac{n^2-m^2}{(2n-1)(2n+1)}} \langle Y_{n-1m}\rangle_0 \right] \\
&- \sqrt{\frac{4\pi}{3}} \langle Y_{nm}\rangle_0 \left[\gamma_3 \langle Y_{10}\rangle_0 - \sqrt{2}\,\mathrm{Re}\{(\gamma_1 - i\gamma_2)\langle Y_{11}\rangle_0\} \right] \Bigg\}
\end{aligned} \tag{2.262}$$

(here $\xi_1 = v\mu_0 M_S H_1 / (kT)$ is the dimensionless probe field parameter). To evaluate the equilibrium averages $\langle Y_{nm}\rangle_0$ occurring in the initial condition Eq. (2.262), we first transform Eq. (2.252) to the homogeneous recurrence relation

$$\mathbf{Q}^-_n \mathbf{R}_{n-1} + \mathbf{Q}_n \mathbf{R}_n + \mathbf{Q}^+_n \mathbf{R}_{n+1} = 0, \tag{2.263}$$

where $\mathbf{Q}_n$ and $\mathbf{Q}^\pm_n$ are defined by Eq. (2.260), and

$$\mathbf{R}_n = \begin{pmatrix} \langle Y_{2n-2n} \rangle_0 \\ \langle Y_{2n-2n+1} \rangle_0 \\ \vdots \\ \langle Y_{2n2n} \rangle_0 \\ \langle Y_{2n-1-2n+1} \rangle_0 \\ \langle Y_{2n-1-2n+2} \rangle_0 \\ \vdots \\ \langle Y_{2n-12n-1} \rangle_0 \end{pmatrix} \text{ with } \mathbf{R}_0 = (\langle Y_{00} \rangle_0) = \left(\frac{1}{\sqrt{4\pi}} \right). \tag{2.264}$$

The exact solution of the recurrence equation, Eq. (2.263), is then by iteration given by the matrix product

$$\mathbf{R}_n = \mathbf{S}_n \mathbf{R}_{n-1} = \frac{1}{\sqrt{4\pi}} \mathbf{S}_n \mathbf{S}_{n-1} \cdots \mathbf{S}_2 \mathbf{S}_1, \tag{2.265}$$

where $\mathbf{S}_n = \mathbf{\Delta}_n(0)\mathbf{Q}_n^-$, and $\mathbf{\Delta}_n(s)$ is defined by Eq. (1.123). Noticing Eq. (2.262), the initial condition vectors $\mathbf{C}_n(0)$ are given by

$$\mathbf{C}_n(0) = \xi_1 \left\{ \hat{\mathbf{K}}_n \mathbf{R}_{n-1} + \mathbf{K}_n \mathbf{R}_n + \hat{\mathbf{K}}_{n+1}^{\dagger} \mathbf{R}_{n+1} \right\}, \tag{2.266}$$

where $\hat{\mathbf{K}}_n$ and $\mathbf{K}_n$ are supermatrices defined by

$$\mathbf{K}_n = \begin{pmatrix} \mathbf{F}_{2n} & \mathbf{D}_{2n} \\ \mathbf{D}_{2n}^H & \mathbf{F}_{2n-1} \end{pmatrix},$$
$$\hat{\mathbf{K}}_n = \begin{pmatrix} \mathbf{0} & \mathbf{0} \\ \mathbf{D}_{2n-1} & \mathbf{0} \end{pmatrix}, \; \hat{\mathbf{K}}_1 = \begin{pmatrix} \mathbf{0} \\ \mathbf{D}_1 \end{pmatrix} \tag{2.267}$$

(the superscript † in Eq. (2.266) denotes the Hermitian conjugate, i.e., the transpose and its complex conjugate). The matrices $\mathbf{K}_n$ and $\hat{\mathbf{K}}_n$ each consist of the *diagonal* submatrix $\mathbf{F}_l$ and a *tridiagonal* submatrix $\mathbf{D}_l$ with respective matrix elements

$$(\mathbf{F}_l)_{nm} = -\sqrt{\frac{4\pi}{3}} \operatorname{Re}\left[\gamma_3 \langle Y_{10} \rangle_0 - \sqrt{2}(\gamma_1 - i\gamma_2) \langle Y_{11} \rangle_0 \right] \delta_{nm},$$
$$(\mathbf{D}_l)_{nm} = \delta_{n-2m} d_{lm-l+1}^- + \delta_{n-1m} d_{lm-l} + \delta_{nm} d_{lm-l-1}^+,$$

where

$$d_{n,m} = \gamma_3 \sqrt{\frac{n^2 - m^2}{4n^2 - 1}},$$

$$d_{n,m}^{-} = -\left(d_{n,-m}^{+}\right)^{*} = -\frac{\left(\gamma_1 + i\gamma_2\right)}{2}\sqrt{\frac{(n+m-1)(n+m)}{4n^2-1}}.$$

Finally, Eq. (2.265) combined with Eq. (2.266) formally yields the initial-value vector as the matrix product

$$\mathbf{C}_n(0) = \frac{\xi_1}{\sqrt{4\pi}}\left[\hat{\mathbf{K}}_n + \left[\mathbf{K}_n + \hat{\mathbf{K}}_{n+1}^{\dagger}\mathbf{S}_{n+1}(0)\right]\mathbf{S}_n(0)\right]\mathbf{S}_{n-1}(0)\cdots\mathbf{S}_1(0). \quad (2.268)$$

The matrix continued fraction solution rendering $\tilde{c}_{lm}(i\omega)$ yields $\tau_{\text{int}}^{\parallel}$ as given by Eq. (2.258) and $\chi_{\parallel}(\omega)$, Eq. (2.256). In like manner, we can evaluate λ_1, consequently the reversal time $\tau = \lambda_1^{-1}$, from the secular equation Eq. (1.150). As usual, all the continued fractions and series involved converge very rapidly; thus, 10–30 downward iterations in the continued fractions, and 10–25 terms of the series, are often enough to arrive at an accuracy of not less than six significant digits in most cases (the number of iterations and the number of terms must be increased with increasing barrier height parameter σ and/or with decreasing friction α). Moreover, by symmetry, the results for the observables are independent of the azimuthal angle of the orientation of the vector $\mathbf{H}_0$. Thus without loss of generality, we can henceforth assume that the vector $\mathbf{H}_0$ lies in the *XZ* plane of the laboratory coordinate system; therefore, on introducing the angle ψ between $\mathbf{H}_0$ and the *Z* axis, regarded as the easy axis of the particle, the direction cosines become $\gamma_1 = \sin\psi$, $\gamma_2 = 0$, and $\gamma_3 = \cos\psi$. The matrix continued fraction algorithm for uniaxial nanomagnets so developed can be adapted to other anisotropy potentials involving spherical harmonics of rank not higher than two, e.g., for *biaxial* and *uniaxial antiferromagnetic nanoparticles* (see Refs. [151,152,157,158]).

Numerical calculations of the reversal time for uniaxial nanomagnets have already been compared in Section 2.3.2 with the asymptotic results yielded by escape-rate theory (see Fig. 2.8). Here to demonstrate that the depletion effect mentioned in Section 2.4.3 again occurs, the integral relaxation time $\tau_{\text{int}}^{\parallel}$ is compared with the reversal time τ in Fig. 2.28. As before for small *h*, both $\tau_{\text{int}}^{\parallel}$ and τ have, as functions of the inverse

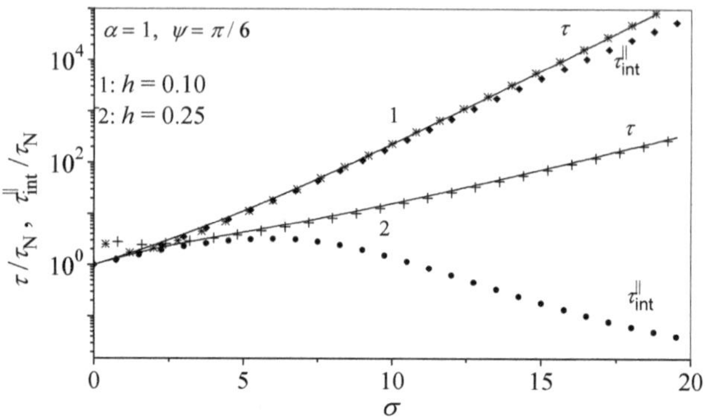

Fig. 2.28. Reversal and integral relaxation times for the oblique field potential vs. the anisotropy (inverse-temperature) parameter σ, compared with the asymptotic escape-rate solution (see Section 2.3.1: asterisks and crosses), $1/\lambda_1$ (Eq. (1.134): solid lines) and the solution rendered by Eq. (2.258) (diamonds and filled circles).

temperature ($\sigma \sim T^{-1}$), Arrhenius behavior (i.e., exponential increase with decreasing temperature). However, marked differences will again occur, if a strong external uniform bias field is applied to the particle. For certain values of h, the Arrhenius behavior of $\tau_{\mathrm{int}}^{\|}$ disappears. Moreover, the effect occurs at a critical value h_c of the ratio h, i.e., bias field parameter/anisotropy barrier-height parameter, which is *far less* than that needed to destroy the bistable nature of the potential [335]. Thus, in the low-temperature limit, the relaxation process is no longer dominated by the slow decay modes associated with the barrier crossing at values of h more than the critical value. The phenomenon, as we have seen, was first discovered (by numerical methods) for uniaxial particles [146], and later (as seen) explained analytically by Garanin [337].

We can also calculate $\chi_{\|}(\omega)$ from Eq. (2.256). $\mathrm{Re}[\chi_{\|}(\omega)]$ and $-\mathrm{Im}[\chi_{\|}(\omega)]$ vs. $\omega\tau_N$ are shown in Figs. 2.29–2.31 over a wide range of frequencies, bias field strength, oblique angle, and damping (the calculations were carried out for $v\mu_0 M_S^2/(kT) = 1$). The results reveal a marked dependence of $\chi_{\|}(\omega)$ on α, and also indicate that *three distinct dispersion peaks* now appear in the spectrum $-\mathrm{Im}[\chi_{\|}(\omega)]$. The characteristic frequency and the half-width of the low-frequency relaxation peak (LRP) are determined by the characteristic frequency $\omega_1 \sim \tau^{-1}$ of the overbarrier relaxation mode. In addition, a far weaker

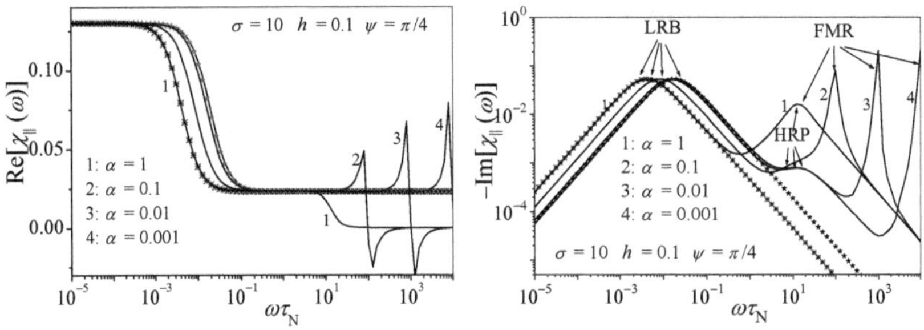

Fig. 2.29. $\mathrm{Re}[\chi_{\|}(\omega)]$ and $-\mathrm{Im}[\chi_{\|}(\omega)]$ vs. $\omega\tau_{\mathrm{N}}$ for various α and for $\sigma = 10$, $h = 0.1$, and $\psi = \pi/4$. Curves 1–4: numerical calculations based on the matrix continued fraction solution. Stars and asterisks: low-frequency approximation, Eq. (2.269).

second relaxation peak appears at high frequencies. This high-frequency relaxation peak (HRP) is due to the intrawell modes. The third FMR peak, a result of the excitation of transverse modes having frequencies close to the precession frequency ω_{pr}, appears only at low damping, and strongly manifests itself in the high-frequency region. (This is just the magnetic analog of the far-infrared absorption peak resulting from inertial librations of electric dipoles in the wells of a mean-field potential, as treated in [97].) As α decreases, the FMR peak shifts to higher frequencies since $\omega_{\mathrm{pr}} \sim 1/(\alpha\tau_{\mathrm{N}})$. Moreover, for $\psi = 0$ or $\psi = \pi$, *the FMR peak vanishes, because the transverse modes then no longer participate in the relaxation process*. The dependence of the linear response on the bias field strength is illustrated in Fig. 2.30. Here, the depletion effect is apparent: at fields above the critical field h_c at which depletion occurs, it is possible to substantially suppress the low-frequency peak, and even cause it to disappear altogether (curves 3 and 3′).

If one is interested solely in the low-frequency ($\omega\tau < 1$) part of $\chi_{\|}(\omega)$ (where the effect of the high-frequency modes may be completely ignored, so that the relaxation at long times may be approximated by a single exponential with the characteristic time $\tau \approx \lambda_1^{-1}$), then the Debye-like formula, viz.,

$$\frac{\chi_{\|}(\omega)}{\chi_{\|}} = \frac{\Delta_1}{1+i\omega\tau} + 1 - \Delta_1, \tag{2.269}$$

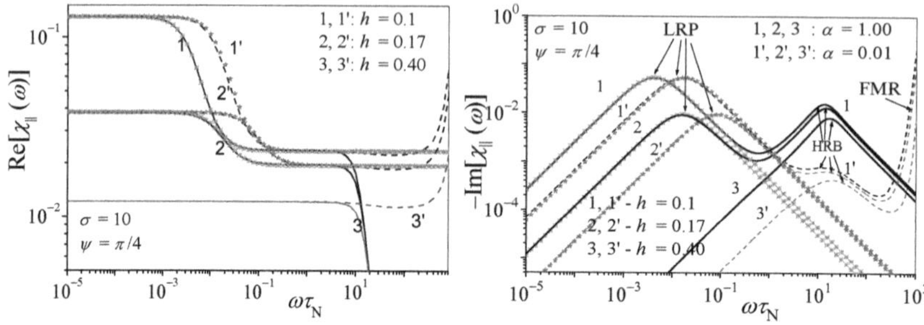

Fig. 2.30. $\mathrm{Re}[\chi_{\|}(\omega)]$ and $-\mathrm{Im}[\chi_{\|}(\omega)]$ vs. $\omega\tau_{\mathrm{N}}$ for $\sigma = 10$, $\psi = \pi/4$, $\alpha = 1.0$ (IHD: solid lines 1, 2, and 3) and $\alpha = 0.01$ (low damping: dashed lines 1′, 2′, and 3′). Lines 1, 1′ ($h = 0.01$); 2, 2′ ($h = 0.17$); and 3, 3′ ($h = 0.4$) are numerical calculations based on the matrix continued fraction solution. Stars and asterisks: Eq. (2.269).

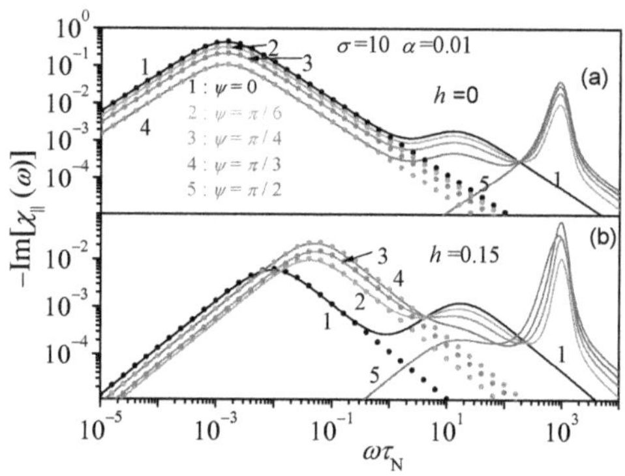

Fig. 2.31. $-\mathrm{Im}[\chi_{\|}(\omega)]$ vs. $\omega\tau_{\mathrm{N}}$ for various values of the angle ψ and (a) $h = 0$ and (b) $h = 0.15$. Solid lines, matrix continued fraction solution; filled circles, Eq. (2.269).

provides an accurate description of the low-frequency part of the spectra (see Figs. 2.29–2.31). The parameters $\chi_{\|}$ and Δ_1 depend on ξ, ψ, and σ, and can be measured experimentally, calculated numerically, and/or estimated theoretically [337]. Figures 2.29–2.31 indicate that the simple Eq. (2.269) satisfactorily determines the low-frequency part of the spectra.

Using matrix continued fractions, one can also calculate related physical quantities, such as temperature variations in the switching field, dynamic magnetic hysteresis, nonlinear dynamic susceptibilities, the signal-to-noise ratio in stochastic resonance, etc.

2.5.2 *Cubic anisotropy*

We now consider the longitudinal magnetization relaxation of single-domain ferromagnetic particles with cubic magnetic anisotropy which has already been treated by Aharoni and Eisenstein and Aharoni using the Fokker–Planck equation in Refs. [360–362], where several low-order eigenvalues were evaluated numerically. The reversal time was also estimated in Refs. [18,33–35,90,354,355]. The effect of a dc bias field on the reversal time τ and on the dynamic susceptibility $\chi_{\|}(\omega)$ of cubic crystals, was studied in Refs. [31] and [363]. Here, the matrix continued fraction method is extended to obtain τ and $\chi_{\|}(\omega)$ for a single-domain particle with cubic anisotropy. We follow the exposition of Refs. [31,33–35].

The cubic anisotropy potential, Eq. (1.15), expressed as a sum of spherical harmonics, is

$$\begin{aligned} E(\vartheta,\varphi) &= A_{40}Y_{40}(\vartheta,\varphi) + A_{44}Y_{44}(\vartheta,\varphi) \\ &\quad + A_{4-4}Y_{4-4}(\vartheta,\varphi) + \text{const}, \end{aligned} \tag{2.270}$$

where

$$A_{40} = -8\sigma\sqrt{\pi}/15, A_{4\pm4} = -4\sigma\sqrt{10\pi/7}/15,$$

$\sigma = vK_c/(4kT)$ is the dimensionless barrier-height parameter, and K_c is the anisotropy constant, which may have either *positive* or *negative* values (see Figs. 1.3(c) and 1.3(d)). Thus, the differential-recurrence relation for the statistical moments $c_{nm}(t) = \langle Y_{nm}\rangle(t) - \langle Y_{nm}\rangle_0$ governing the magnetization relaxation (using the methods of Chapter 1, see also [34,35,108,377]) is

$$\tau_N \dot{c}_{nm}(t) = \sum_{s=-1}^{1}\sum_{r=-4}^{4} e_{nm\,n+rm+4s}c_{n+rm+4s}(t), \tag{2.271}$$

where the explicit equations for $e_{nm\,n'm'}$ are given below. Thus the 27-term differential-recurrence equation, Eq. (2.271), can again be compactly written as the matrix three-term differential-recurrence equation, Eq. (2.200), where the column vectors $\mathbf{C}_n(t)$ are now defined as

$$\mathbf{C}_1(t)=\begin{pmatrix} c_{4-4}(t) \\ c_{40}(t) \\ c_{44}(t) \\ c_{30}(t) \\ c_{20}(t) \\ c_{10}(t) \end{pmatrix}, \mathbf{C}_n(t)=\begin{pmatrix} \mathbf{c}_{4n}(t) \\ \mathbf{c}_{4n-1}(t) \\ \mathbf{c}_{4n-2}(t) \\ \mathbf{c}_{4n-3}(t) \end{pmatrix}, \mathbf{c}_{4n-i}(t)=\begin{pmatrix} c_{4n-i-4(n-1+\delta_{i0})}(t) \\ c_{4n-i-4(n-2+\delta_{i0})}(t) \\ \vdots \\ c_{4n-i4(n-1+\delta_{i0})}(t) \end{pmatrix} \tag{2.272}$$

($i = 0, 1, 2, 3$), while the matrices $\mathbf{Q}_n^{\pm}$, $\mathbf{Q}_n$ are defined as

$$\mathbf{Q}_n^- = \begin{pmatrix} \mathbf{J}_{4n} & \mathbf{0} & \mathbf{0} & \mathbf{0} \\ \mathbf{D}_{4n-1} & \mathbf{J}_{4n-1} & \mathbf{0} & \mathbf{0} \\ \mathbf{P}_{4n-2} & \mathbf{D}_{4n-2} & \mathbf{J}_{4n-2} & \mathbf{0} \\ \mathbf{B}_{4n-3} & \mathbf{P}_{4n-3} & \mathbf{D}_{4n-3} & \mathbf{J}_{4n-3} \end{pmatrix}, \mathbf{Q}_n = \begin{pmatrix} \mathbf{A}_{4n} & \mathbf{B}_{4n} & \mathbf{P}_{4n} & \mathbf{D}_{4n} \\ \mathbf{B}_{4n}^{\mathrm{T}} & \mathbf{A}_{4n-1} & \mathbf{B}_{4n-1} & \mathbf{P}_{4n-1} \\ f_{4n}\mathbf{P}_{4n}^{\mathrm{T}} & \mathbf{B}_{4n-1}^{\mathrm{T}} & \mathbf{A}_{4n-2} & \mathbf{B}_{4n-2} \\ \mathbf{D}_{4n}^{\mathrm{T}} & f_{4n-1}\mathbf{P}_{4n-1}^{\mathrm{T}} & \mathbf{B}_{4n-2}^{\mathrm{T}} & \mathbf{A}_{4n-3} \end{pmatrix},$$

$$\mathbf{Q}_n^+ = \begin{pmatrix} g_{4n+4}\mathbf{J}_{4n+4}^{\mathrm{T}} & \mathbf{D}_{4n+3}^{\mathrm{T}} & f_{4n+2}\mathbf{P}_{4n+2}^{\mathrm{T}} & \mathbf{B}_{4n+3}^{\mathrm{T}} \\ \mathbf{0} & g_{4n+3}\mathbf{J}_{4n+3}^{\mathrm{T}} & \mathbf{D}_{4n+2}^{\mathrm{T}} & f_{4n+1}\mathbf{P}_{4n+1}^{\mathrm{T}} \\ \mathbf{0} & \mathbf{0} & g_{4n+2}\mathbf{J}_{4n+2}^{\mathrm{T}} & \mathbf{D}_{4n+1}^{\mathrm{T}} \\ \mathbf{0} & \mathbf{0} & \mathbf{0} & g_{4n+1}\mathbf{J}_{4n+1}^{\mathrm{T}} \end{pmatrix}, \tag{2.273}$$

where

$$f_n = -\frac{2n-11}{2n+9},$$

$$g_n = -\frac{n-4}{n+1},$$

and the index "T" means transposition. The column vector $\mathbf{C}_n(t)$ has $8n-2$ elements. The dimensions of the matrices $\mathbf{Q}_n, \mathbf{Q}_n^+$, and $\mathbf{Q}_n^-$ are, respectively, $(8n-2)\times(8n-2)$, $(8n-2)\times(8n+6)$ and $(8n-2)\times(8n-10)$. The exception is $\mathbf{Q}_1^-$, which merely degenerates to a column vector of dimension 6. The $\mathbf{Q}_n, \mathbf{Q}_n^{\pm}$ comprise three-diagonal submatrices **A**, **B**, **D**, **J**, and **P** as shown in Eq. (2.273). Evidently two distinct types of submatrices exist, viz., $\mathbf{A}_{4n}$, $\mathbf{A}_{4n-1}$, $\mathbf{A}_{4n-2}$, $\mathbf{A}_{4n-3}$, $\mathbf{B}_{4n-1}$, $\mathbf{B}_{4n-2}$, $\mathbf{B}_{4n-3}$, $\mathbf{D}_{4n-1}$, $\mathbf{P}_{4n-1}$, $\mathbf{P}_{4n-2}$. The first type $\mathbf{X}_{4n-i}$ ($i = 0, 1, 2, 3$) is of generic form

$$\mathbf{X}_{4n-i} = \begin{pmatrix} x_{4n-i,-4(n-1+\delta_{i0})} & x^{+}_{4n-i,-4(n-1+\delta_{i0})} & 0 & \cdots & 0 \\ x^{-}_{4n-i,-4(n-2+\delta_{i0})} & x_{4n-i,-4(n-2+\delta_{i0})} & x^{+}_{4n-i,-4(n-2+\delta_{i0})} & \cdots & 0 \\ 0 & x^{-}_{4n-i,-4(n-3+\delta_{i0})} & x_{4n-i,-4(n-3+\delta_{i0})} & \cdots & 0 \\ \vdots & \vdots & \vdots & \ddots & \vdots \\ 0 & 0 & 0 & \cdots & x_{4n-i,4(n-1+\delta_{i0})} \end{pmatrix}$$

and has dimension $[2(n+\delta_{0i})-1]\times[2(n+\delta_{0i})-1]$, while the second kind $\mathbf{X}_{4n-i}$ $(i=0, 1, 2, 3)$ comprising the submatrices $\mathbf{B}_{4n}$, $\mathbf{D}_{4n}$, $\mathbf{J}_{4n}$, $\mathbf{P}_{4n}$, $\mathbf{D}_{4n-2}$, $\mathbf{D}_{4n-3}$, $\mathbf{J}_{4n-1}$, $\mathbf{J}_{4n-2}$, $\mathbf{J}_{4n-3}$, $\mathbf{P}_{4n-3}$ is defined as

$$\mathbf{X}_{4n-i} = \begin{pmatrix} x^{+}_{4n-i,-4(n-1+\delta_{i0})} & 0 & 0 & \cdots & 0 \\ x_{4n-i,-4(n-2+\delta_{i0})} & x^{+}_{4n-i,-4(n-2+\delta_{i0})} & 0 & \cdots & 0 \\ x^{-}_{4n-i,-4(n-3+\delta_{i0})} & x_{4n-i,-4(n-3+\delta_{i0})} & x^{+}_{4n-i,-4(n-3+\delta_{i0})} & \cdots & 0 \\ \vdots & \vdots & \vdots & \ddots & \vdots \\ 0 & 0 & 0 & \cdots & x^{-}_{4n-i,4(n-1+\delta_{i0})} \end{pmatrix}$$

and has dimension $[2(n+\delta_{0i})-1]\times[2(n+\delta_{0i})-3]$. The matrix elements are formed from the $e_{lml'm'}$ above and are given by (supposing that the elements a_{nm}, b_{nm}, etc. pertain to the matrices **A**, **B**, etc.)

$$a_{nm} = e_{nmnm} = \sigma\frac{9n(n^2-1)(n+2)-15m^2[6n(n+1)-5-7m^2]}{(2n-3)(2n-1)(2n+3)(2n+5)} - \frac{n(n+1)}{2},$$

$$a^{-}_{nm} = a^{+}_{n-m} = e_{nmnm-4}$$

$$= \frac{15\sigma\sqrt{(n+m)(n-m+4)[n^2-(m-3)^2][n^2-(m-2)^2][n^2-(m-1)^2]}}{2(2n-3)(2n-1)(2n+3)(2n+5)},$$

$$b_{nm} = e_{nmn-1m} = \frac{3i\sigma m(3n^2-5-7m^2)}{\alpha(4n^2-9)}\sqrt{\frac{n^2-m^2}{4n^2-1}},\ b^{-}_{nm} = -b^{+}_{n-m} = e_{nmn-1m-4}$$

$$= \frac{3i\sigma}{2\alpha(4n^2-9)}\sqrt{\frac{(n+m)(n+m-4)\prod_{k=1}^{3}[n^2-(m-k)^2]}{4n^2-1}},$$

$$d_{nm} = e_{nmn-3m} = \frac{7i\sigma m}{\alpha(2n-3)(2n-1)}\sqrt{\frac{(n^2-m^2)[(n-1)^2-m^2][(n-2)^2-m^2]}{(2n-5)(2n+1)}},$$

$$d_{nm}^{-} = -d_{n-m}^{+} = e_{nmn-3m-4} = \frac{-i\sigma}{2\alpha(2n-3)(2n-1)}$$

$$\times\sqrt{\frac{(n-m+1)\prod_{k=0}^{6}(n+m-k)}{(2n-5)(2n+1)}},$$

$$p_{nm} = e_{nm\,n-2m} = \frac{\sigma(2n+9)\left(n^2-n-2-7m^2\right)}{(2n-5)(2n-1)(2n+3)}\sqrt{\frac{(n^2-m^2)[(n-1)^2-m^2]}{(2n+1)(2n-3)}},$$

$$p_{nm}^{-} = p_{n-m}^{+} = e_{nmn-2m-4} = -\frac{\sigma(2n+9)}{2(2n-5)(2n-1)(2n+3)}$$

$$\times\sqrt{\frac{(n+m-5)(n+m-4)(n+m-3)(n+m)[n^2-(m-2)^2][n^2-(m-1)^2]}{(2n+1)(2n-3)}},$$

$$j_{nm} = e_{nmn-4m} = \frac{7\sigma(n+1)}{(2n-5)(2n-3)(2n-1)}$$

$$\times\sqrt{\frac{[(n-3)^2-m^2][(n-2)^2-m^2][(n-1)^2-m^2](n^2-m^2)}{(2n-7)(2n+1)}},$$

$$j_{nm}^{-} = j_{n-m}^{+} = e_{nmn-4m-4} = \frac{\sigma(n+1)}{2(2n-5)(2n-3)(2n-1)}\sqrt{\frac{\prod_{k=0}^{7}(n+m-k)}{(2n-7)(2n+1)}}.$$

As before the *exact* solution for the Laplace transform $\tilde{\mathbf{C}}_1(s)$ as formally given by Eq. (2.203) holds and the initial value vectors $\mathbf{C}_n(0)$ may be evaluated using the same continued fractions because the initial value matrix elements $c_{nm}(0) = \langle \cos\vartheta Y_{nm}\rangle_0$ are given by

$$c_{nm}(0) = \sqrt{\frac{(n+1)^2-m^2}{(2n+1)(2n+3)}}\langle Y_{n+1m}\rangle_0 + \sqrt{\frac{n^2-m^2}{4n^2-1}}\langle Y_{n-1m}\rangle_0. \quad (2.274)$$

The recurrence relation for the equilibrium averages $\langle Y_{nm}\rangle_0$, viz.,

$$\sum_{s=-1}^{1}\sum_{r=-4}^{4} e_{nmn+rm+4s}\left\langle Y_{n+rm+4s}\right\rangle_0 = 0, \quad (2.275)$$

may then be written explicitly as the algebraic three-term matrix recurrence relation, Eq. (2.263), where the matrices $\mathbf{Q}_n$ and $\mathbf{Q}_n^{\pm}$ are now defined by Eqs. (2.273) and the vectors $\mathbf{R}_n$ are given by

$$\mathbf{R}_0 = \frac{1}{\sqrt{4\pi}}, \ \mathbf{R}_n = \begin{pmatrix} \mathbf{r}_{4n} \\ \mathbf{r}_{4n-1} \\ \mathbf{r}_{4n-2} \\ \mathbf{r}_{4n-3} \end{pmatrix}, \ \mathbf{r}_{4n-i} = \begin{pmatrix} \left\langle Y_{4n-i-4(n-1+\delta_{i0})} \right\rangle_0 \\ \left\langle Y_{4n-i-4(n-2+\delta_{i0})} \right\rangle_0 \\ \vdots \\ \left\langle Y_{4n-i4(n-1+\delta_{i0})} \right\rangle_0 \end{pmatrix} \tag{2.276}$$

($i = 0, 1, 2, 3$). Thus, the initial value vector $\mathbf{C}_n(0)$ is again rendered by the matrix product (cf. Eq. (2.268))

$$\mathbf{C}_n(0) = \frac{1}{\sqrt{4\pi}}\left[\hat{\mathbf{K}}_n + \left[\mathbf{K}_n + \hat{\mathbf{K}}_{n+1}^T \mathbf{S}_{n+1}\right]\mathbf{S}_n\right]\mathbf{S}_{n-1}\cdots\mathbf{S}_1, \tag{2.277}$$

where $\mathbf{K}_n$ and $\hat{\mathbf{K}}_n$ are the supermatrices defined as

$$\mathbf{K}_n = \begin{pmatrix} \mathbf{0} & \mathbf{U}_{4n} & \mathbf{0} & \mathbf{0} \\ \mathbf{U}_{4n}^{\mathrm{T}} & \mathbf{0} & \mathbf{U}_{4n-1} & \mathbf{0} \\ \mathbf{0} & \mathbf{U}_{4n-1}^{\mathrm{T}} & \mathbf{0} & \mathbf{U}_{4n-2} \\ \mathbf{0} & \mathbf{0} & \mathbf{U}_{4n-2}^{\mathrm{T}} & \mathbf{0} \end{pmatrix}, \ \hat{\mathbf{K}}_n = \begin{pmatrix} \mathbf{0} & \mathbf{0} & \mathbf{0} & \mathbf{0} \\ \mathbf{0} & \mathbf{0} & \mathbf{0} & \mathbf{0} \\ \mathbf{0} & \mathbf{0} & \mathbf{0} & \mathbf{0} \\ \mathbf{U}_{4n-3} & \mathbf{0} & \mathbf{0} & \mathbf{0} \end{pmatrix}.$$

The supermatrices $\mathbf{K}_n$ and $\hat{\mathbf{K}}_n$ are comprised of submatrices $\mathbf{U}_{4n}$ and $\mathbf{U}_{4n-i}$ ($i = 1, 2, 3$). The dimensions of the submatrices $\mathbf{U}_{4n}$ and $\mathbf{U}_{4n-i}$ are, respectively, $(2n + 1) \times (2n - 1)$ and $(2n - 1) \times (2n - 1)$, while their matrix elements are given by

$$\left(\mathbf{U}_{4n}\right)_{pq} = \delta_{p-1q} u_{4n-4n+4q},$$
$$\left(\mathbf{U}_{4n-i}\right)_{pq} = \delta_{pq} u_{4n-i-4(n-q)},$$

where

$$u_{nm} = \sqrt{\frac{n^2 - m^2}{4n^2 - 1}}.$$

Once again, all the continued fractions and series involved in the matrix continued fraction solution converge very rapidly; thus 10–20 downward iterations in calculating the continued fractions, and 11–15 terms in the series, are usually sufficient to arrive at an accuracy of not less than six significant digits.

2.5.2.1 *Complex susceptibility and relaxation times*

Before proceeding, we first summarize the principal results of *linear response theory* as applied on this occasion to nanomagnets with cubic anisotropy. There the decay of the longitudinal component of the magnetization $\langle M_{\|}\rangle(t)$ of a system of noninteracting single-domain ferromagnetic particles with cubic anisotropy, when a small uniform external field $\mathbf{H}_1$ ($\mu_0 v(\mathbf{M}\cdot\mathbf{H}_1) << kT$) applied along the Z axis (which is chosen as the easy axis of the particle for $K > 0$) has been switched off at time $t = 0$, is

$$\langle M_{\|}\rangle(t) = \chi_{\|} H_1 C_{\|}(t). \tag{2.278}$$

Here $\chi_{\|} = v\mu_0 M_S^2 / (3kT)$ is as usual the static susceptibility, and $C_{\|}(t) = c_{10}(t) / c_{10}(0)$ is the normalized ACF of the longitudinal component of the magnetization. The longitudinal susceptibility $\chi_{\|}(\omega)$ is given by Eq. (2.256). Other relevant quantities are again the integral and effective relaxation times $\tau_{\text{int}}^{\|}$ and $\tau_{\text{ef}}^{\|}$, defined as

$$\begin{aligned} \tau_{\text{int}}^{\|} &= \tilde{C}_{\|}(0), \\ \tau_{\text{ef}}^{\|} &= \frac{2\tau_N \langle \cos^2 \vartheta \rangle_0}{1 - \langle \cos^2 \vartheta \rangle_0} = \tau_N. \end{aligned} \tag{2.279}$$

Here, we have also recalled that, for cubic anisotropy $\langle \cos^2 \vartheta \rangle_0 = 1/3$. According to Eq. (2.279), the effective relaxation time is independent of σ. Hence, the high-frequency asymptotic behavior of $\text{Im}[\chi_{\|}(\omega)]$ is also independent of σ, that is,

$$-\chi_{\|}^{-1} \lim_{\omega\tau_N \to \infty} \omega\tau_N \, \text{Im}[\chi_{\|}(\omega)] = \frac{\tau_{\text{ef}}^{\|}}{\tau_N} = 1. \tag{2.280}$$

Having evaluated the Laplace transform $\tilde{c}_{10}(s)$, we have all desired observables. The behavior of $\tau_{\text{int}}^{\|}$ as a function of σ, for various values of α, is shown in Fig. 2.32. Here, the exact results are compared with those evaluated from the asymptotic formulas for the reversal time τ in the low-temperature or high-barrier limit ($|\sigma| >> 1$) for various α. As is clear from Fig. 2.32, $\tau_{\text{int}}^{\|}$ for particles with cubic anisotropy strongly depends on α. As far as physical interpretation is concerned, $\tau_{\text{int}}^{\|}$ is again determined by

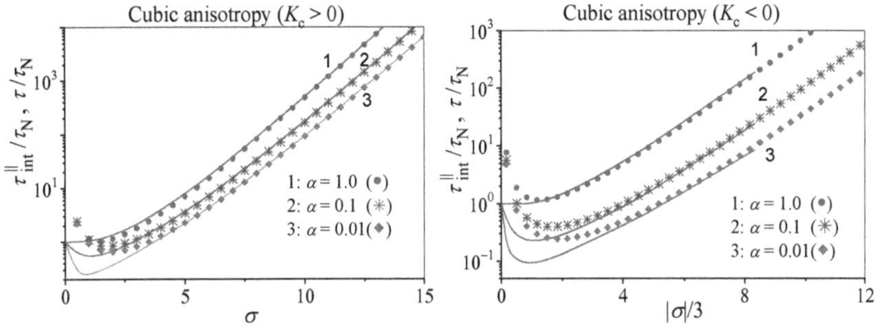

Fig. 2.32. Integral relaxation time $\tau_{int}^{\parallel}/\tau_N$ and reversal time τ/τ_N vs. barrier-height parameter for various values of the damping parameter α. Solid lines: matrix continued fraction solution for $\tau_{int}^{\parallel}/\tau_N$ at $\alpha = 1$ (curve 1); $\alpha = 0.1$ (2); and $\alpha = 0.01$ (3). Symbols: the reversal time calculated from Eqs. (2.116) and (2.120).

the slowest low-frequency relaxation mode, which governs transitions of the magnetization over the barriers separating one potential well from another. The characteristic frequency of this overbarrier relaxation mode is determined by τ^{-1}. Asymptotic formulas for the reversal time for cubic crystals were also obtained in Section 2.3.4, Eqs. (2.116) and (2.120).

The characteristic feature of noncircularly symmetric potentials, namely, *intrinsic dependence* on the damping parameter also appears in the spectra of $\chi_{\parallel}''(\omega)$, as shown in Fig. 2.33. (The calculations were carried out with $v\mu_0 M_S^2 / kT = 1$.) Two peaks in the loss spectrum are visible in these figures. The first (low-frequency) peak as expected is located at frequencies of the order of the average frequency of reorientation of the magnetization vector. However, the characteristic frequency and the half-width are now determined by the two smallest nonvanishing eigenvalues λ_1 and λ_2 [27,28]. For high barriers, λ_1 and λ_2 only differ by a factor of order of 1.5, appearing in the spectrum of $\chi_{\parallel}''(\omega)$ as merely a *single* Lorentzian (recall that to observe two distinct peaks in a sum of two Lorentzians, the characteristic frequencies of each one must differ by at least 5.8 times.) Therefore, the characteristic frequency of the low-frequency peak ω_1 can still be estimated as $\omega_1 \approx \tau^{-1}$, which can be evaluated from the Brown–Kramers escape-rate theory (Section 2.3.4). The second peak is due to the (fast) high-frequency intrawell and transverse relaxation modes. Like the oblique-field configuration, there is an *inherent geometric dependence of* $\chi_{\parallel}(\omega)$ *and* $\tau_{int}^{\parallel}/\tau_N$ *on* α *for particles*

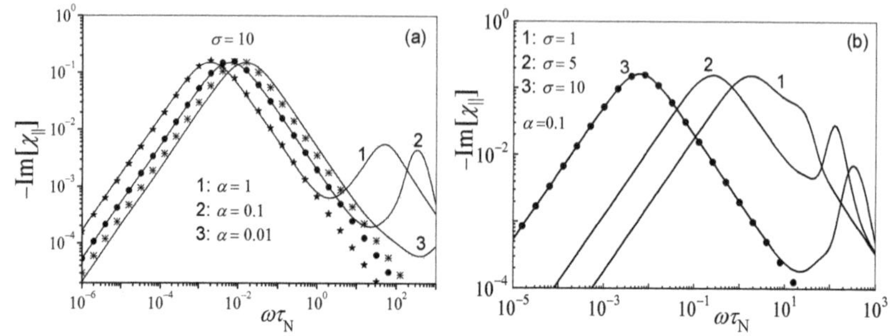

Fig. 2.33. (a) $\chi_{\parallel}''$ vs. $\omega\tau_{\mathrm{N}}$ for $\sigma = 10$ and $\alpha = 1$ (curve 1), $\alpha = 0.1$ (2), and $\alpha = 0.01$ (3). Symbols: Eqs. (2.269) and (2.116). (b) $\chi_{\parallel}''$ vs. $\omega\tau_{\mathrm{N}}$ for $\alpha = 0.1$ and $\sigma = 1$ (1), $\sigma = 5$ (2), and $\sigma = 10$ (3). Filled circles: Eqs. (2.269) and (2.116).

with cubic anisotropy. This dependence again arises from *coupling of the longitudinal and transverse relaxation modes*. A simplified description of the spectra applies in the low-frequency limit, $\omega\tau < 1$. Here, the contribution of the high-frequency modes becomes negligible, so that $\chi_{\parallel}(\omega)$ may be approximated by the single Lorentzian, Eq. (2.269), where the reversal time τ can be calculated from the analytic escape rate, Eqs. (2.116) and (2.120), and the amplitude Δ_1 can be simply evaluated as $\Delta_1 \approx 1 - 1/(4\sigma)$ and $\Delta_1 \approx 1 - 3/(8|\sigma|)$ ($\sigma >> 1$) for positive and negative anisotropy constants, respectively [378]. In Fig. 2.33, the simple Debye Eq. (2.269) describes the low-frequency part of $\chi_{\parallel}''(\omega)$.

2.5.2.2 *Magnetic stochastic resonance in nanomagnets with cubic anisotropy*

The magnetic SR in a *uniaxial particle* (see Section 2.4.4 above), where the bistability arises from a circularly symmetric potential involves only one coordinate ϑ. In contrast, SR in a cubic crystal is governed by *two* coordinates giving rise to *multistable* states. Here we demonstrate that magnetic SR in particles with noncircularly symmetric anisotropy potentials, e.g., cubic anisotropy, has behavior which has little in common with uniaxial anisotropy due to the lowered local symmetry (both ϑ and φ are now involved) of the energyscape near the easy axis direction in the space of orientations. Whereas in uniaxial anisotropy the energyscape is a uniform equatorial ridge (zone) separating two polar minima and has no saddle points, cubic anisotropy now generates *azimuthally nonuniform* energy distributions with numerous saddle points. Such a noncircularly symmetric energyscape leads to a new effect, viz., strong *intrinsic* dependence of SR in superparamagnetic particles with cubic anisotropy on *the damping coefficient* α.

The behavior of the SNR for cubic crystals can be understood in qualitative fashion by observing that in the low-temperature limit, $\sigma >> 1$, the longest relaxation time τ for cubic crystals is rendered by asymptotic formulas via the Kramers-Brown theory [97] for the escape rate. The universal turnover formulas which apply to all damping regimes have been given in Section 2.3.4, Eqs. (2.116) and (2.120). Hence, noting that for cubic crystals $\tau_{\text{int}}^{\|} \approx \tau$ and $\langle \cos^2 \vartheta \rangle_0 = 1/3$, we may easily estimate the SNR in the low-temperature and low-frequency limits as (cf. Eq. (2.228))

$$R = R(0) \approx \frac{\tau_0 \sigma}{3\tau} \approx R_{\text{IHD}}\, e^{\frac{1}{\pi}\int_0^\infty \ln\left(1 - e^{\frac{-8\sqrt{2}\alpha|\sigma|(\lambda^2+1/4)}{9}}\right)\frac{d\lambda}{\lambda^2+1/4}}, \tag{2.281}$$

where $\tau_0 = \tau_{\text{N}} / (\alpha + \alpha^{-1})$ is a damping-independent reference time and R_{IHD} is the SNR for IHD values of damping ($\alpha \geq 1$) given by

$$R_{\text{IHD}} = \frac{2\sqrt{2}\sigma^2 e^{-\sigma}}{3\pi(1+\alpha^2)} \begin{cases} 3e^{-\sigma}\left(\sqrt{9\alpha^2+8}+\alpha\right), & \sigma > 0, \\ e^{-|\sigma|/3}\left(\sqrt{9\alpha^2+8}-\alpha\right), & \sigma < 0. \end{cases} \tag{2.282}$$

The exponential factor in Eq. (2.281) interpolates between the VLD and IHD values of R [181,355], while the exponential temperature dependence of R_{IHD} in Eq. (2.282) is due to the Arrhenius-like dependence of τ (σ and $|\sigma|/3$ are the dimensionless barrier heights for the cubic anisotropy potential [18,97]). Expressions (2.281) and (2.282) yield an accurate simplified description of SR for cubic anisotropy in the high-barrier (low-temperature) limit and are compared with the exact matrix continued fraction solution [35] in Fig. 2.34. A striking feature of magnetic SR for cubic anisotropy is that the SNR curve as initially conjectured *intrinsically depends* on the damping parameter α which is once again a signature of *precession-aided relaxation*, i.e., the intrinsic coupling between the relaxational and precessional modes arising from the noncircularly

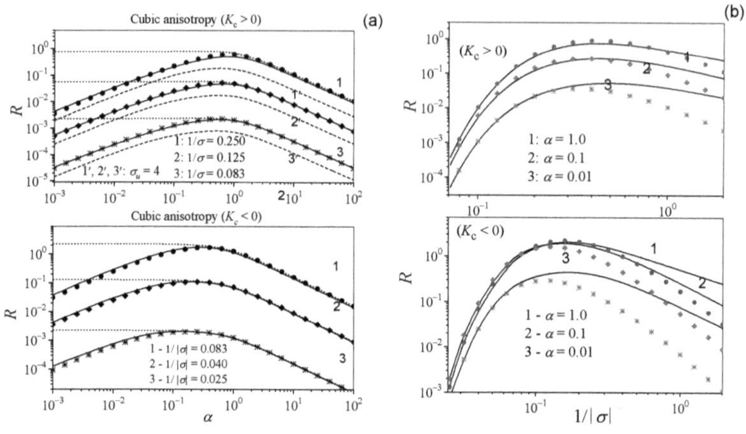

Fig. 2.34. (a) SNR vs. α in the low-temperature limit for $\sigma > 0$ (Fe-type crystals) and $\sigma < 0$ (Ni-type crystals) for various values of $1/\sigma$. Solid lines: matrix continued fraction solution (Section 2.5.2) [35]. Symbols: calculations from Eq. (2.281). Dotted lines: the SNR in the IHD limit given by Eq. (2.282). Dashed lines: the SNR for the mixed anisotropy $\sigma(\sin^4\vartheta\sin^2 2\varphi + \sin^2 2\vartheta) + \sigma_u \sin^2\vartheta$ with the dimensionless uniaxial anisotropy constant $\sigma_u = 4$. (b) SNR vs. $1/\sigma$ for $\sigma > 0$ and $\sigma < 0$ and $\alpha = 1.0$ (1); 0.1 (2); 0.01 (3). Solid lines: matrix continued fraction solution [35]. Symbols: calculations from Eq. (2.281). Dotted lines: the SNR in the IHD limit given by Eq. (2.282).

symmetric potential. Here, at each temperature, there exists a certain α_{max} for which the SNR has a maximum R_{max}. The analytic form of Eqs. (2.282) and (2.281) again allows one to estimate the peak R_{max} and the half-width δR of the damping dependence of R shown in Fig. 2.34: $R_{max} \approx R_{IHD}\big|_{\alpha\to 0}$ and δR are given by $R_{max} \sim 8\pi^{-1}\sigma^2 e^{-\sigma}$, $\delta R \sim (\sqrt{14}+\sqrt{2})/2$ for $\sigma > 0$ and $R_{max} \sim 8\sigma^2 e^{-|\sigma|/3}/(3\pi)$, $\delta R \sim (\sqrt{7}-1)/\sqrt{2}$ for $\sigma < 0$, respectively (δR can be regarded as a solution of the algebraic equation $R_{IHD} = R_{max}/2$).

The *strong intrinsic magnetic mode coupling* so demonstrated for cubic crystals suggests another way of determining the damping coefficient from measurements of SNR in a fine-particle system, namely fitting the theory to the experimental SNR, Eq. (2.224), at a given frequency Ω. Thus, the sole fitting parameter is α allowing its estimation (e.g., by adjusting α_{max} and R_{max}). Moreover, α can also be determined for different T, so yielding its temperature dependence. Such experiments are much more easily accomplished than those for uniaxial particles subject to a symmetry-breaking strong dc external magnetic field [37,38,83] suggesting that the noncircularly symmetric problem studied is of interest in both its fundamental and applied physics aspects. From a fundamental viewpoint a knowledge of $\alpha(T)$ is required to identify various possible mechanisms of orientational reversal of **M**, e.g., macroscopic quantum tunneling [379], since $\alpha(T)$ allows one both to distinguish between and to separate the various relaxation mechanisms. In applied physics, the dependence $\alpha(T)$ is essential for the development of the various procedures generically called thermal-assisted magnetic recording on single grains.

Other types of noncircularly symmetric anisotropies can be treated in like manner. For illustration, we show in Fig. 2.34(a) the SNR for mixed cubic and uniaxial anisotropies evaluated for different values of $\sigma_u/\sigma =$ 1.0, 0.5 and 0.33 and for $\sigma > 0$. Here the effect of the uniaxial anisotropy is important since the parameter σ_u/σ now determines the barrier heights of the (cubic+uniaxial) potential and moreover yields a marked contribution to the decrease of the SNR, $\sim e^{-\sigma-\sigma_u}$, with increasing σ_u/σ.

2.5.2.3 *Effect of a dc bias field*

We now consider magnetization relaxation of nanomagnets with cubic anisotropy subjected to a uniform biasing magnetic field $\mathbf{H}_0$ [31] and [363]. For simplicity, we suppose that $\mathbf{H}_0$ is directed along the easy axis {001} (see Fig. 2.35). The normalized free energy density of a particle with cubic magnetic anisotropy is

$$E(\vartheta,\varphi) = \sigma\left(\sin^4\vartheta\sin^2 2\varphi + \sin^2 2\vartheta - 2h\cos\vartheta\right), \qquad (2.283)$$

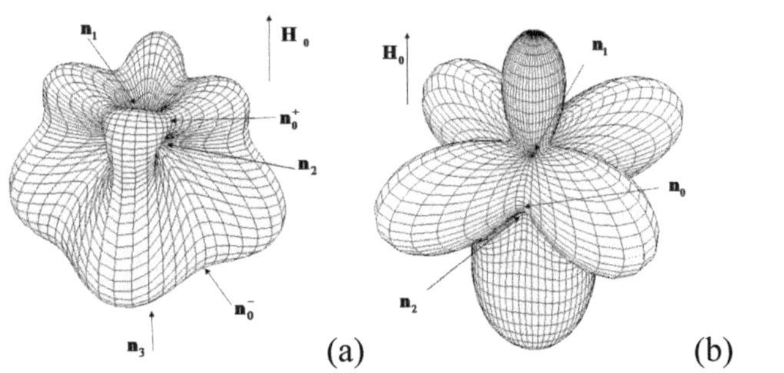

Fig. 2.35. Cubic anisotropy with a dc bias field for (a) $\sigma > 0$ and (b) $\sigma < 0$.

Table 2.6. Properties of the anisotropy potential, Eq. (2.283), for $\sigma > 0$ (for $\sigma < 0$, maxima and minima are interchanged, and the signs of σ and h are changed).

$\{\vartheta_N^p, \varphi_N^p\}$	$E(\vartheta_N^p, \phi_N^p)$
Two minima at $\vartheta_I^{min} = 0$ and $\vartheta_{III}^{min} = \pi$	$\mp 2h\sigma$
Four minima at ϑ_{II}^{min} and $\phi_{II}^{min} = n\pi/2,\ n = 0,1,2,3$	$-\frac{\sigma}{4}\left(h^2 + \frac{1}{16}h^4 + \ldots\right)$
Eight maxima at ϑ_I^{max} and ϑ_{II}^{max} and $\phi^{max} = n\pi/4,\ n = 1,3,5,7$	$\sigma\left(\frac{4}{3} \mp \frac{2h}{\sqrt{3}} + \frac{h^2}{4} \pm \frac{\sqrt{3}h^3}{16} + \frac{3h^4}{32} + \ldots\right)$
Eight saddle points at ϑ_I^{sad} and ϑ_{II}^{sad} and $\phi^{sad} = n\pi/2,\ n = 0,1,2,3$	$\sigma\left(1 \mp \sqrt{2}h + \frac{h^2}{8} \pm \frac{h^3}{32\sqrt{2}} + \frac{h^4}{128} + \ldots\right)$
Four saddle points at ϑ_{III}^{sad} $\phi_{III}^{sad} = n\pi/4,\ n = 1,3,5,7$	$\sigma\left(1 - \frac{h^2}{2} - \frac{3h^4}{16} + \ldots\right)$

$$\begin{Bmatrix} \vartheta_{II}^{min} \\ \vartheta_I^{sad} \end{Bmatrix} = \arccos\left\{2^{-4/3}\,\mathrm{Re}\left[\left(1 \mp i\sqrt{3}\right)\left(h - i\sqrt{-h^2 + 32/27}\right)^{1/3}\right]\right\}$$

$$\begin{Bmatrix} \vartheta_I^{max} \\ \vartheta_{III}^{sad} \end{Bmatrix} = \arccos\left\{\frac{1}{\sqrt[3]{12}}\mathrm{Re}\left[\left(1 \pm i\sqrt{3}\right)\left(h - i\sqrt{-h^2 + 16/81}\right)^{1/3}\right]\right\}$$

$$\vartheta_{II}^{max} = \arccos\left\{-\frac{1}{\sqrt[3]{12}}\left[\left(h - \sqrt{h^2 - 16/81}\right)^{1/3} + \left(h + \sqrt{h^2 - 16/81}\right)^{1/3}\right]\right\}$$

$$\vartheta_{II}^{sad} = \arccos\left\{-2^{-4/3}\left[\left(h - \sqrt{h^2 - 32/27}\right)^{1/3} + \left(h + \sqrt{h^2 - 32/27}\right)^{1/3}\right]\right\}$$

where $h = \xi/(2\sigma)$ is, as usual, the dc field parameter. The properties of the anisotropy, Eq. (2.283), are summarized in Table 2.6.

Here the matrix continued fraction algorithm developed in Section 2.5.2 requires only minor modifications. Indeed, the differential-recurrence equation for the equilibrium correlation functions

$$c_{lm}(t) = \left\langle \cos\vartheta(0) Y_{lm}(t) \right\rangle_0 - \left\langle \cos\vartheta(0) \right\rangle_0 \left\langle Y_{lm} \right\rangle_0 ,$$

governing the dynamics of the magnetization with a dc applied field, is very similar to that without the field, namely Eq. (2.271),

$$\tau_{\mathrm{N}}\dot{c}_{lm}(t)=\sum_{s=-1}^{1}\sum_{r=-4}^{4}e_{lmrs}c_{l+rm+4s}(t). \tag{2.284}$$

Thus out of all the 27 coefficients $e_{nmn+rm\pm 4s}$, there are only three that depend on the field parameter h and which differ from those listed after Eq. (2.271) for $h=0$. They are

$$e_{nmnm}=a_{nm}$$

$$=\sigma\frac{9(n^2-1)n(n+2)-15m^2[6n(n+1)-5-7m^2]}{(2n-3)(2n-1)(2n+3)(2n+5)}-\frac{n(n+1)}{2}+\frac{im\sigma h}{\alpha},$$

$$e_{nmn-1m}=b_{nm}=\sigma\left[h(n+1)+\frac{3im\left(3n^2-5-7m^2\right)}{\alpha(4n^2-9)}\right]\sqrt{\frac{n^2-m^2}{4n^2-1}},$$

$$e_{nmn+1m}=b'_{n+1m}=-\sigma\left[hn-\frac{i3m\left(3n^2+6n-2-7m^2\right)}{\alpha(2n-1)(2n+5)}\right]\sqrt{\frac{(n+1)^2-m^2}{(2n+1)(2n+3)}}.$$

The solution of Eq. (2.284) has been dealt with in the previous section. The only modifications required are in $\mathbf{Q}_n^+$, where the submatrices $\mathbf{A}_{4n}$, $\mathbf{A}_{4n-1}$, $\mathbf{A}_{4n-2}$, $\mathbf{A}_{4n-3}$, $\mathbf{B}_{4n}$, $\mathbf{B}_{4n-1}$, $\mathbf{B}_{4n-2}$, and $\mathbf{B}_{4n-3}$ must be formed from the field-dependent coefficients a_{nm} and b_{nm} [b'_{nm} must be used *only* in the calculation of the submatrix $\mathbf{B}_{4n+3}$ in Eq. (2.273), instead of b_{nm}] and in $\mathbf{Q}_1^-$, which now becomes

$$\mathbf{Q}_1^-=\begin{pmatrix}4\sigma\sqrt{10/63}\\ 8\sigma/3\\ 4\sigma\sqrt{10/63}\\ 0\\ 0\\ 2\sigma h/\sqrt{3}\end{pmatrix}.$$

Having determined $\tilde{c}_{10}(i\omega)$, we have the one-sided Fourier transform of the autocorrelation function $C_{\parallel}(t)$ of the longitudinal component of the magnetization of the particle and its integral relaxation time $\tau_{\mathrm{int}}^{\parallel}$, viz.

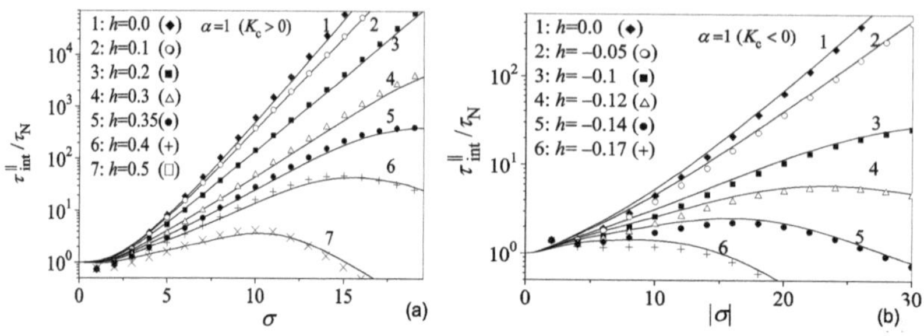

Fig. 2.36. Correlation time $\tau_{\text{int}}^{\|}$ vs. the inverse-temperature parameter σ for various values of the field parameter h with the damping parameter $\alpha = 1$. (a): $\sigma > 0$. (b): $\sigma < 0$. Solid lines: matrix continued fraction solution [31]. Symbols: Eq. (2.288).

$$C_{\|}(t) = \frac{\langle M_{\|}(0)M_{\|}(t)\rangle_0 - \langle M_{\|}(0)\rangle_0^2}{\langle M_{\|}^2(0)\rangle_0 - \langle M_{\|}(0)\rangle_0^2} = \frac{c_{10}(t)}{c_{10}(0)}, \tag{2.285}$$

$$\tau_{\text{int}}^{\|} = \int_0^\infty C_{\|}(t)dt = \frac{\tilde{c}_{10}(0)}{c_{10}(0)}. \tag{2.286}$$

The dynamic susceptibility $\chi_{\|}(\omega)$ is then given by Eq. (2.256).

Without the bias field $\mathbf{H}_0$, the behavior of $\tau_{\text{int}}^{\|}$ as a function of σ for various values of α, is as investigated in the previous section. However, marked differences again occur (see Fig. 2.36) if a strong field is applied to the particle: once more, for certain values of h, the Arrhenius behavior (exponential increase with increasing barrier height) of $\tau_{\text{int}}^{\|}$ disappears. For uniaxial nanomagnets, this effect (as seen) is a natural consequence of the depletion, by the uniform field, of the shallower of the two potential wells involved in the barrier crossing. Thus, at low temperatures, the fast intrawell modes may again come to dominate the relaxation. Furthermore, in Ref. [335] it has been shown that this depletion effect always exists in relaxation in one-dimensional bistable potentials (the magnetization relaxation of uniaxial particles is governed by the one-variable Fokker–Planck equation). Moreover, it has also been asserted that the effect is a general feature of relaxation in biased double-well potentials. However, the relaxation in a cubic potential subjected to a bias field is governed by the *two-variable* Fokker–Planck equation, so that at first glance the results of Ref. [335] do not appear to be directly applicable to this problem.

Nevertheless, only barrier-crossing transitions involving the particular wells at $\vartheta = 0$ and/or at $\vartheta = \pi$, result in *changes of the magnetization direction*, with the azimuthal angle φ *remaining constant*. Thus, the dynamics of these transitions are effectively described by *one* coordinate. Hence, one may also apply the method of Ref. [335] in order to describe them.

For the purpose of discussion, we consider the potential $E(\vartheta, \varphi)$, Eq. (2.283), which at $\varphi = 0$ has a maximum at the saddle point $\vartheta_0 = \vartheta_{\mathrm{I}}^{\mathrm{sad}}$ and two minima at $\vartheta_2 = \vartheta_{\mathrm{I}}^{\mathrm{min}}$ and $\vartheta_1 = \vartheta_{\mathrm{II}}^{\mathrm{min}}$ (see Table 2.6). For simplicity we consider the IHD regime ($\alpha \geq 1$), since the values of h_c are practically independent of α. At low temperatures ($\sigma >> 1$) and IHD damping ($\alpha \geq 1$), $C_{\|}(t)$ from Eq. (2.285) may be approximated by a *sum* of two exponentials, which again correspond to *intrawell* and *overbarrier* relaxation processes, namely

$$C_{\|}(t) \approx \Delta e^{-t/\tau} + \Delta_{\mathrm{well}} e^{-t/\tau_{\mathrm{W}}}, \tag{2.287}$$

where τ_{W} is the "intrawell" relaxation time and so has only a weak temperature dependence, viz.,

$$\tau_{\mathrm{W}} \sim \frac{\tau_{\mathrm{N}}}{2\sigma(2+h)} \quad (\sigma > 0)$$

and

$$\tau_{\mathrm{W}} \sim \frac{3\tau_{\mathrm{N}}}{8|\sigma| + 2\sigma h\sqrt{3})} \quad (\sigma < 0).$$

The reversal time τ has Arrhenius (exponential) temperature dependence, namely, $\tau \sim e^{E_C - E_1}$, with the consequence that $\Delta / \Delta_{\mathrm{well}} \sim e^{E_2 - E_1} << 1$, where $E_i = E(\vartheta_i, 0)$. According to Eqs. (2.286) and (2.287), the integral relaxation time is given by

$$\tau_{\mathrm{int}}^{\|} \approx \tau_{\mathrm{W}} + \tau \frac{\Delta}{\Delta_{\mathrm{well}}}, \tag{2.288}$$

where

$$\tau \frac{\Delta}{\Delta_{\mathrm{well}}} \sim e^{E_C + E_2 - 2E_1}. \tag{2.289}$$

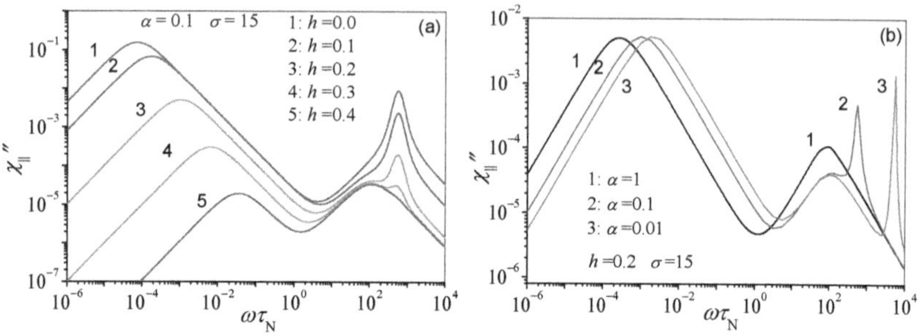

Fig. 2.37. $\chi''_{\parallel}$ vs. $\omega\tau_N$ (a) for $\sigma = 15$, $\alpha = 0.1$, and various h and (b) for $\sigma = 15$, $h = 0.2$, and various α.

However, the argument of the exponential in Eq. (2.289) $E_C + E_2 - 2E_1$ may again *change its sign* at some critical value of h_c. Thus if $h < h_c$, the quantity $\tau\Delta/\Delta_{\text{well}}$ *increases exponentially* as the temperature T decreases, and so once again entirely determines the temperature dependence of the correlation time $\tau_{\text{int}}^{\parallel}$. In contrast, if $h > h_c$ the quantity $\tau\Delta/\Delta_{\text{well}}$ *decreases exponentially* as T decreases, so that $\tau_{\text{int}}^{\parallel}$ no longer has Arrhenius behavior. Therefore, at this critical value of the bias parameter, the relaxation switches as before from being dominated by the slowest overbarrier mode to being dominated by the fast intrawell relaxation modes, because of the depletion of the upper (shallow) wells at low temperatures. Thus, on using the equations for E_1, E_2, and E_C from Table 2.6 for $\sigma > 0$ and $\sigma < 0$, we have, respectively,

$$\tau\frac{\Delta}{\Delta_{\text{well}}} \propto e^{\sigma[1-(2+\sqrt{2}h)]}$$

and

$$\tau\frac{\Delta}{\Delta_{\text{well}}} \propto e^{|\sigma|(1+6\sqrt{3}h)}.$$

Hence, the critical values of h are now given, respectively, by $h_c = 1/(2+\sqrt{2}) \approx 0.3$ and $|h_c| = 1/6\sqrt{3} \approx 0.1$. Clearly, from Fig. 2.35, these estimates agree with the numerical calculations of [363].

The imaginary part $\chi''_{\parallel}(\omega)$ is shown in Fig. 2.37 (where the calculations were carried out with $v\mu_0 M_S^2/(kT) = 1$). In this figure, three peaks appear in the loss spectrum. The first (low-frequency) relaxation peak is again located at frequencies of the order of the average frequency of reorientation of the magnetization vector. The characteristic frequency and half-width in this low-frequency peak are determined by τ. The second, much weaker, relaxation peak is caused by the high-frequency intrawell relaxation modes, and is located at frequencies of the order of τ_W^{-1}. The third (resonance) peak comes from the transverse modes, which give rise to ferromagnetic resonance at frequencies corresponding to the magnetization vector precession frequency $\omega_{pr} = \gamma H_{an}$. With decreasing α, the dimensionless precession frequency $\tau_N \omega_{pr}$ is shifted to higher frequencies as α^{-1} and the half-width decreases $\sim \alpha$. In the IHD regime ($\alpha > 1$), this resonance peak disappears as expected. An intrinsic geometric dependence of $\chi_{\|}(\omega)$ on α, arising from the *coupling of the longitudinal and transverse relaxation modes* always exists even without the bias field.

More information about dc field effects on relaxation processes in cubic crystals is available in Refs. [18,33–35,360,376].

2.6 Nonlinear stationary ac response of nanomagnets

2.6.1 *Uniaxial nanomagnets*

Due to the *large magnitude* of the magnetic dipole moment ($\sim 10^4$–10^5 μ_B) of a magnetic nanoparticle, resulting in a relatively large Zeeman energy even in relatively weak external magnetic fields, the magnetization reversal process has a pronounced field dependence causing nonlinear effects in the dynamic susceptibility and field-induced birefringence, stochastic resonance, dynamic hysteresis, etc. However, as seen previously, the determination of the nonlinear response to an external field represents an extremely difficult task even for dilute systems, because the response now always depends on the *precise nature* of the stimulus; hence there exists no *unique* response function valid for all stimuli, unlike in linear response.

The nonlinear magnetic response of an individual superparamagnetic nanoparticle with thermal agitation can be evaluated, like the linear response, by calculating the relevant statistical averages from the magnetic Langevin equation, Eq. (1.42). Here the free energy density comprises the magnetic anisotropy potential and the Zeeman energy densities, however the latter now arises from external dc and ac magnetic fields $\mathbf{H}(t) = \mathbf{H}_0 + \mathbf{H}\cos\omega t$ each of *arbitrary* strength and *orientation*. The nonlinear ac stationary response has often been calculated for *uniaxial* nanomagnets, either (i) by assuming that the energy of a particle in external fields is much less than the thermal energy kT, meaning that the response may be evaluated via *perturbation theory* (e.g., Refs. [37] and [39]), or (ii) by assuming that the strong external fields are directed along the easy axis of the particle, so that *circular symmetry* is preserved (e.g., Refs. [39,47,153,154]). However, these results are in themselves very restricted since many interesting nonlinear phenomena (such as the *intrinsic* damping dependence of the response, and the *interplay* between precession and thermoactivation) are excluded. The reason being that *in circular symmetry, no dynamical coupling between the longitudinal and transverse (or precessional) modes of motion exists.* In contrast, by discarding the above assumptions, we shall now describe (following Ref. [155]) an exact *non-perturbative* method for the nonlinear magnetization relaxation of superparamagnetic particles, with an *arbitrary* anisotropy potential E in a *strong* ac driving field superimposed on a *strong* dc bias field of *arbitrary* orientation. Moreover, taking as an example *uniaxial nanomagnets*, we shall demonstrate that, for *arbitrary* orientations of the ac and dc bias fields, (thus breaking the circular symmetry), the magnetization dynamics alters substantially, leading to new nonlinear effects which are impossible to treat via perturbation theory.

Thus, we first consider the general situation, where the magnitude of the applied ac field $\mathbf{H}(t)$ is so large that the Zeeman energy of a particle is comparable to, or greater than, kT. One is therefore faced with an *intrinsically* nonlinear problem which, of course, cannot be treated by perturbation theory. Nevertheless, it can still be solved in a general sense, because the infinite hierarchy of averaged differential-recurrence relations for the statistical moments $\langle Y_{lm}\rangle(t)$ still remains of the form

$$\tau_N \frac{d}{dt}\langle Y_{lm}\rangle(t) = \sum_{r,s} e_{l\,m\,l+r\,m+s}(t)\langle Y_{l+r\,m+s}\rangle(t), \tag{2.290}$$

with the difference that the coefficients $e_{l\,m\,l+r\,m+s}(t)$ are now *explicitly* time-dependent. The general solution is obtained by introducing column vectors $\mathbf{c}_n(t)$ $(n = 0, 1, 2, 3, \ldots)$ with elements $c_{lm}(t) = \langle Y_{lm} \rangle(t)$. Equation (2.290) then becomes the forced matrix recurrence equation, viz.,

$$\begin{aligned} \tau_{\mathrm{N}} \frac{d}{dt}\mathbf{c}_n(t) &= \mathbf{q}_n^{-}\mathbf{c}_{n-1}(t) + \mathbf{q}_n\mathbf{c}_n(t) + \mathbf{q}_n^{+}\mathbf{c}_{n+1}(t) \\ &+ \left[\mathbf{p}_n^{-}\mathbf{c}_{n-1}(t) + \mathbf{p}_n\mathbf{c}_n(t) + \mathbf{p}_n^{+}\mathbf{c}_{n+1}(t)\right](e^{i\omega t} + e^{-i\omega t}), \end{aligned} \quad (2.291)$$

where $\mathbf{c}_0 = \langle Y_{00} \rangle = 1/\sqrt{4\pi}$, the matrices $\mathbf{q}_n, \mathbf{q}_n^{\pm}$ and $\mathbf{p}_n, \mathbf{p}_n^{\pm}$ are generated from the coefficients $e_{l\,ml+r\,m+s}(t)$. The explicit form of $\mathbf{c}_n(t)$ and $\mathbf{q}_n, \mathbf{q}_n^{\pm}, \mathbf{p}_n, \mathbf{p}_n^{\pm}$ depends on the *precise* form of the free energy density $E(\vartheta, \varphi)$ chosen and for uniaxial nanomagnets, the $\mathbf{c}_n(t)$ are given by Eq. (2.300) below. Since we are again solely concerned with the stationary ac response, independent of the initial conditions, we require only the steady-state solution of Eq. (2.291). Thus, we may seek all the elements $c_{lm}(t)$ and thus all the column vectors $\mathbf{c}_n(t)$ as Fourier series in the time, namely

$$c_{lm}(t) = \langle Y_{lm} \rangle(t) = \sum_{k=-\infty}^{\infty} c_{lm}^{k} e^{ik\omega t} \quad (2.292)$$

and

$$\mathbf{c}_n(t) = \sum_{k=-\infty}^{\infty} \mathbf{c}_n^{k} e^{ik\omega t}.$$

By substituting this series into Eq. (2.291), we then have algebraic recurrence relations for the corresponding Fourier amplitudes $\mathbf{c}_n^k$, namely

$$\begin{aligned} &\mathbf{q}_n^{-}\mathbf{c}_{n-1}^{k} + (\mathbf{q}_n - ik\tau_{\mathrm{N}}\omega\mathbf{I})\mathbf{c}_n^{k} + \mathbf{q}_n^{+}\mathbf{c}_{n+1}^{k} + \mathbf{p}_n^{-}(\mathbf{c}_{n-1}^{k-1} + \mathbf{c}_{n-1}^{k+1}) \\ &+ \mathbf{p}_n(\mathbf{c}_n^{k-1} + \mathbf{c}_n^{k+1}) + \mathbf{p}_n^{+}(\mathbf{c}_{n+1}^{k-1} + \mathbf{c}_{n+1}^{k+1}) = 0, \end{aligned} \quad (2.293)$$

where $\mathbf{I}$ is the identity matrix. Equation (2.293) can be transformed into the compact recurrence equations

$$\begin{aligned} \mathbf{Q}_1\mathbf{C}_1 + \mathbf{Q}_1^{+}\mathbf{C}_2 &= -\mathbf{Q}_1^{-}\mathbf{C}_0, \\ \mathbf{Q}_n^{-}\mathbf{C}_{n-1} + \mathbf{Q}_n\mathbf{C}_n + \mathbf{Q}_n^{+}\mathbf{C}_{n+1} &= 0 \ (n = 2, 3, \ldots), \end{aligned} \quad (2.294)$$

where the vectors $\mathbf{C}_n$ and the matrices $\mathbf{Q}_n$ and $\mathbf{Q}_n^{\pm}$ are defined as

$$\mathbf{C}_n = \begin{pmatrix} \vdots \\ \mathbf{c}_n^{-2} \\ \mathbf{c}_n^{-1} \\ \mathbf{c}_n^{0} \\ \mathbf{c}_n^{1} \\ \mathbf{c}_n^{2} \\ \vdots \end{pmatrix}, \qquad \begin{aligned} (\mathbf{Q}_n)_{lm} &= \delta_{l-1m}\mathbf{p}_n + \delta_{lm}\mathbf{q}_n(m\omega) + \delta_{l+1m}\mathbf{p}_n, \\ (\mathbf{Q}_n^{\pm})_{lm} &= \delta_{l-1,m}\mathbf{p}_n^{\pm} + \delta_{l,m}\mathbf{q}_n^{\pm} + \delta_{l+1,m}\mathbf{p}_n^{\pm}. \end{aligned} \tag{2.295}$$

Therefore the exact solution of Eq. (2.294) for $\mathbf{C}_1$ can now be given using matrix continued fractions as before, viz.,

$$\mathbf{C}_1 = \mathbf{\Delta}_1 \mathbf{Q}_1^{-} \mathbf{C}_0, \tag{2.296}$$

where the matrix continued fraction $\mathbf{\Delta}_1$ is defined by the recurrence equation

$$\mathbf{\Delta}_n = \left[-\mathbf{Q}_n - \mathbf{Q}_n^{+} \mathbf{\Delta}_{n+1} \mathbf{Q}_{n+1}^{-} \right]^{-1}. \tag{2.297}$$

The vector $\mathbf{C}_1$ contains all the Fourier amplitudes required for the nonlinear stationary response. We emphasize that so far our matrix solution, Eq. (2.296), is valid for an *arbitrary* anisotropy potential.

For illustration, we next specialize the general matrix method to *uniaxial* nanomagnets with the ac and dc bias ac fields applied in *arbitrary* directions, so that the free energy in dimensionless form is

$$E = \sigma \sin^2 \vartheta - \xi_0 \cos\Theta - \xi \cos\omega t \cos\Theta'. \tag{2.298}$$

Here $\xi_0 = v\mu_0 M_S H_0 / (kT)$ and $\xi = v\mu_0 M_S H / (kT)$ are the usual dc and ac external field parameters, Θ and Θ' are the angles between the vectors $\mathbf{H}_0$ and $\mathbf{M}$ and $\mathbf{H}$ and $\mathbf{M}$, respectively, so that

$$\begin{aligned} \cos\Theta &= \gamma_1 \sin\vartheta \cos\varphi + \gamma_2 \sin\vartheta \sin\varphi + \gamma_3 \cos\vartheta, \\ \cos\Theta' &= \gamma_1' \sin\vartheta \cos\varphi + \gamma_2' \sin\vartheta \sin\varphi + \gamma_3' \cos\vartheta, \end{aligned} \tag{2.299}$$

$\gamma_1, \gamma_2, \gamma_3$ and $\gamma_1', \gamma_2', \gamma_3'$ are the direction cosines of the vectors $\mathbf{H}_0$ and $\mathbf{H}$. Thus the recurring vectors $\mathbf{c}_n(t)$ in Eq. (2.291), the Fourier amplitude vectors $\mathbf{c}_n^k$ in Eq. (2.293), the vector $\mathbf{R} = \mathbf{Q}_1^{-} \mathbf{C}_0$ and the various submatrices $\mathbf{q}_n$, $\mathbf{q}_n^{\pm}$, $\mathbf{p}_n$, $\mathbf{p}_n^{\pm}$ in Eq. (2.293) are given by

$$\mathbf{c}_n(t)=\begin{pmatrix}\langle Y_{2n-2n}\rangle(t)\\ \vdots\\ \langle Y_{2n2n}\rangle(t)\\ \langle Y_{2n-1-2n+1}\rangle(t)\\ \vdots\\ \langle Y_{2n-12n-1}\rangle(t)\end{pmatrix},\ \mathbf{c}_n^k=\begin{pmatrix}c^k_{2n-2n}\\ \vdots\\ c^k_{2n2n}\\ c^k_{2n-1-2n+1}\\ \vdots\\ c^k_{2n-12n-1}\end{pmatrix},$$

$$\mathbf{R}=\frac{-1}{\sqrt{4\pi}}\begin{pmatrix}\vdots\\ \mathbf{0}\\ \mathbf{p}_1^-\\ \mathbf{q}_1^-\\ \mathbf{p}_1^-\\ \mathbf{0}\\ \vdots\end{pmatrix},\mathbf{q}_1^-=\frac{1}{\sqrt{6}}\begin{pmatrix}0\\ 0\\ 2\sigma\sqrt{6/5}\\ 0\\ 0\\ (\gamma_1-i\gamma_2)\xi_0\\ \sqrt{2}\gamma_3\xi_0\\ -(\gamma_1+i\gamma_2)\xi_0\end{pmatrix},\mathbf{p}_1^-=\frac{\xi}{2\sqrt{6}}\begin{pmatrix}0\\ 0\\ 0\\ 0\\ 0\\ (\gamma_1-i\gamma_2)\\ \sqrt{2}\gamma_3\\ -(\gamma_1+i\gamma_2)\end{pmatrix}\tag{2.300}$$

$$\mathbf{p}_n^-=\begin{pmatrix}\mathbf{0}&\mathbf{0}\\ \mathbf{b}_{2n-1}&\mathbf{0}\end{pmatrix},\ \mathbf{p}_n=\begin{pmatrix}\mathbf{a}_{2n}&\mathbf{b}_{2n}\\ \mathbf{d}_{2n-1}&\mathbf{a}_{2n-1}\end{pmatrix},\ \mathbf{p}_n^+=\begin{pmatrix}\mathbf{0}&\mathbf{d}_{2n}\\ \mathbf{0}&\mathbf{0}\end{pmatrix},$$
$$\mathbf{q}_n^-=\begin{pmatrix}\mathbf{V}_{2n}&\mathbf{0}\\ \mathbf{W}_{2n-1}&\mathbf{V}_{2n-1}\end{pmatrix},\mathbf{q}_n=\begin{pmatrix}\mathbf{X}_{2n}&\mathbf{W}_{2n}\\ \mathbf{Y}_{2n-1}&\mathbf{X}_{2n-1}\end{pmatrix},\mathbf{q}_n^+=\begin{pmatrix}\mathbf{Z}_{2n}&\mathbf{Y}_{2n}\\ \mathbf{0}&\mathbf{Z}_{2n-1}\end{pmatrix}.\tag{2.301}$$

Here $\mathbf{0}$ denotes zero matrices of the appropriate dimensions. The matrices $\mathbf{q}_n$ and $\mathbf{q}_n^{\pm}$ now comprise the five submatrices $\mathbf{V}_l$, $\mathbf{W}_l$, $\mathbf{X}_l$, $\mathbf{Y}_l$, and $\mathbf{Z}_l$; these also appear in the linear response and are therefore defined by Eq. (2.261). The tridiagonal submatrices $\mathbf{a}_l$, $\mathbf{b}_l$, and $\mathbf{d}_l$ also appearing above have dimensions, respectively, $(2l+1)\times(2l+1)$, $(2l+1)\times(2l+3)$, and $(2l+1)\times(2l-1)$, while their matrix elements are explicitly given by

$$\begin{aligned}(\mathbf{a}_l)_{nm}&=\delta_{n-1m}a^-_{l\,-l+m}+\delta_{nm}a_{l\,-l+m-1}+\delta_{n+1m}a^+_{l\,-l+m-2},\\ (\mathbf{b}_l)_{nm}&=\delta_{nm}b^-_{l\,-l+m-1}+\delta_{n+1m}b_{l\,-l+m-2}+\delta_{n+2m}b^+_{l\,-l+m-3},\\ (\mathbf{d}_l)_{nm}&=\delta_{n-2m}d^-_{l\,-l+m+1}+\delta_{n-1m}d_{l\,-l+m}+\delta_{nm}d^+_{l\,-l+m-1},\end{aligned}\tag{2.302}$$

where $a^-_{nm} = -(a^+_{n-m})^*$, $b^-_{nm} = -(b^+_{n-m})^*$, $d^-_{nm} = -(d^+_{n-m})^*$ and explicitly

$$a_{nm} = -i\frac{m\xi\gamma_3'}{4\alpha},\ a^+_{nm} = -i\frac{\xi(\gamma_1' - i\gamma_2')}{8\alpha}\sqrt{(n+m+1)(n-m)},$$
$$b_{nm} = -\frac{\xi\gamma_3' n}{4}\sqrt{\frac{(n+1)^2 - m^2}{(2n+1)(2n+3)}},\ d_{nm} = \frac{\xi\gamma_3'(n+1)}{4}\sqrt{\frac{n^2 - m^2}{4n^2 - 1}},$$
$$b^+_{nm} = \frac{\xi(\gamma_1' - i\gamma_2')n}{8}\sqrt{\frac{(n+m+1)(n+m+2)}{(2n+1)(2n+3)}},$$
$$d^+_{nm} = \frac{\xi(\gamma_1' - i\gamma_2')(n+1)}{8}\sqrt{\frac{(n-m)(n-m-1)}{(2n+1)(2n-1)}}. \tag{2.303}$$

The Fourier amplitudes c^k_{lm} are determined from the general Eq. (2.296), so that we now have the magnetization expressed as the following Fourier series:

$$M_H(t) = M_S \sum_{k=-\infty}^{\infty} m^k_1(\omega) e^{ik\omega t}, \tag{2.304}$$

where the Fourier coefficients $m^k_1(\omega)$ are rendered via c^k_{lm} as

$$m^k_1(\omega) = \sqrt{2\pi/3}\left[\sqrt{2}\gamma_3' c^k_{10} + (\gamma_1' + i\gamma_2')c^k_{1-1} - (\gamma_1' - i\gamma_2')c^k_{11}\right]. \tag{2.305}$$

We also have $M_H(t)$ for an assembly of *randomly* oriented non-interacting uniaxial nanomagnets (particles) via

$$\overline{M_H}(t) = M_S \sum_{k=-\infty}^{\infty} \overline{m^k_1(\omega)} e^{ik\omega t}, \tag{2.306}$$

where the overbar denotes averaging over the instantaneous orientations of the particle easy axis. This averaging procedure can be accomplished numerically. To implement it we shall assume, from now on, that the field vectors $\mathbf{H}_0$ and $\mathbf{H}$ are parallel. Without loss of generality, we also suppose as before that both $\mathbf{H}_0$ and $\mathbf{H}$ lie in the *XZ* plane of the laboratory coordinate system, so that the direction cosines are simply $\gamma_1 = \gamma_1' = \sin\psi$, $\gamma_2 = \gamma_2' = 0$, and $\gamma_3 = \gamma_3' = \cos\psi$, where ψ is the angle between $\mathbf{H}_0$ and the *Z*-axis taken as the easy axis of the particle. The last assumption is justified, since for $\mathbf{H}_0 \parallel \mathbf{H}$, both the magnetization $M_H(t)$ and its associated Fourier amplitudes $m_1^k(\omega)$ are *independent* of the azimuthal angle due to the circular symmetry about the *Z*-axis. Since the coefficients $m_1^k(\omega)$ now only depend on ψ, in the calculation of the averages for an assembly of particles with *randomly* oriented easy axes using Gaussian quadratures, we simply require [155] (cf. Eq. (2.100))

$$\overline{m_1^k(\omega)} = \int m_1^k(\omega,\psi)\sin\psi d\psi = \frac{\pi}{4}\sum_{i=1}^{n} w_i m_1^k(\omega,\psi_i)\sin\psi_i,$$

where we have used $m_1^1(\omega,\psi) = m_1^1(\omega,\pi-\psi)$ and w_i and ψ_i are given by Eq. (2.101).

As far as the behavior of the Fourier coefficients $m_1^k(\omega)$ as functions of the ac field amplitude ξ is concerned, strong nonlinear effects are expected to occur for $\xi > 1$. For cobalt nanoparticles with mean diameter $a \sim 10$ nm and saturation magnetization $M_S \sim 1.4\,10^6$ A/m, ξ is of order unity for $H \sim 450$ A/m$(\sim 5$ Oe$)$ at T~30 K. Both ac and dc fields of this order are easily attained in nonlinear response measurements of magnetic nanoparticles [27,28]. To illustrate them, we first study the *time-independent* or dc component of the normalized dc magnetization $M_\xi(\omega)$ defined by the mean value over a cycle

$$M_\xi(\omega) = \frac{\omega}{2\pi M_S}\int_0^{2\pi/\omega} M_H(t)\,dt = m_1^0(\omega), \tag{2.307}$$

which by inspection is entirely real and *frequency-dependent*. In other words when combined with a strong ac driving field, the dc component of the magnetization of uniaxial particles $M_\xi(\omega)$ *alters markedly from the linear response; it becomes frequency-dependent and superposition no longer applies*. The subsequent analysis may however be simplified as follows. First observe that in the limits of vanishing ac field, $\xi \to 0$, and/or very high frequency field, $\omega \to \infty$, Eq. (2.307) obviously yields *two limiting* values for the normalized dc magnetization, namely

$$\lim_{\xi\to 0} M_\xi(\omega) = M_0 \text{ and } \lim_{\omega\to\infty} M_\xi(\omega) = M_\xi(\infty), \tag{2.308}$$

respectively. However, both of these fundamental limits are equal, i.e., $M_\xi(\infty) = M_0 = M_0(\sigma,\xi_0,\psi)$, and so each can be simply expressed in terms of the stationary average $\langle\cos\Theta\rangle_0$ as the integral

$$\langle\cos\Theta\rangle_0 = \frac{1}{Z}\int_0^{2\pi}\int_0^{\pi}\cos\Theta e^{\sigma\cos^2\vartheta+\xi_0\cos\Theta}\sin\vartheta d\vartheta d\varphi, \qquad (2.309)$$

where Z is the partition function. Moreover, in the *very low-frequency* limit, $\omega\to 0$, the mean value $M_\xi(0)$ can also be evaluated by using a quasistatic Boltzmann distribution in Eq. (2.309) so that

$$M_\xi(0) = \lim_{\omega\to 0}\frac{\omega}{2\pi}\int_0^{2\pi/\omega} M_0(\sigma,\xi_0+\xi\cos\omega t,\psi)dt. \qquad (2.310)$$

Therefore for $\xi \ll 1$, $M_\xi(0)$ can also be evaluated from the stationary averages as [161]

$$M_\xi(0) - M_\xi(\infty)$$
$$= \frac{\xi^2}{4}\left(\left\langle\cos^3\Theta\right\rangle_0 - 3\left\langle\cos^2\Theta\right\rangle_0\left\langle\cos\Theta\right\rangle_0 + 2\left\langle\cos\Theta\right\rangle_0^3\right) + o\left(\xi^2\right). \qquad (2.311)$$

For purposes of exposition, we first consider circular symmetry, i.e., with both the dc and ac fields applied along the easy axis, we then have $\psi = 0$ and $\Theta = \vartheta$, so that $\left\langle\cos\Theta\right\rangle_0 = \langle\cos\vartheta\rangle_0$ and so Eq. (2.309) takes on the closed form, Eq. (A.48) of Appendix A.1.5.

To facilitate our discussion, plots of both $M_\xi(\omega)$ and the real part of the *fundamental* component of the nonlinear dynamic susceptibility, viz. $2m_1^1/\xi$, as a function of the normalized frequency $\omega\tau_N$ for various anisotropy (or inverse temperature) parameters $\sigma = vK/(kT)$ are shown in Fig. 2.38 for an assembly of *aligned* nanoparticles with the angle $\psi = 0$ (so that circular symmetry prevails). By inspection two *distinc*t low- and high-frequency dispersion regions appear both in the spectrum of $M_\xi(\omega)$ and in the susceptibility $2m_1^1/\xi$ (in the latter the second dispersion region would be visible on the logarithmic scale). Moreover, the low-frequency dispersion of each of the two functions $M_\xi(\omega)$ and $2m_1^1/\xi$ is clearly governed by barrier crossing relaxation modes with the *same* characteristic frequency *indicating that the magnetization reversal time may be directly determined from measurements of the dc response* $M_\xi(\omega)$. In addition, for weak ac fields the characteristic frequency ω_1 of this low-frequency peak may be determined *explicitly* as $\omega_1 = \lambda_1$, i.e., λ_1 of the Fokker–Planck operator associated with the overbarrier relaxation processes. We recall

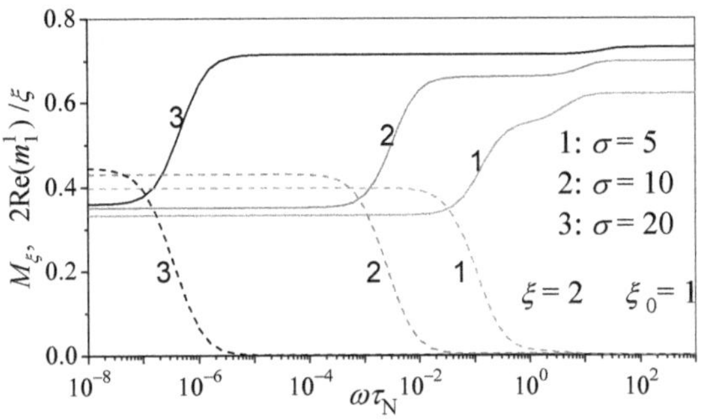

Fig. 2.38. Comparison of the dc magnetization M_ξ (solid lines) and real part of the fundamental component of the nonlinear dynamic susceptibility $2\mathrm{Re}(m_1^1)/\xi$ (dashed lines) vs. normalized frequency $\omega\tau_N$ for $\xi_0 = 1$, $\xi = 2$, $\psi = 0$ (circular symmetry) and various anisotropy parameters $\sigma = vK/(kT)$

that subject to the conditions $\xi << 1$ and $\sigma(1-h)^2 >> 1$, λ_1 for $\psi = 0$ (corresponding to circular symmetry) is given by Brown's asymptotic formula Eq. (2.4). Now at the opposite end of the spectrum, the high-frequency peak is entirely due to "intrawell" relaxation modes. These *individual near-degenerate* high-frequency modes are, however, virtually indistinguishable in the frequency spectrum of $M_\xi(\omega)$ appearing merely as a single high-frequency dispersion again as in the dynamic susceptibility. Thus, the multi-mode spectrum $M_\xi(\omega)$ may in practice merely be approximated [161] by a sum of *two* Lorentzians, viz.,

$$\frac{M_\xi(\omega) - M_\xi(\infty)}{M_\xi(0) - M_\xi(\infty)} \approx \frac{\Delta}{1+(\omega/\omega_1)^2} + \frac{1-\Delta}{1+(\omega/\omega_2)^2}. \tag{2.312}$$

In Eq. (2.312) ω_1 and ω_2 are the characteristic frequencies of the overbarrier relaxation and high-frequency modes, respectively. Both frequencies are closely related to the reversal time and the near-degenerate high-frequency "intrawell" relaxation modes (with an ac external field), while Δ is an adjustable amplitude parameter. The normalized dc magnetization $M_\xi(\omega)$ both as a function of the frequency $\omega\tau_N$ and also as a function of the dc field ξ_0 for various ac driving field amplitudes ξ is shown in Fig. 2.39. As shown in Fig. 2.39(a) and (b), the approximate two-mode Eq. (2.312) accurately fits the exact matrix continued fraction

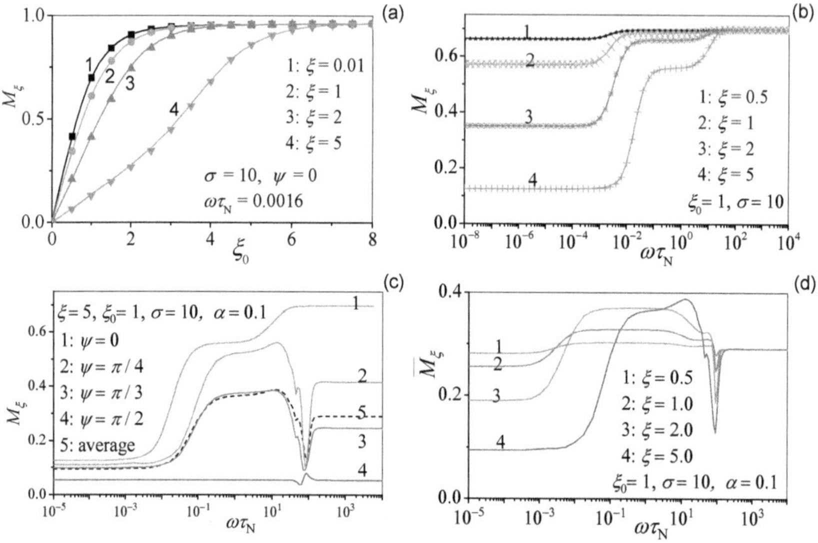

Fig. 2.39. (a) Normalized dc magnetization M_ξ vs. dc field parameter ξ_0 for $\omega\tau_N = 0{,}0016$, $\sigma = 10$, and various values of the ac field ξ displaying monotonic increase to the asymptotic value ($\xi = 0.01$ represents linear response conditions where the ac and dc field responses do not entangle). (b) Dc magnetization M_ξ vs. normalized frequency $\omega\tau_N$ for $\sigma = 10$, $\psi = 0$, $\xi_0 = 1$, and various ac field amplitudes ξ showing pronounced frequency dependence including two distinct dispersion regions caused by the nonlinear combination of the dc and ac responses (the superposition principle no longer applies). Solid lines: matrix continued fraction solution. Symbols: the approximate Eq. (2.312). (c) Dc magnetization M_ξ vs. $\omega\tau_N$ for $\sigma = 10$, $\xi_0 = 1$, $\xi = 5$, and various oblique angles ψ showing pronounced frequency dependence now comprising three distinct dispersion regions. Solid lines (1–4): matrix continued fraction solution. Dashed line 5: the average dc magnetization $\overline{M}_\xi$. Notice the much weaker effect for a purely transverse applied field, $\psi = \pi/2$. (d) Average dc magnetization $\overline{M}_\xi$ vs. $\omega\tau_N$ for $\sigma = 10$, $\alpha = 0.1$, $\xi_0 = 1$, and various ac field amplitudes ξ. Solid lines (1–4): matrix continued fraction solution. Notice the *parametric resonance behavior* in the high-frequency dispersion (a subharmonic weak resonance dispersion at frequencies ~ $\omega_{pr}/2$).

results. In Fig. 2.39, the limiting values of $M_\xi(\omega)$ as $\omega\to\infty$ and as $\omega\to 0$ are clearly given by Eq. (2.309) and Eq. (2.310), respectively.

The results in Fig. 2.39(a) and (b) pertain to both the *dc bias and ac fields applied parallel to the easy axis* of the uniaxial nanoparticle. Thus, the inherent coupling between the longitudinal relaxational and transverse precessional modes implied by the magnetic Langevin equation (1.42) is suppressed due to the circular symmetry. Hence the important precessional and longitudinal mode coupling effects ocurring in noncircularly symmetric potentials cannot be detected if one is so restricted. If, in contrast, the external fields are applied at an oblique angle ψ to the easy axis so breaking the circular symmetry, pronounced precessional and longitudinal mode coupling effects will also occur in the frequency-dependent dc magnetization. These nonlinear frequency-dependent effects due to the loss of circular symmetry essentially comprise a new high-frequency dispersion of resonant character near the FMR angular frequency $\omega_{\rm pr}$ which also exhibits *parametric resonance* appearing at frequencies $\sim \omega_{\rm pr}/2$ along with ψ angular dependence of the dc magnetization curves (see Fig. 2.39(c) and (d)). Both effects again arise from the coupling of the slow thermally activated magnetization reversal mode with the fast precessional modes via the driving ac field acting in conjunction with the dc bias field. Notice that the nonlinear behavior of the frequency-dependent dc component of the magnetization of nanomagnets driven by an ac external field closely resembles that of the frequency-dependent dc component of the dynamic Kerr effect in both polar liquids and liquid crystals insofar as it has a *frequency-dependent dc response*.

Consider now the fundamental component of the nonlinear ac stationary response. For a weak ac field, $\xi \to 0$, $\chi(\omega) = 6m_1^1(\omega)/\xi$ merely defines the normalized linear dynamic susceptibility, concurring with the benchmark linear response solution given in Section 2.5.1, so providing a check on the two independent algorithms used. However, in strong ac fields, $\xi > 1$, pronounced nonlinear effects also occur in the susceptibility (see Fig. 2.40). The low-frequency peak of $-\mathrm{Im}\,[\chi(\omega)]$ *can no longer be approximated by a single Lorentzian.* Furthermore, its characteristic frequency ω_1 (i.e., the angular frequency at which $-\mathrm{Im}\,[\chi(\omega)]$ attains its maximum) as a function of the ac field amplitude strongly depends on whether or not a dc field is applied. For a strong dc bias, $\xi_0 > 1$, the low-frequency peak shifts to lower frequencies; attaining a maximum for $\xi \sim \xi_0$ and thereafter decreases exponentially with increasing ξ, meaning that as the dc field increases, the magnetization reversal time *initially increases* and thereafter, having first attained its maximum at some critical value $\xi \sim \xi_0$, decreases exponentially. On the other hand, for *weak* dc bias,

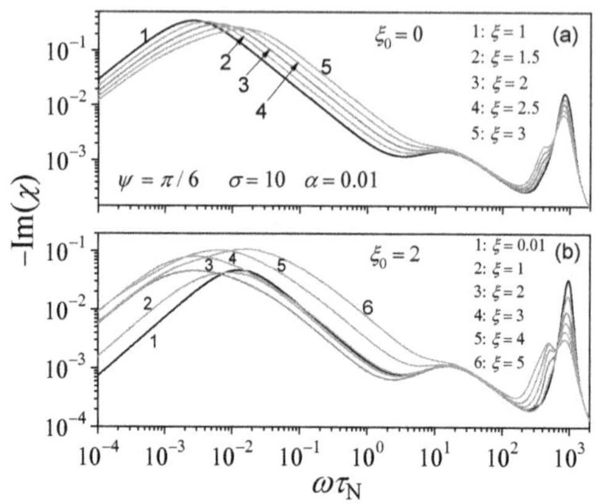

Fig. 2.40. $-\mathrm{Im}[\chi(\omega)]$ vs. $\omega\tau_N$ for various ac and dc field parameters.

$0 \le \xi_0 < 1$, the low-frequency peak shifts monotonically to higher frequencies. Moreover, as seen in Fig. 2.40, the *FMR peak with frequencies close to the precession frequency* ω_{pr} *of the magnetization strongly manifests itself at high frequencies.* As the ac field amplitude increases, this peak decreases and also broadens, showing pronounced nonlinear saturation effects. Besides, a *second weak subharmonic resonance peak*, owing to the *parametric resonance* of the nonlinear oscillatory (precessional) motion of the magnetization $\mathbf{M}(t)$, appears at frequencies $\sim \omega_{pr}/2$. The appearance of *subharmonic resonance* is similar to that already associated with nonlinear oscillators driven by an ac external force [380], e.g., in driven Josephson junctions. The appearance of the FMR peak also indicates that dynamic magnetic hysteresis (DMH) occurs at high frequencies. Thus, the DMH of such particles, arising from a high-frequency periodic signal, may be evaluated together with the damping, angular, and field rate of change dependencies of the coercivity thereby permitting *quantitative analysis of the ultrafast switching of the magnetization.*

The above results concern an *individual particle*, or an assembly of *fully aligned noninteracting particles*. In Fig. 2.41, $\mathrm{Re}[\chi(\omega)]$, $-\mathrm{Im}[\chi(\omega)]$ vs. $\omega\tau_N$ of an assembly of *randomly oriented non-interacting* uniaxial particles are compared with $\mathrm{Re}[\chi(\omega)]$ and $-\mathrm{Im}[\chi(\omega)]$ of an individual particle for various oblique angles ψ. For $\psi = 0$, the FMR peak entirely

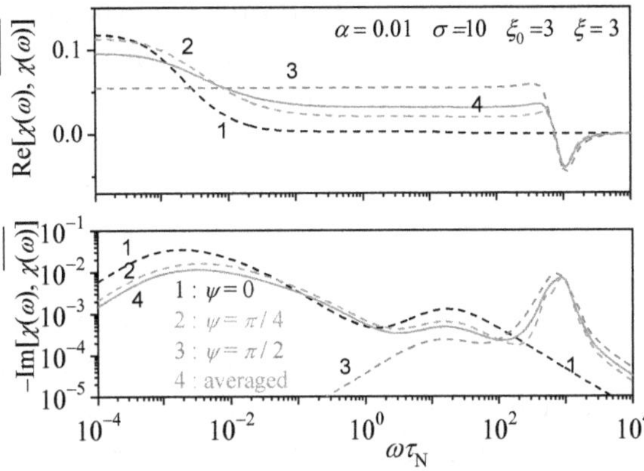

Fig. 2.41. $\mathrm{Re}[\overline{\chi(\omega)}], -\mathrm{Im}[\overline{\chi(\omega)}]$ (solid lines 4) and $\mathrm{Re}[\chi(\omega)], -\mathrm{Im}[\chi(\omega)]$ for $\psi = 0$ (curve 1), $\psi = \pi/4$ (2), and $\psi = \pi/2$ (3)] vs. $\omega\tau_N$ for $\sigma = 10$, $\alpha = 0.01$, $\xi_0 = 3$, and $\xi = 3$. Solid and dashed lines: matrix continued fraction solution.

disappears because the transverse modes no longer take part in the relaxation process. Hence as in the linear response, the commonly used assumption that the strong external fields are directed along the easy axis of the particle (so that *circular symmetry* is preserved) yields an altogether very restricted picture of the nonlinear response of an assembly of uniaxial nanoparticles (e.g., compare curves 1 and 4 in Fig. 2.41).

In summary, we have treated the time-independent but frequency-dependent magnetization for an ensemble of *fully aligned* noninteracting particles, and an ensemble of particles with *randomly oriented easy axes*, driven by strong dc and ac fields using a *nonperturbative* approach. The nonlinear susceptibility displays typical phenomena associated with nonlinear oscillators, etc. In all cases the nonlinear response (with a strong ac field applied in conjunction with a strong dc bias field at an angle to the easy axis of the particle so that the circular symmetry is broken) is very sensitive to both ac field orientation, damping, and frequency. Moreover, *the intrinsic damping dependence* of the ac nonlinear response for the oblique-field configuration serves as a *signature of the coupling between the longitudinal and precessional modes of the magnetization.* Hence, it should be possible to determine the damping coefficient α from measurements of nonlinear response characteristics [37]. This might be accomplished by fitting the theory to the experimental dependence of $\chi(\omega)$

on the angle ψ and the ac and dc bias field strengths. Then the sole fitting parameter is α, which can be determined at different temperatures T, yielding its temperature dependence. All the calculations, since they are valid for ac fields of *arbitrary* strengths and orientations, allow one to both predict and moreover interpret quantitatively, nonlinear phenomena in nanomagnets, where perturbation theory and the assumption that circular symmetry is preserved are no longer valid. Our results suggest that the existing experimental measurements of nonlinear susceptibilities of fine particles, e.g., Refs. [27,28], should be repeated for the frequency-dependent dc component of the magnetization in a strong bias-field configuration because *sui generis* it shows pronounced nonlinear effects overlooked in previous measurements. Similar nonlinear effects can be measured in the dc polarization of polar liquids in superimposed external dc bias and ac electric fields because from a physical point of view, the stochastic magnetization dynamics of single-domain ferromagnetic particles (magnetic dipoles) in magnetic fields is analogous to the rotational Brownian motion of polar molecules (electric dipoles) in electric fields [97]. For simplicity, only uniaxial particles have been treated. Particles with other anisotropies, e.g., cubic, biaxial, etc., can be considered in like manner [160,162]. For practical applications (e.g., in hyperthermia), to include the *polydispersity* of the particles of an actual sample, one must average the nonlinear response functions over an appropriate distribution function of particle volumes (which may also be accomplished numerically using Gaussian quadratures).

2.6.2 *Biaxial nanomagnets*

Biaxial anisotropy generates *azimuthally nonuniform energy distributions with two saddle points* leading to new effects, viz., strong *intrinsic* dependence of the magnetic characteristics on the damping via coupling of the longitudinal and transverse relaxation modes [160] (Fig. 1.3(b)) which cannot be ignored because of the *inherent* noncircular symmetry. The biaxial anisotropy is also essential in constructing a generic model of thermally assisted magnetization reversal subjected to spin-transfer torque (see Section 1.3.3). In contrast to uniaxial particles, the nonlinear ac stationary response strongly depends on the azimuthal direction of the ac field which we now analyze.

First the free energy density E of a single-domain ferromagnetic nanoparticle with biaxial anisotropy in superimposed homogeneous external magnetic dc and ac fields $\mathbf{H}(t) = \mathbf{H}_0 + \mathbf{H}\cos\omega t$ of arbitrary strengths and orientations is

$$E(\vartheta,\varphi,t) = -\sigma\cos^2\vartheta + \Delta\sin^2\vartheta\cos^2\varphi - \left(\xi_0 + \xi\cos\omega t\right)\Theta(\vartheta,\varphi), \quad (2.313)$$

where Δ and σ are the dimensionless anisotropy parameters, ξ_0 and ξ are again the dc and ac external field parameters, respectively, $\Theta(\vartheta, \varphi)$ is given by Eq. (2.299), where $\gamma_1 = \sin\psi\cos\phi$, $\gamma_2 = \sin\psi\sin\phi$, $\gamma_3 = \cos\psi$ are the direction cosines of the ac external field, ψ is the angle between $\mathbf{H}$ and the Z axis assumed as the easy axis of the particle, and ϕ is the azimuthal angle (we again assume that $\mathbf{H}_0$ and $\mathbf{H}$ are parallel).

For zero ac field $\mathbf{H} = 0$, the free energy density is expressed in terms of the Y_{lm} as [90]

$$\begin{aligned} E(\vartheta,\varphi) = {} & \sqrt{\frac{2\pi}{15}}\Delta\left[Y_{22}(\vartheta,\varphi) + Y_{2-2}(\vartheta,\varphi)\right] \\ & -\frac{4}{3}\sqrt{\frac{\pi}{5}}\left(\sigma + \frac{\Delta}{2}\right)Y_{20}(\vartheta,\varphi) - 4\sigma h\sqrt{\frac{\pi}{3}}Y_{10}(\vartheta,\varphi) + \text{const.} \end{aligned} \quad (2.314)$$

Next, we write the differential-recurrence relations for the statistical moments (expectation values of the spherical harmonics) $c_{nm}(t) = \langle Y_{nm}\rangle(t)$ governing the dynamics of the magnetization of single-domain particles with orthorhombic anisotropy with the field $\mathbf{H}_0$ applied along the easy axis of the magnetization:

$$\begin{aligned} \tau_N \frac{d}{dt}c_{nm}(t) = {} & v_{nm}c_{n-2m}(t) + w_{nm}c_{n-1m}(t) + x_{nm}c_{nm}(t) + y_{nm}c_{n+1m}(t) + z_{nm}c_{n+2m}(t) \\ & + v_{nm}^{++}c_{n-2m+2}(t) + w_{nm}^{++}c_{n-1m+2}(t) + x_{nm}^{++}c_{nm+2}(t) + y_{nm}^{++}c_{n+1m+2}(t) + z_{nm}^{++}c_{n+2m+2}(t) \\ & + v_{nm}^{--}c_{n-2m-2}(t) + w_{nm}^{--}c_{n-1m-2}(t) + x_{nm}^{--}c_{nm-2}(t) + y_{nm}^{--}c_{n+1m-2}(t) + z_{nm}^{--}c_{n+2m-2}(t) \\ & + x_{nm}^{+}c_{nm+1}(t) + x_{nm}^{-}c_{nm-1}(t) + y_{nm}^{+}c_{n+1m+1}(t) + y_{nm}^{-}c_{n+1m-1}(t) \\ & + w_{nm}^{+}c_{n-1m+1}(t) + w_{nm}^{-}c_{n-1m-1}(t), \end{aligned} \quad (2.315)$$

where $n \geq 1$, $-n \leq m \leq n$ and the coefficients $v_{nm}, v^{\pm}_{nm}$, etc. are explicitly defined as [151,152]

$$w_{nm} = \left[\frac{\gamma_3 \xi_0 (n+1)}{2} + i\left(\sigma + \frac{\Delta}{2}\right)\frac{m}{\alpha}\right]\sqrt{\frac{n^2 - m^2}{4n^2 - 1}},$$

$$w^{-}_{nm} = -\left(w^{+}_{n-m}\right)^{*} = -(\gamma_1 + i\gamma_2)\xi_0 \frac{n+1}{4}\sqrt{\frac{(n+m-1)(n+m)}{4n^2 - 1}},$$

$$w^{--}_{nm} = -w^{++}_{n-m} = -i\frac{\Delta}{4\alpha}\sqrt{\frac{(n+m-2)(n+m)[n^2 - (m-1)^2]}{4n^2 - 1}},$$

$$v_{nm} = \left(\sigma + \frac{\Delta}{2}\right)\frac{n+1}{2n-1}\sqrt{\frac{[(n-1)^2 - m^2](n^2 - m^2)}{(2n+1)(2n-3)}},$$

$$v^{--}_{nm} = v^{++}_{n-m} = -\frac{\Delta}{4}\frac{n+1}{2n-1}\sqrt{\frac{(n+m-3)(n+m-2)(-1+n+m)(n+m)}{(2n+1)(2n-3)}},$$

$$x_{nm} = -\frac{n(n+1)}{2} + \left(\sigma + \frac{\Delta}{2}\right)\frac{n(n+1) - 3m^2}{(2n-1)(2n+3)} + i\frac{\xi_0 m \gamma_3}{2\alpha},$$

$$x^{-}_{nm} = -\left(x^{+}_{n-m}\right)^{*} = -i\frac{(\gamma_1 + i\gamma_2)\xi_0}{4\alpha}\sqrt{(1+n-m)(n+m)},$$

$$x^{--}_{nm} = x^{++}_{n-m} = \frac{3\Delta}{4}\sqrt{\frac{[(n+1)^2 - (m-1)^2][n^2 - (m-1)^2]}{(2n-1)(2n+3)}},$$

$$y^{--}_{nm} = -y^{++}_{n-m} = i\frac{\Delta}{4\alpha}\sqrt{\frac{(n+1-m)(n+3-m)[(n+1)^2 - (m-1)^2]}{(2n+1)(2n+3)}},$$

$$y_{nm} = -\left[\gamma_3 \xi_0 \frac{n}{2} - i\left(\sigma + \frac{\Delta}{2}\right)\frac{m}{\alpha}\right]\sqrt{\frac{(n+1)^2 - m^2}{(2n+1)(2n+3)}},$$

$$y^{-}_{n,m} = -\left(y^{+}_{n,-m}\right)^{*} = -\frac{n}{4}\xi_0(\gamma_1 + i\gamma_2)\sqrt{\frac{(n+1-m)(n+2-m)}{(2n+1)(2n+3)}},$$

$$z_{nm} = -\left(\sigma + \frac{\Delta}{2}\right)\frac{n}{2n+3}\sqrt{\frac{[(n+1)^2 - m^2][(n+2)^2 - m^2]}{(2n+1)(2n+5)}},$$

$$z^{--}_{nm} = z^{++}_{n-m} = \frac{\Delta}{4}\frac{n}{2n+3}\sqrt{\frac{(n-m+4)(n-m+3)(n-m+2)(n-m+1)}{(2n+1)(2n+5)}}.$$

The stationary ac response is again calculated from the formally exact matrix continued fraction solution, Eqs. (2.294)–(2.297) with the five submatrices $\mathbf{V}_l$, $\mathbf{W}_l$, $\mathbf{X}_l$, $\mathbf{Y}_l$, and $\mathbf{Z}_l$ and the column vectors $\mathbf{p}_1^-$ and $\mathbf{q}_1^-$ now redefined as

$$\begin{aligned}
(\mathbf{V}_l)_{nm} &= \delta_{n-4,m} v_{l,-l+m+3}^- + \delta_{n-2,m} v_{l,-l+m+1}^- + \delta_{n,m} v_{l,-l+m-1}^+,\\
(\mathbf{W})_{nm} &= \delta_{n-3,m} w_{l,-l+m+2}^- + \delta_{n-1,m} w_{l,-l+m}^- + \delta_{n+1,m} w_{l,-l+m-2}^+,\\
(\mathbf{X})_{nm} &= \delta_{n-2,m} x_{l,-l+m+1}^- + \delta_{n,m} x_{l,-l+m-1}^- + \delta_{n+2,m} x_{l,-l+m-3}^+,\\
(\mathbf{Y})_{nm} &= \delta_{n-1,m} y_{l,-l+m}^- + \delta_{n+1,m} y_{l,-l+m-2}^- + \delta_{n+3,m} y_{l,-l+m-4}^+,\\
(\mathbf{Z})_{nm} &= \delta_{n,m} z_{l,-l+m-1}^- + \delta_{n+2,m} z_{l,-l+m-3}^- + \delta_{n+4,m} z_{l,-l+m-5}^+,
\end{aligned}$$

$$\mathbf{q}_1^- = \begin{pmatrix} -\sqrt{3/10}\Delta \\ 0 \\ (2\sigma+\Delta)/\sqrt{5} \\ 0 \\ -\sqrt{3/10}\Delta \\ (\gamma_1 - i\gamma_2)\xi_0/\sqrt{6} \\ \gamma_3\xi_0/\sqrt{3} \\ -(\gamma_1 + i\gamma_2)\xi_0/\sqrt{6} \end{pmatrix}, \quad \mathbf{p}_1^- = \frac{\xi}{2\sqrt{6}} \begin{pmatrix} 0 \\ 0 \\ 0 \\ 0 \\ 0 \\ (\gamma_1 - i\gamma_2) \\ \gamma_3\sqrt{2} \\ -(\gamma_1 + i\gamma_2) \end{pmatrix}. \tag{2.316}$$

The tridiagonal submatrices $\mathbf{a}_l$, $\mathbf{b}_l$, and $\mathbf{d}_l$ coincide with those for uniaxial particles and are given explicitly by Eq. (2.302). For $\Delta = 0$, the above solution reduces to that for uniaxial particles. The relevant amplitudes c_{10}^k and $c_{1\pm1}^k$ yield the nonlinear stationary response of the magnetization as given by Eqs. (2.304) and (2.305).

For a *weak* ac field, $\xi \to 0$, $\chi(\omega) = 6m_1^1(\omega)/\xi$ again represents the normalized linear dynamic susceptibility, the results concurring with the benchmark linear response solution [151,152], indicating that a marked dependence of $\chi(\omega)$ on σ (inverse temperature), Δ (biaxiality), ξ (ac field), ξ_0 (dc field), α (damping), and the angles ψ and ϕ exists. Moreover, three *distinct dispersion regions* once again appear in $\chi(\omega)$ (see curve 1 in Fig. 2.42(b)) and the low-frequency behavior is dominated by the barrier-crossing mode. In addition, a weaker second relaxation peak again appears at high frequencies. This relaxation peak is yet another manifestation of

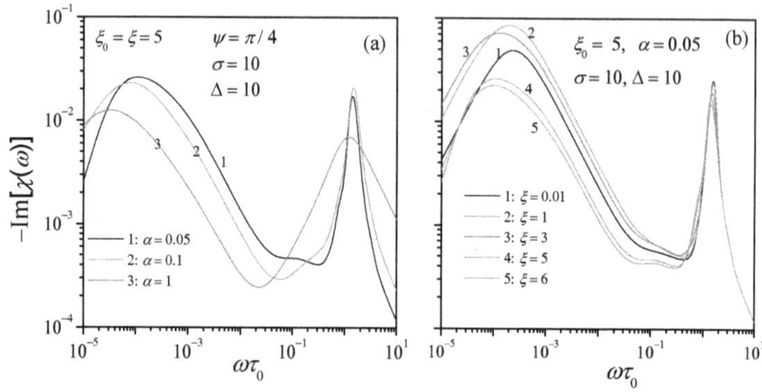

Fig. 2.42. $-\mathrm{Im}\,[\chi(\omega)]$ vs. the dimensionless frequency $\omega\tau_0$ $[\tau_0 = \tau_N(\alpha+\alpha^{-1})^{-1}]$ for barrier parameter $\sigma = 10$, biaxiality parameter $\Delta = 10$, dc field parameter $\xi_0 = 5$, angles $\psi = \pi/4$ and $\phi = 0$, and several values of damping α (a) and ac field amplitude ξ (b).

the near-degenerate "intrawell" modes, which are indistinguishable in the frequency spectrum appearing as a single high-frequency peak and for small dc fields, the amplitude of this peak is far weaker than that of the first peak. However, in strong fields $\xi_0 >> 1$, the former may again dominate the spectrum because as ξ_0 increases, the magnitude of the low-frequency peak decreases. Furthermore, even without the dc field, an inherent geometric dependence of $\chi(\omega)$ on α arising from the *coupling of the longitudinal and transverse relaxation modes* exists. As before this *intrinsic* coupling appears in the dynamical equations of motion, whereby the longitudinal component of the magnetization is entangled with the transverse components. Thus, a third FMR peak in $\chi(\omega)$ appears again due to excitation of transverse modes with characteristic frequencies near the precession frequency of the magnetization which exists only for low damping ($\alpha << 1$) and strongly manifests itself at high frequencies. As α decreases, the half-width of the FMR peak decreases.

In *strong ac fields*, $\xi > 1$, pronounced nonlinear effects again occur (see Figs. 2.42 and 2.43 illustrating the dependence of the nonlinear response on the biaxiality Δ, damping α, angle ψ, dc field ξ_0 and the ac field ξ). The low-frequency peak can no longer be approximated by a *single* Lorentzian indicating that many other modes now contribute to the magnetization reversal process. However, the characteristic frequency ω_{max} of the maximum magnetic loss still yields an effective reversal time

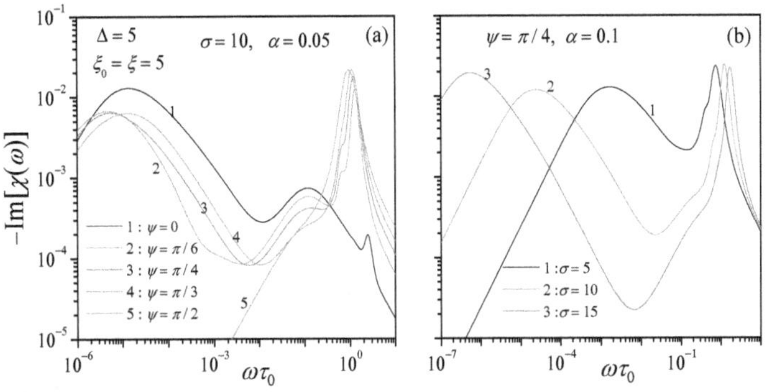

Fig. 2.43. $-\mathrm{Im}[\chi(\omega)]$ vs. $\omega\tau_0$ for the biaxiality parameter $\Delta = 5$, dc field parameter $\xi_0 = 5$, ac field parameter $\xi = 5$, angle $\phi = 0$, and various field angles ψ (a) and anisotropy parameters σ (b).

τ, i.e., $\tau \approx \omega_{\max}^{-1}$. The behavior of $\omega_{\max}$ (and, hence, τ) as a function of the ac field amplitude again depends on whether or not a dc field is applied. For strong dc, $\xi_0 > 1$, the low-frequency peak shifts to lower frequencies attaining a maximum at $\xi \sim \xi_0$ thereafter $\omega_{\max}$ decreases with increasing ξ. In tandem with this the reversal time initially increases and on attaining its maximum at some critical $\xi \sim \xi_0$ then decreases (see Fig. 2.44). In contrast for weak dc bias, $\xi_0 << 1$, the low-frequency peak once again shifts monotonically to higher frequencies. Also (v. Fig. 2.42) with increasing ξ, the magnitude of the FMR peak at the precession frequency ω_{pr} decreases and broadens showing pronounced nonlinear saturation effects as are characteristic of a soft spring. Moreover, for low damping a second weak (subharmonic) resonance peak [owing to *phase locking* of the nonlinear oscillatory (precessional) motion of $\mathbf{M}(t)$] appears at frequencies $\sim \omega_{\mathrm{pr}}/2$ (Fig. 2.43). This *subharmonic generation* is again similar to that existing in nonlinear parametric oscillator systems driven by an ac external force [380]. Nevertheless, the high-frequency behavior ($\omega_{\mathrm{pr}} >> \omega$) remains virtually unchanged. Indeed $-\mathrm{Im}[\chi(\omega)]$ for $\omega_{\mathrm{pr}} >> \omega$ is independent of ξ (see Fig. 2.42). *Moreover, in contrast to uniaxial particles, the FMR peak invariably appears for low damping in the spectrum* $-\mathrm{Im}[\chi(\omega)]$ *at* $\psi = 0$ *because now the transverse modes always intrinsically take part in the longitudinal relaxation process.* In Fig. 2.42

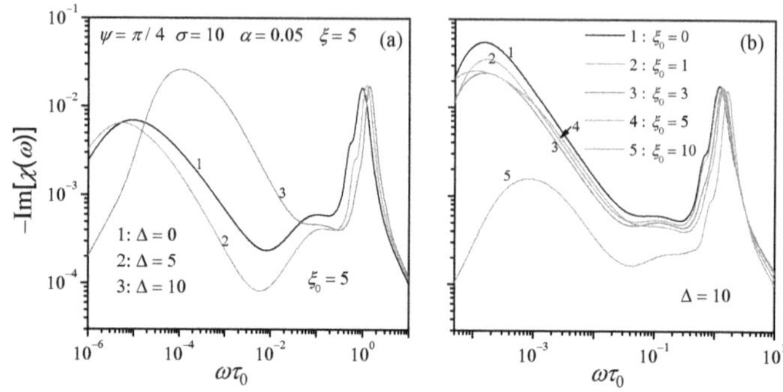

Fig. 2.44. $-\mathrm{Im}\,[\chi(\omega)]$ vs. $\omega\tau_0$ for the barrier parameter $\sigma = 10$, ac field parameter $\xi = 5$, damping $\alpha = 0.05$, angles $\psi = \pi/4$ and $\phi = 0$ and various biaxiality parameters Δ (a) and dc field amplitudes ξ_0 (b).

(b) the height of the first peak (for weak fields) increases with increasing ac amplitude and then reduces at higher ac fields, implying that the changes in the ac field magnitude may be used to *modulate the heat production* (specific power loss) in a magnetic nanoparticle working as a hyperthermia source. In general, however, the most effective way to *tune* the heat production is to vary the magnitude of the dc field [47,153] (see Fig. 2.44(b)). As ξ_0 is increased the magnitude of the low-frequency peak again *decreases due to the depletion of the population in the shallowest potential well.* Finally, the FMR peak is in itself a *signature of the onset of dynamic hysteresis* at very high frequencies, a feature which is entirely absent for high damping.

2.7 Dynamic magnetic hysteresis

2.7.1 *DMH in uniaxial nanomagnets*

The DMH induced in superparamagnetic particles by an external ac field is a useful phenomenon in practice, because it occurs in *magnetodynamic hyperthermia* occasioned by induction heating of nanoparticles [9,12,13, 93,94]. Superparamagnetism, since it plays a crucial role in the magnetization dynamics of nanoparticles, naturally emphasizes the physics of DMH because the *temperature directly influences the remagnetization conditions*, causing changes of loop shape, coercive force and specific power loss.

The theory of DMH in single-domain magnetically *isotropic* particles subjected to thermal fluctuations having been initiated by Ignatchenko and Gekht [41], was later extended to uniaxial superparamagnetic particles with moderate-to-high internal barriers using both approximate and numerically exact methods [42,47–50,60,61,159,160,197,381,382]. Here we shall use Brown's model to study the effect of a dc bias magnetic field on the DMH of superparamagnetic particles subjected to an ac external field sufficient to induce remagnetization and encountered in at least two actual situations. The first concerns particles with *exchange bias* [383, 384], e.g., occurring in oxidized metal nanograins with a ferromagnetic (FM) core and an antiferromagnetic (AFM) oxide outer layer, e.g. Ni–NiO and Co–CoO [357]. Here the bias effect is due to *interfacial exchange coupling* between the FM and AFM phases, and explicitly manifests itself as a shift of the Stoner–Wohlfarth hysteretic curve along the field axis [383,384]. Thus, it is referred to as the *unidirectional exchange anisotropy* with the ensuing displacement of the hysteresis loop yielding the exchange bias field H_b. The exchange bias effect has been studied both in films and on crystal surfaces and is now widely used in spintronics to both pin and modify the switching conditions for the magnetization of FM layers in spin-valve sandwiches; e.g. see Ref. [357]. In FM-AFM core-shell particles, the exchange bias also strongly affects the quasistatic magnetization cycles as well as altering the ferromagnetic resonance conditions [385]. In nanoparticles exhibiting superparamagnetism, where the *ratio of barrier energy to thermal energy is very easily altered by varying the temperature*, the exchange bias field becomes still more important, crucially influencing the magnetization relaxation processes. Yet another use is in intentional biasing of the external ac field both in heat generation experiments and in hyperthermia techniques for controlling the specific power loss in systems of single-domain magnetic particles. As far as theoretical treatment is concerned, modelling the DMH in all the above situations leads to an identical set of equations. Here we follow closely the exposition of Ref. [47].

Therefore, to model DMH, we just consider a uniaxial single-domain ferromagnetic particle subjected to an ac field of amplitude H_1 and a dc field of amplitude H_0 so that the free energy E is simply

$$E = -\sigma(\mathbf{u}\cdot\mathbf{n})^2 - \xi_0(\mathbf{u}\cdot\mathbf{b}) - \xi\cos\omega t(\mathbf{u}\cdot\mathbf{q}). \tag{2.317}$$

Here $\mathbf{q}$ and $\mathbf{n}$ are unit vectors along the external ac field and easy axes, respectively; $\mathbf{u} = \mathbf{M}/M_S$, $\mathbf{b}$ is a unit vector in the dc field direction, $\xi_0 = v\mu_0 M_S H_0/(kT)$ and $\xi = v\mu_0 M_S H/(kT)$ are the dimensionless dc and ac field parameters, $\sigma = vK/(kT)$ is the bulk anisotropy parameter, and K is the bulk anisotropy constant. If *exchange anisotropy* is included, the factor $\mu_0 M_S H_0$ in Eq. (2.317) must be replaced by $K_u S/v$, where the parameter K_u is the surface energy density and S represents the surface of the particle. For simplicity, we suppose that $\mathbf{q}\,||\,\mathbf{n}\,||\,\mathbf{b}$, meaning circular symmetry so that Eq. (2.317) becomes

$$E = \sigma\sin^2\vartheta - (\xi_0 + \xi\cos\omega t)\cos\vartheta + \text{const.} \tag{2.318}$$

The calculation of the ac stationary response for arbitrary orientations of the vectors $\mathbf{q}$ and $\mathbf{b}$ has been given previously in Section 2.6.

Due to the circular symmetry, the azimuthal dependence of the observables may be ignored; hence, the magnetization dynamics are described by relatively simple differential-recurrence equations for the statistical moments (relaxation functions) $f_n(t) = \langle P_n(\cos\vartheta)\rangle(t)$, viz., (cf. Eq. (2.165))

$$\begin{aligned}\tau_N \dot{f}_n(t) &= d_n f_n(t) + c_n f_{n-2}(t) + e_n f_{n+2}(t)\\ &+ (\xi_0 + \xi_1\cos\omega t)a_n\left[f_{n-1}(t) - f_{n+1}(t)\right].\end{aligned} \tag{2.319}$$

Here the coefficients a_n, d_n, c_n, and e_n are defined in Section 2.4 by Eq. (2.167). Since we seek only the stationary ac response, which is independent of the initial conditions, we require only the steady-state solution of Eq. (2.319). Hence, we may seek $f_n(t)$ as a Fourier series in the time, viz.,

$$f_n(t) = \sum_{k=-\infty}^{\infty} F_k^n e^{ik\omega t} = F_0^n + 2\sum_{k=1}^{\infty}\operatorname{Re}\left\{F_k^n e^{ik\omega t}\right\}. \tag{2.320}$$

All the $f_n(t)$ must be real, therefore the Fourier amplitudes F_k^n must satisfy the condition $F_{-k}^n = (F_k^n)^*$, where the asterisk denotes the complex conjugate. By substituting Eq. (2.320) into Eq. (2.319), we then have algebraic recurrence equations for F_k^n, viz.,

$$\begin{aligned}&\xi_0 a_n (F_k^{n-1} - F_k^{n+1}) + \frac{\xi_1 a_n}{2}(F_{k-1}^{n-1} + F_{k+1}^{n-1} - F_{k-1}^{n+1} - F_{k+1}^{n+1}) \\ &\quad + (d_n - i\omega\tau_N k)F_k^n + c_n F_k^{n-2} + e_n F_k^{n+2} = 0.\end{aligned} \tag{2.321}$$

The *exact* solution of this equation can once again be posed as a matrix continued fraction because the *nine-term scalar* recurrence relation of Eq. (2.321) transforms into a *three-term matrix* recurrence equation. We have

$$\mathbf{Q}_1\mathbf{C}_1 + \mathbf{Q}_1^+\mathbf{C}_2 = \mathbf{R} \text{ and } \mathbf{Q}_n^-\mathbf{C}_{n-1} + \mathbf{Q}_n\mathbf{C}_n + \mathbf{Q}_n^+\mathbf{C}_{n+1} = 0, \tag{2.322}$$

where the column vectors $\mathbf{C}_n$ and $\mathbf{R}$ are defined as

$$\mathbf{C}_n = \begin{pmatrix} \mathbf{c}_{2n} \\ \mathbf{c}_{2n-1} \end{pmatrix}, \ \mathbf{R} = \begin{pmatrix} \mathbf{r}_2 \\ \mathbf{r}_1 \end{pmatrix}, \ \mathbf{c}_n = \begin{pmatrix} \vdots \\ F_{-2}^n \\ F_{-1}^n \\ F_0^n \\ F_1^n \\ F_2^n \\ \vdots \end{pmatrix}, \ \mathbf{r}_1 = \begin{pmatrix} \vdots \\ 0 \\ \xi/6 \\ \xi_0/3 \\ \xi/6 \\ 0 \\ \vdots \end{pmatrix}, \ \mathbf{r}_2 = \begin{pmatrix} \vdots \\ 0 \\ 0 \\ 2\sigma/5 \\ 0 \\ 0 \\ \vdots \end{pmatrix},$$

and the supermatrices $\mathbf{Q}_n^-$, $\mathbf{Q}_n$, and $\mathbf{Q}_n^+$ are [47]

$$\mathbf{Q}_n^- = \begin{pmatrix} c_{2n}\mathbf{I} & \mathbf{0} \\ a_{2n-1}\mathbf{Y} & c_{2n-1}\mathbf{I} \end{pmatrix}, \ \mathbf{Q}_n = \begin{pmatrix} \mathbf{Z}_{2n} & a_{2n}\mathbf{Y} \\ -a_{2n-1}\mathbf{Y} & \mathbf{Z}_{2n-1} \end{pmatrix}, \ \mathbf{Q}_n^+ = \begin{pmatrix} e_{2n}\mathbf{I} & -a_{2n}\mathbf{Y} \\ \mathbf{0} & e_{2n-1}\mathbf{I} \end{pmatrix}.$$

Here $\mathbf{I}$ is the identity matrix, the symbols $\mathbf{0}$ designate a zero matrix of infinite dimensions. The matrix elements of the infinite submatrices $\mathbf{Y}$ and $\mathbf{Z}_n$ are

$$(\mathbf{Y})_{rs} = \delta_{rs-1}\xi/2 + \delta_{rs}\xi_0 + \delta_{rs+1}\xi/2, \quad (\mathbf{Z}_n)_{rs} = \delta_{rs}(d_n - i\omega\tau_N s).$$

Equation (2.322) can again be solved for $\mathbf{C}_1$ using matrix continued fractions, viz.,

$$\mathbf{C}_1 = \mathbf{\Delta}_1 \cdot \mathbf{R}, \tag{2.323}$$

where the matrix continued fraction $\mathbf{\Delta}_1$ is defined by the recurrence relation, Eq. (2.297). Clearly, $\mathbf{C}_1$ contains all the Fourier amplitudes of $f_1(t)$, as needed for the nonlinear stationary response. Thus, we have the steady-state magnetization

$$M(t) = n_0 v M_S f_1(t) = M_S \sum_{k=-\infty}^{\infty} m_k^1(\omega) e^{ik\omega t}, \tag{2.324}$$

where $m_k^1(\omega) = n_0 v F_k^1(\omega)$ and n_0 is the number density of the assembly of (non-interacting) particles and ultimately yields the DMH loops, i.e., *parametric plots* of $M(t)$ vs. $H_0 + H\cos\omega t$, where t spans the periodic time of the ac field for a given amplitude H. Therefore, we automatically have the area of the DMH loop $A(\omega)$, i.e.

$$A(\omega) = \oint M(t) dH(t) = -\pi M_S H \,\mathrm{Im}\left(m_1^1\right). \tag{2.325}$$

This expression is simply the *energy loss per particle and per cycle of the ac field*. Moreover, *the power loss per unit volume* $P(\omega)$ (i.e., the specific power loss) is related to the DMH loop area via

$$P(\omega) = \omega A(\omega)/2\pi = 2\omega M_S H A_n(\omega)/\pi, \tag{2.326}$$

where $A_n(\omega) = -(\pi/4)\,\mathrm{Im}\left(m_1^1\right)$ is the loop area normalized with respect to its maximum value $4M_S H$.

Since we treat DMH in a dc bias field, we now introduce additional dimensionless temperature-independent field parameters h_1 and h_0, viz.,

$$h_1 = \xi/(2\sigma) = H/H_A, \quad h_0 = \xi_0/(2\sigma) = H_0/H_A, \tag{2.327}$$

where the bulk anisotropy field H_A is given by $H_A = 2K/\mu_0 M_S$. In the weak ac field regime ($h_1 \ll 1$), linear response theory again applies. Thus, the normalized ac susceptibility $\chi(\omega)/\chi = m_1^1(\omega)/m_1^1(0)$ is given by the simple analytic (two-mode approximation) Eq. (1.150). As seen the two Lorentzian in Eq. (1.150) implies that the overall relaxation process comprises two distinct processes, namely the *slow* thermally activated (overbarrier) process, and the *fast* relaxation in the wells. However, for low frequencies and high barriers between the wells, only the first term in Eq. (1.150) is required. Therefore, the approximation $\Delta \approx 1$ may also be used here. Thus, the steady-state magnetization is (alias the simple Debye formula)

$$M(t) = n_0 v M_S \left[\langle \cos\vartheta \rangle_0 + \chi'(0) H \frac{\cos\omega t + \omega\tau \sin\omega t}{1+\omega^2\tau^2} \right], \quad (2.328)$$

with $\tau = \lambda_1^{-1}$. Writing

$$x(t) = \frac{H(t)}{H} = \cos\omega t \text{ and } y(t) = \frac{1}{\chi'(0)H}\left[\frac{M(t)}{n_0 v M_S} - \langle \cos\vartheta \rangle_0\right] \quad (2.329)$$

and on eliminating the time from these two equations, we have the cartesian equation of an ellipse in the (x, y) plane, namely [47]

$$x^2 + (\omega\tau)^{-2}\{[1+(\omega\tau)^2]y - x\}^2 = 1. \quad (2.330)$$

This ellipse is centered at (0,0) with its major axis inclined to the x axis at an angle $\arctan[2(\omega\tau)^{-2}]/2$. For *large* ac field amplitudes ($h_1 \sim 1$), an analytical formula for $M(t)$ is not available. Nevertheless, the main features of DMH may still be qualitatively understood. Suppose we have the situation characterized by the two conditions: $\sigma >> 1$ (so that superparamagnetism is *weak*) and $\xi \sim \sigma$ (meaning that the ac *field amplitude H is comparable with the internal anisotropy field*). Then, the external ac field besides *saturating* the particle magnetic moment, may also induce its inversion (i.e., *switching between the directions of the easy axis*). Now under *quasistatic* conditions for *zero temperature and for zero bias* $H_0 = 0$, the magnetization simply exhibits the classic Stoner–Wohlfarth hysteresis cycle. However, for *finite* temperatures and frequencies, the magnetization relaxation undergoes pronounced transformations, due to the strong dependence of the magnetic relaxation time τ on both the anisotropy barrier σ and the applied fields ξ and ξ_0. First we discuss the *zero-temperature* behavior; then by imposing a bias field h_0 on a particle in the direction of $\mathbf{n}$ labeled as "up", with ac field amplitude h_1 greater than $1 + h_0$ (i.e., the Stoner–Wohlfarth coercive force plus the bias), it follows that at zero temperature the ac field $h_1(t) = h_1 \cos\omega t$ merely reverses the particle magnetic moment $\mu\mathbf{u} = vM_S\mathbf{u}$ twice per cycle. The result is a standard rectangular quasistatic hysteresis loop $M(h)$ merely shifted to the negative direction along the h axis by an amount h_0

(here $h = h_1 + h_0$). However, if the ac field amplitude is equal to the critical value $h_1 = 1$, the situation becomes *qualitatively different*, because the ac field is now only able to entrain the magnetic moment in the *positive half-cycle*. Once that happens, no further reversal of the magnetic moment can occur in the quasistatic regime, therefore, instead of a loop, we have only the straight-line segment $M = n_0 v M_S$ spanning the h axis over the interval $(-1, 1)$. For smaller h_1, the shape of the loop for an assembly of particles depends on the initial conditions, i.e., the relative numbers of magnetic moments initially directed "up" and "down". However, for any given particle it still remains a straight line (zero area) bounded by the interval (h_1, h_1).

Returning now to *finite* temperatures, where owing to thermal agitation, the particle magnetic moment (*i*) is never completely saturated, and (*ii*) becomes superparamagnetic, meaning that it may now (in contrast to zero temperature) flip between the "up" and "down" states for *sub-critical* fields ($h_1 < 1$). Thus τ for a given anisotropy σ then depends strongly on the *instantaneous* value of the applied field. However, the response $M(t)$ in the stationary state is independent of the initial conditions. Hence it simply represents *retarded phase* forced oscillations resulting in a DMH loop $M(h)$ with shape for given σ and h_0, depending only on the ac field amplitude h_1 and frequency ω. Figure 2.45 shows an example of loop shapes obtained for strong fields by using continued fractions. Notice that σ is a dimensionless inverse-temperature parameter while the ac and dc field amplitudes are expressed in terms of the reduced parameters h_1 and h_0, respectively, as defined in Eq. (2.327). The actual loops shown in Fig. 2.45 correspond to approximately half ($h_1 = 0.45$) and complete nucleation ($h_1 = 1$) Stoner–Wohlfarth coercive force; see the panels *a* and *b*, respectively. Next if we choose $\sigma = 25$ so as to ensure the strong-field limit condition and to simultaneously have τ of the order (of magnitude) of the period $2\pi/\omega$ at $|h| \sim 0.5$, then for time scales lying within the period of oscillation of the field if $|h| > 0.5$, the response of the magnetization is practically instantaneous. Thus, the process exemplified in Fig. 2.46 is exclusively controlled by the magnetization dynamics. Indeed, the effective coercive force is merely determined by the condition that the response time should be less than $2\pi/\omega$. Therefore, a particle

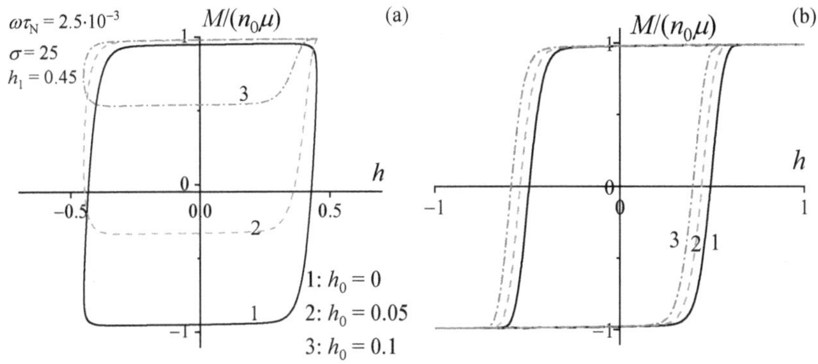

Fig. 2.45. DMH loops vs. the field parameter h for ac field amplitudes $h_1 = 0.45$ (a) and 1.0 (b) with anisotropy and frequency parameters $\sigma = 25$ and $\omega\tau_N = 2.5\cdot10^{-3}$ and $h_0 = 0$ (solid lines), 0.05 (dashed lines), and 0.1 (dot-dashed lines); (Reproduced from Ref. [47] by permission of the American Institute of Physics.)

assembly (characterized by the foregoing parameters) *may completely reverse its magnetization while subjected to a field of only about half that required for quasistatic switching*. Therefore, for *zero* bias, *doubling* of the ac field amplitude will not cause any significant changes in the loop area (see Fig. 2.46). However even a weak dc bias field will now compel the loop (for the ac parameters $h_1 = 0.45$ and $h_1 = 1$) to alter profoundly. See, e.g., Fig. 2.46(b), where alteration of the dc field h_0 does not influence the loop shape at all, merely shifting the loop further along the h axis. In contrast the amplitude h_1 is now sufficient to establish an adiabatic regime near each extremum of the ac field. This regime is characterized by the magnetization and field both oscillating in phase (see the horizontal lines going left- and rightward). In contrast (as Fig. 2.46(a) demonstrates) if h_1 lies outside the range where adiabatic response is possible, even a *slight* perturbation of the external conditions suffices to cause significant alterations in the loops.

To explain bias field effects on the loop area, recall that τ depends on the *instantaneous value* of *the applied field*, hence under the conditions of Fig. 2.46 (a), the *positive peak* of $h(t)$ corresponds to a *total applied field* $h_1 + h_0$, favoring relaxation to *positive values* of M. In contrast the *negative peak* of $h(t)$ amounts only to a total applied field $h_1 - h_0$ which cannot induce sufficiently fast (in comparison to the period $2\pi/\omega$) relaxation to

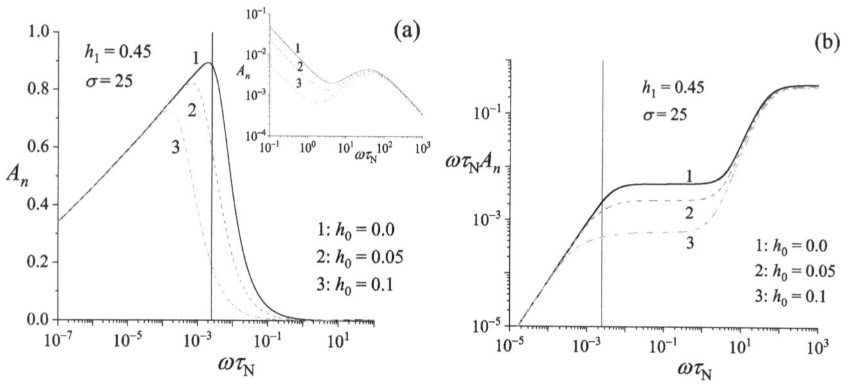

Fig. 2.46. Normalized area of the DMH loop A_n (a) and the product ωA_n (b) vs. $\omega\tau_N$ with varying dc bias field $h_0 = 0$ (1, solid line), 0.05 (2, dashed line), and 0.1 (3, dot-dashed line); anisotropy parameter $\sigma = 25$, ac field amplitude $h_1 = 0.45$; thin vertical lines correspond to the frequency $\omega\tau_N = 2.5{\cdot}10^{-3}$, their intersections with the curves render the areas of the loops shown in Fig. 2.48 (a). (Reproduced from Ref. [47] by permission of the American Institute of Physics.)

negative values of M, thus the "centroid" of the loop slides upwards. Another effect of the foregoing overall "asymmetrical" acceleration of the relaxation is the emergence (in the loop for a dc field with $h_0 = 1$) of a narrow rightward-directed ledge indicating that the adiabatic regime has almost been attained; for still higher bias h_0, the $M(h)$ curve gradually degenerates into a nearly horizontal line.

DC bias field effects on DMH are further illustrated in Figs. 2.47 and 2.48, where alterations in DMH due to varying h_0 are shown in detail as functions of frequency, temperature, and field amplitude. Clearly, the remagnetization time is highly sensitive to variation of these parameters. For example, in a *strong* ac field, the temperature dependence of the relaxation time, $\tau \sim e^{\sigma}$ (accurately accounting for the linear regime) is modified because the strong ac field now drastically reduces the effective response time. Thus, the field strength facilitates remagnetization regimes, *which are unattainable in weak ac fields*. Moreover, under the appropriate conditions, the dc bias component can *efficiently tune* this effect, either by *enhancing or blocking the action of the strong ac field*. In Fig. 2.48, the subtle interplay of the various parameters is illustrated via the frequency

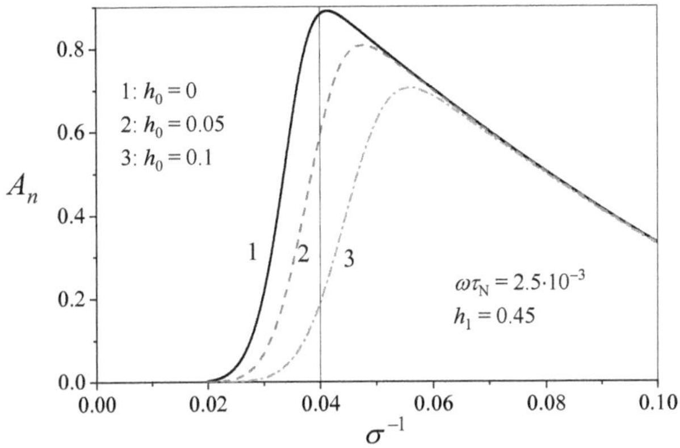

Fig. 2.47. Normalized area of the DMH loop A_n vs. the dimensionless temperature σ^{-1} for various dc bias field parameters $h_0 = 0$ (1), 0.05 (2), 0.1 (3); frequency parameter $\omega\tau_N = 2.5 \cdot 10^{-3}$, ac field amplitude $h_1 \approx 0.45$; thin vertical line corresponds to anisotropy parameter σ=25. (Reproduced from Ref. [47] by permission of the American Institute of Physics.)

dependence of (the normalized loop area) A_n and the specific energy loss P. The main peak of A_n (Fig. 2.48(a)) originates in the *field-induced modifications of the superparamagnetic relaxation time* which are now strongly tuned by the dc bias field. Clearly the curve $A_n(\omega)$ exhibits a much smaller high-frequency peak, originating in the *intrawell forced response* of the magnetic moment. However, including the latter response is essential to correctly describing the *specific* power loss, since $P \propto \omega A_n$; see Eq. (2.326). As Fig. 2.48(b) shows, while the saturation (plateau) for R attained at $\omega\tau_0 \leq 1$ is merely an intermediate one: at still higher frequencies (where the intrawell response is accordingly highly excited) the saturation level of P increases.

The temperature dependence of the DMH is described by the relation $\sigma \propto 1/T$. Figure 2.48 shows that the loop area A_n alters, however, the *tuning action* of the dc bias field is only effective over a certain temperature interval. Thus, for *subcritical ac fields* ($h_1 < 1$), the magnetization relaxation is mainly due to thermal fluctuations so that the magnetic response time retains a strong temperature dependence.

The normalized loop area and the specific power loss as functions of the ac field amplitude are shown in Fig. 2.48. To explain the decrease of

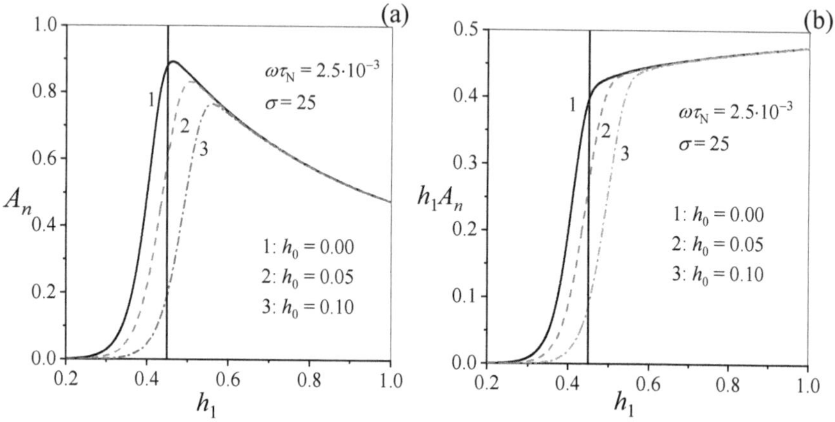

Fig. 2.48. Normalized area of the DMH loop A_n (a) and $h_1 A_n$ (b) vs. the ac field amplitude h_1 with varying bias field parameter $h_0 = 0$ (1, solid line), 0.05 (dashed line), and 0.1 (3, dot-dashed line); anisotropy parameter $\sigma = 25$, frequency parameter $\omega\tau_N = 2.5\cdot10^{-3}$; thin vertical lines correspond to the ac field amplitude $h_1 \approx 0.45$; their intersections with the curves render the areas of the loops of Fig. 2.47 (a). (Reproduced from Ref. [47] by permission of the American Institute of Physics.)

A_n shown in Fig. 2.48(a), recall that, for sufficiently high σ, the loop area A scales as $h_1 A_n \propto \ln h_1$ starting from relatively small values of the field amplitude [162]. Fig. 2.48(b) illustrates this logarithmic increase of A at $h_1 > 0.5$. This slow increase of the actual loop area A compared to the normalized loop area A_n, renders the decrease of A_n as $\propto \ln h_1/h_1$. Clearly from Fig. 2.48(a), the *bias-induced tuning* of the specific power loss is effective in the field interval $0.3 < h_1 < 0.5$.

Finally, we consider the bias-controlled specific power losses in a nanoparticle assembly. Then with a precession time $\tau_0 = 10^{-10}$ s, for example, the condition $\omega\tau_N = 2.5\cdot10^{-3}$ renders a characteristic frequency $f = \omega/2\pi$ ~100 kHz. As the bias control is efficient for $\sigma \approx 25$, then converting this relation to $v = 25\,kT/K$ and setting the anisotropy energy density $K \sim 3\cdot10^4$ J m^{-3} yields a particle volume $v = 3\cdot10^{-24}$ m^3 at room temperature. Hence the loops shown in Fig. 2.45, and the cross-sections in Figs. 2.46–2.48, pertain to particles with a mean diameter 15–20 nm and ferrite nanoparticles of this size are optimal for magnetic hyperthermia purposes. The ac field amplitude H corresponding to the condition $h_1 \approx 0.5$ (equivalent to $H \approx K/\mu_0 M_S$ for a ferrite with magnetization M_S for the given K)

renders $H \approx 5.6 \cdot 10^{-3}$ A / m·(~ 700 Oe), *which could be significantly reduced by simply limiting the amplitude to* $h_1 \approx 0.3$, *where the bias-control is still effective* at a given inverse-temperature parameter σ.

The nonlinear DMH of an assembly of coherently oriented single-domain uniaxial magnetic particles has been treated using the simplest version of the Néel–Brown model. The solution is based on matrix continued fractions, without any *a priori* assumptions regarding the potential barrier height, the temperature or the magnetizing field strength. Under appropriate conditions, a bias dc field of a *small magnitude* (i.e., small compared to the internal anisotropy) can *strongly affect* the DMH loop shape in single-domain particles, so that *via changes in the dc bias field magnitude, one may effectively control the heat production (specific power loss) in a magnetic nanosystem working as a hyperthermia source.* An interesting feature of DMH in *weakly* superparamagnetic particles (σ>>1) is that certain regimes exist, where an ac field of *subcritical strength* ($H \leq H_A$) induces losses differing but little from those with $H = H_A$. Therefore, it may be possible in DMH to *substantially reduce the applied ac field amplitudes, thus diminishing both the undesirable effects of electromagnetic radiation and the power of the set-ups used.* In principle, the model as developed is applicable both to particles with an *intrinsic* bias field (e.g., with a core–shell FM–AFM structure) and to *magnetically uniform* nanoparticles subjected to an external dc bias field. One noteworthy point is that the exchange unidirectional anisotropy (i.e., the source of the internal bias) is itself temperature-dependent so that treating the DMH becomes more challenging. Finally, our treatment supposes that the ac and dc fields are directed along the easy axis. Therefore, the DMH for *arbitrarily* directed fields has also been studied in Ref. [52,159].

2.7.2 *Dynamic magnetic hysteresis in biaxial nanomagnets*

Unlike uniaxial anisotropy, biaxial anisotropy has *intrinsic* azimuthal angle dependence arising from the Δ (biaxiality) term (see Fig. 1.11(b)) leading to intrinsic α dependence of the DMH, i.e., yet another signature of the ubiquitous coupling between the longitudinal and precessional

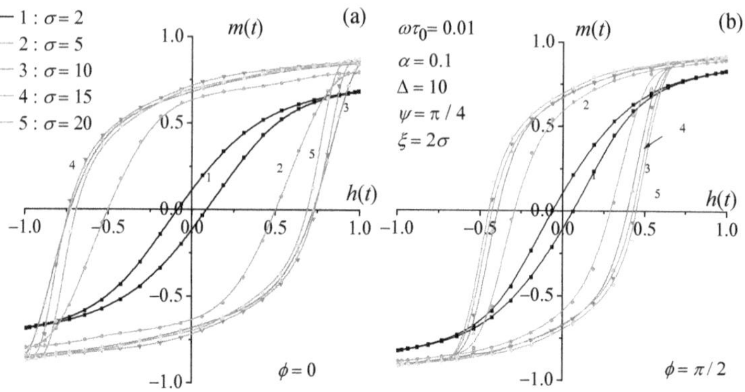

Fig. 2.49. DMH loops for the dimensionless frequency $\omega\tau_0 = 0.01$ for biaxiality parameter Δ=10, damping $\alpha = 0.1$ and various values of the barrier parameter σ. Figures (a) and (b) correspond to orientation of the external fields in the *XZ* (angle $\phi = 0$) or *YZ* ($\phi = \pi/2$) planes, respectively; oblique angle $\psi = \pi/4$ and ac field parameter $\xi = 2\sigma$.

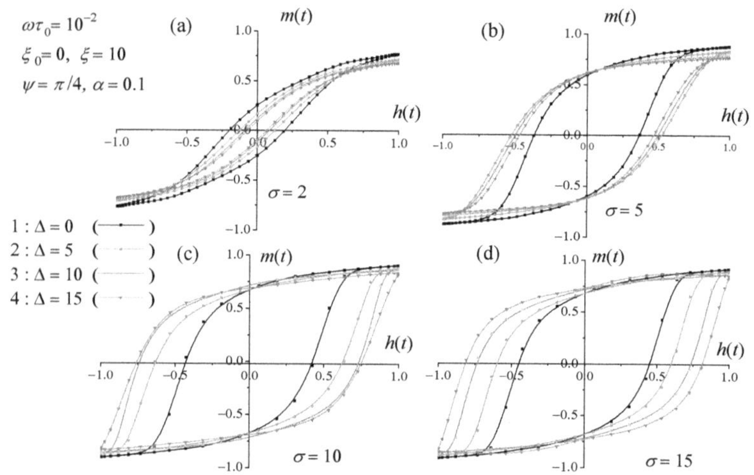

Fig. 2.50. DMH loops for $\omega\tau_0 = 0.01$ for various anisotropy parameters Δ and σ and damping $\alpha = 0.1$, ac and dc field parameters $\xi = 10$, $\xi_0 = 0$, angles $\psi = \pi/4$, $\phi = 0$.

modes of the magnetization. In Figs. 2.49–2.52, the normalized DMH loops, i.e., $m(t) = M(t)/M_S$ vs. $h(t) = H(t)/H = \cos\omega t$, and the normalized loop area A_n (specific power loss) given by the line integral

$$A_n = -\frac{1}{4M_S H}\oint M_H(t)\,dH(t) = -\frac{\pi}{2}\mathrm{Im}(m_1^1) \qquad (2.331)$$

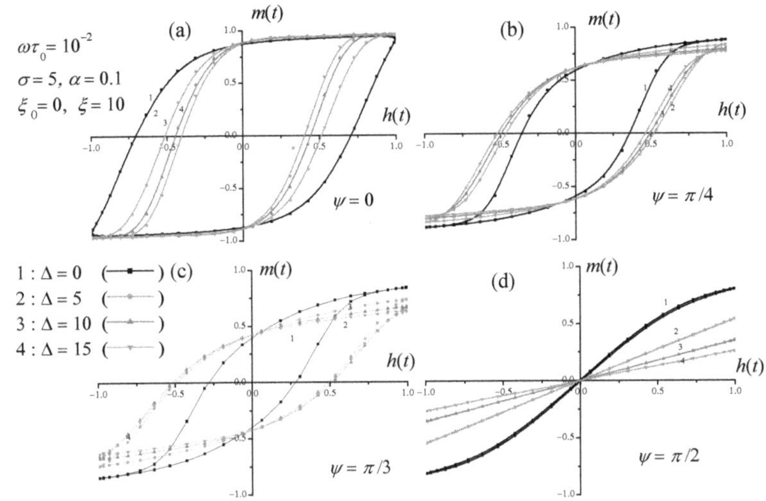

Fig. 2.51. DMH loops for $\omega\tau_0 = 0.01$ with barrier parameter $\sigma = 5$, damping $\alpha = 0.1$, ac field parameter $\xi = 10$, dc field parameter $\xi_0 = 0$ and various biaxiality parameters Δ, $\phi = 0$, and angle $\psi = 0$, $\pi/4$, $\pi/3$ and $\pi/2$.

for various ac amplitudes ξ, damping α and barrier parameters $\sigma \propto 1/T$ are shown. In this section, we use the reduced time scale $\tau_0 = \tau_N/(\alpha + \alpha^{-1})$ so that $\omega\tau_0$ becomes a *damping-independent* dimensionless frequency. Clearly in Fig. 2.49, in marked contrast to uniaxial particles, the DMH now strongly depends on the *azimuthal* direction of the ac field **H** and not just on the polar angle ψ between the easy axis and **H**. Moreover, the coercivity, the remanent magnetization, and the saturation magnetization also strongly depend on the barrier-height (inverse temperature) parameter σ, so that at low frequencies considerable variation in both the size and shape of the loops exists for different σ. For $\omega\tau_0 = 0.01$, $\alpha = 0.1$, $\xi_0 = 0$, and $\sigma \geq 5$, all the loops are *large*, i.e., a large portion of the saturation field remains when the driving field is removed. In contrast for $\sigma \leq 2$, we observe *narrow* loops, i.e., only a small amount of energy is invariably dissipated in repeatedly reversing the magnetization. Figure 2.53 demonstrates that the normalized loop area A_n depends strongly on Δ, ψ, α and the dimensionless frequency $\omega\tau_0$ so that the magnetization relaxation is due to thermal fluctuations. At *very high* frequencies, e.g., $\omega\tau_0 = 1$ (see, Fig. 2.52), *the dynamic hysteresis is due to the resonant*

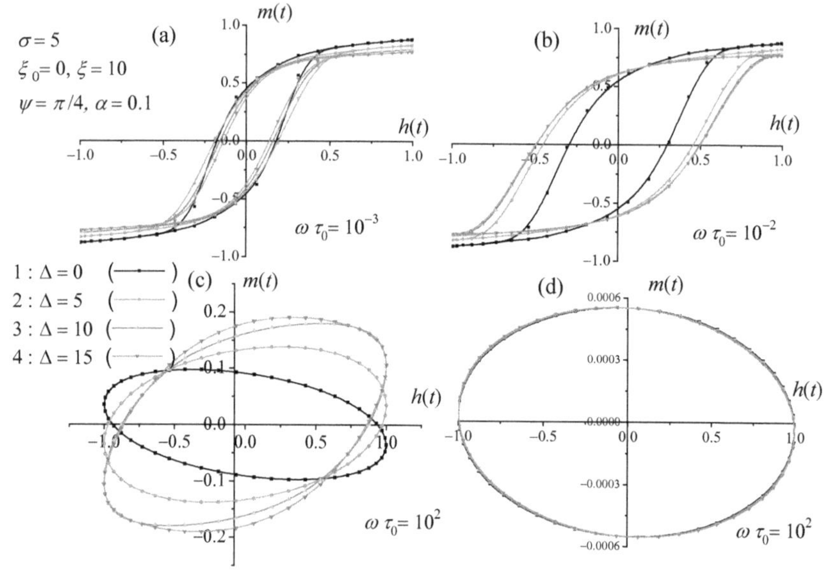

Fig. 2.52. DMH loops with barrier parameter $\sigma = 5$, damping $\alpha = 0.1$, ac field parameter $\xi = 10$, dc field parameter $\xi_0 = 0$, angles $\psi = \pi/4$ and $\phi = 0$ and various biaxiality parameters Δ and dimensionless frequency $\omega\tau_0 = 0.001$, 0.01, 1 and 100.

absorption in the FMR peak. Hence, the DMH arising from a very high-frequency periodic signal may be evaluated, thus permitting quantitative analysis of *ultrafast switching* of the magnetization. Indeed, the rotation of the hysteresis loop exemplified in Fig. 2.52(c), which originates in the sensitivity of the FMR frequency ω_{pr} to the biaxial parameter (implying that the phase shift between $M(t)$ and $H(t)$ may in typical resonant fashion vary rapidly near ω_{pr}) can be used to determine Δ. Clearly, at low frequencies (Fig. 2.51) the DMH loop area increases as the biaxiality increases which is also true if either α or ψ is decreased. The effect is shown in Fig. 2.53 and may become quite pronounced in the superparamagnetic regime $\sigma = 2$. Two hypotheses can be advanced to explain this: (i) that *low damping* implies a longer response time (required to explain the temperature dependence) and (ii) that the precession actually takes place about the direction of the *total field, including thermal contributions*. Thus, this *noise-induced* random precession may hinder the magnetic response, consequently the area is increased.

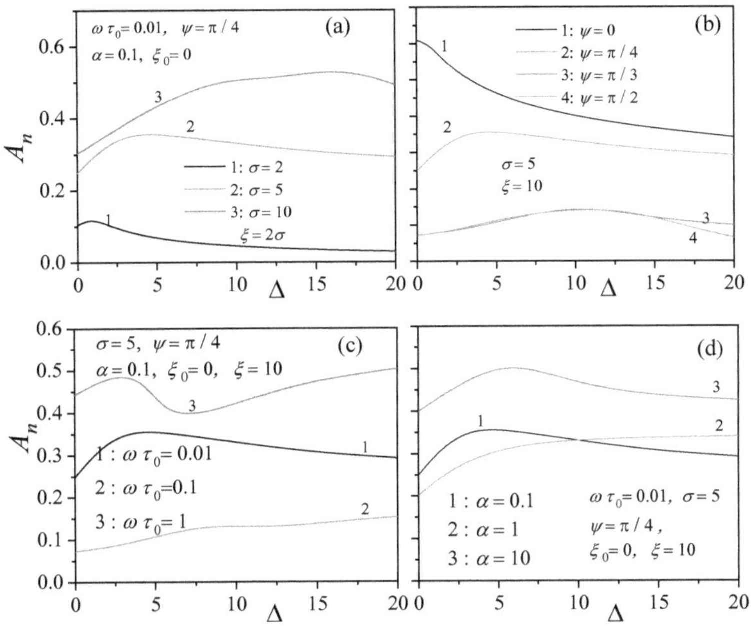

Fig. 2.53. Normalized area of the DMH loop A_n as a function of the biaxiality parameter Δ and various barrier parameters σ (a) and angles ψ (b) for damping $\alpha = 0.1$, dc field parameter $\xi_0 = 0$, frequency $\omega\tau_0 = 0.01$, and $\phi = 0$; and for various frequencies $\omega\tau_0$ (c) and damping α (d) and barrier parameter $\sigma = 5$, ac field parameter $\xi = 10$, dc field parameter $\xi_0 = 0$, angles $\psi = \pi/4$ and $\phi = 0$.

The foregoing calculations of the DMH indicate that a marked dependence of the former on both the biaxiality Δ and the azimuthal angle ϕ exists, again a general feature of noncircularly symmetric systems originating in the coupling of the longitudinal and transverse modes. In other words, the noncircularly symmetric anisotropy generates an azimuthally nonuniform energy distribution leading to strong dependence of the magnetic characteristics on Δ. This arises from the *mode coupling* due to Δ so that as far as DMH is concerned, at low frequencies, the loops exhibit a pronounced damping dependence again due to coupling of the thermally activated magnetization reversal mode with the precessional modes via the time-dependent azimuthal direction of the applied ac field. Therefore, the *frequency dependence of the DMH* for fixed angles ψ and ϕ and temperature is very significant because the *remagnetization time is highly sensitive to the frequency of the applied field.* For example, in a strong ac driving field, the Arrhenius dependence of the relaxation time on temperature $\tau \sim e^{\sigma}$, accurately accounting for the linear regime, is modified since the strong ac field drastically reduces the effective response time. Thus, new remagnetization regimes, which could never be attained for weak ac fields are possible. Furthermore, the dc bias component under the appropriate conditions, *efficiently tunes this effect by either enhancing or blocking the action of the ac field.* The foregoing description of DMH holds over a wide field-frequency-temperature range and avoids drastic oversimplifications such as assuming linear response or high-barrier approximations. Hence, it is a useful tool for both verification of existing solutions and inviting extension to novel applications involving magnetic nanoparticles, e.g., *low-frequency magnetic hyperthermia* and *precise focusing of the heat generated by them.* DMH also has potential applications valuable in magnetic recording because the phenomenon can be used by industry (especially in data storage) to characterize the recording density, signal-to-noise ratio, etc. The description of the angular dependence of the DMH which has been given could also be used to analyze the switching times in nanomagnets allowing for both imperfect orientational distributions of particle easy axes and nonuniformity of the head field inside the recording layer.

2.8 Spin-transfer torque effects

2.8.1 *STT effects in the magnetization reversal*

Thermal fluctuations of nanomagnets driven by spin-polarized currents can be treated via the Landau–Lifshitz–Gilbert equation generalized to include both the random thermal noise field and the Slonczewski STT term (v. Section 1.3.3). The analysis is usually effected using the generic nanopillar model of a spin-torque device which comprises two ferromagnetic strata representing the free and fixed layers and a nonmagnetic conducting spacer all sandwiched between two ohmic contacts (see Fig. 1.12). By solving the corresponding magnetic Langevin equation as in zero STT, the effects of thermal fluctuations and STT on relevant switching characteristics may also be treated over wide ranges of temperature, damping, external magnetic field, and spin-polarized current thereby identifying new STT effects in the thermally assisted magnetization reversal.

The differential-recurrence equation for the $\langle Y_{lm}\rangle(t)$ derived in Chapter 1, Section 1.3.5, is valid for *arbitrary* free energy E and nonconservative potential Φ. We now specialize this general differential-recurrence equation to a particular free energy E which written in the standard form of superimposed easy-plane and in-plane easy-axis anisotropies plus the Zeeman term (cf. Eq. (2.60)) is

$$E(\vartheta,\varphi)=\sigma\left[\delta\cos^2\vartheta-\sin^2\vartheta\cos^2\varphi-2h\cos\Theta(\vartheta,\varphi)\right]. \quad (2.332)$$

Here Θ is the angle between the vectors $\mathbf{H}_0$ and $\mathbf{M}$ given by Eq (2.299), where $\gamma_1=\cos\varphi_\xi\sin\vartheta_\xi$, $\gamma_2=\sin\varphi_\xi\sin\vartheta_\xi$, $\gamma_3=\cos\vartheta_\xi$ are the direction cosines of the applied field (the remaining notation is as explained in Chapter 1, Section 1.3.3). The free energy density, Eq. (2.332), can be written as

$$E(\vartheta,\varphi)=-\frac{2\sigma}{3}\sqrt{\pi}+\sum_{r=1}^{2}\sum_{s=-r}^{r}A_{rs}Y_{rs}(\vartheta,\varphi), \quad (2.333)$$

where the nonzero expansion coefficients A_{rs} are given by

$$\begin{aligned}A_{10}&=-4\sigma h\gamma_3\sqrt{\frac{\pi}{3}},\ A_{1\pm1}=\pm\sigma h(\gamma_1\mp i\gamma_2)\sqrt{\frac{8\pi}{3}},\\ A_{20}&=\sigma(1+2\delta)\sqrt{\frac{4\pi}{45}},\ A_{2\pm2}=-\sigma\sqrt{\frac{2\pi}{15}}.\end{aligned} \quad (2.334)$$

Furthermore, the nonconservative potential Φ, Eq. (1.78) due to the spin-polarized current is [109]

$$\frac{v\Phi}{kT}=\frac{J}{c_P}\ln\left(1+c_P\frac{4\pi}{3}\sum_{m=-1}^{1}Y_{1m}(\vartheta,\varphi)Y_{1m}^*(\vartheta_P,\varphi_P)\right), \quad (2.335)$$

where the unit vector $\mathbf{e}_P$ identifies the magnetization direction in the *fixed layer* (see Fig. 1.12), ϑ_P and φ_P are the spherical polar coordinates of $\mathbf{e}_P$, J is the dimensionless spin-polarized current parameter, taken as positive

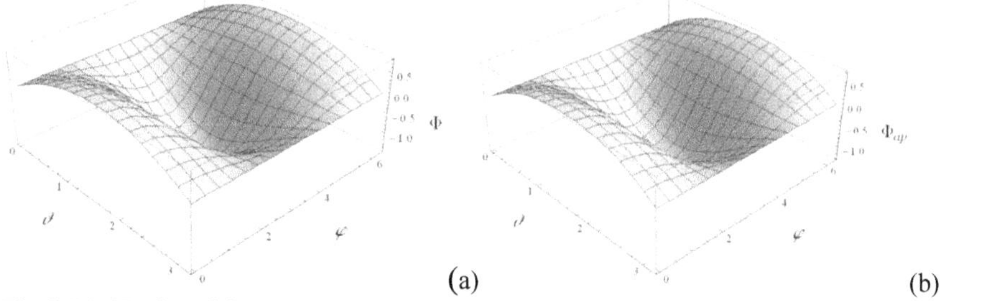

(a) (b)

Fig. 2.54. 3D plot of the non-conservative potentials $\Phi(\vartheta,\varphi)$ Eq. (2.335), (a) and $\Phi_{ap}(\vartheta,\varphi)$ Eq. (2.336), (b) for typical model parameters ($P = 0.3$, $J = 1$, $\varphi_P = 0$ and $\vartheta_P = \pi/2$).

when the electrons flow from the free into the fixed layer and the coefficient c_P, determined by the spin-polarization factor P ($0 < P < 1$), is given by Eq. (1.79). Next by retaining only the two leading terms in the Taylor series expansion $\ln(1+x) = x - x^2/2 + x^3/3 - \ldots$, which converges fairly well for typical model parameters, we have from Eq. (2.335)

$$\frac{v\Phi_{ap}}{kT} = \sum_{r=0}^{2}\sum_{s=-r}^{r} B_{rs}Y_{rs}(\vartheta,\varphi), \tag{2.336}$$

where the expansion coefficients B_{rs} are defined as

$$B_{00} = -\frac{4\pi^{3/2}c_P J}{9}\left[Y_{10}^{*2} - 2Y_{11}^{*}Y_{1-1}^{*}\right], \quad B_{1m} = \frac{4\pi J}{3}Y_{1m}^{*},$$

$$B_{20} = -\frac{8\pi^{3/2}c_P J}{9\sqrt{5}}\left[Y_{10}^{*2} + Y_{11}^{*}Y_{1-1}^{*}\right],$$

$$B_{2\pm1} = -\frac{8\pi^{3/2}c_P J}{3\sqrt{15}}Y_{10}^{*}Y_{1\pm1}^{*}, \quad B_{2,\pm2} = -\frac{8\pi^{3/2}c_P J}{3\sqrt{30}}Y_{1\pm1}^{*2}$$

(henceforth $Y_{lm}^{*} = Y_{lm}^{*}(\vartheta_P,\varphi_P)$). The approximate Eq. (2.336) accurately reproduces all the features of Φ, Eq. (2.335), to within 5% in the majority of cases (see Fig. 2.54) since $c_P^2(\mathbf{u}\cdot\mathbf{e}_P)^2/3 < 0.15$ for $P \leq 0.4$.

The general differential-recurrence Eqs. (1.91) and (1.111), as specialized to Eqs. (2.333) and (2.336), yield in the usual notation the 25-term differential-recurrence relation for the statistical moments $c_{nm}(t) = \langle Y_{nm}\rangle(t)$ governing the magnetization dynamics, viz.

$$\tau_{\mathrm{N}}\frac{d}{dt}c_{nm}(t)=v_{nm}^{--}c_{n-2m-2}(t)+v_{nm}^{-}c_{n-2m-1}(t)+v_{nm}c_{n-2m}(t)+v_{nm}^{+}c_{n-2m+1}(t)$$
$$+v_{nm}^{++}c_{n-2m+2}(t)+w_{nm}^{--}c_{n-1m-2}(t)+w_{nm}^{-}c_{n-1m-1}(t)+w_{nm}c_{n-1m}(t)+w_{nm}^{+}c_{n-1m+1}(t)$$
$$+w_{nm}^{++}c_{n-1m+2}(t)+x_{nm}^{--}c_{nm-2}(t)+x_{nm}^{-}c_{nm-1}(t)+x_{nm}c_{nm}(t)+x_{nm}^{+}c_{nm+1}(t)$$
$$+x_{nm}^{++}c_{nm+2}(t)+y_{nm}^{--}c_{n+1m-2}+y_{nm}^{-}c_{n+1m-1}(t)+y_{nm}c_{n+1m}(t)+y_{nm}^{+}c_{n+1m+1}(t)$$
$$+y_{nm}^{++}c_{n+1m+2}(t)+z_{nm}^{--}c_{n+2m-2}(t)+z_{nm}^{-}c_{n+2m-1}(t)+z_{nm}c_{n+2m}(t)+z_{nm}^{+}c_{n+2m+1}(t)$$
$$+z_{nm}^{++}c_{n+2m+2}(t), \tag{2.337}$$

where the various coefficients x_{nm}, etc. are now given by

$$x_{nm}=-\frac{n(n+1)}{2}+i\frac{m\sigma h\gamma_3}{\alpha}+i\sqrt{\frac{\pi}{3}}mJY_{10}^{*}+\frac{n(n+1)-3m^2}{(2n-1)(2n+3)}$$
$$\times\left[-\sigma\left(\frac{1}{2}+\delta\right)+\frac{2\pi c_P J}{3\alpha}\left(Y_{10}^{*2}+Y_{11}^{*}Y_{1-1}^{*}\right)\right],$$

$$x_{nm}^{\pm\pm}=-\frac{\sqrt{(n\pm m+1)(n\pm m+2)(n\mp m-1)(n\mp m)}}{(2n-1)(2n+3)}\left(\frac{3\sigma}{4}+\frac{\pi c_P J}{\alpha}Y_{1\pm1}^{*2}\right),$$

$$x_{nm}^{\pm}=\sqrt{(1+n\pm m)(n\mp m)}\left\{(\gamma_1\mp i\gamma_2)\frac{i\sigma h}{2\alpha}\right.$$
$$\left.\mp i\sqrt{\frac{\pi}{6}}JY_{1\pm1}^{*}+\frac{\sqrt{2}\pi b_P c_P J(1\pm 2m)}{\alpha(2n-1)(2n+3)}Y_{10}^{*}Y_{1\pm1}^{*}\right\},$$

$$y_{nm}=\sqrt{\frac{(n+1)^2-m^2}{(2n+1)(2n+3)}}\left\{-i\frac{m\sigma}{\alpha}\left(\frac{1}{2}+\delta\right)-n\sigma h\gamma_3+\frac{nJ}{\alpha}\sqrt{\frac{\pi}{3}}Y_{10}^{*}\right.$$
$$\left.-i\frac{2\pi m c_P J}{3}\left(Y_{10}^{*2}+Y_{11}^{*}Y_{1-1}^{*}\right)\right\},$$

$$y_{nm}^{\pm\pm}=\mp i\left(\frac{\sigma}{4\alpha}-\frac{\pi c_P J}{3}Y_{1\pm1}^{*2}\right)\sqrt{\frac{(1+n\pm m)(2+n\pm m)(3+n\pm m)(n\mp m)}{(1+2n)(3+2n)}},$$

$$y_{nm}^{\pm}=\sqrt{\frac{(1+n\pm m)(2+n\pm m)}{(1+2n)(3+2n)}}\left\{\pm\frac{n}{2}\sigma h(\gamma_1\mp i\gamma_2)+\frac{nJ}{\alpha}\sqrt{\frac{\pi}{6}}Y_{1\pm1}^{*}\right.$$
$$\left.\pm i\pi\sqrt{2/9}c_P J(n\mp 2m)Y_{10}^{*}Y_{1\pm1}^{*}\right\},$$

$$w_{nm} = \sqrt{\frac{n^2-m^2}{4n^2-1}}\left\{-\frac{im\sigma}{\alpha}\left(\frac{1}{2}+\delta\right)+(n+1)\sigma h\gamma_3 - \frac{J(n+1)}{\alpha}\sqrt{\frac{\pi}{3}}Y_{10}^{*}\right.$$
$$\left. -i\frac{2m\pi c_P J}{3}\left(Y_{10}^{*2}+Y_{11}^{*}Y_{1-1}^{*}\right)\right\},$$

$$w_{nm}^{\pm\pm} = \pm i\left(\frac{\sigma}{4\alpha}-\frac{\pi c_P J}{3}Y_{1\pm1}^{*2}\right)\sqrt{\frac{(n\mp m-2)(n\mp m-1)(1+n\pm m)(n\mp m)}{4n^2-1}},$$

$$w_{nm}^{\pm} = \sqrt{\frac{(n\mp m)(n\mp m-1)}{4n^2-1}}\left\{\frac{J(n+1)}{\alpha}\sqrt{\frac{\pi}{6}}Y_{1\pm1}^{*}\right.$$
$$\left.\pm\frac{n+1}{2}\sigma h(\gamma_1 \mp i\gamma_2) \pm i\frac{\sqrt{2}\pi c_P J(n+1\pm 2m)}{3}Y_{10}^{*}Y_{1\pm1}^{*}\right\},$$

$$z_{nm} = \frac{n}{2n+3}\sqrt{\frac{[(n+1)^2-m^2][(n+2)^2-m^2]}{(2n+1)(2n+5)}}$$
$$\times\left\{\sigma(\delta+1/2)-\frac{2\pi c_P J}{3\alpha}\left(Y_{10}^{*2}+Y_{11}^{*}Y_{1-1}^{*}\right)\right\},$$

$$z_{nm}^{\pm\pm} = -\left(\frac{\sigma}{4}+\frac{\pi c_P J}{3\alpha}Y_{1\pm1}^{*2}\right)\frac{n}{2n+3}\sqrt{\prod_{k=1}^{4}(n\pm m+k)\frac{1}{(2n+1)(2n+5)}},$$

$$z_{nm}^{\pm} = -\frac{n\sqrt{8}\pi c_P J}{3\alpha(2n+3)}Y_{10}^{*}Y_{1\pm1}^{*}\sqrt{\frac{[(n+1)^2-m^2](n\pm m+2)(n\pm m+3)}{(2n+1)(2n+5)}},$$

$$v_{nm} = \frac{n+1}{2n-1}\sqrt{\frac{[(n-1)^2-m^2](n^2-m^2)}{(2n+1)(2n-3)}}\left\{-\sigma\left(\frac{1}{2}+\delta\right)+\frac{2\pi c_P J}{3\alpha}\left(Y_{10}^{*2}+Y_{11}^{*}Y_{1-1}^{*}\right)\right\},$$

$$v_{nm}^{\pm} = -\frac{\sqrt{8}\pi c_P J}{3\alpha}Y_{10}^{*}Y_{1\pm1}^{*}\frac{n+1}{2n-1}\sqrt{\frac{(n\mp m-2)(n\mp m-1)(n^2-m^2)}{(2n+1)(2n-3)}},$$

$$v_{nm}^{\pm\pm} = \left(\frac{\sigma}{4}+\frac{\pi c_P J}{3\alpha}Y_{1\pm1}^{*2}\right)\frac{n+1}{2n-1}\sqrt{\prod_{k=0}^{3}(n\mp m-k)\frac{1}{(2n+1)(2n-3)}}.$$

The differential-recurrence equations obtained in this section will be used below to analyze the dynamical behavior of the magnetization of nanomagnets. This analysis again involves the determination of the usual characteristic times of the magnetization decay, dynamic susceptibilities (linear and nonlinear), stochastic resonance, dynamic magnetic hysteresis, etc.

Equation (2.337) may once again be solved by using matrix continued fractions because we can transform the *scalar* differential-recurrence relation, Eq. (2.337), into the *tridiagonal vector* differential-recurrence relation, Eq. (2.200), where the column vectors $\mathbf{C}_n(t)$ and the matrices $\mathbf{Q}_n, \mathbf{Q}_n^{\pm}$ are defined by Eqs. (2.259) and (2.260), respectively. The matrices $\mathbf{Q}_n, \mathbf{Q}_n^{\pm}$ themselves comprise five submatrices $\mathbf{V}_l$, $\mathbf{W}_l$, $\mathbf{X}_l$, $\mathbf{Y}_l$, and $\mathbf{Z}_l$. The matrix elements of these five-diagonal submatrices, which are formed from the coefficients occurring in Eq. (2.337), are given by

$$
\begin{aligned}
(\mathbf{V}_l)_{nm} &= \delta_{n-4m} v^{--}_{l,-l+m+3} + \delta_{n-3m} v^{-}_{l,-l+m+2} + \delta_{n-2m} v_{l,-l+m+1} \\
&\quad + \delta_{n-1m} v^{+}_{l,-l+m} + \delta_{nm} v^{++}_{l,-l+m-1}, \\
(\mathbf{W}_l)_{nm} &= \delta_{n-3m} w^{--}_{l,-l+m+2} + \delta_{n-2m} w^{-}_{l,-l+m+1} + \delta_{n-1m} w_{l,-l+m} \\
&\quad + \delta_{nm} w^{+}_{l,-l+m-1} + \delta_{n+1m} w^{++}_{l,-l+m-2}, \\
(\mathbf{X}_l)_{nm} &= \delta_{n-2m} x^{--}_{l,-l+m+1} + \delta_{n-1m} x^{-}_{l,-l+m} + \delta_{nm} x_{l,-l+m-1} \\
&\quad + \delta_{n+1m} x^{+}_{l,-l+m-2} + \delta_{n+2m} x^{++}_{l,-l+m-3}, \\
(\mathbf{Y}_l)_{nm} &= \delta_{n-1m} y^{--}_{l,-l+m} + \delta_{nm} y^{-}_{l,-l+m-1} + \delta_{n+1m} y_{l,-l+m-2} \\
&\quad + \delta_{n+2m} y^{+}_{l,-l+m-3} + \delta_{n+3m} y^{++}_{l,-l+m-4}, \\
(\mathbf{Z}_l)_{nm} &= \delta_{nm} z^{--}_{l,-l+m-1} + \delta_{n+1m} z^{-}_{l,-l+m-2} + \delta_{n+2m} z_{l,-l+m-3} \\
&\quad + \delta_{n+3m} z^{+}_{l,-l+m-4} + \delta_{n+4m} z^{++}_{l,-l+m-5}.
\end{aligned}
$$

The formal exact matrix continued fraction solution of Eq. (2.200) for the Laplace transform $\tilde{\mathbf{C}}_1(s)$ is then given by Eq. (2.203), where the initial condition vectors $\mathbf{C}_n(0)$ can also be evaluated from the continued fractions. The stationary averages $\langle Y_{lm} \rangle_0$, needed to evaluate the $\mathbf{C}_n(0)$, the stationary distribution function W_0, and the magnetization can be calculated via matrix continued fractions as follows. The recurrence relation for the stationary averages $\langle Y_{nm} \rangle_0$ may again be written explicitly as the algebraic three-term matrix recurrence relation, Eq. (2.263), where $\mathbf{Q}_n$ and $\mathbf{Q}_n^{\pm}$ are defined by Eqs. (2.260), and $\mathbf{R}_n$ are given by Eq. (2.264). Thus, the $\mathbf{R}_n$ are again rendered by the matrix product, Eq. (2.265). Using the stationary statistical moments $\langle Y_{1\pm1}\rangle_0$, we have the stationary magnetization in the easy X-direction $\langle u_X\rangle_0$, viz.,

$$\begin{aligned}\langle u_X\rangle_0 &= \langle \sin\vartheta\cos\varphi\rangle_0 \\ &= \sqrt{\frac{2\pi}{3}}\left(\langle Y_{1-1}\rangle_0 - \langle Y_{11}\rangle_0\right)\end{aligned} \tag{2.338}$$

and the *effective potential* E_{ef} via

$$E_{\mathrm{ef}}(\vartheta,\varphi) = -\ln\left[W_0(\vartheta,\varphi)\right], \tag{2.339}$$

where $W_0(\vartheta,\varphi)$ is the *out-of-equilibrium* or *time-independent* stationary distribution given by Eq. (1.89). Furthermore, we also have λ_1^{-1} (yielding the reversal time) from Eq. (1.134).

The anisotropy and spin-polarization parameters will be taken throughout as $D_\| = 0.034$, $\delta = 20$, and $P = 0.3$ Moreover, the applied field $\mathbf{H}_0$ and the unit vector $\mathbf{e}_P$, identifying the magnetization direction in the fixed layer, are assumed to lie in the equatorial or XY plane, i.e., with colatitudes $\vartheta_P = \pi/2$ and $\vartheta_\xi = \pi/2$. Thus the orientations of the vectors $\mathbf{H}_0$ and $\mathbf{e}_P$ in the XY plane are entirely determined by their respective azimuthal angles φ_ξ and φ_P, respectively. Moreover, $\varphi_\xi = \varphi_P = 0$ corresponds to the particular configuration whereby both vectors $\mathbf{H}_0$ and $\mathbf{e}_P$ are directed along the easy X-axis. For $\gamma = 2.2\cdot10^5\ \mathrm{m\ A^{-1}s^{-1}}$, $T = 300$ K, $v \sim 10^{-24}\ \mathrm{m}^3$, $J_e \approx 10^7\ \mathrm{A\ cm^{-2}}$, $J_p \approx 10^9\ \mathrm{A\ cm^{-2}}$, $M_S \approx 1.4\cdot10^6\ \mathrm{A\ m^{-1}}$ (cobalt), and $\alpha = 0.02$, we have the following estimates for the principal parameters of the model $J \approx 0.98$ and $\sigma \approx 20.2$. Furthermore instead of the free diffusion time τ_{N}, we shall use as the normalizing time $\tau_0 = \tau_{\mathrm{N}}/[\sigma(\alpha+\alpha^{-1})] = (2\gamma M_S D_\|)^{-1}$, and the above numerical values yield $\tau_0 \approx 4.8\cdot10^{-11}$ s.

Thus having determined the time-independent stationary distribution $W_0(\vartheta,\varphi)$ via Eq. (1.89), we also have the effective potential E_{ef} from Eq. (2.339). A typical example of the calculations is shown in Figs. 2.55 and 2.56 (cf. Fig. 2.6). Clearly E_{ef} comprises a double-well structure with nonequivalent wells (Fig. 2.55). Moreover, this energyscape, as expected on intuitive grounds, strongly depends on damping, external magnetic field magnitude and orientation, and spin-polarized current. Therefore, by

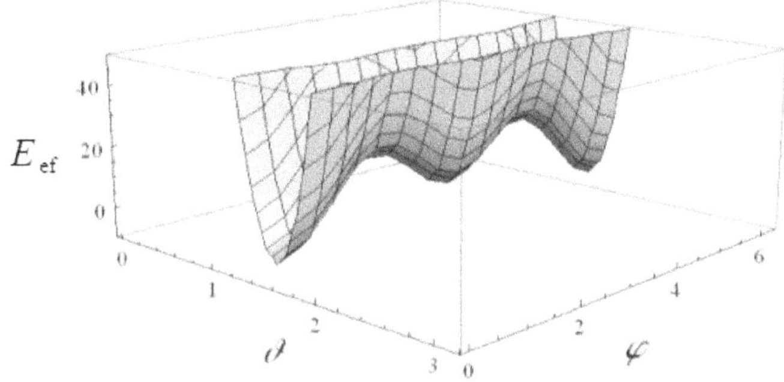

Fig. 2.55. 3D plot of the effective potential $E_{ef}(\vartheta,\varphi)$, Eq. (2.339), near the minima for parameters $P = 0.3$, $J = 1$, $\alpha = 0.2$, $h = 0.15$, $\sigma = 20$, $\delta = 20$, $\varphi_P = 0$, and $\varphi_\xi = 0$.

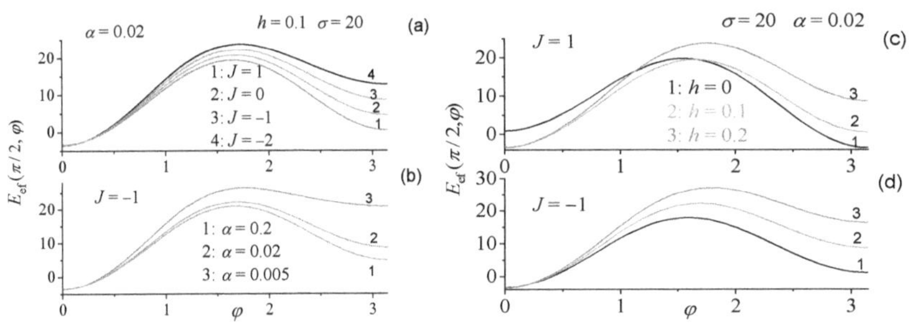

Fig. 2.56. $E_{ef}(\pi/2,\varphi)$ vs the azimuthal angle φ for various spin-polarized current parameters $J = 1, 0, -1, -2$ and $\alpha = 0.02$ (a); for various damping $\alpha = 0.2, 0.02, 0.005$ and $J = -1$ (b); for various external field parameters $h = 0.0, 0.1, 0.2$ and $J = 1$ (c) and $J = -1$ (d) ($\alpha = 0.02$, $\sigma = 20$, $P = 0.3$, $\delta = 20$, $\varphi_P = 0$ and $\varphi_\xi = 0$).

varying the magnitude of the spin-polarized current, damping, etc., one may substantially alter the effective barriers and thus the reversal time τ.

The stationary average $\langle u_X \rangle_0$ is calculated from Eq. (2.338). In Fig. 2.57, we show the consequent dependence of $\langle u_X \rangle_0$ on the spin-polarized current as a family of curves for various damping, external field magnitude (h), orientation (φ_ξ), and the magnetization direction in the fixed layer (φ_P), all regarded as parameters. Clearly the *switching current* J_{SW} (i.e., the *current at which* $\langle u_X \rangle_0$ *changes sign corresponding to reversal of the direction of precession*) strongly depends on these model parameters, viz., α, J, h, σ, φ_ξ, and φ_P. As both α and h increase J_{SW} rapidly increases and may also vary significantly with both the orientation of the external field

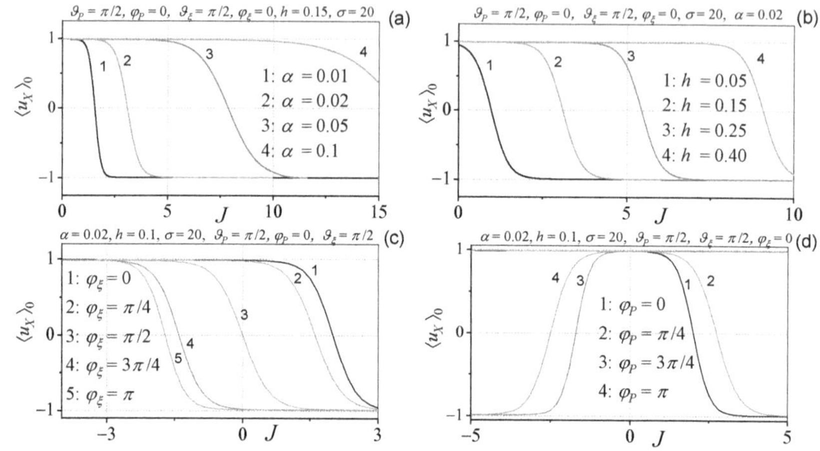

Fig. 2.57. Stationary normalized magnetization $\langle u_X \rangle_0$ vs the dimensionless current parameter J for various damping parameters α (a), external field parameters h (b), external field orientations in the free layer φ_ξ (c), and spin-polarization orientations φ_P (d).

and the direction of the magnetization of the fixed layer. The precise onset of the slope in $\langle u_X \rangle_0$ largely depends on both α and σ: higher values of σ (lower temperatures) and smaller values of α correspond to a more rapid onset of the slope. Moreover, λ_1 of the Fokker–Planck operator (inverse of the reversal time of the direction of precession, i.e., that of the X-component of the magnetization **M**) may be evaluated from the secular Eq. (1.134). Next, via λ_1 we also have the dependence of the magnetization reversal time $\tau = 1/\lambda_1$ on the spin-polarized current, anisotropy parameters, damping, external field magnitude and orientation in the free layer, and the magnetization direction in the fixed layer. In general, a pronounced dependence of τ on all these model parameters exists. Examples are shown in Fig. 2.58. Figure 2.58(a) illustrates the damping dependence of τ for various currents J while Figs. 2.58(b) and 2.58(c) show the temperature dependence of τ for various J and external field h parameters. Clearly by altering J, the ensuing variation of τ may be as much as several orders of magnitude for very low damping, $\alpha \ll 1$ (Fig. 2.58(a)). Furthermore, τ may greatly exceed or, on the other hand, be much less than that for $J = 0$, while the increase or decrease in τ is *entirely governed by the current direction*, i.e., by the sign of J as expected. The

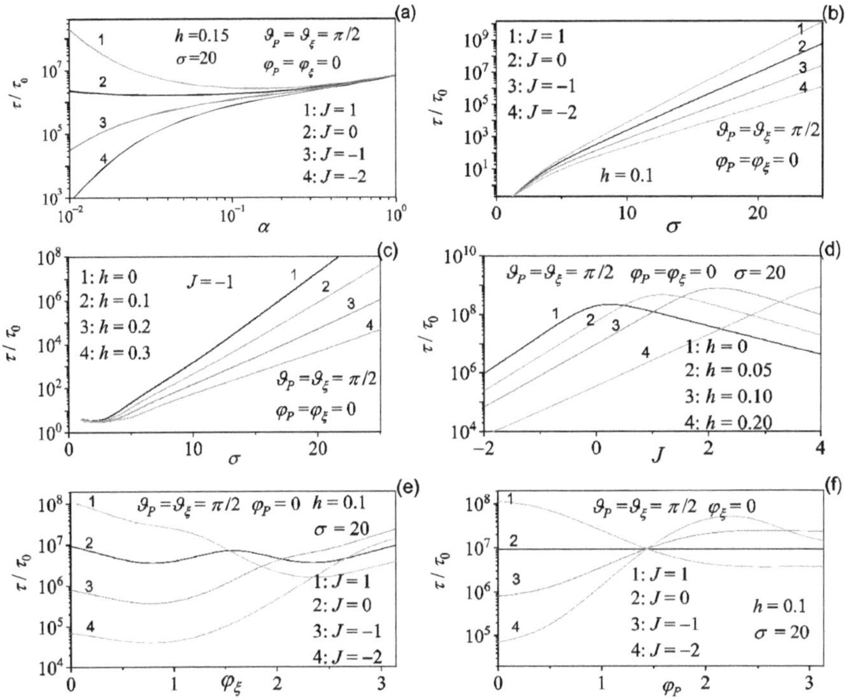

Fig. 2.58 (a) Reversal time τ/τ_0 vs the damping α for various spin-polarized current parameters J. τ/τ_0 vs the inverse-temperature parameter $\sigma \sim T^{-1}$ for (b) various J and (c) various external field parameters h. (d) τ/τ_0 vs J for various h. τ/τ_0 vs the azimuthal angles φ_ξ (e) and φ_P (f) for various J; (α=0.02, P=0.3, δ=20).

temperature dependence of τ can be readily understood via the *effective potential* E_{ef}, Eq. (2.339). Obviously, at high barriers, $\sigma > 5$ this *temperature dependence* has the customary Arrhenius behavior $\tau \sim e^{\Delta E_{\mathrm{ef}}}$, i.e., *exponentially increasing with decreasing temperature*. Furthermore, the slope of $\tau(T^{-1})$ markedly depends on J, h, α, etc. because the barrier height ΔE_{ef} of the shallow well is strongly affected by these. The slope of $\tau(T^{-1})$ significantly decreases with increasing h due to a decrease of the barrier height ΔE_{ef} via the action of the external field. At low barriers, $\tau(T^{-1})$ may deviate considerably from Arrhenius-like behavior. Figure 2.58(d) illustrates the dependence of τ on J for various h. Clearly, as J increases from negative values, *τ exponentially increases at first, attaining a maximum at a critical value of the spin-polarized current and then smoothly switching to exponential decrease as J is further increased* *through positive values*. This dependence of τ on J implies that both kinds of scaling for the switching time suggested in the literature, namely, $\tau \propto e^{-C|J_{SW}-J|}$ and $\tau \propto e^{-C(J_{SW}-J)^2}$ may be realized for $|J_{SW}-J|>1$ and $|J_{SW}-J|<1$, respectively (where C and J_{SW} are parameters depending, in general, on h, σ, δ, α, etc.). Figures 2.58(e) and 2.58(f) exemplify the pronounced dependence of τ on the azimuthal angles of the applied field φ_ξ and the magnetization direction φ_P in the fixed layer for various J which may comprise several orders of magnitude [note that $\tau(\varphi_\xi)=\tau(2\pi-\varphi_\xi)$ and $\tau(\varphi_P)=\tau(2\pi-\varphi_P)$]. *Invariably strong STT effects on the magnetization reversal exist only for low damping $\alpha \le 1$, because the magnitude of these effects in the magnetization reversal is governed by the ratio J/α. In this damping regime the variation of τ with J may be of several orders of magnitude. In contrast for high damping $\alpha \ge 1$ the STT term in the magnetic Langevin equation, Eq. (1.75), does not influence the reversal process at all because it is then negligible compared to the damping and random torques.*

Thus, one may treat STT effects in the thermally assisted magnetization reversal via the evolution equation for the statistical moments yielded by the magnetic Langevin equation, Eq. (1.75), rendering the switching characteristics over wide ranges of temperature, damping, external magnetic field, and spin-polarized current. Variation of that current may alter the reversal time by several orders of magnitude concurring with experimental results. The numerical solutions of the recurrence relations for the relevant statistical moments hold for the most comprehensive formulation of the generic nanopillar model (Fig. 1.12), i.e., for *arbitrary directions* of the external field and spin polarization and for *arbitrary free energy density,* yielding the STT switching characteristics under conditions otherwise inaccessible, thus serving both as a basis for theoretical investigations and interpretation of a broad range of STT experiments. Additionally, the results are essential for the *future development* of both escape-rate theory and stochastic dynamics simulations of the magnetization reversal time in STT systems, representing once more rigorous benchmarks since they obviously constitute exact solutions with which calculations of that time by any other method must comply.

2.8.2 *Nonlinear ac stationary response*

We saw STT effects on certain out-of-equilibrium time- and frequency-independent stationary observables subject to a dc *bias field alone* via the generic nanopillar model (Fig. 1.12) have just been treated by exactly solving the magnetic Langevin equation, Eq. (1.75) via matrix continued fractions. These observables comprise the *stationary* distribution of the magnetization orientations, the effective potential, and the in-plane component of the magnetization of the free layer. These *time- and frequency-independent* observables have been studied over wide ranges of the spin-polarized current, the dissipative coupling to the thermal bath, the anisotropy parameters and the magnitude and orientation of the applied external dc field, which was supposed *constant* in time. Besides these stationary observables, the reversal time of the in-plane component of the magnetization of the free layer has also been evaluated via λ_1^{-1} of the corresponding Fokker–Planck operator again as a function of the parameters mentioned. In the previous section, the external applied (bias) field was supposed *time-independent*, i.e., representing a dc field applied in the infinite past. Hence, the previous results cannot be applied to virtually all *dynamical* aspects of the time-dependent magnetization response (except for the relaxation time). These include magnetization switching of STT devices the line shape of STT nano-oscillators driven by ac external magnetic fields and currents [386–393], stochastic resonance [394–397], etc. As shown experimentally, the magnetization reversal in STT devices driven by superimposed dc and ac currents or by a direct spin-polarized current combined with an ac magnetic field may enable a more efficient STT magnetization reversal in comparison to that achievable by purely dc currents alone. A finding which is of technological interest in improving switching characteristics of magnetic random-access memories [393].

Here to comprehensively investigate STT effects on the dynamical characteristics of the generic nanopillar model (Fig. 1.12) due to an *ac magnetic field of arbitrary strength and frequency*, we generalize the approach developed originally for zero STT in Section 2.6. Thus, we shall have the response characteristics for all frequencies from the very low ones characterizing overbarrier relaxation processes up to the very high ones appropriate to the ferromagnetic resonance (GHz) range using a single model. *A priori* STT effects in the ac stationary response of a nanomagnet inherently pose a more complicated problem than in the time-independent out-of-equilibrium case because the observables are now both *time-* and *frequency-dependent*. However, these difficulties may again be overcome using the matrix continued fraction method as in the nonlinear ac response without STT.

The main thrust of our investigation is the effect of STT on the complex magnetic susceptibility and DMH loops of a nanoscale ferromagnet subjected to superimposed ac and dc bias fields $\mathbf{H}_0 + \mathbf{H}\cos\omega t$ of *arbitrary* strengths and orientations. Here the normalized free energy density $E(\vartheta,\varphi,t)$ of the free layer may conveniently be written as ($\mathbf{H}_0$ and $\mathbf{H}$ are assumed *parallel*)

$$\begin{aligned} E(\vartheta,\varphi,t) &= \sigma\left(\delta\cos^2\vartheta - \sin^2\vartheta\cos^2\varphi\right) \\ &\quad -\left(\xi_0 + \xi\cos\omega t\right)\cos\Theta(\vartheta,\varphi), \end{aligned} \tag{2.340}$$

where $\cos\Theta(\vartheta,\varphi)$ is given by Eq. (2.299); ξ_0 and ξ are the usual dc and ac field parameters, respectively. The magnetization $M_H = M_S\langle\cos\Theta\rangle$ in the direction of the ac driving field $\mathbf{H}\cos\omega t$ may be expressed via the statistical moments $\langle Y_{10}\rangle(t)$ and $\langle Y_{11}\rangle(t)$ as [104]

$$M_H(t) = M_S\sqrt{\frac{4\pi}{3}}\left\{\gamma_3\langle Y_{10}\rangle(t) - \sqrt{2}\,\mathrm{Re}\left[(\gamma_1 - i\gamma_2)\langle Y_{11}\rangle(t)\right]\right\}, \tag{2.341}$$

where $\gamma_1 = \cos\varphi_\xi \sin\vartheta_\xi, \gamma_2 = \sin\varphi_\xi \sin\vartheta_\xi, \gamma_3 = \cos\vartheta_\xi$ are the direction cosines of the applied ac field $\mathbf{H}$. However, like zero STT (see Section 2.6) due to the sinusoidal term in the applied field $\mathbf{H}_0 + \mathbf{H}\cos\omega t$, the stationary ac response of $M_H(t)$ must be written as a Fourier series in the time, Eqs. (2.304) and (2.305), because (with the notable exception of the linear response) all harmonics of the ac field will now be involved. In Eqs. (2.304) and (2.305), the c_{lm}^k are themselves the Fourier coefficients in a Fourier series development (in the time) of the average spherical harmonics, Eq. (2.292). Equation (2.304) includes the linear response as a special case, $\xi \to 0$, whereupon all higher harmonics may be discarded in Eq. (2.304) and only the term $m_1^1(\omega)$ now linear in ξ remains. With the Fourier amplitudes $m_1^k(\omega)$, we have $M_H(t)$ and other related parameters such as the dynamic susceptibilities, etc. This calculation will also yield the DMH loop representing as we have seen a parametric plot of the stationary time-dependent magnetization as a function of the ac field, i.e., $M_H(t)$ vs. $H(t) = H\cos\omega t$, and the normalized area enclosed by the loop A_n, Eq. (2.331).

Now the calculation of the ac stationary response from the magnetic Langevin equation (1.75) can once more be reduced to the solution of an infinite hierarchy of 25-term differential-recurrence relations for the statistical moments $c_{lm} = \langle Y_{lm} \rangle(t)$, Eq. (2.337). Such a hierarchy has been derived in Section 2.8.1 for the nonconservative potential due to a spin-polarized current given by Eq. (2.336) and the conservative biaxial anisotropy plus the Zeeman term due to a spatially uniform dc bias field $\mathbf{H}_0$. In like manner, we can generalize this derivation to our ac case representing the response to a dc bias field *temporally modulated* by an ac field $\mathbf{H}\cos\omega t$. Here the total free energy density E is given by Eq. (2.340) above and the infinite hierarchy of 25-term differential-recurrence relations for $c_{lm} = \langle Y_{lm} \rangle(t)$ is given by Eq. (2.337). Furthermore, the coefficients $x_{nm}(t)$, $y_{nm}(t)$, etc. now have the same form as those in Eq. (2.337) save that they are now *time-dependent*. To keep in step with the notation of Section 2.7, we define reduced fields

$$h_0 = H_0 / (2M_S D_\|) \text{ and } h_1 = H / (2M_S D_\|) \tag{2.342}$$

Thus, in our particular case involving a single ac forcing term, we have

$$h(t) \to h_0 + 2h_1 \cos\omega t \to h_0 + h_1(e^{i\omega t} + e^{-i\omega t}) \tag{2.343}$$

Equation (2.343) then implies that in superimposed ac and dc fields, the coefficients x_{nm}, y_{nm}, w_{nm} in Eq. (2.337) may be written as the *sum of a dc term* and *one oscillating* at the fundamental frequency, viz.

$$\begin{aligned} w_{nm} &= w^0_{nm} + 2w^1_{nm}\cos\omega t, \\ x_{nm} &= x^0_{nm} + 2x^1_{nm}\cos\omega t, \\ y_{nm} &= y^0_{nm} + 2y^1_{nm}\cos\omega t. \end{aligned}$$

The time-independent coefficients $v^0_{nm}, w^0_{nm}, x^0_{nm}, y^0_{nm}, z^0_{nm}$ are as listed explicitly in Section 2.6.2 while the coefficients $w^1_{nm}, x^1_{nm}, y^1_{nm}$ are given by

$$x_{nm}^{1} = i\frac{m\sigma h_1\gamma_3}{\alpha},$$

$$x_{nm}^{\pm1} = \frac{i\sigma h_1}{2\alpha}\sqrt{(1+n\pm m)(n\mp m)}\,(\gamma_1 \mp i\gamma_2),$$

$$y_{nm}^{1} = -\sigma h_1\gamma_3 n\sqrt{\frac{(n+1)^2-m^2}{(2n+1)(2n+3)}},$$

$$y_{nm}^{\pm1} = \pm\frac{n}{2}\sigma h_1(\gamma_1 \mp i\gamma_2)\sqrt{\frac{(1+n\pm m)(2+n\pm m)}{(2n+1)(2n+3)}},$$

$$w_{nm}^{1} = \sigma h_1\gamma_3(n+1)\sqrt{\frac{n^2-m^2}{4n^2-1}},$$

$$w_{nm}^{\pm1} = \pm\sigma h_1(\gamma_1 \mp i\gamma_2)\frac{n+1}{2}\sqrt{\frac{(n\mp m)(n\mp m-1)}{4n^2-1}}.$$

The differential-recurrence relations Eq. (2.337) with time-dependent coefficients can be solved via matrix continued fractions like the zero STT case treated in Section 2.6 with only slight modifications of the algorithm previously used. First by defining vectors $\mathbf{c}_n^k$ via Eq. (2.300) with elements composed of the Fourier amplitudes c_{nm}^k (see Eq. (2.292)), we have a nine-term algebraic recurrence relation for the $\mathbf{c}_n^k$, Eq. (2.293), where the matrices $\mathbf{q}_n, \mathbf{q}_n^{\pm}, \mathbf{p}_n, \mathbf{p}_n^{\pm}$, are now

$$\mathbf{q}_n = \begin{pmatrix} \mathbf{X}_{2n}^0 & \mathbf{W}_{2n}^0 \\ \mathbf{Y}_{2n-1}^0 & \mathbf{X}_{2n-1}^0 \end{pmatrix}, \mathbf{q}_n^{+} = \begin{pmatrix} \mathbf{Z}_{2n} & \mathbf{Y}_{2n}^0 \\ \mathbf{0} & \mathbf{Z}_{2n-1} \end{pmatrix},$$

$$\mathbf{q}_n^{-} = \begin{pmatrix} \mathbf{V}_{2n} & \mathbf{0} \\ \mathbf{W}_{2n-1}^0 & \mathbf{V}_{2n-1} \end{pmatrix}, \mathbf{p}_n = \begin{pmatrix} \mathbf{X}_{2n}^1 & \mathbf{W}_{2n}^1 \\ \mathbf{Y}_{2n-1}^1 & \mathbf{X}_{2n-1}^1 \end{pmatrix},$$

$$\mathbf{p}_n^{+} = \begin{pmatrix} \mathbf{0} & \mathbf{Y}_{2n}^1 \\ \mathbf{0} & \mathbf{0} \end{pmatrix}, \mathbf{p}_n^{-} = \begin{pmatrix} \mathbf{0} & \mathbf{0} \\ \mathbf{W}_{2n-1}^1 & \mathbf{0} \end{pmatrix}.$$

Here $\mathbf{V}_n, \mathbf{Z}_n, \mathbf{W}_n^i, \mathbf{X}_n^i$, and $\mathbf{Y}_n^i$ have virtually the same form as the submatrices $\mathbf{V}_n, \mathbf{Z}_n, \mathbf{W}_n, \mathbf{X}_n$, and $\mathbf{Y}_n$ from Section 2.6.2 defined in terms of the *time-independent* elements $x_{nm} = x_{nm}^0, y_{nm} = y_{nm}^0, w_{nm} = w_{nm}^0$. The few differences which occur are highlighted by the superscript i = 0, 1 in the submatrices $\mathbf{W}_n^i, \mathbf{X}_n^i$, and $\mathbf{Y}_n^i$, so indicating that the elements w_{nm}, x_{nm}, y_{nm}

appearing in them must now be replaced by $w^{j}_{nm}, x^{i}_{nm}, y^{i}_{nm}$, etc., respectively. Next, we introduce supercolumn vectors $\mathbf{C}_n$ via

$$\mathbf{C}_0 = \begin{pmatrix} \vdots \\ 0 \\ 0 \\ \mathbf{c}_0^0 \\ 0 \\ 0 \\ \vdots \end{pmatrix}, \mathbf{C}_n = \begin{pmatrix} \vdots \\ \mathbf{c}_n^{-2} \\ \mathbf{c}_n^{-1} \\ \mathbf{c}_n^{0} \\ \mathbf{c}_n^{1} \\ \mathbf{c}_n^{2} \\ \vdots \end{pmatrix}, \mathbf{Q}_1^{-}\mathbf{C}_0 = \frac{1}{\sqrt{4\pi}}\begin{pmatrix} \vdots \\ \mathbf{0} \\ \mathbf{p}_1^{-} \\ \mathbf{q}_1^{-} \\ \mathbf{p}_1^{-} \\ \mathbf{0} \\ \vdots \end{pmatrix}, \mathbf{p}_1^{-} = \begin{pmatrix} 0 \\ 0 \\ 0 \\ 0 \\ 0 \\ w_{1-1}^{+1} \\ w_{10}^{1} \\ w_{11}^{-1} \end{pmatrix}, \mathbf{q}_1^{-} = \begin{pmatrix} v_{2-2}^{++} \\ v_{2-1}^{+} \\ v_{20} \\ v_{21}^{-} \\ v_{22}^{--} \\ w_{1-1}^{+} \\ w_{10} \\ w_{11}^{-} \end{pmatrix}. \quad (2.344)$$

thereby yielding the tridiagonal algebraic matrix recurrence relation, Eq. (2.294), where the tridiagonal supermatrices $\mathbf{Q}_n$ and $\mathbf{Q}_n^{\pm}$ are defined as

$$\left[\mathbf{Q}_n^{\pm}\right]_{lm} = \delta_{l-1m}\mathbf{p}_n^{\pm} + \delta_{lm}\mathbf{q}_n^{\pm} + \delta_{l+1m}\mathbf{p}_n^{\pm},$$

$$\left[\mathbf{Q}_n\right]_{lm} = \delta_{l-1m}\mathbf{p}_n + \delta_{lm}\left(\mathbf{q}_n - im\tau_{\mathrm{N}}\omega\mathbf{I}\right) + \delta_{l+1m}\mathbf{p}_n.$$

The exact solution of Eq. (2.294) is then given by Eq. (2.296), where the vector $\mathbf{C}_1$ obviously contains the elements c_{10}^{k} and $c_{1\pm1}^{k}$ needed for both the linear and nonlinear ac stationary responses given by Eqs. (2.304) and (2.305).

Next (as explained above), the vectors $\mathbf{H}_0$, $\mathbf{H}$, and $\mathbf{e}_P$ are assumed throughout to lie in the equatorial or *XY* plane with colatitudes $\vartheta_\xi = \pi/2$ and $\vartheta_P = \pi/2$, respectively, so that the orientations of $\mathbf{H}_0$, $\mathbf{H}$, and $\mathbf{e}_P$ are entirely specified by the azimuthal angles of the applied fields φ_ξ and spin -polarization φ_P, respectively. The specific values $\varphi_\xi = \varphi_P = 0$ again correspond to the configuration with the vectors $\mathbf{H}_0$, $\mathbf{H}$, and $\mathbf{e}_P$ all directed along the (easy) *X*-axis. The anisotropy parameter $D_{\|}$, spin-polarization azimuthal angle φ_P, biaxiality parameter δ, spin-polarization factor P, and damping α selected are $D_{\|} = 0.034$, $\varphi_P = 0$, $\delta = 20$, $P = 0.3$, and $\alpha = 0.01$. For a particle volume $v \sim 10^{-24}\,\mathrm{m}^3$, $M_S \approx 1.4\cdot10^6\,\mathrm{Am}^{-1}$ (cobalt), and $T \sim 293$ K, the field parameters ξ_0 and ξ are of the order of unity for

$$H_0, H \sim \frac{kT}{v\mu_0 M_S} \approx 2.3\times10^3\ \mathrm{A\,m}^{-1}.$$

2.8.3 Linear dynamic susceptibility

In the *weak* ac field situation, $\xi \to 0$, all nonlinear effects in the response may be ignored, so that the magnetization $M_H(t)$ is simply given by the linear response

$$M_H(t)/M_S = M_0 + \mathrm{Re}\left\{\chi(\omega)\xi e^{i\omega t}\right\}, \quad (2.345)$$

where

$$M_0 = \langle\cos\Theta\rangle_{st} = m_1^0(\omega) = \frac{\omega}{2\pi M_S}\int_0^{2\pi/\omega} M_H(t)dt$$

is the stationary *time- and frequency-independent magnetization*, $\langle\ \rangle_{st}$ denotes the statistical average, and $\chi(\omega) = 2m_1^1(\omega)/\xi$ is the linear dynamic susceptibility which is of course independent of the ac field strength. The corresponding plots of the real and imaginary parts of the normalized susceptibility $\chi(\omega)/\chi$ vs. $\omega\tau_N$ are shown in Fig. 2.59, where $\chi = \chi(0)$ is the static susceptibility. As in zero STT, analysis and subsequent interpretation of the linear response radically simplifies at low frequencies. The simplification occurs because the low-frequency behavior of $\chi(\omega) = \chi'(\omega) - i\chi''(\omega)$ may then be accurately described by a *single* Lorentzian, Eq. (2.269). In Eq. (2.269), τ is the longest (overbarrier) relaxation time without the ac external field, and Δ_1 is a parameter accounting for the mid- and high-frequency relaxation processes. τ is as usual related to the frequency ω_{max} of the low-frequency peak in the loss spectrum $\chi''(\omega)$, where it attains a maximum, and the half-width $\Delta\omega$ of the spectrum $\chi'(\omega)$ via

$$\tau \approx \omega_{max}^{-1} \approx \Delta\omega^{-1}. \quad (2.346)$$

Since τ is also the magnetization reversal time (effectively the inverse escape rate), it can always be associated with the inverse of the smallest nonvanishing eigenvalue λ_1^{-1} of the Fokker–Planck operator. Comparison of τ as extracted from the spectra $\chi(\omega)$ via Eq. (2.346) with $\tau = \lambda_1^{-1}$ calculated independently via λ_1 of the Fokker–Planck operator shows that both methods yield identical results. Also, variation of the material and geometrical model parameters alters the reversal time τ by several orders of magnitude concurring with experimental results.

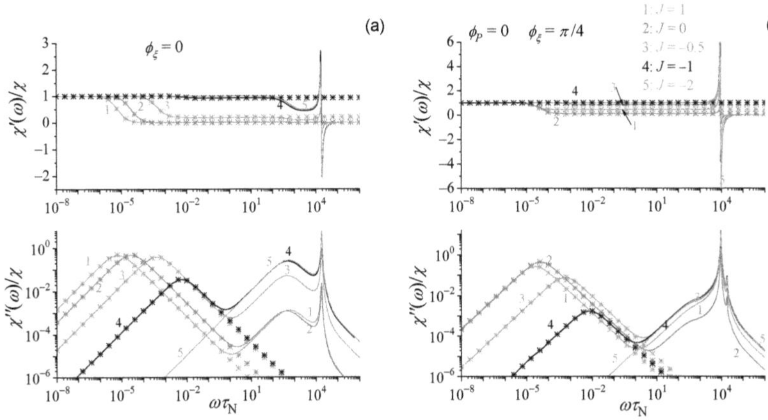

Fig. 2.59. Real and imaginary parts of the normalized linear susceptibility $\chi(\omega)/\chi$ vs. the normalized frequency $\omega\tau_N$ for various spin-polarized current parameters $J = 1, 0, -0.5, -1, -2$ and for various orientations of the applied fields $\varphi_\xi = 0$ (a) and $\varphi_\xi = \pi/4$ (b) with the anisotropy parameter $\sigma = 20$ and the dc field parameter $\xi_0 = 2$. Solid lines: matrix continued fraction solution. Asterisks: Approximate Eq. (2.269) with the reversal time $\tau = \lambda_1^{-1}$ calculated using the independent method based on the smallest nonvanishing eigenvalue of the Fokker-Planck equation.

The main features of the normalized linear susceptibility plots are as follows. For intermediate J, the overall picture is similar to that for $J = 0$, i.e., we observe the low-frequency overbarrier (interwell) relaxation, mid-frequency (intrawell) relaxation, and high-frequency resonance behavior in a biaxial potential. Thus, we again have, in general, three dispersion regions in $\chi'(\omega)$ and three corresponding absorption peaks in the magnetic loss spectrum $\chi''(\omega)$ (see Fig. 2.59). The broad low-frequency peak in $\chi''(\omega)$ corresponds to *slow* reversal of the magnetization vector over the potential barriers and is accurately described by the approximate Eq. (2.269). The most pronounced STT effect is that the decrease of J from large positive values *initially* shifts the low-frequency relaxation peak to *lower* frequencies until the peak frequency ω_{max} reaches a minimum at some intermediate value of J above which the peak is shifted to *higher* frequencies (only the shift to higher frequencies is shown in Fig. 2.59). This minimum frequency peak relates to the particular situation, where the STT has annulled the effect of the dc bias field so that the effective potential has equal well depths corresponding to the maximum relaxation time at a definite value of J_{max}, depicted graphically in Fig. 2.58(d).

Moreover, for high positive or negative J, the magnitude of the low-frequency peak in $\chi''(\omega)$ decreases until it merges with the mid-frequency peak, signifying that the overbarrier relaxation process has been completely extinguished due to the action of STT. Thus, *high magnitude spin-polarized current seems to have virtually the same effect on the magnetization reversal as that of a strong dc bias field in single domain ferromagnetic particles at zero STT*. Here we have seen that at a certain critical value of this field far less than the nucleation field the integral relaxation time (area under the curve of the magnetization decay) *diverges exponentially from the overbarrier relaxation time due to the depletion of the population of the shallowest well of the potential by the dc field*. This event is also signified by the virtual disappearance of the low-frequency peak in the magnetic loss spectrum. Thus, all that now remains are the dynamical processes within the wells. The explanation appears to be that the high positive or negative J reduces the *effective* potential barrier so that the overbarrier time decreases. Regarding the second peak at intermediate frequencies, this is due to *fast near-degenerate exponential decays in the wells* of the effective potential and comprises the familiar longitudinal *intrawell* relaxation. Lastly, we observe a third FMR peak at the Larmor frequency ω_{pr} with origin lying in the magnetization precession in the effective field due to both the anisotropy and applied dc field. Notice that the susceptibility is strongly influenced by the azimuthal angle φ_ξ of the applied ac field. For $\varphi_\xi \neq 0$, *high-frequency resonant harmonic modes are also discernible in the FMR peak generating a comb-like structure with characteristic frequencies* $n\omega_{pr}$, $n = 2, 3, \ldots$ reminiscent of that occurring in inertia-corrected dielectric relaxation of polar molecules at THz frequencies under the influence of a mean field potential [97,398]. This comb-like structure virtually disappears, however, for $\varphi_\xi = 0$. The STT has no effect at all on the mid-frequency and FMR regions with any apparent changes being purely an artifact of the normalization.

2.8.4 *Nonlinear response*

In strong applied ac fields, $\xi > 1$, pronounced frequency-dependent nonlinear effects again occur (see Figs. 2.60 and 2.61 illustrating the

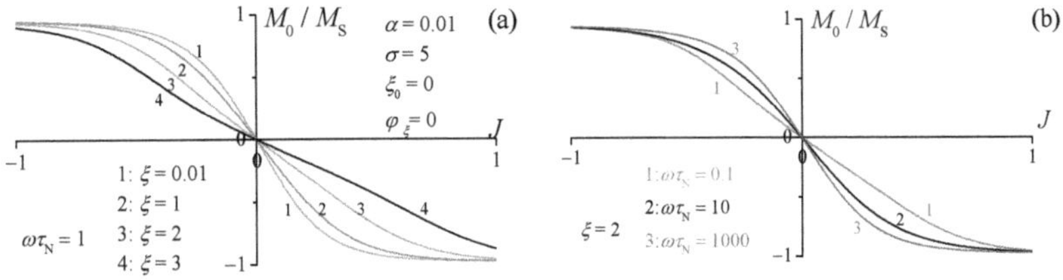

Fig. 2.60. (a) Time-independent (dc) component of the magnetization M_0 / M_S vs. the spin-polarized current parameter J for (a) various ac field amplitudes $\xi = 0.01$, 1, 2, and 3 ($\xi = 0.01$ represents linear response) and $\omega\tau_N = 1$ and for (b) various frequencies $\omega\tau_N = 10^{-1}, 10$, and 10^3 at $\xi = 2$ at $\sigma = 5$, $\varphi_\xi = 0$, and $\xi_0 = 0$. Solid lines: the matrix continued fraction solution.

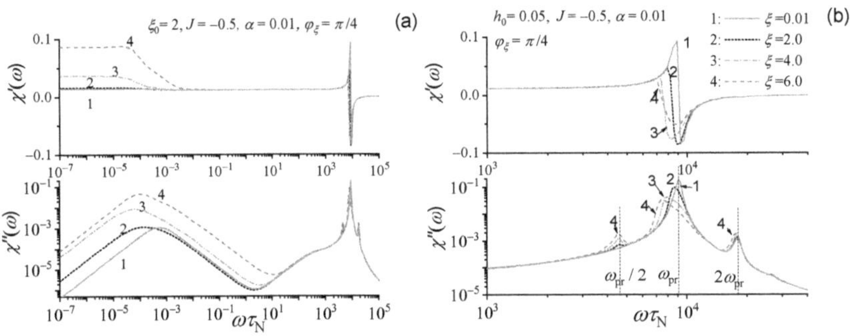

Fig. 2.61. (a) Real $\chi'(\omega)$ and imaginary $\chi''(\omega)$ parts of the nonlinear susceptibility vs. $\omega\tau_N$ for various ac field amplitudes ξ and $J = -0.5$, $\varphi_\xi = \pi/4$, $\sigma = 20$, $\alpha = 0.01$ and $\xi_0 = 2$. Solid and dashed lines: linear and nonlinear response, respectively. (b) The high-frequency parts of the spectra alone.

dependence of the nonlinear response on ξ). In contrast to the linear response, the stationary *time-independent* but now *frequency-dependent* dc magnetization $M_0(\omega) = M_S m_1^0(\omega)$ and the nonlinear dynamic susceptibility $\chi(\omega) = 2m_1^1(\omega)/\xi = \chi'(\omega) - i\chi''(\omega)$ as well as all other higher harmonic responses $m_1^k(\omega)$ with $k > 1$ now strongly depend on ξ. Moreover, for given ξ, all the $m_1^k(\omega)$ also markedly depend on the φ_ξ, ξ_0, σ, δ, α, and J, while the time-independent (dc) component of the magnetization M_0 alters profoundly leading to new nonlinear effects. That component in typical nonlinear fashion becomes *dependent on both the amplitude and frequency of the ac field* (see Fig. 2.60), in sharp contrast to that of nanomagnets in an ac field omitting STT, where *one*

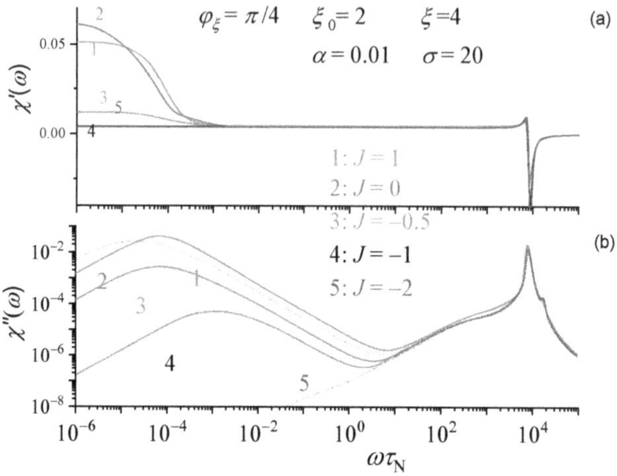

Fig. 2.62. $\chi'(\omega)$ and $\chi''(\omega)$ vs $\omega\tau_N$ for various spin-polarized current parameters $J = 1, 0, -0.5, -1, -2$, and $\sigma = 20$, $\varphi_\xi = \pi/4$, $\alpha = 0.01$, $\xi_0 = 2$, and $\xi = 4$.

must also apply a dc bias in combination with a *strong ac field* to observe the frequency dependence of the dc response. This striking effect being due to *entanglement* of the nonlinear ac and dc responses. However, with STT included the ac field amplitude and frequency dependence of M_0 *always* exists even for zero dc bias field, i.e., $\xi_0 = 0$. Hence, the spin-polarized current seems to have the same effect on $M_0(\omega)$ as that of a dc bias field at zero STT. Furthermore, the dc response in Fig. 2.60 is not an odd function of J due to the nonconservative nature of the potential Φ.

In strong ac fields, it appears that the low-frequency part of $\chi(\omega)$ can no longer be approximated by the single Lorentzian Eq. (2.269). Nevertheless, the Eq. (2.346) may still be used to estimate an effective magnetization reversal time τ as $\tau \approx \Delta\omega^{-1}$. Furthermore, as the ac field strength ξ increases, the magnitude of the low-frequency peak in $\chi''(\omega)$ is enhanced (Fig. 2.61) and also with increasing ξ the overbarrier peak on initially shifting to lower frequencies, attains a minimum frequency, thereafter shifting to higher frequencies. Omitting STT, this minimum frequency peak will occur at $\xi \sim \xi_0$. However, this is no longer true with STT as the latter now acts in combination with the applied field (Fig. 2.62). The reversal time τ may also be evaluated from the spectra of *both* the dc component $M_0(\omega)$ and the higher-order harmonics $m_1^k(\omega)$ with $k > 1$

because the low-frequency parts of these spectra are themselves dominated by overbarrier relaxation processes with the characteristic time τ. As seen in Fig. 2.61(b), again with increasing ξ, the magnitude of the main FMR peak at the precession frequency ω_{pr} decreases and broadens showing pronounced saturation effects. Moreover, a new high-frequency dispersion of *subharmonic resonant* character near the frequency $\sim \omega_{\text{pr}}/2$ again due to *parametric resonance* appears, like that commonly occurring in nonlinear mechanical oscillators driven by an ac external force. Nevertheless, the very high-frequency ($\omega >> \omega_{\text{pr}}$) behavior of the spectrum remains virtually unchanged (see Fig. 2.62). Parametric excitations of a current-biased nanomagnet by a microwave magnetic field were observed by Urazhdin *et al.* [392] amply demonstrating that this phenomenon can be used to determine dynamical properties of nanomagnets.

The nonlinear susceptibility for various J (Fig. 2.62) exhibits many of the same characteristics as the corresponding linear susceptibility, i.e., three dispersion regions in $\chi'(\omega)$ and three corresponding absorption peaks in the magnetic loss spectrum $\chi''(\omega)$ for J of intermediate magnitude, merging of the overbarrier peak with the mid-frequency peak for high positive or negative J, and virtually no STT effect in the mid-frequency and FMR regions. For $J < 0$ the magnitude of the low-frequency overbarrier peak *decreases* and shifts to higher frequencies, while for $J > 0$ its magnitude increases and shifts to higher frequencies.

2.8.5 *Dynamic magnetic hysteresis*

For a *weak* ac field, $\xi \to 0$, the DMH loops [normalized magnetization $m(t) = M_H(t) / M_S$ vs. reduced ac field $h(t) = \cos\omega t$] curves are ellipses with normalized area A_n given by Eq. (2.331); the behavior of $A_n \sim \chi'' \sim -\text{Im}(m_1^1)$ being similar [cf. Eq. (2.331)] to that of the magnetic loss $\chi''(\omega)$. Now for *relatively weak* ac fields, $\xi \approx 1$, the DMH loops still have an ellipsoidal shape implying that only a few harmonics contribute to the nonlinear response. However, in *strong* ac fields, $\xi > 1$, the loop shape alters substantially (see Figs. 2.63 and 2.64). In Fig. 2.63, the loops are plotted for various J and ac amplitudes ξ showing how their form alters as these parameters vary. The pronounced frequency dependence of the

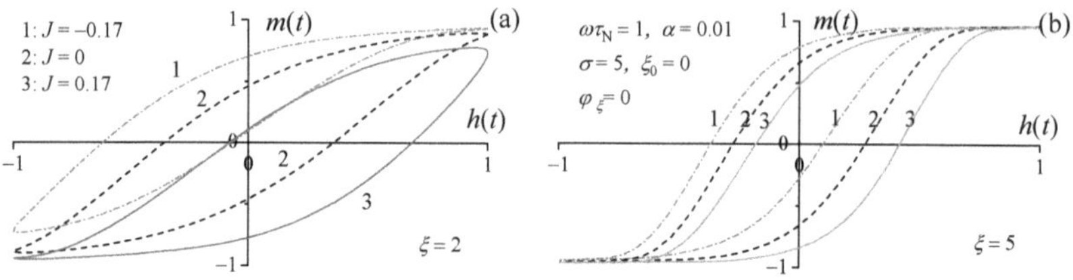

Fig. 2.63. DMH loops for various spin-polarized current parameters $J = -0.17$, 0, 0.17 and (a) $\xi = 2$ and (b) $\xi = 5$ with $\omega\tau_N = 1$, $\varphi_\xi = 0$, $\alpha = 0.01$, $\sigma = 5$ and $\xi_0 = 0$.

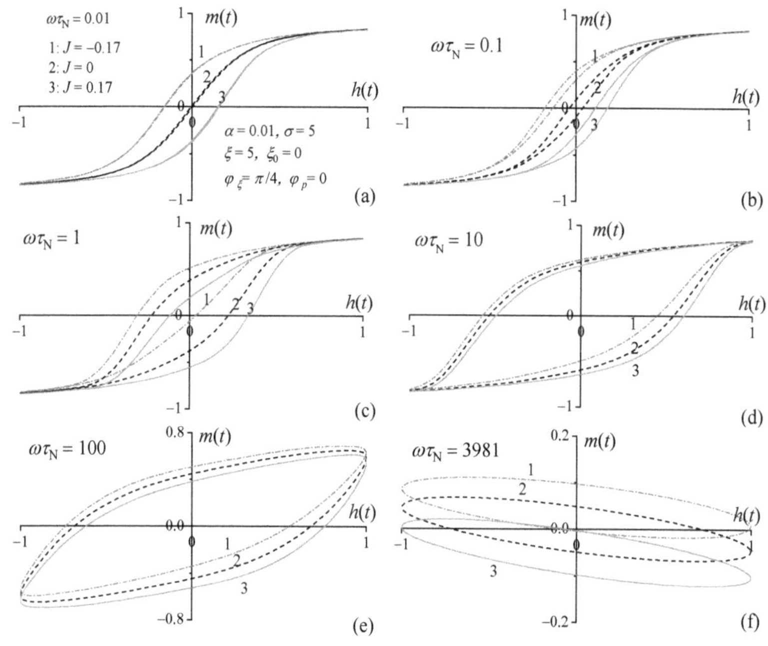

Fig. 2.64. DMH loops for various spin-polarized current parameters $J = -0.17$, 0, 0.17 and frequencies $\omega\tau_N = 10^{-2}$ (a), 10^{-1} (b), 1 (c), 10 (d), 10^2 (e), and 3981 (f) with $\varphi_\xi = \pi/4$, $\varphi_P = 0$, $\alpha = 0.01$, $\xi_0 = 0$, $\xi = 5$, and $\sigma = 5$.

DMH is highlighted in Fig. 2.64. At low frequencies, the field changes are *quasiadiabatic*, so that the magnetization reverses due to the *cooperative* shuttling action of thermal agitation, STT, and ac field. In contrast at high frequencies (see Fig. 2.64(f)), the origin of the DMH again lies in the *resonant* dispersion and absorption in the FMR peak. Here, the phase difference $\Delta\phi$ between $M_H(t)$ and $H(t)$, governing the *loop orientation*, may undergo a pronounced variation in the very high frequency FMR peak

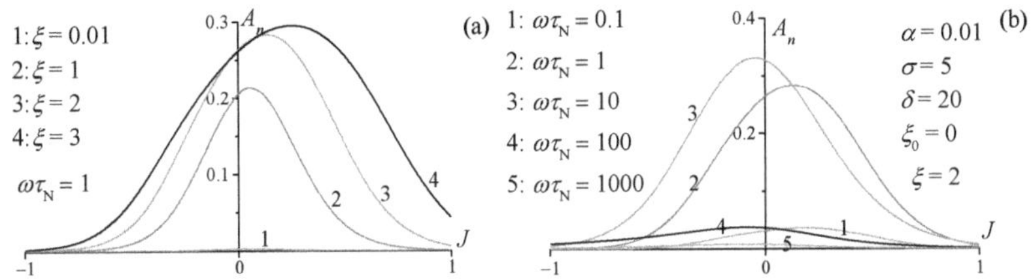

Fig. 2.65. Normalized area of the DMH loop A_n, Eq. (2.331), vs. spin-polarized current parameter J (a) for various ac field amplitudes ξ and $\omega\tau_N = 1$ and (b) for various frequencies $\omega\tau_N$ and $\xi = 2$ with $\varphi_\xi = 0$, $\alpha = 0.01$, $\xi_0 = 0$ and $\sigma = 5$.

as is typical of a resonant process and it may exceed $\pi/2$. Obviously, this large resonant effect does not exist at all for low and intermediate frequencies, where $\Delta\phi$ is always less than $\pi/2$. Thus, at FMR frequencies, DMH occurs due to the *resonant* behavior of the nonlinear response and obviously the switching may be termed "resonant", leading naturally to the novel concept of *resonant switching of the magnetization.* Since resonant DMH occurs at very high (GHz) frequencies, the magnetization switching is for the most part governed by the rate of change of the external driving field. Here the magnetization may be more *advantageously switched, because the field needed to reverse it is then much smaller than the quasistatic coercive force.*

By plotting the normalized area A_n vs. the spin-polarized current (Fig. 2.65), A_n can invariably be represented as a bell-like curve with the height, width, and center of the peak determined by the various parameters. This resembles a plot of A_n vs. the dc bias field strength ξ_0 except that the latter will always have the center of the peak along $\xi_0 = 0$. In Fig. 2.65(a), on increasing the ac field strength ξ, A_n also increases and the range of J, for which a significant DMH loop area exists, broadens. In *strong* ac fields, $\xi > 5$, the normalized area alters substantially (see Fig. 2.65(a)). Nevertheless, A_n is still determined by $-\mathrm{Im}(m_1^1)$ [cf. Eq. (2.331)]. Thus, A_n strongly depends on the frequency ω (see Fig. 2.65(b)). Moreover, on increasing the driving frequency, the normalized area initially increases, reaches a maximum, and then decreases.

Thus, STT effects on the ac stationary forced response of nanoscale ferromagnets driven by an ac magnetic field of arbitrary strength may be treated using the same nonperturbative approach as developed for zero

STT. The method, based on the solution of the differential-recurrence relation for the infinite hierarchy of statistical moments generated by either the Langevin or Fokker–Planck equations as augmented by STT terms, indicates that STT profoundly alters the nonlinear response of a nanomagnet leading to many new effects. Furthermore, the statistical moment approach holds for the *most comprehensive* formulation of the generic nanopillar model (Fig. 1.11), i.e., for *arbitrary directions* of the dc and ac external fields. Thus, it is again possible to treat STT effects on frequency-dependent characteristics under conditions which are otherwise inaccessible. Clearly, at low damping, the stationary response to an ac driving field is very sensitive to both the intensity of the spin-polarized current and the frequency and amplitude of that field again owing to the *intrinsic* coupling between the magnetization precession and its thermally activated reversal. Furthermore, the calculations (valid for ac fields of *arbitrary* strength and orientation) quantify the role played by STT in nonlinear phenomena in nanoscale ferromagnets such as nonlinear stochastic resonance and dynamic magnetic hysteresis, nonlinear ac field effects on the switching-field curves, etc., where perturbation theory is no longer valid. In addition, the method yields the response for *all frequencies of interest* including the very high frequencies covering the ferromagnetic resonance (GHz) range so exemplifying various nonlinear phenomena such as *parametric resonance* and *higher harmonic generation* (which may stimulate new experiments). Hence, the high-frequency linear and nonlinear FMR spectra may explain the line shape of STT nano-oscillators driven by ac external magnetic fields and currents. Likewise, the DMH loops and their area (yielding the Joule heating during the switching process) as well as the calculations of the effective magnetization reversal time via the low-frequency peak of the magnetic loss spectra may be useful for the prediction, modeling, and interpretation of switching processes in recording techniques. Finally, the DMH arising from a *high-frequency* periodic signal may be exactly evaluated permitting quantitative analysis of *ultrafast switching* of the magnetization. Furthermore, the moment method may be similarly generalized to the forced response of a nanoscale ferromagnet driven by an *alternating current*.

2.9 Antiferromagnetic nanoparticles

The physical properties of both ferromagnets and antiferromagnets are drastically modified when their dimensions are reduced to the nanometric scale thereby prompting extensive studies of such nanoparticles. However, the magnetization dynamics of *antiferromagnetic* nanoparticles may differ markedly from those of ferromagnetic ones because of the intrinsic properties of antiferromagnetic materials [88]. Furthermore, their magnetic behavior can be quite different from that observed in the bulk. Examples include enhanced magnetic moment and coercivity, exchange bias, increase in magnetic moment with temperature, decrease in the susceptibility with temperature below the ordering (Néel) temperature T_N and its enhancement compared to that of the bulk [399,400]. Moreover, due to thermal agitation, *like ferromagnetic nanoparticles, antiferromagnetic ones should become superparamagnetic at finite temperatures* (the so-called *superantiferromagnetism* [401–404]). By using Brown's continuous diffusion model, Raikher and Stepanov [399,400] have evaluated both the characteristic relaxation times and the linear dynamic susceptibility of a suspension of antiferromagnetic nanoparticles for dc and ac magnetic fields lying parallel to the easy axis of the particle. Furthermore, their results were extended in Ref. [158] to a dc magnetic field applied at an *arbitrary* angle to the easy axis. Moreover, the magnetization reversal time τ of antiferromagnetic nanoparticles has been evaluated in Ref. [157] via the Kramers escape-rate theory as adapted to magnetic nanoparticles [97,239]; the analytic equations for τ so obtained agree favorably with the numerical solution of the Fokker–Planck equation [157,158]. The primary goal of this section is to give a detailed investigation of the magnetization, nonlinear dynamic magnetic susceptibility, and DMH of antiferromagnetic nanoparticles. Furthermore, we shall demonstrate that in superimposed dc and ac fields $\mathbf{H}_0 + \mathbf{H}(t)$ these magnetic characteristics substantially alter in turn leading to interesting nonlinear effects. As seen before the theoretical treatment of the nonlinear response and associated phenomena inherently poses a complicated mathematical problem because as usual no *unique* response function relating the transient and ac stationary responses, unlike in linear response, exists. However, these difficulties may be overcome by extending the methods already developed in this chapter and in Refs. [37,47,52,155,156,159,381,382,]. Thus, one may again treat the dynamic nonlinear stationary response of antiferromagnetic nanoparticles driven by an ac magnetic field allowing one to evaluate both the linear and nonlinear ac stationary responses over a wide range of damping and from very low to high (GHz) frequencies. Once again if the field $\mathbf{H}_0$ is applied at an arbitrary angle to the easy axis of the particle, a strong intrinsic dependence of magnetic characteristics (such as the reversal time, complex magnetic susceptibility, etc.) on the damping α arising from coupling of the longitudinal and magnetization transverse modes exists. Therefore, the nonlinear response can again be used to determine α [37].

The Néel–Brown model of the thermal fluctuations in ferromagnetic nanoparticles may also be adapted to antiferromagnetic nanoparticles. In an *antiferromagnet* the exchange energy is *negative* so that neighboring spins *alternate* in orientation and there is *no net moment* even though long-range magnetic ordering exists [88]. In *ferrimagnets*, however, a *spontaneous moment* exists because despite the antiferromagnetic coupling, interpenetrating spin-up and spin-down lattices with different moments also exist that do not cancel [88]. The ensuing adaptation of the model is as follows. First observe that an ideal antiferromagnet can be divided into two sublattices, with *equal* and *opposite* magnetic moments and if the numbers N_1 and N_2 of ions in these sublattices are equal, the sample is said to be "compensated". In *fine* antiferromagnetic particles, however, such *total magnetic compensation of the sublattices is impossible* for many reasons (unequal numbers of spins in crystal planes, spin frustration near the surface, lattice defects, etc.) so that $N_1 \neq N_2$ resulting in *effective spontaneous magnetization* of these particles. Furthermore, if we assume that *the ionic magnetic moments m remain the same* and $N_1 > N_2$, then the total moments of the sublattices $m_1 = N_1 m\mathbf{u}$ and $m_2 = -N_2 m\mathbf{u}$ are unequal. Here $\mathbf{u}$ is the *antiferromagnetic vector*, specifying the *decompensation* magnetic moment

$$\boldsymbol{\mu} = \mathbf{m}_1 + \mathbf{m}_2 = \mathbf{u}\mu,$$

where $\mu = m(N_1 - N_2)$. Now in an antiferromagnetic nanoparticle, whereas an external dc magnetic field $\mathbf{H}_0$ applied along $\mathbf{u}$ will not tend to rotate the moments $\mathbf{m}_1$ and $\mathbf{m}_2$, a field *normal* to $\mathbf{u}$ will. Hence, an antiferromagnetic nanoparticle possesses a considerable *transverse* magnetic susceptibility χ_A characterizing the induced magnetic moment of the particle which may be written as

$$\frac{v\chi_A}{2}[\mathbf{H}_0 - \mathbf{u}(\mathbf{u}\cdot\mathbf{H}_0)].$$

Consequently, the magnetic moments of two sublattices of the particle can be written allowing for *decompensation* [399,400] as

$$\mathbf{m}_{1,2} = \frac{1}{2}\mathbf{u}\left[\pm 2vM_S + \mu - v\chi_A\left(\mathbf{H}_0\cdot\mathbf{u}\right)\right], \qquad (2.347)$$

where $M_S = m(N_1 + N_2)/(2v)$ is the bulk sublattice magnetization and v is the particle volume. Specifically due to *superantiferromagnetism*, χ_A exceeds that of a bulk crystal by a factor of 2 or 3 attaining a typical value of order 10^{-4} [16,15,399,400,] while the magnitude of the magnetic moment $\boldsymbol{\mu}$ is roughly $\mu \sim \mu_B(N_1 - N_2)(N_1 + N_2)^{1/2}$ [15,16,399,400], where μ_B is the Bohr magneton. Consequently for a particle with a diameter ~ 10 nm and $N \sim 10^5$–10^6, μ may vary from a few μ_B to a few hundred μ_B [405–410] (that is μ of the same order of magnitude as the magnetization of weak ferromagnets). At temperatures below the Néel temperature T_N if the external magnetic fields are much weaker than the exchange field, one may therefore assume [399,400] that the *decompensation magnetic moment is constant in magnitude*. Thus, the unit moment vector **u** can only rotate yet may not alter its length. Hence as in nanoscale ferromagnets, *the magnetization dynamics of antiferromagnetic nanoparticles subject to thermal agitation can also be described using Brown's diffusion model of a magnetic moment* (classical spin) leading as before to the Fokker–Planck equation, Eq. (1.65) of Chapter 1, for the distribution function $W(\vartheta, \varphi, t)$ of magnetic moment orientations on the unit sphere [17,18]. Furthermore if (as in the simplest case of a uniaxial antiferromagnetic particle) the easy axis of the nanoparticle coincides with the Z-axis of the laboratory coordinate system, then the normalized free energy $E(\vartheta, \varphi, t)$ in superimposed magnetic dc and ac fields $\mathbf{H}(t) = \mathbf{H}_0 + \mathbf{H}_1\cos\omega t$ is [399,400,403]

$$E(\vartheta,\varphi,t) = \sigma \sin^2\vartheta - \xi\cos\omega t\cos\Theta$$
$$-\frac{\zeta}{2}(\xi_0 + \xi\cos\omega t)^2 \sin^2\Theta. \tag{2.348}$$

Here $\sigma = vK/(kT)$, $\xi_0 = \mu_0\mu H_0/(kT)$ and $\xi = \mu_0\mu H_1/(kT)$ are the anisotropy (inverse temperature), dc, and ac applied field parameters, respectively, while $\zeta = vkT\chi_A/(\mu_0\mu^2)$ is the "antiferromagnetic" parameter, K is the anisotropy constant typically of the order of a few 10^4 J/m^3 for ferritin [16], the dc and ac external fields $\mathbf{H}_0$ and $\mathbf{H}_1$ are assumed parallel; thus the angle Θ between the vectors $\mathbf{u}$ and $\mathbf{H}_0$ is given by Eq. (2.299). If $\chi_A = 0$, Eq. (2.348) is just the familiar free energy for uniaxial ferromagnetic nanoparticles including the anisotropy and Zeeman terms. Thus, the last term on the right-hand side of Eq. (2.348) is the contribution due to the *induced moment* which *does not exist at all* for ferromagnetic nanoparticles. However, this term affects the dynamics of the magnetic moment subjected to an ac driving field and thus determines the main features of the nonlinear response of nanoscale antiferromagnets. *Hence even relatively weak external magnetic fields may cause nonlinear effects to appear in the magnetization dynamics so that these differ from those observed in ferromagnetic nanoparticles.* Since the magnetic moment decompensation originates in the applied field the parameters ξ_0 and ξ for antiferromagnetic nanoparticles are about *two orders of magnitude smaller than those for ferromagnetic nanoparticles of comparable size* ~10 nm [399,400]. Therefore, although nonlinear effects (saturation, etc.) in the ac stationary response of antiferromagnetic nanoparticles become really pronounced only for $\xi > 1$, nevertheless, their principal features may appear and, moreover, may be studied for $\xi << 1$.

Once more we seek the solution of the Fokker–Planck equation (1.65) with free energy potential defined by Eq. (2.348), as a series of spherical harmonics $Y_{lm}(\vartheta, \varphi)$ (see Eq. (1.85) of Chapter 1). The calculation of the nonlinear ac stationary response of an antiferromagnetic nanoparticle to an external driving field can then be reduced [97] (*v.* also Chapter 1) to the solution of an infinite hierarchy of 25-term differential-recurrence relations for the statistical moments (as usual the expectation values of the spherical harmonics) $c_{lm}(t) = \langle Y_{lm}\rangle(t)$:

$$\begin{aligned}
\tau_{\mathrm{N}}\frac{d}{dt}c_{nm}(t) &= v_{nm}^{--}c_{n-2m-2}(t)+v_{nm}^{-}c_{n-2m-1}(t)+v_{nm}c_{n-2m}(t)\\
&+v_{nm}^{+}c_{n-2m+1}(t)+v_{nm}^{++}c_{n-2m+2}(t)+w_{nm}^{--}c_{n-1m-2}(t)+w_{nm}^{-}(t)c_{n-1m-1}(t)\\
&+w_{nm}(t)c_{n-1m}(t)+w_{nm}^{+}(t)c_{n-1m+1}(t)+w_{nm}^{++}c_{n-1m+2}(t)+x_{nm}^{--}c_{nm-2}(t)\\
&+x_{nm}^{-}(t)c_{nm-1}(t)+x_{nm}(t)c_{nm}(t)+x_{nm}^{+}(t)c_{nm+1}(t)+x_{nm}^{++}c_{nm+2}(t)\\
&+y_{nm}^{--}c_{n+1m-2}(t)+y_{nm}^{-}(t)c_{n+1m-1}(t)+y_{nm}(t)c_{n+1m}(t)+y_{nm}^{+}(t)c_{n+1m+1}(t)\\
&+y_{nm}^{++}c_{n+1m+2}(t)+z_{nm}^{--}c_{n+2m-2}(t)+z_{nm}^{-}c_{n+2m-1}(t)+z_{nm}c_{n+2m}(t)\\
&+z_{nm}^{+}c_{n+2m+1}(t)+z_{nm}^{++}c_{n+2m+2}(t).
\end{aligned}\tag{2.349}$$

The various time-dependent coefficients w_{nm}, x_{nm}, etc. in Eq. (2.349) may for brevity be written in the form

$$\begin{aligned}
w_{nm} &= w_{nm}^{0}+w_{nm}^{1}(e^{i\omega t}+e^{-i\omega t})+w_{nm}^{2}(e^{2i\omega t}+e^{-2i\omega t}),\\
x_{nm} &= x_{nm}^{0}+x_{nm}^{1}(e^{i\omega t}+e^{-i\omega t})+x_{nm}^{2}(e^{2i\omega t}+e^{-2i\omega t}),\\
y_{nm} &= y_{nm}^{0}+y_{nm}^{1}(e^{i\omega t}+e^{-i\omega t})+y_{nm}^{2}(e^{2i\omega t}+e^{-2i\omega t}), \text{ etc.}
\end{aligned}$$

in which the time-independent coefficients w_{nm}^{j}, x_{nm}^{j}, etc. are explicitly given by

$$x_{nm}^{2}=\frac{\zeta\xi^{2}(1-3\gamma_3^{2})\left[n(n+1)-3m^{2}\right]}{16(2n-1)(2n+3)},\quad x_{nm}^{1}=i\frac{\xi m\gamma_3}{4\alpha}+x_{nm}^{2}\frac{4\xi_0\xi}{\xi^{2}},$$

$$x_{nm}^{0}=-\frac{n(n+1)}{2}+i\frac{\xi_0 m\gamma_3}{2\alpha}+\sigma\frac{n(n+1)-3m^{2}}{(2n-1)(2n+3)}+x_{nm}^{2}\frac{2(2\xi_0^{2}+\xi^{2})}{\xi^{2}},$$

$$x_{nm}^{-2}=(\gamma_1+i\gamma_2)\frac{3(2m-1)\zeta\xi^{2}\gamma_3}{16(2n-1)(2n+3)}\sqrt{(1+n-m)(n+m)},$$

$$x_{nm}^{-1}=x_{nm}^{-2}\frac{4\xi_0\xi}{\xi^{2}}+\frac{i\xi}{8\alpha}(\gamma_1+i\gamma_2)\sqrt{(1+n-m)(n+m)},$$

$$x_{nm}^{-0}=x_{nm}^{-2}\frac{2(2\xi_0^{2}+\xi^{2})}{\xi^{2}}+\frac{i\xi_0}{4\alpha}(\gamma_1+i\gamma_2)\sqrt{(1+n-m)(n+m)},$$

$$x_{nm}^{--0}=\frac{3\zeta(2\xi_0^{2}+\xi^{2})(\gamma_1+i\gamma_2)^{2}}{16(2n-1)(2n+3)}\sqrt{[n^{2}-(m-1)^{2}](2+n-m)(n+m)},$$

$$x_{nm}^{--1} = \frac{2\xi_0\xi}{2\xi_0^2 + \xi^2} x_{nm}^{--0}, \; x_{nm}^{--2} = \frac{\xi^2/2}{2\xi_0^2 + \xi^2} x_{nm}^{--0},$$

$$y_{nm}^{2} = \frac{im\zeta\xi^2(1-3\gamma_3^2)}{16\alpha}\sqrt{\frac{(n+1)^2 - m^2}{(2n+1)(2n+3)}},$$

$$y_{nm}^{1} = -\frac{\xi}{4}\left[n\gamma_3 - \frac{im\zeta\xi_0(1-3\gamma_3^2)}{\alpha}\right]\sqrt{\frac{(n+1)^2 - m^2}{(2n+1)(2n+3)}},$$

$$y_{nm}^{0} = -\left(\frac{\xi_0}{2}\gamma_3 n - \frac{im}{\alpha}\left[\sigma + \frac{\zeta(2\xi_0^2+\xi^2)(1-3\gamma_3^2)}{8}\right]\right)\sqrt{\frac{(n+1)^2 - m^2}{(2n+1)(2n+3)}},$$

$$y_{nm}^{-2} = -\frac{i\zeta\gamma_3\xi^2}{16\alpha}(\gamma_1 + i\gamma_2)(n+2m)\sqrt{\frac{(1+n-m)(2+n-m)}{(1+2n)(3+2n)}},$$

$$y_{nm}^{-1} = \frac{4\xi_0\xi}{\xi^2} y_{nm}^{-2} - \frac{n\xi}{8}(\gamma_1 + i\gamma_2)\sqrt{\frac{(1+n-m)(2+n-m)}{(1+2n)(3+2n)}},$$

$$y_{nm}^{-0} = \frac{2(2\xi_0^2 + \xi^2)}{\xi^2} y_{nm}^{-2} - \frac{n\xi_0}{4}(\gamma_1 + i\gamma_2)\sqrt{\frac{(1+n-m)(2+n-m)}{(1+2n)(3+2n)}},$$

$$y_{nm}^{--0} = -\frac{i\zeta(2\xi_0^2 + \xi^2)(\gamma_1 + i\gamma_2)^2}{16\alpha}\sqrt{\frac{1}{(1+2n)(3+2n)}\prod_{k=0}^{3}(k+n-m)},$$

$$y_{nm}^{--1} = \frac{2\xi_0\xi}{2\xi_0^2 + \xi^2} y_{nm}^{--0}, \; y_{nm}^{--2} = \frac{\xi^2/2}{2\xi_0^2 + \xi^2} y_{nm}^{--0},$$

$$z_{nm}^{2} = \frac{n\zeta\xi^2(3\gamma_3^2 - 1)}{16(2n+3)}\sqrt{\frac{[(n+1)^2 - m^2][(n+2)^2 - m^2]}{(2n+1)(2n+5)}}, \; z_{nm}^{1} = \frac{4\xi_0\xi}{\xi^2} z_{nm}^{2},$$

$$z_{nm}^{0} = \frac{2(2\xi_0^2 + \xi^2)}{\xi^2} z_{nm}^{2} - \frac{\sigma n}{2n+3}\sqrt{\frac{[(n+1)^2 - m^2][(n+2)^2 - m^2]}{(2n+1)(2n+5)}},$$

$$z_{nm}^{-0} = \frac{n(2\xi_0^2 + \xi^2)\zeta(\gamma_1 + i\gamma_2)\gamma_3}{4(2n+3)}\sqrt{\frac{[(n+1)^2 - m^2](n-m+2)(3+n-m)}{(2n+1)(2n+5)}},$$

$$z_{nm}^{--1} = \frac{2\xi_0\xi}{2\xi_0^2 + \xi^2} z_{nm}^{--0}, \; z_{nm}^{--2} = \frac{\xi^2/2}{2\xi_0^2 + \xi^2} z_{nm}^{--0},$$

$$z_{nm}^{-1} = \frac{2\xi_0 \xi}{2\xi_0^2 + \xi^2} z_{nm}^{-0}, \quad z_{nm}^{-2} = \frac{\xi^2 / 2}{2\xi_0^2 + \xi^2} z_{nm}^{-0},$$

$$z_{nm}^{--0} = \frac{n\zeta(2\xi_0^2 + \xi^2)(\gamma_1 + i\gamma_2)^2}{16(2n+3)} \sqrt{\frac{1}{(2n+1)(2n+5)} \prod_{k=1}^{4} (k+n-m)},$$

$$w_{nm}^{2} = \frac{im\zeta\xi^2(1-3\gamma_3^2)}{16\alpha} \sqrt{\frac{n^2 - m^2}{4n^2 - 1}}, \quad w_{nm}^{1} = w_{nm}^{2} \frac{4\xi_0 \xi}{\xi^2} + \frac{\xi\gamma_3(n+1)}{4} \sqrt{\frac{n^2 - m^2}{4n^2 - 1}},$$

$$w_{nm}^{0} = \frac{2(2\xi_0^2 + \xi^2)}{\xi^2} w_{nm}^{2} + \left(\frac{\gamma_3 \xi_0}{2}(n+1) + \frac{im\sigma}{\alpha} \right) \sqrt{\frac{n^2 - m^2}{4n^2 - 1}},$$

$$w_{nm}^{-2} = -\frac{i(1-2m+n)\zeta\xi^2\gamma_3(\gamma_1 + i\gamma_2)}{16\alpha} \sqrt{\frac{(n+m)(n+m-1)}{4n^2 - 1}},$$

$$w_{nm}^{-1} = \frac{4\xi_0 \xi}{\xi^2} w_{nm}^{-2} - \xi \frac{n+1}{8} (\gamma_1 + i\gamma_2) \sqrt{\frac{(n+m)(n+m-1)}{4n^2 - 1}},$$

$$w_{nm}^{-0} = \frac{2(2\xi_0^2 + \xi^2)}{\xi^2} w_{nm}^{-2} - \xi_0 \frac{n+1}{4} (\gamma_1 + i\gamma_2) \sqrt{\frac{(n+m)(n+m-1)}{4n^2 - 1}},$$

$$w_{nm}^{--0} = -\frac{i\zeta(2\xi_0^2 + \xi^2)(\gamma_1 + i\gamma_2)^2}{16\alpha} \sqrt{\frac{(n+m-2)[n^2 - (m-1)^2](n+m)}{4n^2 - 1}},$$

$$w_{nm}^{--1} = \frac{2\xi_0 \xi}{2\xi_0^2 + \xi^2} w_{nm}^{--0}, \quad w_{nm}^{--2} = \frac{\xi^2 / 2}{2\xi_0^2 + \xi^2} w_{nm}^{--0},$$

$$v_{nm}^{2} = \frac{(n+1)\zeta\xi^2(1-3\gamma_3^2)}{16(2n-1)} \sqrt{\frac{[(n-1)^2 - m^2](n^2 - m^2)}{(2n+1)(2n-3)}}, \quad v_{nm}^{1} = v_{nm}^{2} \frac{4\xi_0 \xi}{\xi^2},$$

$$v_{nm}^{0} = \left[\frac{\zeta(2\xi_0^2 + \xi^2)}{8}(1-3\gamma_3^2) + \sigma \right] \frac{n+1}{2n-1} \sqrt{\frac{[(n-1)^2 - m^2](n^2 - m^2)}{(2n+1)(2n-3)}},$$

$$v_{nm}^{-0} = \frac{(n+1)\zeta(2\xi_0^2 + \xi^2)(\gamma_1 + i\gamma_2)\gamma_3}{4(2n-1)} \sqrt{\frac{(n+m-2)(n+m-1)(n^2 - m^2)}{(2n+1)(2n-3)}},$$

$$v_{nm}^{-1} = \frac{2\xi_0 \xi}{2\xi_0^2 + \xi^2} v_{nm}^{-0}, \quad v_{nm}^{-2} = \frac{\xi^2 / 2}{2\xi_0^2 + \xi^2} v_{nm}^{-0},$$

$$v_{nm}^{--0} = -\frac{(n+1)\zeta(2\xi_0^2 + \xi^2)(\gamma_1 + i\gamma_2)^2}{16(2n-1)} \sqrt{\frac{1}{(2n+1)(2n-3)} \prod_{k=0}^{3} (n+m-k)},$$

$$v_{nm}^{--1} = \frac{2\xi_0 \xi}{2\xi_0^2 + \xi^2} v_{nm}^{--0}, \quad v_{nm}^{--2} = \frac{\xi^2 / 2}{2\xi_0^2 + \xi^2} v_{nm}^{--0}.$$

Notice that the remaining coefficients, those with superscripts + and ++ may be calculated according to the replacement rule: $a_{nm}^{+} = (a_{n-m}^{-})^{*}$ and $a_{nm}^{++} = (a_{n-m}^{--})^{*}$ offering a certain amount of simplification.

To find the ac stationary response which is independent of the initial conditions, so that only the steady-state solution of Eq. (2.349) is needed, we expand $c_{lm} = \langle Y_{lm} \rangle(t)$ in a Fourier series, Eq. (2.292), where the Fourier amplitudes c_{lm}^{k} can once again be evaluated using matrix continued fractions. Thus, Eq. (2.349) can be rearranged as the matrix three-term recurrence relations, Eq. (2.294), where on this particular occasion the matrix elements of the supermatrices $\mathbf{Q}_n$ and $\mathbf{Q}_n^{\pm}$ are defined as [155]

$$\left[\mathbf{Q}_n^{\pm}\right]_{lm} = \delta_{l-2m}\mathbf{r}_n^{\pm} + \delta_{l-1m}\mathbf{p}_n^{\pm} + \delta_{lm}\mathbf{q}_n^{\pm} + \delta_{l+1m}\mathbf{p}_n^{\pm} + \delta_{l+2m}\mathbf{r}_n^{\pm},$$

$$\left[\mathbf{Q}_n\right]_{lm} = \delta_{l-2m}\mathbf{r}_n + \delta_{l-1m}\mathbf{p}_n + \delta_{lm}\left(\mathbf{q}_n - im\tau_{\mathrm{N}}\omega\mathbf{I}\right) + \delta_{l+1m}\mathbf{p}_n + \delta_{l+2m}\mathbf{r}_n,$$

while the supermatrices $\mathbf{q}_n, \mathbf{q}_n^{\pm}, \mathbf{p}_n, \mathbf{p}_n^{\pm}, \mathbf{r}_n, \mathbf{r}_n^{\pm}$ are as defined by Eq. (2.301) above and

$$\mathbf{R} = -\frac{1}{\sqrt{4\pi}}\begin{pmatrix} \vdots \\ \mathbf{0} \\ \mathbf{r}_1^{-} \\ \mathbf{p}_1^{-} \\ \mathbf{q}_1^{-} \\ \mathbf{p}_1^{-} \\ \mathbf{r}_1^{-} \\ \mathbf{0} \\ \vdots \end{pmatrix}, \quad \mathbf{q}_1^{-} = \begin{pmatrix} -\sqrt{6/5}\sigma^2\left(2h^2 + h_0^2\right)\zeta\left(\gamma_1 - i\gamma_2\right)^2 \\ -2\sqrt{6/5}\sigma^2\left(2h^2 + h_0^2\right)\zeta\left(\gamma_1 - i\gamma_2\right)\gamma_3 \\ -\frac{2\sigma}{\sqrt{5}}\left[\sigma\left(2h^2 + h_0^2\right)\zeta\left(3\gamma_3^2 - 1\right) - 1\right] \\ 2\sqrt{6/5}\sigma^2\left(2h^2 + h_0^2\right)\zeta\left(\gamma_1 + i\gamma_2\right)\gamma_3 \\ -\sqrt{6/5}\sigma^2\left(2h^2 + h_0^2\right)\zeta\left(\gamma_1 + i\gamma_2\right)^2 \\ \sqrt{2/3}h_0\left(\gamma_1 - i\gamma_2\right)\sigma \\ \frac{2}{\sqrt{3}}h_0\gamma_3\sigma \\ -\sqrt{2/3}h_0\left(\gamma_1 + i\gamma_2\right)\sigma \end{pmatrix},$$

$$\mathbf{p}_1^- = \begin{pmatrix} -2\sqrt{6/5}hh_0\zeta(\gamma_1 - i\gamma_2)^2\sigma^2 \\ -4\sqrt{6/5}hh_0\zeta(\gamma_1 - i\gamma_2)\gamma_3\sigma^2 \\ -4hh_0\zeta(3\gamma_3^2 - 1)\sigma^2/\sqrt{5} \\ 4\sqrt{6/5}hh_0\zeta(\gamma_1 + i\gamma_2)\gamma_3\sigma^2 \\ -2\sqrt{6/5}hh_0\zeta(\gamma_1 + i\gamma_2)^2\sigma^2 \\ \sqrt{2/3}h(\gamma_1 - i\gamma_2)\sigma \\ 2h\gamma_3\sigma/\sqrt{3} \\ -\sqrt{2/3}h(\gamma_1 + i\gamma_2)\sigma \end{pmatrix}, \quad \mathbf{r}_1^- = \begin{pmatrix} -\sqrt{6/5}h^2\zeta(\gamma_1 - i\gamma_2)^2\sigma^2 \\ -2\sqrt{6/5}h^2\zeta(\gamma_1 - i\gamma_2)\gamma_3\sigma^2 \\ -2h^2\zeta(3\gamma_3^2 - 1)\sigma^2/\sqrt{5} \\ 2\sqrt{6/5}h^2\zeta(\gamma_1 + i\gamma_2)\gamma_3\sigma^2 \\ -\sqrt{6/5}h^2\zeta(\gamma_1 + i\gamma_2)^2\sigma^2 \\ 0 \\ 0 \\ 0 \end{pmatrix}.$$

The exact solution is again rendered by the matrix continued fraction, Eq. (2.296), which contains all the Fourier amplitudes c_{lm}^k required to determine the nonlinear stationary response.

Henceforth, we can assume without loss of generality that the vectors $\mathbf{H}_0$ and $\mathbf{H}$ both lie in the XZ-plane of the laboratory coordinate system, i.e., $\varphi = 0$, with the direction cosines $\gamma_1 = \sin\psi, \gamma_2 = 0, \gamma_3 = \cos\psi$. Therefore, the formal expression for the average magnetic moment of the particle in the direction of the ac driving field is

$$\mu_H(t) = \mu\left[\langle\cos\Theta\rangle(t) + \zeta(\xi_0 + \xi\cos\omega t)\langle\sin^2\Theta\rangle(t)\right]. \quad (2.350)$$

Moreover, we also have $\langle\cos\Theta\rangle(t)$ and $\langle\cos^2\Theta\rangle(t)$ expressed in terms of the statistical moments $\langle Y_{1m}\rangle(t)$ and $\langle Y_{2m}\rangle(t)$ as

$$\langle\cos\Theta\rangle(t) = \sqrt{2\pi/3}\left(\sqrt{2}\cos\psi\langle Y_{10}\rangle(t) + \sin\psi\left[\langle Y_{1-1}\rangle(t) - \langle Y_{11}\rangle(t)\right]\right),$$

$$\langle\cos^2\Theta\rangle(t) = \frac{1}{3} + \sqrt{\frac{2\pi}{15}}\Big\{\sqrt{2/3}\left(3\cos^2\psi - 1\right)\langle Y_{20}\rangle(t)$$
$$+\sin 2\psi\left[\langle Y_{2-1}\rangle(t) - \langle Y_{21}\rangle(t)\right] + \sin^2\psi\left[\langle Y_{22}\rangle(t) + \langle Y_{2-2}\rangle(t)\right]\Big\},$$

formally indicating which Fourier amplitudes are explicitly needed in order to calculate $\mu_H(t)$. As before, the average magnetic moment can itself be written as a Fourier series in the time, viz.,

$$\mu_H(t) = \mu\sum_{k=-\infty}^{\infty} m_1^k(\omega)e^{ik\omega t}, \quad (2.351)$$

where $m_1^k(\omega)$ is the amplitude of the kth harmonic component of the nonlinear response, viz.,

$$m_1^k(\omega)=\frac{2\zeta\xi_0}{3}\delta_{0k}+\frac{\zeta\xi}{6}(\delta_{1k}+\delta_{-1k})+\sqrt{\frac{4\pi}{3}}\left[\cos\psi c_{10}^k+\frac{\sin\psi}{\sqrt{2}}\left(c_{1-1}^k-c_{11}^k\right)\right]$$

$$-\zeta\sqrt{\frac{2\pi}{15}}\left\{\sqrt{2/3}\left(3\cos^2\psi-1\right)\left[\xi_0 c_{20}^k+\xi\left(c_{20}^{k-1}+c_{20}^{k+1}\right)/2\right]\right.$$
$$+\sin 2\psi\left[\xi_0\left(c_{2-1}^k-c_{21}^k\right)+\xi\left(c_{2-1}^{k-1}+c_{2-1}^{k+1}-c_{21}^{k-1}-c_{21}^{k+1}\right)/2\right] \tag{2.352}$$
$$\left.+\sin^2\psi\left[\xi_0\left(c_{22}^k+c_{2-2}^k\right)+\xi\left(c_{22}^{k-1}+c_{22}^{k+1}+c_{2-2}^{k-1}+c_{2-2}^{k+1}\right)/2\right]\right\}.$$

Thus by calculating c_{10}^k and $c_{1\pm1}^k$ via Eq. (2.296), we have the desired Fourier amplitudes $m_1^k(\omega)$.

As a first illustration of the nonlinear effects induced by the ac field, we can again study the *time-independent* or dc component of the magnetization denoted by M_ξ which is formally defined by the mean value over a period, viz.,

$$M_\xi=\frac{\omega}{2\pi\mu}\int_0^{2\pi/\omega}\mu_H(t)\,dt=m_1^0(\omega). \tag{2.353}$$

This expression is entirely real and as before yields in the limits of *vanishing ac field*, $\xi\to 0$, the static dc magnetization

$$\lim_{\xi\to 0}M_\xi=M_0=\langle\cos\Theta\rangle_0+\zeta\xi_0\left(1-\langle\cos^2\Theta\rangle_0\right),$$

where the angular brackets $\langle\ \rangle_0$ denote equilibrium ensemble averaging defined as

$$\langle\cos^n\Theta\rangle_0=\frac{1}{Z}\int_0^{2\pi}\int_0^{\pi}\cos^n\Theta e^{-E_0(\vartheta,\varphi)}\sin\vartheta d\vartheta d\varphi,\ Z=\int_0^{2\pi}\int_0^{\pi}e^{-E_0(\vartheta,\varphi)}\sin\vartheta d\vartheta d\varphi,$$

and the free energy $E_0(\vartheta,\varphi)$ is

$$E_0(\vartheta,\varphi)=\sigma\sin^2\vartheta-\xi_0\cos\Theta-\frac{\zeta\xi_0^2}{2}\sin^2\Theta. \tag{2.354}$$

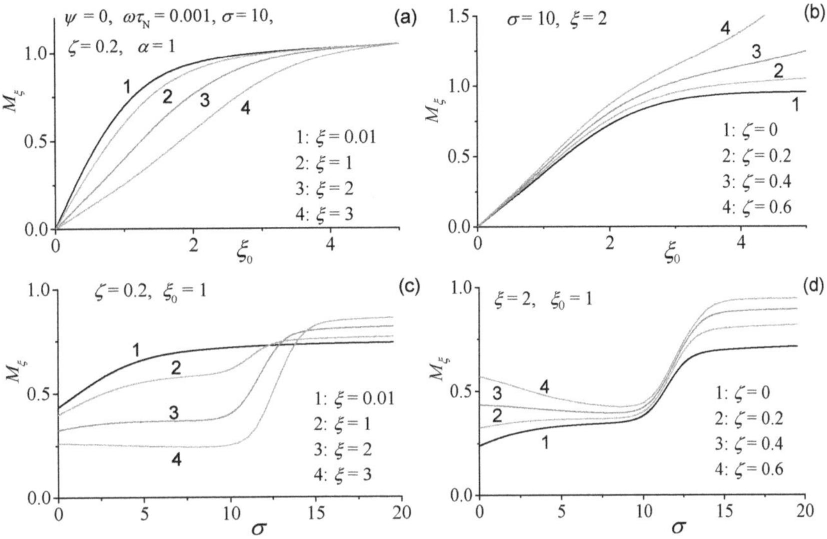

Fig. 2.66. Magnetization M_ξ vs. the dc bias field ((a) and (b)) and anisotropy (inverse temperature) parameters ((c) and (d)) for various ac fields and antiferromagnetic parameters for damping $\alpha = 1$, oblique angle $\psi = 0$, and frequency $\omega\tau_N = 0.001$.

The free energy potential, Eq. (2.354), represents, in general, a bistable structure with two minima separated by potential barriers with a saddle point [158]. Since the vector $\mathbf{H}_0$ lies in the XZ-plane ($\gamma_1 = \sin\psi, \gamma_2 = 0$, $\gamma_3 = \cos\psi$), we have $\cos\Theta = \sin\psi \sin\vartheta \cos\varphi + \cos\psi \cos\vartheta$. Proceeding to the time-dependent situation, we now evaluate the spectrum of the fundamental component of the magnetization defined by

$$\chi(\omega) = \chi'(\omega) - i\chi''(\omega) = \frac{6m_1^1(\omega)}{\xi}, \tag{2.355}$$

representing both the linear and the lowest order nonlinear susceptibility. The amplitude $m_1^1(\omega)$ also yields the normalized area of the DMH loop A_n, which as before is the energy loss per particle and per cycle of the ac field, defined by Eq. (2.331). All other harmonic components $m_1^k(\omega)$ with $k > 1$ may be investigated in like manner.

Briefly returning now to the interpretation of nonlinear effects in the time-independent (i.e., dc) but frequency-dependent magnetization, $M_\xi(\omega)$ is plotted in Fig. 2.66 as a function of the applied dc field and

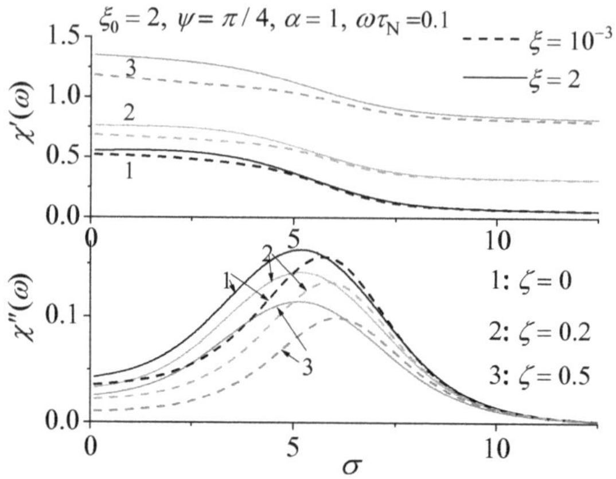

Fig. 2.67. Real and imaginary parts of the dynamic susceptibility vs. $\sigma \sim T^{-1}$ for various antiferromagnetic parameters ζ for dc field parameter $\xi_0 = 2$, frequency $\omega\tau_N = 0.1$, damping $\alpha = 1$, angle $\psi = \pi/4$, and ac field parameters $\xi = 0.001$ (linear response) and $\xi = 2$ (nonlinear response). Dashed and solid lines: the linear and nonlinear responses, respectively.

anisotropy (or the inverse-temperature) parameter σ for various ac field magnitudes ξ and antiferromagnetic parameters ζ. As far as the dynamic susceptibility is concerned the imaginary $\chi''(\omega)$ and the real $\chi'(\omega)$ parts of the susceptibility as functions of σ and the frequency ω are shown in Figs. 2.67 and 2.68, respectively, for various values of ζ both in the linear ($\xi \leq 1$) and nonlinear ($\xi \geq 1$) regimes. (The condition $\xi \to 0$ again corresponds to the linear response, where $\mu_H(t)/H$ is independent of the ac field strength.) All the calculations indicate as expected that a marked dependence of $\chi(\omega)$ on the antiferromagnetic parameter ζ, anisotropy σ, ac field ξ, dc field ξ_0, damping α, and angle ψ exists. Moreover, for the special case of linear response, $\xi << 1$, the results agree with independent numerical calculations [157]. However, like ferromagnetic particles in strong ac fields, where $\xi \geq 1$, pronounced nonlinear effects occur; see Figs. 2.66–2.68 illustrating the dependence of the nonlinear response on the ac field parameter ξ. Again, like ferromagnetic nanoparticles [47,52,155,156, 159], three peaks appear in the spectra of the magnetic loss $\chi''(\omega)$ in the

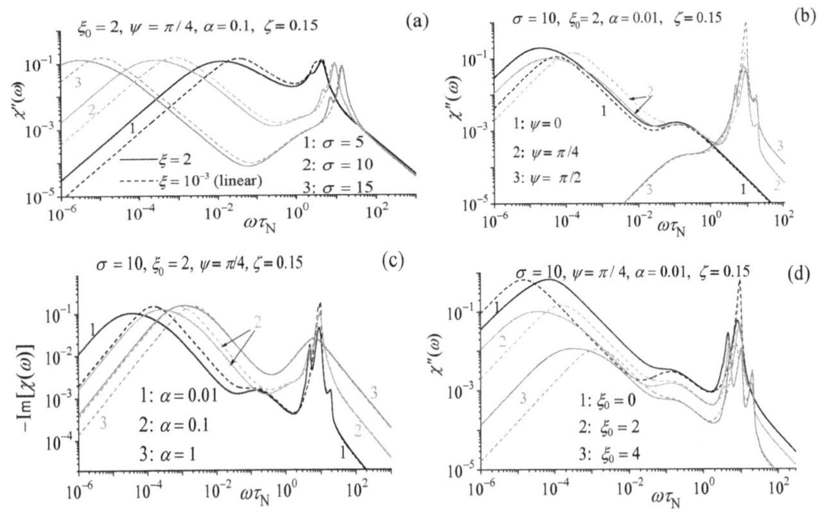

Fig. 2.68. $\chi''(\omega)$ vs. the dimensionless frequency $\omega\tau_N$ for (a) various anisotropy parameters σ, (b) angles ψ, (c) damping α, and (d) dc field parameters ξ_0 for ac field amplitudes $\xi = 0.001$ (linear response: dashed lines) and 2 (nonlinear response: solid lines).

neighborhood of the frequencies, where $\chi'(\omega)$ alters (see Fig. 2.68). The characteristic frequencies of these peaks, i.e., where $\chi''(\omega)$ attains local maxima, are τ^{-1}, $\alpha\omega_{pr}$, and ω_{pr}. As usual τ is the switching (reversal) time of the magnetic moment between two minima of the free energy separated by a potential barrier with a saddle point and $\omega_{pr} = \gamma H_{ef}$ is the precession frequency of the moment in the effective magnetic field $\mathbf{H}_{ef}$. Clearly the low-frequency behavior of both $\chi'(\omega)$ and $\chi''(\omega)$ is dominated by the barrier-crossing relaxation mode as for ferromagnetic particles. Likewise the reversal time of the magnetization τ can be evaluated from the characteristic frequency ω_{max}, where $\chi''(\omega)$ reaches a maximum, and/or the half-width $\Delta\omega$ of the spectrum of $\chi'(\omega)$ via Eq. (2.346).

In weak ac fields, $\xi << 1$, τ can again be associated with λ_1 of the Fokker–Planck operator in Eq. (1.65) [97]. Then comparison of τ as extracted from the spectra $\chi(\omega)$ via Eq. (2.346) with $\tau = 1/\lambda_1$ as calculated using an independent method [97], namely numerical calculation of λ_1 of the Fokker–Planck operator, shows that both results are identical. The calculations also indicate that when the dc field parameter ξ_0 is increased the magnitude of the low-frequency peak decreases. This is again a signature of the familiar depletion of the population in the shallowest potential well of the free-energy density $E_0(\vartheta, \varphi)$, Eq. (2.354), and is marked by the virtual disappearance of the low-frequency peak in the magnetic loss spectrum $\chi''(\omega)$ (see Fig. 2.68(d)). For weak dc bias fields, $\xi_0 < 0.3$, the low-frequency peak shifts monotonically to higher frequencies as the ac field amplitude ξ is increased. In contrast for large $\xi_0 > 1$, the low-frequency peak shifts to lower frequencies reaching a maximum at $\xi \sim \xi_0$ and thereafter decreases rapidly with increasing ξ. In other words, as ξ_0 increases, τ *initially increases* and thereafter having attained its maximum at some critical value $\xi \sim \xi_0$, τ *decreases*. The second relaxation peak in $\chi''(\omega)$ and the corresponding dispersion of $\chi'(\omega)$ appearing at higher frequencies $\sim \alpha\omega_{\mathrm{pr}}$ are again due to the near-degenerate "intrawell" relaxation modes, which individually are virtually indistinguishable in the spectra. For $h_0 < 0.5$, the amplitude of this intermediate peak is far weaker than that of the low-frequency one. Nevertheless, for $h_0 > 0.2$, it may still dominate the spectra because as h_0 increases, the low-frequency peak drastically decreases (see Fig. 2.68(d)). Figure 2.68(d) illustrates the inherent dependence of $\chi(\omega)$ on the damping α via coupling of the longitudinal and transverse relaxation modes [37]. This coupling explicitly appears in the dynamical equation of motion of the magnetic moment $\boldsymbol{\mu}(t)$, where its longitudinal component again gives rise to *entanglement* with the transverse ones. The entanglement of the third antiferromagnetic resonance (AFMR) peak in $\chi''(\omega)$ is due to excitation of transverse modes with characteristic frequencies near the precession frequency ω_{pr} in the effective magnetic field $\mathbf{H}_{\mathrm{ef}}$ (see Fig. 2.68). The AFMR peak like the previous FMR one is evident only for low damping ($\alpha << 1$) and strongly manifests itself at high frequencies $\omega \sim \omega_{\mathrm{pr}}$ (see Fig. 2.68(c); a feature which is absent for $\alpha > 1$). Moreover, as seen in Fig. 2.68, as the ac field ξ increases, the magnitude of the AFMR peak correspondingly decreases and the half-width increases showing pronounced nonlinear effects. In addition, a second weak (now familiar) resonance peak owing to *parametric resonance* of the nonlinear oscillatory (precessional) motion of the magnetic moment $\boldsymbol{\mu}(t)$ appears at frequencies $\sim \omega_{\mathrm{pr}}/2$ (Fig. 2.68(a)), while for very low damping, $\alpha < 0.01$, harmonic resonance peaks with characteristic frequencies $n\,\omega_{\mathrm{pr}}$, $n = 2,3,\ldots$ due to the high-frequency resonant modes which are now discernible in the spectrum (only the peak with $n = 2$ is visible in Figs. 2.68(b)-2.68(d)). These peaks virtually disappear, however, for $\psi = 0$. Both phenomena are a powerful signature of the nonlinear effects invariably existing in nonlinear oscillator systems driven by an ac external stimulus [380].

The reversal time τ can be also estimated using escape-rate theory like that of ferromagnets. The free energy, Eq. (2.354), represents a bistable structure with minima at $\mathbf{n}_A^+$ and $\mathbf{n}_A^-$ separated by a potential barrier with a saddle point at $\mathbf{n}_C$. The saddle point generally lies in the equatorial region, while $\mathbf{n}_A^+$ and $\mathbf{n}_A^-$ lie in the north and south polar regions, respectively. Once more for $\psi \neq 0$, corresponding to a noncircularly symmetric energyscape, we have a strong intrinsic dependence of the reversal time and other magnetic characteristics (such as the complex magnetic susceptibility) on α arising as usual from *coupling of the longitudinal and transverse relaxation modes*. For the antiferromagnetic anisotropy potential, Eq. (2.354), the time τ can be estimated using the turnover Eq. (2.91). First to evaluate τ_{IHD}, we need $c_1^{(p)}$, $c_2^{(p)}$, and E_p, viz.,

$$\begin{aligned} c_1^{(p)} &= 2\sigma[\cos 2\vartheta_p + h\cos(\vartheta_p - \psi) - 2h^2\zeta\sigma\cos 2(\vartheta_p - \psi)], \\ c_2^{(p)} &= 2\sigma[\cos^2\vartheta_p + h\cos(\vartheta_p - \psi) - 2h^2\zeta\sigma\cos^2(\vartheta_p - \psi)], \\ E_p &= \sigma[\sin^2\vartheta_p - 2h\cos(\vartheta_p - \psi) + 2h^2\zeta\sigma\cos^2(\vartheta_p - \psi)]. \end{aligned} \tag{2.356}$$

Here the ϑ_p are the solutions of the equation $\partial_\vartheta E\big|_{\varphi=0} = 0$ determining the stationary points, viz.,

$$x = h\left(\cos\psi - \frac{x\sin\psi}{\sqrt{1-x^2}}\right)\left[1 - 2h\zeta\sigma(x\cos\psi + \sin\psi\sqrt{1-x^2})\right] \tag{2.357}$$

($x = \cos\vartheta$). The roots of this equation, namely, $x_C = \cos\vartheta_C$ and $x_{1,2} = \cos\vartheta_{1,2}$ corresponding to the saddle point and minima of the free-energy density, respectively, can also be written as convergent Taylor series for any desired order of the field parameter h, viz.,

$$\begin{aligned} \cos\vartheta_C &= -h\cos\psi \\ &\times\left[1 - h(2\zeta\sigma - 1)\sin\psi - h^2[\zeta\sigma(1 - 3\cos 2\psi) - \sin^2\psi] + \cdots\right], \\ \cos\vartheta_{1,2} &= \pm 1 \mp (h^2/2)\sin^2\psi + h^3(1 + 2\zeta\sigma)\sin^2\psi\cos\psi + \cdots. \end{aligned} \tag{2.358}$$

The corresponding Taylor series for $\Delta E_{1,2}$, $c_1^{(C)}, c_2^{(C)}$, $\omega_{1,2}$ and ω_C are obtained from Eqs. (2.354), (2.356), and (2.358), so that

$$\Delta E_{1,2} = \sigma\{1 \pm 2h(\cos\psi \mp \sin\psi) + (1 - 2\zeta\sigma\cos 2\psi)h^2$$
$$+\cos\psi\sin\psi\left[(1-4\zeta\sigma)\cos\psi \mp (4\zeta\sigma+1)\sin\psi\right]h^3 + \cdots\},$$
$$\sigma^{-1}c_1^{(C)} = -2 + 2h\sin\psi + [1 + (1+4\zeta\sigma)\cos 2\psi]h^2$$
$$+(5+4\zeta\sigma)\cos^2\psi\sin\psi h^3 + ...,$$
$$\sigma^{-1}c_2^{(C)} = 2h\sin\psi + (1+4\zeta\sigma)\cos^2\psi\sin\psi h^3 - \cdots.$$
$$\omega_{1,2} = \frac{2\sigma\gamma kT}{v\mu_0 M_S}\{1 \pm h\cos\psi - h^2\left[1 + 2\sigma\zeta(6\sigma\zeta - 1)\cos 2\psi\right]/4 \pm \cdots\},$$
$$\omega_C = \frac{2\sigma\gamma kT}{v\mu_0 M_S}\{2 - h\sin\psi + \left[8\zeta\sigma - 3 - (1+8\zeta\sigma)\cos 2\psi\right]h^2/8 - (1/16)$$
$$\times\left[\left(19 + 40\zeta\sigma + 32\zeta^2\sigma^2 + \left(17 + 56\zeta\sigma + 32\zeta^2\sigma^2\right)\cos 2\psi\right)\sin\psi\right]h^3 + \cdots\},$$

To evaluate the actions S_1 and S_2 defined by Eq. (1.304), we now determine the critical trajectories (separatrixes), obtained by solving the equation $E(\vartheta, \varphi) = E_C$. The critical trajectories $\cos\vartheta_{1,2}(\varphi)\big|_{E=E_C}$ can also be written as Taylor series in h, viz.,

$$\cos\vartheta_{1,2}(\varphi)\big|_{E=E_C} = -h\cos\psi\{1 + h(1-2\zeta\sigma)\sin\psi\cos\varphi\}$$
$$\mp 2\sin(\varphi/2)\sqrt{h\sin\psi}\{1 - (h/2)\left[\zeta\sigma + (\zeta\sigma - 1)\cos\varphi\right]\sin\psi\} + \cdots.$$

Thus, we have S_1 and S_2 analytically as

$$S_{1,2} = \frac{8}{3}\sigma\sqrt{h\sin\psi}\,\{6 - (13 + 2\zeta\sigma)h\sin\psi$$
$$+(h^2/40)\left[15 + 4\zeta\sigma(61 - 3\zeta\sigma) + \left(4\zeta\sigma(3\zeta\sigma - 61) - 315\right)\cos 2\psi\right]\}$$
$$\pm 2\pi\sigma h^2\sin 2\psi\{-4 + (3 + 2\zeta\sigma)h\sin\psi\} + \cdots.$$

Having derived explicit equations for $\Delta E_{1,2}$, $S_{1,2}$, $c_1^{(C)}, c_2^{(C)}$, $\omega_{1,2}$ and ω_C, we then have τ from the turnover Eq. (2.91). Moreover, for circular symmetry, $\psi = 0$, where the above equations are no longer valid, τ may be evaluated using the MFPT approach as for uniaxial nanomagnets. Here the normalized free energy of the nanoparticle, Eq. (2.354), may be written in the uniaxial particle form, viz.,

$$E_{\mathrm{ef}}(z) = -\sigma' z^2 - \xi z, \tag{2.359}$$

where $\sigma' = \sigma - \xi^2\zeta/2$ represents an *effective* anisotropy constant, and $z = \cos\vartheta$. Thus, the Fokker–Planck equation again reduces to one in a single space variable. The asymmetric bistable potential E_{ef} has a maximum at $z_{\max} = -\xi/(2\sigma')$ and two nonequivalent minima separated by the barriers $\Delta E_{\pm} = \sigma'(1 \pm \xi/(2\sigma'))^2$. The reversal time can then be evaluated via the MFPT. Hence, by analogy with Eqs. (2.2) and (2.3)

$$\tau^{-1} = \frac{1}{4\tau_{\mathrm{N}}}\left[\left(\int_{-1}^{z_{\max}} \frac{e^{E_{\mathrm{ef}}(z)}}{1-z^2}\int_{-1}^{z} e^{-E_{\mathrm{ef}}(z')}dz'dz\right)^{-1} + \left(\int_{z_{\max}}^{1} \frac{e^{E_{\mathrm{ef}}(z)}}{1-z^2}\int_{z}^{1} e^{-E_{\mathrm{ef}}(z')}dz'dz\right)^{-1}\right]. \tag{2.360}$$

In the high-barrier approximation, $\sigma' >> 1$, asymptotic estimation of Eq. (2.360) using steepest descents then yields Brown's high-energy-barrier asymptotic formula for uniaxial particles subjected to a dc field, with an *effective anisotropy parameter* σ' and the usual field parameter ξ

$$\tau \sim \frac{\tau_{\mathrm{N}}\sqrt{\pi}e^{\sigma'(1-\xi/2\sigma')^2}}{\sigma'^{3/2}[1-(\xi/2\sigma')^2][1-\xi/2\sigma' + (1+\xi/2\sigma')e^{-2\xi}]}. \tag{2.361}$$

If the departures from circular symmetry are small, $h \sin\psi < 0.03$, the circularly symmetric results, Eq. (2.361) may be corrected as before via appropriate interpolation integrals (v. Section 2.3.1).

The DMH loops, i.e., $m(t) = \mu_H(t)/\mu$ vs. $h(t) = H(t)/H = \cos\omega t$, for various antiferromagnetic ζ and inverse-temperature $\sigma \propto 1/T$ parameters are exhibited in Figs. 2.69 and 2.70 for a wide range of the other model parameters (α, ψ, etc.). We again conclude that at finite temperatures due to the thermal motion, the particle magnetic moment is never completely saturated instead wandering between the "up" and "down" states. Moreover, as before the shape of DMH loops for given anisotropy and antiferromagnetic parameters σ, ζ, damping α, oblique angle ψ and dc field ξ_0 depends on the amplitude ξ and frequency ω of the ac field. Figures 2.69(a) and 2.69(b) indicate that the form of the DMH loops strongly depends on both σ and ξ_0. Figure 2.69(b) illustrates that their area

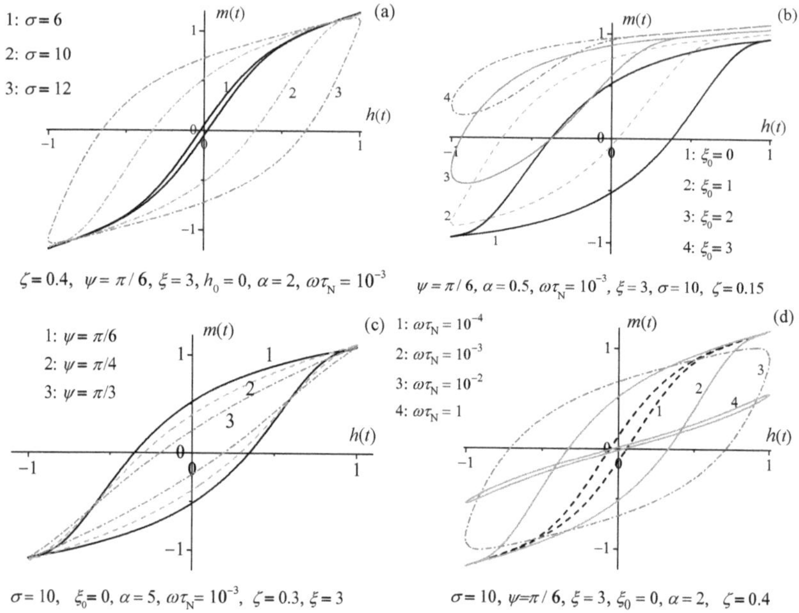

Fig. 2.69. DMH loops for (a) various anisotropy parameters $\sigma = 6, 10, 12$, (b) dc fields $\xi_0 = 0, 1, 2, 3$, (c) angles $\psi = \pi/6, \pi/4, \pi/3$, and (d) frequency $\omega\tau_N = 10^{-4}, 10^{-3}, 10^{-2}, 1$.

decreases with increasing ξ_0. Also, as σ decreases, i.e., with increasing temperature, they become narrow (Fig. 2.69(a)), implying that again only a small amount of energy is dissipated in reversing the magnetization. Figure 2.69(c) illustrates the dependence of their form on ψ while in Fig. 2.69(d), the loops are shown for very low and intermediate frequencies. At low frequencies, $\omega\tau_N \sim 10^{-4} - 10^{-3}$, they enclose large areas while at intermediate frequencies, $\omega\tau_N \sim 1$, their shapes become elliptic with a relatively small area. Moreover, also at low frequencies, where changes of the ac field are quasi-adiabatic, the magnetization dynamics again represent the so-called switching regime in which the magnetization reverses due to the cooperative shuttling action of both thermal agitation and applied field so that the DMH loop area decreases as ζ increases. Both the coercivity, the remanent and the saturation magnetization strongly depend on σ, so that pronounced variations in the form of the loops exist at low frequencies. At high frequencies, $\omega\tau_N >> 1$, DMH is mainly due to the absorption and dispersion associated with the "intrawell" and AFMR modes. Thus, the DMH arising from a high-frequency periodic signal may be calculated, thereby permitting quantitative analysis of *ultrafast* switching of the magnetization in nanoscale antiferromagnets. While at $\omega \sim \omega_{\mathrm{pr}}$, DMH once again occurs due to the *resonant* behavior of the nonlinear response, hence the switching may also be termed "resonant" leading naturally to the concept of *resonant switching of the magnetization* [47,52,156,159]. Since resonant DMH occurs at very high (GHz) frequencies, the magnetization switching is (like the purely ferromagnetic case) mainly governed by the frequency ω of the external driving field and so may be advantageously switched in this instance, because the field needed to reverse it is then much smaller than the quasistatic coercive force. As in ferromagnetism, the phase difference δ between $\mu_H(t)$ and $H(t)$, governing *loop orientation*, may undergo a pronounced variation. In particular, δ may exceed $\pi/2$ as is typical of a resonant process [47,52,156,159]. Obviously, this effect is nonexistent at low and intermediate frequencies, where relaxation processes dominate with the result that δ is always less than $\pi/2$. The form of the DMH loops also alters as the antiferromagnetic parameter ζ varies (see Fig. 2.71). Figure 2.72 shows A_n as a function of ξ for various $\sigma \sim T^{-1}$. From Fig. 2.72, the area of the DMH loop (A_n - specific power loss) strongly depends on temperature insofar as with increasing σ, the loop area initially increases, reaches a maximum, and then decreases. Furthermore, for small σ (high temperatures), the behavior of A_n for both antiferromagnetic and ferromagnetic nanoparticles is very similar however for large $\sigma > 10$ this is not so. For a *weak* $\xi < 1$ ac field, the DMH loops are ellipses with area A_n given by Eq. (2.331); the behavior of $A_n \sim -\mathrm{Im}(m_1^1)$ being similar to that of the magnetic loss $-\mathrm{Im}\,[\chi(\omega)]$ [cf. Eq. (2.355)]. In *strong* ac fields, $\xi > 1$, although the DMH area substantially alters (see Fig. 2.72); nevertheless, A_n is still determined by $-\mathrm{Im}(m_1^1)$ with maxima appearing for very strong fields, $\xi \geq 2$.

Thus, the nonlinear forced ac stationary response of antiferromagnetic nanoparticles in superimposed ac and dc bias external magnetic fields exhibits a *significant dependence on the specific antiferromagnetic* parameter ζ when studied via the continuous diffusion model of Brown [51,53]. Moreover, the nonlinear dynamic susceptibility and DMH in an ac field applied at an angle to the easy axis of the particle (so that the

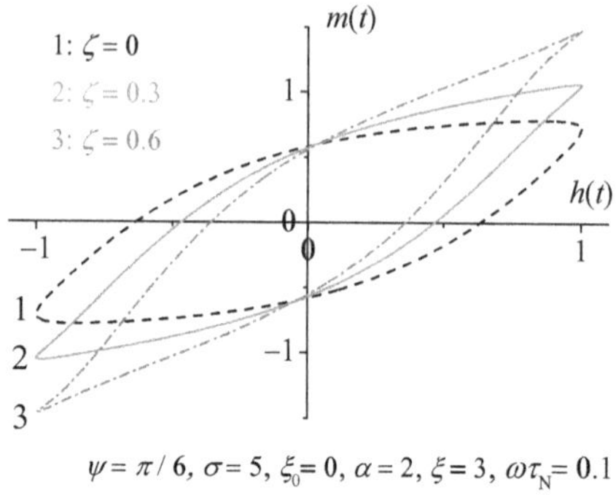

Fig. 2.70. DMH loops for various antiferromagnetic parameters ζ at $\omega\tau_N = 0.1$, $\sigma = 5$, $\alpha = 2$, $\psi = \pi/6$, $\xi = 3$, and $\xi_0 = 0$.

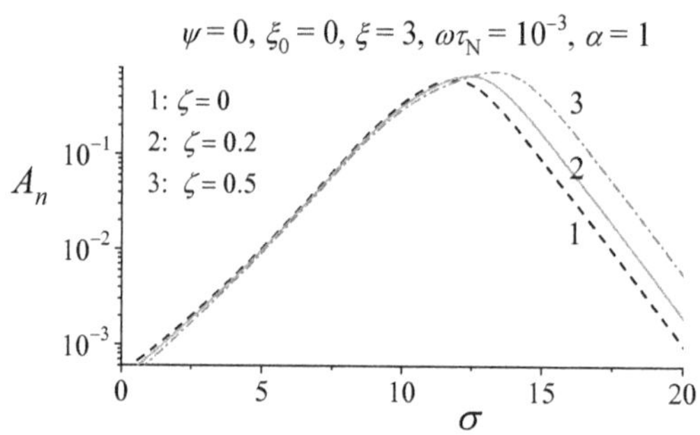

Fig. 2.71. DMH loop area A_n vs. the anisotropy (inverse temperature) parameter $\sigma \sim T^{-1}$ for various antiferromagnetic parameters ζ at $\xi = 3$, $\omega\tau_N = 10^{-3}$, $\alpha = 1$, and $\xi_0 = 0$.

circular symmetry is broken) are both very sensitive to ac field orientation and amplitude, again owing to the coupling (induced by the symmetry breaking driving field) between the precession of the magnetic moment and its thermally activated reversal. Thus, one may explain why the nonlinear ac stationary response of antiferromagnetic nanoparticles is very sensitive to damping. This sensitivity is once again just a signature of the coupling between the longitudinal and precessional modes of the magnetization exactly as that occurring in nanoscale ferromagnets [37,47, 52,155,156,159,381,382]. Moreover, under appropriate conditions a small dc bias field can strongly affect the shape of the DMH loops in antiferromagnetic nanoparticles, implying that by varying the dc bias field

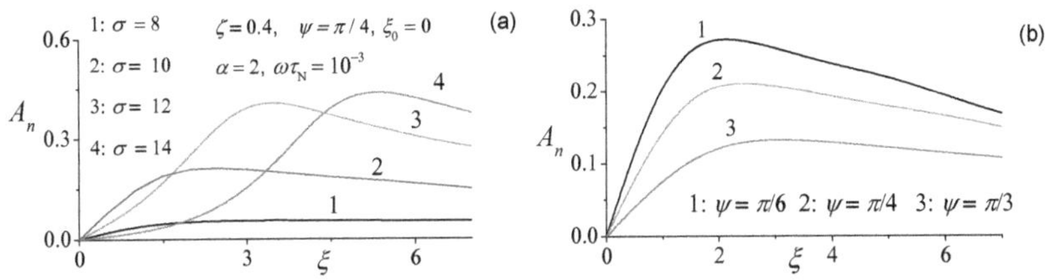

Fig. 2.72. Area of the DMH loop A_n vs. the ac field ξ: (a) for various barrier parameters $\sigma = 8$, 10, 12, 14, antiferromagnetic parameter $\zeta = 0.4$, dc field parameter $\xi_0 = 0$, frequency $\omega\tau_N = 10^{-3}$ and damping $\alpha = 2$; (b) angles $\psi = \pi/6$, $\pi/4$, $\pi/3$.

strength, one may control the heat production (specific power loss) in such nanoparticles. Since the results are valid for ac fields of *arbitrary* strength and orientation, they also provide a rigorous basis for analyzing the nonlinear ac stationary response of antiferromagnetic nanoparticles in strong ac fields, where perturbation theory is invalid. The calculations of the nonlinear response of an individual antiferromagnetic nanoparticle can be generalized as before to the average magnetic moment of an assembly of *randomly oriented* noninteracting antiferromagnetic nanoparticles by averaging over particle easy axis orientations [155].

2.10 Finite barrier correction for the FMR frequency

FMR was predicted in 1935 by Landau and Lifshitz [89]. Later it was independently verified in experiments by Griffiths [411]. As seen FMR arises from the precessional motion of the magnetization **M** of a ferromagnetic material in an external magnetic field **H**. The precession frequency of the magnetization depends on the orientation of the material, the strength of the magnetic field, as well as on the magnetization of the sample. Thermal fluctuations strongly affect the ferromagnetic resonance in nanomagnets (see, e.g., Refs. [89,412–416]). However only a few treatments of their FMR exist. The first one is due to Berger *et al.* [412, 413], who interpreted FMR spectra via the usual Landau–Lifshitz formula [89] (which ignores thermal fluctuations) with a modified anisotropy constant and saturation magnetization. A similar method was suggested by de Biasi *et al.* [414–416]. They modified the well-known Smit-Beljers-Suhl formula for the precession frequency ω_A, viz. [417–419]

$$\omega_A = \frac{\gamma}{\mu_0 M_S \sin\vartheta} \sqrt{\partial_{\vartheta\vartheta} V \partial_{\varphi\varphi} V - \left(\partial_{\vartheta\varphi} V\right)^2}\bigg|_{\vartheta=\vartheta_A;\varphi=\varphi_A}, \qquad (2.362)$$

where φ_A, ϑ_A are the angular coordinates of the minimum A of the free energy density (Eq. (2.362) is just the magnetic analog of the attempt angular frequency ω_A of a point particle in the small oscillation approximation; see Eq. (1.249) of Chapter 1). Furthermore, Noginova *et al.* [420,421] calculated the FMR absorption spectra of nanomagnets in the manner of paramagnetic resonance theory, i.e., as a *superposition of transitions between all allowed states*. The FMR in an assembly of noninteracting magnetic nanoparticles was also studied by Sukhov *et al.* [422], Raikher *et al.* [105,376,423–426], and (as seen) by Kalmykov and Coffey [148] by solving Brown's magnetic Langevin equation.

To evaluate the FMR frequency of a nanomagnet, i.e., the salient parameter of the problem, we consider a magnetic dipole placed in a potential well created by the magnetocrystalline anisotropy $V(\vartheta, \varphi)$. Moreover, we use the fact that for very low damping ($\alpha < 0.01$), where the energy loss per cycle (due to Brownian motion) of a precessing magnetization is very much less than the thermal energy, the energy trajectories diffuse very slowly therefore they differ but little from those of the undamped precessional motion in a well. In other words, the magnetization dynamics in a potential well are *effectively deterministic* for the purpose of calculating the FMR frequency. Consequently, the motion is described by the deterministic Larmor-like equation, viz.

$$\dot{\mathbf{u}}(t) = \gamma\left[\mathbf{H}(t) \times \mathbf{u}(t)\right]. \qquad (2.363)$$

The FMR frequency Ω_r can then be approximately calculated as the precession frequency ω_A at the bottom of the *deepest* potential well, which is given by Eq. (2.362). This equation allows one to accurately estimate Ω_r only in the *zero-temperature limit*, $T \to 0$, corresponding to effectively *infinite* potential barriers between the metastable states. Unfortunately, Eq. (2.362) leads to substantial errors in Ω_r for *shallow* potential wells with *relatively small* potential barriers (~3–5 kT). Moreover, Eq. (2.362) cannot be used for anisotropies with *flat* potential wells, where the factor $(\partial_{\vartheta\vartheta} V \partial_{\varphi\varphi} V - (\partial_{\vartheta\varphi} V)^2)^{1/2}$ is exactly equal to zero. This is so because the paraboloidal approximation for V in the vicinity of the minimum A, which is used in the derivation of Eq. (2.362), is no longer valid. Hence, the *precession frequency of the magnetization depends (in typical nonlinear fashion) on the energy of the magnetic moment*, which we can write in usual nondimensional form as $E(\vartheta,\varphi) = vV(\vartheta,\varphi)/(kT)$, where v is the volume of the nanoparticle, k is Boltzmann's constant, and T is the temperature. The distribution of energy in the well is then given by the Boltzmann law $Z_A^{-1}e^{-E(\vartheta,\varphi)}$, where

$$Z_A = \iint_{E<E_C} e^{-E(\vartheta,\varphi)} \sin\vartheta d\vartheta d\varphi$$

is the well partition function and $E_C = vV(\vartheta_C, \varphi_C)/(kT)$ is a saddle point energy. The precession frequency $\omega_E = 2\pi P_E^{-1}$ is then obtained from the precession period P_E at a given E and so (as seen) is determined by the line integral, Eq. (1.158) of Chapter 1. At the bottom of the well (at energy $E_A = vV(\vartheta_A, \varphi_A)/(kT)$), the precession frequency which is now just $\omega_A \equiv 2\pi P_{E_A}^{-1}$ reduces to Eq. (2.362) in the paraboloidal approximation. In writing the general Eq. (1.158) for P_E, it is assumed [86,165,166] that one may parametrize the instantaneous magnetization direction by the *slow* energy variable E and the *fast* precessional variable running uniformly along a closed Stoner–Wohlfarth orbit [14] of energy E. Thus, only the purely gyromagnetic term without damping and noise torques need be considered in the magnetic Langevin equation (see Refs. [86,165,166] for details). Moreover, in thermal equilibrium, the FMR frequency Ω_r of a nanoparticle may be estimated from the energy-dependent precession frequency ω_E of the magnetization vector in the well when *averaged over all possible precessional trajectories inside the well* so that

$$\Omega_r = \frac{1}{Z_A} \iint_{E<E_C} \omega_{E(\vartheta,\varphi)} e^{-E(\vartheta,\varphi)} \sin\vartheta d\vartheta d\varphi. \tag{2.364}$$

The integral in Eq. (2.364) is calculated via the precession period P_E, Eq. (1.181) of Chapter 1, *along the lossless energy trajectories*, i.e., the Stoner–Wohlfarth orbits. However, cases exist (low potential barriers or flat wells), where the angular frequency Ω_r may differ considerably from the bottom precession frequency as given by Eq. (2.362). Therefore, we must then use Eq. (2.364) to evaluate the effects of thermal fluctuations on the FMR frequency of nanomagnets. The accuracy of the results for Ω_r as obtained from Eq. (2.364) is then ascertained by comparing them with independent calculations of the FMR frequency from the spectra of the transverse component of the dynamic susceptibility tensor of a nanoparticle obtained via the matrix continued fractions described earlier.

2.10.1 *FMR frequency for uniaxial anisotropy with a transverse magnetic field*

First, we use Eq. (2.364) to determine the FMR frequency for a uniaxial nanomagnet subject to a dc magnetic field $\mathbf{H}_0$ applied perpendicular to the easy axis and therefore characterized by the dimensionless free energy

$$E(\vartheta,\varphi)=\sigma(\sin^2\vartheta-2h\sin\vartheta\cos\varphi), \tag{2.365}$$

where $\sigma=vK_u/(kT)$ is as usual the dimensionless anisotropy (inverse temperature) parameter, K_u is the uniaxial anisotropy constant, and $h=\mu_0 M_S H_0/(2K_u)$ is the external field parameter. The potential Eq. (2.365) has (as seen) two *equivalent* wells with minima $E_A=-\sigma h^2$ at $\mathbf{n}_A^+$ and $\mathbf{n}_A^-$ separated by a potential barrier $\Delta E=E_C-E_A=\sigma(1-h)^2$ with a saddle point at $E_C=\sigma(1-2h)$ at $\mathbf{n}_C$ (see Fig. 2.1). The saddle point lies in the equatorial region, while $\mathbf{n}_A^+$ and $\mathbf{n}_A^-$ lie in the north and south polar regions, respectively. In general, $E(\vartheta,\varphi)$ from Eq. (2.365) retains its *bistable* form for $0\le h<h_c$, where $h_c=1$ is the critical value of h, where $E(\vartheta,\varphi)$ loses its bistable character. The details of the numerical treatment of the problem are given in Section 2.5.1 above.

For $E(\vartheta,\varphi)$ as given by Eq. (2.365), the Larmor-like gyromagnetic Eq. (2.363) can be rewritten in terms of the cartesian components (u_X, u_Y, u_Z) of the unit vector $\mathbf{u}=\mathbf{M}/M_S$ as

$$\tau_0\frac{d}{dt}\begin{pmatrix}u_X\\u_Y\\u_Z\end{pmatrix}=\begin{pmatrix}-u_Z u_Y\\u_Z(u_X-h)\\hu_Y\end{pmatrix}, \tag{2.366}$$

where $\tau_0=\mu_0 M_S/(2\gamma K_u)$ is a precession time constant. The line integral Eq. (1.158) for the precession period P_E can be rewritten in terms of the cartesian components (u_X, u_Y, u_Z) as

$$P_E = \oint_E \frac{\dot{u}_X du_X + \dot{u}_Y du_Y + \dot{u}_Z du_Z}{\dot{u}_X^2 + \dot{u}_Y^2 + \dot{u}_Z^2}, \tag{2.367}$$

where (u_X, u_Y, u_Z) are obviously given by the *deterministic* gyromagnetic equation of motion of the magnetization, Eq. (2.366). By substituting Eqs. (2.366) into Eq. (2.367), we have

$$P_E = \tau_0 \oint_E \frac{-u_Z u_Y du_X + (u_Z u_X - hu_Z)du_Y + hu_Y du_Z}{u_Z^2 u_Y^2 + u_Z^2 (u_X - h)^2 + h^2 u_Y^2}. \tag{2.368}$$

To calculate this integral analytically in terms of elliptic integrals, we rearrange it by using the obvious constraints of energy conservation $E = \text{const}$ and normalization $|\mathbf{u}| = 1$, so that

$$\begin{gathered} 1 - \sigma^{-1}E = u_Z^2 + 2hu_X = p_E^2 = \text{const}, \\ u_X^2 + u_Y^2 + u_Z^2 = 1, \end{gathered} \tag{2.369}$$

where we recall that in the well $-h^2 < E/\sigma < 1 - 2h$, so $p_E^2 > 0$ for $0 < h < 1$. Next, we introduce a new function $u(t)$ related to u_X, u_Y, and u_Z via

$$u_X = \frac{p_E^2}{2h}\left(1 - u^2\right),\ u_Y = \sqrt{1 - p_E^2 u^2 - \frac{p_E^4}{4h^2}\left(1 - u^2\right)^2},\ u_Z = p_E u. \tag{2.370}$$

Thus, the line integral, Eq. (2.368), can be expressed in terms of the single variable u. Therefore, on substituting Eq. (2.370) into Eq. (2.368), we immediately have the period of the precessional motion in the well

$$P_E = \frac{2\tau_0}{p_E} \oint_E \frac{du}{\sqrt{\Phi(u)}} = \frac{2\tau_0}{p_E} \oint_E \frac{du}{\sqrt{(e_1 - u)(u - e_2)(u - e_3)(u - e_4)}}, \tag{2.371}$$

where

$$\Phi(u) = \frac{4h^2}{p_E^4} - 1 + \left(2 - \frac{4h^2}{p_E^2}\right)u^2 - u^4 \tag{2.372}$$

and the roots of the fourth-order polynomial $\Phi(u)$ are given by

$$e_{1,2} = -e_{3,4} = \sqrt{1 - 2h^2 p_E^{-2}\left[1 \pm \sqrt{1 + \left(1 - p_E^2\right)h^{-2}}\right]}. \tag{2.373}$$

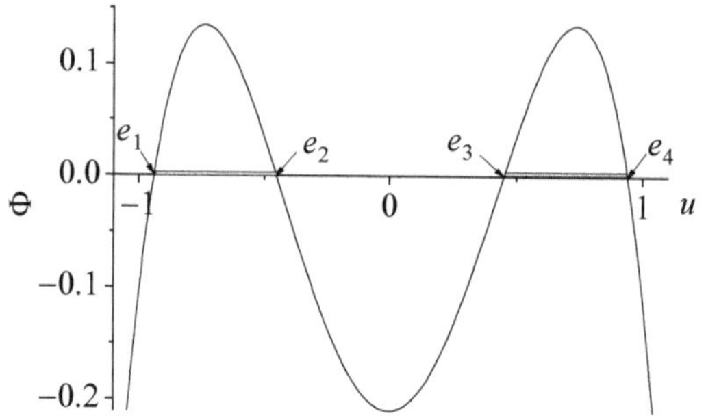

Fig. 2.73. The function $\Phi(u)$, Eq. (2.372), for $h=0.4$ and $p_E=0.95$.

Moreover, for external field parameters $0<h<h_c=1$, $e_1 \le e_2 \le e_3 \le e_4$. Note that in one well u varies in the interval $e_1 \le u \le e_2$ while in the other u varies in the interval $e_3 \le u \le e_4$ (see Fig. 2.73). Noticing that

$$\int_{e_3}^{e_4} \frac{dx}{\sqrt{(e_1-x)(x-e_2)(x-e_3)(x-e_4)}} = \frac{2K(m_E)}{\sqrt{(e_1-e_3)(e_2-e_4)}}, \quad (2.374)$$

where $K(m)$ is the complete elliptic integral of the first kind [290,353] with the modulus

$$m_E = \frac{(e_1-e_2)(e_3-e_4)}{(e_1-e_3)(e_2-e_4)}, \quad (2.375)$$

we then have for the period of the precessional motion

$$P_E = \frac{8\tau_0 K(m_E)}{p_E\sqrt{(e_1-e_3)(e_2-e_4)}}. \quad (2.376)$$

In the *particular* case of zero external field, $h=0$, Eq. (2.376) becomes

$$P_E = \frac{2\pi\tau_0}{\sqrt{1-E/\sigma}}. \quad (2.377)$$

The *averaged* precession frequency Ω_r, Eq.(2.364), as a function of the dc field parameter h is shown in Fig. 2.74 for various σ. Here Ω_r as determined from Eq. (2.364) is compared with independent calculations of the FMR frequency from the FMR absorption spectra via numerical

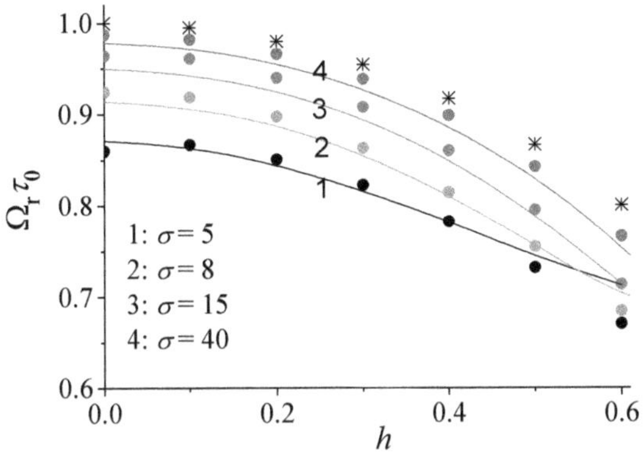

Fig. 2.74. FMR frequency Ω_r vs. the dc field parameter h for various anisotropy (inverse temperature) parameters $\sigma = vK_u/(kT)$. Solid lines: continued fraction solution [148]; filled circles: averaged precession frequency Eqs. (2.364) and (2.376); asterisks: precession frequency at the bottom of the well Eq. (2.378).

solution of the magnetic Langevin equation (see for details Section 2.5.1). Figure 2.74 shows that the approximate Eqs. (2.364) and (2.376) used in combination yield an accurate solution for the FMR frequency both at high (σ< 10) and low (σ< 25) temperatures for a wide range of the external dc field. Moreover, for large σ, i.e., high barriers, the FMR frequency Ω_r is, as expected, accurately approximated by the precession frequency ω_A at the bottom of the well, namely,

$$\omega_A = \tau_0^{-1}\sqrt{1-h^2} \, . \tag{2.378}$$

However, Fig. 2.74 also clearly illustrates that for *low* barriers, $\Delta E < 5$, the *effect of thermal fluctuations on the FMR frequency may exceed* 10 %.

2.10.2 *FMR frequency for biaxial anisotropy*

Next, we consider a biaxial anisotropy potential augmented by the Zeeman term due to an external magnetic field $\mathbf{H}_0$. The dimensionless free energy density $E(\vartheta, \varphi)$ of magnetic nanoparticles taken in standard form including the dc bias field $\mathbf{H}_0$ is then [97,160]

$$E(\vartheta,\varphi) = \sigma\left[\sin^2\vartheta + \delta\sin^2\vartheta\cos^2\varphi - 2h\cos\Theta(\vartheta,\varphi)\right], \tag{2.379}$$

where again $\sigma = vK_u/(kT)$ is the dimensionless anisotropy parameter, K_u is the anisotropy constant, δ is the dimensionless biaxiality parameter accounting for both the magnetizing and magnetocrystalline effects, $h = \mu_0 M_S H_0/(2K_u)$ is as before the external field parameter, and $\Theta(\vartheta, \varphi)$ is given by Eq. (2.299). We recall that biaxial anisotropy may yield an appreciable contribution to the free energy density of magnetic nanoparticles [97]. Furthermore, the bistable potential is used to describe the free energy density of a spheroidal nanoparticle, with the axis of symmetry inclined at a certain angle to the easy anisotropy axis of the particle as well that of elongated particles, where easy- and hard-axis anisotropy terms are present [97]. Moreover, the bistable potential is (as seen) commonly used in spintronic applications [97,137] to represent the free energy density of a nanopillar in the standard form of superimposed easy-plane and in-plane easy axis anisotropies.

For the biaxial anisotropy, Eq. (2.379), the gyromagnetic equation, Eq. (2.363), can again be written for the cartesian components (u_X, u_Y, u_Z) of the unit vector $\mathbf{u} = \mathbf{M}/M_S$ as the system of differential equations:

$$\tau_0 \frac{d}{dt}\begin{pmatrix} u_X \\ u_Y \\ u_Z \end{pmatrix} = \begin{pmatrix} -u_Y u_Z - h\gamma_3 u_Y + h\gamma_2 u_Z \\ (1+\delta)u_Z u_X - h\gamma_1 u_Z + h\gamma_3 u_X \\ -\delta u_X u_Y - h\gamma_2 u_X + h\gamma_1 u_Y \end{pmatrix}. \tag{2.380}$$

We consider now three separate cases:

a): $\mathbf{H}_0 \parallel Z$ $(\gamma_3 = 1, \gamma_1 = \gamma_2 = 0)$, the dc magnetic field $\mathbf{H}_0$ is applied along the easy axis of magnetization;

b): $\mathbf{H}_0 \parallel Y$ $(\gamma_2 = 1, \gamma_1 = \gamma_3 = 0)$, $\mathbf{H}_0$ is applied along along the hard axis of magnetization, and

c): $\mathbf{H}_0 \parallel X$ $(\gamma_1 = 1, \gamma_2 = \gamma_3 = 0)$ (see Fig. 2.75).

The free energy density for each of the three cases becomes

$$\text{a)} \quad E(\vartheta,\varphi) = \sigma\left(\sin^2\vartheta + \delta\sin^2\vartheta\cos^2\varphi - 2h\cos\vartheta\right); \tag{2.381}$$

$$\text{b)} \quad E(\vartheta,\varphi) = \sigma\left(\sin^2\vartheta + \delta\sin^2\vartheta\cos^2\varphi - 2h\sin\vartheta\sin\varphi\right); \tag{2.382}$$

$$\text{c)} \quad E(\vartheta,\varphi) = \sigma\left(\sin^2\vartheta + \delta\sin^2\vartheta\cos^2\varphi - 2h\sin\vartheta\cos\varphi\right). \tag{2.383}$$

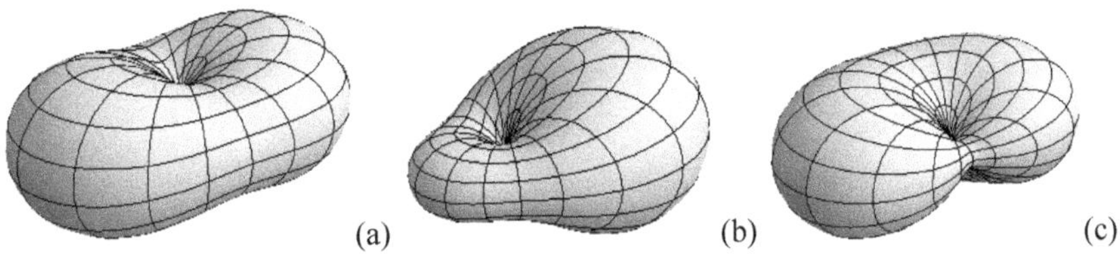

Fig. 2.75. 3D plots of $\sigma^{-1}E(\vartheta,\varphi)$ (a): Eq. (2.381), (b): Eq. (2.383) and (c): Eq. (2.382) for $\delta = 0.6$ and $h = 0.3$.

Table 2.7. The bottom precession frequency.

Potential	Position of minimum	Precession frequency ω_A
$\gamma_3 = 1, \gamma_1 = \gamma_2 = 0$	$\vartheta_A = 0$ and $\vartheta_A = \pi$	$\tau_0^{-1}\sqrt{(1+h)(1+h+\delta)}$
$\gamma_1 = 1, \gamma_2 = \gamma_3 = 0$	$\varphi_A = 0$ and $\sin\vartheta_A = \pm h(1+\delta)^{-1}$	$\tau_0^{-1}\sqrt{1+\delta-h^2(1+\delta)^{-1}}$
$\gamma_2 = 1, \gamma_1 = \gamma_3 = 0$	$\varphi_A = \pi/2$ and $\sin\vartheta_A = \pm h$	$\tau_0^{-1}\sqrt{(1-h^2)(1+\delta)}$

The free energy density $E(\vartheta, \varphi)$ from Eqs. (2.381)–(2.383) has two wells (which may be either nonequivalent or equivalent) and one or two equivalent saddle points (see Fig. 2.75). The oscillation frequencies for the deepest well ω_A, Eq. (2.362), for each of the three potentials (2.381)–(2.383) are given in Table 2.7. The solution of Eq. (2.380) for these three cases can be obtained in the same manner as uniaxial anisotropy [427]. Here we give only the resultant equations for the precessional period:

a) $\gamma_3 = 1, \gamma_1 = \gamma_2 = 0$
$$P_E = \frac{2\tau_0 K(m_E)}{p_E\sqrt{(\delta+1)(1+e_+)(1-e_-)}}, \tag{2.384}$$

where

$$p_E^2 = \frac{\sigma\delta - E}{\sigma(\delta+1)} + \frac{h^2}{(\delta+1)^2} = \left(u_Z + \frac{h}{\delta+1}\right)^2 + \frac{\delta}{\delta+1}u_Y^2,$$
$$e_\pm = \frac{1}{p_E}\left(-\frac{h\delta}{\delta+1} \pm \sqrt{h^2 - \frac{E}{\sigma}}\right), \quad m_E = \frac{(1+e_-)(1-e_+)}{(1+e_+)(1-e_-)}. \tag{2.385}$$

(b) $\gamma_2 = 1, \gamma_1 = \gamma_3 = 0$
$$P_E = \frac{4\tau_0 K(m_E)}{p_E\sqrt{\delta(1+e_+)(1-e_-)}}, \tag{2.386}$$

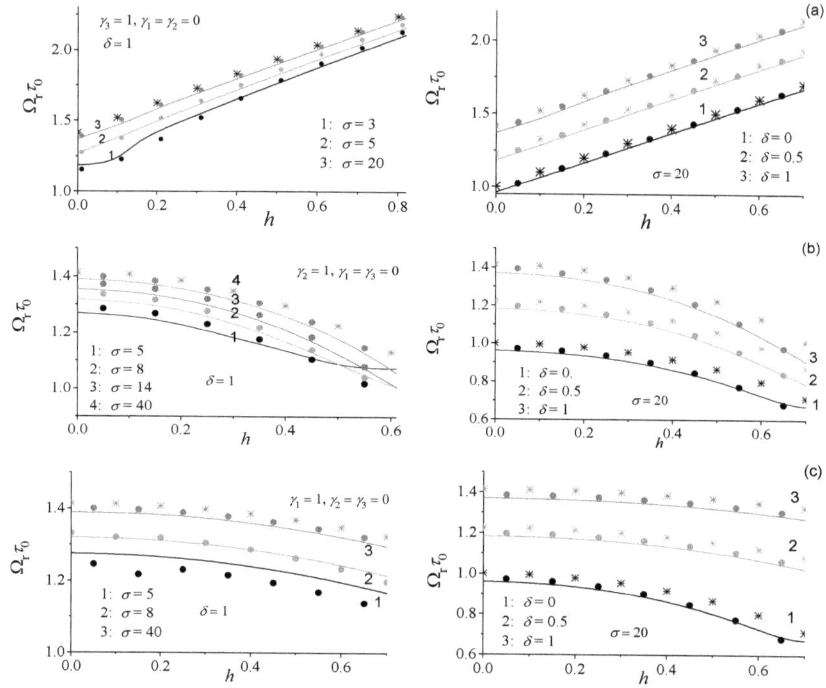

Fig. 2.76. FMR frequency Ω_r vs. the dc field parameter h for $\gamma_3 = 1, \gamma_1 = \gamma_2 = 0$ (a), $\gamma_2 = 1, \gamma_1 = \gamma_3 = 0$ (b), $\gamma_1 = 1, \gamma_2 = \gamma_3 = 0$ (c), and various anisotropy parameters σ and δ. Solid line: the FMR frequency calculated via continued fraction solution [35]; filled circles: averaged precession frequency Eqs. (2.364), (2.384), (2.386), and (2.388); asterisks: precession frequency at the bottom of the deepest well (see Table 2.7).

where

$$p_E^2 = \frac{\delta + 1 - \sigma^{-1}E}{\delta} + \frac{h^2}{\delta^2} = \left(u_Y + \frac{h}{\delta}\right)^2 + \frac{\delta + 1}{\delta} u_Z^2,$$
$$e_\pm = \frac{h(\delta + 1)}{\delta p_E} \pm \frac{\sqrt{h^2 + \sigma^{-1}E}}{p_E}, \quad m_E = \frac{2(e_+ - e_-)}{(1 + e_+)(1 - e_-)}. \tag{2.387}$$

(c) $\gamma_1 = 1, \gamma_2 = \gamma_3 = 0$

$$P_E = \frac{4\tau_0}{p_E\sqrt{\delta}}\left[\frac{K(m_E^{-1})}{\sqrt{2(e_+ - e_-)}} - \frac{K(m_E)}{\sqrt{(1 + e_+)(1 - e_-)}}\right], \tag{2.388}$$

where

$$p_E^2 = \frac{\sigma^{-1}E}{\delta+1} + \frac{h^2}{(\delta+1)^2} = \left(u_X - \frac{h}{\delta+1}\right)^2 + \frac{1}{\delta+1}u_Y^2,$$
$$e_\pm = \frac{h \pm \sqrt{h^2 + \delta(1+\delta)(p_E^2(1+\delta)-1)}}{p_E(1+\delta)\delta}, \quad m_E = \frac{2(e_+ - e_-)}{(1+e_+)(1-e_-)}. \tag{2.389}$$

The FMR frequency Ω_r as a function of the field parameter h for each of the three cases is shown in Fig. 2.76. Here Ω_r as given by Eq. (2.364) is compared with independent calculations of the FMR frequency from the dynamic susceptibility spectra of the nanoparticle obtained via numerical solution of the magnetic Langevin equation (see for details Section 2.6.2). For high barriers, the paraboloidal approximation solution rendered by Eq. (2.362) is accurate while the weighted Ω_r from Eq. (2.364) also yields an accurate solution (see Fig. 2.76). However, for low barriers (small particle and/or high temperatures), the effect of thermal fluctuations on the FMR frequency may again be substantial (more than 10 %).

2.10.3 *Mixed anisotropy: breakdown of the paraboloid approximation*

The origin of Eq. (2.362) lies in the paraboloidal approximation for the free energy density potential $E(\vartheta, \varphi)$ near the bottom of a potential well. However, (as seen) there are certain situations, where the well frequency ω_A is zero. This obviously incorrect result may occur if the paraboloidal approximation fails. We saw that the breakdown of that approximation in the well is encountered, for example, for a potential $E(\vartheta, \varphi)$ with flat well bottoms. An example is *mixed* uniaxial and cubic anisotropy, Eq. (2.123) [365,366,369]. Here for cubic-to-uniaxial anisotropy ratio $\zeta = K_c / K_u = 1$, the anisotropy energy, Eq. (2.123); becomes (see Fig. 2.15c)

$$E(\vartheta,\varphi) = \frac{\sigma}{8}(7+\cos 4\varphi)\sin^4\vartheta, \tag{2.390}$$

where $\sigma = vK_u / (kT)$ is the dimensionless anisotropy parameter and K_u denotes the uniaxial anisotropy constant. This value of ζ is of much interest, since Eq. (2.362) cannot now be used to estimate the FMR frequency because

$$\omega_A = \frac{\gamma}{\mu_0 M_S \sin\vartheta}\sqrt{\partial_{\vartheta\vartheta}V\partial_{\varphi\varphi}V - \left(\partial_{\vartheta\varphi}V\right)^2}\Bigg|_{\vartheta=\vartheta_{\min};\varphi=\varphi_{\min}} \equiv 0. \quad (2.391)$$

For the anisotropy potential, Eq. (2.390), the gyromagnetic Eq. (2.363) can also be written in terms of the cartesian components (u_X, u_Y, u_Z) of the unit vector **u** as

$$\tau_0 \frac{d}{dt}\begin{pmatrix} u_X \\ u_Y \\ u_Z \end{pmatrix} = \begin{pmatrix} -u_Y u_Z(1-u_Z^2+u_Y^2) \\ u_Z u_X(1-u_Z^2+u_X^2) \\ u_X u_Y(u_Y^2-u_X^2) \end{pmatrix}. \quad (2.392)$$

The magnetization trajectories of the precessional dynamics must again obey the two constraints, namely, $|\mathbf{u}| = 1$, and

$$\begin{aligned} p_E^2 &= 1-\sigma^{-1}E \\ &= u_Z^2 + u_X^2u_Y^2 + u_X^2u_Z^2 + u_Y^2u_Z^2 = \text{const}, \end{aligned} \quad (2.393)$$

where $E_A \le E \le E_C$. Two equivalent wells with minima $E_A = 0$ at $\vartheta_A = 0, \pi$ and four saddle points $E_C = 3\sigma/4$ at $\vartheta_C = \pi/2$ and $\varphi_C = \pi n/2, n = 0,1,2,3$ (see Fig. 2.15(c)) exist. Since

$$u_Y^2 u_X^2 = u_Z^4 - 2u_Z^2 + 1 - \sigma^{-1}E$$

and

$$\left(u_X^2 - u_Y^2\right)^2 = -\left(3u_Z^4 - 6u_Z^2 + 3 - 4\sigma^{-1}E\right),$$

we can therefore again introduce a new function $u(t)$ related to $u_X(t), u_Y(t)$, and $u_Z(t)$ via

$$\begin{aligned} u_X &= 2^{-1/2}\sqrt{1-u^2 \mp \sqrt{3(u^2-1)+4\sigma^{-1}E}}, \\ u_Y &= 2^{-1/2}\sqrt{1-u^2 \pm \sqrt{3(u^2-1)+4\sigma^{-1}E}}, \\ u_Z &= u. \end{aligned} \quad (2.394)$$

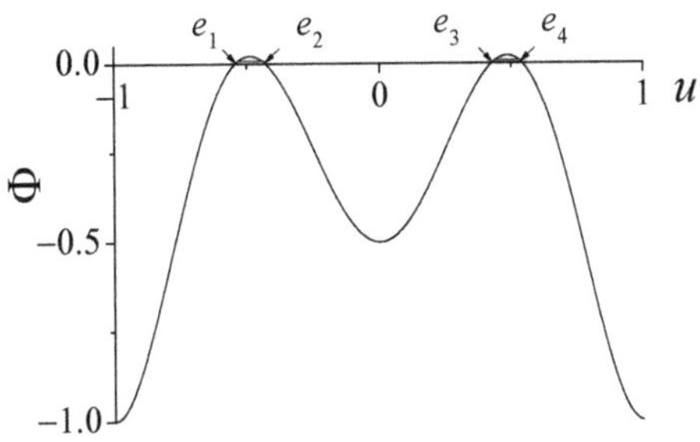

Fig. 2.77. The function $\Phi(u)$, Eq. (2.396), for $E = \sigma/2$.

By substituting $u_X(t), u_Y(t), u_Z(t)$ given by Eq. (2.394) into Eq. (2.367), we have

$$P_E = \tau_0 \oint_E \frac{-u_Y u_Z (u_X^2 + 2u_Y^2) du_X + u_Z u_X (u_Y^2 + 2u_X^2) du_Y + u_X u_Y (u_Y^2 - u_X^2) du_Z}{u_Y^2 u_Z^2 (u_X^2 + 2u_Y^2)^2 + (u_Y^2 + 2u_X^2)^2 u_Z^2 u_X^2 + (u_Y^2 - u_X^2)^2 u_X^2 u_Y^2}$$

$$= -\tau_0 \oint_E \frac{du}{\sqrt{\Phi(u)}}, \tag{2.395}$$

where

$$\Phi(u) = \left[(1-u^2)^2 - \sigma^{-1}E\right]\left[4\sigma^{-1}E - 3(1-u^2)^2\right] \tag{2.396}$$

and u varies in the intervals $e_1 \le u \le e_2$ and $e_3 \le u \le e_4$ (see Fig. 2.77), where

$$e_{1,3} = \mp\sqrt{1 - 2\sqrt{\frac{E}{3\sigma}}}, \quad e_{2,4} = \mp\sqrt{1 - \sqrt{\frac{E}{\sigma}}}. \tag{2.397}$$

are the roots of the polynomial $\Phi(u)$. The sign of the integral is again determined by the direction along the closed trajectory with $E = \text{const}$. Therefore, we have Eq. (2.398) for the precessional period P_E

$$P_E = 2\tau_0 \int_{e_3}^{e_4} \frac{du}{\sqrt{\left[(1-u^2)^2 - \sigma^{-1}E\right]\left[4\sigma^{-1}E - 3(1-u^2)^2\right]}}. \tag{2.398}$$

The FMR frequency as a function of the parameter $\sigma = vK_u/(kT)$ is shown in Fig. 2.78. Clearly the definition of Ω_r via Eq. (2.364) in

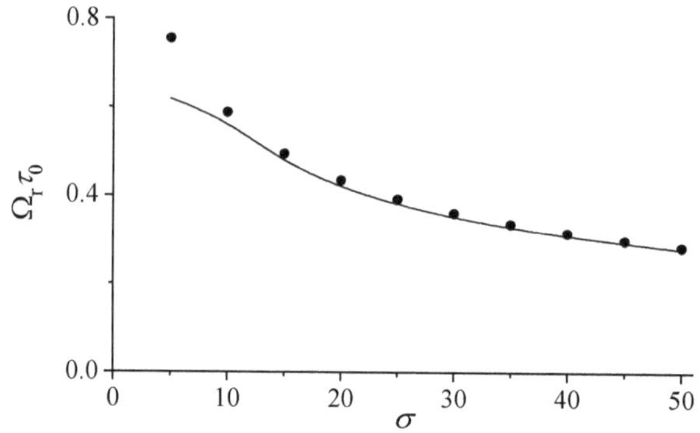

Fig. 2.78. FMR frequency Ω_r vs. the anisotropy (inverse temperature) parameter σ for $\zeta = -1$ (flat wells). Solid line: continued fraction solution [38]; filled circles: averaged precession frequency inside the well Ω_r from Eqs. (2.364) and (2.398).

combination with the closed Eq. (2.398) provides a good approximation to the FMR frequency for $\sigma > 10$.

Finally at low potential barriers, $\sigma < 5$, comparison of Ω_r from Eq. (2.364) with the matrix continued fraction solution becomes rather complicated as the FMR peak in the spectrum of the imaginary part of transverse dynamic susceptibility $\chi_\perp(\omega)$ then becomes asymmetric and very large due to the inhomogeneous broadening.

We have evaluated the effects of thermal fluctuations on the *average* FMR frequency of a magnetic nanoparticle for a variety of anisotropy potentials showing that for an acceptable solution all that is required is a *knowledge of the deterministic dynamics governed by the simple gyromagnetic equation and the Boltzmann equilibrium distribution*. The calculations clearly illustrate that for low barriers, $\Delta E < 5$ (corresponding to small particles and/or high temperatures), the effect of thermal fluctuations on the FMR frequency may be substantial (more than 10 %). In this respect, the determination of the FMR frequency is rather similar to the calculation of the underdamped Kramers' escape rate [166], which is also obtained from the deterministic dynamics of the magnetization. This similarity is unsurprising given that the underdamped escape rate and the FMR frequency are intrinsically linked to each other since they are both aspects of the same phenomenon, namely, the very lightly damped periodic motion in the wells of the magnetocrystalline anisotropy potential. In other words, at low frequencies the effect of slightly damped librational motion of spins with energy near the barrier energy is to give rise to overbarrier relaxation while at high frequencies the effect of the precessional motion deep in the well is to give rise to ferromagnetic resonance. Moreover, the linking of the high-frequency resonant process with the low-frequency overbarrier one is yet another example [428] of the original Kramers concept of oscillations in a potential well of Brownian particles with energy equal to the separatrix energy before escape as both are simply limiting cases of the same family of librational dynamical processes in a well. The salient difference between the inertial Brownian motion case (pertaining to particles and rigid rotators) and the magnetic one is that in the former the low damping process is due to *inertial effects* while in the latter that process arises from geometry.

2.11 Magnetization dynamics of two interacting spins in an external magnetic field

The one-spin approximation described above, has been used to interpret many experiments. Nevertheless, systems of (both metallic and ferrite) particles exist, where deviations in behavior from that predicted by the one-spin approximation have been observed [20]. An understanding of these effects *requires a microscopic theory of the magnetization dynamics, capable of distinguishing and accounting for the various crystallographic local environments inside a nanoparticle and on its surface*. In formulating such a theory, one is invariably faced with complex N-body problems. Thus, before one can even attempt to calculate the magnetization relaxation time of many interacting spins, one needs to understand the effect of interactions on the magnetization relaxation time of the simplest interacting system, conceived of as a pair of spins coupled via exchange and/or dipole-dipole interactions, including the usual magnetocrystalline anisotropy and Zeeman terms. This will be referred to as the *two-spin problem* (TSP). This apparently simple problem poses, in the course of its general formulation, the usual difficulties associated with analyzing the energyscape (location of the minima, maxima, and saddle points of the potential energy) in a system with many degrees of freedom, which is a crucial step in the calculation of the relaxation time whatever the context. The effect of interactions between magnetic moments on the magnetization relaxation time τ has been already treated in several papers. For example, Lyberatos *et al.* [429] and Andersson *et al.* [430] used a Monte-Carlo method to study the dynamics of an assembly of interacting magnetic moments. Berkov and Gorn [431] numerically solved the set of Langevin equations for a system of spherical single-domain particles with uniaxial anisotropy and coupled by dipole-dipole interactions. This calculation was done without any assumptions regarding the strength of the interactions and is focused on the temperature and frequency behavior of the imaginary part of the ac susceptibility of an assembly of such nanoparticles. They also found a crossover-type behavior between assemblies with low and high anisotropy accompanied by a nonmonotonic temperature behavior of the susceptibility peak as a function of temperature and concentration at moderate to low damping. In Ref. [432], a pair of coupled dipoles was treated using numerical Langevin dynamics simulations. Hinzke and Nowak [433,434] compared calculations of thermally activated magnetization reversal in systems of interacting classical magnetic moments obtained both by a Monte-Carlo method and Langevin dynamics simulations. Denisov and Trohidou [435] by solving the Fokker–Planck equation for the distribution function of two-dimensional ensembles of ferromagnetic nanoparticles, derived an evolution equation for the magnetization and found its solution in certain limiting cases. Approximate analytic equations for τ were obtained under particular conditions for assemblies of magnetic moments in Refs. [436–438]. Rodé *et al.* [439] numerically solved the Fokker–Planck equation for two interacting particles and calculated the time decay of the magnetization. Chen *et al.* [440] presented an analytic solution for τ for two identical interacting single domain particles and obtained good agreement with the numerical results using the Fokker–Planck equation. Yoshimori and Korringa [441,442] and Solomon [443] treated relaxational processes of a system of two spins in a magnetic field. Also, many authors (see, e.g., [444–449]) proceeded by using a self-consistent potential in the Fokker–Planck equation to handle the long-range magnetic dipole-dipole interactions. The approach by Zubarev and Iskakova [444] renders, a result similar to that of Zwanzig [450] for a lattice of electric dipoles for the linear response, namely, the effect of the interactions leads to a discrete set of microscopic relaxation mechanisms so that the characteristic time scale increases because of these. Ivanov *et al.* [445–448] have obtained formulas for the static and dynamic susceptibility that compared favorably with numerical simulations and experimental data while the approach [449], based on the nonlinear response of magnetic dipoles to external alternating fields, allows one to evaluate the global time scale in agreement with the predictions of the mean field theory. These investigations have been recently extended by Déjardin *et al.* [451] who derived an analytical formula for the dynamic susceptibility of isotropic dipolar assemblies with long-range interactions using the forced rotational diffusion equation beyond the mean field approximation. A complementary method for treating the magnetization dynamics of a pair of exchange coupled dipoles was developed by Titov *et al.* [452,453], i.e., by averaging the stochastic Landau-Lifshitz-Gilbert equation for the magnetization over its realizations, so reducing the problem to solving a system of linear differential-recurrence relations for the statistical moments (averaged spherical harmonics). The system was then solved in the frequency domain by matrix continued fractions yielding the complete solution of the two-spin problem. The results for the relaxation time for two spins in a uniaxial potential have been compared favorably with Langer's theory of the decay of metastable states [103]. This theory we recall also comprises the generalization of the Kramers calculation of the IHD escape rate for a single degree of freedom system with a separable and additive Hamiltonian to multi-degree of freedom systems with nonseparable Hamiltonians as obtain in magnetic systems. The IHD solution corresponds physically to the situation where the energy loss per cycle of the almost periodic motion of a spin having the saddle point energy is $\geq kT$ (see Chapter 1, Section 1.5.3).

Here, we demonstrate how the hierarchy of differential-recurrence relations for the relaxation functions arise naturally from the coupled vector Langevin equations for the two-spin system thus bypassing the problem of constructing and solving the Fokker–Planck equation entirely. Consider a system of two interacting magnetic dipoles $\boldsymbol{\mu}_p(t)$, $p = 1, 2$,

$$\boldsymbol{\mu}_p = \mu \mathbf{s}_p, \tag{2.399}$$

where $\mathbf{s}_p = \mathbf{i}\sin\vartheta_p\cos\varphi_p + \mathbf{j}\sin\vartheta_p\sin\varphi_p + \mathbf{k}\cos\vartheta_p$ and ϑ_p and φ_p are the respective polar and azimuthal angles, $\mathbf{s}_p$ denotes a unit vector along $\boldsymbol{\mu}_p$ and μ is the nominal value of the dipole moment. For a simple exposition, the two easy axes are taken as parallel to each other and to the applied field, which is parallel to the reference (Z) axis. The dynamics of an individual spin $\mathbf{s}_p(t)$ is governed by the stochastic Landau–Lifshitz–Gilbert equation, i.e., the deterministic Landau–Lifshitz–Gilbert equation augmented by a random field $\mathbf{h}^{(p)}(t)$ [18,97] (cf. Eq. (1.59))

$$\dot{\mathbf{s}}_p = b\mu_0\mu\left(\alpha^{-1}\left[\left(\mathbf{H}_p + \mathbf{h}_p\right)\times\mathbf{s}_p\right] - \left[\mathbf{s}_p\times\left[\mathbf{s}_p\times\left(\mathbf{H}_p + \mathbf{h}_p\right)\right]\right]\right), \tag{2.400}$$

where $b = (2kT\tau_{\mathrm{N}})^{-1}$, $\tau_{\mathrm{N}} = \mu_0\mu(1+\alpha^2)/(2\gamma\alpha kT)$ is a characteristic (free diffusion) time, γ is the gyromagnetic ratio, $\mathbf{H}_p = -kT\mu_0^{-1}\partial_{\boldsymbol{\mu}_p}E$ is the effective magnetic field, α is the dimensionless damping parameter representing the dissipative coupling to the heat bath and the random Gaussian field $\mathbf{h}_p(t)$ has the white noise properties, Eq. (1.43), with the constant $D = kT\alpha(\gamma\mu_0\mu)^{-1}$. The random field describes as before the thermal fluctuations of the spin via the fluctuation-dissipation theorem. The effective magnetic field $\mathbf{H}_p(t)$ acting on the spin consists of the externally applied magnetic fields, the crystalline anisotropy field, and exchange) field produced by the other spin.

Next, we assume as before that the magnitude of an external applied (spatially) uniform dc magnetic field is suddenly altered at time $t = 0$ from $\mathbf{H}^{\mathrm{I}}$ to $\mathbf{H}^{\mathrm{II}}$. We then then study the relaxation starting from an equilibrium state I with the Boltzmann distribution function $W_{\mathrm{I}} = Z_{\mathrm{I}}^{-1}e^{-E_{\mathrm{I}}}$ $(t \le 0)$ to another equilibrium state II with the distribution function $W_{\mathrm{II}} = Z_{\mathrm{II}}^{-1}e^{-E_{\mathrm{II}}}$ $(t \to \infty)$. Here E_i is the free energy, $\beta = 1/kT$ and Z_i $(i = \mathrm{I}, \mathrm{II})$ are the partition functions. Thus, the spin dynamics $\mathbf{s}_p(t)$ immediately following the alteration of the field may be described using the normalized relaxation function

$$f_p(t) = \frac{\langle \mathbf{s}_p\cdot\mathbf{e}_Z\rangle(t) - \langle \mathbf{s}_p\cdot\mathbf{e}_Z\rangle_{\mathrm{II}}}{\langle \mathbf{s}_p\cdot\mathbf{e}_Z\rangle_{\mathrm{I}} - \langle \mathbf{s}_p\cdot\mathbf{e}_Z\rangle_{\mathrm{II}}}, \tag{2.401}$$

where $\mathbf{e}_Z$ is the unit vector along the Z axis, the angular brackets $\langle \mathbf{s}_p\cdot\mathbf{e}_Z\rangle(t)$ are the time-dependent ensemble averages, and the brackets

$\langle \mathbf{s}_p \cdot \mathbf{e}_Z \rangle_i$ designate the *equilibrium* ensemble average in the initial I and final II states,

$$\langle \mathbf{s}_p \cdot \mathbf{e}_Z \rangle_i = \int_0^{2\pi}\int_0^{2\pi}\int_0^{\pi}\int_0^{\pi} (\mathbf{s}_p \cdot \mathbf{e}_Z) W_i \sin\vartheta_2 \sin\vartheta_1 d\vartheta_2 d\vartheta_1 d\varphi_2 d\varphi_1. \quad (2.402)$$

Note that the transient response so formulated is truly *nonlinear* because the change in amplitude $\mathbf{H}_{\mathrm{I}} - \mathbf{H}_{\mathrm{II}}$ of the external dc magnetic field is *arbitrary*. The overall behavior of $f_p(t)$ is characterized by the *integral relaxation time* τ_{int} (see Section 1.3.6). As usual this time is the area under the curve of $f_p(t)$:

$$\tau_{\mathrm{int}} = \int_0^{\infty} f_p(t)dt. \quad (2.403)$$

Again, for external fields less than a certain critical field (see below), τ_{int} yields a close approximation to the greatest relaxation time of the system. The effective relaxation time τ_{ef} may be evaluated by averaging the stochastic Gilbert equations (2.400). Thus, we have for the sharp values

$$\tau_{\mathrm{N}} \frac{d}{dt}\cos\vartheta_p = -\cos\vartheta_p + \frac{\sin\vartheta_p}{2}\frac{\partial E}{\partial \vartheta_p} + \frac{1}{2\alpha}\frac{\partial E}{\partial \varphi_p}, \quad (2.404)$$

where $p = 1, 2$. The first term on the right-hand side of Eq. (2.404) is the noise-induced drift. It follows from Eq. (2.404) that

$$\tau_{\mathrm{N}} \left\langle (\cos\vartheta_1 + \cos\vartheta_2)\frac{d}{dt}(\cos\vartheta_1 + \cos\vartheta_2) \right\rangle_0 =$$

$$-\left\langle (\cos\vartheta_1 + \cos\vartheta_2)\sum_{p=1}^{2}\cos\vartheta_p \left(1 - \frac{\tan\vartheta_p}{2}\frac{\partial E}{\partial \vartheta_p} - \frac{1}{2\alpha}\frac{\partial E}{\partial \varphi_p}\right)\right\rangle_0 \quad (2.405)$$

$$= -\frac{1}{2}\left\langle (\sin^2\vartheta_1 + \sin^2\vartheta_2) \right\rangle_0$$

(here we have used integration by parts). Thus, we have

$$\tau_{\mathrm{ef}} = 2\tau_{\mathrm{N}} \frac{\left\langle (\cos\vartheta_1 + \cos\vartheta_2)^2 \right\rangle_0 - \left\langle \cos\vartheta_1 + \cos\vartheta_2 \right\rangle_0^2}{\left\langle \sin^2\vartheta_1 + \sin^2\vartheta_2 \right\rangle_0}. \tag{2.406}$$

The equilibrium averages (quadruple integrals) in Eq. (2.406) can be evaluated numerically.

Either of the relaxation functions $f_1(t)$ or $f_2(t)$ describe the transient nonlinear response of the longitudinal component of the magnetic dipole moment of the system $m_Z(t) = \mu\langle(\mathbf{s}_1 + \mathbf{s}_2)\cdot\mathbf{e}_Z\rangle(t)$, viz.,

$$m_Z(t) = 2\mu\left[\langle \mathbf{s}_1 \cdot \mathbf{e}_Z\rangle_{\mathrm{II}} + \left(\langle \mathbf{s}_1 \cdot \mathbf{e}_Z\rangle_{\mathrm{I}} - \langle \mathbf{s}_1 \cdot \mathbf{e}_Z\rangle_{\mathrm{II}}\right) f_1(t)\right], \tag{2.407}$$

where $\mu = \mu_0^{-1}\gamma\hbar S$ is the magnetic moment associated with the spin. As usual, a special case is the *linear response* of the two-spin system to infinitesimally small changes in the strength of the strong dc field $\mathbf{H}^{\mathrm{I}}$, i.e., for $\mathbf{H}^{\mathrm{II}} = \mathbf{H}^{\mathrm{I}} - \boldsymbol{\kappa}$ as $\boldsymbol{\kappa} \to 0$. Here the relaxation function $f_p(t)$ as described by Eq. (2.401) coincides with the normalized longitudinal dipole equilibrium correlation function $C_{\parallel}(t)$, that is

$$\lim_{\kappa\to 0} f_p(t) = C_{\parallel}(t) = \frac{\langle m_Z(0) m_Z(t)\rangle_{\mathrm{I}} - \langle m_Z(0)\rangle_{\mathrm{I}}^2}{\langle m_Z^2(0)\rangle_{\mathrm{I}} - \langle m_Z(0)\rangle_{\mathrm{I}}^2}. \tag{2.408}$$

Again according to linear response theory (see Chapter 1, Section 1.3.6), having determined the one-sided Fourier transform $\tilde{C}_{\parallel}(i\omega)$ [the spectrum of $C_{\parallel}(t)$], one has the integral relaxation time in the linear response approximation that is the correlation time $\tau_{\mathrm{int}} = \tilde{C}_{\parallel}(0)$ and the normalized dynamic susceptibility $\chi(\omega) = 1 - i\omega\tilde{C}_{\parallel}(\omega)$. The asymptotic behavior of $\chi(\omega)$ in the extremes of very low and very high frequencies is given by Eq. (1.147). The integral relaxation (correlation) time τ_{int} and effective relaxation time τ_{ef} may equivalently be written as explained earlier in terms of the eigenvalues λ_k of the Fokker–Planck operator L_{FP}; viz.,

$$\tau_{\mathrm{int}} = \sum_k c_k / \lambda_k \quad \text{and} \quad \tau_{\mathrm{ef}} = \left(\sum_k c_k \lambda_k\right)^{-1} \tag{2.409}$$

as defined by the Fokker–Planck equation accompanying Eq. (2.400), viz. $\dot{W} = \mathrm{L}_{\mathrm{FP}} W$, for the joint distribution function $W(\mathbf{s}_1, \mathbf{s}_2, t)$ of the orientations of the spins (see Section 1.3.6). Equation (2.409) emphasizes that the integral relaxation time τ_{int} contains contributions from *all* the eigenvalues λ_k. As usual the smallest nonvanishing eigenvalue λ_1 is associated with the slowest overbarrier or interwell relaxation mode and so with the long-time behavior of $f_p(t)$; the other eigenvalues λ_k characterize high-frequency "intrawell" modes. Recall that to evaluate τ_{int} numerically from Eq. (2.409), all the λ_k and c_k are, in principle, required. However, as repeatedly seen in the low-temperature (high-barrier) limit, $\lambda_1 << |\lambda_k|$ and $c_1 \approx 1 >> c_k$ $(k \neq 1)$, provided the wells of the potential remain approximately equivalent (true for a small external field), one has

$$\tau_{\text{int}} \approx \lambda_1^{-1}. \tag{2.410}$$

Hence, the integral relaxation time for nearly equivalent potential wells in the low-temperature limit closely approximates λ_1^{-1}. By using matrix continued fractions, one can also calculate the smallest nonvanishing eigenvalue λ_1 of the Fokker–Planck operator L_{FP} from the secular equation (1.134).

2.11.1 *Exchange interaction*

We first consider the two-spin problem with *exchange interaction*. This will be accomplished by (i) deriving an exact system of equations for the statistical moments by directly averaging the Landau-Lifshitz-Gilbert equations for the motion of the magnetization augmented by a random field due to the heat bath over its realizations; (ii) solving the resulting hierarchy of moment equations via matrix continued fractions [97,143]. As in the one-spin problem, this method will allow us to accurately evaluate the time decay of the magnetization in all ranges of the damping, exchange coupling, anisotropy and applied field parameters as well as other physical parameters (spectra of the relaxation functions, the complex susceptibility, etc.). Furthermore, in the IHD regime, an asymptotic formula for the greatest relaxation time [454] based on Langer's calculation is naturally expected to apply [181]. In this damping regime, the numerical results for the relaxation time obtained are compared with those of Ref. [454] in order to establish a range of validity for asymptotic calculations of the relaxation time in the IHD range.

The reduced free energy density E of the exchanged-coupled two-spin system in the dc magnetic field may be written as [454]

$$\begin{aligned} E &= -\varsigma(\mathbf{s}_1 \cdot \mathbf{s}_2) - \sum_{p=1,2}\left[\xi_i(\mathbf{e}_Z \cdot \mathbf{s}_p) + \sigma(\mathbf{e}_Z \cdot \mathbf{s}_p)^2\right] \\ &= -\varsigma\left[\cos\vartheta_1 \cos\vartheta_2 + \sin\vartheta_1 \sin\vartheta_2 \cos(\varphi_1 - \varphi_2)\right] \\ &\quad -\xi_\gamma\left(\cos\vartheta_1 + \cos\vartheta_2\right) - \sigma\left(\cos\vartheta_1^2 + \cos\vartheta_2^2\right), \quad i = \mathrm{I,II} \end{aligned} \tag{2.411}$$

where $\xi_i = \mu_0 \mu H^i / (kT)$, σ, and $\varsigma = JS^2/(kT)$ are the dimensionless field, anisotropy, and interaction parameters, respectively, and J is the ferromagnetic exchange coupling constant. The vector stochastic differential equation (2.400) written in spherical polar coordinates leads to the system of two coupled scalar stochastic differential equations

$$\dot{\vartheta}_p = \frac{\mu_0\mu(h_{\vartheta_p} + \alpha^{-1}h_{\varphi_p})}{2kT\tau_\mathrm{N}} - \frac{1}{2\tau_\mathrm{N}}\left[\frac{\partial E}{\partial \vartheta_p} + \frac{1}{\alpha \sin\vartheta_p}\frac{\partial E}{\partial \varphi_p}\right], \tag{2.412}$$

$$\dot{\varphi}_p = \frac{\mu_0\mu(h_{\varphi_p} - \alpha^{-1}h_{\vartheta_p})}{2kT\tau_\mathrm{N}\sin\vartheta_p} - \frac{1}{2\tau_\mathrm{N}}\left[\frac{1}{\sin^2\vartheta}\frac{\partial E}{\partial \varphi_p} - \frac{1}{\alpha\sin\vartheta_p}\frac{\partial E}{\partial \vartheta_p}\right]. \tag{2.413}$$

We now introduce as new variables the product of spherical harmonics

$$M_{l_1 l_2 m}(t) = Y_{l_1 m}[\vartheta_1(t), \varphi_1(t)] Y_{l_2 -m}[\vartheta_2(t), \varphi_2(t)] \tag{2.414}$$

corresponding to a complete set of orthogonal functions characterizing the dynamics of the two exchange-coupled spins. Thus, we have from Eqs. (2.412)–(2.414) the stochastic equation of motion for the functions $M_{l_1 l_2 m}$, namely

$$\frac{d}{dt}M_{l_1 l_2 m} = \sum_{p=1,2} \dot{\vartheta}_p \frac{\partial M_{l_1 l_2 m}}{\partial \vartheta_p} + \dot{\varphi}_p \frac{\partial M_{l_1 l_2 m}}{\partial \varphi_p}. \tag{2.415}$$

Upon averaging the stochastic differential Eq. (2.415) over its realizations in an infinitesimal time (see details in Chapter 1), we obtain

$$\tau_N \dot{M}_{l_1,l_2,m} = \sum_{p=1,2} \frac{1}{4}\left[\left(L^{(p)}\right)^2\left(EM_{l_1l_2m}\right) - E\left(L^{(p)}\right)^2 M_{l_1l_2m} - M_{l_1l_2m}\left(L^{(p)}\right)^2 E\right]$$
$$+\sum_{p=1,2}\frac{i}{4\alpha}\sqrt{\frac{3}{2\pi}}\left\{\left(Y_{1,1}^{(p)}\right)^{-1}\left[\left(L_Z^{(p)}E_+^{(p)}\right)\left(L_+^{(p)}M_{l_1l_2m}\right)-\left(L_+^{(p)}E_+^{(p)}\right)\left(L_Z^{(p)}M_{l_1l_2m}\right)\right]\right.$$
$$\left.+\left(Y_{1-1}^{(p)}\right)^{-1}\left[\left(L_Z^{(p)}E_-^{(p)}\right)\left(L_-^{(p)}M_{l_1l_2m}\right)-\left(L_-^{(p)}E_-^{(p)}\right)\left(L_Z^{(p)}M_{l_1l_2m}\right)\right]\right\}$$
$$-\frac{1}{2}\sum_{p=1,2}\left(L^{(p)}\right)^2 M_{l_1l_2m}, \tag{2.416}$$

where

$$\left(L^{(p)}\right)^2 = -\frac{1}{\sin\vartheta_p}\frac{\partial}{\partial\vartheta_p}\left(\sin\vartheta_p\frac{\partial}{\partial\vartheta_p}\right) - \frac{1}{\sin^2\vartheta_p}\frac{\partial^2}{\partial\varphi_p^2},$$
$$L_Z^{(p)} = -i\frac{\partial}{\partial\varphi_p},\ L_\pm^{(p)} = e^{\pm i\varphi_p}\left(\pm\frac{\partial}{\partial\vartheta_p} + i\cot\vartheta_p\frac{\partial}{\partial\varphi_p}\right)$$

are the orbital angular momentum operators [104]. Here we have used the following representation for the expansion of the normalized free energy E, Eq. (2.411), in terms of spherical harmonics

$$E = -\frac{4\pi}{3}\varsigma\sum_{m=-1}^{1}(-1)^m Y_{1m}(\vartheta_1,\varphi_1)Y_{1m}(\vartheta_2,\varphi_2)$$
$$-\sum_{p=1,2}\left[\xi_{\mathrm{N}}\sqrt{\frac{4\pi}{3}}Y_{10}(\vartheta_p,\varphi_p)+\sigma\frac{4}{3}\sqrt{\frac{\pi}{5}}Y_{20}(\vartheta_p,\varphi_p)\right]+\text{const.} \tag{2.417}$$

Noting the properties of the angular momentum operators (see Appendix A.1.7), Eq. (2.416) can be further transformed to the equation of motion of the sharp values $M_{l_1l_2m}$, viz.,

$$\tau_N \dot{M}_{l_1l_2m} = \sum_{i,j=-2}^{2}\sum_{k=-1}^{1} d_{l_1+il_2+jm+k}^{l_1l_2m} M_{l_1+il_2+jm+k}, \tag{2.418}$$

where the coefficients $d_{l_1+il_2+jm+k}^{l_1l_2m}$ are given below. Taking the averages of $M_{l_1l_2m}$ over the sharp values of ϑ_p and φ_p [97], we now have from Eq. (2.418) the desired set of recurrence relations for the relaxation functions $c_{l_1l_2m}(t) = \langle M_{l_1l_2m}\rangle(t) - \langle M_{l_1l_2m}\rangle_{\mathrm{II}}$, viz.,

$$\tau_N \dot{c}_{l_1 l_2 m} = \sum_{i,j=-2}^{2} \sum_{k=-1}^{1} d_{l_1+il_2+jm+k}^{l_1 l_2 m} c_{l_1+il_2+jm+k}. \tag{2.419}$$

These differential-recurrence relations Eq. (2.419) must be solved subject to the initial conditions $c_{l_1 l_2 m}(0) = \langle M_{l_1 l_2 m} \rangle_{\mathrm{I}} - \langle M_{l_1 l_2 m} \rangle_{\mathrm{II}}$, where the equilibrium averages $\langle M_{l_1 l_2 m} \rangle_i$ ($i = \mathrm{I}, \mathrm{II}$) satisfy the recurrence relation:

$$\sum_{i,j=-2}^{2} \sum_{k=-1}^{1} d_{l_1+i,l_2+j,m+k}^{l_1,l_2,m} \left\langle M_{l_1+i,l_2+j,m+k} \right\rangle_i = 0. \tag{2.420}$$

The equilibrium averages $\langle M_{l_1 l_2 m} \rangle_i$ corresponding to the initial and final states of the system can be evaluated using Eq. (2.420) or from the integral

$$\langle M_{l_1 l_2 m} \rangle_i = \int_0^{2\pi} \int_0^{2\pi} \int_0^{\pi} \int_0^{\pi} M_{l_1 l_2 m} W_i \sin\vartheta_2 \sin\vartheta_1 d\vartheta_2 d\vartheta_1 d\varphi_2 d\varphi_1. \tag{2.421}$$

To proceed in the familiar way, we first introduce the column vectors

$$\mathbf{C}_n(t) = \begin{pmatrix} \mathbf{c}_{2n-10}(t) \\ \vdots \\ \mathbf{c}_{02n-1}(t) \\ \mathbf{c}_{2n0}(t) \\ \vdots \\ \mathbf{c}_{02n}(t) \end{pmatrix}_{4n^2+2n+1}, \quad \mathbf{c}_{nm}(t) = \begin{pmatrix} c_{nm-r}(t) \\ c_{nm-r+1}(t) \\ \vdots \\ c_{nmr}(t) \end{pmatrix}, \tag{2.422}$$

where $r = \min[n, m]$ and $\mathbf{C}_0(t) = 0$. Thus Eq. (2.419) can again be transformed into the archetypal matrix three-term differential-recurrence equation, Eq. (2.200), where the matrices $\mathbf{Q}_n, \mathbf{Q}_n^{\pm}$ are now given by

$$\mathbf{Q}_n^- = \begin{pmatrix} \mathbf{V}_{2n-1} & \mathbf{R}_{2n-1} \\ \mathbf{0} & \mathbf{V}_{2n} \end{pmatrix}, \mathbf{Q}_n = \begin{pmatrix} \mathbf{P}_{2n-1} & \mathbf{S}_{2n-1} \\ \mathbf{R}_{2n} & \mathbf{P}_{2n} \end{pmatrix}, \mathbf{Q}_n^+ = \begin{pmatrix} \mathbf{U}_{2n-1} & \mathbf{0} \\ \mathbf{S}_{2n} & \mathbf{U}_{2n} \end{pmatrix}, \tag{2.423}$$

where

$$\mathbf{U}_m = \begin{pmatrix} \bar{\mathbf{u}}_{m0} & \mathbf{u}_{m0} & \bar{\mathbf{u}}_{0m} & \mathbf{0} & \ddots & \mathbf{0} \\ \mathbf{0} & \bar{\mathbf{u}}_{m-11} & \mathbf{u}_{m-11} & \bar{\mathbf{u}}_{1m-1} & \ddots & \ddots \\ \ddots & \ddots & \ddots & \ddots & \ddots & \mathbf{0} \\ \mathbf{0} & \ddots & \mathbf{0} & \bar{\mathbf{u}}_{0m} & \mathbf{u}_{0m} & \bar{\mathbf{u}}_{m0} \end{pmatrix},$$

$$\mathbf{P}_m = \begin{pmatrix} \mathbf{p}_{m0} & \overline{\mathbf{p}}_{0m} & \ddots & \mathbf{0} \\ \overline{\mathbf{p}}_{m-11} & \mathbf{p}_{m-11} & \ddots & \ddots \\ \ddots & \ddots & \ddots & \overline{\mathbf{p}}_{m-11} \\ \mathbf{0} & \ddots & \overline{\mathbf{p}}_{0m} & \mathbf{p}_{0m} \end{pmatrix}, \mathbf{S}_m = \begin{pmatrix} \mathbf{s}_{m0} & \mathbf{s}^*_{0m} & \mathbf{0} & \ddots & \mathbf{0} \\ \mathbf{0} & \mathbf{s}_{m-11} & \mathbf{s}^*_{1m-1} & \ddots & \ddots \\ \ddots & \ddots & \ddots & \ddots & \mathbf{0} \\ \mathbf{0} & \ddots & \mathbf{0} & \mathbf{s}_{0m} & \mathbf{s}^*_{m0} \end{pmatrix},$$

$$\mathbf{R}_m = \begin{pmatrix} \mathbf{r}^*_{0m} & \mathbf{0} & \ddots & \mathbf{0} \\ \mathbf{r}_{m-11} & \mathbf{r}^*_{1m-1} & \ddots & \ddots \\ \mathbf{0} & \mathbf{r}_{m-22} & \ddots & \mathbf{0} \\ \ddots & \ddots & \ddots & \mathbf{r}^*_{m-11} \\ \mathbf{0} & \ddots & \mathbf{0} & \mathbf{r}_{0m} \end{pmatrix}, \mathbf{V}_m = \begin{pmatrix} \overline{\mathbf{v}}_{0m} & \mathbf{0} & \ddots & \mathbf{0} \\ \mathbf{v}_{m-11} & \overline{\mathbf{v}}_{1m-1} & \ddots & \ddots \\ \overline{\mathbf{v}}_{m-22} & \mathbf{v}_{m-22} & \ddots & \mathbf{0} \\ \mathbf{0} & \overline{\mathbf{v}}_{m-33} & \ddots & \overline{\mathbf{v}}_{m-22} \\ \ddots & \ddots & \ddots & \mathbf{v}_{1m-1} \\ \mathbf{0} & \ddots & \mathbf{0} & \overline{\mathbf{v}}_{0m} \end{pmatrix}.$$

Here the diagonal matrices $\mathbf{p}_{nm}, \mathbf{s}_{nm}, \overline{\mathbf{u}}_{nm}, \overline{\mathbf{v}}_{nm}$ have matrix elements $(\mathbf{x}_{nm})_{ik} = x_{nmi}\delta_{ik}$ while the matrix elements of the tridiagonal matrices $\mathbf{p}^*_{nm}, \mathbf{r}_{nm}, \mathbf{u}_{nm}, \mathbf{v}_{nm}$ are given by $(\mathbf{x}_{nm})_{ik} = x^-_{nmi}\delta_{ik-1} + x_{nmi}\delta_{ik} + x^+_{nmi}\delta_{ik+1}$. Here $\mathbf{x}$ designates one of the submatrices $\mathbf{p}_{nm}, \mathbf{s}_{nm}, \overline{\mathbf{u}}_{nm}, \overline{\mathbf{v}}_{nm}, \mathbf{p}^*_{nm}, \mathbf{r}_{nm}, \mathbf{u}_{nm}, \mathbf{v}_{nm}$. All the submatrices have the same number of rows, namely, $2r + 1$. The number of columns also can be found as $2r_x + 1$, now however each submatrix has its own number r_x, namely

$$\begin{gathered} r_p = \min[n,m], r_{p*} = \min[n+1, m-1], \\ r_s = \min[n+1, m], r_r = \min[n, m-1], \\ r_v = \min[n-1, m-1], r_u = \min[n+1, m+1], \\ r_{v*} = \min[n, m-2], r_{u*} = \min[n+2, m]. \end{gathered}$$

The corresponding matrix elements are

$$d^{l_1 l_2 m}_{l_1 l_2 m} = p_{l_1 l_2 m} = -\sum_{l=l_1,l_2} \left(\frac{1}{2} l(l+1) - \sigma \frac{l(l+1) - 3m^2}{(2l-1)(2l+3)} \right),$$

$$d^{l_1 l_2 m}_{l_1+2 l_2 m} = \overline{u}_{l_1 l_2 m} = -\sigma \frac{l_1}{2l_1+3} \sqrt{\frac{((l_1+1)^2 - m^2)((l_1+2)^2 - m^2)}{(2l_1+1)(2l_1+5)}},$$

$$d^{l_1 l_2 m}_{l_1 l_2+2 m} = \overline{u}_{l_2 l_1 m} = -\sigma \frac{l_2}{2l_2+3} \sqrt{\frac{((l_2+1)^2 - m^2)((l_2+2)^2 - m^2)}{(2l_2+1)(2l_2+5)}},$$

$$d_{l_1-2\,l_2\,m}^{l_1\,l_2\,m}=\overline{v}_{l_2\,l_1\,m}=\sigma\frac{l_1+1}{2l_1-1}\sqrt{\frac{(l_1^2-m^2)((l_1-1)^2-m^2)}{(2l_1+1)(2l_1-3)}},$$

$$d_{l_1\,l_2-2\,m}^{l_1\,l_2\,m}=\overline{v}_{l_1\,l_2\,m}=\sigma\frac{l_2+1}{2l_2-1}\sqrt{\frac{(l_2^2-m^2)((l_2-1)^2-m^2)}{(2l_2+1)(2l_2-3)}},$$

$$d_{l_1+1\,l_2\,m}^{l_1\,l_2\,m}=\left(s_{l_1\,l_2\,m}\right)^*=-\left(\frac{\xi_{\text{II}}}{2}l_1+\frac{i(2\sigma-\varsigma)}{2\alpha}m\right)\sqrt{\frac{(l_1+1)^2-m^2}{4(l_1+1)^2-1}},$$

$$d_{l_1\,l_2+1\,m}^{l_1\,l_2\,m}=s_{l_2\,l_1\,m}=-\left(\frac{\xi_{\text{II}}}{2}l_2-\frac{i(2\sigma-\varsigma)}{2\alpha}m\right)\sqrt{\frac{(l_2+1)^2-m^2}{4(l_2+1)^2-1}},$$

$$d_{l_1-1\,l_2\,m}^{l_1\,l_2\,m}=r_{l_2\,l_1\,m}=\left(\frac{\xi_{\text{II}}}{2}(l_1+1)-\frac{i(2\sigma-\varsigma)}{2\alpha}m\right)\sqrt{\frac{l_1^2-m^2}{4l_1^2-1}},$$

$$d_{l_1\,l_2-1\,m}^{l_1\,l_2\,m}=\left(r_{l_1\,l_2\,m}\right)^*=\left(\frac{\xi_{\text{II}}}{2}(l_2+1)+\frac{i(2\sigma-\varsigma)}{2\alpha}m\right)\sqrt{\frac{l_2^2-m^2}{4l_2^2-1}},$$

$$d_{l_1+1\,l_2+1\,m}^{l_1\,l_2\,m}=u_{l_1\,l_2\,m}=-\frac{1}{2}\varsigma(l_1+l_2)\sqrt{\frac{((l_1+1)^2-m^2)((l_2+1)^2-m^2)}{(2l_1+1)(2l_1+3)(2l_2+1)(2l_2+3)}},$$

$$d_{l_1-1\,l_2-1\,m}^{l_1\,l_2\,m}=v_{l_1\,l_2\,m}=\frac{1}{2}\varsigma(l_1+l_2+2)\sqrt{\frac{(l_1^2-m^2)(l_2^2-m^2)}{(2l_1-1)(2l_1+1)(2l_2-1)(2l_2+1)}},$$

$$d_{l_1+1\,l_2-1\,m}^{l_1\,l_2\,m}=\overline{p}_{l_1\,l_2\,m}=\frac{1}{2}\varsigma(l_2-l_1+1)\sqrt{\frac{((l_1+1)^2-m^2)(l_2^2-m^2)}{(2l_1+1)(2l_1+3)(2l_2-1)(2l_2+1)}},$$

$$d_{l_1-1\,l_2+1\,m}^{l_1\,l_2\,m}=\overline{p}_{l_2\,l_1\,m}=\frac{1}{2}\varsigma(l_1-l_2+1)\sqrt{\frac{((l_2+1)^2-m^2)(l_1^2-m^2)}{(2l_2+1)(2l_2+3)(2l_1-1)(2l_1+1)}},$$

$$d_{l_1+1\,l_2+1\,m\pm1}^{l_1\,l_2\,m}=u_{l_1\,l_2\,m}^{\pm}=\varsigma\frac{l_1+l_2}{4}\sqrt{\frac{(l_1\pm m+1)(l_1\pm m+2)(l_2\pm m+1)(l_2\pm m+2)}{(2l_1+1)(2l_1+3)(2l_2+1)(2l_2+3)}},$$

$$d_{l_1+1\,l_2-1\,m\pm1}^{l_1\,l_2\,m}=\overline{p}_{l_1\,l_2\,m}^{\pm}=\varsigma\frac{l_2-l_1+1}{4}\sqrt{\frac{(l_1\pm m+1)(l_1\pm m+2)(l_2\mp m-1)(l_2\mp m)}{(2l_1+1)(2l_1+3)(2l_2-1)(2l_2+1)}},$$

$$d_{l_1-1\,l_2+1\,m\pm1}^{l_1\,l_2\,m}=\overline{p}_{l_2\,l_1\,m}^{\pm}=\varsigma\frac{l_1-l_2+1}{4}\sqrt{\frac{(l_2\pm m+1)(l_2\pm m+2)(l_1\mp m-1)(l_1\mp m)}{(2l_2+1)(2l_2+3)(2l_1-1)(2l_1+1)}},$$

$$d^{l_1 l_2 m}_{l_1-1 l_2-1 m\pm1} = v^{\pm}_{l_1 l_2 m} = -\varsigma \frac{l_1+l_2+2}{4}\sqrt{\frac{(l_1\mp m-1)(l_1\mp m)(l_2\mp m-1)(l_2\mp m)}{(2l_1-1)(2l_1+1)(2l_2-1)(2l_2+1)}},$$

$$d^{l_1 l_2 m}_{l_1+1 l_2 m\pm1} = \left(s^{\pm}_{l_1 l_2 m}\right)^* = \mp\frac{i\varsigma}{4\alpha}\sqrt{\frac{(l_1\pm m+1)(l_1\pm m+2)(l_2\pm m+1)(l_2\mp m)}{(2l_1+1)(2l_1+3)}},$$

$$d^{l_1 l_2 m}_{l_1 l_2+1 m\pm1} = s^{\pm}_{l_2 l_1 m} = \pm\frac{i\varsigma}{4\alpha}\sqrt{\frac{(l_2\pm m+1)(l_2\pm m+2)(l_1\pm m+1)(l_1\mp m)}{(2l_2+1)(2l_2+3)}},$$

$$d^{l_1 l_2 m}_{l_1 l_2-1 m\pm1} = \left(r^{\pm}_{l_1 l_2 m}\right)^* = \mp\frac{i\varsigma}{4\alpha}\sqrt{\frac{(l_1\pm m+1)(l_1\mp m)(l_2\mp m-1)(l_2\mp m)}{(2l_2-1)(2l_2+1)}},$$

$$d^{l_1 l_2 m}_{l_1-1 l_2 m\pm1} = r^{\pm}_{l_2 l_1 m} = \pm\frac{i\varsigma}{4\alpha}\sqrt{\frac{(l_2\pm m+1)(l_2\mp m)(l_1\mp m-1)(l_1\mp m)}{(2l_1-1)(2l_1+1)}},$$

and

$$d^{l_i l_j m}_{l_i+x l_j+y m\pm1} = (d^{l_j l_i m}_{l_j+y l_i+x m\pm1})^*.$$

By using the general method for solving matrix recurrence equations (see Chapter 1, Section 1.3.5), we have the exact solution, Eq. (2.203), for the Laplace transform $\tilde{\mathbf{C}}_1(s)$. Here the initial value vectors $\mathbf{C}_n(0)$ are calculated in the same manner as before. Thus, we transform Eq. (2.420) to the algebraic matrix recurrence relation

$$\mathbf{Q}_n^-\mathbf{F}^i_{n-1} + \mathbf{Q}_n\mathbf{F}^i_n + \mathbf{Q}_n^+\mathbf{F}^i_{n+1} = \mathbf{0},$$

where

$$\mathbf{F}^i_0 = \frac{1}{4\pi},\ \mathbf{F}^i_n = \begin{pmatrix} \mathbf{f}^i_{2n-10} \\ \mathbf{f}^i_{2n-21} \\ \vdots \\ \mathbf{f}^i_{02n-1} \\ \mathbf{f}^i_{2n0} \\ \mathbf{f}^i_{2n-11} \\ \vdots \\ \mathbf{f}^i_{02n} \end{pmatrix}_{4n^2+2n+1},\ \mathbf{f}^i_{nm} = \begin{pmatrix} M^i_{nm-r} \\ M^i_{nm-r+1} \\ \vdots \\ M^i_{nmr} \end{pmatrix},$$

and $r = \min[n, m]$. The solution of this equation then has the familiar form

$$\mathbf{F}_n^i = \mathbf{\Delta}_n^i(0)\mathbf{Q}_n^-\mathbf{F}_{n-1}^i = \frac{1}{4\pi}\mathbf{\Delta}_n^i(0)\mathbf{Q}_n^-\mathbf{\Delta}_{n-1}^i(0)\mathbf{Q}_n^- \ldots \mathbf{\Delta}_1^i(0)\mathbf{Q}_1^-.$$

Having determined $\tilde{\mathbf{C}}_1(i\omega)$, we have the spectrum of the relaxation function $f_p(t)$:

$$\tilde{f}_p(i\omega) = \frac{\tilde{c}_{100}(i\omega)}{c_{100}(0)} \tag{2.424}$$

consequently the spectrum of the magnetization of the two-spin system [see Eq. (2.407)] and its relaxation time τ_{int} given by the zero frequency limit of $\tilde{f}_p(i\omega)$:

$$\tau_{\text{int}} = \frac{\tilde{c}_{100}(0)}{c_{100}(0)}. \tag{2.425}$$

For linear response, we also have from Eqs. (2.408) the spectrum of the equilibrium correlation function $C_{\|}(t)$, the integral relaxation time $\tau_{\text{int}} = \tilde{C}_{\|}(0)$, as well as the longitudinal complex susceptibility $\chi(\omega) = \chi'(\omega) - i\chi''(\omega) = 1 - i\omega\tilde{C}_{\|}(i\omega)$. The time behavior of $f_p(t), C_{\|}(t)$, and $m_Z(t)$ can be obtained numerically from $\tilde{f}_p(\omega)$ and $\tilde{C}_{\|}(i\omega)$ by inverse Fourier transformation.

The *linear response* relaxation time τ_{int} evaluated from Eqs. (2.424) and (2.425), where the applied field change $\Delta h = h_{\text{I}} - h_{\text{II}}$ is maintained at a small constant value 0.001, is shown in Fig. 2.79. Figure 2.79(a) essentially displays the effect of the exchange coupling for zero reduced bias field on the relaxation time τ_{int} as a function of σ (which may be regarded as the inverse-temperature parameter). It is apparent from Fig. 2.79(a) that the *effect of increasing exchange coupling in the zero bias field situation in linear response is to further increase the relaxation time.* This is entirely in accord with Fig. 2.79(b), where the relaxation time τ_{int} is plotted as a function of the interaction (exchange coupling) parameter ς for reasonably large values of the anisotropy parameter anisotropy σ, $\alpha = 1$ (IHD regime), and for a small bias field with Δh still maintained very small so that we are again considering the linear response with a small bias field. As before the effect of increasing exchange coupling is to

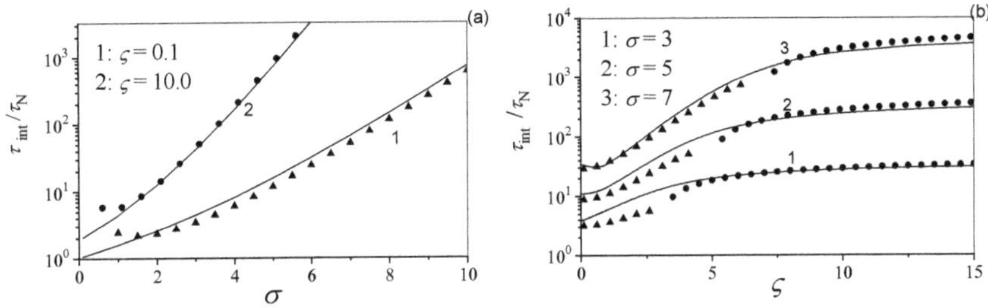

Fig. 2.79. Integral relaxation time τ_{int} vs. the anisotropy (inverse temperature) parameter σ for $\alpha = 1, h_{\text{I}} = 0.001, h_{\text{II}} = 0,$ and two interaction parameters $\varsigma = 0.1$ and 10 (a) and τ_{int} vs. ς for $\alpha = 0.5, h_{\text{I}} = 0.101, h_{\text{II}} = 0.1,$ and various σ (b). The matrix continued fraction solution [solid lines: Eq. (2.425)] is compared with the inverse escape rate rendered by Langer's theory [filled circles and triangles: Eqs. (2.426) and (2.427)].

increase the relaxation time, that is, to raise the potential barrier height. Thus, we infer that the global effect of *increasing exchange coupling* is to *increase the relaxation time*. Although our results are based on the simplistic concept of *two-body interaction* they allow one to qualitatively understand the role played by exchange interaction which is in general to increase the relaxation time over and above that without the interaction.

The longest relaxation time for the two-spin problem in the IHD regime ($\alpha \geq 1$) can also be estimated in semi-analytic fashion in terms of the relaxation rate Γ as [454]

$$\tau \simeq \tau_{\text{N}} \left[\Gamma(h_i) + \Gamma(-h_i) \right]^{-1}, \tag{2.426}$$

$$\Gamma = \begin{cases} 2\Gamma^+\Gamma^-(\Gamma^+ + \Gamma^-)^{-1} & j < 1 - h_i^2, \\ 2\Gamma_{2s} & j > 1 - h_i^2, \end{cases} \tag{2.427}$$

where $j = \varsigma / \sigma$,

$$\Gamma_{2s} = \frac{\sigma^{3/2}(1-h_i^2)(1-h_i)[j+1-h_\gamma]}{2\sqrt{\pi j[j-(1-h_i^2)]}} e^{-2\sigma(1-h_i)^2},$$

$$\Gamma^{\pm} = \frac{\sigma^{3/2}\left|\kappa^{\pm}\right| D^{\pm} P^{\pm}}{\sqrt{\pi j \left| 2R_1^{\pm} R_2^{\pm} + jW^{\pm}(R_1^{\pm} + R_2^{\pm}) \right|}} e^{-\Delta E^{\pm}},$$

$$P^{\pm} = [(1-\cos^2 \vartheta_1^{\pm})(1-\cos^2 \vartheta_2^{\pm})]^{1/4}, R_{1,2}^{\pm} = \cos \vartheta_{1,2}^{\pm}(2\cos \vartheta_{1,2}^{\pm} + h_\gamma) - 1,$$

$$W^{\pm} = P^{\pm} - \frac{j}{2} \pm h_i \sqrt{1 + \frac{(j/2)^2}{1-j}}, \; D^+ = (1-h_i)(j+1-h_i), D^- = 1 - h_i^2 - j,$$

$$\Delta E^{\pm} = -\sigma\left\{ j\sqrt{\left[1 \pm h_i\sqrt{1+(j/2)^2(1-j)^{-1}} - j/2\right]^2 - \left(h_i \pm \sqrt{1-j}\right)^2} \right.$$
$$-2 + 2h_i \pm (2h_i - j) + j\left[1 \pm h_i\sqrt{1+(j/2)^2(1-j)^{-1}} - j/2\right]$$
$$\left. - \left[h_i^2 + j - 1 \pm 2h_i\sqrt{1+(j/2)^2(1-j)^{-1}}\right]\right\}.$$

Here $\cos\vartheta_{1,2}^{\pm}$ corresponds to the position of the two saddle points for each spin as obtained from each of the two equations $\partial_{\vartheta_{1,2}} E = 0$ so that

$$\cos\vartheta_1^{\pm} = \frac{1}{2}\left(-h_i \mp \sqrt{1-j}\right) + \sqrt{\left(h_i \pm \sqrt{1-j}\right)^2 + 2j \mp 4h_i\sqrt{1+\frac{(j/2)^2}{1-j}}},$$
$$\cos\vartheta_2^{\pm} = \frac{1}{2}\left(-h_i \mp \sqrt{1-j} - \sqrt{\left(h_i \pm \sqrt{1-j}\right)^2 + 2j \mp 4h_i\sqrt{1+\frac{(j/2)^2}{1-j}}}\right).$$

The attempt frequencies $\kappa^{\pm}$ are then computed numerically. The relaxation time for the two-spin system so predicted contains *two unconnected branches* corresponding to the two regimes, $j < j_c = 1 - h_i^2$ and $j > j_c$ [454]; these branches are plotted in Fig. 2.79 as shown by filled circles and triangles, respectively. Apparently, a “critical” value of the exchange coupling, $j_c = 1 - h_i^2$ exists separating two regimes with distinct reversal mechanisms. More precisely, for *strong* coupling ($j > j_c$) the two spins reverse their direction coherently through the only saddle point available. On the other hand, for *weak* coupling two saddle points exist so that reversal of the two spins is a two-step process. Moreover, near j_c it was shown that the relaxation rate diverges due to *flattening of the saddle points.* Thus, Langer’s general calculation of the IHD relaxation rate which is formulated in terms of a quadratic expansion of the potential energy about the saddle point, is not applicable at all in this coupling range, hence *leaving the two ranges unconnected* as far as asymptotic calculations are concerned. A comparison of the exact (numerical) integral relaxation time with the asymptotes obtained in Ref. [454] for the IHD situation typified by $\alpha = 1$ using Langer’s theory is given in Fig. 2.79. First of all, these plots show overall good agreement between the asymptotic calculation and those rendered by the matrix continued fraction method, *excluding the low-barrier region* characterized by small σ and exchange coupling values as exemplified by curve 1 in Fig. 2.79(b). The deviation between the numerical and asymptotic solutions for the relaxation time for small σ, or equivalently, low anisotropy energy barriers, is as expected because Langer's theory on which the asymptotic solution is based assumes high energy barriers and thus *well-defined saddle points* and *metastable states*. However, on examination of Fig. 2.79(b), the striking result appears that the *apparent discontinuity* in the asymptotic expression for the relaxation time as a function of the exchange coupling (region between the dots and triangles) referred to in Ref. [454] is not reproduced at all by the numerical solution. Thus, the discontinuity reported there is merely an artifact of the approximation used in the asymptotic calculation, i.e., the Taylor series expansion of the potential energy near the saddle point, as discussed in Ref. [454]. Moreover, the discrepancy between the numerical and asymptotic solutions manifests itself in the vicinity of the critical exchange coupling. There the saddle points become rather flat so that Taylor series expansion of the potential energy near the saddle point is no longer valid. Hence, the numerical solution implies, although a *critical exchange coupling* marking a fundamental change in the energyscape and therefore in the reversal mechanism (i.e., change from a two-step into a one-step process) exists, that *no corresponding singularity in the relaxation time as a function of that coupling appears*. Thus, the reasons for seeking a numerical solution for the relaxation time become quite apparent. Namely, the numerical solution based on matrix continued fractions yields the behavior of the relaxation time both in the region where the magnetic anisotropy energy is comparable to the thermal energy kT and in the critical exchange coupling region where the saddle points flatten. Hence, the numerical solution allows one to accurately delineate the regions where the asymptotic solution no longer applies. In other words, as without exchange coupling, it is now possible to give a range of validity for the asymptotic formula. Another advantage of the numerical solution is that it is of course valid for all α unlike the asymptotic solution from Langer's method, applying for IHD damping only.

Figure 2.80 indicates that for small h_{II}, the dependence of τ_{int} on the anisotropy parameter σ has an activation character leading to *exponential increase* in the relaxation time as the height of the potential barrier σ

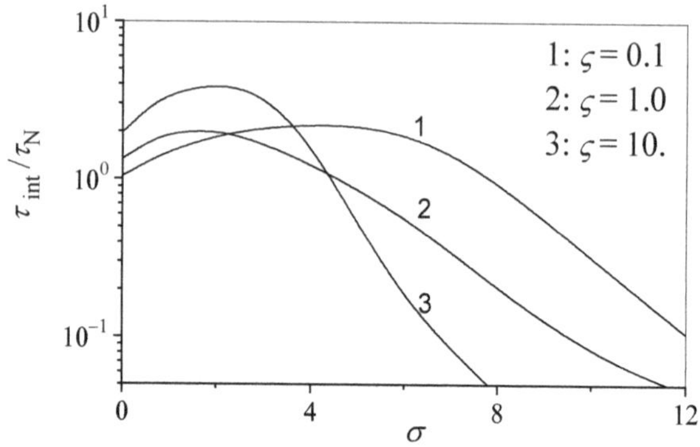

Fig. 2.80. Integral relaxation time $\tau_{\text{int}} / \tau_{\text{N}}$ vs. σ for $\alpha = 1, h_{\text{I}} = 3.001, h_{\text{II}} = 3.0$, and various exchange parameters ς.

increases as predicted by transition state theory. However, as the dc bias field increases (e.g. taking zero exchange coupling described by a bistable potential) the wells of the potential become markedly nonequivalent. Thus, the integral relaxation time τ_{int} can *decrease* with *increasing* σ (see Fig. 2.80). Hence τ_{int} may *differ exponentially* from the inverse Kramers rate. This effect as we have mentioned (see Section 2.4.3) was first reported in Ref. [146] and explained qualitatively in Ref. [337] in an analysis of the linear response of an assembly of noninteracting uniaxial particles in the low-temperature limit and as we recall is due to the depletion of the population of the upper or shallower potential well. An effect which is consequent on the escape of many particles from that well and their subsequent descent to the deeper well from which it is very difficult to escape due to the high energy barrier. Moreover, the depletion effect signifies that if the parameter h_{II} exceeds a certain critical level $h_{\text{c}} \cong 0.17$, τ no longer has an activation character. Consequently, the integral relaxation time no longer provides an accurate approximation to the reversal time of the magnetization. In other words, in this situation contrary to the overbarrier relaxation time, τ_{int} *decreases* as the height of the potential barrier *increases*. Clearly, *the depletion effect also occurs with exchange interactions* as shown in Fig. 2.80. Hence, the integral relaxation time again *exponentially diverges* from the greatest relaxation time or inverse escape rate yielded by Langer's method for values of the reduced field in excess of the critical value h_{c}.

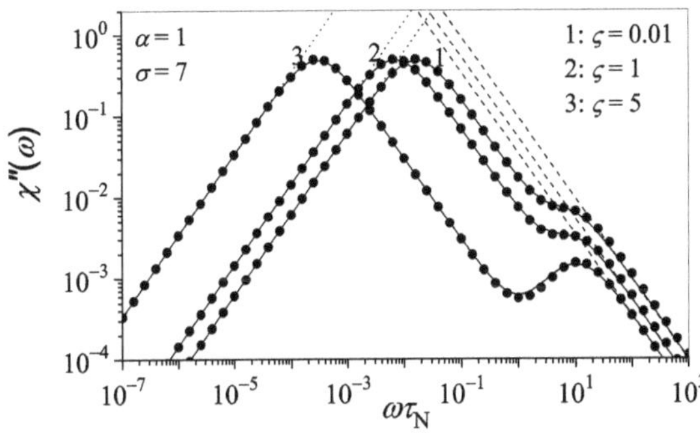

Fig. 2.81. The imaginary part of the complex susceptibility $\chi''(\omega)$ vs. $\omega\tau_N$ for $\alpha=1$, $\sigma=7$, $h_I = 0.001$, $h_{II} = 0$, and various values of the exchange parameter $\zeta = 0.01$, 1.0, and 5. Solid lines: the matrix continued fraction solution; asterisks: the approximate Eq. (1.150); dotted and dashed lines: the low- and high-frequency asymptotes Eq. (1.147), respectively.

The imaginary part $\chi''(\omega)$ of the complex susceptibility for, $\alpha = 1$, $\sigma = 6$, and various exchange parameters ς is shown in Fig. 2.81. Here the usual two peaks appear in the magnetic loss spectrum $\chi''(\omega)$. One dominates the low-frequency part of the spectrum and is due to the *slow* overbarrier relaxation of the spins. The second peak is due to high-frequency "intrawell" modes. Like noninteracting particles, the dynamic susceptibility $\chi(\omega)$ rendered as an infinite set of Lorentzians may be approximated by a sum of two Lorentzians only, Eq. (1.150). The parameters Δ_1 and τ_W in Eq. (1.150) may be determined from $1/\lambda_1$, τ_{int} and τ_{ef} so as to ensure the correct asymptotic behavior of $\chi(\omega)$ in the extremes of very low and high frequencies (see Eq. (1.150)). To fit the spectra in Fig. 2.81 via Eq. (1.150) the only adjustable parameter used is Δ_1. Moreover, since $1 - \Delta_1 << 1$ and $1/\lambda_1 \approx \tau$, the approximate equation $\tau_W \approx (1-\Delta_1)\tau_{ef}$ may also be used (to approximately satisfy the asymptotic behavior of $\chi(\omega)$ in the extremes of very low and very high frequencies, Eq. (1.147)). The other times τ_{int} and τ_{ef} are evaluated from Eqs. (2.425) and (2.406), respectively (the numerical values of τ_{int}, τ_{ef}, and $1-\Delta_1$ are given in the Table 2.8) The low and very high-frequency asymptotes, Eq. (1.147), are also shown for comparison. By inspection of the curves in Fig. 2.81, the low-frequency peak shifts to lower frequencies as ς increases (in accordance with the results of Fig. 2.79).

Table 2.8. Numerical values of parameters in Eq. (1.150) used in Fig. 2.81.

ζ	$\tau_{\mathrm{int}}/\tau_{\mathrm{N}}$, Eq. (2.425)	$\tau_{\mathrm{ef}}/\tau_{\mathrm{N}}$, Eq. (2.406)	$1-\Delta_1$
0.01	61.1	10.52	0.01
1.0	143.8	18.65	0.005
5.0	3348	29.14	0.003

The reader will recall that representing the dynamic susceptibility $\chi(\omega)$ by Eq. (1.150) implies that the long-time behavior of $m_Z(t)$ may be approximated by two exponentials corresponding to overbarrier and "intrawell" relaxation processes (ignoring precessional modes)

$$m_Z(t)/m_Z(0) \approx \Delta_1 e^{-t\lambda_1} + (1-\Delta_1)e^{-t/\tau_W}. \tag{2.428}$$

Hence Eq. (2.428) sensibly approximates the evolution of $m_Z(t)$ in the time domain (for $t > \tau_N$ the contribution of the second term may be ignored).

To summarize, we have discussed here a precise method for the determination of the response functions and relaxation times of the magnetization. This comprises the calculation of the linear and nonlinear transient responses of two interacting spins using the matrix continued fraction technique including nonlinear effects *for all* values of anisotropy and interaction parameters. Simple approximate equations for the dynamic susceptibility and the time behavior of the magnetization are also given. *Hence the ferromagnetic exchange interaction always tends to increase the effective barrier height, that is, to increase the greatest relaxation time.* Moreover, it appears that *the depletion effect, where the integral relaxation time diverges exponentially from the inverse Kramers rate consequent on the application of a strong bias field is further enhanced by exchange interaction*. We emphasize that the asymptotic solution based on Langer's method is valid in the IHD region only, where the energy loss per cycle of a spin with the capability to cross a barrier is much greater than kT. Hence, the asymptotic solution should be generalized to all α and then tested against the numerical solution as done earlier for noninteracting spins. This procedure would ultimately result in an asymptotic formula including the VLD, turnover, and IHD regimes.

2.11.2 *Dipole-dipole coupling*

The overall object here is the effect of *dipole-dipole interaction* on the magnetization relaxation in contrast to the previous section, where *exchange interactions* alone were treated. As a first step, dipole-dipole interactions are examined using the two-spin model representing a system with more than two configurational degrees of freedom and the results are compared with those for exchange interaction. In a simple exposition, we suppose uniaxial anisotropy with the external field applied parallel to the easy Z-axis. The particular advantage of this anisotropy potential is that (although subject to many symmetry restrictions) it ultimately results in a recurrence relation in three recurring indexes only. The overall free energy E_i ($i=\mathrm{I,II}$) including dipole-dipole interaction, anisotropy and Zeeman energies becomes [455]

$$\begin{aligned} E_i &= \varsigma\left[(\mathbf{s}_1\cdot\mathbf{s}_2)-3(\mathbf{u}_r\cdot\mathbf{s}_1)(\mathbf{u}_r\cdot\mathbf{s}_2)\right]-\sum_{p=1,2}\left[\xi_i(\mathbf{e}_Z\cdot\mathbf{s}_p)+\sigma(\mathbf{e}_Z\cdot\mathbf{s}_p)^2\right] \\ &= \varsigma\left[\sin\vartheta_1\sin\vartheta_2\cos(\varphi_1-\varphi_2)-2\cos\vartheta_1\cos\vartheta_2\right] \\ &\quad -\xi_i\left(\cos\vartheta_1+\cos\vartheta_2\right)-\sigma\left(\cos\vartheta_1^2+\cos\vartheta_2^2\right),\quad (i=\mathrm{I,II}), \end{aligned} \tag{2.429}$$

where $\mathbf{e}_Z$ is the unit vector along the (polar) Z axis, $\mathbf{u}_r=\mathbf{r}/r$ is a unit vector with $r=|\mathbf{r}|$ being the distance between the two spins (the vector $\mathbf{r}$ specifying the separation between each dipole is taken as parallel to the easy axis; here, for simplicity, we have assumed that $\mathbf{e}_Z \parallel \mathbf{u}_r$), ξ_i and σ are the dimensionless field and anisotropy parameters, respectively, $\varsigma=\mu_0\mu S^2/(kTr^3)$ is the dipole-dipole interaction parameter. For simplicity the two easy axes of magnetization are supposed parallel to each other and also to the direction of the applied dc field, which in turn is assumed parallel to the reference (Z) axis. Thus, without the dipole-dipole term the task reduces to the well-known problem of the magnetization relaxation in a circularly symmetric anisotropy-Zeeman energy potential (see Section 2.4). It is worthwhile to compare Eq. (2.429) with the free energy of two spins coupled solely by exchange interaction Eq. (2.411). As usual in relaxation problems we require the subsequent behavior of the system of spins starting from an equilibrium state I with Boltzmann distribution function $W_{\mathrm{I}} = Z_{\mathrm{I}}^{-1} e^{-E_{\mathrm{I}}}$ $(t \le 0)$ to another equilibrium state II with the new Boltzmann distribution function $W_{\mathrm{II}} = Z_{\mathrm{II}}^{-1} e^{-E_{\mathrm{II}}}$ $(t \to \infty)$.

For calculation purposes, we represent E_i in terms of spherical harmonics as (cf. Eq. (2.417))

$$E_i = -\frac{4\pi}{6}\varsigma \sum_{m=-1}^{1} (3+(-1)^m) Y_{1m}(\vartheta_1,\varphi_1) Y_{1-m}(\vartheta_2,\varphi_2) - \sum_{p=1,2}\left[\xi_i \sqrt{4\pi/3}\, Y_{10}(\vartheta_p,\varphi_p) + \sigma 4\sqrt{\pi/45}\, Y_{20}(\vartheta_p,\varphi_p)\right] + \text{const.} \tag{2.430}$$

In the manner detailed in the previous section, we have from Eq. (2.418) a set of differential-recurrence relations, Eq. (2.419), for the relaxation functions $c_{l_1 l_2 m}(t) = \langle M_{l_1 l_2 m}\rangle(t) - \langle M_{l_1 l_2 m}\rangle_{\mathrm{II}}$. As only the interaction terms in Eqs. (2.417) and (2.430) are changed, we list below only those coefficients $d_{l_1+i\, l_2+j\, m+k}^{l_1 l_2 m}$ in Eq. (2.419)that differ from the ones listed previously:

$$d_{l_1+1 l_2 m}^{l_1 l_2 m} = \left(s_{l_1 l_2 m}\right)^* = -\left(\frac{\xi_{\mathrm{II}}}{2} l_1 + \frac{i(\sigma-\varsigma)}{\alpha} m\right)\sqrt{\frac{(l_1+1)^2 - m^2}{4(l_1+1)^2 - 1}},$$

$$d_{l_1 l_2+1 m}^{l_1 l_2 m} = s_{l_2 l_1 m} = -\left(\frac{\xi_{\mathrm{II}}}{2} l_2 - \frac{i(\sigma-\varsigma)}{\alpha} m\right)\sqrt{\frac{(l_2+1)^2 - m^2}{4(l_2+1)^2 - 1}},$$

$$d_{l_1-1 l_2 m}^{l_1 l_2 m} = r_{l_2 l_1 m} = \left(\frac{\xi_{\mathrm{II}}}{2}(l_1+1) - \frac{i(\sigma-\varsigma)}{\alpha} m\right)\sqrt{\frac{l_1^2 - m^2}{4l_1^2 - 1}},$$

$$d_{l_1 l_2-1 m}^{l_1 l_2 m} = \left(r_{l_1 l_2 m}\right)^* = \left(\frac{\xi_{\mathrm{II}}}{2}(l_2+1) + \frac{i(\sigma-\varsigma)}{\alpha} m\right)\sqrt{\frac{l_2^2 - m^2}{4l_2^2 - 1}},$$

$$d_{l_1+1 l_2+1 m}^{l_1 l_2 m} = u_{l_1 l_2 m} = -\varsigma(l_1+l_2)\sqrt{\frac{((l_1+1)^2 - m^2)((l_2+1)^2 - m^2)}{(2l_1+1)(2l_1+3)(2l_2+1)(2l_2+3)}},$$

$$d_{l_1-1 l_2-1 m}^{l_1 l_2 m} = v_{l_1 l_2 m} = \varsigma(l_1+l_2+2)\sqrt{\frac{(l_1^2 - m^2)(l_2^2 - m^2)}{(2l_1-1)(2l_1+1)(2l_2-1)(2l_2+1)}},$$

$$d_{l_1+1 l_2-1 m}^{l_1 l_2 m} = \bar{p}_{l_1 l_2 m} = \varsigma(l_2 - l_1 + 1)\sqrt{\frac{((l_1+1)^2 - m^2)(l_2^2 - m^2)}{(2l_1+1)(2l_1+3)(2l_2-1)(2l_2+1)}},$$

$$d_{l_1-1 l_2+1 m}^{l_1 l_2 m} = \bar{p}_{l_2 l_1 m} = \varsigma(l_1 - l_2 + 1)\sqrt{\frac{((l_2+1)^2 - m^2)(l_1^2 - m^2)}{(2l_2+1)(2l_2+3)(2l_1-1)(2l_1+1)}}.$$

The hierarchy of three-index recurrence relations, Eq. (2.419), with the new coefficients $d_{l_1+il_2+jm+k}^{l_1 l_2 m}$ must be solved subject to the initial conditions $c_{l_1 l_2 m}(0) = \langle M_{l_1 l_2 m}\rangle_{\mathrm{I}} - \langle M_{l_1 l_2 m}\rangle_{\mathrm{II}}$. The preceding equilibrium averages $\langle M_{l_1 l_2 m}\rangle_i$ corresponding to initial and final states of the system can be directly evaluated from the integral, Eq. (2.421). Notice that the $\langle M_{l_1 l_2 m}\rangle_i$ themselves satisfy the time-independent recurrence relation, Eq. (2.420). Again, we introduce the vectors Eq. (2.422) indicating that Eq. (2.419) in this instance can be transformed into the tractable tridiagonal vector recurrence relation (admitting of a closed form matrix solution for $\tilde{\mathbf{C}}_1(i\omega)$ in the frequency domain, Eq. (2.203)). The decay response immediately following the alteration of the field may be described via the normalized relaxation function $f_p(t)$, Eq. (2.401).

Having determined $\tilde{\mathbf{C}}_1(i\omega)$, we then have the spectrum $\tilde{f}_p(i\omega) = \tilde{c}_{100}(i\omega)/c_{100}(0)$ of the relaxation function $f_p(t)$ given by Eq. (2.424). The overall behavior of $f_p(t)$ is as before characterized by the *integral relaxation time* $\tau_{\mathrm{int}} = \tilde{c}_{100}(0)/c_{100}(0)$. The functions $f_1(t)$ or $f_2(t)$ and the time τ_{int} describe the transient nonlinear behavior of the longitudinal component of the magnetic moment of the two dipole system $m_Z(t) = \mu\langle(\mathbf{s}_1 + \mathbf{s}_2)\cdot\mathbf{e}_Z\rangle(t)$ given by Eq. (2.407). The *nonlinear response* contains automatically the *linear response* of a two-spin system to infinitesimally small changes in the strength of the strong dc field $\mathbf{H}^{\mathrm{I}}$, i.e., for $\mathbf{H}^{\mathrm{II}} = \mathbf{H}^{\mathrm{I}} - \boldsymbol{\kappa}$ as $\boldsymbol{\kappa} \to 0$, where $\boldsymbol{\kappa}$ is regarded as a small external perturbation. In this instance, $f_p(t)$ from Eq. (2.401) coincides with the longitudinal equilibrium correlation function $C_{\parallel}(t)$, Eq. (2.408). According to linear response theory, having determined the one-sided Fourier transform $\tilde{C}_{\parallel}(i\omega)$, one has τ_{int} which is of course now identical to the correlation time $\tau_{\mathrm{int}} = \tilde{C}_{\parallel}(0)$, the *effective* relaxation time $\tau_{\mathrm{ef}} = -1/\dot{C}_{\parallel}(0)$, and also the normalized dynamic susceptibility $\chi(\omega)$ via $\chi(\omega) = 1 - i\omega\tilde{C}_{\parallel}(i\omega)$. The quantities τ_{int} and τ_{ef} as before characterize the long and short time behavior of $C_{\parallel}(t)$ [97].

The integral relaxation time of the two-spin system with dipole-dipole coupling is plotted in Fig. 2.82 for various model parameters. The corresponding results for exchange interactions are also shown in Fig. 2.82 for comparison. We now compare the results for dipole-dipole interaction with those for exchange interaction for the same value of the interaction parameter. Figure 2.82(a) is a comparison of the relative effect of the

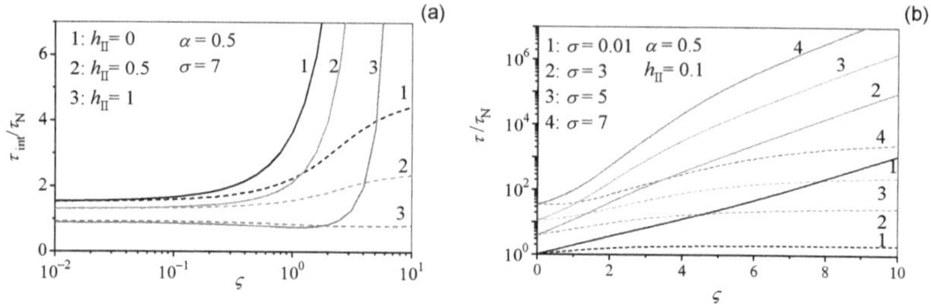

Fig. 2.82. Integral relaxation time τ_{int} / τ_N as a function of the interaction parameter ς (a) for various external field parameters h_{II} subject to the linear response condition $h_I - h_{II} = 0.001$ and (b) for various σ and $h_I = 0.101, h_{II} = 0.1, \alpha = 0.5$. Solid and dashed lines: the matrix continued fraction solution for dipole-dipole and exchange interaction, respectively. Note the very pronounced effect of dipole-dipole interaction which for large ς greatly increases the relaxation time as compared to exchange interaction.

interaction parameter ς on the integral relaxation time in linear response, i.e., the correlation time. Clearly the effect of increasing ς is without doubt to increase the relaxation time. However, *the effect of dipole-dipole coupling is much more pronounced than that of exchange coupling for the same value of* ς. Indeed, increase of ς beyond ~1 causes a rapid increase in τ_{int} for dipole-dipole coupling as compared to exchange coupling. A similar effect also occurs with exchange interaction. However, a much larger value of the exchange interaction parameter ς would be needed relative to that for dipole-dipole coupling. Figure 2.82(b) shows the relaxation time as a function of the interaction strength ς for various anisotropy (or inverse-temperature) parameters. Clearly for both types of interaction the tendency of each is to markedly increase the integral relaxation time with *the enhancement effect being much greater for dipole-dipole rather than exchange interaction*.

Figure 2.83 shows the integral relaxation time τ_{int} and reversal time $\tau = 1/\lambda_1$ vs. the anisotropy (or inverse temperature) parameter σ. Without the external field, i.e., $h_{II} = 0$, the temperature dependence of τ_{int} (just as $\tau = 1/\lambda_1$) has the customary Arrhenius behavior, i.e., exponentially increasing with decreasing temperature (see Fig. 2.83(a)). The slopes of $\tau_{int}(T^{-1})$ and $\tau(T^{-1})$ markedly depend on the interaction parameter ς. Notice that both τ_{int} and $\tau = 1/\lambda_1$ greatly increase as the interaction parameter is raised, with the effect of dipole-dipole interaction again

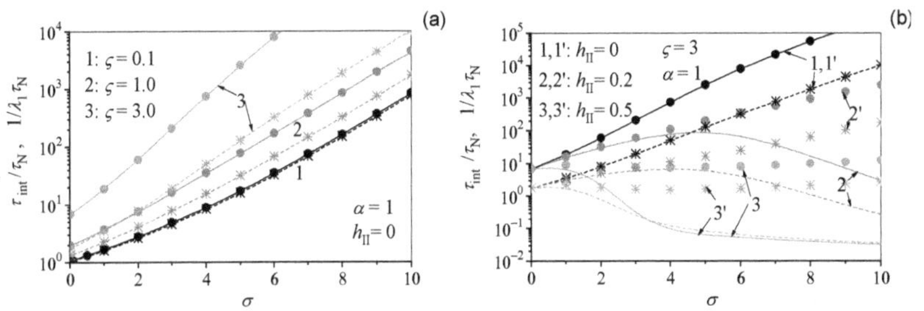

Fig. 2.83. Integral relaxation time τ_{int} / τ_N (solid and dashed lines) and inverse smallest nonvanishing eigenvalue $(\lambda_1\tau_N)^{-1}$ (circles and asterisks) vs. the anisotropy (inverse temperature) parameter σ (a) for various interaction parameters ς and $h_{II} = 0$ and (b) for various external field parameters h_{II} =0, 0.2, 0.5 and $\varsigma = 3$ subject to the linear response condition $h_I - h_{II} = 0.001$. Solid and dashed lines: the matrix continued fraction solution for dipole-dipole and exchange interactions, respectively.

dominating as that parameter increases. At $h_{II} = 0$, the integral relaxation time provides an accurate approximation to the reversal time of the magnetization $\tau = 1/\lambda_1$. However, as the constant bias field increases, so that (taking zero exchange coupling described by a bistable potential as an example) the wells of the interaction potential become markedly nonequivalent then τ_{int} can *decrease* with *increasing* σ (see Fig. 2.83(b)). Thus τ_{int} may *differ exponentially* from the reversal time $\tau = 1/\lambda_1$. This effect was qualitatively explained in Section 2.4.3 for an assembly of noninteracting uniaxial nanomagnets (i.e., for $\varsigma = 0$). In the low-temperature limit, the effect is again due to the depletion of the population of the shallower potential well consequent on the escape of many particles from that well and their subsequent descent to the deeper well where it is very difficult to escape from due to the high energy barrier. Thus, the integral relaxation time τ_{int} can once more deviate considerably from the reversal time of the magnetization $\tau = 1/\lambda_1$ and so is no longer a good approximation to the latter (see Fig. 2.83(b)).

The relative role played by interactions in the behavior of the real $\chi'(\omega)$ and imaginary $\chi''(\omega)$ parts of the dynamic susceptibility is shown in Fig. 2.84(a), while the effect of the external electric field on the dynamic susceptibility is illustrated in Fig. 2.84(b). The spectra $\chi'(\omega)$ and $\chi''(\omega)$ for dipole-dipole interaction are in essence very similar to those for exchange interaction. Furthermore, like noninteracting magnetic dipoles

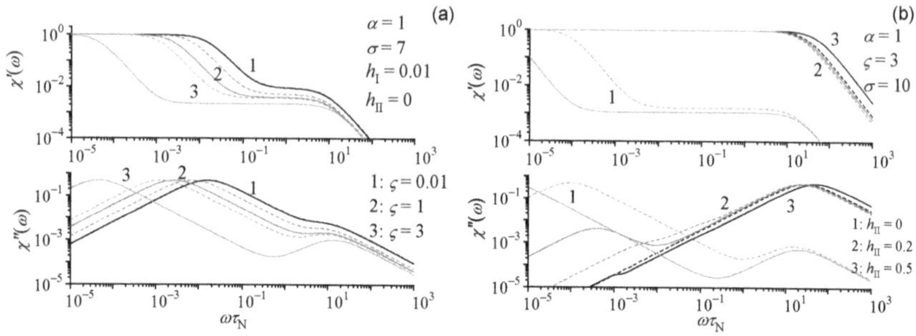

Fig. 2.84. The real $\chi'(\omega)$ and imaginary $\chi''(\omega)$ parts of the complex susceptibility vs. $\omega\tau_{\rm N}$ (a) for $\alpha=1$, $\sigma=7$, $h_{\rm II}=0$, and various interaction parameters $\varsigma = 0.01$, 1.0, and 3 and (b) for various external field parameters $h_{\rm II} = 0$, 0.2, 0.5 and $\alpha=1$, $\sigma= 10$, and $\varsigma = 3$. Solid and dashed lines: the matrix continued fraction solution for dipole-dipole and exchange interactions, respectively (the solid and dashed lines for $\varsigma = 0.01$ lie on top of each other).

(cf. Fig. 2.22), two peaks appear in the spectra of the magnetic loss $\chi''(\omega)$. The characteristic frequencies of these peaks, i.e., where $\chi''(\omega)$ reaches local maxima, are τ^{-1} and $\alpha\omega_{\rm pr}$, where $\omega_{\rm pr}$ is the precession frequency of the magnetic moment in the effective magnetic field. The high-frequency peak is due to the (fast) near-degenerate intrawell modes which are virtually indistinguishable in the frequency spectrum appearing as a single high-frequency peak. At low dc fields, the amplitude of the latter is far weaker than that of the low-frequency one (see Fig. 2.84(b)). However, in a strong dc magnetic field, the high-frequency intrawell modes can dominate the spectrum because as $h_{\rm II}$ increases, the magnitude of the low-frequency peak decreases and may even disappear altogether (curves 3). This is again due to the depletion effect: in strong fields, at some critical value of $h_{\rm II}$, the relaxation switches from being dominated by the slowest barrier-crossing mode to being dominated by the fast intrawell modes. As the low-frequency behavior of $\chi''(\omega)$ is due to the barrier-crossing relaxation mode, the reversal time of the magnetization τ can be evaluated from the characteristic frequency $\omega_{\rm max}$, where $\chi''(\omega)$ reaches a maximum, and/or the half-width $\Delta\omega$ of the spectrum of $\chi'(\omega)$ via Eq. (2.346). The comparison of τ as extracted from the spectra $\chi''(\omega)$ via Eq. (2.346) with $\tau = 1/\lambda_1$ as calculated using an entirely independent method, namely numerical calculation of λ_1 of the Fokker–Planck operator $L_{\rm FP}$ shows that both results are identical. In accordance with the previous figures, this maximum in

$\chi''(\omega)$ moves to lower frequencies for dipole-dipole interactions than for exchange interactions. Once more the dynamic susceptibility $\chi(\omega)$ given as an infinite series of Lorentzians may be approximated by a sum of two Lorentzians only, Eq. (1.150).

Finally, the relative effects of dipole-dipole and exchange interaction have been considered for the most simple case where both easy axes of magnetization are supposed parallel to each other and also to the direction of the applied ac field which is in turn assumed parallel to the reference Z-axis and throughout it has been assumed that the anisotropy potential is represented by the simplest possible uniaxial potential. Thus, our solution is subject to considerable symmetry-based restrictions. Therefore, it can be reasonably argued that such a model cannot provide a reliable description of the many spin relaxation problem. Nevertheless, even our drastically simplified configuration exemplifies the relative roles of dipole-dipole and exchange interactions. The main conclusion to be drawn is that including them causes a marked increase in the reversal time.

2.12 Concluding remarks

We have comprehensively treated Brown's diffusion model of the magnetization relaxation of magnetic nanoparticles, both via the Langevin equation and via the Kramers escape-rate theory as adapted to classical spin systems. Using these two powerful complementary methods, we have both numerical (provided by matrix continued fractions) and simple approximate analytical solutions of diverse problems occurring in superparamagnetism. These include estimation of the characteristic times of the magnetization reversal process, susceptibilities (both linear and nonlinear), switching-field curves and surfaces at finite temperatures, dynamic hysteresis, and magnetic stochastic resonance. Moreover, the infinite hierarchy of differential-recurrence equations for the statistical moments, governing the kinetics of the magnetization of a nanomagnet possessing *arbitrary* free energy density, can always be solved using matrix continued fractions over wide ranges of temperature and damping. The main advantage of this method is that it can be used for low potential barriers (high temperatures), where asymptotic methods like escape-rate theory fail. Alternatively, in model parameter ranges, where it is no longer feasible to use continued fractions, e.g., for very high barriers and very low damping, escape-rate theory provides a powerful tool for studying the relaxation processes in nanomagnets. Concerning the range of applicability of particular results of escape-rate theory, it has been clearly demonstrated that the TST escape rate Γ_A^{TST} yields a lower bound for the reversal time τ in a nanomagnet, and that it renders τ *only* for damping $\alpha \sim 0.1-1$. Therefore, in general, the correction prefactor Λ must be incorporated in the escape rate equation, viz., $\Gamma_A = \Lambda\Gamma_A^{\mathrm{TST}}$. An explicit equation for Γ_A is given by the turnover equation, Eq. (1.306), providing a reliable estimate of τ in all damping ranges. Regarding the single most important problem in nanomagnetism, namely the magnetization reversal due to thermal agitation, accurate analytical formulas for the reversal time of nanomagnets with uniaxial, biaxial, cubic, and mixed anisotropies have been obtained. Furthermore, it has been demonstrated that the low-frequency part of the longitudinal susceptibility of nanomagnets with various anisotropies is accurately described by the Debye-like equation, Eq. (1.151). Therefore in the time domain, the relaxation of the averaged longitudinal component of the magnetization $\langle M_{\parallel}\rangle(t)-\langle M_{\parallel}\rangle_0$ of a nanomagnet may be accurately approximated at long times ($t>>\tau$) by a *single* exponential, namely $\langle M_{\parallel}\rangle(t)-\langle M_{\parallel}\rangle_0 \sim e^{-t/\tau}$. Moreover, all the methods can be applied, with little modification, to thermal agitation in *current-induced magnetization dynamics* in nanomagnets governed by the magnetic Langevin equation, Eq. (1.42). Finally, the relaxation time of the magnetization and other related observables for *two interacting* magnetic moments (macrospins) coupled via exchange and dipole-dipole interactions can also be calculated.

We now proceed to the detailed treatment of spin size effects based on the general Stratonovich formulation of the spin dynamics in phase space introduced in Chapter 1.

Chapter 3

Quantum Effects in the Magnetization Relaxation of Nanomagnets

3.1 Introduction

It will by now have become quite apparent to our readers that the interpretation of spin relaxation experiments poses a fundamental problem of condensed phase physics and chemistry. A problem which if it were solved would provide a well-defined means of extracting information concerning the spin dynamics in gases, liquids and solids effectively acting as a *bridge* between microscopic and macroscopic physics. We may discuss this concept as follows: since the number of spins in a sample roughly corresponds to the number of atoms, on an *atomic* level, nuclear magnetic and related spin resonance experiments, etc. examine the time evolution of the *individual elementary spins* [213,456] of nuclei, electrons, muons, etc., while on *mesoscales* the time evolution of magnetic *molecular clusters* (i.e., spins 15–25 μ_B) exhibiting relatively large quantum effects is relevant to the fabrication of molecular magnets [457]. On *nanoscales* single-domain ferromagnetic particles (giant spins 10^4–10^5 μ_B) with a given orientation of the particle moment and permanent magnetization exist. These have spawned very extensive magnetic recording industries, the particles commonly used being on or near the microsize scale. Also, on a *nanoscale* level, we have magnetic fluids composed of single-domain ferromagnetic particles in a colloidal suspension. Here relaxation experiments detect [97,458] both the Arrhenius or solid-state like (Néel) mechanism [239] of magnetization

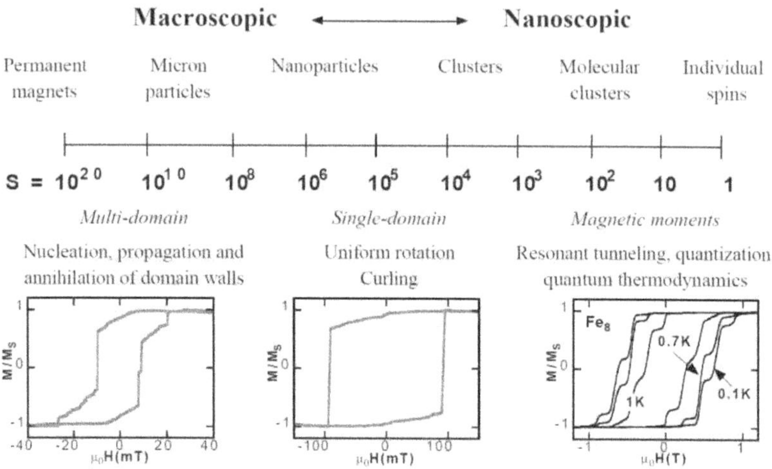

Fig. 3.1. The transition from macroscopic to nanoscopic magnets. The hysteresis loops are typical examples of magnetization reversal via nucleation, propagation and annihilation of domain walls (left), via uniform rotation (middle), and via quantum tunneling (right). Reprinted from W. Wernsdorfer, Adv. Chem. Phys. **118**, 99 (2001) with the permission of John Wiley & Sons.

relaxation which may overcome via thermal agitation, anisotropy potential barriers inside the particle and also the Debye orientational relaxation [110,111,459]. The latter being due to physical Brownian rotation of the suspended particles subjected to thermal agitation. Here quantum effects are expected to be very much smaller. Finally, on the *bulk macroscopic scale* permanent magnets (10^{20} μ_B) exist, i.e., *multi-domain* systems, where magnetization reversal occurs via the *macroscopic* processes of nucleation, propagation and annihilation of domain walls. Thus, we see that a *well-defined spin number scale* ranging from the *bulk macroscopic* down to *individual* atom and spins naturally occurs [20] (see Fig. 3.1), so prompting a quantum treatment of spin relaxations.

On an *atomic level* spin relaxation experiments in nuclear magnetic or electron spin resonance have been usually interpreted via the *phenomenological* Bloch [460] equations (Chapter 1) and their later modifications [213,456]. These describe the relaxation of *elementary* spins subjected to an external magnetic field and interacting with an environment constituting a *heat reservoir* at constant temperature T. Such simple linear equations of motion for the nuclear magnetization were originally proposed on phenomenological grounds. The main assumption is that the effects of the heat bath can be described by *two*-time constants, the so-called relaxation times, which provide a quantitative description for liquid samples [456]. However, microscopic theories of the relaxation in quantum spin systems have been developed by Bloch [461,462], Bloembergen, Purcell, and Pound [222], and others (see, e.g., [223–225]). These microscopic theories have yielded general evolution equations for the density matrix operator of the spin system allowing one, in principle, to evaluate all desired observables such as relaxation times, etc.

On meso- and nanoscales both the behavior of the hysteresis loop and the magnetization reversal time of molecular clusters and nanomagnets are essential for observation of the transition from the macroscopic to the nanoscale level. The reason being that *quantum effects are likely to occur as the spin number decreases* (see Fig. 3.1). In this region, the magnetization in molecular nanomagnets may reverse *both due to thermal agitation and quantum tunneling* as may be observed in the corresponding hysteresis loop [20]. In contrast, in ferromagnetic nanoparticles (originally encountered in the study of paleomagnetism in the context of past reversals of the Earth's magnetic field, where depending on the volume of the particle, the relaxation time may vary from nanoseconds to millions of years), the magnetization reversal is treated *classically* by assuming *uniform* rotation of the magnetization vector as conjectured by Néel [16] and Stoner and Wohlfarth [14]. The relaxation time epochs represent the transition from the Langevin paramagnetic behavior of nanoparticles (superparamagnetism) with no hysteresis involved via the magnetic after effect stage (where the magnetization reversal time is of the order of the time of a measurement) to stable ferromagnetism. There a *given ferromagnetic state* corresponds to one of *many possible such metastable states* in which the magnetization vector is held in a preferred orientation. In the hypothesis of uniform or coherent rotation, the *exchange interactions render all atomic spins collinear* and the magnitude of the magnetization vector is constant in space. This hypothesis should hold for fine (nanometric) magnetic particles, where domain walls cannot form in the sample because it is energetically unfavorable.

The *static* magnetization properties of single-domain particles are, as seen, usually treated via the elegant method of Stoner and Wohlfarth [14]. Their procedure simply consists in *minimizing the free energy of the particle* (i.e., the sum of the Zeeman and anisotropy energies with respect to the polar and azimuthal angles, specifying the orientation of the magnetization) for each value of the applied field. The calculation always leads to hysteresis because in certain field ranges *two or more minima* exist and thermally agitated transitions between them are neglected. We recall that the applied field, at which the magnetization reverses, is called the *switching field* and the *angular dependence of that field with respect to the easy axis of the magnetization* yields the familiar Stoner–Wohlfarth astroids [14,70]. These were originally given for *uniaxial shape anisotropy* only, which is the anisotropy of the magnetostatic energy of the sample induced by its nonspherical shape. The astroids (Chapter 1) represent a *parametric plot of the parallel versus the perpendicular component of the switching field*, which in the uniaxial case is the field that destroys the bistable structure of the free energy. The Stoner–Wohlfarth astroid concept (as seen) was later generalized to *arbitrary* effective anisotropy by Thiaville [71] including any given magneto-crystalline and surface anisotropy. He proposed a geometrical method (Chapter 1) for the calculation of the energy of a particle allowing one to determine the switching field for all values of the applied magnetic field yielding the *critical switching-field surface* analogous to the Stoner–Wohlfarth astroids. Knowledge [20] of the switching-field surface permits one to determine the effective anisotropy of the particle and all related parameters such as the frequencies of oscillation in the wells of the potential, i.e., the ferromagnetic resonance frequency, etc. We reiterate that these *static* calculations all ignore thermal effects on the switching field, i.e., *transitions between the minima of the potential are neglected* so that they are strictly only valid at zero temperature.

As far as the magnetization dynamics of fine particles at *finite* temperatures are concerned, Néel [16] (as seen) determined the magnetization relaxation time (i.e., the time of reversal of the magnetization of the particle, due to thermal agitation over its internal magnetocrystalline anisotropy barrier) from the inverse escape rate over the barriers using transition state theory [85] as specialized to magnetic moments, viz.,

$$\tau \sim f_A^{-1} e^{\Delta E}. \tag{3.1}$$

Here ΔE is the barrier height and f_A is the so-called attempt frequency associated with the frequency of the *gyromagnetic precession* in the effective field of the magnetocrystalline anisotropy. We have also seen that, by varying either the volume or the temperature of the particles, τ can be made to vary from 10^{-9} s to *millions of years* (f_A^{-1} is often taken as small as 10^{-10}–10^{-11} s in practice). The presence of the exponential factor in Eq. (3.1) indicates that, to approach the *zero remanence* (corresponding to thermal equilibrium), a sufficient number of particles (magnetic moments) must be reversed by thermal activation over the energy barrier. However, the Néel treatment, given in detail for uniaxial anisotropy only, is confined to a *discrete* set of orientations for the magnetic moment of the particle. Moreover, the equilibrium distribution is all that is ever required since the disturbance to the Boltzmann distribution in the wells of the magnetocrystalline anisotropy potential due to the escape of the magnetization over the barrier is ignored. Besides the overbarrier relaxation process, Bean and Livingston [8] have also suggested that in a single-domain particle the magnetization may also reverse by *macroscopic quantum tunneling* (macroscopic since a giant spin is involved) through the magnetocrystalline-Zeeman energy potential barrier. In general, the magnetization may reverse by quantum tunneling at *very low temperatures* [175–177], which may be observed in the behavior of the Stoner-Wohlfarth astroids and associated hysteresis loops. Therefore, to distinguish *tunneling* reversal from reversal by *thermal agitation*, systematic ways of introducing quantum effects into the magnetization reversal of nano-mesoscale magnets are required. In other words, *quantum effects in parameters characterizing the decay of metastable states in spin systems*, are the subject of this chapter.

In the context of thermal effects in the magnetization reversal of classical spins, we have seen in Chapter 1 that the original dynamical calculations of Néel for single-domain particles utilize classical TST, whereas in the Néel–Brown model [20,239] (which explicitly treats the system as a gyromagnetic one and which includes nonequilibrium effects due to the loss of magnetization at the barrier) the time evolution of $\mathbf{u}(t) = \mathbf{M}(t)/M_S$ is described by a classical (magnetic) Langevin equation which reads

$$\dot{\mathbf{u}} = \gamma\left[\mathbf{u}\times\left(\mathbf{H}_{ef}+\mathbf{h}\right)\right]-\alpha\left[\mathbf{u}\times\dot{\mathbf{u}}\right]. \tag{3.2}$$

The stochastic differential equation, Eq. (3.2), for the isotropic Brownian motion of the classical spin containing both precessional and aligment terms is then used in Chapter 1 to derive the Fokker–Planck equation accompanying Eq. (3.2), governing the time evolution of the distribution function $W(\vartheta, \varphi, t)$ of magnetization orientations on the surface of a sphere of constant radius M_S. The relevant Fokker–Planck equation is [97,239] (see Chapter 1)

$$\frac{\partial W}{\partial t} = \mathrm{L}_{\mathrm{FP}} W, \tag{3.3}$$

where the Fokker–Planck operator L_{FP} again comprising both precessional and alignment terms is defined by Eq. (1.65). Brown's model [17,18] rooted in a magnetic Langevin equation, Eq. (3.2), allows one (as seen in Chapter 2) to treat the relaxation processes in classical spin systems (single-domain particles, etc.) using powerful computational techniques which have already been developed in the theory of the Brownian motion. These include (described in Chapters 1 and 2) continued fractions, mean first-passage times, escape-rate theory, etc. [97,239]. Thus, we can compare theoretical predictions with experimental data on superparamagnetic relaxation, which plays a fundamental role in information storage, paleomagnetism, biotechnology, etc.

Here we shall demonstrate how the results of calculations of relaxation times, dynamic susceptibilities, etc. for various model systems via the phase-space formalism introduced in Chapter 1 are in complete agreement with those obtained by solving quantum master equations for the density matrix. The results are significant particularly because they elucidate the role played by quantum effects in the various spin relaxation processes in nanomagnets as well as providing a basis for a theory of macroscopic quantum tunneling to which the semiclassical approach is ideally suited.

The contents of this chapter can be summarized as follows. We begin by calculating Wigner functions for various equilibrium model spin systems with *arbitrary* spin number S (e.g., spins in a uniform external field, uniaxial nanomagnets in longitudinal and transverse fields, biaxial and cubic systems). We also calculate quantum corrections to both switching-field curves and to the spin TST reversal time. Thus, we derive and solve phase-space master equations for *circularly symmetric* problems including both noninteracting spins in a dc magnetic field and uniaxial nanomagnets subjected to a dc magnetic field. Hence, we evaluate characteristic relaxation times and dynamic susceptibilities for these spin systems as well as treating nonlinear longitudinal relaxation of spins in superimposed ac and dc magnetic fields and investigating quantum effects in dynamic magnetic hysteresis and stochastic resonance. Next, by using the statistical moment equations for the time-dependent average spherical harmonics for an *arbitrary* spin Hamiltonian (see Chapter 1) we determine the characteristic relaxation times and dynamic susceptibility of a uniaxial nanomagnet subjected to a dc bias field of *arbitrary* orientation with respect to the easy axis. Throughout, we shall compare the results obtained via the spin phase-space formalism with the corresponding classical limit, $S \to \infty$. Finally, to facilitate a better understanding of that formalism and its connection to classical spin relaxation, the appendixes describe the basic properties of spin and polarization operators (Appendix A.2) and spherical harmonics (Appendix A.1.7). Appendix A.3 describes the intricate calculations involved in transforming the reduced density matrix evolution equation into a master equation in phase space for a uniaxial nanomagnet. The definitions and associated derivations of characteristic relaxation times and correlation functions are given in Appendix A.4.

3.2 Equilibrium phase-space distribution functions for spins

We now obtain both analytically and numerically, *equilibrium* time-independent quasiprobability distribution functions for spin systems with various time-independent Hamiltonians $\hat{H}_S$, where the equilibrium density matrix $\hat{\rho}_S$ is given by

$$\hat{\rho}_S = \frac{1}{Z_S} e^{-\beta \hat{H}_S} \tag{3.4}$$

with the partition function $Z_S = \mathrm{Tr}(e^{-\beta \hat{H}_S})$. First recall that according to Eq. (1.543) the density matrix $\hat{\rho}_S$ for an arbitrary Hamiltonian $\hat{H}_S$ can be written as a finite series of polarization operators, viz.,

$$\hat{\rho}_S = \sum_{L=0}^{2S} \sum_{M=-L}^{L} \hat{T}_{LM}^{(S)} \left\langle \hat{T}_{LM}^{\dagger(S)} \right\rangle_0, \tag{3.5}$$

while the general Eq. (1.579) yields the corresponding equilibrium phase-space distributions $W_S^{(s)}(\vartheta,\varphi)$ as a *finite* series of spherical harmonics, viz.,

$$\frac{2S+1}{4\pi} W_S^{(s)}(\vartheta,\varphi) = \sum_{L=0}^{2S} \sum_{M=-L}^{L} Y_{LM}(\vartheta,\varphi) \left\langle Y_{LM}^{*} \right\rangle_0^{(s)}, \tag{3.6}$$

where *averaged* spherical harmonics $\left\langle Y_{LM}^{*} \right\rangle_0^{(s)}$ are related to *averaged* polarization operators $\left\langle \hat{T}_{LM}^{\dagger(S)} \right\rangle_0$ via

$$\left\langle Y_{LM}^{*} \right\rangle_0^{(s)} = \sqrt{\frac{4\pi}{2S+1}} \left(C_{SSL0}^{SS} \right)^{-s} \left\langle \hat{T}_{LM}^{\dagger(S)} \right\rangle_0. \tag{3.7}$$

Hence, $W_S^{(s)}(\vartheta,\varphi)$ can also be written as a finite series of the averaged polarization operators thus it can in principle always be determined (analytically or numerically) for a given Hamiltonian $\hat{H}_S$.

Following Ref. [260,261], we shall evaluate analytically (for small S) or numerically (for large S) from Eqs. (3.6) and (3.7) the equilibrium phase-space distributions $W_S^{(s)}(\vartheta,\varphi)$. To implement this procedure for a *particular Hamiltonian*:

(i) First, we write the density matrix operator $\hat{\rho}_S$ from Eqs. (3.4) and (3.5) for the given Hamiltonian $\hat{H}_S$ expressed either in terms of polarization operators $\hat{T}_{LM}^{(S)}$ or, using Eqs. (A.147) from Appendix A.2, in terms of the spin operators $\hat{S}_i$ $(i = X, Y, Z)$.

(ii) Next, we calculate the averaged polarization operator $\left\langle \hat{T}_{LM}^{\dagger(S)} \right\rangle_0 = \mathrm{Tr}(\hat{\rho}_S \hat{T}_{LM}^{\dagger(S)})$ using the operator expansion method described in Appendix A.2.

(iii) Then, we can write from Eq. (3.7) the expansion (Fourier) coefficients $\left\langle Y_{LM}^{*}\right\rangle_0^{(s)}$ connecting the *average of a spherical harmonic to that of a polarization operator.*

(iv) Thus, we obtain the phase-space distribution $W_S^{(s)}(\vartheta,\varphi)$ for the chosen Hamiltonian $\hat{H}_S$ from the formal finite series solution Eq. (3.6) for any particular S.

All the calculations for a given Hamiltonian $\hat{H}_S$, which are rather tedious, can be accomplished using MATHEMATICA®.

For illustration, we initially evaluate the equilibrium Wigner ($s=0$), Q- ($s=-1$) and P- ($s=1$) phase-space distributions $W_S^{(s)}(\vartheta,\varphi)$ for the very simple case of a spin with spin number S and magnetic moment $\mu=\gamma\hbar S/\mu_0$ in an external dc field $\mathbf{H}_0$ applied along the Z-axis (essentially these distributions correspond to the usual treatment of quantum paramagnetism). Consequently, the spin Hamiltonian $\hat{H}_S$ is just

$$\hat{H}_S=-\xi(\beta S)^{-1}\hat{S}_Z, \tag{3.8}$$

where $\xi=\beta\mu_0\mu H_0$ is the dimensionless external field parameter. Hence, we shall see that the Q-function $W_S^{(-1)}(\vartheta,\varphi)$ alone satisfies the *nonnegativity* condition, viz., $W_S^{(-1)}\geq 0$, required of a true probability density function. The quasiprobability densities $W_S^{(1)}$ and $W_S^{(0)}$ do not satisfy this condition (because they may take on negative values). Thus, in future determinations of a phase-space representation, we shall usually restrict ourselves to the (Q-) function $W_S^{(-1)}(\vartheta,\varphi)$ as all the other functions can be treated in like manner. First, we shall determine the equilibrium Q- ($s=-1$) phase-space distributions for an assembly of noninteracting spins in an external dc field $\mathbf{H}_0$ applied in an *arbitrary* direction in space. Here the spin Hamiltonian is

$$\hat{H}_S=-\xi(\beta S)^{-1}\left(\gamma_X\hat{S}_X+\gamma_Y\hat{S}_Y+\gamma_Z\hat{S}_Z\right), \tag{3.9}$$

where γ_X, γ_Y, γ_Z are the direction cosines of $\mathbf{H}_0$. Next, we shall treat various magnetic anisotropies so establishing one or more preferred orientations of the magnetization of an assembly of spins. For example, we shall consider a *uniaxial* paramagnet in an external magnetic field with *arbitrary* orientation so that

$$\hat{H}_S = -\xi(\beta S)^{-1}\left(\gamma_X \hat{S}_X + \gamma_Y \hat{S}_Y + \gamma_Z \hat{S}_Z\right) - \sigma(\beta S^2)^{-1}\hat{S}_Z^2 \tag{3.10}$$

as well as two *particular* cases of Eq. (3.10), namely, a uniaxial nanomagnet in both a longitudinal and a transverse external field with

$$\hat{H}_S = -\xi(\beta S)^{-1}\hat{S}_Z - \sigma(\beta S^2)^{-1}\hat{S}_Z^2 \tag{3.11}$$

and

$$\hat{H}_S = -\xi(\beta S)^{-1}\hat{S}_X - \sigma(\beta S^2)^{-1}\hat{S}_Z^2, \tag{3.12}$$

respectively, where σ is the dimensionless anisotropy parameter. In the classical limit, $S \to \infty$, the latter Hamiltonian corresponds to a uniaxial nanomagnet with two equivalent ground states of magnetization separated by a magnetocrystalline anisotropy energy barrier (the σ term) with an applied transverse field. The transverse field in the quantum case will *enhance* the tunneling probability. Finally, we shall consider biaxial and cubic-like systems with Hamiltonians

$$\hat{H}_S = -(\beta S^2)^{-1}\left[\sigma \hat{S}_Z^2 + \delta(\hat{S}_X^2 - \hat{S}_Y^2)\right], \tag{3.13}$$

(δ is a dimensionless biaxiality parameter) and

$$\hat{H}_S = -\xi(\beta S)^{-1}\hat{S}_Z - \sigma_c(2\beta S^4)^{-1}\left(\hat{S}_X^4 + \hat{S}_Y^4 + \hat{S}_Z^4\right), \tag{3.14}$$

where σ_c is the dimensionless cubic anisotropy parameter, which may be either positive or negative.

Having determined the equilibrium quasiprobability distributions $W_S^{(s)}(\vartheta,\varphi)$ corresponding to these Hamiltonians $\hat{H}_S$, our second purpose is to calculate the magnetization reversal time via the quantum generalization of TST (previously treated for classical spins by Néel [16]) permitting one to estimate temperature effects in the astroids and hysteresis loops within the limitations imposed by quantum TST (moderate damping, etc.) [97,239]. Finally, we shall calculate the Stoner-Wohlfarth switching-field curves as a function of spin number S for various Hamiltonians $\hat{H}_S$, so generalizing Thiaville's geometrical method [71] to include quantum effects due to *finite* spin number. Thus, one may study the behavior of the astroids in the interesting *magnetic cluster* –

single-domain nanoparticle transition region. In the magnetic context, explicit equations for the equilibrium phase-space distributions (described below) have already been obtained for an assembly of noninteracting spins in a uniform magnetic field [229,230] and for spins in the simplest uniaxial potential of the magnetocrystalline anisotropy and Zeeman energy [260].

3.2.1 *Spins in a uniform external field*

First, we evaluate the equilibrium phase-space distributions for the uniaxial situation pertaining to a spin with spin number S (integer or half-integer) in an external dc field $\mathbf{H}_0$ applied along the Z-axis. The spin Hamiltonian $\hat{H}_S$ is then given by Eq. (3.8), namely, [213]

$$\hat{H}_S = -\gamma\hbar H_0 \hat{S}_Z \tag{3.15}$$

with the eigenenergies $E_m = -\gamma\hbar H_0 m$ with $m = -S, -S+1, \ldots, S,$ the spacing between adjacent energy levels being $\gamma\hbar H_0$. In equilibrium, the phase-space distributions $W_S^{(s)}(\vartheta)$ are independent of the azimuthal angle φ and, according to Eq. (3.6), can be expressed as the series [260]

$$\begin{aligned} W_S^{(s)}(\vartheta) &= \sqrt{\frac{4\pi}{2S+1}} \sum_{L=0}^{2S} \left(C_{SSL0}^{SS}\right)^{-s} Y_{L0}(\vartheta,\varphi) \left\langle \hat{T}_{L0}^{\dagger(S)} \right\rangle_0 \\ &= \frac{1}{Z_S(2S+1)} \sum_{L=0}^{2S} (2L+1) \left(C_{SSL0}^{SS}\right)^{-s} P_L(\cos\vartheta) \sum_{m=-S}^{S} C_{SmL0}^{Sm} e^{\xi m/S}, \end{aligned} \tag{3.16}$$

where $P_L(z)$ are the Legendre polynomials [290], $\xi = \beta\gamma\hbar H_0 S$ is the dimensionless field parameter, and the partition function Z_S is given explicitly by

$$Z_S = \sum_{m=-S}^{S} e^{\xi m/S} = \frac{\sinh\left[\left(S+\frac{1}{2}\right)\xi/S\right]}{\sinh\left(\frac{1}{2}\xi/S\right)}. \tag{3.17}$$

In writing Eq. (3.16), we have used the definition of the spherical harmonic of rank L, Eq. (A.70) from Appendix A.1.7, as well as the existing explicit expression for the matrix elements ρ_{mn} of the density matrix $\hat{\rho}_S$, viz.,

$$\rho_{mn} = \frac{\delta_{mn}}{Z_S} e^{\xi m/S} \tag{3.18}$$

along with Eq. (A.143) from Appendix A.2 for $\hat{T}_{LM}^{\dagger(S)}$ so yielding the equilibrium average polarization operators in Eq. (3.16), viz.,

$$\left\langle \hat{T}_{L0}^{\dagger(S)} \right\rangle_0 = \frac{1}{Z_S}\sqrt{\frac{2L+1}{2S+1}} \sum_{m=-S}^{S} C_{SmL0}^{Sm} e^{\xi m/S}. \tag{3.19}$$

The finite series in Eq. (3.16) for $s=-1$ can be summed (after some algebra, which is best accomplished via MATHEMATICA®). Thus, the Q-function $W_S^{(-1)}(\vartheta)$ can finally be written for arbitrary S in the (previously known) concise closed form [230], viz.,

$$W_S^{(-1)}(\vartheta) = \frac{1}{Z_S}\left[\cosh\frac{\xi}{2S} + \cos\vartheta \sinh\frac{\xi}{2S}\right]^{2S}. \tag{3.20}$$

Moreover, using the general rule for the calculation of the expected value of the spin operator $\hat{S}_Z$ via the corresponding (c-number) function, Eq. (1.589), the average longitudinal component of the spin at equilibrium is (because the integral over φ is 2π)

$$\begin{aligned} \left\langle \hat{S}_Z \right\rangle_0 &= \left(S+\tfrac{1}{2}\right) S \left(\frac{S+1}{S}\right)^{(s+1)/2} \int_0^{\pi} \cos\vartheta W_S^{(-s)}(\vartheta) \sin\vartheta d\vartheta \\ &= \frac{\sqrt{S(S+1)}}{Z_S} \sum_{m=-S}^{S} C_{Sm10}^{Sm} e^{\xi m/S} = \frac{1}{Z_S}\sum_{m=-S}^{S} m e^{\xi m/S} = SB_S(\xi), \end{aligned} \tag{3.21}$$

where $B_S(\xi)$ is the Brillouin function defined by Eq. (1.368) of Chapter 1 [213] and we have utilized the Clebsch–Gordan identity [104]

$$C_{Sm10}^{Sm} = \frac{m}{\sqrt{S(S+1)}}.$$

Obviously, $\left\langle \hat{S}_Z \right\rangle_0$ from Eq. (3.21) has the *all-important feature that it is independent of the representation distribution parameter s*. Notice for future reference that setting $s=1$ in the Weyl symbol Eq. (1.589) means that we are utilizing the Q-distribution, where $s=-1$. Moreover, Eq. (3.21) is in complete agreement with the long-established result (Chapter 1, Eq. (1.367)) for the equilibrium magnetization of an assembly of noninteracting spins in a uniform magnetic field [213,456]. Furthermore, in the classical limit,

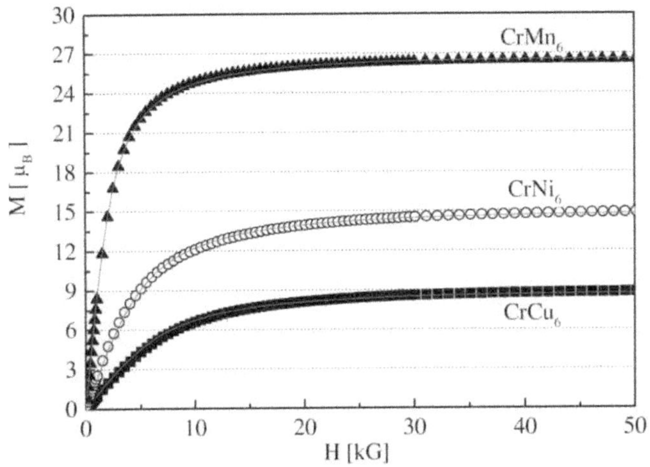

Fig. 3.2. The magnetization per molecule $M = g\mu_B SB_S(\beta\mu_0 g\mu_B SH)$ in units of Bohr magnetons μ_B of $CrCu_6$, $CrNi_6$, and $CrMn_6$ vs. the external applied field H at $T = 2$ K. The experimental data (symbols) are accurately described by the Brillouin functions (solid lines). Reproduced by permission of the American Physical Society from Z. Salman, A. Keren, P. Mendels, V. Marvaud, A. Sciuller, M. Verdaguer, J. S. Lord, and C. Baines, *Phys. Rev. B* **65,** 132403 (2002).

$$S \to \infty \text{ and } \mu = \gamma\hbar S / \mu_0 = \text{const}, \tag{3.22}$$

the distribution $W_S^{(s)}(\vartheta)$ tends to the Boltzmann distribution for classical magnetic dipoles, i.e.

$$(S+1/2)W_S^{(s)}(\vartheta) \to Z_{\text{cl}}^{-1} e^{\xi\cos\vartheta}, \tag{3.23}$$

while the Brillouin function $B_S(\xi)$ tends to the Langevin function $L(\xi) = \coth\xi - 1/\xi$, Eq. (1.370) of Chapter 1. Here Z_{cl} is the classical partition function given by Eq. (1.371) of Chapter 1.

Experimental studies of the magnetization of various *paramagnetic* atoms and molecules indicate that they agree closely with the Brillouin function, Eq. (1.368) of Chapter 1 (see, e.g., Refs. [463,464]). The magnetization per molecule in units of μ_B as a function of the applied field is shown in Fig. 3.2 for the three isotropic high-spin molecules $[Cr\{(CN)Cu(tren)\}_6](ClO_4)_{21}$ ($S = 9/2$), $[Cr\{(CN)Ni(tetren)\}_6](ClO_4)_9$ (S=15/2), $[Cr\{(CN)Mn(tren)\}_6](ClO_4)_{21}$ ($S = 27/2$), which are labelled as $CrCu_6$, $CrNi_6$, and $CrMn_6$, respectively. These consist of clusters of metal

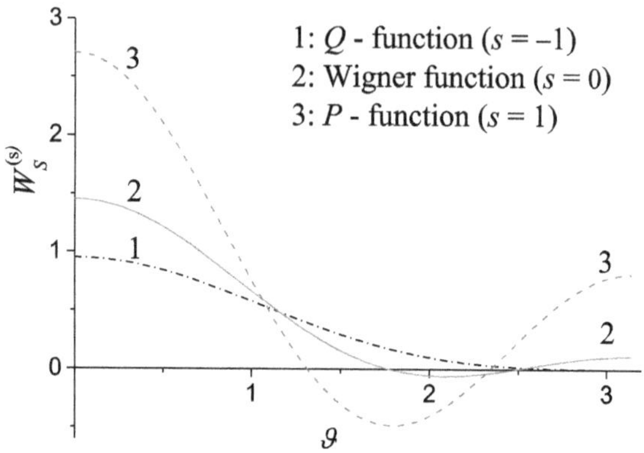

Fig. 3.3. $W_S^{(s)}(\vartheta)$ vs. the polar angle ϑ for $s = 0, \pm 1$, $\xi = 3$ and $S = 1$.

ions ordered in a crystal lattice and coupled only via Heisenberg ferromagnetic or antiferromagnetic interactions between spins inside the molecule. In all cases, M increases with the applied field reaching a saturation value of (9/2) $g\,\mu_B$, (15/2) $g\,\mu_B$ and (27/2) $g\,\mu_B$ in $CrCu_6$, $CrNi_6$, and $CrMn_6$, respectively ($g = 2$ is Landé's factor).

The three phase-space equilibrium distributions $W_S^{(s)}(\vartheta)$ embodied in the finite series representation Eq. (3.16) are shown for comparison purposes in Fig. 3.3 for $s = 0, \pm 1$ so demonstrating that the Q-function $W_S^{(-1)}(\vartheta,\varphi)$ alone satisfies the *nonnegativity* condition, viz., $W_S^{(-1)} \geq 0$, required of a true probability density function. The other quasiprobability distribution functions $W_S^{(1)}$ and $W_S^{(0)}$ violate this condition (because they may take on negative values). Thus for convenience we shall consider $W_S^{(-1)}(\vartheta,\varphi,t)$ only omitting everywhere the superscript (-1) in $W_S^{(-1)}(\vartheta,\varphi,t)$ [nevertheless all results may be easily generalized for $W_S^{(1)}(\vartheta,\varphi,t)$ and $W_S^{(0)}(\vartheta,\varphi,t)$, which can be treated in like manner].

Next we calculate the phase-space Q-function distribution (henceforth this will be given the generic title "Wigner function") for spins in an external uniform field $\mathbf{H}_0$ of an *arbitrary* orientation rather than just applied along the Z axis so that the Hamiltonian is given by Eq. (3.9). This operator can be rewritten in terms of the spherical spin operators $\hat{S}_\mu$ (see Appendix A.2) as

$$\beta \hat{H}_S = -\frac{\xi}{S} \sum_{\mu=-1}^{1} \gamma^{\mu} \hat{S}_{\mu},$$

where in terms of direction cosines $\gamma^{\pm 1} = \mp 2^{-1/2}(\gamma_X \mp i\gamma_Y)$, $\gamma^0 = \gamma_Z$ and the matrix elements of the Hamiltonian operator $\hat{H}_S$ can again be given in closed form, viz.,

$$\left[\beta \hat{H}_S\right]_{m'm} = -\frac{\xi}{S}\left(\gamma^{-1}\left[\hat{S}_{-1}\right]_{m-1m} \delta_{m-1m'} + \gamma^0 m \delta_{mm'} + \gamma^1 \left[\hat{S}_1\right]_{m+1m} \delta_{m+1m'}\right), \quad (3.24)$$

where the matrix elements of the spherical spin operators $\hat{S}_{\pm 1}$ are given by Eq. (A.137) from Appendix A.2, namely,

$$\left[\hat{S}_{\pm 1}\right]_{m\pm 1m} = \mp\sqrt{\frac{1}{2}(S \mp m)(S \pm m + 1)}. \quad (3.25)$$

Furthermore, for small S, the equilibrium density matrix $\hat{\rho}_S = e^{-\beta \hat{H}_S} / Z_S$ can be written in closed form as a finite series of the spin operators, e.g., for $S = 1/2$, $S = 1$, we have

$$\hat{\rho}_{1/2} = \frac{1}{Z_{1/2}}\left(\hat{I}\cosh\xi + 2\sum_{\mu=-1}^{1} \gamma^{\mu} \hat{S}_{\mu} \sinh\xi\right),$$
$$\hat{\rho}_1 = \frac{1}{Z_1}\left[\hat{I} + \sum_{\mu=-1}^{1} \gamma^{\mu} \hat{S}_{\mu} \sinh\xi + 2\left(\sum_{\mu=-1}^{1} \gamma^{\mu} \hat{S}_{\mu}\right)^2 \sinh^2\frac{\xi}{2}\right], \quad (3.26)$$

where $\hat{I}$ is the identity matrix.

The corresponding Q-distribution $W_S(\vartheta, \varphi)$ can then be calculated in the finite series form Eq. (3.6). In turn, this finite series can then be summed (after tedious algebra, which is again best accomplished via MATHEMATICA®) so that the distribution $W_S(\vartheta, \varphi)$ can ultimately be written for arbitrary S in concise closed form [261], viz., [cf. Eq. (3.20)]

$$W_S(\vartheta, \varphi) = \frac{1}{Z_S}\left[\cosh\frac{\xi}{2S} + F(\vartheta, \varphi)\sinh\frac{\xi}{2S}\right]^{2S}, \quad (3.27)$$

where

$$F(\vartheta, \varphi) = \gamma_X \sin\vartheta\cos\varphi + \gamma_Y \sin\vartheta\sin\varphi + \gamma_Z \cos\vartheta \quad (3.28)$$

and

$$Z_S = \frac{2S+1}{4\pi}\int_0^{\pi}\int_0^{2\pi}\left[\cosh\frac{\xi}{2S} + F(\vartheta,\varphi)\sinh\frac{\xi}{2S}\right]^{2S}\sin\vartheta d\vartheta d\varphi$$
$$= \frac{\sinh\left[\left(S+\frac{1}{2}\right)\xi / S\right]}{\sinh\left(\frac{1}{2}\xi / S\right)} \tag{3.29}$$

is the partition function [Eq. (3.29) concurs with Eq. (3.17)]. Moreover, for the three specific cases represented by the following direction cosines:

$$1)\, \gamma_X = 1, \gamma_Y = \gamma_Z = 0;\ 2)\, \gamma_Y = 1, \gamma_X = \gamma_Z = 0;\ 3)\, \gamma_Z = 1, \gamma_X = \gamma_Y = 0;$$

the phase-space distribution, Eq. (3.27), reduces to the equations already derived by Takahashi and Shibata [229,230] and also reproduces Eq. (3.20). The average component of the magnetic moment in the direction of the dc field $\mathbf{H}_0$ is then from Eqs. (3.27) and (3.28) [cf. Eq. (3.21)]

$$\left\langle \hat{\mu}_{H_0} \right\rangle_0 = \mu \frac{(2S+1)(S+1)}{4\pi}\int_0^{\pi}\int_0^{2\pi} F(\vartheta,\varphi) W_S(\vartheta,\varphi)\sin\vartheta d\vartheta d\varphi$$
$$= \mu B_S(\xi), \tag{3.30}$$

where $B_S(x)$ is once again the Brillouin function. In the classical limit, $S \to \infty$, the equilibrium distribution $W_S(\vartheta,\varphi)$ given by Eq. (3.27) tends to the Boltzmann distribution

$$\frac{2S+1}{4\pi} W_S(\vartheta,\varphi) \to \frac{1}{Z_{\mathrm{cl}}} e^{\xi F(\vartheta,\varphi)}, \tag{3.31}$$

while the equilibrium average $\left\langle \hat{\mu}_{H_0} \right\rangle_0$ tends to $\mu L(\xi)$, where $L(\xi)$ is the Langevin function, Eq. (1.370) of Chapter 1, and Z_{cl} is the classical partition function given by Eq. (1.371) of Chapter 1.

Clearly, the above calculations represent both quantum and classical treatments of paramagnetism [213]. Here, quantum effects as identified via the Brillouin function, Eq. (1.368) of Chapter 1, become important at small S when that function *must* be used instead of the Langevin function originating in Eq. (3.31), etc., which is valid in the classical limit.

3.2.2 Uniaxial nanomagnet in an external field

Here we calculate the equilibrium Wigner function $W_S(\vartheta,\varphi)$ for a *uniaxial* nanomagnet of arbitrary spin number S in an external magnetic field of an *arbitrary* orientation. First, we briefly consider the more general case of a spin in an external dc field $\mathbf{H}_0$ with the Hamiltonian operator $\hat{H}_S$ given by Eq. (3.10); then we specialize it to a longitudinal field. In the general case, the matrix elements of $\hat{H}_S$ can again be given via the matrix elements of the spherical spin operators $\hat{S}_\mu$ ($\mu = 0, \pm 1$) and $\hat{S}_0^2$, viz. [cf. Eq. (3.24) with superimposed anisotropy term],

$$\left[\beta\hat{H}_S\right]_{m'm} = \left[-\frac{\xi}{S}\sum_{\mu=-1}^{1}\gamma^\mu\hat{S}_\mu - \frac{\sigma}{S^2}\hat{S}_0^2\right]_{m'm} = -\left(m^2\frac{\sigma}{S^2} + \gamma^0 m\frac{\xi}{S}\right)\delta_{mm'} \\ -\frac{\xi}{S}\left(\gamma^{-1}\left[\hat{S}_{-1}\right]_{m-1m}\delta_{m-1m'} + \gamma^1\left[\hat{S}_1\right]_{m+1m}\delta_{m+1m'}\right), \tag{3.32}$$

where the matrix elements $[\hat{S}_{\pm 1}]_{m\pm 1m}$ are defined by Eq. (3.25). The corresponding phase-space distribution $W_S(\vartheta,\varphi)$ can then be calculated in the finite series form of Eqs. (3.6) and (3.7). Now motivated by the form of the classical potential, the results of the calculation of an "effective" free energy potential defined by $E_{ef}(\vartheta,\varphi) = -\ln W_S(\vartheta,\varphi)$ are shown in Fig. 3.4 for various values of S. The *effective potential* $E_{ef}(\vartheta, \varphi)$ has two nonequivalent minima (the minimum at $\vartheta = \pi$ is masked in these plots) and one saddle point in the plane $\varphi = 0$; the potential shape and barrier heights strongly depend on the spin number S. Moreover, in the classical limit, $S \to \infty$, $E_{ef}(\vartheta, \varphi)$ tends to the normalized classical free energy $E_{cl}(\vartheta, \varphi)$ given by

$$E_{cl}(\vartheta,\varphi) = -\sigma\left\{\cos^2\vartheta + 2h\left[\left(\gamma_X\cos\varphi + \gamma_Y\sin\varphi\right)\sin\vartheta + \gamma_Z\cos\vartheta\right]\right\}, \tag{3.33}$$

which is also shown in Fig. 3.4 for comparison.

However, the general treatment above considerably simplifies for a *longitudinal* field (ubiquitous in magnetic applications) so that $\gamma_X = \gamma_Y = 0$, $\gamma_Z = 1$ [see Eq. (3.11)]. Here the density matrix $\hat{\rho}_S$ is *diagonal* with matrix elements $\rho_{mm'}$ given explicitly by the simple closed form [228,329,330]

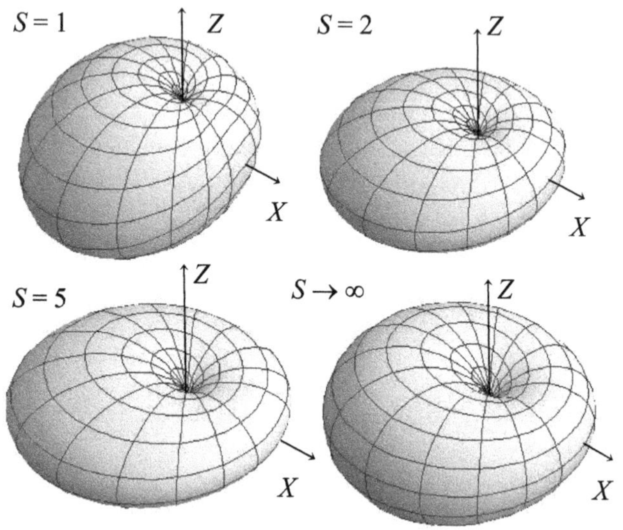

Fig. 3.4. 3D plot of the "effective" free energy potential $E_{ef}(\vartheta, \varphi)$ for a uniaxial nanomagnet in an external field for various values of $S = 1, 2, 5$, and $S \to \infty$ [classical limit; Eq. (3.33)] and the parameters $\sigma = 5$, $\xi = 2$, $\gamma_Z = 1/2$, $\gamma_Y = 0$, $\gamma_X = \sqrt{3}/2$ (i.e., the field $\mathbf{H}_0$ is in the XZ-plane and directed at an angle $\pi/3$ to the Z-axis).

$$\rho_{mm'} = \frac{\delta_{mm'}}{Z_S} e^{\frac{\sigma m^2}{S^2} + \frac{\xi m}{S}}, \quad Z_S = \sum_{m=-S}^{S} e^{\frac{\sigma m^2}{S^2} + \frac{\xi m}{S}}. \tag{3.34}$$

The matrix elements $\rho_{mm'}$, Eq. (3.34), can then in turn be used to evaluate the averages $\left\langle \hat{T}_{LM}^{\dagger(S)} \right\rangle_0$ in Eq. (3.7) in closed form as [cf. Eq. (3.19)]

$$\left\langle \hat{T}_{LM}^{\dagger(S)} \right\rangle_0 = \frac{\delta_{M0}}{Z_S} \sqrt{\frac{2L+1}{2S+1}} \sum_{m=-S}^{S} C_{SmLM}^{Sm} e^{\frac{\sigma m^2}{S^2} + \frac{\xi m}{S}}. \tag{3.35}$$

Furthermore, due to the (circular) symmetry about the Z-axis, the phase-space distribution function $W_S(\vartheta, \varphi) \equiv W_S(\vartheta)$ is now independent of the azimuthal angle φ so that the Wigner function Eq. (3.6) for this important problem merely reduces to the finite series of Legendre polynomials [260]

$$W_S(\vartheta) = \sum_{L=0}^{2S} \frac{2L+1}{2S+1} \langle P_L \rangle_0 P_L(\cos\vartheta). \tag{3.36}$$

Here $\langle P_L \rangle_0$ are the equilibrium averages of the Legendre polynomials $P_L(\cos\vartheta)$ given by

$$\begin{aligned}\langle P_L \rangle_0 &= (S+1/2)\int_0^{\pi} P_L(\cos\vartheta)W_S(\vartheta)\sin\vartheta d\vartheta \\ &= \frac{(2S)!\sqrt{2S+1}}{Z_S\sqrt{(2S-L)!(2S+L+1)!}}\sum_{m=-S}^{S} C_{SmL0}^{Sm} e^{\frac{\sigma m^2}{S^2}+\frac{\xi m}{S}}\end{aligned} \tag{3.37}$$

and we have used Eq. (3.35) and also the definition of $P_L(\cos\vartheta)$ Eq. (A.35) from Appendix A.1.5. The equilibrium statistical moment $\langle P_1 \rangle_0$ then yields the average longitudinal component of the spin as

$$\langle \hat{S}_Z \rangle_0 = (S+1)\langle P_1 \rangle_0 = (S+1)\langle \cos\vartheta \rangle_0 = \frac{1}{Z_S}\sum_{m=-S}^{S} m e^{\frac{\sigma m^2}{S^2}+\frac{\xi m}{S}}, \tag{3.38}$$

which concurs with the well-known result for the equilibrium magnetization for arbitrary S [228,329,330]. Here we have used the Weyl symbol of the operator $\hat{S}_Z$ given by Eq. (1.589) with $s = 1$. From the explicit expressions for the several $P_L(\cos\vartheta)$ [290] in Eq. (3.36), we then have explicit trigonometric forms for the distribution functions, e.g., for $S = 1/2, 1, 3/2, 2$, etc. [260]

$$W_{1/2}(\vartheta) = \frac{e^{\sigma}}{Z_{1/2}} f_{\xi}(\vartheta), \quad W_1(\vartheta) = \frac{e^{\sigma}}{Z_1}\left[\left(f_{\xi}^{(1)}(\vartheta)\right)^2 + \frac{1}{2}\left(e^{-\sigma}-1\right)\sin^2\vartheta\right],$$

$$W_{3/2}(\vartheta) = \frac{e^{9\sigma/4}}{Z_{3/2}}\left[\left(f_{\xi}^{(3/2)}(\vartheta)\right)^3 + \frac{3}{4}\left(e^{-8\sigma/9}-1\right)f_{\xi}^{(3/2)}(\vartheta)\sin^2\vartheta\right],$$

$$\begin{aligned}W_2(\vartheta) = \frac{e^{\sigma}}{Z_2}\Big[&\left(f_{\xi}^{(2)}(\vartheta)\right)^4 + \left(e^{-3\sigma/4}-1\right)\left(f_{\xi}^{(2)}(\vartheta)\right)^2\sin^2\vartheta \\ &+ \frac{1}{8}\left(3e^{-\sigma} - 4e^{-3\sigma/4} + 1\right)\sin^4\vartheta\Big],\end{aligned}$$

etc., where the function $f_{\xi}^{(S)}(\vartheta)$ is given by

$$f_{\xi}^{(S)}(\vartheta) = \cosh\frac{\xi}{2S} + \cos\vartheta\sinh\frac{\xi}{2S}. \tag{3.39}$$

Moreover, for *any S*, the trigonometric series in $[f_\xi^{(S)}(\vartheta)]^{2(S-m)}\sin^{2m}\vartheta$ for the distribution $W_S(\vartheta)$ can be rewritten in compact general form as

$$W_S(\vartheta)=\frac{e^{\sigma}}{Z_S}\sum_{n=0}^{[S]}b_n\left(f_\xi^{(S)}(\vartheta)\right)^{2(S-n)}\sin^{2n}\vartheta, \tag{3.40}$$

where [*S*] means the whole part of *S* and the leading coefficients b_n are

$$b_0=1,\ b_1=\frac{S}{2}\left[e^{-(2S-1)\sigma/S^2}-1\right],\ \text{etc.}$$

The distribution $W_S(\vartheta)$ given by Eq. (3.36) is shown in Fig. 3.5 as a function of the polar angle ϑ. The maxima of $W_S(\vartheta)$ occur at $\vartheta=0$ and $\vartheta=\pi$, where

$$W_S(0)=e^{\xi+\sigma}/Z_S \text{ and } W_S(\pi)=e^{-\xi+\sigma}/Z_S, \tag{3.41}$$

respectively, meaning classically speaking that the spins are concentrated at the bottom of the wells, where the minima of the potential energy occur. In the classical limit, $S\to\infty$, $W_S(\vartheta)$ from Eq. (3.36) tends to the usual Boltzmann distribution for a uniaxial nanomagnet in a longitudinal field, i.e.,

$$\begin{gathered}(S+1/2)W_S(\vartheta)\to\frac{1}{Z_{\text{cl}}}e^{\xi\cos\vartheta+\sigma\cos^2\vartheta},\\ Z_{\text{cl}}=\int_0^{\pi}e^{\xi\cos\vartheta+\sigma\cos^2\vartheta}\sin\vartheta d\vartheta,\end{gathered} \tag{3.42}$$

where Z_{cl} is the classical partition function. Clearly from Fig. 3.5 (a), the deviations of the quantum distribution $(S+1/2)W_S(\vartheta)$ from the classical Boltzmann distribution Eq. (3.42) become pronounced only for *small* spin numbers $S<10$ while as S increases, the distribution $(S+1/2)W_S(\vartheta)$ tends to the classical expression Eq. (3.42) (e.g., for $S=20$, the differences between the two distributions Eqs. (3.40) and (3.42) do not exceed 10 percent; see the curve 5 in Fig. 3.5 (a)). Due to the biasing effect of the external field, the maxima are unequal in height. Moreover, in the low temperature limit, the dynamics of the spin in the vicinity of the maxima $\vartheta=0$ and $\vartheta=\pi$, *represent precession in the effective magnetic field* with characteristic angular frequencies ω^+ and ω^-, viz.,

$$\omega^{\pm} = \frac{1}{\beta\hbar}\left[\pm\frac{\xi}{S} + (2S-1)\frac{\sigma}{S^2}\right],$$

so that the general distribution $W_S(\vartheta)$ can be explicitly approximated by

$$W_S(\vartheta) \approx \begin{cases} Z_S^{-1} e^{-\sigma(1-S^{-1})}\left(f^{(S)}_{\beta\hbar\omega^+}(\vartheta)\right)^{2S}, & (\vartheta \le 1) \\ Z_S^{-1} e^{-\sigma(1-S^{-1})}\left(f^{(S)}_{\beta\hbar\omega^-}(\vartheta)\right)^{2S}, & (\pi - \vartheta \le 1) \end{cases} \tag{3.43}$$

For $\sigma = 0$, i.e., for *a spin in a uniform external magnetic field*, when the Hamiltonian simply becomes $\beta\hat{H}_S = -\xi\hat{S}_Z / S$, Eq. (3.36) reduces to our previous result Eq. (3.20), viz.,

$$W_S(\vartheta) = \left(f_\xi^{(S)}(\vartheta)\right)^{2S} / Z_S, \tag{3.44}$$

where Z_S is defined by Eq. (3.17). As may be seen in Fig. 3.5 (b), the "oscillator" function f from Eq. (3.44) describes with a very high degree of accuracy the behavior of $W_S(\vartheta)$ near $\vartheta = 0$ and $\vartheta = \pi$ as expected from intuitive reasoning. The above distribution Eq. (3.44) represents a *quantum analog of the Boltzmann distribution, Eq. (3.23), for classical magnetic dipoles* $\boldsymbol{\mu}$ precessing in the uniform magnetic field $\mathbf{H}_0$ with precession angular frequency $\omega_0 = \gamma H_0$.

For another particular case, viz., $\xi = 0$, i.e., *for a uniaxial spin system alone* with Hamiltonian $\beta\hat{H}_S = -\sigma\hat{S}_Z^2 / S^2$, the equilibrium phase-space distribution $W_S(\vartheta)$ from Eq. (3.36) simplifies to

$$W_S(\vartheta) = \frac{1}{Z_S}\sum_{L=0}^{2S}\frac{(2S)!(2L+1)P_L(\cos\vartheta)}{\sqrt{(2S+1)(2S-L)!(2S+L+1)!}}\sum_{m=-S}^{S} C^{Sm}_{SmL0} e^{\sigma m^2/S^2}, \tag{3.45}$$

where the partition function

$$Z_S = \sum_{m=-S}^{S} e^{\sigma m^2/S^2}. \tag{3.46}$$

Near $\vartheta = 0$ and $\vartheta = \pi/2$, the leading terms of the series expansion of the equilibrium distribution $W(\vartheta)$ from Eq. (3.45) (i.e., in $\sin^2\vartheta$ and $\cos^2\vartheta$, respectively) are [260]

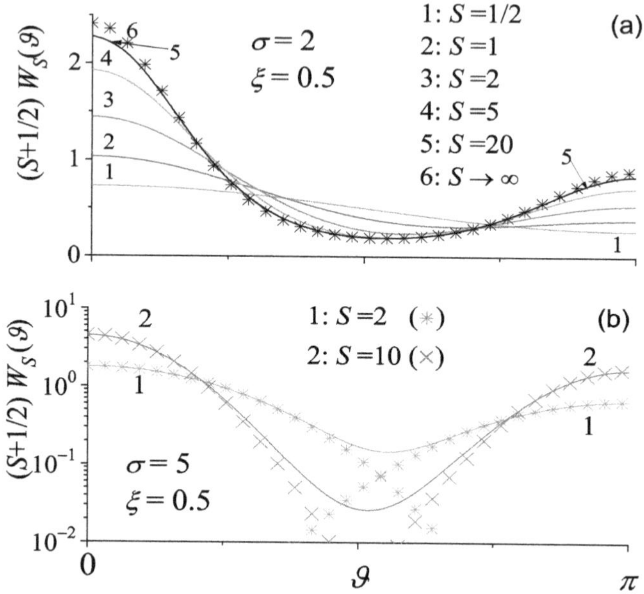

Fig. 3.5. (a) $(S+1/2)W_S(\vartheta)$ vs. ϑ for $\sigma = 2$, $\xi = 0.5$, and various values of S including the classical limit, $S \to \infty$ (∗), Eq. (3.42). (b) The distribution $(S+1/2)W_S(\vartheta)$ (solid lines) for $\sigma = 5$, $\xi = 0.5$, and $S = 2$ and 10. Crosses (×) and asterisks (∗): Eq. (3.43).

$$W_S(\vartheta) = W_S(0)\left\{1 + \frac{S}{2}\left[e^{-(2S-1)\sigma/S^2} - 1\right]\sin^2\vartheta + \ldots\right\}, \tag{3.47}$$

$$W_S(\vartheta) = W_S(\pi/2)\left[1 + A\cos^2\vartheta + \ldots\right], \tag{3.48}$$

where

$$W_S(0) = \frac{e^{\sigma}}{Z_S}, \quad W_S(\pi/2) = \frac{(2S)!}{2^{2S}Z_S}\sum_{m=-S}^{S}\frac{e^{\sigma m^2/S^2}}{(S+m)!(S-m)!},$$
$$A = \frac{\displaystyle\sum_{m=-S+1}^{S-1} e^{m^2\sigma/S^2}\,\frac{e^{(1-2m)\sigma/S^2} + e^{(1+2m)\sigma/S^2} - 2}{(S-1+m)!(S-1-m)!}}{\displaystyle 2\sum_{m=-S}^{S}\frac{e^{m^2\sigma/S^2}}{(S+m)!(S-m)!}}. \tag{3.49}$$

3.2.3 *Uniaxial nanomagnet in a transverse field*

As a further example, we calculate the Wigner function of a uniaxial nanomagnet in a *transverse* external field with Hamiltonian $\hat{H}_S$ given by Eq. (3.12) otherwise known as the Lipkin–Meshkov Hamiltonian [465,466]. For small *S*, the density matrix $\hat{\rho}_S = e^{-\beta\hat{H}_S}/Z_S$ can again be calculated in closed form using MATHEMATICA®. For consecutive $S = 1/2$, $S = 1$, etc., we have in terms of the spin operators [261]

$$\hat{\rho}_{1/2} = \frac{e^{\sigma}}{Z_{1/2}}\left[\hat{I}\cosh\xi + 2\hat{S}_X\sinh\xi\right],$$

$$\hat{\rho}_1 = \frac{1}{Z_1}\Big[(e^{\sigma} - \sigma A)\hat{I} + A(\xi\hat{S}_X + \sigma\hat{S}_Z^2) \tag{3.50}$$

$$+\left(e^{\sigma/2}\cosh\sqrt{\xi^2 + \sigma^2/4} - e^{\sigma} + \sigma A/2\right)\hat{S}_X^2\Big],$$

etc., where

$$Z_{1/2} = 2e^{\sigma}\cosh\xi,\ \ Z_1 = e^{\sigma} + 2e^{\sigma/2}\cosh\sqrt{\xi^2 + \sigma^2/4},$$

$$A = \frac{e^{\sigma/2}}{\sqrt{\xi^2 + \sigma^2/4}}\sinh\sqrt{\xi^2 + \sigma^2/4}.$$

The corresponding equations for the individual phase-space distributions $W_S(\vartheta,\varphi)$ which can be obtained from the general expressions Eqs. (3.6) and (3.7) are given by [261]

$$W_{1/2}(\vartheta,\varphi) = \frac{e^{\sigma}}{Z_{1/2}}\left[\cosh\xi + \sinh\xi\sin\vartheta\cos\varphi\right],$$

$$W_1(\vartheta,\varphi) = \frac{1}{2Z_1}\Big\{e^{\sigma}\left(1 - \sin^2\vartheta\cos^2\varphi\right) + e^{\sigma/2}\cosh\sqrt{\xi^2 + \sigma^2/4}$$

$$\times\left(1 + \sin^2\vartheta\cos^2\varphi\right) + \frac{\sigma A}{2}\left[\cos 2\vartheta + \sin^2\vartheta\cos^2\varphi + \frac{4\xi}{\sigma}\sin\vartheta\cos\varphi\right]\Big\},$$

etc. However, once again as the spin number *S* increases, the analytical equations for the Wigner function $W_S(\vartheta,\varphi)$ rapidly become more and more complicated and thus rather impractical to use instead $W_S(\vartheta,\varphi)$ for

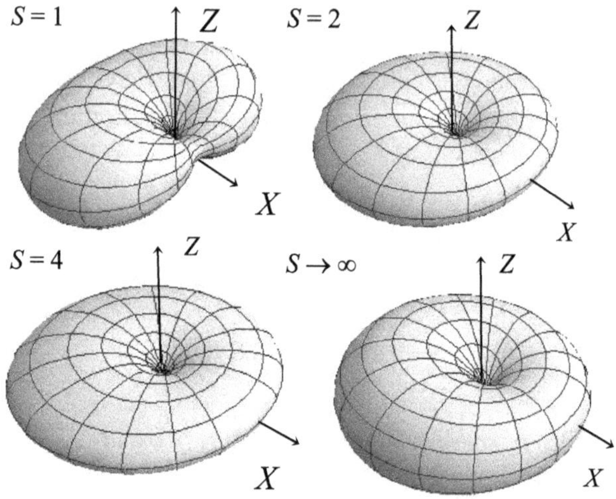

Fig. 3.6. 3D plot of the effective potential $E_{\mathrm{ef}}(\vartheta, \varphi)$ for σ=10, h =0.1 and S = 1, 2, 4, and $S\to\infty$ (classical limit).

given spin may always be calculated much faster numerically from the general series expression Eq. (3.6).

Calculations of the "effective" free energy $E_{\mathrm{ef}}(\vartheta,\varphi) = -\ln W_S(\vartheta,\varphi)$ are shown in Fig. 3.6 for various values of S and σ = 10 with $h = \xi/(2\sigma) = 0.1$. In the classical limit, $S\to\infty$ the *effective* free energy function $E_{\mathrm{ef}}(\vartheta, \varphi)$ becomes the classical free energy $E_{\mathrm{cl}}(\vartheta,\varphi)$ given by

$$E_{\mathrm{cl}}(\vartheta,\varphi) = -\sigma\left(\cos^2\vartheta + 2h\cos\varphi\sin\vartheta\right), \tag{3.51}$$

which is also shown in Fig. 3.6 for comparison. The "effective" potential $E_{\mathrm{ef}}(\vartheta, \varphi)$, like the classical free energy $E_{\mathrm{cl}}(\vartheta, \varphi)$, has two equivalent minima and one saddle point in the plane $\varphi = 0$ at $\vartheta = \pi/2$ the potential characteristics (such as the shape and barrier heights) strongly depend on S, e.g., the smallest barrier height increases with increasing S from 0 (at S = 1/2) to its classical value $\sigma(1-h^2)$.

3.2.4 *Biaxial anisotropy*

We now calculate the Wigner function of a *biaxial* Hamiltonian $\hat{H}_S$, Eq. (3.13). Here the density matrix $\hat{\rho}_S = e^{-\beta\hat{H}_S}/Z_S$ can again be calculated in

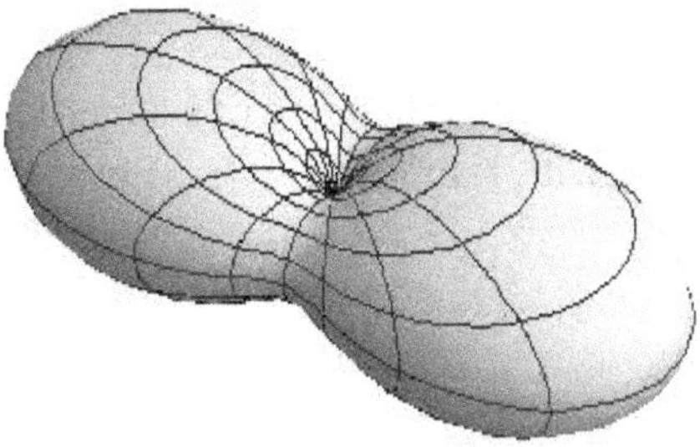

Fig. 3.7. 3D plot of $E_{ef}(\vartheta, \varphi)$ for $S = 2$, $\sigma = 5$ and $\delta = 2$.

simple closed form for small S using MATHEMATICA®. For example, for $S = 1/2$, $S = 1$, etc., we have in terms of the spin operators [261]

$$\hat{\rho}_{1/2} = \frac{e^{\sigma}}{Z_{1/2}}\hat{I}, \; \hat{\rho}_1 = \frac{1}{Z_1}\left[\hat{I} + (e^{\sigma}\cosh\delta - 1)\hat{S}_Z^2 - e^{\sigma}\sinh\delta(\hat{S}_X^2 - \hat{S}_Y^2)\right], \quad (3.52)$$

etc., where $Z_{1/2} = 2e^{\sigma}$ and $Z_1 = 1 + 2e^{\sigma}\cosh\delta$. The corresponding equations for the phase-space distributions $W_S(\vartheta,\varphi)$ are [261]

$$\begin{aligned} W_{1/2}(\vartheta,\varphi) &= 1/2, \\ W_1(\vartheta,\varphi) &= \frac{1}{2Z_1}\left[\sin^2\vartheta - e^{\sigma}\sinh\delta\sin^2\vartheta\cos 2\varphi\right. \\ &\quad \left. + e^{\sigma}\cosh\delta\left(1 + \cos^2\vartheta\right)\right], \end{aligned} \quad (3.53)$$

etc. As before with increasing S, $W_S(\vartheta,\varphi)$ may always be calculated numerically from the general series expression Eq. (3.6).

The "effective" potential $E_{ef}(\vartheta,\varphi) = -\ln W_S(\vartheta,\varphi)$ is shown in Fig. 3.7 for $S = 2$, $\sigma = 5$, and $\delta = 2$. In the classical limit, $S \to \infty$, $E_{ef}(\vartheta, \varphi)$ again tends to the classical free energy $E_{cl}(\vartheta, \varphi)$ given by

$$E_{cl}(\vartheta,\varphi) = -\sigma\cos^2\vartheta + \delta\cos 2\varphi\sin^2\vartheta.$$

The "effective" potential $E_{ef}(\vartheta, \varphi)$ [like $E_{cl}(\vartheta, \varphi)$] has two equivalent minima and two saddle points in the plane XZ at $\vartheta = \pi/2$; potential characteristics (such as the shape and barrier heights) again strongly depend on S, e.g. the barrier height once again increases with increasing S from 0 (at $S = 1/2$) to its classical value σ.

3.2.5 *Cubic anisotropy*

Finally, we calculate the Wigner function of a cubic anisotropy free energy subjected to a dc field with Hamiltonian $\hat{H}_S$ Eq. (3.14). For small S, the density matrix $\hat{\rho}_S = e^{-\beta\hat{H}_S}/Z_S$ can again be evaluated in closed form using MATHEMATICA®, e.g. for the first few spin numbers $S = 1/2$, $S = 1$, $3/2$, and 2 we have [261]

$$\hat{\rho}_{1/2} = \frac{e^{3\sigma_c/2}}{Z_{1/2}}\left(\hat{I}\cosh\xi + 2\hat{S}_Z\sinh\xi\right),$$

$$\hat{\rho}_1 = \frac{e^{\sigma_c}}{Z_1}\left[\hat{I} + \sinh\xi\hat{S}_Z + 2\sinh^2\frac{\xi}{2}\hat{S}_Z^2\right],$$

$$\hat{\rho}_{3/2} = \frac{e^{41\sigma_c/54}}{Z_{3/2}}\left\{\frac{1}{8}\left(9\cosh\frac{\xi}{3} - \cosh\xi\right)\hat{I} + \frac{1}{12}\left(27\sinh\frac{\xi}{3} - \sinh\xi\right)\hat{S}_Z\right.$$
$$\left. + 2\cosh\frac{\xi}{3}\sinh^2\frac{\xi}{3}\hat{S}_Z^2 + \frac{4}{3}\sinh^3\frac{\xi}{3}\hat{S}_Z^3\right\},$$

$$\hat{\rho}_2 = \frac{1}{Z_2}\left\{\left(e^{3\sigma_c/4} - \frac{3\sigma_c}{4\xi}R\right)\hat{I} + \frac{1}{6}\left(8e^{9\sigma_c/16}\sinh\frac{\xi}{2} - R\right)\hat{S}_Z\right.$$
$$+\frac{1}{12}\left(16e^{9\sigma_c/16}\cosh\frac{\xi}{2} - 15e^{3\sigma_c/4} - P - \frac{93\sigma_c}{32\xi}R\right)\hat{S}_Z^2 - \frac{1}{6}\left(2e^{9\sigma_c/16}\sinh\frac{\xi}{2} - R\right)\hat{S}_Z^3$$
$$\left. + \frac{1}{12}\left(3e^{3\sigma_c/4} - 4e^{9\sigma_c/16}\cosh\frac{\xi}{2} + P - \frac{9\sigma_c}{32\xi}R\right)\hat{S}_Z^4 + \frac{\sigma_c}{32\xi}R\left(\hat{S}_X^4 + \hat{S}_Y^4\right)\right\},$$

where

$$Z_{1/2} = 2e^{3\sigma_c/2}\cosh\xi,\ Z_1 = e^{\sigma_c}(1 + 2\cosh\xi),$$

$$Z_{3/2} = 4e^{\frac{41}{54}\sigma_c}\cosh\frac{\xi}{3}\cosh\frac{2\xi}{3},\ Z_2 = e^{\frac{3}{4}\sigma_c} + 2e^{\frac{9}{16}\sigma_c}\cosh\frac{\xi}{2} + 2P,$$

$$P = e^{\frac{21}{32}\sigma_c}\cosh\sqrt{\xi^2 + \left(\frac{3\sigma_c}{32}\right)^2},\ R = \frac{\xi e^{\frac{21}{32}\sigma_c}}{\sqrt{\xi^2 + (3\sigma_c/32)^2}}\sinh\sqrt{\xi^2 + \left(\frac{3\sigma_c}{32}\right)^2}.$$

The corresponding equations for the Wigner function $W_S(\vartheta,\varphi)$ are [261]

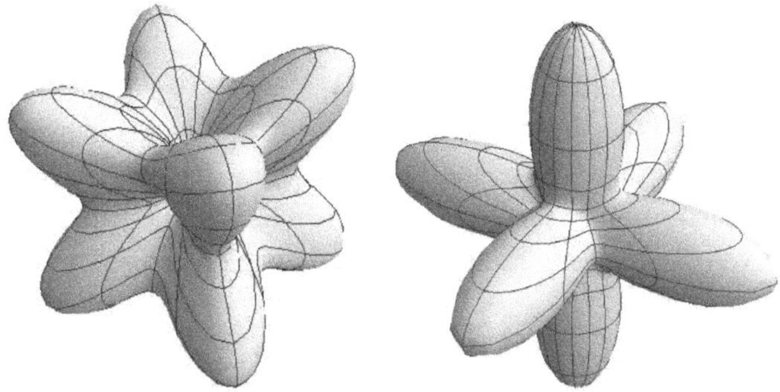

Fig. 3.8. 3D plot of $E_{ef}(\vartheta, \varphi)$ for positive (left) and negative (right) cubic anisotropies for $S = 4$ and $\xi = 0$.

$$W_{1/2}(\vartheta,\varphi) = Z_{1/2}^{-1} e^{\frac{3}{2}\sigma_c}\left[\cosh(\xi/2) + \sinh(\xi/2)\cos\vartheta\right],$$

$$W_1(\vartheta,\varphi) = Z_1^{-1} e^{\sigma_c}\left[\cosh(\xi/2) + \sinh(\xi/2)\cos\vartheta\right]^2,$$

$$W_{3/2}(\vartheta,\varphi) = Z_{3/2}^{-1} e^{\frac{41}{54}\sigma_c}\left[\cosh(\xi/2) + \sinh(\xi/2)\cos\vartheta\right]^3,$$

$$W_2(\vartheta,\varphi) = \frac{1}{8Z_2}\Big\{2e^{\frac{9}{16}\sigma_c}\sin^2\vartheta\left[4\cos\vartheta\sinh(\xi/2) + (3+\cos 2\vartheta)\cosh(\xi/2)\right]$$
$$+3e^{3\sigma_c/4}\sin^4\vartheta + P(8\cos^2\vartheta + \sin^4\vartheta) +$$
$$4R\left(\cos\vartheta + \cos^3\vartheta + \frac{3\sigma}{128\xi}\cos 4\varphi \sin^4\vartheta\right)\Big\}. \tag{3.54}$$

For larger S, $W_S(\vartheta,\varphi)$ may always be calculated numerically from the general series expression Eq. (3.6).

The normalized "effective" potential $E_{ef}(\vartheta,\varphi) = -\ln W_S(\vartheta,\varphi)$ is shown in Fig. 3.8 for $\sigma_c = \pm 8$, $\xi = 0$ and $S = 4$. The potential characteristics (such as the shape and barrier heights) again strongly depend on S. In the classical limit, $S \to \infty$, that potential once more tends to the classical free energy $E_{cl}(\vartheta, \varphi)$ given by (cf. Eq. (1.15))

$$E_{cl}(\vartheta,\varphi) = \frac{\sigma_c}{4}\left(\sin^2 2\vartheta + \sin^4\vartheta\sin^2 2\varphi\right) - \xi\cos\vartheta.$$

For positive anisotropy constant $\sigma_c > 0$ the cubic potential has 6 minima (wells), 8 maxima, and 12 saddle points. For $\sigma_c < 0$ the maxima and minima are interchanged.

Clearly the Wigner-Stratonovich transformation yields in principle the equilibrium phase-space distribution via its finite series representation for any given anisotropy free energy. These results may be used to estimate the spin reversal time from transition state theory as was accomplished for particles in Section 1.8.3.

3.2.6 *TST reversal time*

As we have seen in Section 1.8.3, TST because it is based on *equilibrium distributions* affords the simplest possible description of quantum corrections to the thermally activated escape rate. Now in applying TST to classical spins (i.e., classical magnetic moments $\boldsymbol{\mu}$) to determine the escape rate (for arbitrary free energy) due to thermal agitation from one metastable orientation say A to another metastable orientation say B, we suppose that the free energy $E_{\mathrm{cl}}(\vartheta, \varphi)$ has a multistable structure. Such a structure has minima at $\mathbf{n}_A$ and $\mathbf{n}_B$ separated by a potential barrier with a saddle point at $\mathbf{n}_C$. In the high barrier approximation, as far as TST is concerned, the classical escape rate Γ_{cl} may be estimated via the flux-over-population [85] (see also Chapter 1, Section 1.5.1)

$$\Gamma_{\mathrm{cl}} = I_C^{\mathrm{cl}} / Z_A^{\mathrm{cl}}, \tag{3.55}$$

where the well partition function Z_A^{cl} and the total current I_C^{cl} of the (spin) representative points at the saddle point C are, respectively,

$$Z_A^{\mathrm{cl}} \approx \iint_{well} e^{-E_{\mathrm{cl}}(\vartheta,\varphi)} \sin\vartheta d\vartheta d\varphi \tag{3.56}$$

and

$$I_C^{\mathrm{cl}} \approx \iint_{\mathrm{saddle}} J_C(\vartheta,\varphi) \sin\vartheta d\vartheta d\varphi \tag{3.57}$$

[$J_C(\vartheta,\varphi)$ is the current density near the saddle point C]. Near the metastable minimum $\mathbf{n}_A$, the spin precesses about a *uniform* "effective" field which may be represented as the gradient of a potential, viz.,

$$\mathbf{H}_A = -\frac{1}{\beta\mu_0}\frac{\partial E_{\rm cl}}{\partial \boldsymbol{\mu}}, \tag{3.58}$$

so that the equation of motion of the magnetic moment $\boldsymbol{\mu}$ is

$$\frac{d\boldsymbol{\mu}}{dt} = -\gamma[\boldsymbol{\mu}\times\mathbf{H}_A]. \tag{3.59}$$

Then to evaluate the classical partition function $Z_A^{\rm cl}$ and the total current $I_C^{\rm cl}$, we simply suppose as in Chapter 1 that the free energy $E_{\rm cl}$ near the minimum $\mathbf{n}_A$ and the saddle point $\mathbf{n}_C$ can be approximated by the first two terms of its Taylor expansion, viz., [18,97,239]

$$E_{\rm cl}(u_1^{(A)},u_2^{(A)}) \approx E_{cl}(\mathbf{n}_A) + \frac{1}{2}\left[c_1^{(A)}\left(u_1^{(A)}\right)^2 + c_2^{(A)}\left(u_2^{(A)}\right)^2\right] \tag{3.60}$$

and

$$E_{\rm cl}(u_1^{(C)},u_2^{(C)}) \approx E_{cl}(\mathbf{n}_C) + \frac{1}{2}\left[c_1^{(C)}\left(u_1^{(C)}\right)^2 + c_2^{(C)}\left(u_2^{(C)}\right)^2\right], \tag{3.61}$$

where $(u_1^{(A)},u_2^{(A)},u_3^{(A)})$ and $(u_1^{(C)},u_2^{(C)},u_3^{(C)})$ denote the direction cosines of the magnetic moment vector $\boldsymbol{\mu}$ near $\mathbf{n}_A$ and $\mathbf{n}_C$, $c_1^{(A)} = \partial^2 E_{cl}/\partial u_1^{(A)2} > 0$ and $c_2^{(A)} = \partial^2 E_{cl}/\partial u_2^{(A)2} > 0$, i.e., the well has the form of an *elliptic* paraboloid, while $c_1^{(C)} = \partial^2 E_{cl}/\partial u_1^{(C)2} < 0$ and $c_2^{(C)} = \partial^2 E_{cl}/\partial u_2^{(C)2} > 0$, i.e., the saddle has the form of an *hyperbolic* paraboloid. Hence, in order to estimate $Z_A^{\rm cl}$ from Eq. (3.56) in the high barrier limit, we have via Gaussian integrals using the Taylor expansion Eq. (3.60) the well partition function

$$\begin{aligned} Z_A^{\rm cl} &\approx \iint_{\rm well} e^{-E_{\rm cl}(u_1^{(A)},u_2^{(A)})}du_1^{(A)}du_2^{(A)} \approx \int_{-\infty}^{\infty}\int_{-\infty}^{\infty} e^{-E_{\rm cl}(u_1^{(A)},u_2^{(A)})}du_1^{(A)}du_2^{(A)} \\ &= \frac{2\pi e^{-E_{\rm cl}(\mathbf{n}_A)}}{\sqrt{c_1^{(A)}c_2^{(A)}}} = \frac{2\pi\gamma e^{-E_{\rm cl}(\mathbf{n}_A)}}{\beta\mu\mu_0\omega_A}, \end{aligned} \tag{3.62}$$

where

$$\omega_A = \frac{\gamma}{\beta\mu_0\mu}\sqrt{c_1^{(A)}c_2^{(A)}} \tag{3.63}$$

is the well (precession) frequency playing the role of the attempt angular frequency in TST [85]. Here the limits of integration may as usual be formally extended to ± infinity without significant error since the spins are almost all sited at the well bottom $\mathbf{n}_A$. The total current I_C^{cl} of representative points at the saddle point $\mathbf{n}_C$ may then be estimated as follows. Having supposed that the saddle region has the shape of a hyperbolic paraboloid we further suppose that the u_1-axis of the local coordinate system at the saddle point $\mathbf{n}_C$ *lies in the same direction as the current density* $\mathbf{J}_C$ *over the saddle.* Now in TST, the Boltzmann distribution $\sim e^{-E_{\text{cl}}}$ holds *everywhere* so that the current density J_C at the saddle point C is given by [18]

$$\begin{aligned} J_C(u_1^{(C)}, u_2^{(C)}) &= -\frac{\gamma}{\beta\mu_0\mu}\delta\left(u_1^{(C)}\right)\frac{\partial E_{\text{cl}}}{\partial u_2^{(C)}} e^{-E_{\text{cl}}\left(u_1^{(C)}, u_2^{(C)}\right)} \\ &= \frac{\gamma}{\beta\mu_0\mu}\delta\left(u_1^{(C)}\right)\frac{\partial}{\partial u_2^{(C)}} e^{-E_{\text{cl}}\left(u_1^{(C)}, u_2^{(C)}\right)}, \end{aligned} \tag{3.64}$$

where $\delta(u)$ is the Dirac-delta function (see Appendix A.1.10). Thus, we must have for the current at the saddle point

$$I_C^{\text{cl}} \approx \iint_{\text{saddle}} J_C(u_1^{(C)}, u_2^{(C)}) du_1^{(C)} du_2^{(C)} = \frac{\gamma}{\mu\mu_0\beta} e^{-E_{\text{cl}}(\mathbf{n}_C)}. \tag{3.65}$$

Hence, using Eqs. (3.62) and (3.65), the flux-over-population Eq. (3.55) yields the *classical* TST formula for spins

$$\Gamma_{\text{cl}} = \frac{\omega_A}{2\pi} e^{-\Delta E_{\text{cl}}}, \tag{3.66}$$

where

$$\Delta E_{\text{cl}} = E_{\text{cl}}(\mathbf{n}_C) - E_{\text{cl}}(\mathbf{n}_A) \tag{3.67}$$

is the potential barrier height which always requires a detailed knowledge of the energy landscape.

In like manner, the *quantum* escape rate Γ for a spin from a metastable orientation A to another metastable orientation B via the saddle point C as determined by quantum TST may be rendered by an equation similar to the classical Eq. (3.55), viz.,

$$\Gamma = \frac{I_C}{Z_A}. \tag{3.68}$$

However, the well quantum partition function Z_A and the total current over the saddle point I_C *must now be evaluated using the equilibrium phase-space distribution function* $W_S(\vartheta,\varphi)$ *of the spin system with the quantum spin Hamiltonian* $\hat{H}_S$ instead of the classical Boltzmann distribution $\sim e^{-E_{\mathrm{cl}}(\vartheta,\varphi)}$. Nevertheless, the dynamics of a spin $\hat{\mathbf{S}}$ still comprise steady precession of angular frequency $\boldsymbol{\omega}_A^S = \gamma \mathbf{H}_A^S$ in the *effective* magnetic field $\mathbf{H}_A^S$ in the well *near* the metastable minimum $\mathbf{n}_A$. Thus the spin Hamiltonian $\hat{H}_S$ may be approximated by the simple equation $\hat{H}_S \approx -\hbar(\boldsymbol{\omega}_A^S \cdot \hat{\mathbf{S}})$, so that the quantum dynamics of the spin $\hat{\mathbf{S}}$ obey the Larmor equation [2,37] [cf. Eq. (3.59)]

$$\frac{d}{dt}\hat{\mathbf{S}} = \frac{i}{\hbar}\left[\hat{H}_S, \hat{\mathbf{S}}\right] = -\gamma\left[\hat{\mathbf{S}} \times \mathbf{H}_A^S\right]. \tag{3.69}$$

Furthermore, near the metastable minimum $\mathbf{n}_A$, the phase-space distribution W_S can be approximated simply by the Zeeman energy distribution for a spin in a *uniform* "effective" field $\mathbf{H}_A^S$ given by Eq. (3.27), viz., [cf. Eq. (3.43) for the uniaxial nanomagnet]

$$W_S(\vartheta,\varphi) \approx W_S(\mathbf{n}_A)e^{-S\xi_A}\left[\cosh\frac{\xi_A}{2} + F_A(\vartheta,\varphi)\sinh\frac{\xi_A}{2}\right]^{2S}, \tag{3.70}$$

where $\xi_A = \beta\hbar\left|\boldsymbol{\omega}_A^S\right|$ in accordance with Eq. (3.27),

$$F_A(\vartheta,\varphi) = \left(\gamma_{X_A}\cos\varphi + \gamma_{Y_A}\sin\varphi\right)\sin\vartheta + \gamma_{Z_A}\cos\vartheta,$$

and γ_X, γ_Y, γ_Z are the direction cosines of the "effective" field $\mathbf{H}_A^S$ at the minimum $\mathbf{n}_A$. This effective field distribution approximation Eq. (3.70) is just the *rotational analog of the harmonic oscillator distribution approximation for the well dynamics in the particle case* (see Chapter 1, Section 1.9.3). Like the classical case, the precession frequency $\boldsymbol{\omega}_A^S$ can be estimated from the well angular frequency Eq. (3.63), however, the coefficients $c_1^{(A)}$ and $c_2^{(A)}$ must now be determined from the truncated *Taylor series expansion of the Weyl symbol* H_S of the spin Hamiltonian $\hat{H}_S$, viz.,

$$\beta H_S = \beta H_S(\mathbf{n}_A) + \frac{1}{2}\left[c_1^{(A)}\left(u_1^{(A)}\right)^2 + c_2^{(A)}\left(u_2^{(A)}\right)^2\right]. \tag{3.71}$$

Then by using the partition function for spins in a uniform dc field, Eq. (3.29), we can approximate the well partition function as

$$\begin{aligned} Z_A &\approx W_S(\mathbf{n}_A)e^{-S\xi_A}\int_{\text{well}}\left[\cosh\frac{\xi_A}{2} + \sinh\frac{\xi_A}{2}F_A(\vartheta,\varphi)\right]^{2S}\sin\vartheta d\vartheta d\varphi \\ &= 4\pi W_S(\mathbf{n}_A)e^{-S\xi_A}\frac{\sinh[(S+1/2)\xi_A]}{(2S+1)\sinh(\xi_A/2)} = \frac{2\pi[1-e^{-(2S+1)\xi_A}]}{(S+1/2)(1-e^{-\xi_A})}W_S(\mathbf{n}_A). \end{aligned} \tag{3.72}$$

Finally, the spin current I_C from the well may be estimated in the manner of the classical Eq. (3.65) by defining the current density J_C at the saddle point $\mathbf{n}_C$ of the "effective" potential $E_{\text{ef}}(u_1^{(C)}, u_2^{(C)}) = -\ln W_S(u_1^{(C)}, u_2^{(C)})$. Thus, we have for the spin current [cf. Eq. (3.65)]

$$I_C \approx \frac{W_S(\mathbf{n}_C)}{\beta\hbar S}, \tag{3.73}$$

since the magnetic moment is now given by $\mu = \gamma\hbar S/\mu_0$. Hence, we obtain the quantum TST escape rate as determined by Eqs. (3.68), (3.72), and (3.73), viz.,

$$\Gamma = \frac{(S+1/2)(1-e^{-\xi_A})}{2\pi\beta\hbar S[1-e^{-(2S+1)\xi_A}]}\frac{W_S(\mathbf{n}_C)}{W_S(\mathbf{n}_A)}. \tag{3.74}$$

To compare this equation with the classical TST Eq. (3.66), we rewrite it in the form of the quantum TST Eq. (3.66) as done in Chapter 1 for particles, viz.,

$$\Gamma = \frac{\omega_A}{2\pi}\Xi_S e^{-\Delta E_{\text{cl}}} = \Xi_S\Gamma_{\text{cl}}, \tag{3.75}$$

where

$$\Xi_S = \frac{\omega_A^S(S+1/2)\left(1-e^{-\xi_A}\right)}{\omega_A\xi_A S\left[1-e^{-(2S+1)\xi_A}\right]}\frac{W_S(\mathbf{n}_C)}{W_S(\mathbf{n}_A)}e^{\Delta E_{\text{cl}}} \tag{3.76}$$

represents the *quantum correction* factor which strongly depends on the spin number S yielding $\Xi_S \to 1$ and $\Gamma \to \Gamma_{\text{cl}}$ in the classical limit, $S \to \infty$.

We study as a particular example of the explicit calculation of Ξ_S a uniaxial nanomagnet with Hamiltonian $\beta\hat{H}_S = -\sigma\hat{S}_Z^2 / S^2$, where the Weyl symbol $\beta H_S(\vartheta)$ and equilibrium phase-space distribution $W_S(\vartheta)$ are given by Eqs. (1.596) and (3.36), respectively. Here $W_S(\mathbf{n}_A) = W_S(\vartheta = 0)$ and $W_S(\mathbf{n}_C) = W_S(\vartheta = \pi/2)$ are given by Eq. (3.49). In the classical limit, we have

$$\omega_A = \frac{2\sigma\gamma}{\mu_0\mu\beta} \text{ and } \Delta E_{\text{cl}} = \sigma. \tag{3.77}$$

Moreover, for our uniaxial nanomagnet $\omega_A^S = \omega_A\left[1-(2S)^{-1}\right]$, so that the quantum correction factor Ξ_S is explicitly given by

$$\Xi_S = \frac{(S+1/2)[1-e^{-\sigma(2S-1)/S^2}]}{\sigma 2^{2S+1}[1-e^{-\sigma(4S^2-1)/S^2}]}\sum_{m=-S}^{S}\frac{(2S)!e^{\sigma m^2/S^2}}{(S+m)!(S-m)!}. \tag{3.78}$$

For this particular case in the limit $S \to \infty$, Eq. (3.75) reduces to the well-known classical TST (Néel) formula for a *uniaxial nanomagnet*, viz.,

$$\Gamma_{\text{cl}} = \left(2\pi\tau_0\right)^{-1} e^{-\sigma}, \tag{3.79}$$

where $\tau_0 = 1/\omega_A$ is a normalizing time constant. The normalized inverse escape rate $(\tau_0\Gamma)^{-1}$ as a function of the inverse-temperature parameter $\sigma \sim 1/T$ is shown in Fig. 3.9 for various values of S. Clearly, the *qualitative behavior of the quantum escape rates* Γ *for finite* S *strongly deviates from the Arrhenius behavior of the classical escape rates* Γ_{cl} *at low temperatures*. This difference is due to *the tunneling effect*.

Like the classical case (having evaluated the escape rate Γ for a given anisotropy) we also have the reversal time at finite temperatures. By *equating the reversal time to the measuring time of a switching time experiment* one may determine the switching-field curves at *finite* temperatures in the manner of the classical theory [97]. Although TST always implies that the dissipation to the bath does not affect the escape rate at all, nevertheless, the results should still apply in a *wide range of dissipation*. The latter may be defined as wide enough to ensure that thermal noise is sufficiently strong to thermalize the escaping system yet not so wide as to disturb the thermal equilibrium in the well, i.e., an equilibrium distribution still prevails everywhere including the saddle

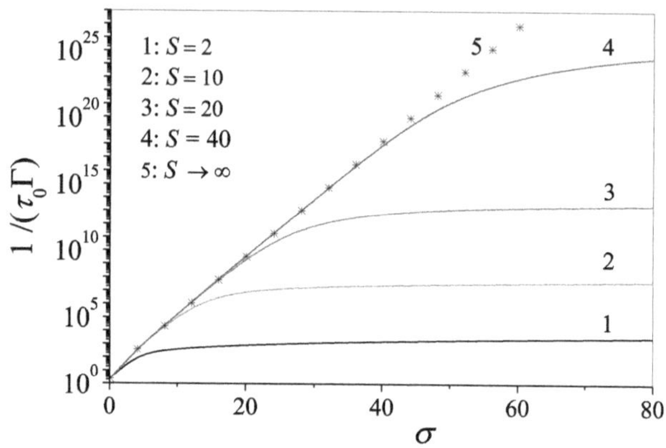

Fig. 3.9. Normalized inverse escape rate $(\tau_0\Gamma)^{-1}$ as a function of the inverse-temperature parameter $\sigma \sim 1/T$ for various values of $S = 2$, 10, 20, 40, and $S \to \infty$ (classical limit).

point region. In classical Kramers escape-rate theory [239], this represents the so-called *intermediate damping* case.

We shall now demonstrate how the phase-space representation for a given spin Hamiltonian may be used to calculate switching-field curves and/or surfaces as a function of spin number S at zero temperature.

3.2.7 *Switching-field curves*

We recall that the first calculation of the magnetization reversal of single-domain ferromagnetic particles with uniaxial anisotropy subjected to an applied field was made by Stoner and Wohlfarth [14] with the hypothesis of *coherent rotation* of the magnetization and *zero temperature* so that *thermally induced switching between the potential minima is ignored*. In the simplest uniaxial anisotropy as considered by them, the magnetization reversal consequently occurs at that value of the applied field (called the *switching* or *nucleation* field) which destroys the bistable nature of the potential. The parametric plot of the parallel vs. the perpendicular component of the switching field then yields the famous astroids. As seen, Thiaville [71] later developed a universal geometrical method for the calculation of the energy of a particle allowing one to determine the switching field for all values of the applied magnetic field thereby yielding the critical switching-field surface analogous to the Stoner-Wohlfarth astroids (see Chapter 1, Section 1.6). This surface, as it generalizes the critical curves of the 2D problem of Stoner and Wohlfarth [14], is called the *limit of metastability surface*. In the general approach to the calculation of switching curves via the geometrical method of Thiaville [71], these curves or surfaces may be constructed for particles with *arbitrary* anisotropy at zero temperature. By fitting experimental switching-field curves and surfaces, one can then determine the free energy of a nanomagnet and the corresponding anisotropy constants (see Fig. 1.19).

To generalize Thiaville's geometrical method [71] (and the Stoner-Wohlfarth calculations) to include quantum effects in switching-field curves and surfaces of a spin system with a model spin Hamiltonian $\hat{H}_S$, we must first determine the *Weyl symbol* $H_S(\vartheta,\varphi)$ corresponding to $\hat{H}_S$, which is defined by the bijective map onto phase space given by Eq. (1.595). Then one may, in principle, again calculate the switching fields using Thiaville's method [71] (see Eqs. (1.322)–(1.325)). Numerous examples of the calculation of switching-field surfaces in 3D space and the 2D critical curves for various *classical* free energy densities have been given in Chapter 1, Section 1.6 and, e.g., in Refs. [70,71,97,239]. Here, following [261], we calculate 2D critical curves for the uniaxial, biaxial, cubic, and mixed Hamiltonians as defined by Eqs. (1.591)–(1.594). The Weyl symbols $H_S^{\mathrm{un}}(\vartheta,\varphi)$, $H_S^{\mathrm{bi}}(\vartheta,\varphi), H_S^{\mathrm{cub}}(\vartheta,\varphi)$, and $H_S^{\mathrm{mix}}(\vartheta,\varphi)$ of these four model Hamiltonians are given by Eqs. (1.596)–(1.599), which bear an obvious resemblance to the corresponding classical free energy densities (see Chapters 1 and 2). If we further suppose that a uniform external magnetic field **H** is applied in the *x*–*z* plane, the Zeeman term operator $-(\xi / S)\left(\sin\psi \hat{S}_X - \cos\psi \hat{S}_Z\right)$ just transforms to the simple phase-space expression $-\xi\cos(\vartheta-\psi)$, where ψ is the angle between the applied field **H** and the *Z* axis. Thus, the switching fields $\mathbf{h}_{\mathrm{un}}$, $\mathbf{h}_{\mathrm{bi}}$, and $\mathbf{h}_{\mathrm{cub}}$ in the *x*–*z* plane (i.e., for $\varphi = 0$) can be directly calculated from the Weyl symbols Eqs. (1.596)–(1.599) yielding

$$\begin{aligned} \mathbf{h}_{\mathrm{un}} &= Q_{\mathrm{un}}\mathbf{h}_{\mathrm{un}}^{\mathrm{cl}}, \\ \mathbf{h}_{\mathrm{bi}} &= Q_{\mathrm{bi}}\mathbf{h}_{\mathrm{bi}}^{\mathrm{cl}}, \\ \mathbf{h}_{\mathrm{cub}} &= Q_{\mathrm{cub}}\mathbf{h}_{\mathrm{cub}}^{\mathrm{cl}}, \end{aligned} \tag{3.80}$$

where

$$Q_{\text{un}} = Q_{\text{bi}} = 1-(2S)^{-1},\ Q_{\text{cub}} = (S-1/2)(S-1)(S-3/2)S^{-3} \quad (3.81)$$

are the quantum correction factors to the corresponding classical switching fields $\mathbf{h}_{\text{un}}^{\text{cl}}$, $\mathbf{h}_{\text{bi}}^{\text{cl}}$, and $\mathbf{h}_{\text{cub}}^{\text{cl}}$ in the *x-z* plane given by the known equations [97,239] (see Chapter 1, Section 1.6.1)

$$\begin{aligned}\mathbf{h}_{\text{un}}^{\text{cl}} &= \left(\sin^3\vartheta, -\cos^3\vartheta\right),\\ \mathbf{h}_{\text{cub}}^{\text{cl}} &= \left(\sin^3\vartheta(3\cos 2\vartheta+2), \cos^3\vartheta(3\cos 2\vartheta-2)\right),\\ \mathbf{h}_{\text{bi}}^{\text{cl}} &= \left(\sin\vartheta[2\Delta+f(\vartheta)], \cos\vartheta[2-f(\vartheta)]\right),\end{aligned} \quad (3.82)$$

where $\Delta = \delta/\sigma$ and $f(\vartheta) = (1+\Delta)\sin^2\vartheta + \left|2\Delta-(1+\Delta)\sin^2\vartheta\right|$. For *mixed* anisotropy, however, the corresponding equation for the switching field $\mathbf{h}_{\text{mix}}$ is much more complicated and therefore must be calculated numerically. The parametric plots of the parallel h_Z vs. the perpendicular h_X component of the switching field for the above spin systems are shown in Fig. 3.10. In general, the figure indicates that the switching field amplitudes *increase* markedly with *increasing S* all the while tending to their classical limiting values as $S \to \infty$ corresponding to *diminishing tunneling effects* as that mechanism is gradually shut off.

We emphasize that the above calculations because they are entirely based on the phase-space representation of the Hamiltonian operator *ignore thermal effects* as in the original Stoner–Wohlfarth and Thiaville calculations. To account for these, it is necessary to estimate the temperature dependence of the spin reversal time, which may be accomplished, e.g., using the quantum TST, which we have described in Section 3.2.6. This will again only involve the quantum equilibrium phase-space distributions, which we have calculated in the preceding sections.

We have just shown how the phase-space method generates equilibrium distribution functions in the configuration space of polar angles (ϑ, φ) for spin systems in the equilibrium state described by the equilibrium distribution $\hat{\rho}_S = e^{-\beta\hat{H}_S}/Z_S$. As seen the Wigner function may be represented in all cases as a finite series of spherical harmonics like the corresponding classical orientational distribution and transparently reduces to the usual Fourier series Eq. (1.538) in the classical limit, $S \to \infty$

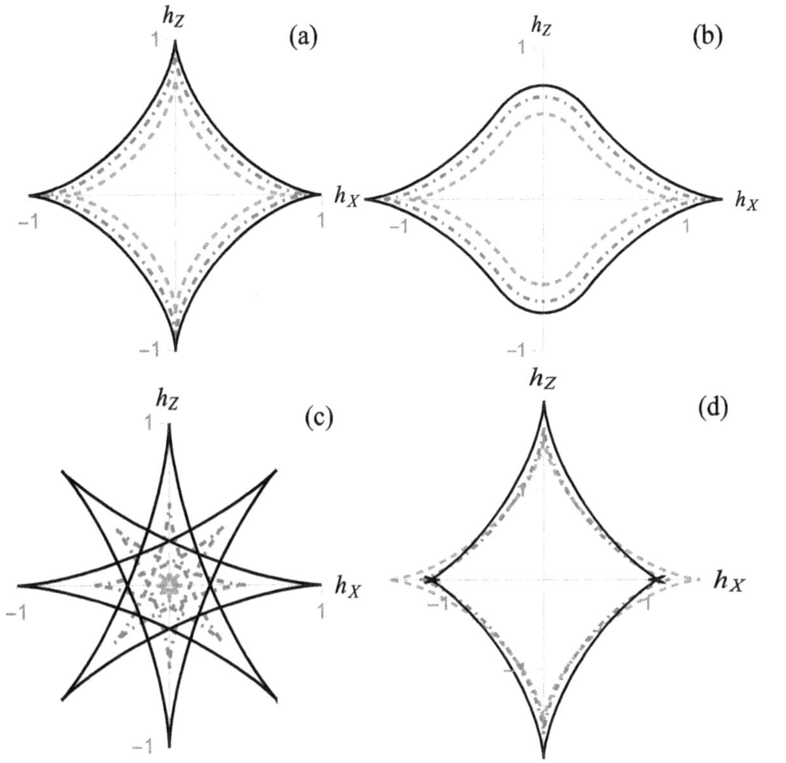

Fig. 3.10. Spin dependence of switching field curves for uniaxial (a), biaxial at $\sigma/\delta = 0.25$ (b), cubic (c), and mixed at $\sigma_1/\sigma_2 = 0.5$ and $\chi = 0$ (d) anisotropies for $S = 2$ (dashed lines), 5 (dash-dotted lines), and $S \to \infty$ (solid lines; classical limit).

Moreover, relevant quantum mechanical averages (such as the magnetization) may be calculated in a manner analogous to the corresponding classical averages using the Weyl symbol of the appropriate quantum operator [see the general mapping Eq. (1.581)]. The resulting Wigner functions can now be used to determine the spin dependence of the switching fields and hysteresis curves and may also be applied to other problems requiring only a knowledge of equilibrium distributions. This conclusion is significant particularly from an experimental point of view as *the transition between magnetic molecular cluster and single-domain ferromagnetic nanoparticle behavior is essentially demarcated via the hysteresis loops and the corresponding switching-field curves* [20]. Furthermore, these Wigner functions are important, in the interpretation of quantum tunneling phenomena in ferromagnetic nanoparticles and molecular magnets (see, e.g., [20]) and also in the investigation of the crossover region between reversal of the magnetization of these particles by thermal agitation and reversal by macroscopic quantum tunneling. For instance, by analogy with Néel's classical calculation [16] the simplest description of quantum effects in the magnetization reversal time of a nanoparticle is provided by the inverse escape rate from the wells of the magnetocrystalline and external field potential as rendered by quantum TST. Thus, the TST rate provides an important benchmark for both analytical calculations of the escape rate, which account for dissipation using quantum rate theory and for the numerical results obtained from the appropriate quantum master equation as well as allowing one to incorporate thermal effects in the switching fields. As seen TST ignores the disturbance to the equilibrium distribution in the wells created by the loss of the magnetization due to escape over the barrier and so involves the equilibrium distribution only in so far as that is assumed to prevail everywhere. Nevertheless, via *TST as corrected for quantum effects [e.g., Eq.* (3.75)*] which stems from the phase-space representation it is possible to predict the temperature dependence of the switching fields and corresponding hysteresis loops* within the limitations imposed by that theory. Therefore, the results should be relevant to experiments seeking evidence for macroscopic quantum tunneling where the *temperature dependence* of the loops is crucial as they are used [20] to differentiate tunneling from thermal agitation behavior. The equilibrium quantum distribution is also essential in the inclusion of *nonequilibrium* effects in the quantum escape rate. For example, a master equation describing the time evolution of the quasiprobability density in the representation space is required in generalizing the classical escape-rate calculations initiated by Kramers [84] for point particles and by Brown [17,18] for single-domain ferromagnetic particles using the Fokker–Planck equation.

In Section 3.3, we shall show that a knowledge of the equilibrium phase-space distribution is also important in two other fundamental aspects, i.e., nonequilibrium phenomena involving master equations in the phase-space representation. The first is in formulating the initial conditions for their solution as the appropriate quantum equilibrium distribution, which must now play the role of the Boltzmann distribution in the corresponding classical problem. Secondly, the equilibrium quantum distribution (as seen for point particles in Chapter 1) plays a vital role in the determination of the diffusion coefficients in a quantum master equation because this distribution must be the stationary solution of that equation. This fact, analogous to Einstein and Smoluchowski's imposition of the Maxwell–Boltzmann distribution as the stationary solution of the Fokker–Planck equation in order to determine drift and diffusion coefficients, will also allow one to calculate these coefficients in the quantum case. This is illustrated for the case of a spin in a uniform field in Section 3.3.5 (see also [256,257]) indicating clearly how all the solution methods developed for the classical Fokker–Planck equation apply to the quantum case just as the corresponding solutions for particles [272] (see Section 1.8.4). However, the calculation of the drift and diffusion coefficients for single variable potentials is much simpler than the general case since only the single polar angle ϑ is involved rather than the two angles ϑ and φ. The restriction to a single variable leads to further mathematical simplifications, since the quantum master equation now has essentially the same form as the classical Fokker–Planck equation in the single coordinate ϑ implying that formulas for the mean first-passage time, integral relaxation time, etc., obtained by quadratures may be directly carried over to the quantum case. This is not so for potentials, where two variables are involved now posing a problem like that encountered in solving the Wigner problem for particles in classical phase space as we have shown in Chapter 1.

3.3 Master equation in phase space for circularly symmetric systems

3.3.1 *Master equation for a uniaxial nanomagnet subjected to a dc magnetic field*

We shall now apply, as an illustrative example, the phase-space method to a *uniaxial* nanomagnet of arbitrary spin number S in an external dc magnetic field $\mathbf{H}_0$ applied along the Z-axis, i.e., the easy axis, where the Hamiltonian operator $\hat{H}_S$ has the form of Eq. (3.11), namely,

$$\beta\hat{H}_S = -\frac{\sigma}{S^2}\hat{S}_Z^2 - \frac{\xi}{S}\hat{S}_Z. \tag{3.83}$$

This Hamiltonian comprises a uniaxial anisotropy term $-\sigma\hat{S}_Z^2 / S^2$ plus the Zeeman coupling to the external field $-\xi\hat{S}_Z / S$, constituting a generic model for relaxation phenomena in uniaxial spin systems such as molecular magnets, nanoclusters, etc. (see, e.g., Refs. [457] and [226] and references cited therein). In the standard basis of spin functions $|S, m\rangle$ (see Appendix A.2), which describe states with definite spin S and spin projection m onto the Z-axis, i.e., $\hat{S}_Z|S,m\rangle = m|S,m\rangle$, the Hamiltonian $\hat{H}_S$, Eq. (3.83), has an energy spectrum with a double-well structure and two minima at $m = \pm S$ separated by a potential barrier. Notice that in strong bias fields, $\xi > \sigma(2S-1)/S$, the barrier entirely disappears. Now recall that generally speaking, spin reversal can take place either by thermal activation or by tunneling or a combination of both. The tunneling may occur from one side of the barrier to the other *between resonant, equal-energy states coupled by transverse fields* or *high-order anisotropy terms* [226–228]. Garanin [226] and García-Palacios and Zueco [227,228] by using the *spin density matrix in the second order of perturbation theory in the spin bath coupling* have studied the longitudinal relaxation of quantum uniaxial nanomagnets with the Hamiltonian Eq. (3.83). In other words, they gave a concise treatment of the spin dynamics by *directly proceeding from the quantum Hubbard operator representation* of the evolution equation for the spin density matrix. This intrinsically circularly symmetric problem has also been treated via the phase-space method in Ref. [259] and may be summarized as follows.

Using the collision operator in the symmetrized Hubbard form (1.430) as written for the particular Hamiltonian given by Eq. (3.83), we have from the general reduced density matrix evolution Eq. (1.411) the evolution equation for the reduced density matrix of a uniaxial nanomagnet [259]

$$\frac{\partial\hat{\rho}_S}{\partial t} - \frac{i}{\hbar\beta}\left\{\frac{\sigma}{S^2}\left[\hat{S}_0^2, \hat{\rho}_S\right] + \frac{\xi}{S}\left[\hat{S}_0, \hat{\rho}_S\right]\right\} = \mathrm{St}(\hat{\rho}_S), \tag{3.84}$$

where the collision kernel operator $\mathrm{St}(\hat{\rho}_S)$, characterizing the spin-bath interaction, is given by

$$\mathrm{St}(\hat{\rho}_S)=D_0\left(\left[\hat{S}_0,\hat{\rho}_S\hat{S}_0\right]+\left[\hat{S}_0\hat{\rho}_S,\hat{S}_0\right]\right)$$
$$-D_{-1}\left(\left[\hat{S}_{-1},\hat{\rho}_S e^{-\frac{\sigma}{2S^2}\hat{S}_0^2-\frac{\xi}{2S}\hat{S}_0}\hat{S}_1 e^{\frac{\sigma}{2S^2}\hat{S}_0^2+\frac{\xi}{2S}\hat{S}_0}\right]+\left[e^{\frac{\sigma}{2S^2}\hat{S}_0^2+\frac{\xi}{2S}\hat{S}_0}\hat{S}_1 e^{-\frac{\sigma}{2S^2}\hat{S}_0^2-\frac{\xi}{2S}\hat{S}_0}\hat{\rho}_S,\hat{S}_{-1}\right]\right)$$
$$-D_1\left(\left[\hat{S}_1,\hat{\rho}_S e^{-\frac{\sigma}{2S^2}\hat{S}_0^2-\frac{\xi}{2S}\hat{S}_0}\hat{S}_{-1} e^{\frac{\sigma}{2S^2}\hat{S}_0^2+\frac{\xi}{2S}\hat{S}_0}\right]+\left[e^{\frac{\sigma}{2S^2}\hat{S}_0^2+\frac{\xi}{2S}\hat{S}_0}\hat{S}_{-1} e^{-\frac{\sigma}{2S^2}\hat{S}_0^2-\frac{\xi}{2S}\hat{S}_0}\hat{\rho}_S,\hat{S}_1\right]\right).$$
$$(3.85)$$

Because of the following operator relations permitting the interchanges:

$$e^{\frac{\sigma}{2S^2}\hat{S}_0^2+\frac{\xi}{2S}\hat{S}_0}\hat{S}_{\pm1}e^{-\frac{\sigma}{2S^2}\hat{S}_0^2-\frac{\xi}{2S}\hat{S}_0}=e^{-\frac{\sigma}{2S^2}\pm\frac{\xi}{2S}}e^{\pm\frac{\sigma}{S^2}\hat{S}_0}\hat{S}_{\pm1}, \qquad (3.86)$$

$$\hat{S}_{\pm1}e^{\mp\frac{\sigma}{S^2}\hat{S}_0}=e^{\frac{\sigma}{S^2}}e^{\mp\frac{\sigma}{S^2}\hat{S}_0}\hat{S}_{\pm1}, \qquad (3.87)$$

we have from Eq. (3.85) a much simplified form of $\mathrm{St}(\hat{\rho}_S)$, viz.,

$$\mathrm{St}(\hat{\rho}_S)=D_{\parallel}\left(\left[\hat{S}_0,\hat{\rho}_S\hat{S}_0\right]+\left[\hat{S}_0\hat{\rho}_S,\hat{S}_0\right]\right)$$
$$-2D_{\perp}\left\{e^{\frac{\sigma}{2S^2}+\frac{\xi}{2S}}\left(\left[\hat{S}_1 e^{\frac{\sigma}{S^2}\hat{S}_0}\hat{\rho}_S,\hat{S}_{-1}\right]+\left[\hat{S}_1,\hat{\rho}_S e^{\frac{\sigma}{S^2}\hat{S}_0}\hat{S}_{-1}\right]\right)\right. \qquad (3.88)$$
$$\left.+e^{\frac{\sigma}{2S^2}-\frac{\xi}{2S}}\left(\left[\hat{S}_{-1},\hat{\rho}_S e^{-\frac{\sigma}{S^2}\hat{S}_0}\hat{S}_1\right]+\left[\hat{S}_{-1}e^{-\frac{\sigma}{S^2}\hat{S}_0}\hat{\rho}_S,\hat{S}_1\right]\right)\right\},$$

where we have introduced the new notation $D_{\perp}=D_{\pm1}/2$ and $D_{\parallel}=D_0$ for the diffusion coefficients. Equation (3.84) describes the evolution of our spin system in contact with the thermal bath at temperature T. Thus one important property of the collision kernel operators is satisfied by $\mathrm{St}(\hat{\rho}_S)$ as given by Eq. (3.88), namely that the equilibrium spin density matrix $\hat{\rho}_{\mathrm{eq}}=e^{-\beta\hat{H}_S}/\mathrm{Tr}\{e^{-\beta\hat{H}_S}\}$ renders the collision kernel equal to zero, i.e., $\mathrm{St}(\hat{\rho}_{\mathrm{eq}})=0$. Conditions for the validity of the reduced density matrix evolution Eq. (3.84) have been discussed in detail in Section 1.8.3.

We now proceed to the phase-space representation of the Hilbert space Eq. (3.84). This is accomplished by rewriting that equation as the *inverse map of a Weyl symbol* (see also [183]). By substituting the density matrix $\hat{\rho}_S$ so rendered into the reduced density matrix evolution Eq. (3.84), we have the inverse map (dropping the representation parameter s)

$$\int\left(\hat{w}\frac{\partial}{\partial t}W_S - W_S\left\{\frac{i}{\hbar\beta}\left(\frac{\sigma}{S^2}\left[\hat{S}_0^2,\hat{w}\right]+\frac{\xi}{S}\left[\hat{S}_0,\hat{w}\right]\right)+\mathrm{St}\left(\hat{w}\right)\right\}\right)d\Omega=0, \quad (3.89)$$

where $d\Omega = \sin\vartheta\, d\vartheta\, d\varphi$. As it stands, the *formal* inverse map Eq. (3.89) does not have the standard (Weyl symbol) form, Eq. (1.581), of the usual inverse Wigner–Stratonovich map with kernel $\hat{w}(\vartheta,\varphi)$. Therefore to facilitate this objective we must first transform the various commutators occurring in the integrand of Eq. (3.89) into the phase-space representation. These commutators will then appear as configuration space differential operators acting on the Wigner-Stratonovich kernel $\hat{w}(\vartheta,\varphi)$ [cf. Eq. (A.175) *et seq.* in Appendix A.3]. This procedure (involving lengthy operator manipulations for each commutator occurring in Eq. (3.89) as fully described in Appendix A.3) will then allow one to express the integrand above in the standard phase-space form demanded by Eq. (1.582). In this way, we will ultimately have the master equation for the spin phase-space distribution $W_S(\vartheta,\varphi,t)$, viz.,

$$\begin{aligned}
&\frac{\partial}{\partial t}W_S - \frac{\sigma}{\hbar\beta S^2}\left(2S\cos\vartheta - \sin\vartheta\frac{\partial}{\partial\vartheta}+\frac{S\xi}{\sigma}\right)\frac{\partial}{\partial\varphi}W_S \\
&= D_\parallel\frac{\partial^2}{\partial^2\varphi}W_S + D_\perp\frac{\cot\vartheta}{\sin\vartheta}\left[\cos\vartheta\frac{\partial^2}{\partial\varphi^2}R_+^{\prime(S)}W_S+\frac{\partial^2}{\partial\varphi^2}R_-^{\prime(S)}W_S\right] \\
&+\frac{D_\perp}{\sin\vartheta}\left[\frac{\partial}{\partial\vartheta}\sin\vartheta\left(\frac{\partial}{\partial\vartheta}R_+^{\prime(S)}W_S+\cos\vartheta\frac{\partial}{\partial\vartheta}R_-^{\prime(S)}W_S+2S\sin\vartheta R_-^{\prime(S)}W_S\right)\right] \\
&+D_\perp\left[\left(\sin\vartheta\frac{\partial}{\partial\vartheta}-2S\cos\vartheta\right)\frac{\partial}{\partial\varphi}R_+^{\prime\prime(S)}W_S-\frac{\partial}{\partial\varphi}R_-^{\prime\prime(S)}W_S\right], \quad (3.90)
\end{aligned}$$

where the phase-space differential operators $R_\pm^{(S)} = R_\pm^{\prime(S)} + iR_\pm^{\prime\prime(S)}$ are defined in Appendix A.3. The left-hand side of Eq. (3.90) is just the quantum analog of the classical Liouville equation for a uniaxial nanomagnet, while the collision operator given by the right-hand side of Eq. (3.90) is the quantum analog of the Fokker–Planck operator for classical spins, Eq. (1.223) of Chapter 1. In summary, the master Eq. (3.90) follows from the equation of motion of the reduced density matrix Eq. (3.84) written as the standard form Eq. (1.581) of the inverse Wigner–Stratonovich map of a Weyl symbol (see Appendix A.3). *Everywhere the interactions between the spin and the heat bath are taken small enough to use the weak coupling limit and the correlation time characterizing the bath is taken short enough to regard the stochastic process originating in the bath as Markovian.*

In *longitudinal* relaxation, when the azimuthal angle dependence of W_S may be ignored, the Liouville term vanishes in Eq. (3.90) and the corresponding phase-space evolution equation for $W_S(z=\cos\vartheta,t)$ then simplifies to the single variable equation

$$\frac{\partial W_S}{\partial t}=\frac{\partial}{\partial z}\left(D^{(2)}W_S+D^{(1)}W_S\right), \tag{3.91}$$

where

$$D^{(1)}=2D_\perp S(1-z^2)R_-^{\prime(S)},$$
$$D^{(2)}=D_\perp(1-z^2)\left(\frac{\partial}{\partial z}R_+^{\prime(S)}+z\frac{\partial}{\partial z}R_-^{\prime(S)}\right). \tag{3.92}$$

The phase-space master Eq. (3.91) is then *formally* similar to the single spatial variable Fokker–Planck equation for the orientation distribution function $W(z, t)$

$$\frac{\partial W}{\partial t}=D_\perp\frac{\partial}{\partial z}\left[(1-z^2)\left(\frac{\partial W}{\partial z}+W\frac{\partial E}{\partial z}\right)\right] \tag{3.93}$$

describing rotational diffusion of a classical spin in a potential [97]

$$E=-\sigma\cos^2\vartheta-\xi\cos\vartheta \tag{3.94}$$

and in the classical limit, Eq. (3.91) reduces after lengthy manipulations to the latter equation as we shall demonstrate in Appendix A.3.

One of the major computational difficulties associated with the deceptively simple phase-space master Eq. (3.90) now appears. By inspection of that equation we see that even for circular symmetry *high order spin number dependent differential operators occur apart from two notable exceptions*. The first of these comprises *noninteracting* spins in a uniform field where all the higher order derivatives in the operators $R_\pm^{(S)}$

vanish. Thus they become closed transcendental functions (see for example Eq. (3.113) below). Here we will have differential-recurrence relations (see Eq. (3.124) *et seq.* below) which are essentially similar to those occurring in the corresponding classical problem (see, for example, Eq. (3.137) *et seq.* below). The other exception is that without any potential whatsoever, i.e., when $\sigma = 0$ and $\xi = 0$, the differential operators merely reduce to $R_+^{(S)} = 1$ and $R_-^{(S)} = 0$, respectively (see Appendix A.3), so that Eq. (3.90) becomes (setting $D_{\|} = D_{\perp}$)

$$\frac{\partial W_S}{\partial t} = D_{\perp}\left[\frac{1}{\sin\vartheta}\frac{\partial}{\partial\vartheta}\left(\sin\vartheta\frac{\partial W_S}{\partial\vartheta}\right) + \frac{1}{\sin^2\vartheta}\frac{\partial^2 W_S}{\partial\varphi^2}\right]. \tag{3.95}$$

Equation (3.95) corresponds to the classical Fokker–Planck equation [97] in the single coordinate ϑ for the orientational distribution function of free magnetic dipole moments on the unit sphere. Hence, like the free quantum translational Brownian motion, the phase-space distribution W_S of free quantum spins obeys the classical Fokker–Planck equation Eq. (3.95) for the rotational diffusion of free classical spins [97].

Thus it would appear that the phase-space master Eq. (3.90) is in general of limited practical use. However, this does not preclude one from deriving differential-recurrence relations for observables via the density matrix evolution equation itself by using *the one-to-one correspondence between the averages of polarization operators in Hilbert space and the averages of the spherical harmonics* as we now describe. This will automatically yield the recurrence relations (v. Eq. (3.100) below) for the averages of the polarization operators which may then be mapped onto averages of the spherical harmonics via Eq. (3.103) below. This procedure is of course just a special case of the formal one outlined in Section 1.9.8 [Eq. (1.600) *et seq.*]

3.3.2 *Differential-recurrence relations for the statistical moments*

Recalling the last paragraph of the previous section the formal solutions of the density matrix evolution Eq. (3.84) corresponding to the uniaxial Hamiltonian (3.83) and the corresponding phase-space Eq. (3.90) for *arbitrary S* may be written as the operator series

$$\hat{\rho}_S(t) = \hat{\rho}_{\text{eq}} + \sum_{L=1}^{2S} c_L(t)\hat{T}_{L0}^{(S)} \tag{3.96}$$

and the corresponding series of spherical harmonics

$$W_S(\vartheta,\varphi,t) = W_S^{\text{eq}}(\vartheta) + \sum_{L=1}^{2S} b_L(t)Y_{L0}(\vartheta,\varphi), \tag{3.97}$$

respectively. The equilibrium phase-space distribution $W_S^{\text{eq}}(\vartheta)$ is the stationary solution of both Eqs. (3.90) and (3.91). We emphasize that $W_S^{\text{eq}}(\vartheta)$ corresponds to the equilibrium spin density matrix $\hat{\rho}_{\text{eq}}$ and is defined by Eq. (1.581), that is

$$W_S^{\text{eq}}(\vartheta) = \text{Tr}\left\{\hat{\rho}_{\text{eq}}\hat{w}(\vartheta)\right\}. \tag{3.98}$$

The distribution $W_S^{\text{eq}}(\vartheta)$ defined by the Wigner-Stratonovich map Eq. (3.98) has already been calculated in Section 3.2.2 and is given by the finite series of Legendre polynomials Eq. (3.36) while the collision kernel of Eq. (3.90) satisfies $\text{St}(W_S^{\text{eq}}) = 0$, i.e., the distribution W_S^{eq} defined by Eq. (3.36) is indeed the stationary solution of the phase-space master Eqs. (3.90) and (3.91). The coefficients $c_L(t)$ and $b_L(t)$ (corresponding to the statistical moments) are, in turn, the averages of the polarization operators $\hat{T}_{L0}^{(S)}$ and the spherical harmonics Y_{L0}, respectively, viz.,

$$\begin{aligned} c_L(t) &= \left\langle \hat{T}_{L0}^{(S)} \right\rangle(t) - \left\langle \hat{T}_{L0}^{(S)} \right\rangle_0, \\ b_L(t) &= \left\langle Y_{L0} \right\rangle(t) - \left\langle Y_{L0} \right\rangle_0. \end{aligned} \tag{3.99}$$

By substituting the polarization operator expansion Eq. (3.96) and the spherical harmonic expansion Eq. (3.97) into the density matrix evolution Eq. (3.84) and the phase-space Eq. (3.90), respectively, we have in each case a *finite* hierarchy of differential-recurrence equations for the statistical moments (in contrast to the classical case, where the corresponding hierarchy is *infinite*).

Since *either* the polarization operator series Eq. (3.96) or the finite series of the spherical harmonics (3.97) will yield similar hierarchies for the statistical moments (see Section 1.9.8), we shall at first describe the derivation of their recurrence relations using the density matrix. This is accomplished by first substituting the operator expansion Eq. (3.96) into the explicit evolution Eq. (3.84). Next we use Eq. (A.182) from Appendix A.3 for the expansion of the matrix exponents $e^{\sigma/2S^2 \pm \xi/2S} e^{\pm\sigma S^{-2}\hat{S}_0}$ in terms of the polarization operators $\hat{T}_{l0}^{(S)}$ and the product formula Eq. (A.154) from Appendix A.2 for the operators $\hat{T}_{LM}^{(S)}$ which allows products of these to be expressed as a sum. In this way, we ultimately have using the Hilbert space representation a hierarchy of multi-term differential-recurrence equations for the relaxation functions $c_L(t)$ in Eq. (3.96), namely,

$$\tau_{\mathrm{N}} \frac{d}{dt} c_L(t) = \sum_{L'=1}^{2S} g_{LL'}^{(S)} c_{L'}(t). \tag{3.100}$$

Here $\tau_{\mathrm{N}} = (2D_{\perp})^{-1}$ is the characteristic (free diffusion) time and we have the following expression for the coefficients $g_{LL'}^{(S)}$ which are defined in the usual way by the average

$$g_{LL'}^{(S)} = -\frac{L(L+1)}{2} e^{\frac{\sigma}{2S^2}} \mathrm{Tr}\left\{ \hat{T}_{L'0}^{(S)} \left[\cosh\left(\frac{\sigma}{S^2}\hat{S}_0 + \frac{\xi}{2S} \right) \hat{T}_{L0}^{(S)} + \sinh\left(\frac{\sigma}{S^2}\hat{S}_0 + \frac{\xi}{2S} \right) \right.\right.$$
$$\left.\left. \times \left(\frac{\sqrt{(2S-L)(2S+L+2)}}{\sqrt{(2L+3)(2L+1)}} \hat{T}_{L+10}^{(S)} - \frac{\sqrt{(2S-L+1)(2S+L+1)}}{\sqrt{(2L-1)(2L+1)}} \hat{T}_{L-10}^{(S)} \right) \right] \right\}, \tag{3.101}$$

where $\cosh(\hat{A})$ and $\sinh(\hat{A})$ appearing in Eq. (3.101) are matrix functions. Likewise, in the phase-space representation, we have the relevant system of differential-recurrence equations for the relaxation functions $b_L(t)$ from the general recurrence relation Eq. (1.604), the matrix elements Eq. (1.603) and the particular Eq. (3.100), viz.,

$$\tau_{\mathrm{N}} \frac{d}{dt} b_L(t) = \sum_{L'=1}^{2S} p_{L'L}^{(S)} b_{L'}(t), \tag{3.102}$$

where

$$p_{L'L}^{(S)} = \sqrt{\frac{(2S-L')!(2S+L'+1)!}{(2S-L)!(2S+L+1)!}} g_{L'L}^{(S)}. \tag{3.103}$$

Alternatively starting with the phase-space method, the recurrence relation Eqs. (3.102) could be derived by directly substituting the spherical harmonic expansion Eq. (3.97) into the phase-space evolution Eq. (3.90) and then using the recurrence relations, etc., of the spherical harmonics, viz., Eqs. (A.81)–(A.83). However, in general very detailed manipulations would be involved for the reasons we have outlined. It should be mentioned that both of the *equilibrium* averages $\langle \hat{T}_{L0}^{(S)} \rangle_0$ and $\langle Y_{L0} \rangle_0$ satisfy similar time-independent recurrence relations, viz.,

$$\sum_{L'=0}^{2S} g_{L,L'}^{(S)} \langle \hat{T}_{L'0}^{(S)} \rangle_0 = 0 \text{ and } \sum_{L'=0}^{2S} p_{L',L}^{(S)} \langle Y_{L'0} \rangle_0 = 0. \tag{3.104}$$

The resulting system of Eq. (3.101) and/or (3.102), which we have just derived using the operator series Eq. (3.96), can be solved by either *direct* matrix diagonalization which involves calculating the eigenvalues and eigenvectors of the system matrix (see Section 1.8.4) or by the *computationally efficient* (matrix) continued fraction method [97,143]. As will be seen below, the solutions can be obtained both for the transient and ac stationary (linear and nonlinear) responses of spins in magnetic fields.

In the *limiting case of zero anisotropy* $\sigma = 0$, the explicit Eq. (3.101) can be further simplified by once again using the general formula for the product of polarization operators in terms of the Clebsch-Gordan series, i.e., Eq. (A.154) of Appendix A.2, thereby yielding the closed expression

$$g_{LL'}^{(S)} = -\frac{1}{2}L(L+1)\left[\delta_{L'L}\cosh\frac{\xi}{2S} + \sinh\frac{\xi}{2S}\right.$$
$$\left.\times\left(\delta_{L'L+1}\sqrt{\frac{(2S-L)(2S+L+2)}{(2L+3)(2L+1)}} - \delta_{L'L-1}\sqrt{\frac{(2S-L+1)(2S+L+1)}{(2L-1)(2L+1)}}\right)\right]$$

so that with the replacement

$$c_L(t) \to (-1)^L \frac{\sqrt{(2S-L)!(2S+L+1)!(2L+1)}}{4\pi(2S)!} f_L(t), \tag{3.105}$$

we obviously have from Eq. (3.100) the simple three-term ordinary differential-recurrence relation for the quantum after-effect functions:

$$\tau_N \frac{d}{dt} f_L(t) = q_L f_L(t) + q_L^- f_{L-1}(t) + q_L^+ f_{L+1}(t), \tag{3.106}$$

with

$$q_L = -\frac{L(L+1)}{2}\cosh\frac{\xi}{2S},$$
$$q_L^{\pm} = \pm\frac{L(L+1)}{2(2L+1)}\left(2S \pm L + \frac{3}{2} \pm \frac{1}{2}\right)\sinh\frac{\xi}{2S}. \tag{3.107}$$

The quantum relaxation function $f_L(t)$ as defined by Eq. (3.105) for a given S corresponds to $f_L(t) = \langle P_L \rangle(t) - \langle P_L \rangle_0$, where the $P_L(z)$ are the Legendre polynomials [290]. This limiting case exactly corresponds to the spin relaxation in a uniform field which will be comprehensively treated in Section 3.3.3.

Returning to the general case $\sigma \neq 0$, in the *classical limit*, $S \to \infty$, the Hamiltonian Eq. (3.83) corresponds to the classical free energy Eq. (3.94) for a classical uniaxial nanomagnet subjected to a uniform longitudinal field while the relatively complicated quantum differential-recurrence relation Eq. (3.100) reduces in that limit to the usual five-term one, Eq. (2.165) of Chapter 2, for the *classical* relaxation function

$$f_L(t) = \langle P_L(\cos\vartheta)\rangle(t) - \langle P_L(\cos\vartheta)\rangle_0. \tag{3.108}$$

This classical problem has been treated in detail in Section 2.5.1 of Chapter 2 (see also Refs. [17,23,97,146,239,334,336–339]). For zero anisotropy, i.e., $\sigma = 0$, we also have from Eq. (3.100) the known result for a classical spin in a uniform field [97,467–471]

$$\tau_N \frac{d}{dt} f_L(t) + \frac{L(L+1)}{2} f_L(t) = \frac{\xi L(L+1)}{2(2L+1)}\left[f_{L-1}(t) - f_{L+1}(t)\right], \tag{3.109}$$

the exact solution has been given in Chapter 7 of Ref. [97].

We have indicated (see Appendix A.3 for details) how one may derive a master equation for the evolution of the phase-space quasiprobability distribution for a uniaxial nanomagnet in contact with a heat bath at temperature T. This is accomplished by first expressing the reduced density matrix master equation in Hilbert space in terms of an inverse Wigner–Stratonovich transformation according to Eq. (3.89). The various commutators in the integrand of Eq. (3.89) involving the spin operators may then be converted into phase-space differential operators. This is in turn accomplished using the orthogonality and recurrence properties of the polarization operators and the corresponding spherical harmonics to ultimately yield via the standard (Weyl symbol) form of the inverse transformation Eqs. (1.577) and (1.578) the desired master equation for the distribution function in the phase space of the polar angles. Despite the *superficial resemblance* of the quantum diffusion Eqs. (3.90) and (3.91) (governing the behavior of the phase-space distribution) to the Fokker–Planck equations for classical spins, the problem is actually much more complicated. The difficulty lies in the collision kernel which *involves powers of differential operators up to the spin number S considered*, only simplifying for large spin numbers ($S \to \infty$) when the high order derivatives occurring in the operators may be ignored.

We have illustrated the phase-space representation of spin relaxation by treating a uniaxial nanomagnet in a uniform magnetic field of arbitrary strength directed along the easy axis, thereby realizing that only a master equation in configuration space akin to the Fokker–Planck equation for classical spins is involved. Hence for spins (just as particles), the existing classical solution methods [97,143] also apply in the quantum case indeed suggesting new closed form quantum results via classical ones. The magnetization, dynamic susceptibility, characteristic relaxation times, etc., for the uniaxial system may now be evaluated. Notice that the spin relaxation of this uniaxial system has already been treated using the quantum Hubbard operator representation of the evolution equation for the spin density matrix [226–228]. Moreover, as shown in Ref. [256,257] for spins in an external field alone both the phase-space and density matrix methods yield results in outwardly very different forms. Nevertheless, the numerical values from both methods for the same physical quantities (such as relaxation times and susceptibility) coincide thereby establishing a vital corollary between the phase-space and the density matrix methods. Thus, the phase-space representation, because it is closely allied to the classical one, besides being complementary to the operator one, transparently illustrates how quantum distributions reduce to the classical ones. The analysis is carried out via the finite Fourier series representation embodied in the Wigner–Stratonovich map as we have illustrated for a uniaxial potential. It may be extended in the appropriate limits to other systems such as biaxial, cubic, etc. However, the difficulties (e.g., the operator form of the diffusion coefficients in the master equation) encountered in our treatment of single space variable potentials are indicative of the even greater ones which would be faced when generalizing the phase-space representation to such potentials, where the Liouville term never vanishes.

We will now treat in more detail the linear and nonlinear longitudinal relaxation for the model parameters $\sigma = 0$ and $\xi \neq 0$ corresponding as just seen to a spin in an external dc magnetic field $\mathbf{H}_0$ directed along the *Z*-axis. Then we shall consider in Section 3.3.9 the general case $\sigma \neq 0$ and $\xi \neq 0$, i.e., a uniaxial nanomagnet in an external magnetic field.

3.3.3 *Spin relaxation in a dc magnetic field*

For *noninteracting spins* in an external dc magnetic field $\mathbf{H}_0$ directed along the *Z*-axis, the Hamiltonian $\hat{H}_S$ is simply Eq. (3.83) with $\sigma = 0$

$$\hat{H}_S = -\hbar\omega_0 \hat{S}_Z, \tag{3.110}$$

where $\omega_0 = \xi / (\beta\hbar S)$ is the precession (Larmor) frequency. We shall now determine both the *exact* and approximate solutions yielding the averaged longitudinal component of a spin $\langle \hat{S}_Z \rangle (t)$ for arbitrary *S*. Furthermore, we shall show how the solution of the corresponding classical problem [97,471] carries over into the quantum domain and how the *exact* solution for the integral relaxation time due to an *arbitrarily strong* sudden change in the uniform field may be obtained. We remark that the original treatment of this transient response problem via the master equation in phase space was first given by Shibata *et al.* [229–231] and was further developed by Kalmykov *et al.* [256,257]. In the *linear response approximation,* the solution reduces to that previously given by Garanin [226] however using the spin density matrix in the second order of perturbation theory in the spin bath coupling. That result was later rederived by García-Palacios and Zueco [227,228] who (again using the density matrix solution) considered the linear response of the longitudinal relaxation of a spin for arbitrary *S*.

Expanding on the introductory paragraph above following [256,257], we analyze the transient relaxation dynamics of a spin $\hat{\mathbf{S}}$ in an external dc magnetic field $\mathbf{H}_0$ directed along the *Z-axis* and a random field $\mathbf{h}(t)$ characterizing the collision damping (due to the heat bath) incurred by the precessional motion of the spin. For the simple Hamiltonian given by Eq. (3.110), the evolution equation for the reduced density matrix is simply

$$\frac{\partial \hat{\rho}_S}{\partial t} - i\omega_0 \left[\hat{S}_0, \hat{\rho}_S \right] = \mathrm{St}(\hat{\rho}_S). \tag{3.111}$$

Equation (3.111) merely represents the *particular* case $\sigma = 0$ of the more general Eq. (3.84) for a uniaxial nanomagnet subjected to a uniform dc magnetic field treated in the previous section. For the Hamiltonian $\hat{H}_S$ Eq. (3.110), the collision kernel operator $\mathrm{St}(\hat{\rho}_S)$ [Eq. (3.88) with $\sigma = 0$] becomes

$$\begin{aligned} \mathrm{St}(\hat{\rho}_S) &= D_{\|}\left(\left[\hat{S}_0, \hat{\rho}_S \hat{S}_0\right] + \left[\hat{S}_0 \hat{\rho}_S, \hat{S}_0\right]\right) \\ &-2D_{\perp}\left\{e^{-\beta\hbar\omega_0/2}\left(\left[\hat{S}_{-1}, \hat{\rho}_S \hat{S}_1\right] + \left[\hat{S}_{-1}\hat{\rho}_S, \hat{S}_1\right]\right)\right. \\ &\left.+e^{\beta\hbar\omega_0/2}\left(\left[\hat{S}_1\hat{\rho}_S, \hat{S}_{-1}\right] + \left[\hat{S}_1, \hat{\rho}_S \hat{S}_{-1}\right]\right)\right\}. \end{aligned} \tag{3.112}$$

Now, in the phase-space master Eq. (3.90) for the *particular* case $\sigma = 0$ *corresponding to a spin in a uniform field*, all the higher order derivatives will disappear so that the operators $R_{\pm}^{(S)}$ just become the closed transcendental expressions

$$R_{+}^{(S)} = \cosh(\beta\hbar\omega_0/2) \text{ and } R_{-}^{(S)} = \sinh(\beta\hbar\omega_0/2).$$

Then the master Eq. (3.90) takes on a much simpler form, namely,

$$\begin{aligned} \frac{\partial W_S}{\partial t} &= \omega_0 \frac{\partial W_S}{\partial \varphi} + D_{\|}\frac{\partial^2 W_S}{\partial \varphi^2} \\ &+\frac{D_{\perp}\sinh\left(\frac{1}{2}\beta\hbar\omega_0\right)}{\sin\vartheta}\left\{\cot\vartheta\left[\cos\vartheta\coth\left(\tfrac{1}{2}\beta\hbar\omega_0\right)+1\right]\frac{\partial^2 W_S}{\partial\varphi^2}\right. \\ &\left.+\frac{\partial}{\partial\vartheta}\left[\sin\vartheta\left(\coth\left(\tfrac{1}{2}\beta\hbar\omega_0\right)+\cos\vartheta\right)\frac{\partial W_S}{\partial\vartheta} + 2S\sin^2\vartheta W_S\right]\right\}. \end{aligned} \tag{3.113}$$

Next by introducing the *renormalization* of the diffusion coefficients $D_{\perp}e^{-\beta\hbar\omega_0/2} \to D_{\perp}$, Eqs. (3.112) and (3.113) yield the result previously obtained for a *nonsymmetrized* form of the collision kernel operator [229–231]. This master equation describing the time evolution of $W_S(\vartheta,\varphi,t)$ again has essentially the same form as the corresponding Fokker–Planck equation for the distribution function $W(\vartheta,\varphi,t)$ of classical spin orientations in the configuration space [229]

$$\frac{\partial W}{\partial t} = \omega_0 \frac{\partial W}{\partial \varphi} + \frac{D_\perp}{\sin^2\vartheta}\frac{\partial^2 W}{\partial \varphi^2} + \frac{D_\perp}{\sin\vartheta}\frac{\partial}{\partial \vartheta}\left[\sin\vartheta\left(\frac{\partial W}{\partial \vartheta} + \xi \sin\vartheta W\right)\right] \tag{3.114}$$

($\xi = \beta\hbar\omega_0 S$). Equation (3.113) serves as the simplest example of the phase-space method for open spin systems. In fact, it is just the rotational analog of the quantum translational harmonic oscillator treated using the Wigner function by Agarwal [314]. In this instance, the evolution Eq. (1.522) for the Wigner distribution $W(q,p,t)$ in the phase space of positions and momenta has the same mathematical form as the Fokker–Planck equation for the classical oscillator (see Section 1.9.4). For purely *longitudinal* relaxation, where the distribution function W_S is independent of the azimuth, the Liouville term in the evolution equation vanishes and Eq. (3.113) reduces to an equation very similar to that governing the longitudinal relaxation of classical spins in a dc magnetic field [97]

$$\frac{\partial W_S}{\partial t} = \frac{D_\perp \sinh(\beta\hbar\omega_0/2)}{\sin\vartheta} \times \frac{\partial}{\partial\vartheta}\left\{\left[\sin\vartheta\left(\cos\vartheta + \coth(\beta\hbar\omega_0/2)\right)\right]\frac{\partial W_S}{\partial\vartheta} + 2S\sin^2\vartheta W_S\right\}. \tag{3.115}$$

This simplification arises naturally and is to be expected on intuitive grounds because we recall that *precession of a spin in a uniform field is effectively the rotational analog of the translational harmonic oscillator*.

Returning to the more general Eq. (3.113), that equation applies in the *narrowing limit case* in which the correlation time τ_c of the random field $\mathbf{h}(t)$ acting on the spin satisfies the condition $\gamma H\tau_c \ll 1$, where H is the averaged amplitude of the random magnetic field. The left-hand side of Eq. (3.113) is the quantum analog of the Liouville equation for a spin which now is the same as the classical case for particles with quadratic Hamiltonians, while the right-hand side (collision kernel) characterizes the interaction of the spin with the thermal bath at temperature T. The remaining conditions for the validity of Eq. (3.113) have already been discussed. Furthermore in the case of purely longitudinal relaxation, Eq. (3.113) may be plausibly derived by postulating (like in the phase-space treatment of the quantum translational Brownian motion, see Section 1.9.4) a master equation for the Wigner function W_S with collision terms given by a Kramers–Moyal expansion truncated at the second term. The various drift and diffusion coefficients in the truncated expansion may then be calculated by requiring that the equilibrium Wigner distribution function W_{eq}, corresponding to the equilibrium spin density matrix $\hat{\rho}_{\text{eq}} = e^{-\beta \hat{H}_S} / \text{Tr}(e^{-\beta \hat{H}_S})$, renders the collision kernel equal to zero (see Section 3.3.5).

3.3.4 *Quantum analog of the magnetic Langevin equation*

The spin relaxation described by the master equation (3.113) may also be equivalently described using a quantum analog of the magnetic Langevin equation with multiplicative noise. For consistency we use the Stratonovich definition [101] of such equations as usual constituting the mathematical idealization of the spin relaxation process [102]. Thus, it is unnecessary to transform them to Itô stochastic differential equations (e.g., [102]). Moreover, one can then use conventional calculus [97,102]. The governing Langevin equations comprise the two stochastic equations of motion for the variables ϑ and φ corresponding to the phase-space master Eq. (3.113) and are in the Stratonovich interpretation [97,262]

$$\begin{aligned}\dot{\vartheta}(t) = {} & D_\perp \cot \vartheta(t)\left(Q_1^S(t) - \sqrt{Q_1^S(t) Q_2^S(t)} \right) \\ & -2D_\perp (S + 1/4) \sin \vartheta(t) \sinh \frac{\beta \hbar \omega_0}{2} + \sqrt{Q_1^S(t)} \left[h_\vartheta(t) + \alpha^{-1} h_\varphi(t) \right],\end{aligned} \tag{3.116}$$

$$\dot{\varphi}(t) = -\omega_0 + \sqrt{Q_2^S(t)} \csc \vartheta(t) \left[h_\varphi(t) - \alpha^{-1} h_\vartheta(t) \right], \tag{3.117}$$

where

$$Q_1^S(t) = \frac{D_\perp \beta}{\eta(1+\alpha^{-2})}\left(\cosh\frac{\beta\hbar\omega_0}{2} + \cos\vartheta(t)\sinh\frac{\beta\hbar\omega_0}{2}\right),$$

$$Q_2^S(t) = \frac{D_\perp \beta}{\eta(1+\alpha^{-2})} \tag{3.118}$$

$$\times\left\{\cos^2\vartheta(t)\cosh\frac{\beta\hbar\omega_0}{2} + \cos\vartheta(t)\sinh\frac{\beta\hbar\omega_0}{2} + \frac{D_\parallel}{D_\perp}\sin^2\vartheta(t)\right\},$$

and the components $h_\vartheta(t)$, $h_\varphi(t)$ of the random field $\mathbf{h}(t)$ in the spherical coordinate system or basis are expressed in terms of their components $h_X(t)$, $h_Y(t)$, $h_Z(t)$ in the cartesian basis by Eq. (1.61) of Chapter 1 with

$$\overline{h_i(t)} = 0,$$
$$\overline{h_i(t)h_j(t')} = 2\eta\beta^{-1}\delta_{ij}\delta(t-t'). \tag{3.119}$$

Here the indexes $i, j = 1, 2, 3$ in Kronecker's delta δ_{ij} correspond to the cartesian axes X, Y, Z of the laboratory coordinate system $OXYZ$, and $\alpha = \gamma\eta\mu$ is a dimensionless dissipation (damping) parameter, η is a "friction" coefficient, and the overbar means the statistical average over the realizations of the random field. In the isotropic diffusion ($D_\perp = D_\parallel$) and classical ($S \to \infty$) limit, the Langevin Eqs. (3.116) and (3.117) reduce to those for *isotropic* diffusion of a classical spin in spherical coordinates [97], namely, (cf. Eq. (1.62) of Chapter 1)

$$\dot{\vartheta}(t) = D_\perp\left\{-\xi\sin\vartheta(t) + \sqrt{\frac{D_\perp\beta}{\eta(1+\alpha^{-2})}}\left[h_\vartheta(t) + \alpha^{-1}h_\varphi(t)\right]\right\},$$
$$\dot{\varphi}(t) = -\omega_0 + \sqrt{\frac{D_\perp\beta}{\eta(1+\alpha^{-2})}}\csc\vartheta(t)\left[h_\varphi(t) - \alpha^{-1}h_\vartheta(t)\right]. \tag{3.120}$$

To show that the Langevin equations (3.116) and (3.117) are equivalent to the master equation (3.113), we recall that by choosing Langevin equations for a set of two stochastic variables $\{\xi_1(t), \xi_2(t)\} = \{\vartheta(t), \varphi(t)\}$ as (see Section 1.3.2 of Chapter 1)

$$\dot{\xi}_i(t) = H_i[\xi_1(t), \xi_2(t)] + G_{ij}[\xi_1(t), \xi_2(t)]h_j(t), \tag{3.121}$$

($i = 1, 2, j = 1, 2, 3$) and subsequently interpreting them as Stratonovich stochastic differential equations, then the averaged equations for the drift, D_i, and diffusion, D_{ij}, coefficients are [97,143]

$$D_i = H_i(x_1, x_2, t) + \eta\beta^{-1} G_{kj}(x_1, x_2) \frac{\partial}{\partial x_k} G_{ij}(x_1, x_2),$$
$$D_{ij} = \eta\beta^{-1} G_{ik}(x_1, x_2) G_{jk}(x_1, x_2). \quad (3.122)$$

Moreover, due to Eqs. (3.121) and (3.122), we have from Eqs. (3.116) and (3.117) the explicit drift and diffusion coefficients

$$D_1 = -D_\perp \left[(2S+1)\sin\vartheta \sinh\frac{\beta\hbar\omega_0}{2} - \cot\vartheta\left(\cosh\frac{\beta\hbar\omega_0}{2} + \cos\vartheta\sinh\frac{\beta\hbar\omega_0}{2}\right)\right],$$

$$D_2 = -\omega_0, \; D_{11} = D_\perp\left(\cosh\frac{\beta\hbar\omega_0}{2} + \cos\vartheta\sinh\frac{\beta\hbar\omega_0}{2}\right),$$

$$D_{22} = D_\parallel + \frac{D_\perp}{\sin^2\vartheta}\left(\cos^2\vartheta\cosh\frac{\beta\hbar\omega_0}{2} + \cos\vartheta\frac{\beta\hbar\omega_0}{2}\right).$$

The Fokker–Planck equation for the probability density function $P(\vartheta,\varphi,t) = \sin\vartheta W_S(\vartheta,\varphi,t)$ corresponding to Eqs. (3.116) and (3.117) is

$$\frac{\partial P}{\partial t} = -\frac{\partial}{\partial\vartheta}(D_1 P) - \frac{\partial}{\partial\varphi}(D_2 P) + \frac{\partial^2}{\partial\vartheta^2}(D_{11} P) + \frac{\partial^2}{\partial\varphi^2}(D_{22} P). \quad (3.123)$$

Equation (3.123) ultimately reduces to the master equation (3.113) for the phase-space distribution function $W_S(\vartheta,\varphi,t)$.

Moreover via the Langevin equations (3.116) and (3.117) for the stochastic variables $\vartheta(t)$ and $\varphi(t)$, we also have the Langevin equation for the evolution of the spherical harmonics $Y_{lm}(\vartheta,\varphi)$ rendered as [97]

$$\dot{Y}_{lm}[\vartheta(t),\varphi(t)] = \dot{\vartheta}(t)\frac{\partial}{\partial\vartheta}Y_{lm}[\vartheta(t),\varphi(t)] + \dot{\varphi}(t)\frac{\partial}{\partial\varphi}Y_{lm}[\vartheta(t),\varphi(t)],$$

where $\dot{\vartheta}(t)$ and $\dot{\varphi}(t)$ are given by Eqs. (3.116) and (3.117), respectively. Then by averaging the Langevin equation for $Y_{lm}(\vartheta,\varphi)$ over its realizations as described in Chapter 1, and using the recursion relations for the spherical harmonics (see Appendix A.1.7), we ultimately have a *closed* system of differential-recurrence equations for the statistical moments (averaged spherical harmonics), namely,

$$\frac{d}{dt}\langle Y_{lm}\rangle(t) = q_{lm}^{-}\langle Y_{l-1m}\rangle(t) + q_{lm}\langle Y_{lm}\rangle(t) + q_{lm}^{+}\langle Y_{l+1m}\rangle(t), \quad (3.124)$$

where $0 \le l \le 2S$,

$$q_{lm} = -D_{\parallel}m^2 + im\omega_0 - D_{\perp}\left[l(l+1) - m^2\right]\cosh\frac{\beta\hbar\omega_0}{2},$$

$$q_{lm}^{-} = D_{\perp}\left(l - 2S - 1\right)\sqrt{\frac{l^2 - m^2}{4l^2 - 1}}\sinh\frac{\beta\hbar\omega_0}{2},$$

$$q_{lm}^{+} = D_{\perp}\left(l + 2S + 2\right)\sqrt{\frac{(l+1)^2 - m^2}{(2l+1)(2l+3)}}\sinh\frac{\beta\hbar\omega_0}{2}.$$

Here the number of recurrent equations is *finite* namely $2S + 1$ because $\langle Y_{Lm}\rangle(t) = 0$ for $L > 2S$ which is the main difference from the corresponding classical hierarchy of differential-recurrence equations for the moments, where the number of equations is infinite. Thus, the Langevin and master equation treatments are now equivalent and yield the same results.

Here we have illustrated how a phase-space Langevin equation may be written by considering the simplest possible yet physically meaningful problem, viz., the relaxation of a spin of *arbitrary* number S in a uniform magnetic field of *arbitrary* strength directed along the Z-axis. We emphasize that the Langevin equations are written down from *a priori* knowledge of the master equation *unlike the classical case* where they are always written down independently of the Fokker–Planck equation. Consequently, the results of each method in the classical case only coincide due to the Gaussian white noise properties of the random field, particularly Isserlis's (or Wick's) theorem [97] is satisfied. This theorem allows multiple time correlations of Gaussian random variables to be expressed as products of two-time ones thereby leading directly to the correspondence between the Langevin and Fokker–Planck equations in the classical case.

3.3.5 *Exact solution of the master equation for longitudinal relaxation*

Having formulated the relevant evolution equations we shall now explicitly treat transient *nonlinear* spin relaxation by directly using the phase-space master equation because in this particular case it takes on the Fokker–Planck form [see Eq. (3.125)]. To accomplish this, we suppose that the magnitude of an externally uniform dc magnetic field is suddenly altered at time $t = 0$ from $\mathbf{H}_{\mathrm{I}}$ to a new value $\mathbf{H}_{\mathrm{II}}$ (the fields $\mathbf{H}_{\mathrm{I}}$ and $\mathbf{H}_{\mathrm{II}}$ are assumed to be applied parallel to the Z axis of the laboratory coordinate system). Thus we study as in the classical case [97], the transient longitudinal relaxation of a system of noninteracting spins starting from an equilibrium state I say with the initial distribution function $W_{\mathrm{eq}}^{\mathrm{I}}$ $(t \le 0)$ to a new equilibrium state II say with the final distribution function $W_{\mathrm{eq}}^{\mathrm{II}}$ $(t \to \infty)$. Here the longitudinal component of the spin operator relaxes from the initial equilibrium value $\langle \hat{S}_Z \rangle_{\mathrm{I}}$ to the final equilibrium value $\langle \hat{S}_Z \rangle_{\mathrm{II}}$, the intervening transient being described by an appropriate relaxation function $\langle \hat{S}_Z \rangle(t) - \langle \hat{S}_Z \rangle_{\mathrm{II}}$ (see Fig. 3.11). The transient response so formulated is truly *nonlinear* because the change in amplitude $H_{\mathrm{I}} - H_{\mathrm{II}}$ of the external dc magnetic field is *arbitrary* (the linear response can be regarded as the limit $|H_{\mathrm{I}} - H_{\mathrm{II}}| \to 0$). Here, the azimuthal angle dependence of the distribution function W_S may be ignored. Thus, the master equation (3.115) becomes the single variable Fokker–Planck equation

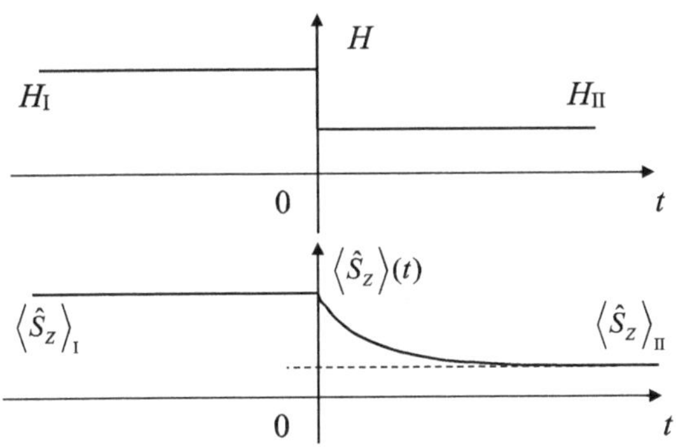

Fig. 3.11. Schematic representation of the nonlinear transient response.

$$\frac{\partial W_S}{\partial t} = \frac{\partial}{\partial z}\left(D_2(z)\frac{\partial W_S}{\partial z} + D_1(z)W_S \right), \tag{3.125}$$

where $z = \cos\vartheta$,

$$D_1(z) = \frac{S(1-z^2)}{\tau_N}\sinh\frac{\xi}{2S},$$
$$D_2(z) = \frac{1-z^2}{2\tau_N}\left(\cosh\frac{\xi}{2S} + z\sinh\frac{\xi}{2S} \right), \tag{3.126}$$

$\tau_N = 1/(2D_\perp)$ is the characteristic time of the free rotational "diffusion" of the spin, and the dimensionless field parameter ξ is defined as

$$\xi = \beta\mu_0\mu H_{\text{II}}. \tag{3.127}$$

Hitherto the explicit expressions Eq. (3.126) for $D_1(z)$ and $D_2(z)$ for spins subjected to a dc magnetic field $\mathbf{H}_0$ have been obtained (as we have just seen) by starting from the evolution equation for the density matrix $\hat{\rho}_S$ giving rise to lengthy calculations. However, these expressions can also be obtained in far simpler fashion merely by knowing the functional form of the master Eq. (3.125) and the equilibrium phase-space distribution $W_{eq}^{\xi}(z)$ for spins. To illustrate this, we shall again select the extension to the semiclassical case of the *Ansatz* of the imposition of a Boltzmann distribution originally used by Einstein, Smoluchowski, Langevin, and Kramers to determine drift and diffusion coefficients in the classical Brownian motion. We have already used this idea for the quantum translational Brownian motion in Section 1.9 and in Ref. [272].

Thus to determine $D_1(z)$ and $D_2(z)$ in Eq. (3.125) explicitly, we first recall that the equilibrium distribution $W_{eq}^{\xi}(z)$ given by Eq. (3.27) with $\gamma_X = \gamma_Y = 0$, viz.,

$$W_{eq}^{\xi}(z) = \frac{\sinh\left(\frac{1}{2}\xi/S\right)}{\sinh\left[\left(S+\frac{1}{2}\right)\xi/S\right]}\left(\cosh\frac{\xi}{2S} + z\sinh\frac{\xi}{2S} \right)^{2S} \tag{3.128}$$

must also be the equilibrium solution of the generic master Eq. (3.125), i.e., it must satisfy

$$\frac{\partial}{\partial z}\left(D_2(z)\frac{\partial}{\partial z}W^{\xi}_{\text{eq}}(z) + D_1(z)W^{\xi}_{\text{eq}}(z) \right) = 0. \qquad (3.129)$$

Now one is at liberty to seek $D_1(z)$ and $D_2(z)$ in series form as

$$\begin{aligned} D_1(z) &= \left(1-z^2\right)\left(a_0^S + a_1^S z + a_2^S z^2 + \cdots\right), \\ D_2(z) &= \left(1-z^2\right)\left(b_0^S + b_1^S z + b_2^S z^2 + \cdots\right). \end{aligned} \qquad (3.130)$$

By substituting Eq. (3.130) into Eq. (3.129), then if $W^{\xi}_{\text{eq}}(z)$ from Eq. (3.128) is to satisfy Eq. (3.129), only the coefficients a_0^S, b_0^S, and b_1^S can be nonzero so that $D_1(z)$ and $D_2(z)$ are given by

$$D_1(z) = 2Sb_0^S\left(1-z^2\right)\tanh\frac{\xi}{2S},$$

$$D_2(z) = b_0^S\left(1-z^2\right)\left(1 + z\tanh\frac{\xi}{2S}\right).$$

To determine the normalizing coefficient b_0^S, we use the fluctuation-dissipation theorem [163] along with the additional requirements that in the classical limit $S \to \infty$, $D_1(z)$ and $D_2(z)$ must reduce to their classical counterparts for the rotational Brownian motion of a classical spin [97], viz., $D_1(z) \to \xi(2\tau_N)^{-1}(1-z^2)$ and $D_2(z) \to (2\tau_N)^{-1}(1-z^2)$. Thus, we obtain $b_0^S = (2\tau_N)^{-1}\cosh(\xi/2S)$ so that $D_1(z)$ and $D_2(z)$ are given by the closed form Eq. (3.126). Now in the derivation of $D_1(z)$ and $D_2(z)$ we have imposed the stationary solution of the master equation as the equilibrium phase-space distribution Eq. (3.128) corresponding to the equilibrium density matrix $\hat{\rho}_{\text{eq}}$ given by Eq. (3.18) describing the system in *thermal equilibrium without coupling* to the thermal bath. However, from the theory of open quantum systems [321], the equilibrium state in general may deviate from the equilibrium density matrix $\hat{\rho}_{\text{eq}}$; the latter describes the thermal equilibrium of the system in the *weak coupling* and *high temperature* limits only. A detailed discussion is given, e.g., by Geva *et al.* [472]. Nevertheless, the imposition of the phase-space distribution Eq. (3.128) as the equilibrium solution of Eq. (3.129) so yielding $D_1(z)$ and $D_2(z)$, once again appears to be the exact analog of the *Ansatz* used by Gross and Lebowitz [319] in their formulation of quantum kinetic models of impulsive collisions. According to [319], for a system with a

Hamiltonian $\hat{H}_S$, the equation governing the time behavior of the density matrix $\hat{\rho}_S$ is Eq. (1.411), where the collision kernel operator $\mathrm{St}(\hat{\rho}_S)$ satisfies the condition $\mathrm{St}(\hat{\rho}_{\mathrm{eq}})=0$. Equation (3.129) is entirely analogous to this condition. Moreover, as we have already seen in Section 1.8.3, the condition $\mathrm{St}(\hat{\rho}_{\mathrm{eq}})=0$ was also used by Redfield [224] to determine the matrix elements of the relaxation operator in his theory of quantum relaxation processes.

The time-dependent solution of the evolution Eq. (3.125) is obtained like in the classical problem by expanding the distribution function $W_S(z,t)$ in Legendre polynomials $P_n(z)$

$$W_S(z,t)=W_{\mathrm{eq}}^{\xi}(z)+\sum_{n=1}^{2S}(n+1/2)P_n(z)f_n(t), \tag{3.131}$$

where $f_n(t)=\langle P_n\rangle(t)-\langle P_n\rangle_0^{\xi}$ are the relaxation functions, the angular brackets $\langle\ \rangle(t)$ and $\langle\ \rangle_0^{\xi}$ designate statistical averaging defined as

$$\langle P_n\rangle(t)=\left(S+\tfrac{1}{2}\right)\int_{-1}^{1}P_n(z)W_S(z,t)dz \tag{3.132}$$

and

$$\langle P_n\rangle_0^{\xi}=\left(S+\tfrac{1}{2}\right)\int_{-1}^{1}P_n(z)W_{\mathrm{eq}}^{\xi}(z)dz. \tag{3.133}$$

The equilibrium distribution $W_{\mathrm{eq}}^{\xi}(z)$ can be defined as [cf. Eq. (3.44)]

$$W_{\mathrm{eq}}^{\xi}(z)=\sum_{n=0}^{2S}(n+1/2)P_n(z)\langle P_n\rangle_0^{\xi}. \tag{3.134}$$

Equations (3.133) and (3.134) yield the *equilibrium* average of $P_1(z)$, viz.,

$$\langle P_1\rangle_0^{\xi}=\left(S+\tfrac{1}{2}\right)\int_{-1}^{1}zW_{\mathrm{eq}}^{\xi}(z)dz=\frac{S}{S+1}B_S(\xi), \tag{3.135}$$

where $B_S(\xi)$ is the Brillouin function defined by Eq. (1.368) of Chapter 1. Substituting Eq. (3.131) into Eq. (3.125) and using the orthogonality and recurrence properties of the $P_n(z)$, we have as in the classical case [97] a differential-recurrence relation for the relaxation functions $f_n(t)$, viz., [cf. the three-term recurrence relation Eq. (3.106)]

$$\tau_n \dot{f}_n(t) = q_n^- f_{n-1}(t) + q_n f_n(t) + q_n^+ f_{n+1}(t), \tag{3.136}$$

where $1 \le n \le 2S$,

$$f_0(t) = f_{2S+1}(t) = 0, \;\; \tau_n = \frac{2\tau_{\rm N}}{n(n+1)},$$

$$q_n = -\cosh\frac{\xi}{2S}, \;\; q_n^{\pm} = \mp\frac{2S \pm n + (3\pm 1)/2}{(2n+1)}\sinh\frac{\xi}{2S}.$$

Since the initial value of the distribution function is $W_S(z,0) = W_{\rm eq}^{\xi+\delta}(z)$, where the transient parameter $\delta = \beta\mu_0\mu\left(H_{\rm II} - H_{\rm I}\right)$, the initial values of the relaxation functions are $f_n(0) = \left\langle P_n \right\rangle_0^{\xi+\delta} - \left\langle P_n \right\rangle_0^{\xi}$. Equation (3.136) is merely a quantum counterpart of the recurrence relation Eq. (3.109) for a classical spin and has been solved in Refs. [243,244] for the *particular* spin numbers S = 1/2, 1, and 3/2.

Using the one-sided Fourier transform, we have from Eq. (3.136)

$$\left(i\omega\tau_n - q_n\right)\tilde{f}_n(i\omega) - q_n^- \tilde{f}_{n-1}(i\omega) - q_n^+ \tilde{f}_{n+1}(i\omega) = \tau_n f_n(0). \tag{3.137}$$

The inhomogeneous algebraic three-term recurrence Eq. (3.137) can be solved exactly for the one-sided Fourier transform of the relaxation function $\tilde{f}_1(i\omega)$ using continued fractions like the corresponding classical problem (see for details the general solution of three-term recurrence relations given in Ref. [97], Chapter 2) yielding the exact solution for the transient quantum nonlinear longitudinal relaxation for *arbitrary S*:

$$\begin{aligned}\tilde{f}_1(i\omega) &= \tau_{\rm N}\,{\rm csch}\frac{\xi}{2S}\sum_{n=1}^{2S}\frac{f_n(0)}{n(n+1)(S+1)}\prod_{k=1}^{n}\frac{q_{k-1}^+}{q_k^-}\Delta_k(\omega,\xi)\\ &= \tau_{\rm N}\,{\rm csch}\frac{\xi}{2S}\sum_{n=1}^{2S}(-1)^{n+1}\frac{(2n+1)(2S+n+1)!(2S-n)!}{n(n+1)\left(S+1\right)(2S+1)!(2S)!}f_n(0)\prod_{k=1}^{n}\Delta_k(\omega,\xi).\end{aligned} \tag{3.138}$$

Here the *finite* continued fraction $\Delta_n(\omega,\xi)$ is defined by the two-term recurrence relation

$$\Delta_n(\omega,\xi) = \frac{q_n^-}{i\omega\tau_n - q_n - q_n^+\Delta_{n+1}(\omega,\xi)}$$

with $\Delta_{2S+1}(\omega, \xi) = 0$, moreover, we have the product in the closed form

$$\prod_{k=1}^{n}\frac{q_{k-1}^{+}}{q_k^{-}}=(-1)^{n+1}\frac{(2n+1)(2S+n+1)!(2S-n)!}{(2S+1)!(2S)!}.$$

The equilibrium averages $\langle P_n\rangle_0^{\xi}$, Eq. (3.133), can also be evaluated explicitly in terms of the continued fraction $\Delta_n(0,\xi)$ since $\langle P_n\rangle_0^{\xi}$ satisfies the three-term recurrence relation

$$q_n^{-}\langle P_{n-1}\rangle_0^{\xi}+q_n\langle P_n\rangle_0^{\xi}+q_n^{+}\langle P_{n+1}\rangle_0^{\xi}=0, \tag{3.139}$$

so that $\Delta_n(0,\xi)=\langle P_n\rangle_0^{\xi}/\langle P_{n-1}\rangle_0^{\xi}$. Consequently, we have

$$\langle P_n\rangle_0^{\xi}=\prod_{k=1}^{n}\Delta_k(0,\xi). \tag{3.140}$$

Equation (3.138) is the exact solution for the one-sided Fourier transform of the nonlinear relaxation function $f_1(t)$ in terms of continued fractions. Having determined $f_1(t)$, various *transient* nonlinear responses of the longitudinal component of the magnetic moment may always be evaluated because in terms of averages of spin operators

$$\langle\hat{S}_Z\rangle(t)-\langle\hat{S}_Z\rangle_{\mathrm{II}}=(S+1)f_1(t), \tag{3.141}$$

where in terms of the Brillouin function the final equilibrium value is

$$\langle\hat{S}_Z\rangle_{\mathrm{II}}=(S+1)\langle P_1\rangle_0^{\xi}=SB_S(\xi).$$

We mention the rise, decay, and rapidly reversing field transient responses, where the general relaxation equation (3.138) can often be simplified. For example, to treat the rise transient we suppose that a *strong* dc field $\mathbf{H}_{\mathrm{II}}$ is suddenly switched on at time $t=0$ (so that $\mathbf{H}_{\mathrm{I}}=0$). Thus we require the nonlinear relaxation behavior of a system of spins starting from an equilibrium state I with the *isotropic* distribution function $W_{\mathrm{eq}}^{0}=(2S+1)^{-1}$ ($t\le 0$) to another equilibrium state II with the *final* distribution function $W_{\mathrm{eq}}^{\mathrm{II}}=W_{\mathrm{eq}}^{\xi}(z)$ ($t\to\infty$). Using Eq. (3.140), Eq. (3.138) simplifies to

$$\tilde{f}_1(i\omega)=\frac{i}{\omega}\left[\Delta_1(0,\xi)-\Delta_1(\omega,\xi)\right]=\frac{i}{\omega}\left[\langle P_1\rangle_0^{\xi}-\Delta_1(\omega,\xi)\right], \tag{3.142}$$

thus, yielding $\tilde{f}_1(i\omega)$ for the rise transient.

3.3.6 *Nonlinear longitudinal relaxation time*

The overall transient behavior of the relaxation function $f_1(t)$ is characterized by the integral relaxation time [97] (see Appendix A.4)

$$\tau_{\text{int}} = \frac{1}{f_1(0)} \int_0^\infty f_1(t)dt = \frac{\tilde{f}_1(0)}{f_1(0)} \tag{3.143}$$

[i.e., the area under the normalized relaxation function $f_1(t)/f_1(0)$] which can be evaluated in series form from the zero frequency limit of Eq. (3.138) and (3.140) yielding [97]

$$\frac{\tau_{\text{int}}}{\tau_{\text{N}}} = \operatorname{csch}\frac{\xi}{2S} \sum_{n=1}^{2S} \frac{(-1)^{n+1}(2n+1)(2S+n+1)!(2S-n)!\,f_n(0)\left\langle P_n \right\rangle_0^\xi}{n(n+1)\left(S+1\right)(2S+1)!(2S)!\,f_1(0)}. \tag{3.144}$$

Moreover, the latter series result can also be written in an equivalent integral form because the master Eq. (3.125) is actually just a single variable Fokker–Planck equation, which may be integrated by quadratures. Indeed, for any system with dynamics governed by a single variable Fokker–Planck equation, e.g., Eq. (3.125), the integral relaxation time τ_{int}, characterizing the nonlinear relaxation behavior of $f_1(t)$, can always be obtained in integral form in terms of the equilibrium distribution and the diffusion coefficient $D_2(z)$ only (see Appendix A.4 for details) [97]. Hence, with Eqs. (3.125) and (A.232) from Appendix A.4, we have like the classical case [97] an *exact* integral expression for τ_{int}, viz.,

$$\tau_{\text{int}} = \frac{1}{\left(S+\frac{1}{2}\right) f_1(0)} \int_{-1}^{1} \frac{\Phi(z)\Psi(z)}{D_2(z) W_{\text{eq}}^\xi(z)} dz, \tag{3.145}$$

where

$$\Psi(z) = \left(S+\tfrac{1}{2}\right) \int_{-1}^{z} \left(x - \left\langle P_1 \right\rangle_0^\xi\right) W_{\text{eq}}^\xi(x) dx,$$

$$\Phi(z) = \left(S+\tfrac{1}{2}\right) \int_{-1}^{z} \left[W_{\text{eq}}^{\xi+\delta}(x) - W_{\text{eq}}^\xi(x)\right] dx,$$

$$f_1(0) = \left\langle P_1 \right\rangle_0^{\xi+\delta} - \left\langle P_1 \right\rangle_0^\xi = \frac{S}{S+1}\left[B_S(\xi+\delta) - B_S(\xi)\right].$$

For the limiting case $S = 1/2$, τ_{int} is independent of the parameter δ and is given by the closed expression

$$\tau_{\text{int}} = \frac{\tau_{\text{N}}}{\cosh\xi}, \tag{3.146}$$

while in the classical limit $S \to \infty$, we have the integral form

$$\tau_{\text{int}} = \frac{\tau_{\text{N}}\xi \operatorname{csch}\xi}{\langle P_1\rangle_0^{\xi+\delta} - \langle P_1\rangle_0^{\xi}} \int_{-1}^{1} \frac{\phi(z)\psi(z)e^{-\xi z}}{1-z^2} dz, \tag{3.147}$$

where

$$\langle P_1\rangle^{\xi} = \coth\xi - \xi^{-1},$$

$$\phi(z) = \int_{-1}^{z} \left[e^{\xi z'} - e^{(\xi+\delta)z'}\right] dz', \; \psi(z) = \int_{-1}^{z} \left(\cos z' - \langle P_1\rangle^{\xi}\right) e^{\xi z'} dz'.$$

agreeing entirely with the established classical result ([97], Chapter 7).

Numerical calculations show that both the series expression Eq. (3.144) and the integral Eq. (3.145) yield identical results. Thus τ_{int} for various nonlinear transient responses (such as the rise, decay, and rapidly reversing field transients) may be easily evaluated from Eq. (3.145). The normalized relaxation time $\tau_{\text{int}}/\tau_{\text{N}}$ from Eq. (3.145) is plotted in Fig. 3.12 for various values of the transient strength δ, the field strength parameter ξ and spin number S. The figure indicates that the relaxation time *decreases* with *increasing field strength* ξ with a strong dependence on *both* S and the transient strength δ. This particular nonlinear effect *comprising accelerated relaxation in the external field* also exists for classical dipoles [97]. An explanation may be given as follows. Without the field ($\xi = 0$), the relaxation time of the spin is just the *free diffusion relaxation time* τ_{N} so that $\tau_{\text{int}} = \tau_{\text{N}}$. However, in a *strong* field ($\xi >> 1$) and $S >> 1$, that relaxation time is determined by the *damped diffusion* of the spin in the field $\mathbf{H}_{\text{II}}$ and the characteristic frequency is now the frequency of the spin oscillation about $\mathbf{H}_{\text{II}}$ (in the vicinity of $z = 0$). Thus τ_{int} is then of the order of $\sim[2D_1(0)]^{-1} = \tau_{\text{N}}/\xi$ so that

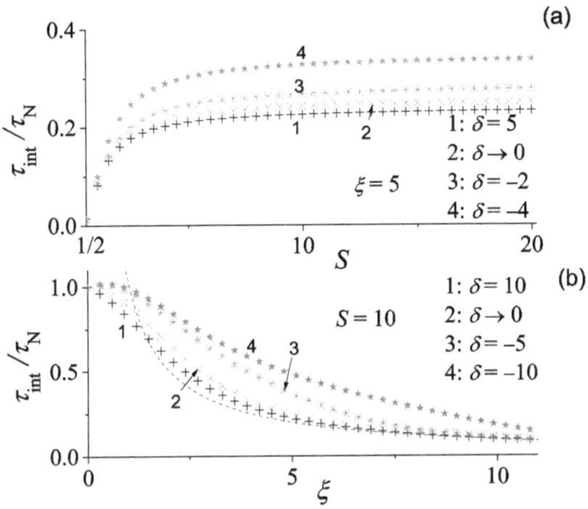

Fig. 3.12. Normalized relaxation time τ_{int}/τ_N from Eq. (3.145) as a function of S (a) and ξ (b) for various values of δ (symbols). Dashed line: Eq. (3.150).

$$\tau_{int} \sim \tau_N \xi^{-1}. \tag{3.148}$$

This asymptotic formula may be used to estimate τ_{int} for $\xi >> 1$ and $\delta > 0$ and $|\delta| << \xi$. However, for $\delta \sim -\xi$ a more accurate form ula is

$$\tau_{int} \sim \frac{\tau_N}{\xi - 1 - \xi(\xi + \delta)}. \tag{3.149}$$

The influence of the transient parameter δ entering into the relaxation time owing to the initial distribution function $W_{eq}^{\xi+\delta}$ is more pronounced for field strengths $\xi \sim 2 \div 7$ (see Fig. 3.12). The enhanced dependence of τ_{int} on δ for negative values of δ can be understood because these cases correspond to rise and rapidly reversing transients, where the initial and final distributions differ considerably. As far as the *spin dependence* of τ_{int} for $\xi >> 1$, $\delta > 0$ and $\delta << \xi$ is concerned a simple asymptotic formula for τ_{int} is (see Fig. 3.12) just

$$\tau_{int} \sim \frac{1}{2D_1(0)} = \frac{\tau_N}{2S} \operatorname{csch} \frac{\xi}{2S}. \tag{3.150}$$

3.3.7 *Linear response*

By using the above results, we may also evaluate the *linear response* of a spin system to *infinitesimally small* changes in the magnitude of the dc field $\mathbf{H}_{\text{II}}$. This is of interest as the corresponding integral relaxation time now becomes the correlation time, which has been previously evaluated [226–228] from the spin density matrix. Thus we again suppose that the uniform dc field $\mathbf{H}_{\text{II}}$ is directed along the Z-axis of the laboratory coordinate system and that a *small* probing field $\mathbf{H}_{\text{I}}$ having been applied to the assembly of noninteracting spins in the distant past ($t = -\infty$) so that equilibrium conditions obtain at time $t = 0$, is suddenly switched off at $t = 0$ Here the normalized relaxation function $f_1(t)/f_1(0)$ reduces to the longitudinal equilibrium correlation function $C(t) = \lim_{\delta \to 0} f_1(t)/f_1(0)$, that is [163,473]

$$C(t) = \frac{1}{\beta S^2 \chi_0} \left\langle \int_0^\beta \left[\hat{S}_Z(-i\lambda\hbar) - \left\langle \hat{S}_Z \right\rangle_{\text{II}} \right] \left[\hat{S}_Z(t) - \left\langle \hat{S}_Z \right\rangle_{\text{II}} \right] d\lambda \right\rangle_{\text{II}}, \quad (3.151)$$

where χ_0 is the normalized static susceptibility defined in terms of the Brillouin function as

$$\chi_0 = \frac{\partial}{\partial \xi} B_S(\xi) = \frac{1}{4S^2} \left[\operatorname{csch}^2\left(\frac{\xi}{2S} \right) - (2S+1)^2 \operatorname{csch}^2\left(\frac{2S+1}{2S} \xi \right) \right]. \quad (3.152)$$

According to linear response theory the dynamic susceptibility $\chi(\omega) = \chi'(\omega) - i\chi''(\omega)$ is defined as [163] (cf. Eq. (1.136) of Chapter 1)

$$\frac{\chi(\omega)}{\chi_0} = 1 - i\omega \int_0^\infty C(t) e^{-i\omega t} dt = 1 - i\omega \tilde{C}(i\omega), \quad (3.153)$$

where $\tilde{C}(i\omega)$ is the one-sided Fourier transform of $C(t)$. However, in linear response, the integral relaxation time, in other words, the correlation time $\tau_{\text{int}}\big|_{\delta \to 0} = \tau_{\text{cor}} = \tilde{C}(0)$ of $C(t)$, follows from the general nonlinear result, Eq. (3.145), in the limit of a very small transient strength parameter $\delta \to \infty$ and is given by

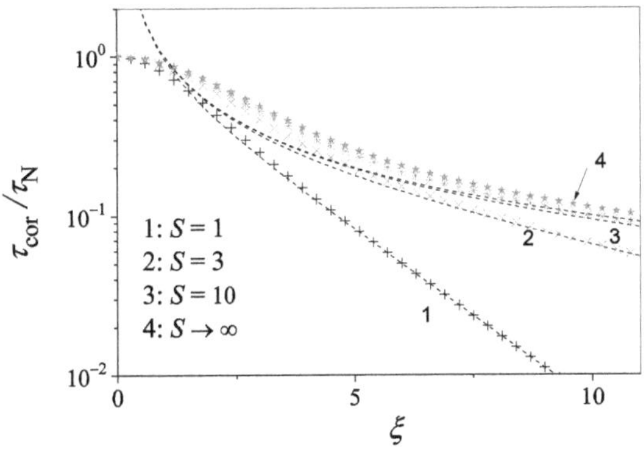

Fig. 3.13. Normalized correlation time τ_{cor}/τ_N of $C_{\parallel}(t)$ from Eq. (3.154) as a function of ξ for various S (symbols). Dashed lines: Eq. (3.150).

$$\tau_{cor} = \frac{(S+1)\left(S+\frac{1}{2}\right)}{\chi_0 S}\int_{-1}^{1}\frac{\int_{-1}^{z}\frac{\partial}{\partial\xi}W_{eq}^{\xi}(x)dx\int_{-1}^{z}\left(y-\frac{SB_S(\xi)}{S+1}\right)W_{eq}^{\xi}(y)dy}{D_2(z)W_{eq}^{\xi}(z)}dz, \quad (3.154)$$

where

$$\frac{\partial W_{eq}^{\xi}}{\partial\xi} = \frac{\mathrm{csch}(\xi/2S)-(2S+1)\mathrm{csch}(\xi+\xi/2S)(\cosh\xi - z\sinh\xi)}{2S\left[\cosh(\xi/2S)+z\sinh(\xi/2S)\right]}W_{eq}^{\xi}.$$

For the limiting case $S = 1/2$, τ_{cor} is merely equal to τ_{int} as yielded by the closed expression Eq. (3.146), while in the classical limit $S \to \infty$, Eq. (3.154) becomes

$$\tau_{cor} = \frac{\tau_N \xi\,\mathrm{csch}\,\xi}{1+\xi^{-2}-\coth^2\xi}\int_{-1}^{1}\left[z-\coth\xi + e^{-\xi(1+z)}(1+\coth\xi)\right]^2\frac{e^{\xi z}dz}{1-z^2} \quad (3.155)$$

concurring with the known result for classical spins ([97], Chapter 7; see also Eq. (2.184)). As far as the spin number dependence of τ_{cor} for $\xi >> 1$ is concerned, an asymptotic formula for $\tau_{cor} = \tau_{int}\big|_{\delta\to 0}$ is given by Eq. (3.150). Generally, τ_{cor} varies smoothly from *power law-like behavior* ($\tau_{cor} \sim \tau_N/\xi$) as $S \to \infty$ to *exponential decrease* ($\tau_{cor} = \tau_N\,\mathrm{csch}\,\xi$ for $S = 1/2$ and is plotted from the exact Eq. (3.154) in Fig. 3.13 as a function of ξ for various S; the asymptotic Eq. (3.150) is also shown for comparison.

We remarked above that the linear response has already been studied by García-Palacios and Zueco [227] using the spin density matrix. They also gave an explicit expression for the linear response integral relaxation time, which was first given by Garanin [226]. He derived his formula pertaining to a *uniaxial nanomagnet in a uniform field* with the Hamiltonian $\hat{H}_S = -\hbar\omega_0\hat{S}_Z - D\hat{S}_Z^2$, which is also valid in the limit $D \to 0$, corresponding to our case. By applying his method [226] to the symmetrized form of the collision kernel given by Eq. (3.229) below, the corresponding equations are [see for details Section 3.3.11 below, Eq. (3.201)]

$$\tau_{\text{int}} = \frac{2\tau_{\text{N}}}{\langle \hat{S}_Z \rangle_{\text{I}} - \langle \hat{S}_Z \rangle_{\text{II}}} \sum_{k=1-S}^{S} \frac{\sum_{m=k}^{S} \left(\rho_m^{\text{I}} - \rho_m^{\text{II}}\right) \sum_{m'=k}^{S} \left(m' - \langle \hat{S}_Z \rangle_{\text{II}}\right) \rho_{m'}^{\text{II}}}{[S(S+1) - k(k-1)]\sqrt{\rho_k^{\text{II}} \rho_{k-1}^{\text{II}}}}, \tag{3.156}$$

$$\tau_{\text{cor}} = \frac{2\tau_{\text{N}}}{S^2 \chi_0} \sum_{k=1-S}^{S} \frac{\left(\sum_{m=k}^{S} \left(m - \langle \hat{S}_Z \rangle_{\text{II}}\right) \rho_m^{\text{II}}\right)^2}{[S(S+1) - k(k-1)]\sqrt{\rho_k^{\text{II}} \rho_{k-1}^{\text{II}}}}, \tag{3.157}$$

where

$$\chi_0 = S^{-2}\left(\sum_{m=-S}^{S} m^2 \rho_m^{\text{II}} - \langle \hat{S}_Z \rangle_{\text{II}}^2\right),$$
$$\langle \hat{S}_Z \rangle_i = \sum_{m=-S}^{S} m\rho_m^i, \ \rho_n^i = e^{\xi_i n/S}\left(\sum_{m=-S}^{S} e^{\xi_i m/S}\right)^{-1}. \tag{3.158}$$

Although the integral and series expressions Eqs. (3.145), (3.154) and (3.156), (3.157), respectively, have outwardly very different forms, nevertheless numerical calculation shows that they yield *identical* results thereby establishing a vital corollary between the phase-space and density matrix methods.

3.3.8 *Single-mode approximation*

Although the exact continued fraction solution, Eq. (3.138), is effective in numerical calculations, it has one significant drawback; namely, the qualitative behavior of the system is not at all obvious in a physical sense.

Thus, to gain such an understanding of the relaxation process, we shall use the single mode approximation suggested by Shibata *et al.* [243,244] and Kalmykov *et al.* [97,256,257] for the relaxation of quantum and classical spins. We first recall that the spectrum $\tilde{f}_1(i\omega)$ from Eq. (3.138) on Fourier inversion, indicates that the time behavior of the relaxation function $f_1(t)$ in general comprises $2S$ exponentials

$$f_1(t) = f_1(0)\sum_{k=1}^{2S} c_k e^{-\lambda_k t}, \tag{3.159}$$

where the λ_k are the eigenvalues of the tridiagonal system matrix **A** with the matrix elements

$$(\mathbf{A})_{qp} = \delta_{pq+1} q_p^- + \delta_{pq} q_p + \delta_{pq-1} q_p^+.$$

In the frequency domain, the corresponding spectrum $\tilde{f}_1(i\omega)$ is thus the series of $2S$ Lorentzians

$$\tilde{f}_1(i\omega) = f_1(0)\sum_{k=1}^{2S} \frac{c_k}{\lambda_k + i\omega}. \tag{3.160}$$

According to Eq. (3.160), the finite number of relaxation modes (corresponding to the eigenvalues λ_k) each contribute to the spectrum $\tilde{f}_1(i\omega)$. However, as we shall see below, these *near-degenerate* individual modes are indistinguishable in the spectrum $\tilde{f}_1(i\omega)$ appearing merely as a single band suggesting once more that $\tilde{f}_1(i\omega)$ may be approximated by the *single* Lorentzian

$$\frac{\tilde{f}_1(i\omega)}{f_1(0)} \approx \frac{\tau_{\text{int}}}{1 + i\omega\tau_{\text{int}}}, \tag{3.161}$$

where τ_{int} is given by Eq. (3.145). In the time domain, the single-mode approximation Eq. (3.161) amounts to the *Ansatz* that the relaxation function $f_1(t)$ as determined by Eq. (3.159) (comprising $2S$ exponentials) may be approximated by a *single* one, viz.,

$$f_1(t) \approx f_1(0) e^{-t/\tau_{\text{int}}}. \tag{3.162}$$

García-Palacios and Zueco [227,228] have also used the single mode approximation to evaluate the *linear* response of an isotropic spin system.

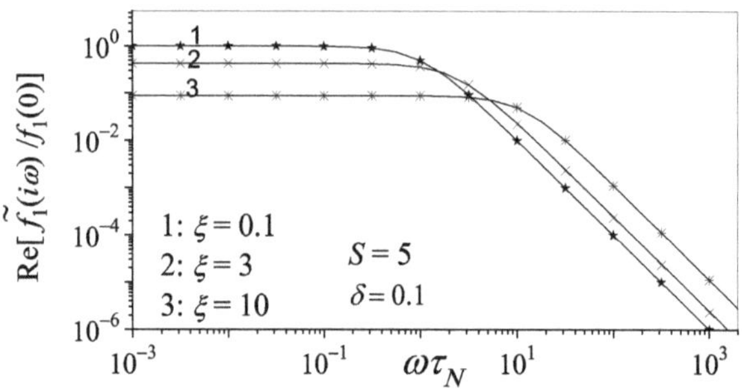

Fig. 3.14. The real parts of the normalized spectra $\tilde{f}_1(i\omega)/f_1(0)$ vs. the normalized frequency $\omega\tau_N$ evaluated from the exact continued fraction solution [Eq. (3.138): solid lines] for $S = 5$, $\delta = 0.1$, and various ξ compared with those calculated from the single Lorentzian approximation Eq. (3.161) (symbols).

In linear response, Eqs. (3.161) and (3.162) can be reformulated for the susceptibility $\chi(\omega)$ and correlation function $C(t)$ as the single exponential

$$C(t) \approx e^{-t/\tau_{\mathrm{cor}}} \tag{3.163}$$

and

$$\chi(\omega) \approx \frac{\chi_0}{1 + i\omega\tau_{\mathrm{cor}}}, \tag{3.164}$$

where τ_{cor} is given by Eq. (3.157).

To test the single mode approximation, we plot in Figs. 3.14 and 3.15 the real parts of the normalized spectra $\tilde{f}_1(i\omega)/f_1(0)$ as calculated both from the exact continued fraction solution [Eq. (3.138): solid lines] and that resulting from Eq. (3.161). Clearly, no practical difference exists between the continued fraction solution and the single mode one [the maximum deviation between the corresponding curves does not exceed a few percent]. Like the classical case ([97], Chapter 7), the single mode approximation is accurate because the finite number ($2S$) of relaxation modes are *near-degenerate* appearing merely as a *single* high-frequency band in the spectrum. Thus, they may be effectively approximated by a single mode, i.e., both the linear and nonlinear longitudinal relaxation for all S is accurately described by the Bloch equation

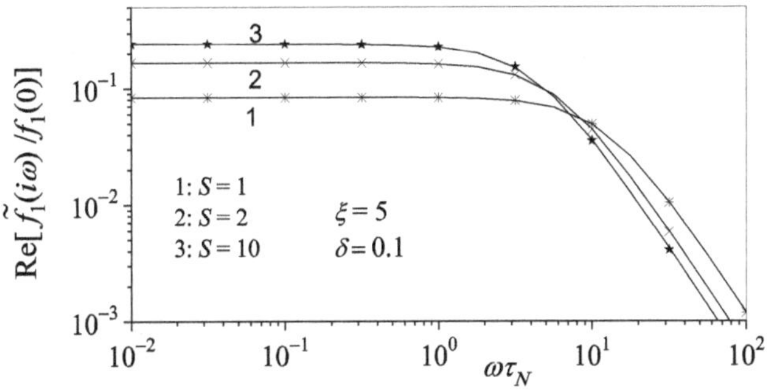

Fig. 3.15. $\tilde{f}_1(i\omega)/f_1(0)$ vs. $\omega\tau_N$ evaluated from the exact continued fraction solution [Eq. (3.138): solid lines] for $\xi = 5$, $\delta = 0.1$ and various S compared with those calculated from the single Lorentzian approximation Eq. (3.161) (symbols).

$$\frac{d}{dt}\left\langle \hat{S}_Z \right\rangle(t) + \frac{1}{T_{\|}}\left(\left\langle \hat{S}_Z \right\rangle(t) - \left\langle \hat{S}_Z \right\rangle_{\mathrm{II}}\right) = 0, \tag{3.165}$$

where $T_{\|} = \tau_{\mathrm{int}}$ is the longitudinal relaxation time and $\left\langle \hat{S}_Z \right\rangle_{\mathrm{II}} = SB_S(\xi)$ is the equilibrium average of the operator $\hat{S}_Z$.

We have treated nonlinear spin relaxation of noninteracting spins using phase-space quasiprobability density evolution equations obtained via the extension of Wigner's formulation of quantum mechanics to *open* systems. The calculations show that in particular limiting cases [e.g., the correlation time Eq. (3.154)] *the results reduce to established ones obtained using the evolution equation for the density matrix in the second order of perturbation theory in the spin-bath coupling*. Thus, we have an important check on the validity of our approach by demonstrating the equivalence of the two methods. Both exact (continued fraction) and approximate (single mode) solutions are given. The continued fraction solution yields the dependence of the longitudinal spin relaxation on the spin number S in closed form. Moreover, this solution is dominated by a *single* exponential having as time constant the integral relaxation time τ_{int}, which strongly depends on both S and the field strength for arbitrary S. Hence, *an accurate description in terms of a Bloch equation* (3.165) *holds even for the nonlinear response of a giant spin*.

Thus, we have explicitly demonstrated for noninteracting spins in an external magnetic field that the existing methods of solution of the classical Fokker–Planck equation (continued fractions, which can be evaluated by iterating a simple algorithm, integral representation of relaxation times, etc.) seamlessly carry over to the quantum case. Again, the methods suggest new closed form quantum results via the corresponding classical ones. An example being the quantum integral relaxation time, Eq. (3.145). We have illustrated the phase-space method via the rudimentary problem of the longitudinal relaxation of a spin in a uniform magnetic field of arbitrary strength directed along the Z-axis [the relaxation of the transverse components of the magnetization can be treated in like manner using the master Eq. (3.113) and the associated quantum recurrence Eq. (3.124)]. This problem is the simplest example of the phase-space method for spins and may be considered as the *rotational analog of the Agarwal problem for the translational harmonic oscillator model* [272,314] described by the master equation Eq. (1.522). Like translational oscillators, the phase-space master Eq. (3.125) for spins has essentially the form of a Fokker–Planck equation. This is not, however, true in general, e.g., for noncircularly symmetric magnetocrystalline anisotropy and external field potentials, where the corresponding master equation may assume a very complicated form indeed. Nevertheless, the simple noninteracting spins problem indicates how one may treat the influence of spin number S on the relaxation behavior using the phase-space method. A factor which is both essential in the formation of magnetic clusters and for nanomagnets in relation to the quest for macroscopic quantum tunneling.

3.3.9 *Longitudinal relaxation of uniaxial nanomagnets*

In contrast to the phase-space mapping used above, we shall now use the density matrix method. We shall consider a uniaxial nanomagnet of arbitrary spin number S in an external *constant* magnetic field $\mathbf{H}_0$ applied along the Z axis, i.e., the axis of symmetry, with the Hamiltonian $\hat{H}_S$ defined by Eq. (3.83) with $\sigma \neq 0$ and $\xi \neq 0$. On this occasion however instead of the relatively complicated phase-space master Eq. (3.91), we shall use the density matrix method in the form based on the correspondence between the averages of polarization operators and averages of spherical harmonics as described in Section 1.8.4 (see also Section 1.3.4). However, as seen, Garanin and García–Palacios *et al.* [226–228] have also treated a uniaxial nanomagnet in a uniform longitudinal field via the quantum Hubbard operator representation of the evolution equation for the spin density matrix. In the Hamiltonian Eq. (3.83), the diagonal terms of the density matrix *decouple* from the nondiagonal ones. Hence, only the former contribute to the time evolution of the longitudinal component of the spin operator so facilitating a simple treatment. Thus, to describe the longitudinal relaxation of a uniaxial nanomagnet, in which case only the *diagonal* terms of the density matrix are involved, the evolution Eq. (3.84) with the collision kernel given by (3.88) simplifies to

$$\frac{\partial \hat{\rho}_S}{\partial t} = -\frac{1}{\tau_{\mathrm{N}}}\left\{ e^{-\frac{\sigma}{2S^2}-\frac{\xi}{2S}}[\hat{S}_{-1}e^{-\frac{\sigma}{S^2}\hat{S}_0}\hat{\rho}_S, \hat{S}_{+1}] + e^{\frac{\sigma}{2S^2}+\frac{\xi}{2S}}[\hat{S}_{+1}e^{\frac{\sigma}{S^2}\hat{S}_0}\hat{\rho}_S, \hat{S}_{-1}] \right\}. \quad (3.166)$$

This simplified method will be described in detail in Section 3.3.10. Meantime the associated evolution equation for the phase-space distribution function $W_S(\cos\vartheta, t)$ corresponding to the reduced density matrix evolution Eq. (3.84) is then given by Eq. (3.91), namely,

$$\frac{\partial W_S}{\partial t} = \frac{1}{\tau_{\mathrm{N}}}\frac{\partial}{\partial z}\left(\frac{1-z^2}{2}\left[\frac{\partial}{\partial z}(R_+'^{(S)}W_S) + z\frac{\partial}{\partial z}(R_-'^{(S)}W_S) + 2SR_-'^{(S)}W_S\right]\right) \quad (3.167)$$

(because the azimuthal angle φ dependence of W_S may be ignored in longitudinal relaxation). In the classical limit, Eq. (3.167) further reduces to the Fokker–Planck equation for a classical uniaxial nanomagnet in a dc magnetic field, Eq. (3.93) [17,97,239].

In general, the solution of Eq. (3.167) for an arbitrary S poses a formidable computational problem. Therefore, we shall now proceed with the density matrix method (however, utilizing only the *equivalence* between the averages of polarization operators and averages of spherical harmonics in this instance), we suppose as before that the magnitude of an external uniform dc magnetic field is suddenly altered at time $t = 0$ from $\mathbf{H}_{\mathrm{I}}$ to $\mathbf{H}_{\mathrm{II}}$ (the magnetic fields $\mathbf{H}_{\mathrm{I}}$ and $\mathbf{H}_{\mathrm{II}}$ are applied parallel to the Z axis of the laboratory coordinate system). Thus we study as in the classical case

[97], the nonlinear transient longitudinal relaxation of a system of spins starting from an equilibrium state I with density matrix $\hat{\rho}_{\text{eq}}^{\text{I}}$ $(t \leq 0)$ say to a new equilibrium state II with density matrix $\hat{\rho}_{\text{eq}}^{\text{II}}$ $(t \to \infty)$ say, see Fig. 3.11. Simultaneously the longitudinal component of the spin $\langle \hat{S}_Z \rangle(t)$ relaxes from the equilibrium value $\langle \hat{S}_Z \rangle_{\text{I}}$ to the new value $\langle \hat{S}_Z \rangle_{\text{II}}$, the ensuing transient response being described by the relaxation function $\langle \hat{S}_Z \rangle(t) - \langle \hat{S}_Z \rangle_{\text{II}}$. The transient response so formulated is again truly *nonlinear* because the *change* in amplitude $H_{\text{I}} - H_{\text{II}}$ of the external dc magnetic field is now *arbitrary* (the linear response is just the particular case $|H_{\text{I}} - H_{\text{II}}| \to 0$). The equilibrium phase-space distributions W_{eq}^{I} and $W_{\text{eq}}^{\text{II}}$ corresponding to the equilibrium spin density matrices $\hat{\rho}_{\text{eq}}^{\text{I}}$ and $\hat{\rho}_{\text{eq}}^{\text{II}}$ comprise the appropriate stationary (time-independent) solutions of Eq. (3.167). These distributions have been extensively studied in Section 3.2 and are given by Eq. (3.36), viz.,

$$(S+1/2)W_{\text{eq}}^{i}(\vartheta) = \sum_{L=0}^{2S} (L+1/2)\langle P_L \rangle_i P_L(\cos\vartheta), \tag{3.168}$$

where $i = \text{I, II}$ and $\langle P_L \rangle_i$ are the equilibrium averages of the Legendre polynomials P_L defined by Eq. (3.37).

As far as the *transient response* is concerned, according to the multi-term differential-recurrence relation Eq. (3.100) for the relaxation functions $c_L(t) = \langle \hat{T}_{L0}^{(S)} \rangle(t) - \langle \hat{T}_{L0}^{(S)} \rangle_{\text{II}}$ in terms of polarization operators, the behavior of any selected $c_L(t)$ is coupled to that of all the others so forming as usual a *finite* hierarchy of the averages of operators (because the index L ranges only between 0 and $2S$). The solution of such a multi-term recurrence relation may always be obtained (as we saw in Chapter 1) by rewriting it as a first-order linear matrix differential equation with constant coefficients, Eq. (1.112), where the column vector

$$\mathbf{C}(t) = \begin{pmatrix} c_1(t) \\ c_2(t) \\ \vdots \\ c_{2S}(t) \end{pmatrix}. \tag{3.169}$$

contains just $2S$ rows (the index L ranges between 1 and $2S$) and the $2S \times 2S$ system matrix $\mathbf{A}$ has the matrix elements defined as

$$(\mathbf{A})_{nm} = -\tau_{\mathrm{N}}^{-1} g_{nm}^{S} \tag{3.170}$$

with g_{nm}^{S} given by Eq. (3.101). For example, for $S = 1$, the system matrix **A** merely takes on the simple two by two form

$$\mathbf{A} = \frac{e^{-\sigma/2}}{\tau_{\mathrm{N}}} \begin{pmatrix} \cosh(\xi_{\mathrm{II}}/2) & \frac{1}{\sqrt{3}}\left(2e^{\sigma}-1\right)\sinh(\xi_{\mathrm{II}}/2) \\ -\sqrt{3}\sinh(\xi_{\mathrm{II}}/2) & \left(2e^{\sigma}+1\right)\cosh(\xi_{\mathrm{II}}/2) \end{pmatrix}. \tag{3.171}$$

The initial conditions for the relaxation functions $c_L(t)$ are

$$c_L(0) = \left\langle \hat{T}_{L0}^{(S)} \right\rangle_{\mathrm{I}} - \left\langle \hat{T}_{L0}^{(S)} \right\rangle_{\mathrm{II}}. \tag{3.172}$$

In general, the solution of the homogeneous matrix Eq. (1.112), is given by Eq. (1.113) [142], namely,

$$\mathbf{C}(t) = e^{-\mathbf{A}t}\mathbf{C}(0) = \mathbf{U}e^{-\mathbf{\Lambda}t}\mathbf{U}^{-1}\mathbf{C}(0), \tag{3.173}$$

where $\mathbf{\Lambda}$ is a diagonal matrix composed of the eigenvalues $\lambda_1, \lambda_1, \ldots, \lambda_{2S}$ of the system matrix **A** and **U** is a right eigenvector matrix composed of all the eigenvectors of **A**. *All* the λ_k are real and positive. The one-sided Fourier transform of Eq. (1.112) yields $\tilde{\mathbf{C}}(i\omega) = (\mathbf{A} + i\omega\mathbf{I})^{-1}\mathbf{C}(0)$. The function $\tilde{c}_1(0)$ is the first row of the column vector $\tilde{\mathbf{C}}(0)$ which is given by $\tilde{\mathbf{C}}(0) = \mathbf{A}^{-1}\mathbf{C}(0)$. The formal matrix solution, Eq. (3.173), will then yield the longitudinal relaxation function $\langle \hat{S}_Z \rangle(t) - \langle \hat{S}_Z \rangle_{\mathrm{II}}$ [cf. Eq. (A.147) of Appendix A.2], viz.,

$$\left\langle \hat{S}_Z \right\rangle(t) - \left\langle \hat{S}_Z \right\rangle_{\mathrm{II}} = \sqrt{\frac{S(S+1)(2S+1)}{3}}\, c_1(t), \tag{3.174}$$

and its spectrum as well as the effective and integral relaxation times from their definitions (see Appendix A.4 and Ref. [97])

$$\tau_{\mathrm{ef}} = -\frac{c_1(0)}{\dot{c}_1(0)},$$
$$\tau_{\mathrm{int}} = \frac{1}{\left\langle \hat{S}_Z \right\rangle_{\mathrm{I}} - \left\langle \hat{S}_Z \right\rangle_{\mathrm{II}}} \int_0^t \left[\left\langle \hat{S}_Z \right\rangle(t) - \left\langle \hat{S}_Z \right\rangle_{\mathrm{II}} \right] dt = \frac{\tilde{c}_1(0)}{c_1(0)}. \tag{3.175}$$

In accordance with the matrix Eq. (3.173), the relaxation function $c_1(t)$ and the associated effective and integral relaxation times are given by [97]

$$c_1(t) = \sum_{k=1}^{2S} u_{1k} r_k e^{-\lambda_k t}, \quad \tau_{\text{ef}} = \frac{\sum_{k=1}^{2S} u_{1k} r_k}{\sum_{k=1}^{2S} u_{1k} r_k \lambda_k}, \quad \tau_{\text{int}} = \frac{\sum_{k=1}^{2S} u_{1k} r_k \lambda_k^{-1}}{\sum_{k=1}^{2S} u_{1k} r_k}, \tag{3.176}$$

where the u_{1k} are the matrix elements of the eigenvector matrix $\mathbf{U}$ as defined above and r_k are those of the associated column vector $\mathbf{U}^{-1}\mathbf{C}(0)$. As usual, both the integral and effective relaxation times each contain contributions from *all* the eigenvalues λ_k and so they characterize the overall relaxation behavior, while the inverse of the smallest nonvanishing eigenvalue λ_1 characterizes the spin reversal time. Furthermore, because the influence of the high-frequency relaxation modes on the low-frequency relaxation may often be ignored, λ_1 usually provides adequate information concerning the low-frequency dynamics of the system (see Section 3.3.11).

Obviously, the matrix method also allows us to evaluate the *linear response* of a spin system due to infinitesimally small changes in the magnitude of the dc field, evaluated in [226–228] via the spin density matrix. Thus we again suppose that the uniform dc field $\mathbf{H}_{\text{II}}$ is directed along the Z axis of the laboratory coordinate system and that a small probing field $\mathbf{H}_{\text{I}}$ $(\mathbf{H}_{\text{I}} \parallel \mathbf{H}_{\text{II}})$ having been applied to the assembly of spins in the distant past $(t = -\infty)$ so that equilibrium conditions obtain by the time $t = 0$, is then switched off. The only difference lies in the initial conditions. Instead of the *general* Eq. (3.172) pertaining to the transient response of *arbitrary strength*, in *linear response*, $\xi_{\text{I}} - \xi_{\text{II}} = \varepsilon << 1$, they simply become

$$\begin{aligned} c_L(0) &= \left\langle \hat{T}_{L0}^{(S)} \right\rangle(0) - \left\langle \hat{T}_{L0}^{(S)} \right\rangle_{\text{II}} \\ &\approx \frac{\varepsilon}{\beta} \operatorname{Tr}\left\{ \hat{\rho}_{\text{eq}}^{\text{II}} \int_0^\beta \hat{S}_0(-i\lambda\hbar) d\lambda \left(\hat{T}_{L0}^{(S)} - \left\langle \hat{T}_{L0}^{(S)} \right\rangle_{\text{II}} \hat{I}^{(S)} \right) \right\}. \end{aligned} \tag{3.177}$$

Here we have used the following identity concerning an exponential function of two operators [163]

$$e^{\beta(\hat{a}+\hat{b})} = e^{\beta\hat{a}}\left(1+\int_0^\beta e^{-\lambda\hat{a}}\hat{b}e^{\lambda(\hat{a}+\hat{b})}d\lambda\right). \tag{3.178}$$

Furthermore, $c_1(t)/c_1(0)$ reduces to the normalized longitudinal correlation function $C(t) = \lim_{\varepsilon\to 0} c_1(t)/c_1(0)$ given by [163,473] [cf. Eq. (3.151)]

$$C(t) = \frac{1}{\beta S^2\chi_0}\left\langle \int_0^\beta \left[\hat{S}_Z(-i\lambda\hbar) - \left\langle \hat{S}_Z\right\rangle_{\text{II}}\right]\left[\hat{S}_Z(t) - \left\langle \hat{S}_Z\right\rangle_{\text{II}}\right]d\lambda\right\rangle_{\text{II}}, \tag{3.179}$$

where

$$\begin{aligned}\chi_0 &= \frac{1}{\beta S^2}\left\langle \int_0^\beta \left[\hat{S}_Z(-i\lambda\hbar) - \left\langle \hat{S}_Z\right\rangle_{\text{II}}\right]\left[\hat{S}_Z(0) - \left\langle \hat{S}_Z\right\rangle_{\text{II}}\right]d\lambda\right\rangle_{\text{II}} \\ &\approx S^{-2}\left(\left\langle \hat{S}_Z^2\right\rangle_{\text{II}} - \left\langle \hat{S}_Z\right\rangle_{\text{II}}^2\right)\end{aligned} \tag{3.180}$$

is the normalized static susceptibility. Then we have the dynamic susceptibility $\chi(\omega) = \chi'(\omega) - i\chi''(\omega)$ defined by Eq. (3.153) [163]. We have also as in the classical case [97] the integral and effective relaxation times $\tau_{\text{cor}} = \tilde{C}(0)$ and $\tau_{\text{ef}} = -1/\dot{C}(0)$, which represent the characteristic times governing the overall and short-time behavior of $C(t)$.

According to the formal definitions embodied in Eqs. (3.153) and (3.176), the normalized dynamic susceptibility is once more a finite sum of Lorentzians, viz.,

$$\frac{\chi(\omega)}{\chi_0} = \sum_{p=1}^{2S}\frac{c_p}{1+i\omega/\lambda_p}, \tag{3.181}$$

where

$$c_p = u_{1p}r_p\left(\sum_{m=1}^{2S}u_{1m}r_m\right)^{-1} \text{ and } \sum_{p=1}^{2S}c_p = 1.$$

Moreover, in the low- ($\omega\to 0$) and high- ($\omega\to\infty$) frequency limits, the behavior of the dynamic susceptibility can be easily discerned as in the classical case [97]. We have from Eqs. (3.176) and (3.181) for the limits $\omega\to 0$ and for $\omega\to\infty$ respectively (cf. Eq. (1.147) of Chapter 1)

$$\chi(\omega) \approx \chi_0\left(1 - i\omega\tau_{\text{cor}} + \ldots\right),\ \omega\to 0, \tag{3.182}$$

$$\chi(\omega) \sim \chi_0 \left(i\omega\tau_{\mathrm{ef}}\right)^{-1} + \ldots, \quad \omega \to \infty. \tag{3.183}$$

Furthermore, the equilibrium averages $\left\langle \hat{S}_Z \right\rangle_{\mathrm{I}}$, $\left\langle \hat{S}_Z \right\rangle_{\mathrm{II}}$, and $\left\langle \hat{S}_Z^2 \right\rangle_{\mathrm{II}}$ can all be expressed in terms of either the density matrix elements or the phase-space distribution as

$$\begin{gathered}
\left\langle \hat{S}_Z \right\rangle_i = \sum_{m=-S}^{S} m\rho_m^i, \quad \left\langle \hat{S}_Z^2 \right\rangle_i = \sum_{m=-S}^{S} m^2 \rho_m^i, \\
\rho_m^i = e^{\xi_i m/S + \sigma m^2/S^2} \left(\sum_{n=-S}^{S} e^{\xi_i n/S + \sigma n^2/S^2} \right)^{-1}, \\
\left\langle \hat{S}_Z \right\rangle_i = \left(S + \tfrac{1}{2}\right)(S+1) \int_{-1}^{1} z W_S^i(z) dz, \\
\left\langle \hat{S}_Z^2 \right\rangle_{\mathrm{II}} = \left(S + \tfrac{1}{2}\right)(S+1) \int_{-1}^{1} \left[(S + \tfrac{3}{2}) z^2 - \tfrac{1}{2} \right] W_S^{\mathrm{II}}(z) dz,
\end{gathered} \tag{3.184}$$

because the Weyl symbols of the operators $\hat{S}_Z$ and $\hat{S}_Z^2$ are, respectively,

$$S_Z = (S+1)\cos\vartheta \text{ and } S_Z^2 = (S+1)\left[(S + \tfrac{3}{2})\cos^2\vartheta - \tfrac{1}{2} \right].$$

Thus one can calculate numerically and, in essence, without using the phase-space master Eq. (3.167) at all, the integral (τ_{ef} and τ_{cor}), effective (τ_{ef}), and longest ($\tau = 1/\lambda_1$) relaxation times as well as the dynamic susceptibility $\chi(\omega)$ for a uniaxial nanomagnet. Moreover, all these observables can be calculated analytically via Eq. (3.184).

3.3.10 *Analytic equations for the characteristic relaxation times and dynamic susceptibility*

Although in general the method of determining observables based on the correspondence between averages of polarization operators discussed in Sections 1.3.4 and 1.8.4 circumvents the complicated phase-space evolution equation yet another much simpler method of treating circularly symmetric problems, which also yields the characteristic relaxation times in closed form, exists. The reason being that for the Hamiltonian given by Eq. (3.83), the diagonal elements of the density matrix *decouple* from the nondiagonal ones. Hence, *only the former* contribute to the time evolution so forming the basis of our simple treatment involving scalar recurrence relations only which we now explain. To appreciate this, we first transform the reduced density-matrix evolution Eq. (3.166) into an evolution equation for its *individual* matrix elements. Thus, we have *directly* from Eq. (3.166) a scalar three-term differential-recurrence equation for the *diagonal* matrix elements $\rho_m = \rho_{mm}$ of the density matrix, viz.,

$$\tau_{\mathrm{N}} \frac{d}{dt} \rho_m(t) = q_m^- \rho_{m-1}(t) + q_m \rho_m(t) + q_m^+ \rho_{m+1}(t),\ t > 0, \qquad (3.185)$$

where $m = -S, -S+1, \ldots, S$, and

$$q_m = -a_m^- e^{-(2m-1)\frac{\sigma}{2S^2} - \frac{\xi_{\mathrm{II}}}{2S}} - a_m^+ e^{(2m+1)\frac{\sigma}{2S^2} + \frac{\xi_{\mathrm{II}}}{2S}},\ q_m^{\pm} = a_m^{\pm} e^{\mp(2m\pm1)\frac{\sigma}{2S^2} \mp \frac{\xi_{\mathrm{II}}}{2S}},$$

$$a_m^{\pm} = -S_{mm\pm1}^{\mp1} S_{m\pm1m}^{\pm1} = \frac{(S \mp m)(S \pm m + 1)}{2}, \qquad (3.186)$$

$$S_{m\pm1m}^{\pm1} = \mp\sqrt{\frac{(S \mp m)(S \pm m + 1)}{2}}.$$

Here $S_{m\pm1m}^{\pm1}$ [cf. Eq. (A.137) from Appendix A.2] are merely the matrix elements of the spherical spin operators $\hat{S}_{\pm1}$. Equation (3.185) is also accompanied by the initial condition $\rho_m(0) = \rho_m^{\mathrm{I}}$. Moreover, due to our usual *Ansatz* that the equilibrium spin density matrix $\hat{\rho}_{\mathrm{eq}}$ must render the collision kernel zero, substitution of the final equilibrium matrix element ρ_m^{II} given by Eq. (3.184) into the right-hand side of Eq. (3.185) requires that

$$q_m^- \rho_{m-1}^{\mathrm{II}} + q_m \rho_m^{\mathrm{II}} + q_m^+ \rho_{m+1}^{\mathrm{II}} = 0. \qquad (3.187)$$

Consequently ρ_m^{II} is the stationary solution of Eq. (3.185). Next, to determine the integral relaxation time, Eq. (3.175), we introduce the set of relaxation functions $f_m(t) = \rho_m(t) - \rho_m^{\mathrm{II}}$. Then the $f_m(t)$ also satisfy the recurrence Eq. (3.185) with the initial conditions $f_m(0) = \rho_m^{\mathrm{I}} - \rho_m^{\mathrm{II}}$. Furthermore since

$$\langle \hat{S}_Z \rangle(t) - \langle \hat{S}_Z \rangle_{\mathrm{II}} = \sum_{m=-S}^{S} m f_m(t) \text{ and } \langle \hat{S}_Z \rangle(0) - \langle \hat{S}_Z \rangle_{\mathrm{II}} = \langle \hat{S}_Z \rangle_{\mathrm{I}} - \langle \hat{S}_Z \rangle_{\mathrm{II}},$$

the one-sided Fourier transform $\tilde{c}_1(i\omega)/c_1(0)$ of the normalized relaxation function $c_1(t)/c_1(0)$ is

$$\frac{\tilde{c}_1(i\omega)}{c_1(0)} = \frac{1}{\langle \hat{S}_Z \rangle_{\mathrm{I}} - \langle \hat{S}_Z \rangle_{\mathrm{II}}} \sum_{m=-S}^{S} m\tilde{f}_m(i\omega). \tag{3.188}$$

Thus, the integral relaxation time τ_{int} is (as usual) by definition [cf. Eq. (3.175)]

$$\tau_{\mathrm{int}} = \frac{1}{\langle \hat{S}_Z \rangle_{\mathrm{I}} - \langle \hat{S}_Z \rangle_{\mathrm{II}}} \sum_{m=-S}^{S} m\tilde{f}_m(0), \tag{3.189}$$

where the $\langle \hat{S}_Z \rangle_i$ are defined by Eq. (3.184).

The spectrum $\tilde{c}_1(i\omega)/c_1(0)$ and the integral relaxation time τ_{int} can now be calculated analytically using continued fractions, which starting from Eq. (3.185), we may describe as follows. For convenience, we first introduce a new index n defined as $n = m + S$. Thus the differential-recurrence Eq. (3.185) can then be rearranged as one for the relaxation functions $f_n(t)$ as defined above, viz.,

$$\tau_{\mathrm{N}} \frac{df_n}{dt} = p_n^- f_{n-1} + p_n f_n + p_n^+ f_{n+1}, \tag{3.190}$$

where the new coefficients $p_n^\pm$ and p_n are [cf. Eq. (3.186)]

$$p_n = -\frac{n(2S-n+1)}{2} e^{-(2n-2S+1)\frac{\sigma}{2S^2} - \frac{\xi_{\mathrm{II}}}{2S}} - \frac{(n+1)(2S-n)}{2} e^{(2n-2S+1)\frac{\sigma}{2S^2} + \frac{\xi_{\mathrm{II}}}{2S}},$$

$$p_n^+ = \frac{(2S-n)(n+1)}{2} e^{-(2n-2S-1)\frac{\sigma}{2S^2} - \frac{\xi_{\mathrm{II}}}{2S}},$$

$$p_n^- = \frac{n(2S-n+1)}{2} e^{(2n-2S-1)\frac{\sigma}{2S^2} + \frac{\xi_{\mathrm{II}}}{2S}}.$$

Consequently, the new recurrence relation Eq. (3.190) can now be written in the *homogeneous* matrix form

$$\tau_{\mathrm{N}} \dot{\mathbf{F}}(t) = \mathbf{\Pi}\mathbf{F}(t), \tag{3.191}$$

where the column vector $\mathbf{F}(t)$ and the tridiagonal system matrix $\mathbf{\Pi}$ are

$$\mathbf{F}(t) = \begin{pmatrix} f_0(t) \\ f_1(t) \\ \vdots \\ f_{2S}(t) \end{pmatrix}, \quad \mathbf{\Pi} = \begin{pmatrix} p_0 & p_0^+ & 0 & \cdots & 0 \\ p_1^- & p_1 & p_1^+ & \cdots & \vdots \\ \vdots & \vdots & \vdots & \ddots & p_{2S-1}^+ \\ 0 & \cdots & 0 & p_{2S}^- & p_{2S} \end{pmatrix}. \tag{3.192}$$

The tridiagonal system matrix $\mathbf{\Pi}$ has exactly the *same eigenvalues* as the actual system matrix $\mathbf{A}$ given by Eq. (3.170) save that it possesses an *additional* zero eigenvalue $\lambda_0 = 0$ corresponding to the thermal equilibrium state. Clearly, the matrix recurrence Eq. (3.191) could again be solved numerically by the matrix methods described in Chapter 1. Instead, we prefer to obtain the *exact analytic solution* in terms of scalar continued fractions. Applying the general method of solution of inhomogeneous three-term recurrence relations to the Fourier-Laplace transform of the scalar Eq. (3.190) ([97], Section 2.7.3), we then have the solution

$$\tilde{f}_n(i\omega) = \Delta_n(\omega) p_n^- \tilde{f}_{n-1}(i\omega) + \frac{\tau_{\mathrm{N}}}{p_{n-1}^+} \sum_{l=n}^{2S} \prod_{k=n}^{l} \left(p_{k-1}^+ \Delta_k(\omega) \right) f_l(0), \tag{3.193}$$

where $\Delta_n(\omega)$ are the continued fractions defined by the recurrence equation

$$\Delta_n(\omega) = \frac{1}{i\omega\tau_{\mathrm{N}} - p_n - p_n^+ p_{n+1}^- \Delta_{n+1}(\omega)}$$

with $0 \le n \le 2S$ and $\Delta_{2S+1}(\omega) = 0$. The spectrum $\tilde{c}_1(i\omega)/c_1(0)$ from the definition Eq. (3.188) is then given by

$$\frac{\tilde{c}_1(i\omega)}{c_1(0)} = \frac{1}{\langle \hat{S}_Z \rangle_{\mathrm{I}} - \langle \hat{S}_Z \rangle_{\mathrm{II}}} \sum_{n=1}^{2S} n \tilde{f}_n(i\omega). \tag{3.194}$$

For $\omega = 0$, Eq. (3.193) further simplifies yielding the recurrence equation

$$\tilde{f}_n(0) = d_n \tilde{f}_{n-1}(0) + r_n, \tag{3.195}$$

where the various coefficients are

$$d_n = e^{(2n-2S-1)\frac{\sigma}{S^2} + \frac{\xi_{\mathrm{II}}}{S}}, \quad r_n = \frac{2\tau_{\mathrm{N}} e^{(2n-2S-1)\frac{\sigma}{2S^2} + \frac{\xi_{\mathrm{II}}}{2S}}}{n(2S-n+1)} \sum_{l=n}^{2S} f_l(0)$$

because $\Delta_n(0)=(p_{n-1}^{+})^{-1}$. However, the modified three-term recurrence equations (3.190) are no longer *linearly independent*, because the determinant of the matrix $\mathbf{\Pi}$ generated from Eq. (3.190) is zero ($\det\mathbf{\Pi}=0$). Thus, all the subsequent $\tilde{f}_n(0)$ needed to calculate the integral relaxation time from Eq. (3.194) *et seq*. can be determined only in terms of $\tilde{f}_0(0)$. Nevertheless, in order to calculate $\tilde{f}_0(0)$, we can now utilize the *normalization properties* of the density matrix, namely,

$$\sum_{m=-S}^{S} f_m(t)=\sum_{m=-S}^{S}\left(\rho_m(t)-\rho_m^{\mathrm{II}}\right)=0,$$

so that

$$\sum_{m=-S}^{S}\tilde{f}_m(0)=0. \tag{3.196}$$

Because of Eqs. (3.195) and (3.196), we now have the identity

$$\begin{aligned}
&\tilde{f}_0(0)+\sum_{n=1}^{2S}\left(d_n\tilde{f}_{n-1}(0)+r_n\right)\\
&=\tilde{f}_0(0)+d_1\tilde{f}_0(0)+r_1+\sum_{n=2}^{2S}\left(d_nd_{n-1}\tilde{f}_{n-2}(0)+d_nr_{n-1}+r_n\right)\\
&=\tilde{f}_0(0)+\tilde{f}_0(0)\sum_{n=1}^{2S}\prod_{l=1}^{n}d_l+\sum_{n=1}^{2S}\frac{r_n}{d_n}\sum_{k=n}^{2S}\prod_{l=n}^{k}d_l=0,
\end{aligned} \tag{3.197}$$

where the various products are explicitly given by

$$\frac{1}{d_n}\prod_{l=n}^{k}d_l=e^{[(k-S)^2-(n-S)^2]\frac{\sigma}{S^2}+(k-n)\frac{\xi_{\mathrm{II}}}{S}},\quad \prod_{l=1}^{n}d_l=e^{\xi_{\mathrm{II}}-\sigma}e^{(n-S)^2\frac{\sigma}{S^2}+(n-S)\frac{\xi_{\mathrm{II}}}{S}}.$$

Since

$$e^{(2k-1)\frac{\sigma}{2S^2}+\frac{\xi_{\mathrm{II}}}{2S}}=\sqrt{\rho_k^{\mathrm{II}}/\,\rho_{k-1}^{\mathrm{II}}}, \tag{3.198}$$

then Eq. (3.197) ultimately yields a closed form expression for $\tilde{f}_0(0)$, viz.,

$$\tilde{f}_0(0)=-\frac{2\tau_{\mathrm{N}}e^{\sigma-\xi_{\mathrm{II}}}}{Z}\sum_{k=1-S}^{S}\frac{\sum_{m=k}^{S}\left(\rho_m^{\mathrm{I}}-\rho_m^{\mathrm{II}}\right)\sum_{j=k}^{S}\rho_j^{\mathrm{II}}}{[S(S+1)-k(k-1)]\sqrt{\rho_k^{\mathrm{II}}\rho_{k-1}^{\mathrm{II}}}}. \tag{3.199}$$

Thus, we have

$$\sum_{n=1}^{2S} n\tilde{f}_n(0) = \tilde{f}_0(0)e^{\xi_{\mathrm{II}}-\sigma}\left\langle \hat{S}_Z \right\rangle_{\mathrm{II}}$$
$$+\sum_{n=1}^{2S} r_n e^{-(n-S)^2\frac{\sigma}{S^2}-(n-S)\frac{\xi_{\mathrm{II}}}{S}} \sum_{k=n}^{2S} k e^{(k-S)^2\frac{\sigma}{S^2}+(k-S)\frac{\xi_{\mathrm{II}}}{S}}. \tag{3.200}$$

Hence, by substituting Eqs. (3.199) and (3.200) into the definition of τ_{int}, viz., Eq. (3.189), we finally have the *integral relaxation time rendered in explicit series form as*

$$\tau_{\mathrm{int}} = \frac{1}{\left\langle \hat{S}_Z \right\rangle_{\mathrm{I}} - \left\langle \hat{S}_Z \right\rangle_{\mathrm{II}}} \sum_{n=1}^{2S} n\tilde{f}_n(0)$$
$$= \frac{2\tau_{\mathrm{N}}}{\left\langle \hat{S}_Z \right\rangle_{\mathrm{I}} - \left\langle \hat{S}_Z \right\rangle_{\mathrm{II}}} \sum_{k=1-S}^{S} \frac{\sum_{m=k}^{S}\left(\rho_m^{\mathrm{I}} - \rho_m^{\mathrm{II}}\right)\sum_{j=k}^{S}\left(j-\left\langle \hat{S}_Z \right\rangle_{\mathrm{II}}\right)\rho_j^{\mathrm{II}}}{\left[S(S+1)-k(k-1)\right]\sqrt{\rho_k^{\mathrm{II}}\rho_{k-1}^{\mathrm{II}}}}. \tag{3.201}$$

Both the eigensolution given by the *formal* Eq. (3.176) and the *explicit* Eq. (3.201) as determined from the definition of the integral relaxation time via the zero frequency limit of the normalized relaxation function $\tilde{c}_1(i\omega)/\mathrm{c}_1(0)$ yield exactly the same numerical result. Thus τ_{int} for various nonlinear transient responses (such as the rise, decay, and rapidly reversing field transients) may be easily evaluated from the explicit Eq. (3.201). Equation (3.201) is also valid for an *arbitrary* diagonal operator $\hat{H}_S(\hat{S}_Z)$, because the *precise form* of its elements is involved only in the *equilibrium matrix elements* of the density operator ρ_m^{II} and in the averages $\left\langle \hat{S}_Z \right\rangle_{\mathrm{I}}$ and $\left\langle \hat{S}_Z \right\rangle_{\mathrm{II}}$. Furthermore, it is useful to recall that in the classical limit, $S \to \infty$, the integral relaxation time τ_{int} of the longitudinal relaxation function $c_1(t) = \langle \cos\vartheta \rangle(t) - \langle \cos\vartheta \rangle_{\mathrm{II}}$ of a *classical uniaxial nanomagnet* with a free energy density $E(\vartheta) = -\sigma\cos^2\vartheta - \xi_{\mathrm{II}}\cos\vartheta$ is given by [97] (see Appendix A.4 for details)

$$\tau_{\mathrm{int}} = \frac{2\tau_{\mathrm{N}}}{\langle \cos\vartheta \rangle_{\mathrm{I}} - \langle \cos\vartheta \rangle_{\mathrm{II}}} \int_{-1}^{1} \frac{\Phi(z)\Psi(z)e^{-\sigma z^2-\xi_{\mathrm{II}}z}}{1-z^2}\, dz, \tag{3.202}$$

where

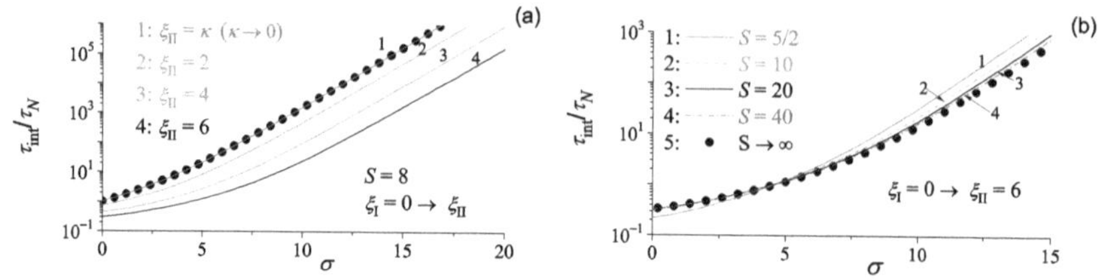

Fig. 3.16. Nonlinear integral relaxation time τ_{int}/τ_N for the rise transient response as a function of the barrier parameter σ (a) for $S = 8$, $\xi_I = 0$ and various $\xi_{II} = 0.01, 2, 4, 6$ (circles: Eq. (3.207) below) and (b) for $\xi_I = 0$ and $\xi_{II} = 6$ and various values of spin $S = 5/2, 10, 20, 40$, and $S \to \infty$ (circles: Eq. (3.202)). Solid lines: calculations from Eq. (3.201).

$$\Psi(z) = \int_{-1}^{z} \left(x - \langle \cos \rangle_{II}\right) e^{\sigma(x^2 + 2h_{II}x)} dx$$

$$= \frac{1}{2\sigma}\left[e^{\sigma(z^2+2h_{II}z)} - e^{\sigma(1-2h_{II})}\right] - e^{\sigma(1-h_{II}^2)} \frac{\sqrt{\pi}\sinh(2\sigma h_{II})}{2\sqrt{\sigma^3} Z_{II}}$$

$$\times \left\{ \operatorname{erfi}[(z+h_{II})\sqrt{\sigma}] + \operatorname{erfi}[(1-h_{II})\sqrt{\sigma}] \right\},$$

$$\Phi(z) = \int_{-1}^{z} \left[W_I(x) - W_{II}(x)\right] dx$$

$$= \frac{\pi^{1/2} e^{-\sigma h_{II}^2}}{2\sigma^{1/2} Z_{II}} \left\{ \operatorname{erfi}[(z+h_{II})\sqrt{\sigma}] + \operatorname{erfi}[(1-h_{II})\sqrt{\sigma}] \right\}$$

$$- \frac{\pi^{1/2} e^{-\sigma h_I^2}}{2\sigma^{1/2} Z_I} \left\{ \operatorname{erfi}[(z+h_I)\sqrt{\sigma}] + \operatorname{erfi}[(1-h_I)\sqrt{\sigma}] \right\},$$

$\langle \cos \vartheta \rangle_i$ and Z_i are defined by Eqs. (A.48) and (A.50) from Appendix A.1.5, respectively, and $h_i = \xi_i/(2\sigma)$ and erfi (x) is the error function of imaginary argument [290] (see Eq. (A.19) from Appendix A.1.3).

The nonlinear relaxation time τ_{int} for the rise transient response as a function of the anisotropy parameter σ and the spin number S is plotted in Fig. 3.16, displaying a pronounced dependence of that time on the field (ξ_{II}), anisotropy (σ), and spin (S) parameters; in particular, that time decreases with increasing field strength ξ_{II}. It is apparent from Fig. 3.16 that for large S, the quantum solutions reduce to the corresponding

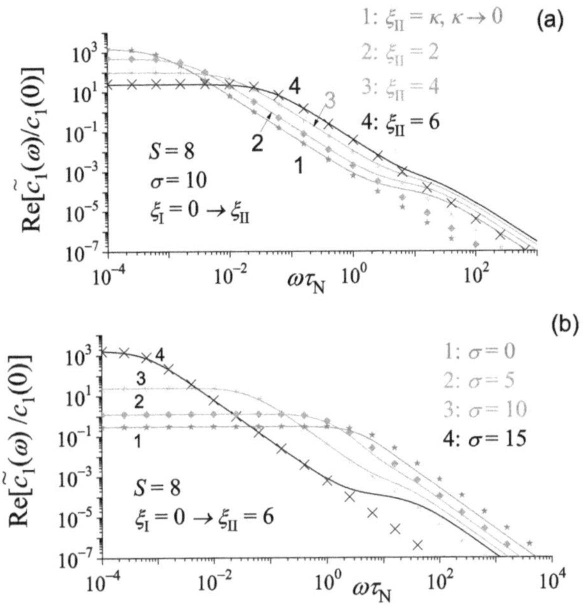

Fig. 3.17. Real part of the spectrum $\tilde{c}_1(i\omega)/c_1(0)$ vs. the normalized frequency $\omega\tau_N$ for the rise transient response (a) for $S = 8$, $\sigma = 10$, $\xi_I = 0$, and various $\xi_{II} = \kappa \to 0, 2, 4, 6$ and (b) for $\xi_I = 0$, $\xi_{II} = 6$, $S = 8$, and various anisotropy parameters σ. Solid lines: calculations from Eq. (3.173); stars: from the single-mode approximation Eq. (3.203).

classical ones. Typical values of S for the quantum-classical crossover are ~20–40. *The smaller the anisotropy σ the smaller the S value required for convergence of the quantum equations to the classical ones.*

In Fig. 3.17, we have plotted the real part of the normalized relaxation function spectrum $\tilde{c}_1(i\omega)/c_1(0)$ vs. the normalized frequency $\omega\tau_N$ for the rise transient response, $\xi_I = 0 \to \xi_{II} \neq 0$. Like the classical case [97], knowledge of λ_1 alone is enough to accurately predict the low-frequency part of $\tilde{c}_1(i\omega)/c_1(0)$ as well as the *long-time behavior* of the relaxation function $c_1(t)/c_1(0)$. Here the single-mode approximation

$$\frac{\tilde{c}_1(i\omega)}{c_1(0)} \approx \frac{\tau_{int}}{1 + i\omega/\lambda_1}, \tag{3.203}$$

where τ_{int} is given by Eq. (3.201) and λ_1 is the smallest nonvanishing eigenvalue of the system matrix **A** defined by Eq. (3.170), is shown for comparison indicating that τ_{int} and λ_1 in combination comprehensively describe the low-frequency behavior of the spectrum $\tilde{c}_1(i\omega)/c_1(0)$ as in the classical case. In the time domain, the single-mode approximation Eq. (3.203) amounts to assuming that the relaxation function $c_1(t)$ as determined by the formal Eq. (3.176) (comprising $2S$ exponentials) may be approximated for $t>0$ by a *single* exponential. Consequently, the longtime relaxation behavior of a uniaxial nanomagnet may once again be accurately approximated by a *single* exponential with relaxation time $T_1=1/\lambda_1$ and thus is again governed by the Bloch Eq. (3.165).

We may also evaluate the *linear response* of a uniaxial nanomagnet corresponding to infinitesimally small changes in the magnitude of the dc field, so that the *integral relaxation time now becomes merely the correlation time*. In linear response, i.e., considering transient relaxation between the states I and II with respective Hamiltonians

$$\hat{H}_S^{\text{I}}=\sigma S^{-2}\hat{S}_Z^2+(\xi_{\text{II}}+\varepsilon)S^{-1}\hat{S}_Z \text{ and } \hat{H}_S^{\text{II}}=\sigma S^{-2}\hat{S}_Z^2+\xi_{\text{II}}S^{-1}\hat{S}_Z,$$

where ε is a small external field parameter, the initial conditions $f_m(0)$ and $\langle\hat{S}_Z\rangle_{\text{I}}-\langle\hat{S}_Z\rangle_{\text{II}}$ reduce to

$$f_m(0)=\frac{1}{Z_S^{\text{I}}}e^{\frac{\sigma}{S^2}m^2+\frac{\xi_{\text{II}}+\varepsilon}{S}m}-\frac{1}{Z_S^{\text{II}}}e^{\frac{\sigma}{S^2}m^2+\frac{\xi_{\text{II}}}{S}m}\approx\frac{\varepsilon}{S}\left(m-\langle\hat{S}_Z\rangle_{\text{II}}\right)\rho_m^{\text{II}}, \quad (3.204)$$

$$\langle\hat{S}_Z\rangle_{\text{I}}-\langle\hat{S}_Z\rangle_{\text{II}}\approx\varepsilon S\chi_0, \quad (3.205)$$

where

$$\chi_0=S^{-2}\left(\left\langle\hat{S}_Z^2\right\rangle_{\text{II}}-\left\langle\hat{S}_Z\right\rangle_{\text{II}}^2\right)=S^{-2}\left[\sum_{m=-S}^{S}m^2\rho_m^{\text{II}}-\left(\sum_{m=-S}^{S}m\rho_m^{\text{II}}\right)^2\right]. \quad (3.206)$$

Thus, in the limit $\varepsilon\to 0$, the general Eq. (3.201) τ_{int}, yields the *correlation time* τ_{cor} in the explicit series form

$$\tau_{\text{cor}}=\frac{2\tau_{\text{N}}}{\chi_0 S^2}\sum_{k=1-S}^{S}\frac{\left(\sum_{m=k}^{S}\left(m-\left\langle\hat{S}_Z\right\rangle_{\text{II}}\right)\rho_m^{\text{II}}\right)^2}{\left[S(S+1)-k(k-1)\right]\sqrt{\rho_k^{\text{II}}\rho_{k-1}^{\text{II}}}}. \quad (3.207)$$

Equation (3.207) in the limit $S \to \infty$ yields numerical results concurring with the classical ones, Eq. (2.182) of Chapter 2.

Furthermore, we can also calculate the effective relaxation time τ_{ef} given by Eq. (A.216) from Appendix A.4. Thus, τ_{ef} yielding precise information on the initial decay of $c_1(t)$ in the time domain and defined as usual by [97]

$$\tau_{\mathrm{ef}} = -\frac{c_1(0)}{\dot{c}_1(0)} = -\frac{S\partial_\varepsilon \left\langle \hat{S}_Z \right\rangle_{\mathrm{I}}\Big|_{\varepsilon=0}}{\sum\limits_{m=-S}^{S} m\dot{f}_m(0)} = -\frac{\chi_0 S^2}{\sum\limits_{m=-S}^{S} m\dot{f}_m(0)} \tag{3.208}$$

is given by

$$\tau_{\mathrm{ef}} = \frac{2\chi_0 S^2 \tau_{\mathrm{N}}}{\sum\limits_{k=1-S}^{S} [S(S+1)-k(k-1)]\sqrt{\rho_k^{\mathrm{II}}\rho_{k-1}^{\mathrm{II}}}}, \tag{3.209}$$

where we have used Eqs. (3.204) and also the closed expression

$$\sum_{m=-S}^{S} m\dot{f}_m(0) = \sum_{n=1}^{2S} n\left[p_n^- f_{n-1}(0) + p_n f_n(0) + p_n^+ f_{n+1}(0)\right]$$

$$= \sum_{n=1}^{2S}\left[(n-1)p_{n-1}^+\rho_n^{\mathrm{II}} - np_n^-\rho_{n-1}^{\mathrm{II}}\right] = \frac{1}{2\tau_{\mathrm{N}}}\sum_{k=1-S}^{S}[S(S+1)-k(k-1)]\sqrt{\rho_k^{\mathrm{II}}\rho_{k-1}^{\mathrm{II}}}.$$

Finally, the longest relaxation time τ, which is associated with the spin reversal time, can be calculated via the smallest *nonvanishing* eigenvalue λ_1 of the system matrix $\tau_{\mathrm{N}}^{-1}\mathbf{\Pi}$ given by Eq. (3.192) as $\tau = 1/\lambda_1$ from the deterministic equation

$$\det\left(\tau_{\mathrm{N}}^{-1}\mathbf{\Pi} - \lambda\mathbf{I}\right) = 0. \tag{3.210}$$

This time may also be expressed in closed form by recalling that the left-hand side of Eq. (3.210) merely represents a polynomial equation of the order $2S + 1$, viz.,

$$\left(k_{2S+1}\lambda^{2S} + k_{2S}\lambda^{2S-1} + \ldots + k_2\lambda + k_1\right)\lambda = 0, \tag{3.211}$$

where

$$k_1 = -\sum_{i=0}^{2S} M_i^i, \; k_2 = \sum_{i=0}^{2S-1} \sum_{j=i+1}^{2S} M_{ij}^{ij}, \tag{3.212}$$

and so on and we have noticed that det $\mathbf{\Pi} = 0$. Here the $M_i^{i'}$ are the first minors of the matrix $\mathbf{\Pi}$, which are in turn the determinants of square matrices as reduced from $\tau_{\mathrm{N}}^{-1}\mathbf{\Pi}$ by removing the ith row and the i'th column of $\tau_{\mathrm{N}}^{-1}\mathbf{\Pi}$ while the $M_{ij}^{i'j'}$ are the minors of the matrix $\tau_{\mathrm{N}}^{-1}\mathbf{\Pi}$, which are the determinants of the square matrix reduced from $\tau_{\mathrm{N}}^{-1}\mathbf{\Pi}$ by removing two (the ith and the jth) of its rows and two (the i'th and the j'th) columns. Consequently, the smallest *nonvanishing* eigenvalue λ_1 can be readily evaluated numerically from Eq. (3.210), e.g., using MATHEMATICA®. However, (in the high barrier approximation, when $\lambda_1 \ll 1$) it can also be evaluated analytically by neglecting all higher powers of λ^n with $n > 2$ in Eq. (3.211). Thus, we have

$$\lambda_1 \approx -k_1 / k_2. \tag{3.213}$$

However, Eq. (3.213) can be equivalently written in matrix form as

$$\lambda_1 \approx \frac{\mathrm{Tr}(\mathbf{M}^{(1)})}{\mathrm{Tr}(\mathbf{M}^{(2)})}, \tag{3.214}$$

where $\mathbf{M}^{(1)}$ is the matrix formed from all the first minors:

$$\mathbf{M}^{(1)} = \begin{pmatrix} M_{2S}^{2S} & M_{2S}^{2S-1} & \cdots & M_{2S}^{0} \\ M_{2S-1}^{2S} & M_{2S-1}^{2S-1} & \cdots & M_{2S-1}^{0} \\ \vdots & \vdots & \ddots & \vdots \\ M_{0}^{2S} & M_{0}^{2S-1} & \cdots & M_{0}^{0} \end{pmatrix}, \tag{3.215}$$

and the matrix $\mathbf{M}^{(2)}$ contains all the $M_{ij}^{i'j'}$ minors:

$$\mathbf{M}^{(2)} = \begin{pmatrix} M_{2S,2S-1}^{2S,2S-1} & M_{2S,2S-1}^{2S,2S-2} & \cdots & M_{2S,2S-1}^{0,0} \\ M_{2S,2S-2}^{2S,2S-1} & M_{2S,2S-2}^{2S,2S-2} & \cdots & M_{2S,2S-2}^{0,0} \\ \vdots & \vdots & \ddots & \vdots \\ M_{0,0}^{2S,2S-1} & M_{0,0}^{2S,2S-2} & \cdots & M_{0,0}^{0,0} \end{pmatrix}. \tag{3.216}$$

The matrices $\mathbf{M}^{(1)}$ and $\mathbf{M}^{(2)}$ have dimensions, respectively, $n \times n$ and $n(n-1)/2 \times n(n-1)/2$, where $n = 2S + 1$. Furthermore, the ordering of the elements in the matrix $\mathbf{M}^{(2)}$ is such that reading across or down the final matrix, the successive lists of positions appear in lexicographic order. Then Tr ($\mathbf{M}^{(1)}$) and Tr ($\mathbf{M}^{(2)}$) can be calculated analytically as

$$\begin{aligned}\mathrm{Tr}\left(\mathbf{M}^{(1)}\right)&=\frac{(-1)^{2S}}{\tau_{\mathrm{N}}^{2S}}\sum_{i=0}^{2S}\prod_{s=1}^{i}p_s^-\prod_{r=i}^{2S-1}p_r^+=\frac{(2S)!}{2^{2S}\tau_{\mathrm{N}}^{2S}}\sum_{i=0}^{2S}e^{(i^2-2iS)\frac{\sigma}{S^2}+(i-S)\frac{\xi_{\mathrm{II}}}{S}}\\&=\frac{(2S)!}{2^{2S}\tau_{\mathrm{N}}^{2S}}\sum_{k=-S}^{S}e^{(k^2-S^2)\frac{\sigma}{S^2}+k\frac{\xi_{\mathrm{II}}}{S}}=\frac{(2S)!e^{-\sigma}}{2^{2S}\tau_{\mathrm{N}}^{2S}}Z_S\end{aligned}\tag{3.217}$$

and in like manner

$$\begin{aligned}\mathrm{Tr}\left(\mathbf{M}^{(2)}\right)&=\frac{(-1)^{2S+1}}{\tau_{\mathrm{N}}^{2S-1}}\sum_{i=0}^{2S-1}\sum_{j=i+1}^{2S}\left(\prod_{s=1}^{i}p_s^-\prod_{r=j}^{2S-1}p_r^+\sum_{m=1}^{j-i}\prod_{u=j+2-m}^{j}p_u^-\prod_{v=i}^{j-m-1}p_v^+\right)\\&=\frac{(2S)!e^{-\sigma}}{2^{2S-1}\tau_{\mathrm{N}}^{2S-1}}\sum_{k=-S}^{S-1}\sum_{n=k+1}^{S}\sum_{m=1}^{n-k}\frac{e^{[2k^2-2n-1+2m(2n-m+1)]\frac{\sigma}{2S^2}+(2k+2m-1)\frac{\xi_{\mathrm{II}}}{2S}}}{(S+n-m+1)(S-n+m)}.\end{aligned}\tag{3.218}$$

Here we have noticed that $\mathbf{\Pi}_{m=a}^{b}p_m^{\pm}=1$ if $b<a$. Thus, in the high barrier approximation, the longest relaxation time $\tau\approx 1/\lambda_1$ is given by the following approximate equation, which is the desired result,

$$\tau\approx\frac{2\tau_N}{Z_S}\sum_{k=-S}^{S-1}\sum_{n=k+1}^{S}\sum_{m=1}^{n-k}\frac{e^{[2k^2-2n-1+2m(2n-m+1)]\frac{\sigma}{2S^2}+(2k+2m-1)\frac{\xi_{\mathrm{II}}}{2S}}}{(S+n-m+1)(S-n+m)}.\tag{3.219}$$

We have already mentioned that the linear response has been previously studied by Garanin [226] and García-Palacios and Zueco [227] using the spin density matrix whereby they also gave analytic expressions for τ_{cor}, τ_{ef} and $\tau\approx 1/\lambda_1$ for more general models of linear and bilinear spin-bath interactions with superohmic damping. Using Garanin's method [226], the longest relaxation time can be found in a similar manner

$$\tau=\frac{2\tau_{\mathrm{N}}}{\chi_\Delta}\sum_{k=1-S}^{S}\frac{\left(\sum_{m=k}^{S}\left(m-\langle\hat{S}_Z\rangle_{\mathrm{II}}\right)\rho_m^{\mathrm{II}}\right)\left(\sum_{m=-S}^{k-1}[\Delta-\mathrm{sgn}(m-m_b)]\rho_m^{\mathrm{II}}\right)}{[S(S+1)-k(k-1)]\sqrt{\rho_k^{\mathrm{II}}\rho_{k-1}^{\mathrm{II}}}},\tag{3.220}$$

where m_b is the quantum number corresponding to the top of the barrier,

$$\Delta = \sum_{m=-S}^{S} \operatorname{sgn}(m - m_b)\rho_m^{\mathrm{II}}$$

and

$$\chi_\Delta = \sum_{m=-S}^{S} m \operatorname{sgn}(m - m_b)\rho_m^{\mathrm{II}} - \left(\sum_{m=-S}^{S} m\rho_m^{\mathrm{II}}\right)\left(\sum_{m=-S}^{S} \operatorname{sgn}(m - m_b)\rho_m^{\mathrm{II}}\right).$$

Furthermore, via the replacement $k + 1 \to k$ and then via

$$\sum_{m=k}^{S}\left(m - \left\langle \hat{S}_Z \right\rangle_{\mathrm{II}}\right)\rho_m^{\mathrm{II}} = -\sum_{m=-S}^{k-1}\left(m - \left\langle \hat{S}_Z \right\rangle_{\mathrm{II}}\right)\rho_m^{\mathrm{II}},$$

one can rearrange Eq. (3.220) as

$$\tau = \frac{2\tau_{\mathrm{N}}}{\chi_\Delta}\sum_{k=-S}^{S-1}\frac{\left(\sum_{m=-S}^{k}\left(m - \left\langle \hat{S}_Z \right\rangle_{\mathrm{II}}\right)\rho_m^{\mathrm{II}}\right)\left(\sum_{m=-S}^{k}\left[\operatorname{sgn}(m - m_b) - \Delta\right]\rho_m^{\mathrm{II}}\right)}{[S(S+1) - k(k+1)]\sqrt{\rho_k^{\mathrm{II}}\rho_{k+1}^{\mathrm{II}}}}. \quad (3.221)$$

For $\xi_{\mathrm{II}} < \sigma$ the relative deviation of τ as given by Eq. (3.221) from λ_1^{-1} as calculated numerically does not exceed 1%.

All the foregoing expressions have been derived via the density matrix method. However, they can also be obtained using the phase-space formalism thereby exemplifying how they may reduce to the classical expressions. For example, τ_{ef} as rendered by Eq. (3.209) can be written as

$$\tau_{\mathrm{ef}} = 2\tau_{\mathrm{N}} e^{\xi/(2S)} \frac{\left\langle \hat{S}_Z^2 \right\rangle_{\mathrm{II}} - \left\langle \hat{S}_Z \right\rangle_{\mathrm{II}}^2}{\left\langle \hat{\mathbf{S}}^2 - \hat{S}_Z^2 + \hat{S}_Z \right\rangle_{\mathrm{II}}}, \quad (3.222)$$

where $\left\langle \hat{S}_Z \right\rangle_{\mathrm{II}}$ and $\left\langle \hat{S}_Z^2 \right\rangle_{\mathrm{II}}$ can also be given via the phase-space Eq. (3.184) and the denominator is given by the phase-space average

$$\left\langle \hat{\mathbf{S}}^2 - \hat{S}_Z^2 + \hat{S}_Z \right\rangle_{\mathrm{II}} = (S+1)(S+\tfrac{1}{2})\int_{-1}^{1}\left[S(1-z^2) + \tfrac{1}{2} + z - \tfrac{3}{2}z^2\right]W_S^{\mathrm{II}}(z)dz, \quad (3.223)$$

on recalling that the Weyl symbol of the operator $\hat{\mathbf{S}}^2 - \hat{S}_Z^2 + \hat{S}_Z$ is just

$$\hat{\mathbf{S}}^2 - \hat{S}_Z^2 + \hat{S}_Z \to (S+1)\left[S\left(1 - \cos^2\vartheta\right) + \cos\vartheta - \frac{3}{2}\cos^2\vartheta + \frac{1}{2}\right]. \quad (3.224)$$

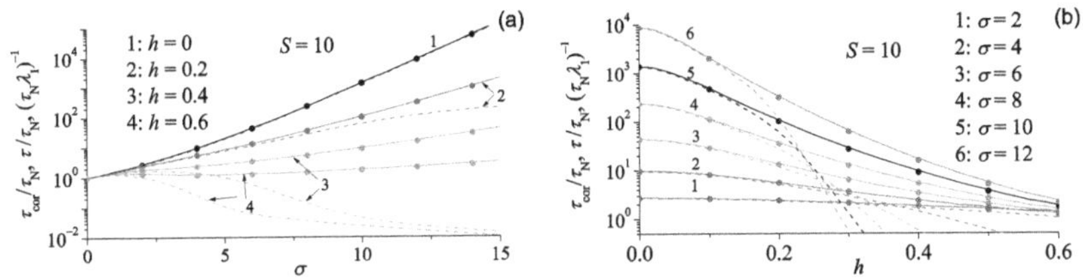

Fig. 3.18. The correlation time τ_{cor} [Eq. (3.207); dashed lines], the inverse of the smallest eigenvalue λ_1 (filled circles) and its approximation τ [Eq. (3.220); solid line] (a) as a function of the barrier height parameter σ for various field parameters $h = \xi_{II}/(2\sigma)$ and (b) as a function of the field parameter h for various σ ($S = 10$).

Clearly, Eq. (3.222) is non other than a quantum analog of the long established equation for the longitudinal effective relaxation time τ_{ef} of classical macrospins, viz. [97,239],

$$\tau_{ef} = 2\tau_N \frac{\left\langle \cos^2 \vartheta \right\rangle_{II} - \left\langle \cos \vartheta \right\rangle_{II}^2}{1 - \left\langle \cos^2 \vartheta \right\rangle_{II}}, \tag{3.225}$$

where $\left\langle \cos \vartheta \right\rangle_{II} = \left\langle P_1 \right\rangle_{II}$ and $\langle \cos^2 \vartheta \rangle_{II} = (2\langle P_2 \rangle_{II} + 1)/3$ are defined via Eqs. (A.48) and (A.49) of Appendix A.1.5. Furthermore, the correlation time τ_{cor} is given by Eq. (2.183) of Chapter 2 [23,97,239,337].

For linear response, the correlation time τ_{cor} and overbarrier time λ_1^{-1}, are plotted in Fig. 3.18 as a function of the field parameter $h = \xi_{II}/(2\sigma)$ and the barrier height parameter σ for $S = 10$. Like the classical case, the behavior of τ_{cor} and λ_1^{-1} is similar only for *small* external fields. In a strong external field, $h \geq 0.2$, λ_1^{-1} can diverge exponentially from τ_{cor}. We recall that this divergence effect for a classical uniaxial nanomagnet was discovered (as seen) numerically by Coffey *et al.* [146] and later explained quantitatively by Garanin [337] (as discussed in Chapters 1 and 2). He showed analytically (as seen) that the contribution of relaxation modes other than the overbarrier one to the integral relaxation time becomes significant for high external fields due to population depletion of the shallower of the two potential wells of a bistable potential under the action of an external applied field. The field is far less than that needed to destroy the bistable nature of the potential [97]. Furthermore, τ_{cor} and λ_1^{-1}, are also plotted in Fig. 3.19 as a function of S for various values of ξ_{II} and σ.

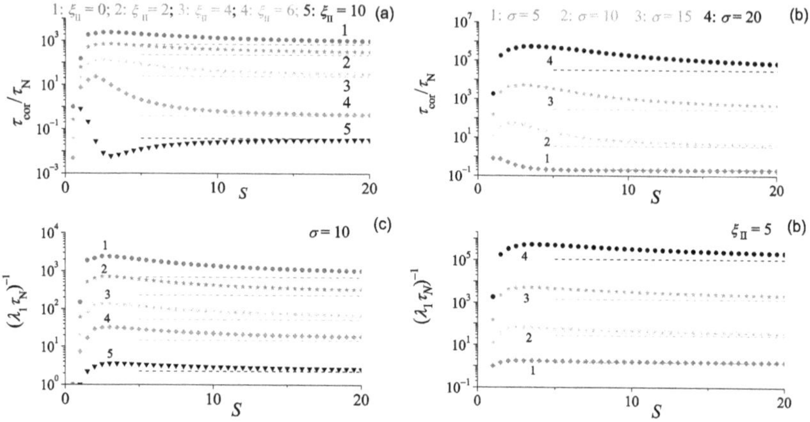

Fig. 3.19. Correlation time τ_{cor}/τ_N (a), (b) and $(\lambda_1\tau_N)^{-1}$ (c), (d) vs. the spin number S for various field parameters ξ_{II} for anisotropy barrier parameter $\sigma = 10$ and for different σ and field parameters $\xi_{II} = 5$. Dashed lines: classical limit, $S \to \infty$.

Clearly *even for relatively small* $S \sim 20$, the quantum formulas are very close to the classical ones.

In Figs. 3.20 and 3.21, we show the real and imaginary parts of the dynamic susceptibility $\chi(\omega)$. Clearly, like the classical case (Chapter 2, Section 2.4.2), two bands appear in the spectrum of the imaginary part – Im $[\chi(\omega)]$. The low-frequency band is due to the slowest "overbarrier" relaxation mode and can be described by a *single* Lorentzian. The characteristic frequency and the half-width of this band are determined by λ_1. The high-frequency band of –Im $[\chi(\omega)]$ as before is due to high-frequency well modes corresponding to the near-degenerate eigenvalues λ_k ($k \geq 2$). These individual "intrawell" modes are again indistinguishable in the spectrum of –Im $[\chi(\omega)]$ appearing merely as a single high-frequency Lorentzian band. Thus, again as in the classical case [97] (see Chapter 1, Section 1.3.6), we may describe the behavior of $\chi(\omega)$ via a two-mode approximation, i.e., by supposing that it is represented by a sum of two Lorentzians, Eq. (1.150) of Chapter 1, viz.,

$$\frac{\chi(\omega)}{\chi_0} \approx \frac{\Delta_1}{1+i\omega\tau} + \frac{1-\Delta_1}{1+i\omega\tau_W}, \tag{3.226}$$

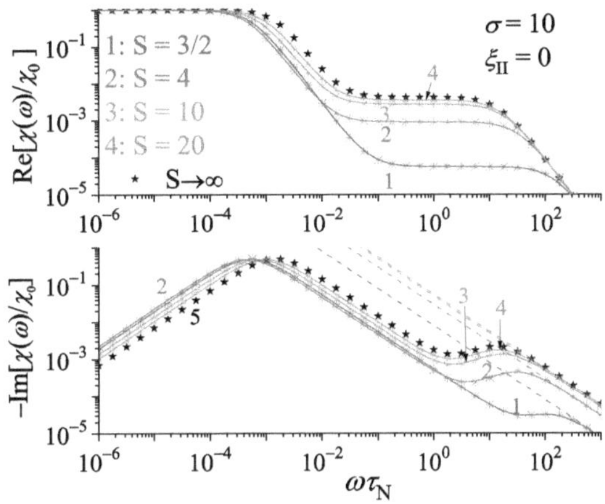

Fig. 3.20. The normalized susceptibility $\chi(\omega)/\chi_0$, Eq. (3.181), vs. $\omega\tau_N$ for barrier parameter $\sigma = 10$, uniform field parameter $\xi_{II} = 0$ (*symmetrical wells*), and various spin numbers $S = 3/2$, 4, 10, 20. Asterisks: the two-mode approximation, Eq. (3.226) below. Straight dashed lines: the high-frequency asymptote, as rendered by Eqs. (3.183) and the effective relaxation time Eq. (3.209). Stars: the corresponding classical limit, $S \to \infty$.

where τ_W is a characteristic relaxation time of the *near-degenerate high-frequency well modes* and Δ_1 in the present context is a parameter characterizing the contribution of these high-frequency modes to the dynamic susceptibility defined via τ_{cor}, τ_{ef} and $\tau \approx 1/\lambda_1$ by Eq. (1.149) of Chapter 1 [338,339]. Here τ_{cor}, τ_{ef} and $\tau \approx 1/\lambda_1$ are given by the quantum Eqs. (3.207), (3.209), and (3.221), respectively. The parameters Δ_1 and τ_W in Eq. (3.226) have been determined by imposing the condition that the two-mode approximation Eq. (3.226) must asymptotically obey both Eqs. (3.182) and (3.183). The longest relaxation time τ is once again related to the frequency ω_{max} of the low-frequency peak in the magnetic loss spectrum $-\text{Im}[\chi(\omega)]$, where it attains a maximum, and/or the half-width $\Delta\omega$ of the spectrum of the real part of the susceptibility $\text{Re}[\chi(\omega)]$ via

$$\tau \approx \omega_{max}^{-1} \approx \Delta\omega^{-1}. \tag{3.227}$$

In the low-frequency region ($\omega\tau \leq 1$), where the effect of the high-frequency modes may be safely ignored, $\chi(\omega)$ may be approximated by Eq. (1.151) of Chapter 1 (a single-mode approximation). We remark that

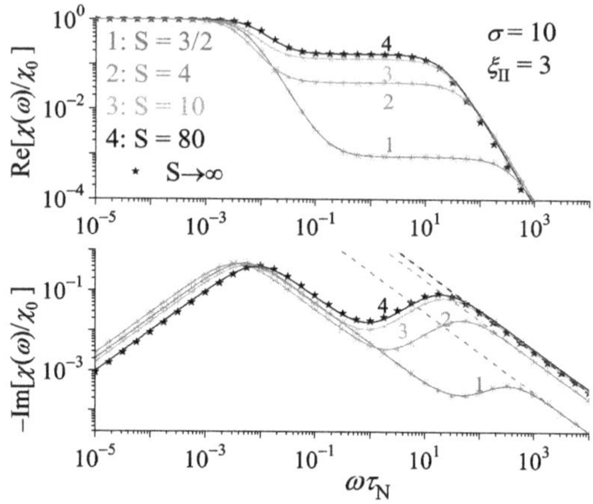

Fig. 3.21. The normalized susceptibility $\chi(\omega)/\chi_0$, Eq. (3.181), vs normalized frequency $\omega\tau_N$ for barrier parameter $\sigma=10$, the uniform field parameter $\xi_{II}=3$, and various spin numbers $S=3/2$, 4, 10, 80. Asterisks: the two-mode approximation, Eq. (3.226). Straight dashed lines: the high-frequency asymptote as rendered by Eq. (3.183) and the effective relaxation time Eq. (3.209). Stars: the classical limit, $S\to\infty$, as rendered by Eq. (3.183) and the effective relaxation time Eq. (3.225).

García-Palacios and Zueco [228] have also shown that the two-mode approximation which was originally developed for classical systems [97,338,339] accurately describes the linear response of quantum uniaxial nanomagnets at all frequencies of interest.

To illustrate the accuracy of the two-mode approximation for the quantum behavior, we plot in Figs. 3.20 and 3.21 the real and imaginary parts of $\chi(\omega)$ as calculated from the matrix solution, Eq. (3.181), representing a finite sum of Lorentzians and the approximate two-mode Eq. (3.226) for zero dc field, $\xi_{II}=0$ (symmetrical wells) and for nonzero dc field, $\xi_{II}=3$ (asymmetrical wells). It is apparent from Figs. 3.20 and 3.21 that at low frequencies no *practical difference exists between the numerical solution and the two-mode approximation* (the maximum relative deviation between the corresponding curves does not exceed a few percent). In the classical limit, $S\to\infty$, the respective Hamiltonians, Eqs. (3.83) or (3.228) below, correspond to the classical free energy $E(\vartheta)=-\sigma\cos^2\vartheta-\xi_{II}\cos\vartheta$. Here both τ_{cor} and τ_{ef} can be expressed in closed form, viz., Eqs. (2.182) and (2.183) of Chapter 2. The classical limit is also shown in Figs. 3.20 and 3.21 for comparison.

We have studied here the transient nonlinear longitudinal relaxation of a *quantum uniaxial nanomagnet* of arbitrary spin *S* in the *high temperature* and *weak spin-bath coupling* limit. The principal result is that one may once again *determine the transition from quantum elementary spin relaxation to the classical superparamagnetic relaxation pertaining to a giant classical spin as a function of the spin number S*. Furthermore, one may accurately estimate the value of *S* (typically in the range 20–40) wherein the crossover to classical superparamagnetic behavior takes place. Thus, one may assign a *range of validity as a function of the spin number S to the classical Néel–Brown treatment of magnetic nanoparticles* with the simplest uniaxial magnetocrystalline anisotropy and Zeeman energy given above. The exact continued fraction solution based on the *diagonal elements* of the density matrix yields in closed form the dependence of the longitudinal spin relaxation function on *S*, which is dominated by a single exponential with time constant the longest relaxation time λ_1^{-1} Thus, *a simple description of the long time behavior of the longitudinal relaxation function as a Bloch equation* (3.165) *again holds for the nonlinear response of a quantum nanomagnet for arbitrary spin S*. In linear response, the approach so developed reproduces (with some modifications due to the symmetrized collision kernel used) the results previously obtained by Garanin [226] and García-Palacios and Zueco [227,228].

We have treated the longitudinal relaxation in two superficially distinct ways, viz., the phase-space formalism embodied in the Wigner-Stratonovich bijective mapping and the density matrix. *The high temperature and weak spin-bath coupling limit being understood in each case*. Thus, we have provided a vital check on the validity of both methods by explicitly demonstrating their equivalence. Again, we emphasize that a very useful feature of the phase-space representation is that existing powerful computational techniques for the Fokker–Planck equation may be extended to the quantum domain which also suggest new closed form quantum results via corresponding classical ones. For example, the integral and effective relaxation times, Eqs. (3.201) and (3.222), are clearly quantum analogs of the corresponding classical expressions Eqs. (3.202) and (3.225).

Next, we study the nonlinear ac stationary responses of quantum nanomagnets by generalizing the known solutions for classical spins driven by a strong ac field so as to treat quantum effects in their ac nonlinear response (see Chapter 2). The investigation is prompted by the fact that it has been shown experimentally [475] for the molecular magnet Mn_{12} with $S = 10$ that the nonlinear susceptibility of quantum nanomagnets differs from that of classical spins.

3.3.11 *Nonlinear longitudinal relaxation in superimposed ac and dc magnetic fields*

The spin reversal process in quantum systems with finite spin number S has a strong field-dependence causing nonlinear effects in the dynamic susceptibility [263], stochastic resonance [220,221], etc. We recall that the nonlinear response to an external field always poses an extremely difficult problem because the response now depends on the *precise nature* of the stimulus [97]. Thus, no *unique* response function valid for all stimuli exists unlike in linear response. These difficulties are all compounded in quantum systems so that the literature available on them is relatively sparse. Nevertheless, the nonlinear longitudinal relaxation of a quantum nanomagnet arising from a sudden change in the magnitude of a strong external dc field was treated in Section 3.3.10 using an evolution equation for the reduced density matrix. The solution of the evolution equation was then written as a finite series of the polarization operators, where the coefficients of the series (statistical averages of the polarization operators) were found from differential-recurrence relations. Moreover, it was shown that the matrix solution simplifies for uniaxial Hamiltonians Eq. (3.228) because the *diagonal terms of the density matrix decouple from the nondiagonal ones so that only the former partake in the time evolution.* We now show how this technique can be applied to the nonlinear dynamic magnetic susceptibility of a quantum nanomagnet with arbitrary S in superimposed ac and dc magnetic uniform fields amounting to the calculation of the nonlinear ac stationary response of the nanomagnet to a strong ac field in the presence of thermal agitation. Calculations of the *nonlinear* ac response of quantum uniaxial nanomagnets have hitherto been made via perturbation theory (e.g., Refs. [474–476]) by supposing that the potential energy of a spin in external magnetic fields is less than the thermal energy so that a small parameter exists. In the response to an ac field of *arbitrary strength*, however, such small parameters do not exist. The approach used is, in some respects, analogous to that of Refs. [477,478] for nonlinear dielectric relaxation behavior of polar molecules in a strong ac electric field and in Refs. [153,156,159] as well as of Chapter 2 for the nonlinear magnetization relaxation of magnetic nanoparticles in superimposed ac and dc magnetic fields. The difference as usual is that for a finite spin number S the solution of the evolution equation is rendered as a *finite* sum of spherical harmonics in contrast to the classical case, where the solution of the evolution equation involves an infinite sum of them. We shall demonstrate that our quantum results in the classical limit, $S \to \infty$, correspond with those of Ref. [153]. Moreover, for small values of the ac applied fields (linear response) they agree with the results of Section 3.3.10 calculated via the switch-off of a small longitudinal uniform field.

As an explicit example, we consider a uniaxial nanomagnet of arbitrary S subjected to both a *uniform* external magnetic field $\mathbf{H}_0$ and to an ac external field $\mathbf{H}(t)$ applied along the Z axis, i.e., the axis of symmetry. The Hamiltonian $\hat{H}_S$ has the uniaxial form [cf. Eq. (3.83)]

$$\hat{H}_S(t) = -\sigma\left(\beta S^2\right)^{-1} \hat{S}_Z^2 - (\xi_0 + \xi\cos\omega t)\left(\beta S\right)^{-1} \hat{S}_Z, \qquad (3.228)$$

where σ is again the dimensionless anisotropy parameter, ξ_0 and ξ are the dc bias and ac field parameters, respectively. This Hamiltonian comprises a uniaxial anisotropy term plus the Zeeman term, representing as before a generic model for quantum spin relaxation phenomena in molecular magnets, nanoclusters, etc.

The density matrix evolution equation describing the longitudinal relaxation of a uniaxial nanomagnet with the time-dependent Hamiltonian defined by Eq. (3.228) is similar to Eq. (3.166) and is given by

$$\tau_{\mathrm{N}} \frac{\partial \hat{\rho}_S}{\partial t} = -e^{-\frac{\sigma}{2S^2} - \frac{\xi_0 + \xi\cos\omega t}{2S}} [\hat{S}_{-1} e^{-\frac{\sigma}{S^2}\hat{S}_0} \hat{\rho}_S, \hat{S}_{+1}] - e^{\frac{\sigma}{2S^2} + \frac{\xi_0 + \xi\cos\omega t}{2S}} [\hat{S}_{+1} e^{\frac{\sigma}{S^2}\hat{S}_0} \hat{\rho}_S, \hat{S}_{-1}]. \qquad (3.229)$$

For our purposes use of the *symmetrized form* of collision kernel Eq. (3.229) is very significant as then it allows a correct description of the harmonics of spectral moments in the nonlinear response when an ac stimulus is imposed, namely the *absence* of the even harmonics for symmetric double-well potentials. Here the magnitude of the ac field ξ is supposed to be so large that the energy of a spin is either comparable to or higher than the thermal energy kT, i.e., $\xi \geq 1$, so that one is always faced with an intrinsically nonlinear problem which is solved as follows.

We recall that for a uniaxial Hamiltonian such as that given by Eq. (3.228), the transformation of the evolution equation (3.229) for the density matrix $\hat{\rho}_S$ into scalar differential-recurrence equations for its *individual* matrix elements can be radically simplified because the diagonal entries of $\hat{\rho}_S$ decouple from the nondiagonal ones. Hence, only the former contribute to the longitudinal spin relaxation. As before, we have from Eq. (3.229) the following three-term differential-recurrence equation for the *diagonal* entries $\rho_m = \rho_{mm}$ [cf. Eq. (3.185)]

$$\tau_N \frac{d}{dt}\rho_m(t) = q_m^-(t)\rho_{m-1}(t) + q_m(t)\rho_m(t) + q_m^+(t)\rho_{m+1}(t), \quad (3.230)$$

where $m = -S, -S+1, \ldots, S$, $\tau_N = (2D_\perp)^{-1}$ is the characteristic (free diffusion) time and now the *time-dependent* coefficients are

$$q_m(t) = -a_m^- e^{-(2m-1)\frac{\sigma}{2S^2} - \frac{\xi_0 + \xi\cos\omega t}{2S}} - a_m^+ e^{(2m+1)\frac{\sigma}{2S^2} + \frac{\xi_0 + \xi\cos\omega t}{2S}},$$
$$q_m^\pm(t) = a_m^\pm e^{\mp(2m\pm1)\frac{\sigma}{2S^2} \mp \frac{\xi_0 + \xi\cos\omega t}{2S}}. \quad (3.231)$$

Since we are solely concerned with the ac response corresponding to the stationary state, which is independent of the initial conditions, in calculating the longitudinal component of the magnetization defined as

$$\langle \hat{S}_Z \rangle(t) = \sum_{m=-S}^{S} m\rho_m(t), \quad (3.232)$$

we may seek both the *diagonal elements* $\rho_m(t)$ and $\langle \hat{S}_Z \rangle(t)$ as the Fourier series, viz.,

$$\rho_m(t) = \sum_{k=-\infty}^{\infty} \rho_m^k e^{ik\omega t},$$
$$\langle \hat{S}_Z \rangle(t) = \sum_{k=-\infty}^{\infty} S_Z^k e^{ik\omega t}, \quad (3.233)$$

where the amplitudes $S_Z^k(\omega)$ are themselves given by the finite series

$$S_Z^k(\omega) = \sum_{m=-S}^{S} m\rho_m^k \,. \tag{3.234}$$

Next, the *time-dependent coefficients* $q_m(t)$ and $q_m^{\pm}(t)$ in Eqs. (3.231) can also be expanded in a Fourier series using the known Fourier–Bessel expansion [290]

$$e^{\pm\frac{\xi}{2S}\cos\omega t} = \sum_{k=-\infty}^{\infty} I_k\left(\pm\frac{\xi}{2S}\right)e^{ik\omega t}, \tag{3.235}$$

where $I_k(z)$ are the modified Bessel functions of the first kind [290] (see Appendix A.1.6). Thus by direct substitution of Eq. (3.233) and the Fourier–Bessel expansion for the time-dependent coefficients $q_m(t)$ and $q_m^{\pm}(t)$ into Eq. (3.230), we obtain a recurrence relation between the various Fourier amplitudes ρ_m^k, viz.,

$$\begin{aligned} ik\omega\tau_{\mathrm{N}}\rho_m^k = \sum_{k'=-\infty}^{\infty} \Bigg\{ & a_m^- e^{\frac{\sigma(2m-1)}{2S^2}+\frac{\xi_0}{2S}} I_{k-k'}\left(\frac{\xi}{2S}\right)\rho_{m-1}^{k'} \\ & + a_m^+ e^{-\frac{\sigma(2m+1)}{2S^2}-\frac{\xi_0}{2S}} I_{k-k'}\left(-\frac{\xi}{2S}\right)\rho_{m+1}^{k'} \\ - & \left[a_m^- e^{-\frac{\sigma(2m-1)}{2S^2}-\frac{\xi_0}{2S}} I_{k-k'}\left(-\frac{\xi}{2S}\right) + a_m^+ e^{\frac{\sigma(2m+1)}{2S^2}+\frac{\xi_0}{2S}} I_{k-k'}\left(\frac{\xi}{2S}\right)\right]\rho_m^{k'} \Bigg\}. \end{aligned} \tag{3.236}$$

The recurrence relation Eq. (3.236) can be solved exactly for ρ_m^k via matrix continued fractions as follows. By introd ucing the column vector

$$\boldsymbol{\rho}_n = \begin{pmatrix} \vdots \\ \rho_n^{-1} \\ \rho_n^{0} \\ \rho_n^{1} \\ \vdots \end{pmatrix}, \tag{3.237}$$

($n = m + S$), we have the following matrix three-term recurrence equation between the $\boldsymbol{\rho}_n$, namely,

$$\mathbf{Q}_n^-\boldsymbol{\rho}_{n-1} + \mathbf{Q}_n\boldsymbol{\rho}_n + \mathbf{Q}_n^+\boldsymbol{\rho}_{n+1} = \mathbf{0}, \tag{3.238}$$

where the matrix elements of the matrices $\mathbf{Q}_n$ and $\mathbf{Q}_n^\pm$ are given by

$$\begin{aligned}\left[\mathbf{Q}_n\right]_{kk'} &= -i\omega\tau_{\mathrm{N}} k\delta_{kk'} - a_n^+ e^{(2n-2S+1)\frac{\sigma}{S^2}+\frac{\xi_0}{S}} I_{k-k'}(\xi/2S) \\ &\quad - a_n^- e^{-(2n-2S-1)\frac{\sigma}{2S^2}-\frac{\xi_0}{2S}} I_{k-k'}(-\xi/2S), \\ \left[\mathbf{Q}_n^\pm\right]_{k,k'} &= a_n^\pm e^{\mp(2n-2S\pm1)\frac{\sigma}{2S^2}\mp\frac{\xi_0}{2S}} I_{k-k'}(\mp\xi/2S).\end{aligned}$$

Using the general method of solving three-term recurrence relations [97,143], all higher order column vectors $\boldsymbol{\rho}_n$ defined by Eq. (3.237) can be expressed in terms of the *lowest order* column vector $\boldsymbol{\rho}_0$ as the product:

$$\boldsymbol{\rho}_n = \mathbf{S}_n\mathbf{S}_{n-1}\cdots\mathbf{S}_1\boldsymbol{\rho}_0, \tag{3.239}$$

where the $\mathbf{S}_m$ are *finite* matrix continued fractions defined by the matrix recurrence relation

$$\mathbf{S}_m = \left[-\mathbf{Q}_m + \mathbf{Q}_m^+\mathbf{S}_{m+1}\right]^{-1}\mathbf{Q}_m^-. \tag{3.240}$$

However, the zero-order column vector $\boldsymbol{\rho}_0$ itself can in this case be found from the normalization condition for the density matrix elements, viz.,

$$\sum_{n=0}^{2S}\rho_n(t) = \sum_{k=-\infty}^{\infty}\sum_{n=0}^{2S}\rho_n^k(\omega)e^{i\omega kt} = 1 \tag{3.241}$$

thereby immediately yielding an inhomogeneous equation for $\boldsymbol{\rho}_0$, viz.,

$$\sum_{n=0}^{2S}\boldsymbol{\rho}_n = \left(\mathbf{I} + \mathbf{S}_1 + \mathbf{S}_2\mathbf{S}_1 + \ldots + \mathbf{S}_{2S}\ldots\mathbf{S}_2\mathbf{S}_1\right)\boldsymbol{\rho}_0 = \mathbf{v}, \tag{3.242}$$

where $\mathbf{I}$ is the unit matrix and the infinite column vector $\mathbf{v}$ has only the single nonvanishing element $v_k = \delta_{k0}$, $-\infty < k < \infty$. Consequently, we have

$$\boldsymbol{\rho}_0 = \left(\mathbf{I} + \mathbf{S}_1 + \mathbf{S}_2\mathbf{S}_1 + \ldots + \mathbf{S}_{2S}\ldots\mathbf{S}_2\mathbf{S}_1\right)^{-1}\mathbf{v}. \tag{3.243}$$

Having calculated the zero-order column vector $\boldsymbol{\rho}_0$, we can then determine via Eq. (3.239) all the other column vectors $\boldsymbol{\rho}_n$ and thus evaluate all the

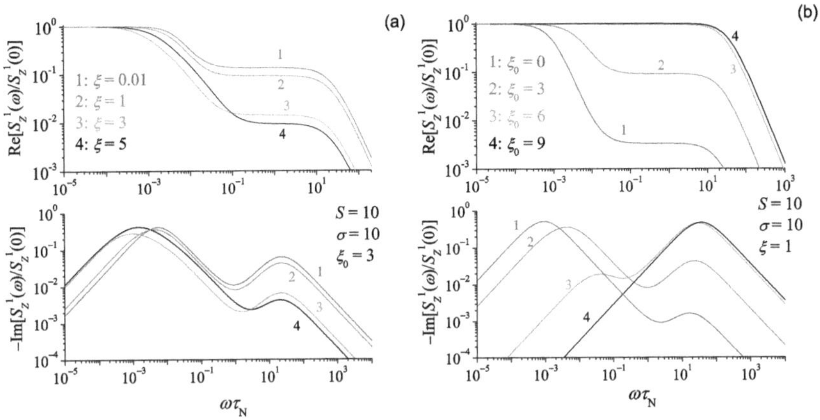

Fig. 3.22. The real and imaginary parts of the normalized fundamental component $S_Z^1(\omega)/S_Z^1(0)$ vs. normalized frequency $\omega\tau_N$ (a) for various applied ac fields ξ and uniform field parameter $\xi_0 = 3$ and (b) for various dc field parameters ξ_0 and $\xi = 1$; spin number $S = 10$ and anisotropy parameter $\sigma = 10$.

$S_Z^k(\omega)$ from Eq. (3.234) thus yielding the nonlinear stationary ac response of a uniaxial nanomagnet.

Initially, we treat the *frequency-dependent* fundamental component of the magnetization $S_Z^1(\omega)$. For a weak ac field, $\xi \to 0$, the normalized fundamental component $S_Z^1(\omega)/S_Z^1(0)$ yields the normalized linear dynamic susceptibility, viz.,

$$\frac{S_Z^1(\omega)}{S_Z^1(0)} \to \frac{\chi(\omega)}{\chi_0}, \tag{3.244}$$

concurring with the linear response solution. In *strong* ac fields, $\xi > 1$, as in linear response, two distinct absorption bands again appear in the spectrum of $-\mathrm{Im}[S_Z^1(\omega)]$ so that two accompanying dispersion regions occur in the spectrum of $\mathrm{Re}[S_Z^1(\omega)]$. However, due to the pronounced nonlinear effects, the low-frequency parts of $\mathrm{Re}[S_Z^1(\omega)]$ and $-\mathrm{Im}[S_Z^1(\omega)]$ may no longer be approximated by a *single* Lorentzian (see Fig. 3.22). Nevertheless, the frequency $\omega_{\max}$ of the maximum loss and/or the half-width $\Delta\omega$ of the spectrum of $\mathrm{Re}[S_Z^1(\omega)]$ may still be used to estimate an effective reversal time τ as defined in Eq. (3.227). The behavior of the low-frequency peak of $-\mathrm{Im}[S_Z^1(\omega)]$ as *a function of the ac field amplitude* again crucially depends on whether or not a dc field is applied. For strong

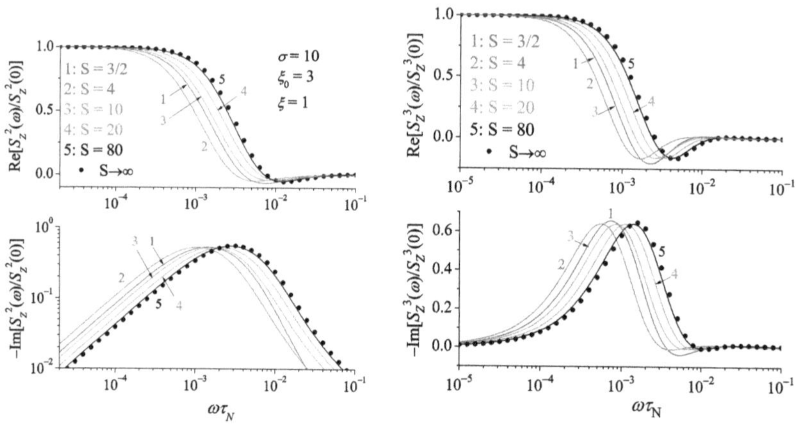

Fig. 3.23. The real and imaginary parts of the normalized second and third harmonic components $S_Z^2(\omega)/S_Z^2(0)$ and $S_Z^3(\omega)/S_Z^3(0)$ of the nonlinear response vs. normalized frequency $\omega\tau_N$ for barrier parameter $\sigma = 10$, uniform field parameter $\xi_0 = 3$ (*nonsymmetrical wells*), ac field parameter $\xi = 1$ and various spin numbers S. Filled circles: the classical limit.

dc bias, $\xi_0 > 1$ (see Fig. 3.22), the low-frequency peak shifts to *lower frequencies reaching a maximum at* $\xi \sim \xi_0$ *thereafter shifting to higher frequencies with increasing* ξ_0. In other words, as the dc field increases, the reversal time of the spin *initially increases* and having attained its maximum at some critical value $\xi \sim \xi_0$ thereafter decreases. For *small* dc bias, $\xi_0 < 0.5$, the low-frequency peak shifts monotonically to higher frequencies with increasing ξ. This behavior is very similar to that observed in the classical case [153,156,159].

Now a striking feature of the nonlinear response is that the effective reversal time may also be evaluated from either the spectrum of the (now) frequency-dependent dc component $S_Z^0(\omega)$ (only for nonzero dc bias, $\xi_0 \neq 0$) or those of the higher order harmonics $S_Z^k(\omega)$ with $k > 1$ because the low-frequency parts of these spectra are themselves, like the spectra of the fundamental, dominated by overbarrier relaxation processes. For illustration, the real and imaginary parts of the normalized second- and third-harmonic components $S_Z^2(\omega)$ and $S_Z^3(\omega)$ are shown in Fig. 3.23. Like the fundamental the behavior of both $-\mathrm{Im}\left[S_Z^2(\omega)\right]$ and $-\mathrm{Im}\left[S_Z^3(\omega)\right]$ depends on whether or not a dc field is applied. For *weak* dc bias $\xi_0 < 0.5$, the low-frequency peak shifts monotonically to higher frequencies. For *strong* dc bias, $\xi_0 > 1$, on the other hand the low-frequency peak shifts to lower frequencies reaching a maximum at $\xi \sim \xi_0$ thereafter decreasing with increasing ξ.

Thus, in the nonlinear relaxation of a uniaxial quantum nanomagnet with arbitrary spin number S subjected to superimposed ac and dc magnetic fields in the high temperature and weak spin-bath coupling limit, we may determine once again the *transition from elementary spin relaxation to that pertaining to a giant spin as a function of the spin number S*. Here only uniaxial nanomagnets have been treated. However, multiaxial anisotropies (cubic, biaxial, etc.) can be considered in like manner.

3.3.12 *Dynamic magnetic hysteresis*

We recall that nanoparticle magnetism has many novel applications, particularly in the (applied) area of information storage [10,11] and in medicine, e.g., in hyperthermia occasioned by induction heating of nanoparticles [12,13]. Moreover, the DMH induced in nanomagnets by an external ac field constitutes a topic of special interest which we now study in the quantum case. Here the temperature directly influences the remagnetization conditions, in turn strongly affecting the effective rates, so altering the loop shape, coercive force, and specific power loss in nanomagnets as described in Chapter 2. Our approach as applied to quantum spins is analogous to that of Refs. [47,159] for DMH of single-domain ferromagnetic particles, where perturbation theory cannot be used. For purposes of exposition, we take a uniaxial nanomagnet of arbitrary spin number S subjected to a uniform external magnetic field $\mathbf{H}_0$ and an ac external field $\mathbf{H}(t)$ applied *along the Z axis*, i.e., the axis of symmetry, with the Hamiltonian $\hat{H}_S$ given by Eq. (3.228) above. Again, our results will coincide with the classical ones in the limit $S \to \infty$.

The stationary ac response of the longitudinal component $\langle \hat{S}_Z \rangle(t)$ was treated in the previous Section 3.3.11. Now, recall that the DMH loop simply represents a parametric plot of the normalized magnetization as a function of the ac field, i.e.,

$$m(t) = \langle \hat{S}_Z \rangle (t) / S \text{ vs. } h(t) = H(t)/H = \cos \omega t,$$

and like the classical case [44–47,52–54,197,382], the normalized area of the DMH loop A_n (which, we recall, is the energy loss per particle and per cycle of the ac field) is given by the exact equation

$$A_n = \frac{1}{4H} \oint m(t) dH(t) = -\frac{\pi}{2S} \mathrm{Im}\left[S_Z^1(\omega) \right]. \tag{3.245}$$

In Figs. 3.24–3.30, we show the effects of ac and dc bias magnetic fields on the DMH loops for a uniaxial nanomagnet with arbitrary S. For a *weak* ac field, $\xi \to 0$, and low frequencies, $\omega\tau \le 1$, the DMH loops are ellipses with normalized area A_n given by Eq. (3.245); the behavior of $A_n \sim -\mathrm{Im}\left[S_Z^1(\omega) \right] \sim \chi''(\omega)$ being similar [cf. Eq. (3.245)] to that of $\chi''(\omega)$ (see Figs. 3.20 and 3.21). Indeed, the two-mode approximation for the susceptibility given by Eq. (3.226) implies that the overall relaxation process consists of two distinct entities, namely, the *slow* thermally activated overbarrier (or interwell) process and the *fast* intrawell relaxation in the wells. At *low* frequencies and with *large* barriers between the wells as well as for *weak* dc bias fields, $\xi_0/(2\sigma) << 1$, the approximation $\Delta_1 \approx 0$ may always be used in Eq. (3.226) so that the *normalized* magnetization $m(t) = \langle \hat{S}_Z \rangle (t)/S$ is given by the simple linear response formula [47] (cf. Eq. (2.328) of Chapter 2)

$$m(t) = \frac{1}{S} \langle \hat{S}_Z \rangle_0 + \frac{\chi_0 \xi}{S} \frac{\cos \omega t + \omega\tau \sin \omega t}{1 + \omega^2 \tau^2}. \tag{3.246}$$

If we introduce new variables x and y defined as

$$x = h(t) = \cos \omega t$$

and

$$y = (\chi_0 \xi)^{-1} \left(Sm(t) - \langle \hat{S}_Z \rangle_0 \right),$$

we then can conclude from Eq. (3.246) that in the linear response approximation, the low-frequency DMH loop in the (x, y) plane is still an ellipse, namely, Eq. (2.329) [47].

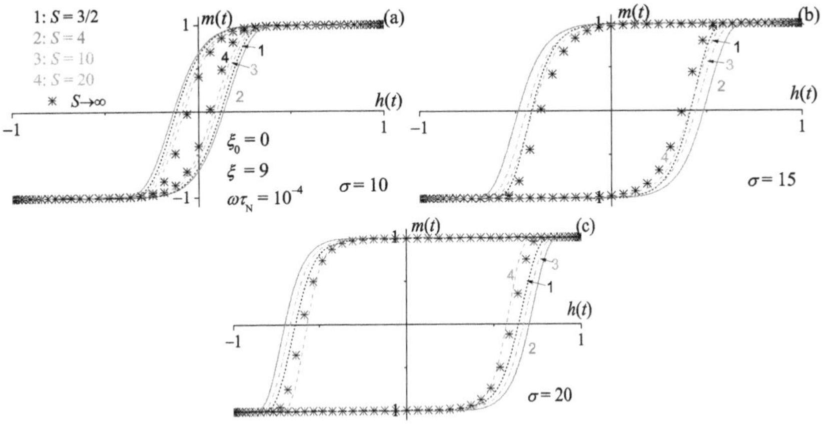

Fig. 3.24. DMH loops for various anisotropy parameters $\sigma = 10$ (a), 15 (b), 20 (c) and various spin numbers $S = 3/2$ (1: short-dashed lines), 4 (2: solid lines), 10 (3: dashed-dotted lines), 20 (4: dashed lines), and ∞ (asterisks) at $\omega\tau_N = 10^{-4}$, $\xi_0 = 0$, and $\xi = 9$.

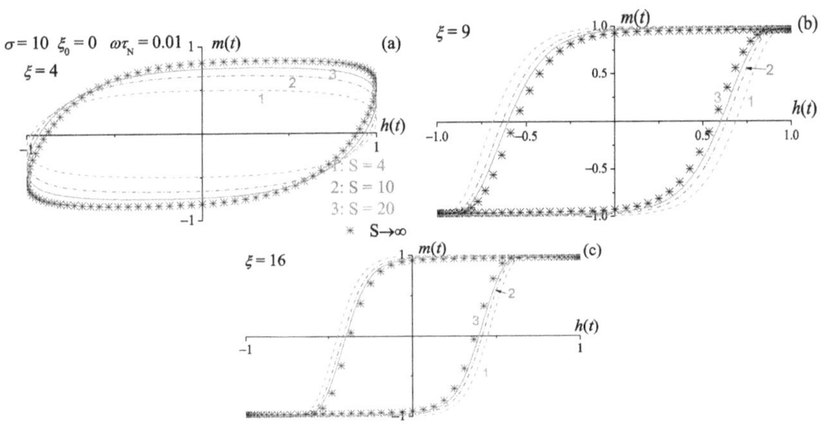

Fig. 3.25. DMH loops for various ac external field parameters $\xi = 4$ (a), 9 (b), 16 (c) and various spin numbers $S = 4$ (1: dashed lines), 10 (2: dashed-dotted lines), 20 (3: solid lines), and ∞ (asterisks) at $\sigma = 10$, $\xi_0 = 0$, and $\omega\tau_N = 10^{-2}$.

For *relatively weak* ac fields, $0.5 < \xi \leq 1$, the DMH loops still retain an approximately ellipsoidal shape implying as in Chapter 2 that only a few harmonics contribute to the weakly nonlinear response. In contrast, in *strong* ac fields, $\xi > 1$, the shape markedly alters and so the normalized area A_n now displays a pronounced dependence on the frequency ω and

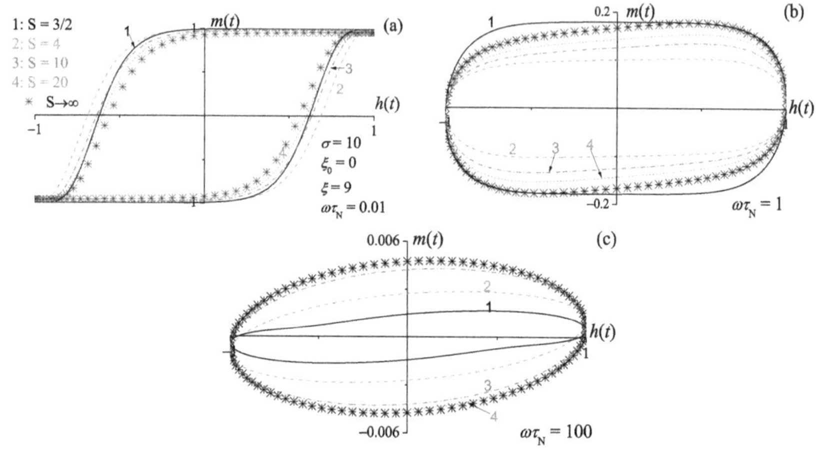

Fig. 3.26. DMH loops for various frequencies $\omega\tau_N = 10^{-2}$ (a), 1 (b), 10^2 (c) and various spin numbers $S = 3/2$ (1: solid lines), 4 (2: dashed lines), 10 (3: dashed-dotted lines), 20 (4: short-dashed lines), and ∞ (asterisks) at $\sigma = 10$, $\xi_0 = 0$, and $\xi = 9$.

the ac and dc bias field amplitudes ξ and ξ_0 as well as on the anisotropy parameter σ and the spin number S (see Figs. 3.24–3.27). In this strong field regime, the external ac field can saturate the magnetic moment as well as inducing its inversion (i.e., switching between the directions of the easy axis). In Figs. 3.24 and 3.25, we plot the DMH loops for various S and anisotropy (σ) and ac field (ξ) parameters exemplifying how their shapes (and consequently their areas) alter as these parameters vary. Clearly, the remagnetization time is highly sensitive to any variation of the parameters. For example, with a strong ac driving field, the Arrhenius dependence of the reversal time on temperature, i.e., $\log(\tau/\tau_N) \propto 1/T$, which accurately accounts for the linear response regime, is modified because the strong ac field intervenes, as in the classical case, to *drastically reduce the effective response time of the nanomagne*t. Thus, the nonlinear behavior again facilitates re-magnetization regimes, never attainable with weak ac fields. The reason being that the dc bias component under the appropriate conditions *efficiently tunes* this effect by either *enhancing* or *blocking* the action of the strong ac field. The *pronounced* frequency dependence of the loops is highlighted in Fig. 3.26 for various S. At low frequencies, the field changes are quasiadiabatic with

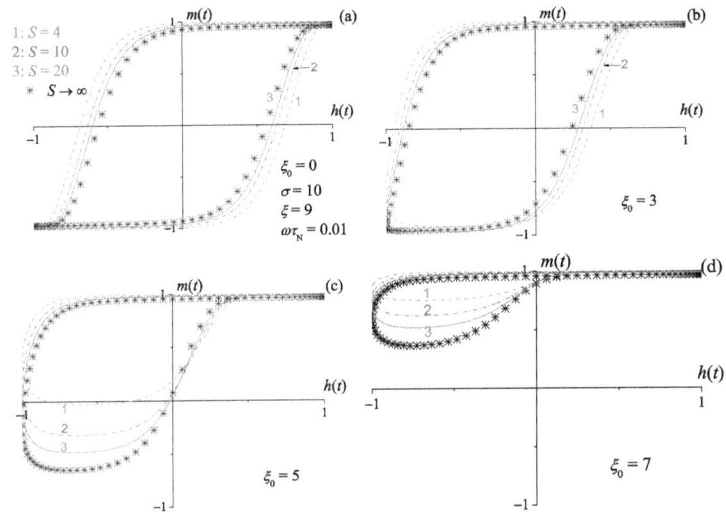

Fig. 3.27. DMH loops for various dc bias field parameters $\xi_0 = 0$ (a), 3 (b), 5 (c), 7 (d) and various spin numbers $S = 4$ (1: dashed lines), 10 (2: dashed-dotted lines), 20 (3: solid lines), and ∞ (asterisks) at $\xi = 9$, $\sigma = 10$, and $\omega\tau_N = 10^{-2}$.

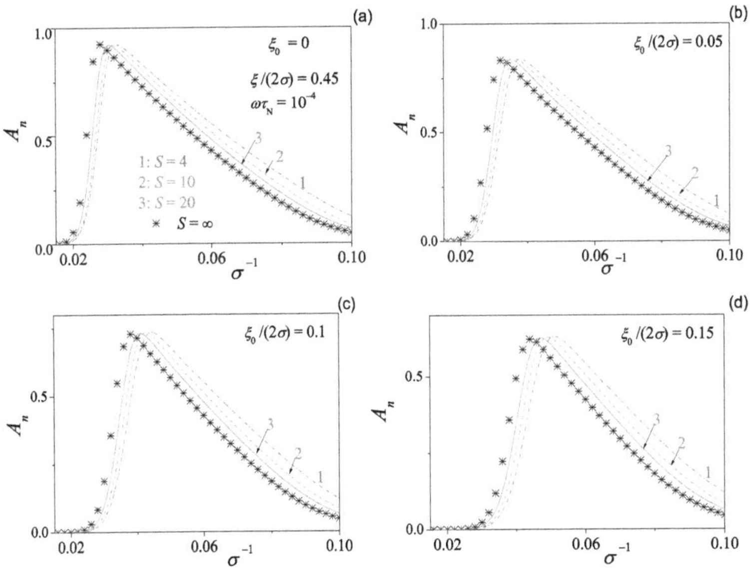

Fig. 3.28. Normalized area of the DMH loop A_n vs. the dimensionless temperature σ^{-1} under variation of the dc bias field parameter $h_0 = \xi_0/(2\sigma) = 0$ (a), 0.05 (b), 0.1 (c), and 0.15 (d) for various spin numbers $S = 4$ (dashed-dotted lines), 10 (dashed lines), 20 (solid lines), and ∞ (asterisks) at the frequency $\omega\tau_N = 10^{-4}$ and ac field amplitude $\xi/(2\sigma) = 0.45$.

the result that the magnetization reverses due to the *cooperative* shuttling action of thermal agitation combined with the ac field. The dc bias field effects on the DMH are illustrated in Fig. 3.27 showing the changes in the DMH caused by varying ξ_0 for several spin numbers S. To understand the effect of the dc bias field on the loop area, we recall that once again the magnetic relaxation time depends on the *instantaneous value* of the applied field. Under the conditions of Fig. 3.27, the *positive limiting* (saturation) value $m(t) \to 1$ corresponds to a total instantaneous field $H_0 + H$, thus *favoring* the magnetization relaxation to the positive saturation value $m(t) \to 1$. However, for *negative* $h(t)$, the total field $H_0 - H$ is much weaker and so cannot induce relaxation to the negative saturation value $m(t) \to -1$ at all. Therefore, the "centroid" of the loop shifts upwards. In the classical limit, $S \to \infty$, our results concur with those [159] for classical uniaxial nanomagnets treated in Chapter 2.

The temperature dependence of the DMH is governed by the dimensionless anisotropy (inverse-temperature) parameter $\sigma \propto 1/T$. The normalized DMH area A_n as a function of σ^{-1} is shown in Fig. 3.28 for various S showing that the *tuning* action of the dc bias field described above is effective over a certain temperature interval. This conclusion once again indicates that the relaxation of the magnetization is mostly caused by thermal fluctuations, implying that the magnetic response time still retains a corresponding strong temperature dependence. The normalized area as a function of the frequency ω and ac field parameter $\xi/(2\sigma)$ is shown in Figs. 3.29 and 3.30, respectively. Invariably, A_n can be represented as a nonmonotonic curve with a maximum in the position which is determined by S as well as by the other model parameters. The peak in A_n (Fig. 3.29) is caused by the field-induced modifications of the reversal time as strongly tuned by the dc bias field. As in Fig. 3.29, variation of the dc field strength shifts the frequency, where the maximum is attained, by several orders of magnitude. The normalized loop area portrayed in Fig. 3.30 illustrates the dependence of A_n on the ac field amplitude, which is similar to that of classical nanomagnets.

The DMH in uniaxial nanomagnets has been treated above without any *a priori* assumptions regarding the potential barrier height, temperature, the magnetizing field strength and/or spin number S. In general, it appears that given appropriate conditions a small bias dc field (in comparison with

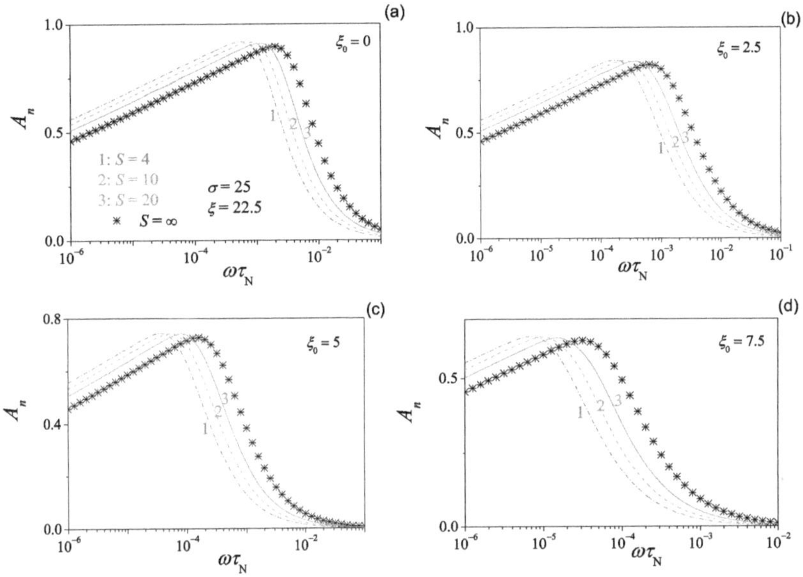

Fig. 3.29. Normalized area of the DMH loop A_n vs. the dimensionless frequency $\omega\tau_N$ under variation of the dc bias field $\xi_0 = 0$ (a), 2.5 (b), 5 (c), and 7.5 (d) for various spin numbers $S = 4$ (dashed-dotted lines), 10 (dashed lines), 20 (solid lines), and ∞ (asterisks); the anisotropy parameter $\sigma = 25$ and ac field parameter $h = 0.45$.

the internal anisotropy field) can profoundly affect the shape of the DMH loops in nanomagnets accompanied by a strong dependence on the spin number S.

3.3.13 *Quantum effects in stochastic resonance*

In contrast to the classical case treated in Chapter 2, Section 2.4.4, in magnetic SR of nanomagnets with smaller spin numbers $S \sim 10$–100 both quantum effects and quantum-classical crossover appear. Here the spin reversal is either due to thermal activation alone or tunneling alone or a combination of both. The quantum effects are not the same as those in the SR for translational Brownian motion (see, e.g., [479–484] and references cited therein) because despite some analogies the quantum spin dynamics differ from those of Brownian particles owing to the different symmetries

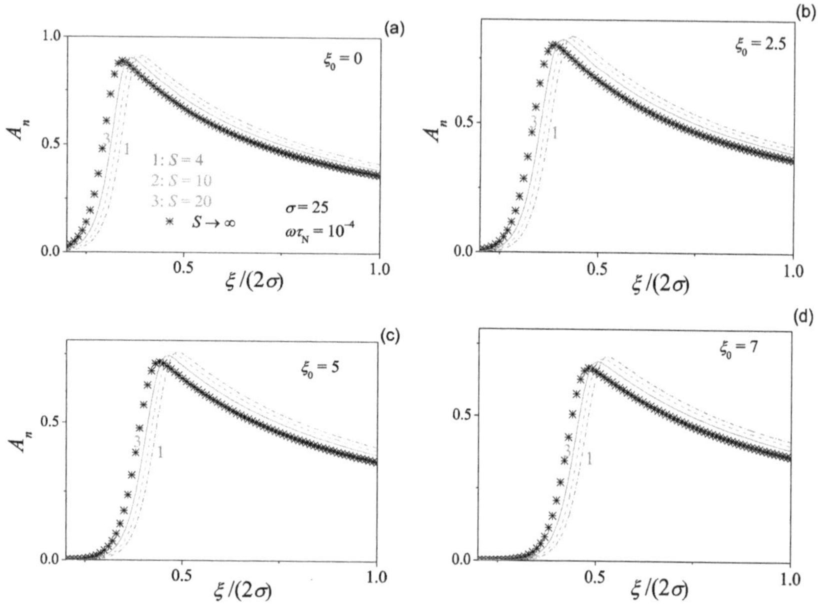

Fig. 3.30. Normalized area of the DMH loop A_n as a function of ac field amplitude $h = \xi/(2\sigma)$ under variation of the dc bias field parameter $\xi_0 = 0$ (a), 2.5 (b), 5 (c), and 7 (d) for various spin numbers $S = 4$ (dashed-dotted lines), 10 (dashed lines), 20 (solid lines), and ∞ (asterisks); the anisotropy parameter $\sigma = 25$ and the frequency $\omega\tau_N = 10^{-4}$.

of the rotational and translational groups. Here we shall treat quantum effects in the SR for spin systems modeled by a uniaxial nanomagnet of arbitrary S in superimposed dc and ac external uniform magnetic fields $\mathbf{H}_0+\mathbf{H}\cos\Omega t$ applied along the Z axis, i.e., the axis of symmetry, so that the time-dependent Hamiltonian $\hat{H}_S$ is [cf. Eq. (3.228)]

$$\beta\hat{H}_S(t) = -\frac{\sigma}{S^2}\hat{S}_Z^2 - \frac{\xi_0 + \xi\cos\Omega t}{S}\hat{S}_Z, \tag{3.247}$$

where σ is the anisotropy constant, ξ_0 and ξ are the external dc and ac magnetic field parameters. The longitudinal relaxation of uniaxial nanomagnets interacting with a thermal bath has been treated in Section 3.3.9 via the respective evolution equations for the reduced density matrix and phase-space distribution function using the methods already available for classical spins. In the large spin limit, we also saw in Section 3.3.9 that the quantum solutions reduce to those of the Fokker–Planck equation for a classical uniaxial nanomagnet while for linear response and finite S, the results agree with those predicted by the solutions of Garanin [226] and García-Palacios and Zueco [227,228]. Here we apply these findings to the SNR for uniaxial quantum nanomagnets.

We saw in Section 3.3.9 that the normalized longitudinal dynamic susceptibility of a quantum nanomagnet is defined in linear response as

$$\frac{\chi(\Omega)}{\chi_0} = 1 - i\Omega\int_0^\infty C(t)e^{-i\Omega t}dt, \tag{3.248}$$

where $C(t)$ is the normalized equilibrium correlation function defined by Eq. (3.179), and

$$\chi_0 = \frac{1}{S^2}\left[\sum_{m=-S}^{S} m^2\rho_m - \left(\sum_{m=-S}^{S} m\rho_m\right)^2\right],$$
$$\rho_m = \frac{e^{\sigma\left(m^2/S^2+2hm/S\right)}}{\sum_{m=-S}^{S} e^{\sigma\left(m^2/S^2+2hm/S\right)}}. \tag{3.249}$$

For a uniaxial nanomagnet, both $C(t)$ and $\chi(\Omega)$ have been calculated in Section 3.3.11 In particular, we recall that $C(t)$ may be written as the finite discrete set of relaxation modes, namely, [cf. Eq. (3.159)]

$$C(t) = \sum_{k=1}^{2S} c_k e^{-\lambda_k t}, \tag{3.250}$$

where λ_k are the eigenvalues of the system matrix **A**, Eq. (3.170), with the replacement $\xi_{\rm II} \to \xi_0$. Consequently, Eqs. (3.248) and (3.250) allow us to *formally* write $\chi(\Omega)$ as the finite discrete set of Lorentzians [cf. Eq. (3.181)]

$$\frac{\chi(\Omega)}{\chi_0} = \sum_{k=1}^{2S} \frac{c_k}{1+i\Omega/\lambda_k}. \tag{3.251}$$

The asymptotic behavior of $\chi(\Omega)$ in the extremes of very low and very high frequencies is given as before by Eqs. (3.182) and (3.183), where $\tau_{\rm cor}$

is the integral relaxation time and τ_{ef} is the effective relaxation time defined in terms of the eigenvalues λ_k as

$$\tau_{cor} = \sum_{k=1}^{2S} c_k / \lambda_k \text{ and } \tau_{ef} = \left(\sum_{k=1}^{2S} c_k \lambda_k \right)^{-1}. \tag{3.252}$$

Furthermore, the smallest nonvanishing eigenvalue λ_1 may as usual be associated with the long-time behavior of $C(t) \sim e^{t/\tau}$, $t >> \tau = 1/\lambda_1$, which is characterized by the longest relaxation (or the reversal) time τ. Now in Section 3.3.11 above, it has been shown that all these times are given by simple analytic formulas, Eqs. (3.207), (3.209), and (3.221), respectively.

As seen in Section 3.3.11, two distinct bands appear in the spectrum of the imaginary part $\chi''(\Omega)$ of the susceptibility for a uniaxial quantum nanomagnet. The low-frequency band is due to the slowest "interwell" relaxation mode. The characteristic frequency and the half-width of this band are determined by λ_1 associated with the long-time behavior of $C(t) \sim e^{-t/\tau}$, $t >> \tau$. The high-frequency band in $\chi''(\Omega)$ is due to the near-degenerate high-frequency modes corresponding to the eigenvalues $\lambda_k >> \lambda_1$ $(2S \geq k \geq 2)$. Thus, if one is interested solely in the low-frequency region $(\Omega\tau \leq 1)$, where their effect may be ignored, $\chi(\Omega)$ may be approximated as the single Lorentzian, Eq. (1.151) of Chapter 1 [220,221]. Next, we recall that magnetic SR may be described using linear response theory yielding for the signal-to-noise ratio (SNR) [62–64,97]

$$\mathrm{SNR} = \frac{\beta^2 \mu_0 \mu N \pi \Omega H^2 |\chi(\Omega)|^2}{2\chi''(\Omega)}, \tag{3.253}$$

where N is the number of nanomagnets per unit volume. By analogy with the SNR for a classical nanomagnet, Eq. (2.226) of Chapter 2, Eq. (3.253) can be written as

$$\mathrm{SNR} = \frac{\pi\beta^2 \mu_0 \mu N \chi_0 H^2}{2\tau_N \sigma} R_\Omega, \tag{3.254}$$

where the dimensionless SNR factor R_Ω is given by

$$R_\Omega = \frac{\sigma\tau_N \Omega \left|\chi(\Omega)\right|^2}{\chi''(\Omega)}. \tag{3.255}$$

Thus, as in the classical case the relevant quantity is R_Ω. In general, R_Ω, besides the obvious dependence on the noise intensity (temperature), the constant (bias) field strength parameter h, and the frequency of the exciting field Ω now depends on the spin number S. In the adiabatic limit, $\Omega \to 0$, with Eqs. (3.251), (3.252), and the correlation time Eq. (3.207), the SNR factor Eq. (3.255) simplifies yielding

$$R_0 = \frac{\tau_N \sigma \chi_0}{\tau_{\text{cor}}} = \frac{\dfrac{\sigma}{2S^2}\left[\sum_{m=-S}^{S} m^2 \rho_m - \left(\sum_{m=-S}^{S} m\rho_m\right)^2\right]^2}{\sum_{k=1-S}^{S}\left(\sum_{m=k}^{S}\left(m - \sum_{n=-S}^{S} n\rho_n\right)\rho_m\right)^2 \dfrac{1}{\left[S(S+1)-k(k-1)\right]\sqrt{\rho_k \rho_{k-1}}}}. \tag{3.256}$$

On the contrary, in the opposite very high-frequency limit, $\Omega \to \infty$, Eq. (3.255) reduces to

$$\begin{aligned} R_{\Omega\to\infty} &= \frac{\tau_N \sigma \chi_0}{\tau_{\text{ef}}} \\ &= \frac{\sigma}{2S^2} \sum_{m=1-S}^{S} \left[S(S+1) - m(m-1)\right]\sqrt{\rho_m \rho_{m-1}} \end{aligned} \tag{3.257}$$

with Eqs. (3.225), (3.251), and (3.252). In the classical limit, $S \to \infty$, the normalized $\chi(\Omega)$ is also given by Eq. (3.251) [cf. Eq. (2.225) of Chapter 2], where $\chi_0 = \left\langle \cos^2 \vartheta \right\rangle_0 - \left\langle \cos\vartheta \right\rangle_0^2$. The classical counterparts of the quantum Eqs. (3.256) and (3.257) are simply Eq. (2.228) of Chapter 2, where the classical relaxation times τ_{ef} and τ_{cor} are given by Eqs. (3.225) and Eq. (2.182) of Chapter 2.

The SNR factor R_0 in the adiabatic limit $\Omega = 0$ as a function of the dimensionless temperature parameter σ^{-1} is shown in Fig. 3.31 for various spin numbers S and field parameters h. Usually the maximum of R_0 is attained in the range $\sigma^{-1} \sim 0.3$–0.5 (corresponding to $T \sim 30$ K for the molecular magnet Mn_{12} acetate with $S = 10$). Moreover, that maximum shifts to higher temperatures with increasing h because the bias field

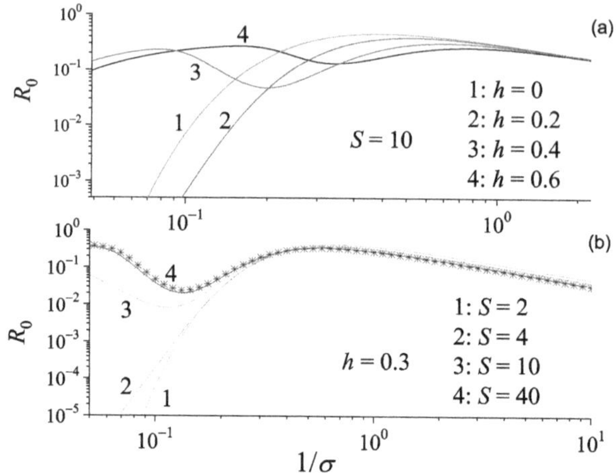

Fig. 3.31. SNR R_0 as a function of dimensionless temperature σ^{-1} for nonzero values of the dc field (a) for various values of h and $S = 10$ and (b) for various values of S and $h = 0.3$ in the adiabatic limit $\Omega = 0$. Solid lines: exact solution. Asterisks: classical limit $S \to \infty$.

radically alters the temperature dependence of the static susceptibility. In a nonzero bias field, the effect of saturation of the magnetization is crucial causing R_0 to tend to zero at zero temperature. Although in the low-temperature limit, $\sigma^{-1} \to 0$, $R_0 \to 0$ for *both small* S and h; nevertheless, as long as S increases at *finite* h or as h increases at *finite* S, $R_0 \to$ constant (see Fig. 3.31). This phenomenon is due to the temperature dependence of τ_{cor}, which causes the latter to progressively lose its Arrhenius character with increasing h. In general, the quantum effects can lead to both *amplification and attenuation* of the SNR. The frequency-dependent SNR R_Ω vs. the dimensionless temperature parameter σ^{-1} is shown in Fig. 3.32 for various spin numbers S and finite values of the forcing frequency Ω. Clearly this figure exemplifies *the quantum effects via a pronounced deviation of the quantum SNR curves from the corresponding classical ones* (up to several orders of magnitude at low temperatures). Now, the SNR as a function of the dimensionless forcing frequency $\Omega\tau_N$ is also presented in Fig. 3.33. Here the SNR monotonically increases from its low-frequency limit Eq. (3.256) to its plateau value given by Eq. (3.257).

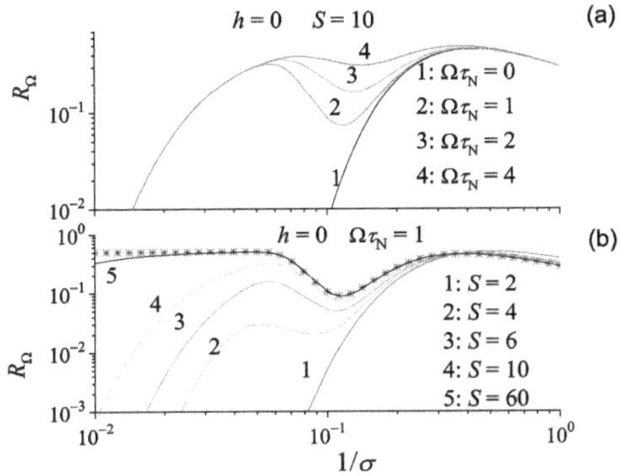

Fig. 3.32. SNR R_Ω vs. the dimensionless temperature σ^{-1} (a) for various normalized frequencies $\Omega\tau_N$ and $S = 10$ and (b) for various S and $\Omega\tau_N = 1$ and $h = 0$. Solid lines: exact solution Eqs. (3.251), (3.252), and (3.255); asterisks: classical limit, $S \to \infty$.

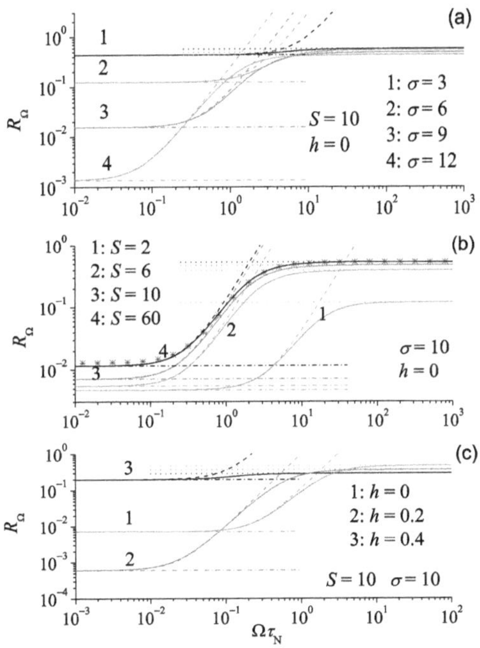

Fig. 3.33. SNR R_Ω vs. the normalized frequency $\Omega\tau_N$ (a) for various σ, $S = 10$, and $h = 0$, (b) for various S, $\sigma = 10$, and $h = 0$, and (c) for various h, $S = 10$ and $\sigma = 10$. Solid lines: exact solution. Dashed lines: the low-frequency asymptote, Eqs. (3.255) and (1.151) of Chapter 1. Dashed-dotted lines: the adiabatic limit, $\Omega = 0$, Eq. (3.256). Dotted lines: the high-frequency limit Eq. (3.257).

We have studied the magnetic SR of a quantum uniaxial nanomagnet of arbitrary spin S in the high temperature and weak spin-bath coupling limit. The principal result is that one may determine *the transition from the SR corresponding to quantum elementary spin relaxation to that pertaining to a giant spin as a function of the spin size S*. Hence, one may accurately estimate the value of S (typically in the range 20–40) at which the crossover from quantum to classical behavior takes place. Thus, one may assign a range of validity to the classical Néel–Brown treatment of a nanomagnet with the simplest uniaxial anisotropy and Zeeman energy. The relatively elementary calculation outlined above is also fundamental towards gaining an understanding of the SR of spin systems characterized by more general Hamiltonians. The extension to particular spin systems such as biaxial, cubic, etc. would also allow one to include spin size effects in important technological applications of magnetic relaxation such as the magnetization reversal time, the switching-field and hysteresis curves, etc.

3.4 Master equation in phase space for noncircularly symmetric systems

3.4.1 *Uniaxial nanomagnet subjected to a dc bias field of arbitrary orientation*

Clearly the analogy between the quantum and classical formulations provided by the phase-space representation via the Wigner–Stratonovich map will also enable the powerful statistical moment method for classical Fokker–Planck equations [97,143] for an *arbitrary* Hamiltonian (which may be expanded in spherical harmonics) to be carried over into the quantum domain. Here our objective is to illustrate this method for the *formal* phase-space master equation (1.600) pertaining to a Hamiltonian $\hat{H}_S$ as accomplished for the Fokker–Planck equation (3.3) for classical spins for *arbitrary* magnetocrystalline anisotropy-Zeeman energy potentials [97,141,377]. We shall illustrate how the magnetization and its relaxation times may be evaluated in the *linear response approximation* and how the solution of the corresponding classical problem [97,239] carries over into the quantum domain. However, in view of the formidable difficulties associated with both the derivation and the solution of phase-space master equations, we will again proceed indirectly using the matrix method of Section 1.3.5 for the calculation of averages of the spherical harmonics. Proceeding in this way, the *explicit* solution may be ultimately written for an *arbitrary* spin Hamiltonian $\hat{H}_S$ as a *finite* series of spherical harmonics analogous to the (infinite) Fourier series representation of the classical case governed by the Fokker–Planck equation (3.3). Therefore, the expansion coefficients, i.e., the statistical averages of the spherical harmonics, may formally be determined as before from a differential-recurrence relation yielding the stochastic spin dynamics for arbitrary spin number *S*. For large *S*, the differential-recurrence relation reduces to that generated by the Fokker–Planck equation. Thus, the spin dynamics may once again be treated in a manner transparently linking to the classical representations, thereby providing quantum corrections to classical averages.

First recall that in accordance with the Wigner–Stratonovich map, the *formal* solution of the phase-space master equation for arbitrary $\hat{H}_S$ may be written as a *finite* linear combination of the spherical harmonics $Y_{LM}(\vartheta, \varphi)$ embodied in the phase-space distribution Eq. (1.579). As shown in Section 1.9.8, by substituting Eq. (1.579) into the master equation Eq. (1.600), we then *formally* have a *finite* set of differential-recurrence relations for the statistical moments $\langle Y_{LM} \rangle(t)$, Eq. (1.604). However, we saw that they can be solved by direct matrix diagonalization, involving the calculation of the eigenvalues and eigenvectors of the system matrix. Thus, it is evident that it is relatively easy to formally generalize the phase-space approach to problems for an *arbitrary* spin Hamiltonian in a manner exactly analogous to that given for the classical Fokker–Planck equation (3.3) in Ref. [377].

Here we shall take as an illustrative example a uniaxial nanomagnet of arbitrary *S* in an external dc magnetic field $\mathbf{H}_0$ applied at an *arbitrary angle* to the *Z*-axis (i.e., the easy axis). Without loss of generality, we may suppose as before that $\mathbf{H}_0$ lies in the *XZ* plane of the laboratory coordinate system so that the spin Hamiltonian $\hat{H}_S$ has the form

$$\beta \hat{H}_S = -\frac{\sigma}{S^2}\hat{S}_Z^2 - \frac{\xi}{S}\left(\cos\psi \hat{S}_Z + \sin\psi \hat{S}_X\right), \qquad (3.258)$$

where ψ is the angle between the dc field $\mathbf{H}_0$ and the Z axis taken as the easy axis of the nanomagnet.

For $s=-1$ (i.e., for the Q-function), the hierarchy of moment equations, Eq. (1.604), can conveniently be rewritten as one for the relaxation functions $c_{L,M}(t)=\langle Y_{LM}\rangle(t)-\langle Y_{LM}\rangle_0$ as

$$\frac{d}{dt}c_{L,M}(t)=\sum_{L',M'}p'_{L'M';LM}c_{L',M'}(t), \tag{3.259}$$

where

$$p'_{L'M';LM}=(-1)^{M-M'}\sqrt{\frac{(2S-L')!(2S+L'+1)!}{(2S-L)!(2S+L+1)!}}g_{L'-M';L-M} \tag{3.260}$$

with the expansion coefficients $g_{L'M';LM}$ given by Eq. (1.446). In the classical limit, $S\to\infty$, Eq. (3.259) reduces to the classical hierarchy treated in Section 2.5.1 [97,141,377]. Next by using MATHEMATICA® for the matrix algebra with the polarization and spin operators as defined in Appendix A.2, the expansion coefficients $p'_{L'M';LM}$ in Eq. (3.260) can then be evaluated from the symmetrized Hubbard form of the collision operator Eq. (1.430) and the spin Hamiltonian, Eq. (3.258). Moreover, the equilibrium averages $\langle Y_{LM}\rangle_0$ themselves satisfy the homogeneous recurrence relation

$$\sum_{L',M'}p'_{L'M';LM}\langle Y_{LM}\rangle_0=0. \tag{3.261}$$

Here the angular brackets $\langle\ \rangle_0$ designate the equilibrium average defined by

$$\langle Y_{LM}\rangle_0=\frac{2S+1}{4\pi}\int_0^{\pi}\int_0^{2\pi}Y_{LM}(\vartheta,\varphi)W_S^{\mathrm{eq}}(\vartheta,\varphi)\sin\vartheta d\vartheta d\varphi, \tag{3.262}$$

where $W_S^{\mathrm{eq}}(\vartheta,\varphi)$ is the equilibrium quasiprobability distribution which can then be evaluated explicitly as described in Section 3.2.2. Alternatively, $\langle Y_{LM}\rangle_0$ can be evaluated directly via the average polarization operators as

$$\langle Y_{LM}\rangle_0 = \sqrt{\frac{2S+1}{4\pi}} C^{SS}_{SSL0} \langle \hat{T}^{(S)}_{LM}\rangle_0 = \sqrt{\frac{2S+1}{4\pi}} C^{SS}_{SSL0} \mathrm{Tr}\left(\hat{\rho}_{\mathrm{eq}} \hat{T}^{(S)}_{LM}\right), \quad (3.263)$$

where the equilibrium density matrix $\hat{\rho}_{\mathrm{eq}}$ is given explicitly by

$$\hat{\rho}_{\mathrm{eq}} = \frac{e^{\sigma \hat{S}_Z^2/S^2 + \xi\left(\cos\psi \hat{S}_Z + \sin\psi \hat{S}_X\right)/S}}{\mathrm{Tr}\{e^{\sigma \hat{S}_Z^2/S^2 + \xi\left(\cos\psi \hat{S}_Z + \sin\psi \hat{S}_X\right)/S}\}}. \quad (3.264)$$

We note that for $\psi = 0$ (uniaxial anisotropy) the hierarchy defined by Eq. (3.259) reduces to the particular recurrence relation Eq. (3.102) already derived in Section 3.3.2. For another particular case, viz., an individual spin in an external uniform magnetic field $\mathbf{H}_0$ directed along the Z-axis of the laboratory coordinate system, where $\beta\hat{H}_S = -\xi\hat{S}_Z / S$, Eq. (3.259) also yields the very simple *three-term* differential-recurrence relation Eq. (3.106) for the relaxation functions $c_{LM}(t)$, which *decouple* for different m and, as shown in Section 3.3.2, can be *exactly* solved using continued fractions [256,257].

By solving Eq. (3.259) for the one-sided Fourier transforms of $c_{10}(t)$ and $c_{1\pm1}(t)$ as described in detail in Refs. [32,38,97,145], we can determine all relevant observables (see, e.g., Figs. 3.34–3.37 below). In the classical limit, $S \to \infty$ and $\hbar S \to$ const, the hierarchy Eq. (3.259) associated with Eq. (3.258) reduces to the differential-recurrence equations for a classical uniaxial nanomagnet subjected to a dc bias field of arbitrary orientation (see Refs. [32,38,47,108,145,156,159,207,342–346] for details). Here the number of recurring equations is *infinite* ($S \to \infty$) again constituting the principal difference between the hierarchies for classical and quantum spins; in the latter case, the number of equations is *finite*. Actually, in the classical limit, $S \to \infty$, the Hamiltonian (3.258) corresponds to a free energy E given by Eq. (2.6) of Chapter 2. To describe the stochastic dynamics of a classical spin (magnetic moment), we have used Gilbert's equation for the motion of the magnetic moment augmented by a random field, Eq. (3.2), thus the problem reduces to the solution of the *infinite* hierarchy of moment equations Eq. (2.251) of Chapter 2 for the relaxation functions $c_{l,m}(t) = \langle Y_{lm}\rangle(t) - \langle Y_{lm}\rangle_0$ [32,38,97,108,145,342–344]. For $S \to \infty$, the hierarchy of quantum relaxation equations (3.259) reduces to the classical Eq. (2.251) so that all the quantum results agree with the classical ones. We shall now compare the predictions of the classical model with those of the quantum one for finite S.

3.4.2 *Characteristic relaxation times and dynamic susceptibility*

We evaluate the *linear response* of the uniaxial nanomagnet with the Hamiltonian Eq. (3.258) due to infinitesimally small changes in the magnitude of the dc field. Thus, we suppose as usual that a small probing field $\mathbf{H}$ $(\mathbf{H} \parallel \mathbf{H}_0,\ \beta\mu_0\mu H \ll 1)$ having been applied to the nanomagnet in the distant past ($t = -\infty$) so that equilibrium conditions obtain at time $t = 0$, is switched off at $t = 0$. By solving the quantum hierarchy Eq. (3.259) for $c_{10}(t)$ and $c_{1\pm1}(t)$, we then have all relevant quantities, namely the integral, effective and reversal times of the magnetization along with the dynamic susceptibility, DMH loops, etc. as dictated by linear response theory. Here the decay of the component $\langle\hat{\mu}_H\rangle(t)$ of the magnetic moment in the direction of a probe field $\mathbf{H}$, when $\mathbf{H}$ has been switched off at time $t = 0$, is given by

$$\langle\hat{\mu}_H\rangle(t) - \langle\hat{\mu}_H\rangle_0 = \beta\mu_0\mu^2 N\chi_0 HC(t). \tag{3.265}$$

In Eq. (3.265), $C(t)$ is the normalized relaxation function defined in this particular instance as

$$C(t) = \frac{\sqrt{2}\cos\psi c_{10}(t) + \sin\psi[c_{1-1}(t) - c_{11}(t)]}{\sqrt{2}\cos\psi c_{10}(0) + \sin\psi[c_{1-1}(0) - c_{11}(0)]}, \tag{3.266}$$

and χ_0 is the static magnetic susceptibility, given by

$$\chi_0 = \sqrt{\frac{2\pi}{3}}\left\{\sqrt{2}\cos\psi c_{10}(0) + \sin\psi[c_{1-1}(0) - c_{11}(0)]\right\}\frac{\gamma\hbar(S+1)}{\mu_0 S^2 H}. \tag{3.267}$$

Furthermore, in writing Eqs. (3.265)–(3.267), we have again used the correspondence rules of spin operators and c-numbers [cf. the Weyl symbols Eqs. (1.587) and (1.589) for $s = 1$], viz.,

$$\hat{S}_X \to \sqrt{\frac{2\pi}{3}}(S+1)\left(Y_{1-1}(\vartheta,\varphi)-Y_{11}(\vartheta,\varphi)\right)=(S+1)\sin\vartheta\cos\varphi$$

and

$$\hat{S}_Z \to \sqrt{\frac{4\pi}{3}}(S+1)Y_{10}(\vartheta,\varphi)=(S+1)\cos\vartheta.$$

The initial conditions $c_{LM}(0)$ in Eqs. (3.266) and (3.267) are

$$\begin{aligned} c_{LM}(0) &= \lim_{\xi_1\to 0}\xi_1^{-1}\left(\langle Y_{LM}\rangle_{\xi_1}-\langle Y_{LM}\rangle_0\right) \\ &= \sqrt{\frac{2S+1}{4\pi}}C^{SS}_{SSL0}\lim_{\xi_1\to 0}\xi_1^{-1}\left(\langle \hat{T}^{(S)}_{LM}\rangle_{\xi_1}-\langle \hat{T}^{(S)}_{LM}\rangle_0\right). \end{aligned} \tag{3.268}$$

Moreover, in calculating the equilibrium averages $\langle \hat{T}^{(S)}_{LM}\rangle_{\xi_1}$, we have also used the equilibrium Boltzmann distribution function ρ_{ξ_1} for the Hamiltonian Eq. (3.258), namely

$$\rho_{\xi_1} = \frac{e^{\frac{\sigma}{S^2}\hat{S}_Z^2+\frac{\xi+\xi_1}{S}\left(\cos\psi\hat{S}_Z+\sin\psi\hat{S}_X\right)}}{\mathrm{Tr}\{e^{\frac{\sigma}{S^2}\hat{S}_Z^2+\frac{\xi+\xi_1}{S}\left(\cos\psi\hat{S}_Z+\sin\psi\hat{S}_X\right)}\}}.$$

The corresponding dynamic susceptibility $\chi(\omega)=\chi'(\omega)-i\chi''(\omega)$ is given via the one-sided Fourier transform $\tilde{C}(i\omega)$ by Eq. (3.153). Thus, we have the integral relaxation time $\tau_{\mathrm{cor}}=\tilde{C}(0)$ (in linear response the correlation time of $C(t)$) and the effective relaxation time $\tau_{\mathrm{ef}}=-1/\dot{C}(0)$. In the frequency domain, these characteristic times determine the low- and high-frequency behavior of the dynamic susceptibility $\chi(\omega)$ via Eqs. (3.182) and (3.183), respectively.

The other relevant quantity is, of course, the inverse of the smallest nonvanishing eigenvalue λ_1 of the transition matrix of the system [Eq. (3.270)] which we recall is the time constant associated with the *long time behavior* of the correlation function $C(t)$ comprising the slowest (lowest frequency) relaxation mode. Thus, λ_1^{-1} may again be associated with the spin reversal time. Furthermore, because the influence of the high-frequency relaxation modes on the low-frequency relaxation may often be ignored, λ_1 again provides more or less complete information concerning the low-frequency dynamics of the system and may be extracted from the eigenvalues of the transition matrix **A** given by Eq. (3.270) below. In the low temperature limit and a weak external dc field, the relation between the various time constants defined above is

$$\lambda_1^{-1} > \tau_{\mathrm{cor}} >> \tau_{\mathrm{N}} >> \tau_{\mathrm{ef}} \,. \tag{3.269}$$

Thus, to determine the magnetization kinetics, we require [cf. the response function Eq. (3.266)] the one-sided Fourier transforms of $c_{10}(t)$ and $c_{1\pm1}(t)$. According to the differential-recurrence Eq. (3.259), these relaxation functions are once again coupled to all the others so forming (unlike the classical case) a *finite* hierarchy of averages (because the index L in Eq. (3.259) ranges only between 0 and $2S$). Once more, the solution of such a multi-term recurrence relation may always be obtained by rewriting it as a first-order linear matrix differential equation like Eq. (1.112) for a column vector $\mathbf{C}(t)$ given by Eq. (1.449) of Chapter 1 with a transition supermatrix **A** of dimension $4S\,(S+1) \times 4S\,(S+1)$ with matrix elements

$$\begin{gathered}(\mathbf{A})_{LL'} = \mathbf{G}^{S}_{LL'}, \\ \left(\mathbf{G}^{S}_{LL'}\right)_{MM'} = -p_{L'M';LM}\,.\end{gathered} \tag{3.270}$$

Having solved the matrix differential Eq. (1.112) as briefly illustrated in Section 3.3.9, we have the relaxation functions $c_{10}(t)$ and $c_{1\pm1}(t)$, their spectra, and all desired observables such as the characteristic times τ_{cor}, τ_{ef}, $\tau = 1/\lambda_1$, and the dynamic susceptibility $\chi(\omega)$. For simplicity, we shall suppose that the diffusion coefficients are given by $D_1 = D_{-1} = 2D_0 = 2D$ (i.e., isotropic diffusion). To compare with the classical case, we simply write $\tau_{\mathrm{N}} = (2D)^{-1}$ and $\alpha = \hbar\, S\beta D$.

The relaxation time τ_{cor} and λ_1^{-1} (the longest relaxation time) as a function of the oblique angle ψ, the anisotropy parameter σ, and the dimensionless damping α are plotted in Figs. 3.34 and 3.35, respectively, for various spin numbers S and field parameters $h = \xi/(2\sigma)$. The classical solutions [32,38,97,145] corresponding to $S \to \infty$ are also shown for comparison. Both τ_{cor} and $\tau = 1/\lambda_1$ exhibit a pronounced dependence on the oblique angle ψ, the field h, anisotropy σ, damping α, and spin number S parameters. It is apparent from Figs. 3.34 and 3.35 that for large S, the quantum solutions reduce to the corresponding classical ones while in

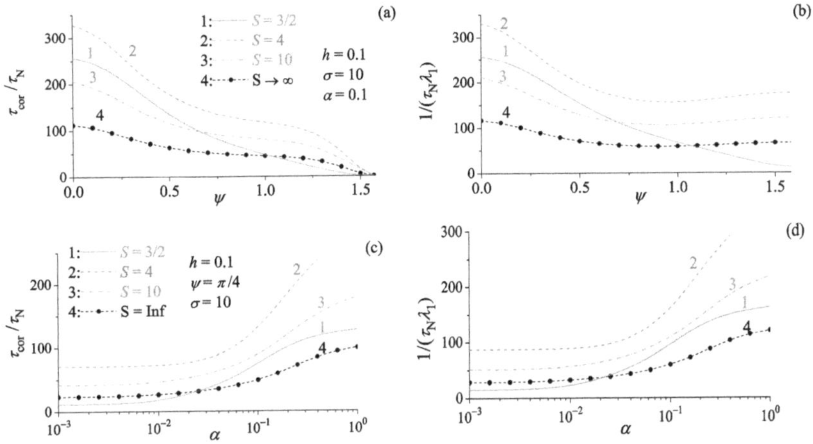

Fig. 3.34. Correlation time τ_{cor} (a), (c) and overbarrier time $1/\lambda_1$ (b), (d) vs. the oblique angle ψ for various spin numbers S and dimensionless damping $\alpha = 0.1$ ((a) and (b)) and vs. dimensionless damping α for various spin numbers S and oblique angle $\psi = \pi/4$, anisotropy parameter $\sigma = 10$, and field parameter $h = 0.1$ ((c) and (d)).

contrast they differ markedly from each other for small S. *Typical values of S for the quantum-classical crossover are* ~20–40. The smaller the anisotropy σ the smaller the S value required for convergence of the quantum results to the classical ones. However, the *intrinsic* damping (α) dependence of these characteristic relaxation times for the oblique field configuration shown in Fig. 3.34 represents coupling between the longitudinal and precessional modes of the magnetization. Hence, it should once more be possible to determine α by fitting the theory to the experimental dependence of the reversal time on the angle ψ and dc bias field strength. The reason being that the sole fitting parameter is α, which can be determined at different temperatures T, exposing its temperature dependence. As before, the behavior of τ_{cor} and $\tau = 1/\lambda_1$ is similar only for small external fields, $h \ll 1$; in a strong external field, $h > 0.2$, τ can *diverge exponentially from* τ_{cor} *as for classical spins* (see Fig. 3.35). This effect was discovered numerically for classical spins by Coffey *et al.* [146] and later interpreted quantitatively by Garanin [337] (see Chapter 1, Section 2.4.3 for details).

In Figs. 3.36 and 3.37, we have plotted the real and imaginary parts of $\chi(\omega)/\chi_0$ vs. normalized frequency ω/D for various model parameter

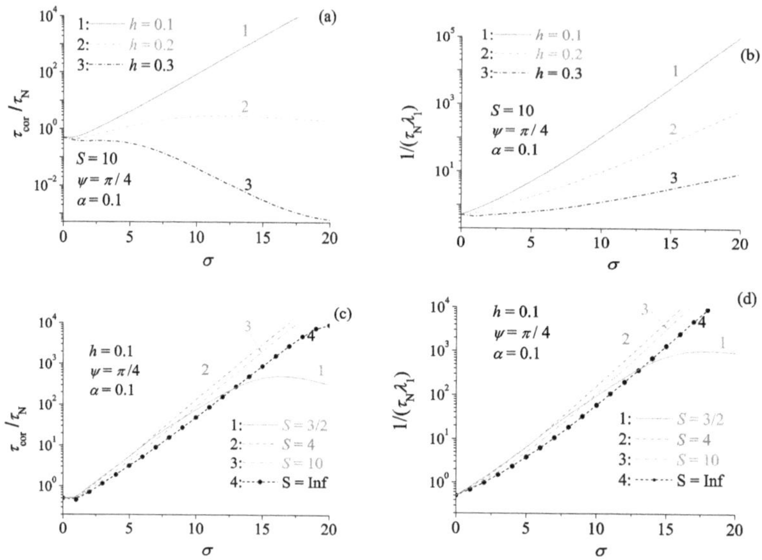

Fig. 3.35. Correlation time τ_{cor} and overbarrier time $1/\lambda_1$ vs. the anisotropy parameter σ for various field parameters h and spin number $S = 10$, (a) and (b), for various S with field parameter $h = 0.1$, (c) and (d) (dimensionless damping $\alpha = 0.1$ and oblique angle $\psi = \pi/4$).

values. Clearly the usual three bands associated with lack of symmetry now appear in $\chi''(\omega)$ two of which are like those in circular symmetry while a third resonance band appears due to high-frequency precession of the spin in the effective field. The low-frequency relaxation band is as before due to the slowest relaxation mode, where the characteristic frequency and bandwidth are determined by λ_1. Like the classical case, λ_1 is sufficient to accurately predict the behavior of the low-frequency part of $\chi(\omega)$ as well as the long-time behavior of $C(t)$. Thus, if again one is interested solely in the low-frequency region ($\omega/\lambda_1 \leq 1$), where the effect of the high-frequency modes may be ignored, $\chi(\omega)$ may be again approximated by the single Lorentzian, Eq. (1.151) of Chapter 1, which implies that $C(t)$ may be approximated for $t > 0$ by a *single* exponential with relaxation time $T_{\|} = 1/\lambda_1$. It is apparent from Figs. 3.36 and 3.37 since the influence of the high-frequency relaxation modes on the low-frequency relaxation may be ignored, that the simple Lorentzian formula, Eq. (1.151) of Chapter 1, again accurately describes the low-frequency

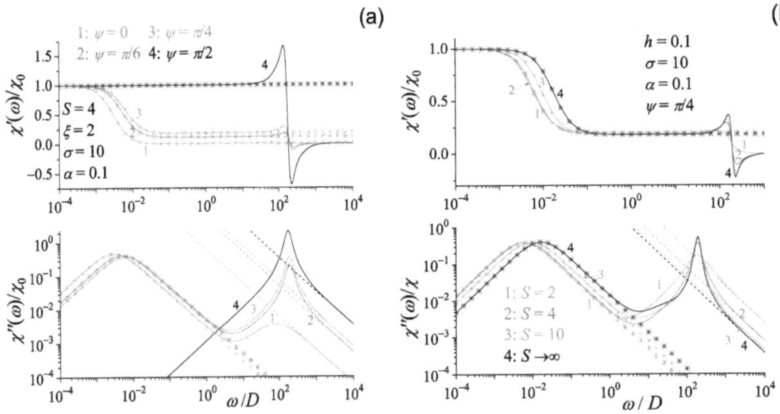

Fig. 3.36. Real and imaginary parts of the dynamic susceptibility $\chi(\omega)$ vs. the normalized frequency ω/D (a) for various oblique angles ψ and field parameter $h = 0.1$, spin $S = 4$, and (b) for various spin numbers S and $h = 0.1$, $\psi = \pi/4$ (damping $\alpha = 0.1$ and anisotropy parameter $\sigma = 10$). Asterisks: the single Lorentzian approximation, Eq. (1.151) of Chapter 1, while the straight dashed lines are the high-frequency asymptotes, Eq. (3.183).

dynamics. The second far weaker high-frequency relaxation band in $\chi''(\omega)$ is once more due to high-frequency longitudinal "intrawell" modes. The individual "intrawell" modes are indistinguishable in the spectrum of $\chi''(\omega)$ appearing merely as a single high-frequency Lorentzian band. This "intrawell" relaxation band is more pronounced when the external field direction coincides with the easy axis, i.e., for $\psi = 0$. However, in general, it is masked by the third sharp resonance band due to *excitation of transverse modes having frequencies close to the precession frequency of the spin* which strongly manifests itself at high frequencies. For $\psi = 0$, the resonance peak entirely disappears because the transverse modes are no longer excited. In contrast it is most pronounced when $\psi = \pi/2$.

In this section, we have solved the differential-recurrence Eq. (3.259) for the evolution of the statistical moments (average spherical harmonics) for the spin Hamiltonian given by Eq. (3.258). For purposes of illustration, the analysis was carried out for a uniaxial nanomagnet subjected to a dc external field applied at an *arbitrary* angle to the easy axis. We have evaluated the characteristic relaxation times along with the linear dynamic susceptibility via obvious generalizations of the methods previously used for classical spins [32,38,97,108,145,342–344]. Thus, the phase-space

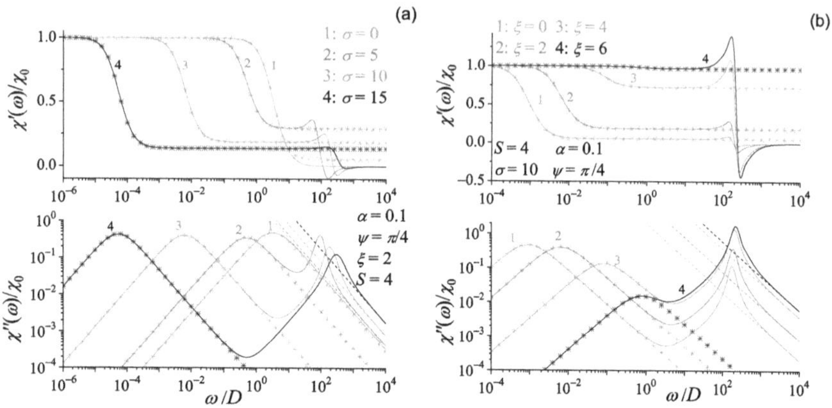

Fig. 3.37. Real and imaginary parts of $\chi(\omega)$ vs. the normalized frequency ω/D (a) for various anisotropy parameters σ and field parameter $\xi = 2$ and (b) for various field parameters ξ and anisotropy parameter $\sigma = 10$ (spin number $S = 4$, damping $\alpha = 0.1$, and oblique angle $\psi = \pi/4$). Asterisks are the single Lorentzian approximation, Eq. (1.151) of Chapter 1, while the straight dashed lines are the high-frequency asymptotes, Eq. (3.183).

representation (because it is closely allied to the classical one) again transparently illustrates how quantum distributions reduce to the classical ones. When the direction of the external fields coincides with the easy axis, i.e., for $\psi = 0$, our method reproduces the results for nanomagnets subjected to a longitudinal field previously obtained in Section 3.3.10. The method may also be extended to other multi-well systems such as biaxial, cubic, mixed, etc. Furthermore, the technique can be generalized to time-dependent Hamiltonians as in Section 3.3.11, so that we can also determine quantum effects in the nonlinear ac stationary response of quantum nanomagnets in the manner of the uniaxial case.

To summarize, we have treated numerous illustrative examples of spin relaxation problems using Wigner's phase-space formulation of quantum mechanics of spins. The one-to-one correspondence between the quantum state in the Hilbert space and a real representation space function first envisaged for the closed system in the spin context by Stratonovich [273], formally represents the quantum mechanics of spins as a statistical theory in the representation space of polar angles (ϑ, φ) which has a clear classical meaning. This procedure effectively generalizes the results of Wigner [6] who represented the quantum mechanics of a particle with

Hamiltonian $\hat{H} = \hat{p}^2 / 2m + V(\hat{q})$ in Hilbert space as a statistical theory in a classically meaningful phase space with the canonical variables position and momentum (q, p). However, Stratonovich [273] proceeded by introducing a quasiprobability density (Wigner) function on the sphere, defined as the linear invertible bijective map onto the representation space comprised of the trace of the product of the system density matrix and the irreducible tensor operators, the analysis being carried out via the finite series in spherical harmonics embodied in the bijective Wigner–Stratonovich map. Hence, the average value of a quantum spin operator may be calculated via its Weyl symbol just as the corresponding classical function in the representation space of polar angles (ϑ, φ). This goal is achieved essentially because the *polarization operators transform under rotation in the same way as the spherical harmonics*. Thus, the Stratonovich representation for spins [273] is very well suited to the development of semiclassical methods of treatment of spin relaxation phenomena allowing one to obtain quantum corrections in a manner closely analogous to the classical case.

Another merit of the phase-space formalism as applied to spin relaxation problems is that only master equations for the phase-space distributions akin to Fokker–Planck equations for the evolution of classical phase-space distributions in configuration space are involved so that operators are, for the most part, unnecessary. The *explicit* solution of these equations can be expanded for an *arbitrary* spin Hamiltonian in a *finite* series of spherical harmonics Eq. (1.579) essentially as the classical case save that only a finite number of spherical harmonics is involved. The expansion coefficients (statistical moments or averages of the spherical harmonics which are obviously by virtue of the Wigner–Stratonovich map the averages of the polarization operators) may be determined from a differential-recurrence relation Eq. (1.601) in a manner similar to the classical case. Although the form of the phase-space master equation for spins is in general very complicated, we can circumvent the problem of determining differential-recurrence relations by directly using the one-to-one correspondence between averages of polarization operators and those of spherical harmonics as outlined in Section 1.9.8. Moreover, we have described this procedure via several illustrative examples. Thus, *we still have a method of treating the spin relaxation for arbitrary spin number S which is closely allied to the classical one even though a phase-space master equation may not be explicitly involved.* In the classical limit, the quantum differential-recurrence relation reduces to that yielded by the classical Fokker–Planck equation for *arbitrary* magnetocrystalline – Zeeman energy potentials. Furthermore, the phase-space representation via the Weyl symbol of the relevant spin operator suggests how powerful computation techniques developed for Fokker–Planck equations (matrix continued fractions, mean first-passage time, integral representation of relaxation times, etc. [97,143]) may be transparently extended to the quantum domain *inter alia* suggesting new closed form quantum results via the corresponding classical ones. A specific example is the determination of the quantum integral relaxation time for a spin in a uniform magnetic field of arbitrary strength directed along the Z-axis, Eq. (3.145). Therefore, having solved the phase-space master equation, one can then, in principle, evaluate all desired observables via obvious generalizations of the methods previously used for classical spins [97,239]. These include the magnetization itself, the magnetization reversal time, the linear and nonlinear dynamic susceptibilities, the temperature dependence of the switching fields, the dynamic hysteresis loops etc. Hence, *one can study the transition of the relaxation behavior from that of an elementary spin to molecular magnets ($S \sim 10$), to nanoclusters ($S \sim 100$), and to classical nanomagnets ($S \geq 1000$).* Thus, all *quantum effects in the spin relaxation phenomena can be treated in a manner linking directly to the classical representations*. Furthermore, by treating a variety of spin relaxation problems, it has been amply demonstrated that although the density matrix and phase-space methods may yield results in *outwardly very different forms*, nevertheless, both yield *identical* numerical values for the *same* physical quantities (such as relaxation times and susceptibility). Hence, a vital corollary between each method has been established thereby demonstrating that they are essentially equivalent while simultaneously providing an important check on the validity of the phase-space method. Thus, the spin phase-space representation, because it is closely allied to the classical representation, besides being complementary to the operator one, transparently illustrates how quantum phase-space distributions reduce to the classical ones in the limit $S \to \infty$.

General Conclusions and Summary

We began by underlining in Chapter 1 the general principles of the classical and quantum Brownian motion of spins. Starting with classical magnetic dipoles (macrospins), *formally exact* solutions for the observables (magnetization, dynamic susceptibility, etc.) were obtained for *arbitrary* magnetocrystalline anisotropy free energy density. This was accomplished either via the magnetic Langevin equation describing the stochastic motion of the magnetic moment in the configuration space of orientations (ϑ, φ) or its associated diffusion (Fokker–Planck) equation for the surface probability density $W(\vartheta, \varphi, t)$ of magnetic moment orientations on the unit sphere. The procedure was then continued by using the theory of angular momentum to reduce the calculation of the observables, namely, the statistical moments, i.e. the averages $\langle Y_{LM} \rangle(t)$ of the spherical harmonics $Y_{LM}(\vartheta, \varphi)$ and *a fortiori* their characteristic relaxation times, to the solution of matrix three-term recurrence relations. These relations were then formally solved in closed operator form in the frequency domain in terms of sums of products of matrix continued fractions so constituting a very rapidly convergent computational algorithm for a given potential as emphasized by Risken [143] in his excellent monograph *The Fokker–Planck equation*.

In Chapter 2, the magnetization relaxation of magnetic nanoparticles, both via the Langevin equation and via the Kramers escape-rate theory as adapted to classical spin systems has been treated in depth. We have shown that the infinite hierarchy of differential-recurrence equations for the statistical moments, governing the kinetics of the magnetization of a nanomagnet possessing *arbitrary* free energy density, can always be solved using matrix continued fractions over wide ranges of temperature and damping. The main advantage of this exact method is that it can be used for low potential barriers (high temperatures), where asymptotic methods like escape-rate theory are not applicable. Alternatively, in model parameter ranges, where it is no longer feasible to use continued fractions, e.g., for very high barriers and very low damping, escape-rate theory provides a powerful tool for studying the relaxation processes in nanomagnets. Using these two powerful complementary methods, one has both exact (in terms of matrix continued fractions) and simple approximate analytical solutions of diverse problems occurring in superparamagnetism. These include estimation of the characteristic times of the magnetization reversal process, susceptibilities (both linear and nonlinear), switching-field curves and surfaces at finite temperatures, dynamic hysteresis, and magnetic stochastic resonance. The overall purpose of the calculations was to highlight the limitations of asymptotic expressions for characteristic times, etc. as well as providing *benchmark solutions* with which the results of Langevin dynamics simulations and all other such methods must comply. We also wished to emphasize that symmetry issues are all-important. For example, *lack of circular symmetry* yields prefactors of escape rates deviating by orders of magnitude from those predicted by simple transition state theory which is valid only for intermediate damping. In fact, TST merely provides an *upper bound* for the escape rate and, hence, a *lower bound* for the reversal time. Moreover, it has been demonstrated that lack of symmetry also gives rise to effects such as *precession aided* contributions to both the reversal and other characteristic times accompanied by ferromagnetic resonance at high frequencies. While if the nonlinear response to imposed ac and dc fields is considered, typical nonlinear effects well known in the dynamics of periodically forced nonlinear systems such as *parametric and subharmonic resonance* at high frequencies and frequency dependent dc responses also occur (*v.* [218]). Lack of circular symmetry also causes the three regions of escape-rate theory, first identified by Kramers for Brownian particles with *separable* and *additive* Hamiltonians, to likewise emerge for classical spins governed by *non-separable* and *non-additive* Hamiltonians. A fact which in our opinion has been very poorly understood in the literature leading to serious underexploitation and lack of full appreciation of the fundamental work of Brown particularly concerning the low damping commonplace in macrospins. His contribution based on the stochastic Landau–Lifshitz–Gilbert equation used as Langevin equation sets classical macrospin stochastic dynamics firmly within the framework of the original theory of the Brownian motion as formulated by Einstein, Smoluchowski, Langevin, Perrin, Debye, Klein, Ornstein, Fürth, Kramers, etc. (see for example, *Selected papers on noise and stochastic processes*, edited by N. Wax, Dover, New York, 1954; A. Einstein, *Investigations on the Theory of the Brownian Motion*, R.H. Fürth (ed.), Methuen, London, 1926, reprinted by Dover, New York, 1954). In tandem with calculations of the statistical moments and their characteristic times, we have used them in Chapter 2 to treat dynamic magnetic hysteresis, magnetic stochastic resonance, generalizations of the Stoner–Wohlfarth calculation of astroids and switching fields and the stochastic dynamics of two macrospins interacting via dipole-dipole and exchange interactions. The latter calculation serving as a preliminary indicator of the relative role played by such interactions in the magnetization reversal.

It has also been demonstrated that the magnetization dynamics of a nanoparticle with *noncircularly symmetric* free energy density exhibit *universal* behavior irrespective of the precise nature of the free energy density considered. This behavior is universal in the sense that the magnetic loss spectra always exhibit (i) a broad low-frequency relaxation peak, caused by *slow* noise-activated escape of the magnetic dipoles over the potential barriers, and so is attributable to an activation process, (ii) a second relaxation peak at higher frequencies arising from *fast* near-degenerate intrawell relaxation modes, and (iii) sharp resonant absorption peaks at very high frequencies. The origin of the resonance peaks lies in the precessions of the magnetic dipole in the wells. Yet another universal effect in the magnetization relaxation in nanoparticles is the *suppression* of the low-frequency (overbarrier) relaxation mode, via the application of a *relatively weak external dc magnetic field*. This effect occurs because of depopulation of the shallow well, so that the overall behavior is mainly dictated by the fast intrawell modes in the deep well; the overbarrier process being relegated to a relatively minor role. Other universal effects are the *dynamic magnetic hysteresis* and *stochastic resonance* arising in an ac driving field. The exact numerical solutions for the smallest nonvanishing eigenvalue of the Fokker–Planck operator for a given system have been verified by comparing them with asymptotic expressions for the magnetization reversal (or greatest relaxation) time for various magnetocrystalline anisotropy-Zeeman energies. The numerical solutions have also been used to verify simple two-Lorentzian closed form approximations to the complex magnetic susceptibility $\chi(\omega)$ at low- and mid- frequencies. The first or slow Lorentzian accounts for the low-frequency overbarrier (Néel–Brown) relaxation process while the second accounts for the near-degenerate fast exponential decays in the potential wells. Moreover, the entire theory has then been modified to include STT effects.

STT effects in the thermally assisted magnetization reversal have been treated via the evolution equation for the statistical moments rendering stationary and nonstationary characteristics for wide ranges of temperature, damping, external magnetic field, and spin-polarized current. The method, based on the solution of the differential-recurrence relation for the infinite hierarchy of statistical moments generated by either the Langevin or Fokker-Planck equations as augmented by STT terms, indicates that STT profoundly alters the magnetization dynamics of a nanomagnet leading to a host of new effects. Furthermore, the statistical moment approach holds for the *most comprehensive* formulation of the generic nanopillar model (Fig. 1.12), i.e., for *arbitrary orientations* of the dc and ac external fields allowing one to treat STT effects under conditions which are otherwise inaccessible. Moreover, the calculations, since they are valid for ac fields of *arbitrary* strength and orientation, quantify the role played by STT in nonlinear phenomena in nanoscale ferromagnets such as nonlinear stochastic resonance and dynamic magnetic hysteresis, nonlinear ac field effects on the switching-field curves, etc., where perturbation theory is no longer valid. In addition, the moment method yields the response at *all frequencies of interest* including the very high frequencies covering the ferromagnetic resonance (GHz) range exemplifying various nonlinear phenomena such as *parametric resonance* and *higher harmonic generation* (which we hope will stimulate new experiments). The high-frequency linear and nonlinear FMR spectra may be suitable for explaining the line shape of STT nano-oscillators driven by ac external magnetic fields and currents. Likewise, the DMH loops and their area (yielding the Joule heating during the switching process) as well as the calculations of the effective magnetization reversal time via the low-frequency band of the magnetic loss spectra may be useful for the prediction, modeling, and interpretation of switching processes in recording techniques. Furthermore, the DMH arising from a *high-frequency* periodic signal may be exactly evaluated permitting quantitative analysis of *ultrafast switching* of the magnetization. Hence, the results may serve both as a basis for theoretical investigations and interpretation of a broad range of STT experiments. Additionally, they are essential for the future development of both escape-rate theory and stochastic dynamics simulations of the magnetization reversal time in STT systems, representing *rigorous benchmark solutions* with which calculations of that time by any other method *must comply*.

Regarding the investigation of quantum effects due to finite spin number S rather than the more or less infinite spin number ($\sim 10^4 - 10^5$) characterizing macrospins, we started with a detailed derivation via the projection operator formalism of the reduced spin density matrix evolution equation, then we proceeded to the high temperature and short correlation time limit of the Markovian approximation. This constitutes a tractable approximation to the quantum behavior which we believe is consistent with finding semi-classical corrections. Next, we recalled that the density matrix of particles with spin S may be represented by a $(2S+1)\times(2S+1)$ square matrix and that in order to describe spin states of such particles the polarization operators $\hat{T}_{LM}^{(S)}$ may be used. These are the Hilbert space correspondents of the spherical harmonics $Y_{LM}(\vartheta,\varphi)$, constituting an orthonormal basis in the Hilbert space of spin matrix operators. Hence by orthogonality, the density matrix may be expressed as a finite series of these operators so that the expansion coefficients are simply their expectation values. Thus, one has via the averages of the spin operators the formal extension of the statistical moment method to the quantum domain so reducing the general solution of the quantum relaxation problem for arbitrary spin Hamiltonians to the task of solving a linear homogeneous matrix differential equation of finite dimensions determined by S. Next mindful that the density matrix method outlined above has hitherto constituted the most common treatment of spin relaxation, we have recalled since we are interested in a classical-like approach, that spins interacting with the bath can also be treated via the relevant quasiprobability distribution function $W_S(\vartheta,\varphi,t)$ in the phase space of the polar and azimuthal angles. Hence in Chapter 1, we at some length gently introduce the reader to the main concepts of the Wigner quasiprobability distribution, which was originally meant to be a reformulation of Schrodinger's wave mechanics which describes quantum mechanics in configuration space q, using the concept of a quasiprobability distribution in the classically meaningful phase space (q,p). Thus, it was shown how observables and matrix elements may be calculated from phase-space integrals of c-number functions via the Weyl correspondence taking as example the simple case of particles with separable and additive Hamiltonians moving in a one-dimensional extension. As an example, we showed how the Wigner–Moyal evolution equation obtained for an arbitrary potential $V(q)$ may be modified so as to include dissipation thereby regaining Agarwal's result [314] for the Brownian motion of a quantum harmonic oscillator in the weak-coupling limit. Proceeding, we then introduced Stratonovich's [273] general discussion of *c*-number quasiprobability distributions for quantum systems in a representation space based on the symmetry properties of the underlying groups (e.g., the SU(2) group for rotations). It was then shown how his direct mapping of a quantum operator $\hat{A}$ in Hilbert space onto the configuration space of orientations (ϑ, φ) and vice versa via a *bijective* transformation $\hat{w}(\vartheta,\varphi)$ depending on the point (ϑ, φ) as a parameter may be used to construct spin quasiprobability distributions on the classically meaningful representation space (ϑ, φ). This powerful and far-reaching bijective operator formalism allows one to calculate the Weyl symbols of spin operators and to express quasiprobability distributions for spins in classical-like fashion as finite Fourier-Laplace series spanning the spin number S via the fundamental relations underpinning the entire theory, viz. Eq. (1.547). Thus one has quasiprobability distributions for spins by identifying averages of the spherical harmonics $Y_{LM}(\vartheta,\varphi)$ as averages of the polarization operators $\hat{T}_{LM}^{(S)}$ via the spin density matrix $\hat{\rho}_S$. Furthermore, since we are interested in *semiclassical corrections* we have considered the SU (2) *coherent or minimum uncertainty* states. Therefore, taking the *Q*-function, comprising the diagonal matrix elements of the density operator $\hat{\rho}_S$ in the spin coherent state representation as an example, we have the classical-like Eq. (1.568) and likewise in the general case of *Q*, *P*, and Wigner spin quasiprobability distributions in the representation space (ϑ, φ) the compact form Eq. (1.579) (for which we use the generic term, Wigner function). In order to illustrate the Wigner–Stratonovich mapping, we explicitly evaluated the Weyl symbol of the spin operator $\hat{\mathbf{S}}$ showing that the magnetic moment operator $\hat{\boldsymbol{\mu}}$ has essentially the form of the magnetic moment vector $\boldsymbol{\mu}$ for a classical spin. As further examples, we evaluated the Weyl symbols for commonly used magnetocrystalline anisotropy Hamiltonians showing that the representation space results which of course explicitly depend on S, resemble the classical free energies of the corresponding magnetocrystalline anisotropies. It was also demonstrated that having calculated the averages of a polarization operator from the reduced matrix evolution Eqs. (1.411) and (1.447), then automatically one has the phase-space distribution without solving the phase-space evolution Eq. (1.600) and vice versa. A fact which is later used in Chapter 3, which we now summarize.

In Chapter 3, building on the fundamentals of quantum mechanics in a classically meaningful representation space established in Chapter 1, we have calculated the equilibrium Wigner distribution for spins in a uniform longitudinal field. The Weyl symbol of the Z component of the spin operator was then used to calculate the average longitudinal component of a spin at equilibrium in terms of the Brillouin function showing that it tends to the Langevin function as $S \to \infty$. Next, the calculation was extended to spins in an external uniform field of *arbitrary* orientation so constituting a *noncircularly symmetric* problem, obtaining the valuable closed expression Eq. (3.27) valid for arbitrary S. Proceeding, we calculated the Wigner function for a uniaxial magnet in a longitudinal field showing that one may determine that function for arbitrary S (cf. Eq. (3.36) *et. seq.* and the discussion following Eq. (3.37) and Fig. 3.5 showing that pronounced quantum effects arise at small S). Next, the Wigner functions for uniaxial, biaxial, and cubic anisotropies were all calculated. In these for the most part noncircularly symmetric problems, the Wigner functions rapidly became more and more complicated as S increases as is evident in Sections 3.2.3 *et seq.* However, they may always be calculated numerically from the general series expression Eq. (3.6). Finally, a feature of all the phase-space potentials derived from the Weyl symbol of a given Hamiltonian is that their characteristics such as shape and barrier heights all depend strongly on S. Furthermore, mindful that TST because it is based on equilibrium distributions affords the simplest possible description of quantum corrections to the thermally activated escape rate, a general expression representing the quantum correction factor to the classical TST rate, which depends very strongly on S, has been obtained. Moreover, it has been shown that the qualitative behavior of the quantum escape rate strongly deviates from the Arrhenius one of the classical escape rate at low temperatures. A difference which is due to the *tunneling* effect. In general, one may conclude that the TST results are very relevant to experiments seeking evidence for *macroscopic quantum tunneling*, where the temperature dependence of the hysteresis loops is crucial as they are used to differentiate the former behavior from that of thermal agitation as discussed in Section 3.2.7. As far as switching fields without thermal agitation are concerned, their amplitudes increase markedly with increasing S and tend to their classical limiting values as $S \to \infty$ corresponding to diminishing tunneling effects.

Turning to non-equilibrium behavior, the phase-space method was used to treat to a uniaxial nanomagnet of arbitrary S in a uniform field applied along the easy axis. We accomplished this by writing the general reduced density matrix Eq. (1.442) in Hilbert space as the evolution equation for the reduced density matrix of a uniaxial nanomagnet. Then we rendered that equation after many intricate manipulations as the standard form Eq. (1.626) of the inverse Wigner–Stratonovich map of a Weyl symbol. The phase-space master Eq. (3.93) for the orientational distribution for longitudinal relaxation so obtained is *formally* similar to a single-variable Fokker–Planck equation. However, we saw that even for circular symmetry, high order spin number-dependent differential operators occur apart from two exceptions: (i) noninteracting spins in a uniform field; (ii) with zero Hamiltonian. Hence apart from these two relatively trivial examples we conclude that the phase-space master Eq. (3.90) is in general of limited practical use. However, that restriction does not preclude one from deriving differential-recurrence relations for the observables by using the one-to-one correspondence between averages of polarization operators in Hilbert space and averages of the spherical harmonics as already alluded to. Therefore we have illustrated this indirect procedure via Eq. (3.96)–(3.109) and have confined our use of the phase-space method to case (i) above. This situation of zero anisotropy Eq. (3.114) is just the rotational analog of the quantum translational harmonic oscillator treated using the Wigner function by Agarwal [314]. Thus, having formulated the relevant evolution equations we explicitly treated transient nonlinear spin relaxation by directly using the phase-space master equation because as we have stated in this instance it takes on *exact* Fokker–Planck form. Hence we have explicitly evaluated the characteristic relaxation times exactly and have also given analytic approximations to them, showing as exemplified by Figs. (3.13)–(3.14) that a pronounced S dependence exists. Moreover, the linear response relaxation time yields an identical result to that previously obtained by the spin density matrix method. Furthermore, one may conclude that both the linear and nonlinear relaxation for all S is *accurately described by a Bloch equation* holding even for the nonlinear response of a macrospin. In contrast to the phase-space method, we have in Section 3.3.9 returned to the spin density matrix one and shown how formally exact expressions for the characteristic relaxation times for a uniaxial nanomagnet with an external uniform field applied along the easy axis may be obtained. Next even though we have circumvented the phase-space evolution equation, it has been shown in Section 3.3.10 how a much simpler method of treating circularly symmetric problems still exists via the spin density matrix equation since taking the above problem as an example, *the diagonal elements of the density matrix decouple from the nondiagonal ones*. This property leads to explicit expressions in series form for all the characteristic times and proves that a simple description of the long-time behavior of the longitudinal relaxation function as a Bloch equation again holds for the nonlinear response of a quantum nanomagnet for arbitrary spin S (cf. Figs. 3.20–3.21). The diagonal-element method was also used in Section 3.3.11 which is concerned with nonlinear longitudinal relaxation in superimposed ac and dc fields, and the results of this investigation are shown in Figs. 3.22 and 3.23. Moreover, as shown in Section 3.12 these results may be applied to dynamic hysteresis in uniaxial nanomagnets without any assumptions regarding potential barrier height, temperature, magnetizing field strength or S. The treatment is exemplified by Figs. 3.26–3.30. The series expressions involving the diagonal matrix elements etc. described above have also been used to describe quantum effects in magnetic stochastic resonance. The principal result is that one may determine the marked transition from the stochastic resonance corresponding to quantum spin relaxation to that pertaining to a macrospin as a function of S. Finally, in Section 3.4 we have again demonstrated how the elegant and far-reaching one-to-one correspondence between the averages of spherical harmonics and polarization operators may be used to tackle the important noncircularly symmetric problem of a uniaxial nanomagnet in a dc bias field of arbitrary orientation.

It is our hope that this necessarily brief summary will both guide the reader through the maze of intricate calculations and will put some of the ideas underlying magnetic nanoparticle relaxation into context ever mindful of the quotation from Pierre-Simon, marquis de Laplace with which we began our lengthy account of thermal fluctuations and relaxation processes in nanomagnets.

Appendixes

Special functions, operators, etc.

A.1. Relevant special functions and associated formulas

A.1.1. Gamma and related functions

The gamma (factorial) function $\Gamma(z)$ is single valued and analytic in the entire complex plane, save at the points $z = -n$ ($n = 0,1,2,\ldots$), where it possesses a single pole with residue $(-1)^n/n!$, and is defined as

$$\Gamma(z) = \int_0^\infty t^{z-1} e^{-t} dt \qquad (\mathrm{Re}\, z > 0). \tag{A.1}$$

It satisfies the recurrence and reflection formulas [290]

$$\Gamma(z+1) = z\Gamma(z) = z! = z(z-1)!, \tag{A.2}$$

$$\Gamma(z)\Gamma(1-z) = -z\Gamma(z)\Gamma(-z) = \pi \csc(\pi z). \tag{A.3}$$

The incomplete gamma function $\gamma(a,z)$ is given by [290]

$$\gamma(a,z) = \int_0^z t^{a-1} e^{-t} dt. \tag{A.4}$$

A.1.2. Confluent hypergeometric (Kummer) functions

The confluent hypergeometric function $M(a,b,z)$ is the solution of Kummer's equation, which is regular at $z = 0$ [290]

$$z\frac{d^2 w}{dz^2} + (b - z)\frac{dw}{dz} - aw = 0 \tag{A.5}$$

and may be defined either via the integral representation

$$M(a,b,z)=\frac{\Gamma(b)}{\Gamma(b-a)\Gamma(a)}\int_0^1 e^{zt}t^{a-1}\left(1-t\right)^{b-a-1}dt \tag{A.6}$$

or by the convergent series

$$M\left(a,b,z\right)=\frac{\Gamma(b)}{\Gamma(a)}\sum_{n=0}^{\infty}\frac{\Gamma(n+a)}{\Gamma(n+b)}\frac{z^n}{n!}. \tag{A.7}$$

The asymptotic expansion of the confluent hypergeometric function $M(a,b,z)$ at $z >> 1$ is given by [290]

$$M\left(a,b,z\right)
=e^z z^{a-b}\frac{\Gamma(b)}{\Gamma(a)}\left[\sum_{n=0}^{N-1}\frac{\Gamma(b-a+n)\Gamma(1-a+n)}{\Gamma(b-a)\Gamma(1-a)z^n n!}+O\left(z^{-N}\right)\right]. \tag{A.8}$$

Furthermore, $M(a, b, z)$ has the following properties

$$M(0,b,z)=1, \tag{A.9}$$

$$M(a,b,z)=e^z M(b-a,b,-z), \tag{A.10}$$

and we also have an integral representation for the product [485]:

$$M(a,b,z)M(a,b,-z)
=\frac{\Gamma^2(b)z^{1-b}}{\Gamma(a)\Gamma(b-a)}\int_{-\infty}^{\infty}I_{b-1}(z\,\mathrm{sech}\,t)\,\mathrm{sech}\,t\,e^{(b-2a)t}dt, \tag{A.11}$$

where $\mathrm{Re}[a] > 0$, $\mathrm{Re}[b-a] > 0$ and $I_{b-1}(z)$ is the modified Bessel function of the first kind of order $b-1$ (see Appendix A.1.6). Also, $M(a, b, z)$ satisfies the recurrence relations

$$M(a,b-1,z)-M(a,b,z)=\frac{az}{b(b-1)}M(a+1,b+1,z), \tag{A.12}$$

$$M(a,b,z)-M(a-1,b,z)=\frac{z}{b}M(a,b+1,z), \tag{A.13}$$

which may be rearranged as a continued fraction ([486], p. 347), viz.,

$$\frac{M(a+1,b+1,z)}{M(a,b,z)}=\cfrac{1}{1-\cfrac{\frac{(b-a)}{b(b+1)}z}{1+\frac{(a+1)z}{(b+1)(b+2)}\frac{M(a+2,b+3,z)}{M(a+1,b+2,z)}}}$$

$$=\cfrac{1}{1-\cfrac{\frac{(b-a)}{b(b+1)}z}{1+\cfrac{\frac{(a+1)}{(b+1)(b+2)}z}{1-\cfrac{\frac{(b-a+1)}{(b+2)(b+3)}z}{1+\cfrac{\frac{(a+2)}{(b+3)(b+4)}z}{1-\ddots}}}}} \tag{A.14}$$

A.1.3. Error function

The error function erf(z) is defined as [290]

$$\operatorname{erf}(z)=\frac{2}{\sqrt{\pi}}\int_0^z e^{-t^2}dt=\frac{2z}{\sqrt{\pi}}M\left(\frac{1}{2},\frac{3}{2},-z^2\right). \tag{A.15}$$

Its Taylor series and the asymptotic ($z\to\infty$) expansion are given by [290]

$$\operatorname{erf}(z)=\frac{2}{\sqrt{\pi}}\sum_{n=0}^{\infty}\frac{(-1)^n z^{2n+1}}{n!(2n+1)}, \tag{A.16}$$

$$\operatorname{erf}(z)\sim 1-\frac{e^{-z^2}}{\pi}\left[\sum_{n=0}^{N}(-1)^n\frac{\Gamma(n+1/2)}{z^{2n+1}}+O\left(z^{-2N-3}\right)\right]. \tag{A.17}$$

The complementary error function erfc(z) = 1 – erf(z) and the error function of imaginary argument erfi(z) = – $i$erf(iz) are given by

$$\operatorname{erfc}(z)=\frac{2}{\sqrt{\pi}}\int_z^{\infty}e^{-t^2}dt \tag{A.18}$$

and

$$\mathrm{erfi}(z) = \frac{2}{\sqrt{\pi}} \int_0^z e^{t^2} dt. \tag{A.19}$$

A.1.4. Hermite and Laguerre polynomials

The Hermite polynomial $H_n(z)$ of order n satisfies the differential equation

$$\frac{d^2w}{dz^2} - 2z\frac{dw}{dz} + 2nw = 0. \tag{A.20}$$

It can be defined either by Rodrigues' formula [487]

$$H_n(z) = (-1)^n e^{z^2} \frac{d^n}{dz^n} e^{-z^2}, \tag{A.21}$$

or explicitly in the series form

$$\begin{aligned} H_n(z) &= n! \sum_{k=0}^{[n/2]} \frac{(-1)^k (2z)^{n-2k}}{k!(n-2k)!} \\ &= (2z)^n - n(n-1)(2z)^{n-2} + \cdots + \frac{(-1)^{n/2} n!}{(n/2)!} \end{aligned} \tag{A.22}$$

(for the odd-order polynomials, the last term in the right-hand side of Eq. (A.22) is equal to zero). The first three Hermite polynomials are [290]

$$H_0(z) = 1, \ H_1(z) = 2z, \text{ and } H_2(z) = 4z^2 - 2. \tag{A.23}$$

The Hermite polynomials satisfy the recurrence equations [290]

$$\frac{d}{dz} H_n(z) = 2nH_{n-1}(z), \tag{A.24}$$

$$H_{n+1}(z) = 2zH_n(z) - 2nH_{n-1}(z), \tag{A.25}$$

and the orthogonality relation

$$\int_{-\infty}^{\infty} e^{-z^2} H_n(z) H_m(z) dz = \sqrt{\pi}\, 2^n n! \delta_{mn}, \tag{A.26}$$

where δ_{nm} denotes Kronecker's delta.

The Laguerre polynomial $L_n(z)$ of order n satisfies the differential equation

$$z\frac{d^2w}{dz^2}+(1-z)\frac{dw}{dz}+nw=0. \tag{A.27}$$

It can be defined either by Rodrigues' formula [487]

$$L_n(z)=\frac{e^z}{n!}\frac{d^n}{dz^n}\left(e^{-z}z^n\right), \tag{A.28}$$

or explicitly in the series form

$$L_n(z)=n!\sum_{k=0}^{n}\frac{(-1)^k(z)^k}{(k!)^2(n-k)!}. \tag{A.29}$$

The first three Laguerre polynomials are [290]

$$L_0(z)=1,\ \ L_1(z)=1-z,\ \text{and}\ L_2(z)=\frac{1}{2}\left(z^2-4z+2\right). \tag{A.30}$$

The Laguerre polynomials satisfy the recurrence equations [290]

$$z\frac{d}{dz}L_n(z)=n\left[L_{n-1}(z)-L_{n-1}(z)\right], \tag{A.31}$$

$$(n+1)L_{n+1}(z)=(2n+1-z)L_n(z)-nL_{n-1}(z), \tag{A.32}$$

and the orthogonality relation

$$\int_0^\infty e^{-z}L_n(z)L_m(z)dz=\delta_{mn}. \tag{A.33}$$

A.1.5. Legendre polynomials

The Legendre polynomials $P_n(z)$ of integer order $n = 0, 1, 2, \ldots$ are solutions of the differential equations, which are regular at $z = 0$ [290]

$$(1-z^2)\frac{d^2w}{dz^2}-2z\frac{dw}{dz}+n(n+1)w=0 \tag{A.34}$$

and are defined as [104]

$$P_n(z)=\frac{1}{2^n n!}\frac{d^n}{dz^n}\left(z^2-1\right)^n. \tag{A.35}$$

In particular, Eq. (A.35) yields

$$\begin{gathered}P_0(z)=1,\ P_1(z)=z,\\ P_2(z)=\frac{1}{2}\left(3z^2-1\right),\ P_3(z)=\frac{1}{2}\left(5z^3-3z\right),\ \text{etc.}\end{gathered} \tag{A.36}$$

The Legendre polynomials satisfy the recurrence relations [290]

$$(2n+1)zP_n(z)=(n+1)P_{n+1}(z)+nP_{n-1}(z), \tag{A.37}$$

$$(z^2-1)\frac{d}{dz}P_n(z)=n\left[zP_n(z)-P_{n-1}(z)\right]=\frac{n(n+1)}{2n+1}\left[P_{n+1}(z)-P_{n-1}(z)\right], \tag{A.38}$$

and the orthogonality relation

$$\int_{-1}^{1}P_n(z)P_{n'}(z)dz=\frac{2\delta_{nn'}}{2n+1}. \tag{A.39}$$

Useful integrals involving the $P_n(z)$ are listed below [146, 370, 471]:

$$\begin{aligned}&\int_{-1}^{1}P_1(z)e^{\xi z+\sigma z^2}dz\\ &=\frac{e^{\sigma}\sinh\xi}{\sigma}-\frac{\xi e^{-\frac{\xi^2}{4\sigma}}}{4\sigma}\sqrt{\frac{\pi}{\sigma}}\left[\operatorname{erfi}\left(\frac{2\sigma+\xi}{2\sqrt{\sigma}}\right)+\operatorname{erfi}\left(\frac{2\sigma-\xi}{2\sqrt{\sigma}}\right)\right],\end{aligned} \tag{A.40}$$

$$\begin{aligned}&\int_{-1}^{1}P_2(z)e^{\xi z+\sigma z^2}dz=\frac{3e^{\sigma}}{2\sigma}\left(\cosh\xi-\frac{\xi}{2\sigma}\sinh\xi\right)\\ &+\frac{e^{-\frac{\xi^2}{4\sigma}}}{2}\sqrt{\frac{\pi}{\sigma}}\left(\frac{3\xi^2}{8\sigma^2}-\frac{3}{4\sigma}-\frac{1}{2}\right)\left[\operatorname{erfi}\left(\frac{2\sigma+\xi}{2\sqrt{\sigma}}\right)+\operatorname{erfi}\left(\frac{2\sigma-\xi}{2\sqrt{\sigma}}\right)\right],\end{aligned} \tag{A.41}$$

$$\int_{-1}^{1}P_n(z)e^{\xi z}dz=\sqrt{\frac{2\pi}{\xi}}I_{n+1/2}(\xi), \tag{A.42}$$

$$\int_{-1}^{1}P_{2n}(z)e^{\sigma z^2}dz=\frac{\sigma^n\Gamma\left(n+\frac{1}{2}\right)}{\Gamma\left(2n+\frac{3}{2}\right)}M\left(n+\tfrac{1}{2},2n+\tfrac{3}{2},\sigma\right), \tag{A.43}$$

$$\int_{-1}^{1} P_1(z)P_{2n+1}(z)e^{\sigma z^2}\,dz = \frac{\sigma^n\Gamma\left(n+\frac{3}{2}\right)}{\Gamma\left(2n+\frac{5}{2}\right)} M\left(n+\tfrac{3}{2},2n+\tfrac{5}{2},\sigma\right), \quad \text{(A.44)}$$

where $I_{n+1/2}(z)$ is the modified spherical Bessel function of the first kind of order $n + 1/2$ (see Appendix A.1.6), $M(a, b, z)$ is the confluent hypergeometric function (see Appendix A.1.2), and erfi (z) is the error function of imaginary argument (see Appendix A.1.3). The above integrals allow us to evaluate analytically the equilibrium averages for various models treated in the book:

$$\left\langle P_n\right\rangle_0 = \int_{-1}^{1} P_n(z)e^{\xi z}dz \Big/ \int_{-1}^{1} e^{\xi z}dz = \frac{I_{n+1/2}(\xi)}{I_{1/2}(\xi)}, \quad \text{(A.45)}$$

$$\left\langle P_{2n}\right\rangle_0 = \frac{1}{Z_{0,\sigma}}\int_{-1}^{1} P_{2n}(z)e^{\sigma z^2}dz = \frac{\sigma^n\Gamma\left(n+\frac{1}{2}\right)M\left(n+\frac{1}{2},2n+\frac{3}{2},\sigma\right)}{2\Gamma\left(2n+\frac{3}{2}\right)M\left(\frac{1}{2},\frac{3}{2},\sigma\right)}, \quad \text{(A.46)}$$

$$\left\langle P_1P_{2n+1}\right\rangle_0 = \frac{1}{Z_{0,\sigma}}\int_{-1}^{1} zP_{2n+1}(z)e^{\sigma z^2}dz = \sigma^n\frac{\Gamma\left(n+\frac{3}{2}\right)M\left(n+\frac{3}{2},2n+\frac{5}{2},\sigma\right)}{2\Gamma\left(2n+\frac{5}{2}\right)M\left(\frac{1}{2},\frac{3}{2},\sigma\right)}, \quad \text{(A.47)}$$

$$\left\langle P_1\right\rangle_0 = \frac{1}{Z_{\xi,\sigma}}\int_{-1}^{1} P_1(z)e^{\xi z+\sigma z^2}dz = \frac{e^{\sigma}}{\sigma Z_{\xi,\sigma}}\sinh\xi - \frac{\xi}{2\sigma}e^{-\frac{\xi^2}{4\sigma}}, \quad \text{(A.48)}$$

$$\begin{aligned}\left\langle P_2\right\rangle_0 &= \frac{1}{Z_{\xi,\sigma}}\int_{-1}^{1} P_2(z)e^{\xi z+\sigma z^2}dz \\ &= \frac{3e^{\sigma}}{2\sigma Z_{\xi,\sigma}}\left(\cosh\xi - \frac{\xi\sinh\xi}{2\sigma}\right) + \frac{3\xi^2}{8\sigma^2} - \frac{3}{4\sigma} - \frac{1}{2},\end{aligned} \quad \text{(A.49)}$$

where the partition function $Z_{\xi,\sigma}$ is given by

$$Z_{\xi,\sigma} = \int_{-1}^{1} e^{\xi z+\sigma z^2}dz = \frac{1}{2}\sqrt{\frac{\pi}{\sigma}}\left[\operatorname{erfi}\left(\frac{2\sigma+\xi}{2\sqrt{\sigma}}\right) + \operatorname{erfi}\left(\frac{2\sigma-\xi}{2\sqrt{\sigma}}\right)\right]. \quad \text{(A.50)}$$

A.1.6. Modified Bessel functions of the first and second kinds

The modified Bessel function of the first kind $I_v(z)$ of order v satisfies the differential equation [290]

$$z^2\frac{d^2w}{dz^2} + z\frac{dw}{dz} - \left(z^2+v^2\right)w = 0. \quad \text{(A.51)}$$

The $I_\nu(z)$ are regular at $z = 0$ and have the following properties

$$I_\nu(ze^{im\pi}) = e^{im\pi\nu} I_\nu(z),$$
$$\frac{d}{dz}\left[z^{\pm\nu} I_\nu(z)\right] = z^{\pm\nu} I_{\nu\mp 1}(z). \tag{A.52}$$

They may be also defined via Kummer's function $M(a, b, z)$ (see Appendix A.1.2) as [290]

$$I_\nu(z) = \frac{e^{-z}}{\Gamma(z)}\left(\frac{z}{2}\right)^\nu M\left(\nu + \frac{1}{2}, 2\nu + 1, z\right). \tag{A.53}$$

The Taylor expansion of $I_\nu(z)$ is given by [290]

$$I_\nu(z) = \left(\frac{z}{2}\right)^\nu \sum_{n=0}^{\infty} \frac{(-1)^n (z^2/4)^n}{n!\Gamma(\nu + n + 1)}, \tag{A.54}$$

while the asymptotic expansion of $I_\nu(z)$ for large z is [290]

$$I_\nu(z) \sim \frac{e^z}{\sqrt{2\pi z}}\left(1 - \frac{4\nu^2 - 1}{8z} + \ldots\right)\left(|\arg z| < \frac{\pi}{2}\right). \tag{A.55}$$

The $I_\nu(z)$ also satisfy the recurrence equation [290]

$$I_{\nu-1}(z) - I_{\nu+1}(z) = \frac{2\nu}{z} I_\nu(z), \tag{A.56}$$

which may be written as the continued fraction:

$$\frac{I_\nu(z)}{I_{\nu-1}(z)} = \frac{1}{\dfrac{2\nu}{z} + \dfrac{I_{\nu+1}(z)}{I_\nu(z)}} = \frac{1}{\dfrac{2\nu}{z} + \dfrac{1}{\dfrac{2(\nu+1)}{z} + \dfrac{1}{\dfrac{2(\nu+2)}{z} + \ddots}}}. \tag{A.57}$$

If $\nu = \pm n$, where n is an integer, $I_n(z)$ is an entire function of z and may be defined using the integral representations [290]

$$I_n(z) = \frac{1}{\pi}\int_0^\pi e^{\pm in\theta + z\cos\theta} d\theta = \frac{1}{\pi}\int_0^\pi \cos n\theta\, e^{z\cos\theta} d\theta. \tag{A.58}$$

If $v = n + 1/2$, where n is an integer, $I_{n+1/2}(z)$ are called the modified spherical Bessel functions of the first kind. For $n = 0$, 1, and 2, the $I_{n+1/2}(z)$ are given by [290]

$$\sqrt{\frac{\pi}{2z}} I_{1/2}(z) = \frac{\sinh z}{z}, \tag{A.59}$$

$$\sqrt{\frac{\pi}{2z}} I_{3/2}(z) = \frac{\cosh z}{z} - \frac{\sinh z}{z^2}, \tag{A.60}$$

$$\sqrt{\frac{\pi}{2z}} I_{5/2}(z) = \frac{3}{z^2}\cosh z + \left(\frac{3}{z^2} + \frac{1}{z}\right)\sinh z, \tag{A.61}$$

etc. Useful series involving the $I_n(z)$ are [290,488]

$$\sqrt{\frac{\pi}{2z}} \sum_{n=0}^{\infty} (\pm 1)^n (2n+1) I_{n+1/2}(z) P_n(\cos\theta) = e^{\pm z\cos\theta}, \tag{A.62}$$

$$\sum_{n=0}^{\infty} (-1)^n \frac{(n+3/4)\Gamma(n+1/2)}{\Gamma(n+2)} I_{2n+3/2}(z) = \frac{\cosh z - 1}{\sqrt{2z}}. \tag{A.63}$$

The modified Bessel function of the second kind $K_v(z)$ of order v also satisfies the differential equation (A.51) being its independent solution, which is not regular at z = 0. It is related to $I_v(z)$ via [290]

$$K_v(z) = \frac{\pi}{2} \frac{I_{-v}(z) - I_v(z)}{\sin v\pi}. \tag{A.64}$$

When v is fixed and $z \to 0$ [290]

$$K_v(z) \sim \frac{\Gamma(v)}{2}\left(\frac{z}{2}\right)^{-v} \quad (\operatorname{Re} v > 0), \qquad K_0(z) \sim -\ln z, \tag{A.65}$$

while the asymptotic expansion of $K_v(z)$ for large z is [290]

$$K_v(z) \sim e^{-z}\sqrt{\frac{\pi}{2z}}\left[1 + \frac{4v^2 - 1}{8z} + \ldots\right] \left(|\arg z| < \frac{\pi}{2}\right). \tag{A.66}$$

A.1.7. Spherical harmonics

A spherical harmonic $Y_{lm}(\vartheta, \varphi)$ is a complex function of two arguments, namely, the colatitude $0 \le \vartheta \le \pi$ and the azimuth $0 \le \varphi \le 2\pi$ and may be defined as [104]

$$Y_{lm}(\vartheta,\varphi) = \sqrt{\frac{(2l+1)(l-m)!}{4\pi(l+m)!}} e^{im\varphi} P_l^m(\cos\vartheta), \qquad \text{(A.67)}$$

$$Y_{l-m} = (-1)^m Y_{lm}^*, \qquad \text{(A.68)}$$

where $P_l^m(x)$ are the associated Legendre functions defined as [104]

$$P_l^m(\cos\vartheta) = \frac{(-1)^m}{2^l l!}(\sin\vartheta)^m \frac{d^{l+m}}{(d\cos\vartheta)^{l+m}}\left(\cos^2\vartheta - 1\right)^l \qquad \text{(A.69)}$$

with $m = l, l-1, \ldots, -l$, and the asterisk denotes the complex conjugate. For $m = 0$, $Y_{l0}(\vartheta, \varphi)$ is given by

$$Y_{l0}(\vartheta,\varphi) = \sqrt{\frac{2l+1}{4\pi}} P_l(\cos\vartheta), \qquad \text{(A.70)}$$

where $P_l(\cos\vartheta)$ is the Legendre polynomial of order l [104,290]. In particular, Eq. (A.67) yields

$$Y_{10}(\vartheta,\varphi) = \sqrt{\frac{3}{4\pi}} \cos\vartheta,\ Y_{1\pm1}(\vartheta,\varphi) = \mp\sqrt{\frac{3}{8\pi}} e^{\pm i\varphi} \sin\vartheta, \text{ etc.} \qquad \text{(A.71)}$$

In quantum mechanics, the $Y_{lm}(\vartheta, \varphi)$ play an important role describing the distribution of particles which move in a spherically symmetric field with the orbital angular momentum l and projection onto the quantization axis m [104]. The $Y_{lm}(\vartheta, \varphi)$ are the eigenfunctions of the square of the angular momentum operator $\hat{\mathbf{L}}$ and its projection $\hat{L}_Z$ onto the Z-axis, namely,

$$\hat{\mathbf{L}}^2 Y_{lm}(\vartheta,\varphi) = l(l+1) Y_{lm}(\vartheta,\varphi), \qquad \text{(A.72)}$$

$$\hat{L}_Z Y_{lm}(\vartheta,\varphi) = m Y_{lm}(\vartheta,\varphi), \qquad \text{(A.73)}$$

where

$$\hat{\mathbf{L}}^2 = -\frac{1}{\sin\vartheta}\frac{\partial}{\partial\vartheta}\left(\sin\vartheta\frac{\partial}{\partial\vartheta}\right) - \frac{1}{\sin^2\vartheta}\frac{\partial^2}{\partial\varphi^2},$$
$$\hat{L}_Z = -i\frac{\partial}{\partial\varphi}. \tag{A.74}$$

The $Y_{lm}(\vartheta, \varphi)$, which are just the components of some irreducible tensor of rank l, may formally be defined by the commutation relations [104]

$$\left[\hat{L}_\mu, Y_{lm}(\vartheta,\varphi)\right] = \sqrt{l(l+1)}C_{lm1\mu}^{lm+\mu}Y_{lm+\mu}(\vartheta,\varphi) \tag{A.75}$$

yielding three separate relations ($\mu = 0, \pm 1$), viz., [88,104]

$$\hat{L}_0 Y_{lm} = mY_{lm},$$
$$\hat{L}_{\pm 1} Y_{lm} = \mp\sqrt{\frac{l(l+1) - m(m\pm 1)}{2}}Y_{lm\pm 1}, \tag{A.76}$$

where $C_{\alpha a b \beta}^{c\gamma}$ are called the Clebsch–Gordan coefficients [104] and $\hat{L}_\mu$ are the components of the orbital angular momentum operator $\hat{\mathbf{L}}$ in the spherical basis defined by [104]

$$\hat{L}_0 = -i\frac{\partial}{\partial\varphi}, \quad \hat{L}_{\pm 1} = -\frac{e^{\pm i\varphi}}{\sqrt{2}}\left(\frac{\partial}{\partial\vartheta} \pm i\cot\vartheta\frac{\partial}{\partial\varphi}\right). \tag{A.77}$$

The angular momentum operators $\hat{L}_\mu$ satisfy the same commutation relations, Eq. (A.156) from Appendix A.2, as the spherical components $\hat{S}_\mu$ of the spin operator $\hat{\mathbf{S}}$ (see Appendix A.2). Various definitions of the $C_{\alpha a b \beta}^{c\gamma}$ the Clebsch–Gordan coefficients are available, e.g., in Ref. [104]. The built-in function Clebsch-Gordan[$\{a, \alpha\},\{b, \beta\},\{c, \gamma\}$] of MATHEMATICA® facilitates the calculation of these coefficients.

The completeness relation for the $Y_{lm}(\vartheta,\varphi)$ is as follows [104]

$$\sum_{l=0}^{\infty}\sum_{m=-l}^{l} Y_{lm}^*(\vartheta,\varphi)Y_{lm}(\vartheta',\varphi') = \delta(\varphi - \varphi')\delta(\cos\vartheta - \cos\vartheta'),$$

while their normalization and orthogonality relations are [104]

$$\int_0^{2\pi}\int_0^{\pi} Y_{lm}(\vartheta,\varphi)Y_{l'm'}^*(\vartheta,\varphi)\sin\vartheta d\vartheta d\varphi = \delta_{ll'}\delta_{mm'}. \tag{A.78}$$

Thus, an arbitrary function $f(\vartheta, \varphi)$ defined on the interval $0 \le \vartheta \le \pi$ and $0 \le \varphi \le 2\pi$ (the unit sphere) which satisfies the square integrability condition

$$\int_0^{2\pi}\int_0^{\pi} |f(\vartheta,\varphi)|^2 \sin\vartheta d\vartheta d\varphi < \infty$$

can be expanded in a series of the spherical harmonics as [104]

$$f(\vartheta,\varphi) = \sum_{l=0}^{\infty}\sum_{m=-l}^{l} a_{lm} Y_{lm}(\vartheta,\varphi), \tag{A.79}$$

where the expansion coefficients a_{lm} are defined by

$$a_{lm} = \int_0^{2\pi}\int_0^{\pi} f(\vartheta,\varphi) Y_{lm}^*(\vartheta,\varphi) \sin\vartheta d\vartheta d\varphi.$$

A product of two spherical harmonics may be expanded in the Clebsch–Gordan series as [104]

$$Y_{l_1 m_1} Y_{l_2 m_2} = \sum_{L,M} \sqrt{\frac{(2l_1+1)(2l_2+1)}{4\pi(2L+1)}} C_{l_1 0 l_2 0}^{L0} C_{l_1 m_1 l_2 m_2}^{LM} Y_{LM}. \tag{A.80}$$

Some useful recurrence relations are [104]

$$\cos\vartheta Y_{lm} = \sqrt{\frac{(l+1)^2 - m^2}{(2l+3)(2l+1)}} Y_{l+1m} - \sqrt{\frac{l^2 - m^2}{(2l+1)(2l-1)}} Y_{l-1m}, \tag{A.81}$$

$$\sin\vartheta \frac{\partial Y_{lm}}{\partial \vartheta} = l\sqrt{\frac{(l+1)^2 - m^2}{(2l+3)(2l+1)}} Y_{l+1m} - (l+1)\sqrt{\frac{l^2 - m^2}{(2l+1)(2l-1)}} Y_{l-1m}. \tag{A.82}$$

The second derivative of $Y_{lm}(\vartheta,\varphi)$ is given by [104]

$$\frac{\partial^2 Y_{lm}}{\partial \vartheta^2} = \left(\frac{m^2}{\sin^2\vartheta} - l(l+1)\right) Y_{lm} - \cot\vartheta \frac{\partial Y_{lm}}{\partial \vartheta}. \tag{A.83}$$

One of the known trigonometric identities for the $Y_{lm}(\vartheta, \varphi)$ which has been used in the main text [cf. Eq. (1.561)] is [104]

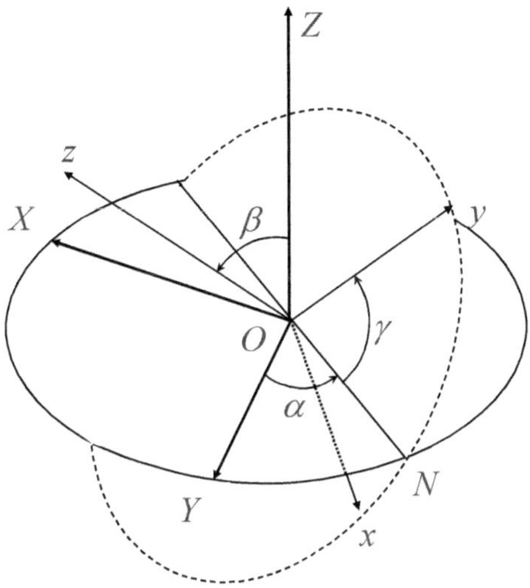

Fig. A.1. The Euler angles relating the laboratory *OXYZ* and the rotating (body-fixed) *Oxyz* coordinate systems.

$$Y_{LM}^{*}(\vartheta,\varphi)=\frac{\cos^{4S}(\vartheta/2)}{C_{SSL0}^{SS}}$$
$$\times\sqrt{\frac{2L+1}{4\pi}}\sum_{m=-S}^{S}\frac{(2S)!C_{Sm-MLM}^{Sm}\left(\tan(\vartheta/2)\right)^{2S-2m+M}e^{-iM\varphi}}{\sqrt{(S+m-M)!(S-m+M)!(S+m)!(S-m)!}}. \tag{A.84}$$

Because the spherical harmonics $Y_{lm}(\vartheta,\varphi)$ are components of some irreducible tensor of rank l, under *arbitrary* rotation of the coordinate system described by the Euler angles α, β, γ, which specify the rotation of a rigid body (see Fig. A.1), the $Y_{lm}(\vartheta,\varphi)$ are transformed according to the rule [104] [cf. Eq. (A.163)]

$$Y_{lm'}(\vartheta',\varphi')=\sum_{m}Y_{lm}(\vartheta,\varphi)D_{mm'}^{l}(\alpha,\beta,\gamma). \tag{A.85}$$

where ϑ, φ and ϑ', φ' are the polar angles of the position vector in the original and final coordinate systems and $D_{mm'}^{l}(\alpha,\beta,\gamma)$ are the Wigner D functions.

A.1.8. Wigner's D-functions

Wigner's *D*-functions $D^J_{MM'}(\alpha,\beta,\gamma)$, being functions of the Euler angles α, β, γ are defined as [104]

$$D^J_{MM'}(\alpha,\beta,\gamma) = e^{-iM\alpha} d^J_{MM'}(\beta) e^{-iM'\gamma}, \tag{A.86}$$

where $d^J_{MM'}(\beta)$ is a real function having various explicit forms, for example, [104]

$$\begin{aligned} d^J_{MM'}(\beta) &= \frac{(-1)^{J-M'}}{2^J}\sqrt{\frac{(J+M)!(1-\cos\beta)^{M'-M}(1+\cos\beta)^{-M'-M}}{(J-M)!(J+M')!(J-M')!}} \\ &\times \frac{d^{J-M}}{(d\cos\beta)^{J-M}}\left[(1-\cos\beta)^{J-M'}(1-\cos\beta)^{J+M'}\right]. \end{aligned} \tag{A.87}$$

In particular, we have [104]

$$D^j_{00}(\alpha,\beta,\gamma) = P_j(\cos\beta), \tag{A.88}$$

$$D^l_{0-m}(\alpha,\beta,\gamma) = \sqrt{\frac{4\pi}{2l+1}} Y_{lm}(\beta,\gamma). \tag{A.89}$$

Wigner's *D*-functions $D^l_{mm'}(\alpha,\beta,\gamma)$ are complex. They are defined in the domain $0 \le \alpha \le 2\pi$, $0 \le \beta \le \pi$, $0 \le \gamma \le 2\pi$ and may be also defined as the matrix elements of the rotational operator $\widehat{D}(\alpha,\beta,\gamma)$ given by [104]

$$\widehat{D}(\alpha,\beta,\gamma) = e^{-i\alpha\hat{J}_x} e^{-i\beta\hat{J}_y} e^{-i\gamma\hat{J}_z}. \tag{A.90}$$

Here $\hat{J}_x, \hat{J}_y, \hat{J}_z$, are the projections of the angular momentum operator $\hat{\mathbf{J}}$ onto the axes of the rotating (body-fixed) coordinate system defined via its spherical components $\hat{J}^{\pm 1}$ and $\hat{J}^0$ as

$$\hat{J}_x = \frac{1}{\sqrt{2}}(\hat{J}^{-1} - \hat{J}^{+1}),\ \hat{J}_y = -\frac{i}{\sqrt{2}}(\hat{J}^{-1} + \hat{J}^{+1}),\ \hat{J}_z = \hat{J}^0, \tag{A.91}$$

where

$$\hat{J}^{\pm 1} = \frac{i}{\sqrt{2}} e^{\mp i\gamma}\left[\pm\cot\beta\frac{\partial}{\partial\gamma} + i\frac{\partial}{\partial\beta} \mp \frac{1}{\sin\beta}\frac{\partial}{\partial\alpha}\right],\ \hat{J}^0 = -i\frac{\partial}{\partial\gamma}. \tag{A.92}$$

The D-functions represent the wave functions of a rigid symmetric top being eigenfunctions of the three operators

$$\hat{J}_z = -i\frac{\partial}{\partial\alpha},\ \hat{J}_Z = -i\frac{\partial}{\partial\gamma}, \tag{A.93}$$

$$\hat{\mathbf{J}}^2 = -\frac{\partial^2}{\partial\beta^2} - \cot\beta\frac{\partial}{\partial\beta} - \frac{1}{\sin^2\beta}\left(\frac{\partial^2}{\partial\alpha^2} - 2\cos\beta\frac{\partial^2}{\partial\alpha\partial\gamma} + \frac{\partial^2}{\partial\gamma^2}\right), \tag{A.94}$$

where $\hat{J}_Z$ is the projection of the angular momentum operator $\hat{\mathbf{J}}$ onto the Z-axis of the laboratory coordinate system and they represent the wave functions ψ_{JMK} of a rigid symmetric top because

$$\psi_{JMK}(\alpha,\beta,\gamma) = i^J\sqrt{\frac{2J+1}{8\pi^2}}D^J_{KM}(\alpha,\beta,\gamma),$$

where J, K, and M are the corresponding quantum numbers.

Using known properties of the angular momentum operators and Wigner's D-functions, we also have [104]

$$\hat{\mathbf{J}}^2 D^j_{nm} = j(j+1)D^j_{nm},\ \ \hat{J}^2_z D^j_{nm} = m^2 D^j_{nm}, \tag{A.95}$$

$$\hat{J}^0 D^j_{nm} = -mD^j_{nm},\ \ \hat{J}^{\pm1} D^j_{nm} = -\sqrt{j(j+1)}C^{jm\pm1}_{jm1\pm1}D^j_{nm\pm1}. \tag{A.96}$$

The orthogonality condition for the D-functions is [104]

$$\int_0^{2\pi}\int_0^{2\pi}\int_0^{\pi} D^{j_2\,*}_{l_2l'_2}(\alpha,\beta,\gamma)D^{j_1}_{l_1l'_1}(\alpha,\beta,\gamma)\sin\beta d\beta d\alpha d\gamma = \frac{8\pi^2}{2j_2+1}\delta_{j_1j_2}\delta_{l_1l_2}\delta_{l'_1l'_2}, \tag{A.97}$$

where $j_1 + j_2$ is an integer,

$$D^{j\,*}_{nm} = (-1)^{n-m}D^j_{-n-m} \tag{A.98}$$

and the asterisk denotes the complex conjugate. Just as products of spherical harmonics $Y_{l_1m_1}Y_{l_2m_2}$, a product of two Wigner functions $D^{j_1}_{l_1m_1}D^{j_2}_{l_2m_2}$ may be expanded in the Clebsch–Gordan series as [104] (cf. Eq. (A.80))

$$D^{j_1}_{l_1m_1}D^{j_2}_{l_2m_2} = \sum_{j=|j_1-j_2|}^{j_1+j_2} C^{jl_1+l_2}_{j_1l_1j_2l_2}C^{jm_1+m_2}_{j_1m_1j_2m_2}D^j_{l_1+l_2m_1+m_2}. \tag{A.99}$$

A.1.9. Elliptic integrals and functions

By introducing a parameter m called the modulus and the argument u given by

$$u = \int_0^{\varphi} \frac{d\theta}{\sqrt{1 - m\sin^2\theta}}, \tag{A.100}$$

where the angle φ is called the *amplitude* $\varphi = \mathrm{am}(u \mid m)$, the Jacobi elliptic functions $\mathrm{sn}(u \mid m)$, $\mathrm{cn}(u \mid m)$ and $\mathrm{dn}(u \mid m)$ can be defined as [290]

$$\mathrm{sn}(u|m) = \sin\varphi,\quad \mathrm{cn}(u|m) = \cos\varphi,\ \text{and}\ \mathrm{dn}(u|m) = \sqrt{1 - m\sin^2\varphi}. \tag{A.101}$$

The functions $\mathrm{sn}(u \mid m)$, $\mathrm{cn}(u \mid m)$ and $\mathrm{dn}(u \mid m)$ are doubly periodic meromorphic functions. They satisfy the functional relations

$$\mathrm{sn}^2(u|m) + \mathrm{cn}^2(u|m) = 1,\ \mathrm{dn}^2(u|m) + m\,\mathrm{sn}^2(u|m) = 1, \tag{A.102}$$

and are the solutions of the following differential equations [487]

$$\frac{d}{du}\mathrm{sn}(u|m) = \sqrt{\left[1 - \mathrm{sn}^2(u|m)\right]\left[1 - m\mathrm{sn}^2(u|m)\right]}, \tag{A.103}$$

$$\frac{d}{du}\mathrm{cn}(u|m) = -\sqrt{\left[1 - \mathrm{cn}^2(u|m)\right]\left[1 - m + m\mathrm{cn}^2(u|m)\right]}, \tag{A.104}$$

$$\frac{d}{du}\mathrm{dn}(u|m) = -\sqrt{\left[1 - \mathrm{dn}^2(u|m)\right]\left[m - 1 + \mathrm{dn}^2(u|m)\right]}. \tag{A.105}$$

The elliptic integrals of the first, second, and third kinds are defined as [290]

$$F(\varphi|m) = \int_0^{\varphi} \frac{d\theta}{\sqrt{1 - m\sin^2\theta}} = F(u|m) = u, \tag{A.106}$$

$$E(\varphi|m) = \int_0^{\varphi} \sqrt{1 - m\sin^2\theta}\, d\theta = E(u|m) = \int_0^{u} \mathrm{dn}^2(w|m)dw, \tag{A.107}$$

$$\begin{aligned}\Pi(n;\varphi|m) &= \int_0^{\varphi} \frac{d\theta}{(1 - n\sin^2\theta)\sqrt{1 - m\sin^2\theta}} \\ &= \Pi(n;u|m) = \int_0^{u} \left(1 - n\,\mathrm{sn}^2(w|m)\right)^{-1} dw,\end{aligned} \tag{A.108}$$

while the *complete* elliptic integrals of the first, second, and third kinds are, respectively, given by [290]

$$K(m)=K=F\left(\frac{\pi}{2}\middle|m\right)=\int_0^{\pi/2}\frac{d\theta}{\sqrt{1-m\sin^2\theta}}, \tag{A.109}$$

$$E(m)=E=E\left(\frac{\pi}{2}\middle|m\right)=\int_0^{\pi/2}\sqrt{1-m\sin^2\theta}d\theta, \tag{A.110}$$

$$\Pi\left(n\middle|m\right)=\Pi\left(n;\frac{\pi}{2}\middle|m\right)=\int_0^{\pi/2}\frac{d\theta}{(1-n\sin^2\theta)\sqrt{1-m\sin^2\theta}}. \tag{A.111}$$

Integrals involving the elliptic function $\mathrm{sn}(u\,|\,m)$ of the following form

$$\frac{1}{4K(m)}\int_0^{4K(m)}\left[\frac{a-\mathrm{sn}^2(u\,|\,m)}{a+\mathrm{sn}^2(u\,|\,m)}\right]^n du, \tag{A.112}$$

can be expressed in terms of the complete elliptic integrals using the table of integrals from Section 1.17.1 of Ref. [350] as

$$\frac{1}{4K(m)}\int_0^{4K(m)}\frac{a-\mathrm{sn}^2(u\,|\,m)}{a+\mathrm{sn}^2(u\,|\,m)}du=2aI(1)-1, \tag{A.113}$$

$$\frac{1}{4K(m)}\int_0^{4K(m)}\left[\frac{a-\mathrm{sn}^2(u\,|\,m)}{a+\mathrm{sn}^2(u\,|\,m)}\right]^2 du=1-4aI(1)+4a^2I(2), \tag{A.114}$$

and so on, where

$$I(n)=\frac{1}{4K(m)}\int_0^{4K(m)}\frac{du}{[a+\mathrm{sn}^2(u\,|\,m)]^n}, \tag{A.115}$$

$$I(1)=\frac{\Pi(-a^{-1}\,|\,m)}{aK(m)}, \tag{A.116}$$

$$\begin{aligned}I(2)=\frac{1}{2a(1+a)}\Bigg[&-1+\frac{E(m)}{(1+am)K(m)}\\&+\frac{1+3a^2m+2a(m+1)}{a(1+am)}\frac{\Pi(-a^{-1}\,|\,m)}{K(m)}\Bigg],\end{aligned} \tag{A.117}$$

$$I(3)=\frac{3+9a^2m+6a(1+m)}{4a(1+a)(1+am)}I(2)-\frac{1+m+3am}{2a(1+a)(1+am)}I(1)+\frac{m}{4a(1+a)(1+am)}, \tag{A.118}$$

$$I(4)=\frac{6+16a^2m+11a(1+m)}{6a(1+a)(1+am)}I(3)+\frac{3+57a^2m+22a(1+m)}{24a^2(1+a)(1+am)}I(2)+\frac{2(1+m+9am)I(1)-m}{24a^2(1+a)(1+am)}, \quad \text{etc.} \tag{A.119}$$

A.1.10. Dirac-delta and Heaviside functions

The Dirac-delta function $\delta(t)$ is a generalized function or distribution, which is defined to have the properties [489]

$$\delta(-t)=\delta(t),\ \delta(t)=0,\ t\neq 0,\ \int_{-\infty}^{\infty}\delta(t)dt=1, \tag{A.120}$$

$$\int_a^b \delta(t-c)f(t)dt=f(c)\ \left(a<c<b\right), \tag{A.121}$$

$$\int_a^b \delta(t-c)f(t)dt=0\ \left(c>b \text{ or } c<a\right), \tag{A.122}$$

$$\int_a^b \delta(t-a)f(t)dt=\frac{1}{2}f(a),\quad \int_a^b \delta(b-t)f(t)dt=\frac{1}{2}f(b), \tag{A.123}$$

where $f(t)$ is any well-behaved function.

The delta function can also be expressed via the Fourier transform as

$$\delta(t)=\frac{1}{2\pi}\int_{-\infty}^{\infty}e^{\pm i\omega t}d\omega. \tag{A.124}$$

The unit-step or Heaviside function $U(t)$ can be defined as

$$U(t) = \begin{cases} 0, & t < 0, \\ 1, & t > 0. \end{cases} \tag{A.125}$$

The unit-step and delta functions are related via

$$U(t) = \int_{-\infty}^{t} \delta(t')dt'. \tag{A.126}$$

A.2. Spin and polarization operators

The spin operator $\hat{\mathbf{S}}$ is Hermitian, i.e. $\hat{\mathbf{S}}^{\dagger} = \hat{\mathbf{S}}$, and is usually represented by a set of three (since the spin vector $\hat{\mathbf{S}}$ has three components) square $(2S+1)\times(2S+1)$ matrices with S being the spin number [104]. For the *cartesian* components of $\hat{\mathbf{S}}$, namely, the operators $\hat{S}_X$, $\hat{S}_Y$, and $\hat{S}_Z$, the Hermitian property takes on the form $\hat{S}_i^{\dagger} = \hat{S}_i$ $(i = X, Y, Z)$, while for the *spherical* components $\hat{S}_{\pm 1}$ and $\hat{S}_0$ that property becomes $\hat{S}_{\mu}^{\dagger} = (-1)^{\mu}\hat{S}_{\mu}$. The relations between the cartesian and spherical components of $\hat{\mathbf{S}}$ are given by [104]

$$\begin{aligned} &\hat{S}_{+1} = -\frac{1}{\sqrt{2}}\left(\hat{S}_X + i\hat{S}_Y\right),\ \hat{S}_X = \frac{1}{\sqrt{2}}\left(\hat{S}_{-1} - \hat{S}_{+1}\right), \\ &\hat{S}_{-1} = \frac{1}{\sqrt{2}}\left(\hat{S}_X - i\hat{S}_Y\right),\ \hat{S}_Y = \frac{i}{\sqrt{2}}\left(\hat{S}_{-1} + \hat{S}_{+1}\right), \\ &\hat{S}_0 = \hat{S}_Z,\ \hat{S}_Z = \hat{S}_0. \end{aligned} \tag{A.127}$$

Polarization states of particles are described by spin functions $\chi_{Sm} = |S, m\rangle$, which depend on the spin variable σ with σ being the spin projection on the Z-axis. This variable takes on $2S+1$ values, namely, $\sigma = -S, -S+1, \ldots, S-1, S$. The dependence of the spin functions $\chi_{Sm}(\sigma)$ on σ is given by $\chi_{Sm}(\sigma) = \delta_{m\sigma}$ [104]. Moreover, the spin functions χ_{Sm} are eigenfunctions of the spin operators $\hat{\mathbf{S}}^2$ and $\hat{S}_Z$, viz., [104]

$$\hat{\mathbf{S}}^2\chi_{Sm} = S(S+1)\chi_{Sm}, \tag{A.128}$$

$$\hat{S}_Z\chi_{Sm} = m\chi_{Sm} \tag{A.129}$$

with $m = -S, -S+1, \ldots, S-1, S$. The spin functions each have $2S+1$ components and describe polarization states of a particle with definite spin

S and spin projection m onto the Z-axis. They can also be rewritten as column matrices [104]

$$\chi_{SS} = \begin{pmatrix} 1 \\ 0 \\ \vdots \\ 0 \end{pmatrix}, \chi_{SS-1} = \begin{pmatrix} 0 \\ 1 \\ \vdots \\ 0 \end{pmatrix}, \ldots, \chi_{S-S} = \begin{pmatrix} 0 \\ 0 \\ \vdots \\ 1 \end{pmatrix}. \tag{A.130}$$

Furthermore, the vector representation shows that the spin functions χ_{Sm} constitute a complete set of functions with the orthonormality and completeness conditions

$$\chi^{\dagger}_{Sm'}\chi_{Sm} = \delta_{mm'}, \tag{A.131}$$

$$\sum_{m=-S}^{S} \chi_{Sm}\chi^{\dagger}_{Sm} = \hat{I}, \tag{A.132}$$

where $\hat{I}$ is the unit $(2S+1)\times(2S+1)$ matrix.

The spherical components of the spin operator $\hat{\mathbf{S}}$ may be expressed using the Clebsch–Gordan coefficients in terms of the spin functions as [104]

$$\hat{S}_{\mu} = \sqrt{S(S+1)} \sum_{m,m'} C^{Sm}_{Sm'1\mu}\chi_{Sm}\chi^{\dagger}_{Sm'}, \quad (\mu = 0, \pm 1). \tag{A.133}$$

The operators $\hat{S}_{\mu}$ also satisfy the following commutation relation [104]

$$\left[\hat{S}_{\mu}, \hat{S}_{\nu}\right] = -\sqrt{2}C^{1\lambda}_{1\mu1\nu}\hat{S}_{\lambda}, \left[\hat{\mathbf{S}}^2, \hat{S}_{\mu}\right] = 0, \quad (\mu, \nu, \lambda = 0, \pm 1). \tag{A.134}$$

The matrix elements of $\hat{S}_{\mu}$ are given by [104]

$$\left[\hat{S}_{\mu}\right]_{m'm} = \chi^{\dagger}_{Sm'}\hat{S}_{\mu}\chi_{Sm} = \sqrt{S(S+1)}C^{Sm'}_{Sm1\mu} \tag{A.135}$$

with the only nonvanishing elements being

$$\left[\hat{S}_{0}\right]_{mm} = m, \tag{A.136}$$

$$\left[\hat{S}_{\pm 1}\right]_{m\pm 1m} = \mp\frac{1}{\sqrt{2}}\sqrt{(S\mp m)(S\pm m+1)}. \tag{A.137}$$

The cartesian and spherical components of $\hat{\mathbf{S}}$ are related by Eq. (A.127). For example, in the particular case $S = 1/2$, the operators $\hat{S}_i$ $(i = X, Y, Z)$ and $\hat{S}_\mu$ $(\mu = 0, \pm 1)$ are square 2×2 matrices given by [104]

$$\hat{S}_X = \frac{1}{2}\begin{pmatrix} 0 & 1 \\ 1 & 0 \end{pmatrix}, \quad \hat{S}_Y = \frac{i}{2}\begin{pmatrix} 0 & -1 \\ 1 & 0 \end{pmatrix}, \quad \hat{S}_Z = \frac{1}{2}\begin{pmatrix} 1 & 0 \\ 0 & -1 \end{pmatrix}, \tag{A.138}$$

$$\hat{S}_{+1} = \frac{1}{\sqrt{2}}\begin{pmatrix} 0 & -1 \\ 0 & 0 \end{pmatrix}, \quad \hat{S}_0 = \frac{1}{2}\begin{pmatrix} 1 & 0 \\ 0 & -1 \end{pmatrix}, \quad \hat{S}_{-1} = \frac{1}{\sqrt{2}}\begin{pmatrix} 0 & 0 \\ 1 & 0 \end{pmatrix}, \tag{A.139}$$

while for $S = 1$, these operators are the square 3×3 matrices given by [104]

$$\hat{S}_X = \frac{1}{\sqrt{2}}\begin{pmatrix} 0 & 1 & 0 \\ 1 & 0 & 1 \\ 0 & 1 & 0 \end{pmatrix}, \quad \hat{S}_Y = \frac{i}{\sqrt{2}}\begin{pmatrix} 0 & -1 & 0 \\ 1 & 0 & -1 \\ 0 & 1 & 0 \end{pmatrix}, \quad \hat{S}_Z = \begin{pmatrix} 1 & 0 & 0 \\ 0 & 0 & 0 \\ 0 & 0 & -1 \end{pmatrix}, \tag{A.140}$$

$$\hat{S}_{+1} = \begin{pmatrix} 0 & -1 & 0 \\ 0 & 0 & -1 \\ 0 & 0 & 0 \end{pmatrix}, \quad \hat{S}_0 = \begin{pmatrix} 1 & 0 & 0 \\ 0 & 0 & 0 \\ 0 & 0 & -1 \end{pmatrix}, \quad \hat{S}_{-1} = \begin{pmatrix} 0 & 0 & 0 \\ 1 & 0 & 0 \\ 0 & 1 & 0 \end{pmatrix}. \tag{A.141}$$

In order to describe a polarization (spin) state of a particle, the polarization operators are often used. These operators denoted by $\hat{T}^{(S)}_{LM}$, where $L = 0, 1, \ldots, 2S$ and $M = -L, -L+1, \ldots, L$, are matrices, which act on spin functions. However, the explicit form of $\hat{T}^{(S)}_{LM}$ depends on the representation chosen for the spin functions. The matrix elements $T_{m'm} = \left[\hat{T}^{(S)}_{LM}\right]_{m'm}$ of the polarization operator $\hat{T}^{(S)}_{LM}$ which has the explicit form

$$\hat{T}^{(S)}_{LM} = \begin{pmatrix} T_{SS} & \cdots & T_{S-S} \\ \vdots & \ddots & \vdots \\ T_{-SS} & \cdots & T_{-S-S} \end{pmatrix} \tag{A.142}$$

are related to those of the spin functions in the spherical basis representation via [104], viz.,

$$T_{m'm} = \chi^{\dagger}_{Sm'} \hat{T}^{(S)}_{LM} \chi_{Sm} = \sqrt{\frac{2L+1}{2S+1}} C^{Sm'}_{SmLM}. \tag{A.143}$$

For example, the operator $\hat{T}_{00}^{(S)}$ is proportional to the unit (2S+1)×(2S+1) matrix

$$\hat{T}_{00}^{(S)} = \frac{1}{\sqrt{2S+1}}\hat{I}, \tag{A.144}$$

while for $L=1$ the operators $\hat{T}_{LM}^{(S)}$ are proportional to the spherical components of the spin operator

$$\hat{T}_{1\mu}^{(S)} = \frac{1}{\sqrt{2}a}\hat{S}_{\mu}, \tag{A.145}$$

where

$$a = \sqrt{\frac{S(S+1)(2S+1)}{6}}. \tag{A.146}$$

Likewise, the cartesian components of $\hat{\mathbf{S}}$ given by Eq. (A.127) may also be expressed via the polarization operators $\hat{T}_{10}^{(S)}$ and $\hat{T}_{1\pm1}^{(S)}$ as

$$\hat{S}_X = a\left(\hat{T}_{1-1}^{(S)} - \hat{T}_{11}^{(S)}\right),\ \hat{S}_Y = ia\left(\hat{T}_{1-1}^{(S)} + \hat{T}_{11}^{(S)}\right),\ \hat{S}_Z = \sqrt{2}a\hat{T}_{10}^{(S)}. \tag{A.147}$$

The polarization operators $\hat{T}_{LM}^{(S)}$ are also normalized and satisfy the relations [104]

$$\hat{T}_{LM}^{\dagger(S)} = (-1)^M \hat{T}_{L-M}^{(S)} \tag{A.148}$$

and

$$\mathrm{Tr}\left(\hat{T}_{L,M}^{(S)}\right) = \sqrt{2S+1}\delta_{L0}\delta_{M0}, \tag{A.149}$$

i.e., all $\hat{T}_{LM}^{(S)}$ have zero trace except $\hat{T}_{00}^{(S)}$. Furthermore, the operators $\hat{T}_{LM}^{(S)}$ also constitute an orthonormal basis in the space of (2S+1)×(2S+1) matrices with S integer or half-integer. Hence, it follows that an *arbitrary* square (2S+1)×(2S+1) matrix operator $\hat{A}$ may be expanded as a series of the polarization operators [104]

$$\hat{A} = \sum_{L=0}^{2S}\sum_{M=-L}^{L} A_{LM}\hat{T}_{LM}^{(S)}, \tag{A.150}$$

where the expansion coefficients A_{LM} are given by

$$A_{LM} = \mathrm{Tr}\left(\hat{T}_{LM}^{\dagger(S)} \hat{A}\right). \tag{A.151}$$

If the matrix $\hat{A}$ is Hermitian ($\hat{A}^\dagger = \hat{A}$) then

$$A_{LM}^{*} = (-1)^{M} A_{L-M}. \tag{A.152}$$

Moreover, the matrix products of the spin functions χ_{Sm} and $\chi_{Sm'}^{\dagger}$ [as arranged in the form of Eq. (A.130)] are themselves square $(2S+1)\times(2S+1)$ matrices and, therefore, may also be expanded as a finite series of polarization operators, viz., [104]

$$\chi_{Sm}\chi_{Sm'}^{\dagger} = \sum_{L} \sqrt{\frac{2L+1}{2S+1}} C_{Sm'LM}^{Sm} \hat{T}_{LM}^{(S)}. \tag{A.153}$$

Furthermore, products of two polarization operators may be written in the form of the Clebsch–Gordan series [104]

$$\hat{T}_{L_1M_1}^{(S)} \hat{T}_{L_2M_2}^{(S)} = \sqrt{(2L_1+1)(2L_2+1)} \sum_{L} (-1)^{2S+L} \begin{Bmatrix} L_1 L_2 L \\ S S S \end{Bmatrix} C_{L_1M_1L_2M_2}^{LM} \hat{T}_{LM}^{(S)}, \tag{A.154}$$

where $\begin{Bmatrix} L_1 L_2 L \\ S S S \end{Bmatrix}$ is Wigner's 6*j*-symbol [104]. The polarization operators also satisfy the commutation relations [104]

$$\left[\hat{T}_{L_1M_1}^{(S)}, \hat{T}_{L_2M_2}^{(S)}\right] = \sqrt{(2L_1+1)(2L_2+1)} \times \sum_{L} (-1)^{2S+L} \left[1-(-1)^{L_1+L_2+L}\right] \begin{Bmatrix} L_1 L_2 L \\ S S S \end{Bmatrix} C_{L_1M_1L_2M_2}^{LM} \hat{T}_{LM}^{(S)}. \tag{A.155}$$

Equation (A.155) then automatically yields the commutation relation for the spherical components of the spin operator $\hat{S}_\mu$ and polarization operator $\hat{T}_{LM}^{(S)}$, namely, [104]

$$[\hat{S}_\mu, \hat{T}_{LM}^{(S)}] = \sqrt{L(L+1)} C_{LM1\mu}^{LM+\mu} \hat{T}_{LM+\mu}^{(S)} \tag{A.156}$$

or

$$[\hat{S}_0, \hat{T}_{LM}^{(S)}] = \sqrt{L(L+1)} C_{LM10}^{LM} \hat{T}_{LM}^{(S)} = M\hat{T}_{LM}^{(S)}, \tag{A.157}$$

$$[\hat{S}_{\pm1},\hat{T}^{(S)}_{LM}]=\sqrt{L(L+1)}C^{LM\pm1}_{LM1\pm1}\hat{T}^{(S)}_{LM\pm1}$$
$$=\mp\sqrt{\frac{L(L+1)-M(M\pm1)}{2}}\hat{T}^{(S)}_{LM\pm1}. \quad \text{(A.158)}$$

Moreover, traces of products of the polarization operators are given by [104]

$$\mathrm{Tr}\left(\hat{T}^{(S)}_{L_1M_1}\hat{T}^{(S)}_{L_2M_2}\right)=(-1)^{M_1}\delta_{L_1L_2}\delta_{M_1,-M_2}, \quad \text{(A.159)}$$

$$\mathrm{Tr}\left(\hat{T}^{\dagger(S)}_{LM}\hat{T}^{(S)}_{L'M'}\right)=\delta_{LL'}\delta_{MM'}, \quad \text{(A.160)}$$

$$\mathrm{Tr}\left(\hat{T}^{(S)}_{L_1M_1}\hat{T}^{(S)}_{L_2M_2}\hat{T}^{(S)}_{L_3M_3}\right)$$
$$=(-1)^{2S+L_3+M_3}\sqrt{(2L_1+1)(2L_2+1)}C^{L_3-M_3}_{L_1M_1L_2M_2}\begin{Bmatrix}L_1L_2L_3\\ SSS\end{Bmatrix}. \quad \text{(A.161)}$$

Finally, under rotation of a coordinate system defined by the Euler angles α,β,γ (see Fig. A.1), the spin basis functions $\chi_{Sm'}$ are transformed to new axes by the rotation operator $\hat{D}^S(\alpha,\beta,\gamma)=e^{-i\alpha\hat{S}_X}e^{-i\beta\hat{S}_Y}e^{-i\gamma\hat{S}_Z}$ given by [104]

$$\hat{D}^S(\alpha,\beta,\gamma)=\sum_{L,M,m,m'}\frac{2L+1}{2S+1}C^{Sm}_{Sm'LM}D^S_{mm'}(\alpha,\beta,\gamma)\hat{T}^{(S)}_{LM}, \quad \text{(A.162)}$$

yielding

$$\chi'_{Sm'}=\hat{D}^S(\alpha,\beta,\gamma)\chi_{Sm'}=\sum_m D^S_{mm'}(\alpha,\beta,\gamma)\chi_{Sm}, \quad \text{(A.163)}$$

where $\chi'_{Sm'}$ describe quantum states with definite spin S and spin projection m' on the new Z'-axis and $D^S_{mm'}(\alpha,\beta,\gamma)$ are the Wigner D functions. Similarly, the polarization operators $\hat{T}^{(S)}_{LM}$, which are irreducible tensors of rank L, are transformed by the operator $\hat{D}^L(\alpha,\beta,\gamma)$ given by Eq. (A.162) with $S=L$.

A.3. Derivation of the master equation for a uniaxial paramagnet subjected to a dc magnetic field

In order to find the phase-space representation of the density matrix evolution equation Eq. (3.110), we must transform the following integrands in Eq. (3.115) into the phase-space representation, viz.,

$$\frac{\sigma}{S^2}\left[\hat{S}_0^2,\hat{w}\right]+\frac{\xi}{S}\left[\hat{S}_0,\hat{w}\right], \tag{A.164}$$

$$\left[\hat{S}_0,\hat{w}\hat{S}_0\right]+\left[\hat{S}_0\hat{w},\hat{S}_0\right], \tag{A.165}$$

$$\left[\hat{S}_1,e^{\frac{\sigma}{S^2}\hat{S}_0}\hat{w}+\left[\hat{S}_1,\hat{w}e^{\frac{\sigma}{S^2}\hat{S}_0}\hat{S}_{-1}\right],\right. \tag{A.166}$$

$$\left[\hat{S}_{-1},\hat{w}e^{-\frac{\sigma}{S^2}\hat{S}_0}\hat{S}_1\right]+\left[\hat{S}_{-1},e^{-\frac{\sigma}{S^2}\hat{S}_0}\hat{w},\hat{S}_1\right]. \tag{A.167}$$

We start with the commutation relation $[\hat{S}_0^2,\hat{w}]$ in Eq. (A.164) and its associated differential operator in configuration space which we have to determine. To accomplish this, we observe that we have from the polarization operator expansion Eq. (1.631) of the Wigner-Stratonovich kernel $\hat{w}$, the following commutation relation indicated by the Liouville term of the integrand of Eq. (3.115), viz.,

$$\left[\hat{S}_0^2,\hat{w}\right]=\sqrt{\frac{4\pi}{2S+1}}\sum_{L=0}^{2S}\sum_{M=-L}^{L}\left(C_{SSL0}^{SS}\right)^{-1}Y_{LM}^*\left[\hat{S}_0^2,\hat{T}_{LM}^{(S)}\right]. \tag{A.168}$$

However, the spin operator $\hat{S}_0^2$ must first be written in terms of the polarization operators $\hat{T}_{LM}^{(S)}$ by using Eqs. (A.146) and (A.154), namely,

$$\hat{S}_0^2=\frac{\sqrt{S(S+1)(2S+1)}}{3}\left(\frac{\sqrt{(2S-1)(2S+3)}}{\sqrt{5}}T_{20}^{(S)}+\sqrt{S(S+1)}T_{00}^{(S)}\right). \tag{A.169}$$

Now the polarization operator $T_{00}^{(S)}$ is given by Eq. (A.144) and is proportional to the unit matrix. However, the commutator of any polarization operator with the unit matrix is zero. Thus the last term in Eq. (A.169) can be discarded. Next, we can use Eq. (A.155) regarding commutators of polarization operators from Appendix A.2 to get for the following commutator

$$\left[\hat{T}_{20}^{(S)},\hat{T}_{LM}^{(S)}\right]=2\sqrt{5(2L+1)}(-1)^{2S+L+1}$$
$$\times\left(\begin{Bmatrix}2LL+1\\SSS\end{Bmatrix}C_{2,0,L,M}^{L+1,M}\hat{T}_{L+1,M}^{(S)}+\begin{Bmatrix}2LL-1\\SSS\end{Bmatrix}C_{20LM}^{L-1M}\hat{T}_{L-1M}^{(S)}\right)$$

whence

$$\left[\hat{S}_0^2, \hat{T}_{LM}^{(S)}\right] = M\sqrt{\frac{\left[(L+1)^2 - M^2\right]\left[(2S+1)^2 - (L+1)^2\right]}{(2L+3)(2L+1)}}\hat{T}_{L+1M}^{(S)} + M\sqrt{\frac{(L^2 - M^2)\left[(2S+1)^2 - L^2\right]}{(2L-1)(2L+1)}}\hat{T}_{L-1M}^{(S)}. \tag{A.170}$$

Therefore, we have from Eqs. (A.168)-(A.170) the desired commutator in terms of polarization operators, viz.,

$$\begin{aligned}\left[\hat{S}_0^2, \hat{w}\right] = &\sqrt{\frac{4\pi}{2S+1}}\sum_{L=0}^{2S}\sum_{M=-L}^{L}\left(C_{SSL0}^{SS}\right)^{-1} MY_{LM}^{*}\\ &\times\left\{\sqrt{\frac{\left[(L+1)^2 - M^2\right]\left[(2S+1)^2 - (L+1)^2\right]}{(2L+3)(2L+1)}}\hat{T}_{L+1M}^{(S)}\right.\\ &\left.+\sqrt{\frac{(L^2 - M^2)\left[(2S+1)^2 - L^2\right]}{(2L-1)(2L+1)}}\hat{T}_{L-1M}^{(S)}\right\}.\end{aligned} \tag{A.171}$$

Next by means of the replacement $L\pm 1 \to L$ in Eq. (A.171) and subsequently using the explicit expression for the Clebsch–Gordan coefficients C_{SSL0}^{SS} from Eq. (1.603), we then have

$$\begin{aligned}\int W_S\left[\hat{S}_0^2, \hat{w}\right]d\Omega = &\sqrt{\frac{4\pi}{2S+1}}\int W_S\sum_{L=0}^{2S}\left(C_{SSL0}^{SS}\right)^{-1}\sum_{M=-L}^{L}\hat{T}_{LM}^{(S)}M\\ &\times\left[L\sqrt{\frac{(L+1)^2 - M^2}{(2L+3)(2L+1)}}Y_{L+1M}^{*} - (L+1)\sqrt{\frac{L^2 - M^2}{4L^2 - 1}}Y_{L-1M}^{*}\right.\\ &\left.+2(S+1)\left(\sqrt{\frac{(L+1)^2 - M^2}{(2L+3)(2L+1)}}Y_{L+1M}^{*} + \sqrt{\frac{L^2 - M^2}{4L^2 - 1}}Y_{L-1M}^{*}\right)\right]d\Omega.\end{aligned} \tag{A.172}$$

In Eq. (A.172), the terms containing the spherical harmonics of odd rank Y_{2S+1M} are omitted because they vanish on averaging (due to the orthogonality relations and because the quasidistribution function W_S contains only the Y_{LM} up to order $L = 2S$). Next by using the recursion relations of the Y_{LM}, Eqs. (A.81) and (A.82) from Appendix A.1.7, we then have the simplified expression

$$\int W_S\left[\hat{S}_0^2,\hat{w}\right]d\Omega=\sqrt{\frac{4\pi}{2S+1}}\int W_S\sum_{L=0}^{2S}\left(C_{SSL0}^{SS}\right)^{-1}\sum_{M=-L}^{L}\hat{T}_{LM}^{(S)}M$$
$$\times\left[\sin\vartheta\frac{\partial}{\partial\vartheta}Y_{LM}^{*}+2(S+1)\cos\vartheta Y_{LM}^{*}\right]d\Omega. \tag{A.173}$$

Thus, via the general prescription, Eq. (1.588), we obtain by inspection the closed form

$$\int W_S\left[\hat{S}_0^2,\hat{w}\right]d\Omega=i\int W_S\left[\sin\vartheta\frac{\partial}{\partial\vartheta}+2(S+1)\cos\vartheta\right]\frac{\partial}{\partial\varphi}\hat{w}d\Omega, \tag{A.174}$$

i.e., we have found in the *configuration representation* (ϑ,φ) that the analog of the commutator $[\hat{S}_0^2,\hat{w}]$ is the differential operator

$$i\left[\sin\vartheta\partial_\vartheta+2(S+1)\cos\vartheta\right]\partial_\varphi\hat{w}.$$

Finally, using integration by parts in Eq. (A.174) in order to render it in the standard form of an inverse Wigner-Stratonovich transformation Eq. (1.626), we have the desired transformation via the inverse map of a Weyl symbol

$$\int W_S\left[\hat{S}_0^2,\hat{w}\right]d\Omega=i\int\hat{w}\left[\frac{1}{\sin\vartheta}\frac{\partial}{\partial\vartheta}\sin^2\vartheta-2(S+1)\cos\vartheta\right]\frac{\partial W_S}{\partial\varphi}d\Omega. \tag{A.175}$$

The above derivation has been given in detail merely as an illustration of how the inverse Wigner-Stratonovich map Eq. (1.626) ultimately leads to the phase-space representation of the density matrix evolution equation via the integrand of Eq. (A.175) and associated equations, which follow.

Next we have the following commutation relation indicated by the second Liouville term $\left[\hat{S}_0,\hat{w}_S\right]$ in Eq. (A.164), viz.,

$$\begin{aligned}\left[\hat{S}_0,\hat{w}\right]&=\sqrt{\frac{4\pi}{2S+1}}\sum_{L=0}^{2S}\sum_{M=-L}^{L}\left(C_{SSL0}^{SS}\right)^{-1}Y_{LM}^{*}\left[\hat{S}_0,\hat{T}_{LM}^{(S)}\right]\\&=\sqrt{\frac{4\pi}{2S+1}}\sum_{L=0}^{2S}\sum_{M=-L}^{L}M\left(C_{SSL0}^{SS}\right)^{-1}Y_{LM}^{*}\hat{T}_{LM}^{(S)}\\&=i\sqrt{\frac{4\pi}{2S+1}}\sum_{L=0}^{2S}\sum_{M=-L}^{L}\left(C_{SSL0}^{SS}\right)^{-1}\frac{\partial Y_{LM}^{*}}{\partial\varphi}\hat{T}_{LM}^{(S)}.\end{aligned} \tag{A.176}$$

Here have we used Eqs. (1.631), (A.127), (A.147), (A.155), (A.73), and (1.588). Thus we have from Eq. (A.176)

$$\left[\hat{S}_0, \hat{w}\right] = i\frac{\partial \hat{w}}{\partial \varphi}. \tag{A.177}$$

Equations (A.176) and (A.177) (again via integration by parts) then yield the Liouville (deterministic) part of the master equation for W_S via the inverse map of a Weyl symbol

$$\begin{aligned}&\int W_S\left\{\frac{\sigma}{S^2}\left[\hat{S}_0^2, \hat{w}\right] + \frac{\xi}{S}\left[\hat{S}_0, \hat{w}\right]\right\} d\Omega \\ &= i\frac{\sigma}{S^2}\int \hat{w}\left(2S\cos\vartheta - \sin\vartheta\frac{\partial}{\partial\vartheta} - \frac{S\xi}{\sigma}\right)\frac{\partial W_S}{\partial\varphi} d\Omega.\end{aligned} \tag{A.178}$$

Next, we consider the commutator $\left[\hat{S}_0\hat{w}, \hat{S}_0\right] + \left[\hat{S}_0, \hat{w}\hat{S}_0\right]$ and determine its phase-space representation. We have

$$\begin{aligned}\left[\hat{S}_0\hat{w}, \hat{S}_0\right] + \left[\hat{S}_0, \hat{w}\hat{S}_0\right] &= -\sqrt{\frac{4\pi}{2S+1}}\sum_{L=0}^{2S}\sum_{M=-L}^{L} M^2 \left(C_{SSL0}^{SS}\right)^{-1} Y_{LM}^{*}\hat{T}_{LM}^{(S)} \\ &= \sqrt{\frac{4\pi}{2S+1}}\sum_{L=0}^{2S}\sum_{M=-L}^{L} \left(C_{SSL0}^{SS}\right)^{-1} \frac{\partial^2 Y_{LM}^{*}}{\partial\varphi^2}\hat{T}_{LM}^{(S)} = \frac{\partial^2 \hat{w}}{\partial\varphi^2}\end{aligned} \tag{A.179}$$

so that integrating by parts, we again obtain the standard form of the inverse map of a Weyl symbol, viz.,

$$\int W_S\left[\hat{S}_0\hat{w}, \hat{S}_0\right] + \left[\hat{S}_0, \hat{w}\hat{S}_0\right] d\Omega = \int \hat{w}\frac{\partial^2 W_S}{\partial\varphi^2} d\Omega. \tag{A.180}$$

Now we consider the remaining commutators (which are more difficult to treat), namely, in the collision operator $\mathrm{St}(\hat{w})$

$$\begin{aligned}&e^{\frac{\sigma}{2S^2}\pm\frac{\xi}{2S}}\left(\left[\hat{S}_{\pm1}e^{\pm\frac{\sigma}{S^2}\hat{S}_0}\hat{w}, \hat{S}_{\mp1}\right] + \left[\hat{S}_{\pm1}\hat{w}e^{\pm\frac{\sigma}{S^2}\hat{S}_0}\hat{S}_{\mp1}\right]\right) \\ &= \left[\hat{S}_{\pm1}\hat{P}_{\pm}^{(S)}\hat{w}, \hat{S}_{\mp1}\right] + \left[\hat{S}_{\pm1}, \hat{w}\hat{P}_{\pm}^{(S)}\hat{S}_{\mp1}\right],\end{aligned} \tag{A.181}$$

where for convenience we have introduced the matrix exponential operators $\hat{P}_{\pm}^{(S)} = e^{\frac{\sigma}{2S^2}\pm\frac{\xi}{2S}}e^{\pm\frac{\sigma}{S^2}\hat{S}_0}$ in Eq. (A.181). However, they too can be expanded as a series of the polarization operators (see Eq. (A.160) *et seq.*) so that

$$\hat{P}_{\pm}^{(S)} = e^{\frac{\sigma}{2S^2} \pm \frac{\xi}{2S}} \sum_{l=0}^{2S} a_l^{\pm} \hat{T}_{l0}^{(S)}, \tag{A.182}$$

where the scalar expansion coefficients $a_l^{\pm}$ can be found using the orthogonality property of the polarization operators, Eq. (A.160), and their matrix elements in terms of the Clebsch–Gordan coefficients, Eq. (A.143). The expansion coefficients are then as usual given by the trace

$$a_l^{\pm} = \mathrm{Tr}\left(e^{\pm\frac{\sigma}{S^2}\hat{S}_0} \hat{T}_{l0}^{(S)} \right) = \sqrt{\frac{2l+1}{2S+1}} \sum_{m=-S}^{S} C_{Sml0}^{Sm} e^{\pm\frac{\sigma}{S^2}m}. \tag{A.183}$$

However, Eq. (A.183) may be further simplified for $l = 0$ using $C_{Sm00}^{Sm} = 1$ to yield the closed form

$$a_0^{\pm} = \sqrt{\frac{2l+1}{2S+1}} \sum_{m=-S}^{S} e^{\pm\frac{\sigma}{S^2}m} = \frac{\sinh[\sigma S^{-2}(S+1/2)]}{\sqrt{2S+1}\sinh(\sigma S^{-2}/2)}. \tag{A.184}$$

All the higher order expansion coefficients may now be found because differential-recurrence relations which allow one to determine $a_l^{\pm}$ may be derived as follows. We first use Eq. (A.154) concerning products of polarization operators to write

$$\begin{aligned} \hat{T}_{10}^{(S)} \hat{T}_{l-10}^{(S)} &= \frac{l\sqrt{3(2S-l+1)(2S+l+1)}}{2\sqrt{(2l+1)(2l-1)S(S+1)(2S+1)}} \hat{T}_{l0}^{(S)} \\ &+ \frac{(l-1)\sqrt{3(2S-l+2)(2S+l)}}{2\sqrt{(2l-3)(2l-1)S(S+1)(2S+1)}} \hat{T}_{l-20}^{(S)}. \end{aligned} \tag{A.185}$$

Next by substituting $\hat{T}_{l0}^{(S)}$ as extracted from Eq. (A.185) into the left hand side of Eq. (A.183), we have

$$\begin{aligned} a_l^{\pm} = \mathrm{Tr}\left(e^{\pm\frac{\sigma}{S^2}\hat{S}_0} \hat{T}_{l0}^{(S)} \right) &= \frac{2\sqrt{(2l+1)(2l-1)S(S+1)(2S+1)}}{l\sqrt{3(2S-l+1)(2S+l+1)}} \mathrm{Tr}\left(e^{\pm\frac{\sigma}{S^2}\hat{S}_0} \hat{T}_{10}^{(S)} \hat{T}_{l-10}^{(S)} \right) \\ &- \frac{(l-1)\sqrt{(2l+1)(2S-l+2)(2S+l)}}{l\sqrt{(2l-3)(2S-l+1)(2S+l+1)}} \mathrm{Tr}\left(e^{\pm\frac{\sigma}{S^2}\hat{S}_0} \hat{T}_{l-20}^{(S)} \right). \end{aligned} \tag{A.186}$$

Since

$$e^{\pm\frac{\sigma}{S^2}\hat{S}_0}\hat{T}_{10}^{(S)}\hat{T}_{l-10}^{(S)} = \pm S\sqrt{\frac{3S}{(S+1)(2S+1)}}\frac{\partial}{\partial\sigma}\left(e^{\pm\frac{\sigma}{S^2}\hat{S}_0}\hat{T}_{l-10}^{(S)}\right), \quad \text{(A.187)}$$

therefore, we have the desired differential-recurrence relation for the expansion coefficients, viz.,

$$\begin{aligned} a_l^+ = &\frac{2S^2}{l}\sqrt{\frac{4l^2-1}{(2S+1)^2-l^2}}\frac{\partial a_{l-1}^+}{\partial\sigma} \\ &+\frac{l-1}{l}\sqrt{\frac{(2l+1)[(2S+1)^2-(l-1)^2]}{(2l-3)[(2S+1)^2-l^2]}}a_{l-2}^+ \end{aligned} \quad \text{(A.188)}$$

with

$$a_l^+ = (-1)^l a_l^- . \quad \text{(A.189)}$$

Next mindful of the matrix exponential operator $e^{\pm\sigma S^{-2}\hat{S}_0}$ embodied in Eq. (A.182) and prompted by that equation we may regard the polarization operator $\hat{T}_{l0}^{(S)}$ as a (matrix) operator acting on the transformation kernel $\hat{w}$ and consequently may denote the corresponding differential operator by $P_l^{(S)}$ and its associated form $\overline{P}_l^{(S)}$. Both of these differential operators are defined in obvious fashion via

$$\int W_S\hat{T}_{l,0}^{(S)}\hat{w}d\Omega = \int W_S P_l^{(S)}\hat{w}d\Omega = \int \hat{w}\overline{P}_l^{(S)}W_S d\Omega . \quad \text{(A.190)}$$

The last term in Eq. (A.190), which serves to define the differential operator $\overline{P}_l^{(S)}$, has now the desired form, viz., an inverse Wigner-Stratonovich map. In Eq. (A.190), the differential operator $\overline{P}_l^{(S)}$ is obtained from the differential operator $P_l^{(S)}$ using integration by parts in the middle term. The operator expansion defined by Eq. (A.182) involving $\hat{P}_\pm^{(S)}$ now allows one to express both of the phase-space differential operators $P_\pm^{(S)}$ and $\overline{P}_\pm^{(S)}$ (as suggested by that equation) in terms of the phase-space differential operators $P_l^{(S)}$ and $\overline{P}_l^{(S)}$ [as defined in Eq. (A.190)] corresponding to the polarization operators $\hat{T}_{l,0}^{(S)}$ as

$$\begin{pmatrix} P_\pm^{(S)} \\ \overline{P}_\pm^{(S)} \end{pmatrix} = e^{\frac{\sigma}{2S^2}\pm\frac{\xi}{2S}}\sum_{l=0}^{2S} a_l^\pm \begin{pmatrix} P_l^{(S)} \\ \overline{P}_l^{(S)} \end{pmatrix} \quad \text{(A.191)}$$

with $a_l^{\pm}$ defined by the recurrence relations Eqs. (A.188) and (A.189). The differential operators $P_l^{(S)}$ and $\bar{P}_l^{(S)}$ on the right-hand side of Eq. (A.191) may now be determined by upward iteration. In order to find them explicitly we first recall the expression for the product of the polarization operators embodied in Eq. (A.185), which may be rearranged as the upward operator recurrence equation

$$\hat{T}_{l0}^{(S)} = A_{l-1}\hat{T}_{10}^{(S)}\hat{T}_{l-10}^{(S)} - A_{l-2}\hat{T}_{l-20}^{(S)}, \tag{A.192}$$

with coefficients A_{l-1} and A_{l-2} given by

$$\begin{aligned} A_{l-1} &= \frac{2}{l}\sqrt{\frac{S(S+1)(2S+1)(4l^2-1)}{3[(2S+1)^2-l^2]}}, \\ A_{l-2} &= \frac{l-1}{l}\sqrt{\frac{(2l+1)[(2S+1)^2-(l-1)^2]}{(2l-3)[(2S+1)^2-l^2]}}. \end{aligned} \tag{A.193}$$

However, by virtue of the correspondence expressed in Eq. (A.190) the phase-space differential operators $P_l^{(S)}$ and $\bar{P}_l^{(S)}$ must also satisfy a similar recurrence equation, viz.,

$$\begin{pmatrix} P_l^{(S)} \\ \bar{P}_l^{(S)} \end{pmatrix} = A_{l-1}\begin{pmatrix} P_{l-1}^{(S)}P_1^{(S)} \\ \bar{P}_1^{(S)}\bar{P}_{l-1}^{(S)} \end{pmatrix} - A_{l-2}\begin{pmatrix} P_{l-2}^{(S)} \\ \bar{P}_{l-2}^{(S)} \end{pmatrix} \tag{A.194}$$

so that we have explicit forms for the first two members of the hierarchy of operator recurrence relations, namely

$$P_0^{(S)} = \frac{1}{\sqrt{2S+1}}\bar{P}_0^{(S)} \tag{A.195}$$

and

$$\begin{pmatrix} P_1^{(S)} \\ \bar{P}_1^{(S)} \end{pmatrix} = \frac{1}{2}\sqrt{\frac{3}{S(S+1)(2S+1)}}\left[(2S+1\pm 1)\cos\vartheta \pm \sin\vartheta\frac{\partial}{\partial\vartheta} \pm i\frac{\partial}{\partial\varphi}\right]. \tag{A.196}$$

Thus, it is now obvious that in general *differential operators of arbitrary order will be involved.* To establish the second member Eq. (A.196), we made the following steps

$$\hat{T}_{10}^{(S)}\hat{w} = \sqrt{\frac{4\pi}{2S+1}}\sum_{L=0}^{2S}\sum_{M=-L}^{L}\left(C_{SSL0}^{SS}\right)^{-1} Y_{LM}^{*}\frac{1}{2}\sqrt{\frac{3}{S(S+1)(2S+1)}}\Bigg[M\hat{T}_{LM}^{(S)}$$
$$+\frac{\sqrt{(L-M+1)(L+M+1)(2S-L)(2S+L+2)}}{\sqrt{(2L+3)(2L+1)}}\hat{T}_{L+1M}^{(S)} \tag{A.197}$$
$$+\frac{\sqrt{(L-M)(L+M)(2S-L+1)(2S+L+1)}}{\sqrt{(2L-1)(2L+1)}}\hat{T}_{L-1M}^{(S)}\Bigg].$$

Next by the replacement $L \pm 1 \to L$ in Eq. (A.197), and then using the explicit expression for the Clebsch–Gordan coefficients C_{SSL0}^{SS} from Eq. (1.603) we have

$$\hat{T}_{10}^{(S)}\hat{w} = \sqrt{\frac{4\pi}{2S+1}}\sum_{L=0}^{2S}\sum_{M=-L}^{L}\left(C_{SSL0}^{SS}\right)^{-1}\hat{T}_{LM}^{(S)}\sqrt{\frac{3}{S(S+1)(2S+1)}}$$
$$\times\left[\frac{M}{2}Y_{LM}^{*}+\frac{2S-L+1}{2}\sqrt{\frac{(L-M)(L+M)}{(2L+1)(2L-1)}}Y_{L-1M}^{*}\right.$$
$$\left.+\frac{2S+L+2}{2}\sqrt{\frac{(L-M+1)(L+M+1)}{(2L+1)(2L+3)}}Y_{L+1M}^{*}\right] \tag{A.198}$$
$$=\sqrt{\frac{3}{S(S+1)(2S+1)}}\left[(S+1)\cos\vartheta+\frac{1}{2}\sin\vartheta\frac{\partial}{\partial\vartheta}+\frac{i}{2}\frac{\partial}{\partial\varphi}\right]\hat{w}$$
$$= P_{1}^{(S)}\hat{w}.$$

Consequently by substituting Eq. (A.198) into the defining Eq. (A.190) and integrating by parts we have Eq. (A.196) above.

Returning to the phase-space representation of the commutators in Eq. (A.181), the matrix exponential operators $\hat{P}_{\pm}^{(S)}$ act on the *polarization operators* in the expansion of transformation kernel $\hat{w}$ while its phase-space correspondents $P_{\pm}^{(S)}$ from Eq. (A.191), in contrast, represent sets of differential operators acting on the *spherical harmonics* in $\hat{w}$ leading in the end to the same result, viz.,.

$$\int W_S[\hat{S}_{\pm 1}\hat{P}_{\pm}^{(S)}\hat{w},\hat{S}_{\mp 1}]d\Omega = \int \overline{P}_{\pm}^{(S)}W_S[\hat{S}_{\pm 1}\hat{w},\hat{S}_{\mp 1}]d\Omega. \tag{A.199}$$

Now because the commutator $[\hat{S}_{\pm 1}\hat{P}_{\pm}^{(S)}\hat{w},\hat{S}_{\mp 1}]$ gives the differential operator, which is complex conjugate to that corresponding to

$[\hat{S}_{\pm1}, \hat{P}_{\pm}^{(S)}\hat{w}\hat{S}_{\mp1}]$ in the right-hand side of Eq. (A.181), we need to consider below only the commutator $[\hat{S}_{\pm1}\hat{P}_{\pm}^{(S)}\hat{w}, \hat{S}_{\mp1}]$. We have

$$[\hat{S}_{\pm1}\hat{w}, \hat{S}_{\mp1}] = \sqrt{\frac{4\pi}{2S+1}} \sum_{L=0}^{2S} \sum_{M=-L}^{L} \left(C_{SSL0}^{SS}\right)^{-1} Y_{LM}^{*}(\vartheta, \varphi) \left[\hat{S}_{\pm1}\hat{T}_{LM}^{(S)}, \hat{S}_{\mp1}\right], \quad \text{(A.200)}$$

where

$$\left[\hat{S}_{\pm1}\hat{T}_{LM}^{(S)}, \hat{S}_{\mp1}\right] = -\frac{1}{4}\left[(L \mp M)(L \pm M + 1)\hat{T}_{LM}^{(S)}\right.$$
$$\mp \frac{(L \pm M + 2)\sqrt{(L+M+1)(L-M+1)(2S-L)(2S+L+2)}}{\sqrt{(2L+3)(2L+1)}}\hat{T}_{L+1M}^{(S)} \quad \text{(A.201)}$$
$$\left.\pm \frac{(L \mp M - 1)\sqrt{(L+M)(L-M)(2S-L+1)(2S+L+1)}}{\sqrt{(2L-1)(2L+1)}}\hat{T}_{L-1M}^{(S)}\right].$$

Next via the replacement $L \pm 1 \to L$ in Eq. (A.171) and then using the explicit expression for the Clebsch–Gordan coefficients C_{SSL0}^{SS} from Eq. (1.603), we have

$$[\hat{S}_{\pm1}\hat{w}, \hat{S}_{\mp1}] = -\sqrt{\frac{\pi}{4(2S+1)}}$$
$$\times \sum_{L=0}^{2S} \sum_{M=-L}^{L} \left(C_{SSL0}^{SS}\right)^{-1} \hat{T}_{LM}^{(S)} \left[(L \mp M)(L \pm M + 1)Y_{LM}^{*}\right.$$
$$\mp \frac{(2S+L+2)(L \pm M + 1)\sqrt{(L+M+1)(L-M+1)}}{\sqrt{(2L-1)(2L+1)}}Y_{L-1M}^{*} \quad \text{(A.202)}$$
$$\left.\pm \frac{(2S-L+1)(L \mp M)\sqrt{(L+M)(L-M)}}{\sqrt{(2L+3)(2L+1)}}Y_{L+1M}^{*}\right].$$

We can ultimately write Eq. (A.202) in differential form using properties of the angular momentum operators $\hat{L}_{\mu}$ and $\hat{\mathbf{L}}^2$ (see Appendix A.1.7), viz.,

$$\hat{\mathbf{L}}^2 Y_{LM}^{*} = L(L+1)Y_{LM}^{*}, \quad \text{(A.203)}$$

$$\hat{L}_0 Y_{LM}^{*} = -MY_{LM}^{*}, \quad \text{(A.204)}$$

$$\hat{L}_{\pm 1} Y_{LM}^* = \pm\sqrt{\frac{L(L+1)-M(M\mp 1)}{2}} Y_{LM\mp 1}^*, \tag{A.205}$$

$$\hat{L}_{\pm 1}\hat{L}_{\mp 1} Y_{LM}^* = -\frac{1}{2}(L\mp M)(L\pm M+1) Y_{LM}^*. \tag{A.206}$$

Thus we can rearrange the commutator given by Eq. (A.200) as the differential form

$$\begin{aligned}
[\hat{S}_{\pm 1}\hat{w}, \hat{S}_{\mp 1}] &= \sqrt{\frac{\pi}{2S+1}} \sum_{L=0}^{2S} \sum_{M=-L}^{L} \left(C_{SSL0}^{SS}\right)^{-1} \hat{T}_{LM}^{(S)} \\
&\times \left[\frac{1}{2}\left(\frac{\partial^2}{\partial \vartheta^2} + \cot\vartheta \frac{\partial}{\partial \vartheta} \pm i \frac{\partial}{\partial \varphi} + \cot^2\vartheta \frac{\partial^2}{\partial \varphi^2} \right) \right. \\
&+ i(S+1)\cos\vartheta \frac{\partial}{\partial \varphi} - \left(S+\frac{1}{2}\right)\sin\vartheta \frac{\partial}{\partial \vartheta} + \frac{i}{2}\sin\vartheta \frac{\partial}{\partial \vartheta}\frac{\partial}{\partial \varphi} \\
&\left. - \frac{1}{2}\cos\vartheta \left[\frac{1}{\sin\theta}\frac{\partial}{\partial \vartheta}\left(\sin\vartheta \frac{\partial}{\partial \vartheta}\right) + \frac{1}{\sin^2\vartheta}\frac{\partial^2}{\partial \varphi^2} \right] \right] Y_{LM}^*.
\end{aligned} \tag{A.207}$$

By substituting Eq. (A.207) into Eq. (A.199) and integrating the latter by parts, we finally have that equation rendered as *the inverse map of a Weyl symbol*, viz.,

$$\begin{aligned}
&\int W_S[\hat{S}_{\pm 1}\hat{P}_{\pm}^{(S)}\hat{w}, \hat{S}_{\mp 1}] d\Omega \\
&= -\frac{1}{4}\int \hat{w} \left\{ \frac{1}{\sin\vartheta}\left[\frac{\partial}{\partial \vartheta}\left(\sin\vartheta (1\pm\cos\vartheta)\frac{\partial}{\partial \vartheta} \pm 2S\sin^2\vartheta \right) \right.\right. \\
&\left.\left. + \cot\vartheta\left(\cos\vartheta \pm 1\right)\frac{\partial^2}{\partial \varphi^2} \right] + i\left(\sin\vartheta \frac{\partial}{\partial \vartheta} - 2S\cos\vartheta \mp 1 \right)\frac{\partial}{\partial \varphi} \right\} \bar{P}_{\pm}^{(S)} W_S d\Omega.
\end{aligned} \tag{A.208}$$

In the derivation of Eq. (A.208), the recurrence properties of the spherical harmonics, Eqs. (A.80)–(A.83), have been used. However, because the commutator $[\hat{S}_{\pm 1}\hat{P}_{\pm}^{(S)}\hat{w}, \hat{S}_{\mp 1}]$ yields the differential operator complex conjugate to that corresponding to $[\hat{S}_{\pm 1}, \hat{P}_{\pm}^{(S)}\hat{w}\hat{S}_{\mp 1}]$, we have once again the standard inverse Wigner–Stratonovich map of a Weyl symbol via

$$\int W_S\left([\hat{S}_{\pm 1}\hat{P}_{\pm}^{(S)}\hat{w},\hat{S}_{\mp 1}]+[\hat{S}_{\pm 1},\hat{w}\hat{P}_{\pm}^{(S)}\hat{S}_{\mp 1}]\right)d\Omega$$

$$=e^{\frac{\sigma}{2S^2}\pm\frac{\xi}{2S}}\int W_S\left([\hat{S}_{\pm 1}e^{\pm\sigma S^{-2}\hat{S}_0}\hat{w},\hat{S}_{\mp 1}]+[\hat{S}_{\pm 1},\hat{w}e^{\pm\sigma S^{-2}\hat{S}_0}\hat{S}_{\mp 1}]\right)d\Omega$$

$$=-\frac{1}{2}\int\hat{w}\,\mathrm{Re}\left\{\left[\frac{1}{\sin\vartheta}\frac{\partial}{\partial\vartheta}\left(\sin\vartheta(1\pm\cos\vartheta)\frac{\partial}{\partial\vartheta}\pm 2S\sin^2\vartheta\right)\right.\right. \quad \text{(A.209)}$$

$$+\frac{1}{\sin\vartheta}\cot\vartheta(\cos\vartheta\pm 1)\frac{\partial^2}{\partial\varphi^2}$$

$$\left.\left.+i\left(\sin\vartheta\frac{\partial}{\partial\vartheta}-2S\cos\vartheta\mp 1\right)\frac{\partial}{\partial\varphi}\right]\overline{P}_{\pm}^{(S)}\right\}W_S d\Omega.$$

Then due to Eq. (3.91) of Chapter 3, we have from the Weyl symbols of Eqs. (A.178), (A.180), and (A.209) the master equation, Eq. (3.92), for the phase-space distribution $W_S(\vartheta,\varphi,t)$, viz.,

$$\frac{\partial W_S}{\partial t}-\frac{\sigma}{\hbar\beta S^2}\left(2S\cos\vartheta-\sin\vartheta\frac{\partial}{\partial\vartheta}-\frac{S\xi}{\sigma}\right)\frac{\partial W_S}{\partial\varphi}=\mathrm{St}\{W_S\}, \quad \text{(A.210)}$$

where the collision kernel $\mathrm{St}\{W_S\}$ is of the form

$$\mathrm{St}\{W_S\}=D_{\|}\frac{\partial^2}{\partial^2\varphi}W_S+D_{\perp}\frac{\cot\vartheta}{\sin\vartheta}\left(\cos\vartheta\frac{\partial^2}{\partial\varphi^2}R_{+}^{\prime(S)}+\frac{\partial^2}{\partial\varphi^2}R_{-}^{\prime(S)}\right)W_S$$

$$+\frac{D_{\perp}}{\sin\vartheta}\left[\frac{\partial}{\partial\vartheta}\sin\vartheta\left(\frac{\partial}{\partial\vartheta}R_{+}^{\prime(S)}+\cos\vartheta\frac{\partial}{\partial\vartheta}R_{-}^{\prime(S)}+2S\sin\vartheta R_{-}^{\prime(S)}\right)\right]W_S \quad \text{(A.211)}$$

$$+D_{\perp}\left[\left(\sin\vartheta\frac{\partial}{\partial\vartheta}-2S\cos\vartheta\right)\frac{\partial}{\partial\varphi}R_{+}^{\prime\prime(S)}-\frac{\partial}{\partial\varphi}R_{-}^{\prime\prime(S)}\right]W_S$$

with the phase-space differential operators $R_{\pm}^{(S)}$ (generally involving differential operators of *arbitrary* order [cf. Eq. (A.191)]) defined as

$$R_{\pm}^{(S)}=R_{\pm}^{\prime(S)}+iR_{\pm}^{\prime\prime(S)}=\overline{P}_{+}^{(S)}\pm\overline{P}_{-}^{(S)}. \quad \text{(A.212)}$$

In the classical limit, we see that since all the terms involving derivatives in the differential operator expansion [Eq. (A.191) *et seq.*] now vanish, the operator $\overline{P}_l^{(S)}$ simply reduces to the Legendre polynomial of order l

$$\overline{P}_l^{(S>>1)} \to Y_{l0}(\vartheta,\varphi)\sqrt{\frac{4\pi}{2S+1}}. \tag{A.213}$$

Thus in this limit by substituting the Legendre polynomials into the operator series Eq. (A.191), we must then have

$$\begin{aligned}\overline{P}_{\pm}^{(S\to\infty)} &= e^{\frac{\sigma}{2S^2}\pm\frac{\xi}{2S}}\sqrt{\frac{4\pi}{2S+1}}\sum_{l=0}^{2S} a_l^{\pm} Y_{l0}(\vartheta,\varphi)\\ &= e^{\frac{\sigma}{2S^2}\pm\frac{\xi}{2S}} F_{\pm}^{(0)}(\vartheta,\varphi),\end{aligned}$$

in which the series

$$F_{\pm}^{(s)}(\vartheta,\varphi) = \sqrt{\frac{4\pi}{2S+1}}\sum_{l=0}^{2S}\left(C_{SSL0}^{SS}\right)^{-s} a_l^{\pm} Y_{l0}(\vartheta,\varphi)$$

represents the Weyl symbol of the Hilbert space operator $e^{\pm\sigma\hat{S}_0/S^2}$. For large S,

$$F_{\pm}^{(s)}(\vartheta,\varphi) \to e^{\pm(\sigma/S)\cos\vartheta}$$

so that proceeding to the limit $S\to\infty$, we have

$$\lim_{S\to\infty} S\left(e^{[\sigma\cos\vartheta+(\sigma/S+\xi)/2]/S} - e^{[-\sigma\cos\vartheta+(\sigma/S-\xi)/2]/S}\right) = \xi + 2\sigma\cos\vartheta$$

and

$$\lim_{S\to\infty} e^{(\xi+\sigma/S+2\sigma\cos\vartheta)/S} = 1. \tag{A.214}$$

Hence it follows that in the classical limit, $S\to\infty$, Eq. (A.210) reduces to the corresponding classical Fokker–Planck equation, Eq. (1.64) for isotropic rotational diffusion of a magnetic dipole in the uniaxial potential [97] (if $D_\perp = D_\|$).

The master equation (A.210) for the evolution of the phase-space quasi-probability distribution $W_S(\vartheta,\varphi,t)$ for a uniaxial spin system in contact with a heat bath at temperature T was derived in the weak coupling limit, i.e., it was supposed that the correlation time characterizing the bath is so short that the stochastic process originating from it is Markovian so that one may assume frequency independent damping. This has been accomplished by expressing the reduced density matrix master equation (3.84) in terms of the inverse Wigner–Stratonovich transformation. In order to achieve this objective, various commutators involving the spin operators occurring in the integrand of Eq. (3.89) have then been evaluated by means of the orthogonality and recurrence properties of the polarization operators and the corresponding spherical harmonics to yield their analogs in phase space. Thus, we have expressed the master equation as a partial differential equation for the distribution function in the phase space of the polar angles. Despite the superficial resemblance of the quantum diffusion Eq. (A.210) to the corresponding classical Fokker-Planck equation for a classical spin in a uniaxial potential Eq. (A.214), it is in reality much more complicated. The difficulties arise because Eq. (A.210) involves the complicated differential operators $R_{\pm}^{(S)}$ only simplifying for large spin numbers ($S\to\infty$) when the higher order derivatives, as indicated by Eq. (A.212), occurring in the operators $R_{\pm}^{(S)}$ may be ignored. Consequently, it is usually much easier to use the density matrix formulation where only the diagonal terms partake in the time evolution.

A.4. Characteristic times of relaxation and correlation functions

We have seen by solving the differential-recurrence equations for the statistical moments, how one can evaluate the characteristic times of the relaxation and/or correlation functions $C_i(t)$ ($i=X, Y, Z$) of the longitudinal and transverse components of spin operators (or vectors in the classical case). To characterize the overall time behavior of $C_i(t)$, we may formally introduce (see Ref. [97]) the integral relaxation time τ_{int}^i, viz.,

$$\tau_{\text{int}}^i = \frac{1}{C_i(0)}\int_0^\infty C_i(t)dt, \tag{A.215}$$

which is just the area under the decay curve of $C_i(t)$. Yet another time constant characterizing the time behavior of $C_i(t)$ is the effective relaxation time τ_{ef}^i defined by

$$\tau_{\text{ef}}^i = -\frac{C_i(0)}{\dot{C}_i(0)} \tag{A.216}$$

(yielding precise information on the initial decay of $C_i(t)$ in the time domain). For spin systems with dynamics governed by Fokker-Planck equations, the times τ^i_{int} and τ^i_{ef} may equivalently be defined using the eigenvalues (λ^i_k) of the Fokker–Planck operator because the normalized relaxation function $C_i(t)/C_i(0)$ may formally be written as (Ref. [97], Chapter 2)

$$\frac{C_i(t)}{C_i(0)} = \sum_k c^i_k e^{-\lambda^i_k t}, \tag{A.217}$$

so that, from Eqs. (A.215), (A.216) and(A.217), we formally have

$$\tau^i_{\mathrm{int}} = \sum_k c^i_k / \lambda^i_k \tag{A.218}$$

and

$$\tau^i_{\mathrm{ef}} = \sum_k \lambda_k c^i_k . \tag{A.219}$$

The relaxation times τ^i_{int} and τ^i_{ef} each contain contributions from *all* the eigenvalues λ^i_k. Therefore in general, in order to evaluate both $C_i(t)$, τ^i_{int}, and τ^i_{ef} numerically, a knowledge of each individual λ_k and c^i_k is required. However, in the low temperature (high barrier) limit, for the *longitudinal* relaxation of the magnetization, $\lambda^Z_1 << \left|\lambda^Z_k\right|$ and $c^Z_1 \approx 1 >> c^Z_k$ $(k \neq 1)$ provided the wells of the potential remain equivalent or nearly equivalent, the approximation $\tau^Z_{\mathrm{int}} \approx 1/\lambda^Z_1$ is valid. In other words, the inverse of the smallest nonvanishing eigenvalue λ^Z_1 closely approximates the longitudinal relaxation time τ^Z_{int} in the low temperature limit for zero or very weak external fields. Furthermore, in the *longitudinal* relaxation of the magnetization, the smallest nonvanishing eigenvalue(s) λ^Z_1 of the Fokker–Planck operator characterizes the long-time behavior of

$$\left\langle \hat{S}_Z \right\rangle(t) - \left\langle \hat{S}_Z \right\rangle_{eq} \sim C_{\|}(t) \sim e^{-\lambda^Z_1 t} = e^{-t/\tau}, \quad t >> \tau. \tag{A.220}$$

Thus, it may be associated with the longest relaxation (reversal) time of the magnetization. In order to evaluate the reversal time τ numerically, we note that the recurrence equations for the statistical moments may always be written in matrix form as

$$\dot{\mathbf{X}}(t) = \mathbf{A}\mathbf{X}(t), \tag{A.221}$$

where **A** is the system matrix and $\mathbf{X}(t)$ is an infinite column vector formed from the statistical moments. The τ may then be determined from the smallest nonvanishing root of the characteristic equation

$$\det(\lambda \mathbf{I} - \mathbf{A}) = 0 \tag{A.222}$$

by selecting a sufficiently large number of equations. The general solution of Eq. (A.221) is determined by successively increasing the size of **A** until convergence is attained.

The effective relaxation time τ_{ef}^{i} can also be expressed in terms of equilibrium averages because both $C_i(0)$ and $\dot{C}_i(0)$ in Eq. (A.216) are given in terms of those averages. For example, for the uniaxial potential V given by Eq. (2.1) of Chapter 2, the effective longitudinal relaxation time $\tau_{\mathrm{ef}}^{\|}$ is expressed via Eq. (2.183) of Chapter 2 in terms of the the equilibrium averages of the Legendre polynomials $\langle P_1 \rangle_0$ and $\langle P_2 \rangle_0$, which are given by Eqs. (A.48)–(A.50) in Appendix A.1.5. The integral relaxation times τ_{int}^{i} can also be calculated via the one-sided Fourier transform of the appropriate correlation function $\tilde{C}_i(i\omega)$ as

$$\tau_{\mathrm{int}}^{i} = \tilde{C}_i(0) / C_i(0).$$

Here we may evaluate the reversal time τ via the one-sided Fourier transform of the correlation function $\tilde{C}_Z(i\omega)$ as follows. We consider the long-time behavior of $C_Z(t)$ which is dominated by an exponential, viz.,

$$C_Z(t) \approx C_0 e^{-t/\tau}. \tag{A.223}$$

Then the longest relaxation time τ can then be extracted from $\tilde{C}_Z(-i\omega)$ (by eliminating C_0) as [97]

$$\tau = \lim_{\omega \to 0} \frac{\tilde{C}_Z(i\omega) - C_Z(0)}{i\omega \tilde{C}_Z(i\omega)}. \tag{A.224}$$

In practical applications, such as to magnetization reversal, matrix continued fractions due to their rapid convergence are much better suited to numerical calculations than standard direct matrix inversion based on the matrix representation, Eq. (A.221), of the infinite system of linear differential-recurrence relations for the averaged spherical harmonics.

In the general case, the integral relaxation time can only be evaluated numerically. However, for systems with dynamics governed by single-variable Fokker–Planck equations it can be calculated analytically for both linear and nonlinear transient responses.

Here, we first derive following Ref. [97], an exact analytic equation for the *nonlinear transient response* relaxation time of a system governed by a one-dimensional Fokker–Planck equation for the probability distribution function $W(z,t)$ of a single variable z, viz.,

$$\frac{\partial W}{\partial t} = \mathrm{L}_{\mathrm{FP}} W, \tag{A.225}$$

where the Fokker–Planck operator L_{FP} may be represented as [97,143]

$$\mathrm{L}_{\mathrm{FP}} W = \frac{\partial}{\partial z}\left[D_2(z)\left(\frac{\partial W}{\partial z} + D_1(z)W\right)\right]. \tag{A.226}$$

Here $D_1(z)$ and $D_2(z)$ are the coordinate dependent coefficients and z is defined in the range $-1 \le z \le 1$). Moreover we assume that the relaxation dynamics of spins obey the single-variable Fokker–Planck equation, Eq. (A.225). Suppose that at time $t=0$, the external field H is suddenly altered from H_{I} to H_{II} (see Fig. 3.12). We are interested in the relaxation of the system starting from an equilibrium (stationary) state I with the distribution function $W_{\mathrm{I}}(z)$, which evolves under the action of the stimulus of arbitrary strength to another equilibrium (stationary) state II with the distribution function $W_{\mathrm{II}}(z)$. This problem is intrinsically nonlinear, because *changes in the magnitude of the potential are arbitrary*. Thus, the concept of relaxation functions and relaxation times must now be used, rather than correlation functions and correlation times.

Following [335], we may define the relaxation function $f_A(t)$ of a dynamical variable A by

$$f_A(t) = \begin{cases} \langle A\rangle(t) - \langle A\rangle_{\mathrm{II}}, & (t>0), \\ \langle A\rangle_{\mathrm{I}} - \langle A\rangle_{\mathrm{II}}, & (t \le 0), \end{cases} \tag{A.227}$$

where $\langle A\rangle(t)$ is the time-dependent average and $\langle A\rangle_{\mathrm{I}}$ and $\langle A\rangle_{\mathrm{II}}$ are equilibrium (stationary) averages defined as

$$\langle A\rangle(t)=\int_{-1}^{1}A(z)W(z,t)dz, \tag{A.228}$$

$$\langle A\rangle_i=\int_{-1}^{1}A(z)W_i(z)dz\ \ (i=\mathrm{I},\mathrm{II}). \tag{A.229}$$

Our goal is to evaluate the integral relaxation time τ_{int} of $f_A(t)$, which is defined as

$$\begin{aligned}\tau_{\mathrm{int}}&=\frac{1}{f_A(0)}\int_0^\infty f_A(t)dt\\&=\frac{1}{f_A(0)}\lim_{s\to 0}\int_0^\infty e^{-st}f_A(t)dt=\frac{\tilde{f}_A(0)}{f_A(0)},\end{aligned} \tag{A.230}$$

where $\tilde{f}_A(s)$ is the Laplace transform of $f_A(t)$. The relaxation time, Eq. (A.230), may be written as

$$\tau_{\mathrm{int}}=\frac{1}{\langle A\rangle_{\mathrm{I}}-\langle A\rangle_{\mathrm{II}}}\int_{-1}^{1}[A(z)-\langle A\rangle_{\mathrm{II}}]\tilde{W}(z,0)dz, \tag{A.231}$$

where

$$\tilde{W}(z,0)=\lim_{s\to 0}\tilde{W}(z,s)$$

and

$$\tilde{W}(z,s)=\int_0^\infty W(z,t)e^{-st}dt.$$

The function $\tilde{W}(z,0)$ in Eq. (A.225) can be calculated analytically via the final-value theorem of Laplace transformation [290], namely,

$$\lim_{s\to 0}s\tilde{W}(z,s)=\lim_{t\to\infty}W(z,t)=W_{\mathrm{II}}(z).$$

Thus, we obtain, from the Fokker–Planck equation, Eq. (A.225), for $t>0$, the ordinary differential equation

$$W_{\mathrm{II}}(z)-W_{\mathrm{I}}(z)=\frac{d}{dz}\left[D_2(z)\left(\frac{d}{dz}\tilde{W}(z,0)+D_1(z)\tilde{W}(z,0)\right)\right]. \quad \text{(A.232)}$$

The particular solution of Eq. (A.232) is

$$\tilde{W}(z,0)=W_{\mathrm{II}}(z)\int_{-1}^{1}\frac{\Phi(y)dy}{D_2(y)W_{\mathrm{II}}(y)}, \quad \text{(A.233)}$$

where

$$\Phi(y)=\int_{-1}^{y}[W_{\mathrm{II}}(z)-W_{\mathrm{I}}(z)]dz \quad \text{(A.234)}$$

and $W_{\mathrm{II}}(z)$ is the stationary solution of the equation

$$\frac{d}{dz}\left[D_2(z)\left(\frac{d}{dz}W_{\mathrm{II}}(z)+D_1(z)W_{\mathrm{II}}(z)\right)\right]=0. \quad \text{(A.235)}$$

Hence, using the definitions Eq. (A.231) and Eq. (A.233), we have

$$\tau_{\mathrm{int}}=\frac{1}{\langle A\rangle_{\mathrm{I}}-\langle A\rangle_{\mathrm{II}}}\int_{-1}^{1}[A(x)-\langle A\rangle_{\mathrm{II}}]W_{\mathrm{II}}(x)\int_{-1}^{x}\frac{\Phi(y)dy}{D_2(y)W_{\mathrm{II}}(y)}dx$$

or, by integration by parts, [335]

$$\tau_{\mathrm{int}}=\frac{1}{\langle A\rangle_{\mathrm{II}}-\langle A\rangle_{\mathrm{I}}}\int_{-1}^{1}\frac{\Phi(x)\Psi(x)}{D_2(x)W_{\mathrm{II}}(x)}dx, \quad \text{(A.236)}$$

where for convenience we have written

$$\Psi(x)=\int_{x_{-1}}^{x}[A(y)-\langle A\rangle_{\mathrm{II}}]W_{\mathrm{II}}(y)dy. \quad \text{(A.237)}$$

Equation (A.236) is an *exact* equation for the *nonlinear transient response relaxation time*. Examples of applications of Eq. (A.236) to nonlinear response problems have been given in Chapter 3, Sections 3.3.6 and 3.3.10 and in Ref. [97].

If we now suppose that the change in the magnitude of the external field H from H_{I} to H_{II} is very small, i.e., $|H_{\mathrm{I}}-H_{\mathrm{II}}|\to 0$, the problem becomes intrinsically linear, because *changes in the magnitude of the potential are insignificant*. Thus, linear response theory and the concept of correlation functions and correlation times rather than relaxation functions and relaxation times can now be used. The *equilibrium* (stationary) correlation function $C_{AB}(t)$ is defined by

$$\begin{aligned} C_{AB}(t) &= \langle A[z(0)]B[z(t)]\rangle_0 - \langle A\rangle_0 \langle B\rangle_0 \\ &= \int_{-1}^{1} [A(z) - \langle A\rangle_0] e^{L_{FP}t} [B(z) - \langle B\rangle_0] W_0(z) dz. \end{aligned} \tag{A.238}$$

Here W_0 is the stationary (equilibrium) distribution function satisfying $L_{FP}W_0 = 0$, the symbols $\langle\ \rangle$ and $\langle\ \rangle_0$ designate the statistical averages over W and W_0, respectively. Now, the salient feature of one-dimensional systems is that an *exact integral formula* for the correlation time τ_{cor}

$$\tau_{cor} = \frac{1}{C_{AB}(0)} \int_0^{\infty} C_{AB}(t) dt \tag{A.239}$$

(defined as the area under the curve of the normalized correlation function $C_{AB}(t)$) wexists because the relevant Fokker–Planck equation (A.225) may be integrated by quadratures as with the nonlinear responses [97,143]. The details of the derivation are given in Refs. [97,143] so we merely quote the final analytic expression for τ_{cor}, viz.,

$$\tau_{cor} = \frac{1}{C_{AB}(0)} \int_{-1}^{1} \frac{\psi_A(x)\psi_B(x) dx}{D_2(x) W_0(x)}, \tag{A.240}$$

where

$$\psi_A(x) = \int_{-1}^{x} [A(y) - \langle A\rangle_0] W_0(y) dy, \tag{A.241}$$

$$\psi_B(x) = \int_{-1}^{x} [B(y) - \langle B\rangle_0] W_0(y) dy. \tag{A.242}$$

For $A = B$, Eq. (A.240) reduces to

$$\tau_{\text{cor}} = \frac{1}{\langle A^2 \rangle_0 - \langle A \rangle_0^2} \int_{-1}^{1} \frac{\psi_A^2(x)dx}{D_2(x)W_0(x)}. \tag{A.243}$$

The relaxation time in integral form, Eq. (A.243), was first given by Szabo [490], and later reproduced by many other authors under different guises [23,97,335,337,491,492].

For the uniaxial potential E given by Eq. (2.1) of Chapter 2, the longitudinal magnetization relaxation in a magnetic nanoparticle is described by the *single-variable* Fokker–Planck equation, Eq. (1.223), so that the integral relaxation time $\tau_{\text{int}}^{\|} \equiv \tau_{\text{cor}}$ is expressed in closed-integral form, Eq. (A.243), yielding Eq. (2.182) of Chapter 2.

References

1. D.C. Mattis, The Theory of Magnetism Made Simple: an Introduction to Physical Concepts and to Some Useful Mathematical Methods, London: World Scientific, 2006.
2. J. M. D. Coey, Magnetism and Magnetic Materials, Cambridge: Cambridge University Press, 2010.
3. D. C. Cassidy, Uncertainty: The Life and Science of Werner Heisenberg, New York: W.H. Freeman, 1992.
4. W. Heisenberg, Die Physikalischen Prinzipien der Quantentheorie, Wiesbaden: Wiesbadener Graphische Betriebe GmbH, 1958.
5. A. Messiah, Quantum mechanics, New York: Dover Publications, 1999.
6. E. P. Wigner, "On the quantum correction for thermodynamic equilibrium," *Phys. Rev.*, vol. 40, p. 749, 1932.
7. W.F. Brown, Jr., Micromagnetics, New York: Wiley Interscience, 1963.
8. C.P. Bean and J.D. Livingston, "Superparamagnetism," *J. Appl. Phys.*, vol. 30, p. S120, 1959.
9. Y. Liu, D.J. Sellmyer, and D. Shindo, Eds., Handbook of Advanced Magnetic Materials, vol. I and II, New York: Springer, 2006.
10. C.D. Mee, The Physics of Magnetic Recording, Amsterdam: North Holland, 1986.
11. A. P. Guimarães, Principles of Nanomagnetism, Berlin: Springer, 2009.
12. Q. A. Pankhurst, N. K. T. Thanh, S. K. Jones, et al., "Progress in applications of magnetic nanoparticles in biomedicine," *J. Phys. D: Appl. Phys.*, vol. 42, p. 224001, 2009.
13. S. Gubin, Ed., Magnetic Nanoparticles, New York: Wiley, 2009.
14. E. C. Stoner and E. P. Wohlfarth, "A mechanism of magnetic hysteresis in heterogeneous alloys," *Philos. Trans. R. Soc. London, Ser. A,* vol. 240, p. 599, 1948.
15. L. Néel, "Influence des fluctuations thermiques sur l'aimantation de grains ferromagnétiques très fins," *C. R. Acad. Sci. Paris,* vol. 228, p. 664, 1949.
16. L. Néel, "Théorie du traînage magnétique des ferromagnétiques en grains fins avec applications aux terres cuites," *Ann. Geophys.*, vol. 5, p. 99, 1949.
17. W. F. Brown, Jr., "Thermal fluctuations of a single-domain particle," *Phys. Rev.*, vol. 130, p. 1677, 1963.

18. W. F. Brown, Jr., "Thermal fluctuations of fine ferromagnetic particles," *IEEE Trans. Magn.,* vol. 15, p. 1196, 1979.
19. D.G. Rancourt, "Magnetism of Earth, planetary, and environmental nanomaterials," *Rev. Mineral. Geochem.,* vol. 44, p. 217, 2001.
20. W. Wernsdorfer, "Classical and quantum magnetization reversal studied in nanometer-sized particles and clusters," *Adv. Chem. Phys.,* vol. 118, p. 99, 2001.
21. D. Kumar and S. Dattagupta, "Non-equilibrium susceptibility of superparamagnetic particles," *J. Phys. C: Solid State Phys.,* vol. 16, p. 3779, 1983.
22. G. S. Agarwal, S. Dattagupta, and K. P. N. Murthy, "Relaxation behaviour of single-domain magnetic particles," *J. Phys. C: Solid State Phys.,* vol. 17, p. 6869, 1984.
23. D. A. Garanin, V. V. Ischenko, and L. V. Panina, "Dynamics of an ensemble of single-domain magnetic particles," *Theor. Math. Phys.,* vol. 82, p. 169, 1990.
24. L. Bessais, L. B. Jaffel, and J. L. Dormann, "Relaxation time of fine magnetic particles in uniaxial symmetry," *Phys. Rev. B,* vol. 45, p. 7805, 1992.
25. W. T. Coffey, P. J. Cregg, and Yu. P. Kalmykov, "On the theory of Debye and Néel relaxation of single domain ferromagnetic particles," *Adv. Chem. Phys.,* vol. 83, p. 263, 1993.
26. E. K. Sadykov and A. G. Isavnin, "Theory of dynamic magnetic susceptibility of uniaxial superparamagnetic particles," *Phys. Solid State,* vol. 38, p. 1160, 1996.
27. T. Bitoh, K. Ohba, M. Takamatsu, et al., "Linear and nonlinear susceptibilities of ferromagnetic fine particles in Cu97Co3 alloy," *J. Phys. Soc. Jpn.,* vol. 64, p. 1311, 1995.
28. T. Bitoh, K. Ohba, M. Takamatsu, et al., "Comparative study of linear and nonlinear susceptibilities of fine-particle and spin-glass systems: quantitative analysis based on the superparamagnetic blocking model," *J. Magn. Magn. Mater.,* vol. 154, p. 59, 1996.
29. Yu. L. Raikher and V. I. Stepanov, "Linear and cubic dynamic susceptibilities of superparamagnetic fine particles," *Phys. Rev. B,* vol. 55, p. 15005, 1997.
30. Yu. L. Raikher and V. I. Stepanov, "Linear and nonlinear superparamagnetic relaxation at high anisotropy barriers," *Phys. Rev. B,* vol. 66, p. 21440, 2002.
31. Yu. P. Kalmykov, "Longitudinal dynamic susceptibility and relaxation time of superparamagnetic particles with cubic anisotropy: Effect of a biasing magnetic field," *Phys. Rev. B,* vol. 61, p. 6205, 2000.
32. Yu. P. Kalmykov and S. V. Titov, "Complex magnetic susceptibility of uniaxial superparamagnetic particles in a strong static magnetic field," *Phys. Solid State,* vol. 40, p. 1492, 1998.
33. Yu. P. Kalmykov and S. V. Titov, "Longitudinal complex magnetic susceptibility of superparamagnetic particles with cubic anisotropy," *Phys. Solid State,* vol. 40, p. 1721, 1998.
34. Yu. P. Kalmykov and S. V. Titov, "Longitudinal dynamic susceptibility of superparamagnetic particles with cubic anisotropy," *J. Exp. Theor. Phys.,* vol. 88, p. 58, 1999.

35. Yu. P. Kalmykov, S. V. Titov, and W. T. Coffey, "Longitudinal complex magnetic susceptibility and relaxation time of superparamagnetic particles with cubic magnetic anisotropy," *Phys. Rev. B,* vol. 58, p. 3267, 1998.
36. M. Respaud, M. Goiran, J. M. Broto, et al., "Dynamical properties of non-interacting Co nanoparticles," *Europhys. Lett.,* vol. 47, p. 122, 1999.
37. J. L. García-Palacios and P. Svedlindh, "Large nonlinear dynamical response of superparamagnets: Interplay between precession and thermoactivation in the stochastic Landau-Lifshitz equation," *Phys. Rev. Lett.,* vol. 85, p. 3724, 2000.
38. W. T. Coffey, D. S. F. Crothers, Yu. P. Kalmykov,et al., "Precessional effects in the linear dynamic susceptibility of uniaxial superparamagnets: Dependence of the ac response on the dissipation parameter," *Phys. Rev. B,* vol. 64, p. 012411, 2001.
39. Yu. L. Raikher and V. I. Stepanov, "Nonlinear dynamic susceptibilities and field-induced birefringence in magnetic particle assemblies," *Adv. Chem. Phys.,* vol. 129, p. 419, 2004.
40. P. E. Jonsson, "Superparamagnetism and spin glass dynamics of interacting magnetic nanoparticle systems," *Adv. Chem. Phys.,* vol. 128, p. 191, 2004.
41. V. A. Ignatchenko and R. S. Gekht, "Dynamic hysteresis of a superparamagnet," *Sov. Phys. JETP,* vol. 40, p. 750, 1975.
42. J. J. Lu, J. H. Huang, and I. Klik, "Field orientations and sweep rate effects on magnetic switching of Stoner–Wohlfarth particles," *J. Appl. Phys.,* vol. 76, p. 1726, 1994.
43. I. Klik and Y. D. Yao, "Hysteresis and limiting cycles in a high frequency ac field," *J. Appl. Phys.,* vol. 89, p. 7457, 2001.
44. Yu. L. Raikher, V. I. Stepanov, and R. Perzynski, "Dynamic hysteresis of a superparamagnetic nanoparticle," *Physica B,* vol. 343, p. 262, 2004.
45. Yu. L. Raikher and V. I. Stepanov, "Dynamic hysteresis of a superparamagnetic nanoparticle at low-to-intermediate frequencies," *J. Magn. Magn. Mater.,* vol. 300, p. e311, 2006.
46. Yu. L. Raikher and V. I. Stepanov, "Absorption of AC field energy in a suspension of magnetic dipoles," *J. Magn. Magn. Mater.,* vol. 320, p. 2692, 2008.
47. P. M. Déjardin, Yu. P. Kalmykov, B. E. Kashevsky, et al., "Effect of a dc bias field on the dynamic hysteresis of single-domain ferromagnetic particles," *J. Appl. Phys.,* vol. 107, p. 073914, 2010.
48. V. Franco and A. Conde, "Thermal effects in a Stoner–Wohlfarth model and their influence on magnetic anisotropy determination," *J. Magn. Magn. Mater.,* vol. 278, p. 28, 2004.
49. C. Tannous and J. Gieraltowski, "A Stoner–Wohlfarth model redux: Dynamic and statistical properties," *Physica B,* vol. 403, p. 3578, 2008.
50. N. A. Usov and Yu. B. Grebenshchikov, "Hysteresis loops of an assembly of superparamagnetic nanoparticles with uniaxial anisotropy," *J. Appl. Phys.,* vol. 106, p. 023917, 2009.

51. N. A. Usov and B. Ya. Liubimov, "Dynamics of magnetic nanoparticle in a viscous liquid: Application to magnetic nanoparticle hyperthermia," *J. Appl. Phys.,* vol. 112, p. 023901, 2012.
52. I. S. Poperechny, Yu. L. Raikher, and V. I. Stepanov, "Dynamic magnetic hysteresis in single-domain particles with uniaxial anisotropy," *Phys. Rev. B,* vol. 82, p. 174423, 2010.
53. J. Carrey, B. Mehdaoui, and M. Respaud, "Simple models for dynamic hysteresis loop calculations of magnetic single-domain nanoparticles: Application to magnetic hyperthermia optimization," *J. Appl. Phys.,* vol. 109, p. 083921, 2011.
54. B. Mehdaoui, J. Carrey, M. Stadler, et al., "Influence of a transverse static magnetic field on the magnetic hyperthermia properties and high-frequency hysteresis loops of ferromagnetic FeCo nanoparticles," *Appl. Phys. Lett.,* vol. 100, p. 052403, 2012.
55. L. Gammaitoni, P. Hänggi, P. Jung et al., "Stochastic resonance," *Rev. Mod. Phys.,* vol. 70, p. 223, 1998.
56. A. N. Grigorenko, V. I. Konov, and P. I. Nikitin, "Magnetostochastic resonance," *JETP Lett.,* vol. 52, p. 593, 1990.
57. A. N. Grigorenko, P. I. Nikitin, A. N. Slavin, et al., "Experimental observation of magnetostochastic resonance," *J. Appl. Phys.,* vol. 76, p. 6335, 1994.
58. A. N. Grigorenko, P. I. Nikitin, and G. V. Roschepkin, "Frequency mixing phenomena in a bistable system," *J. Appl. Phys.,* vol. 79, p. 6113, 1996.
59. N. Grigorenko, P. I. Nikitin, and G. V. Roschepkin, "Frequency mixing in a bistable system in the presence of noise," *J. Exp. Theor. Phys.,* vol. 85, p. 343, 1997.
60. E. K. Sadykov, "Stochastic resonance in finely dispersed magnets," *Sov. Phys. Solid State,* vol. 33, p. 1862, 1991.
61. E. K. Sadykov, "Stochastic resonance in small-particle magnetics. I. Radiospectroscopic study," *J. Phys. Condens. Matter,* vol. 4, p. 3295, 1992.
62. Yu. L. Raikher and V. I. Stepanov, "Stochastic resonance in single-domain particles," *J. Phys. Condens. Matter,* vol. 6, p. 4137, 1994.
63. Yu. L. Raikher and V. I. Stepanov, "Stochastic resonance and phase shifts in superparamagnetic particles," *Phys. Rev. B,* vol. 52, p. 3493, 1995.
64. Yu. L. Raikher, V. I. Stepanov, A. N. Grigorenko, et al., "Nonlinear magnetic stochastic resonance: Noise-strength–constant-force diagrams," *Phys. Rev. E,* vol. 56, p. 6400, 1997.
65. Yu. L. Raikher and V. I. Stepanov, "Noise- and force-induced resonances in noisy rotary oscillations of classical spins," *Phys. Rev. Lett.,* vol. 86, p. 1923, 2001.
66. P. C. Fannin and Yu. L. Raikher, "The estimation of the signal-to-noise ratio of a nanoparticle," *J. Phys. D: Appl. Phys.,* vol. 34, p. 1612, 2001.
67. P. C. Fannin, Yu. L. Raikher, A. T. Giannitsis, et al., "Investigation of the influence which material parameters have on the signal-to-noise ratio of nanoparticles," *J. Magn. Magn. Mater.,* vol. 252, p. 114, 2002.
68. Yu. L. Raikher, V. I. Stepanov, and P. C. Fannin, "Stochastic resonance in a superparamagnetic particle," *J. Magn. Magn. Mater.,* vol. 258–259, p. 369, 2003.

69. Yu. P. Kalmykov, Yu. L. Raikher, W. T. Coffey, et al., "Precession-aided magnetic stochastic resonance in ferromagnetic nanoparticles with cubic anisotropy," *Phys. Rev. B,* vol. 71, p. 012415, 2005.
70. G. Bertotti, Hysteresis in Magnetism, San Diego: Academic, 1998.
71. A. Thiaville, "Coherent rotation of magnetization in three dimensions: A geometrical approach," *Phys. Rev. B,* vol. 61, p. 12221, 2000.
72. M. Jamet, W. Wernsdorfer, C. Thirion, et al., "Magnetic anisotropy of a single cobalt nanocluster," *Phys. Rev. Lett.,* vol. 86, p. 4676, 2001.
73. M. Jamet, W. Wernsdorfer, C. Thirion, et al., "Magnetic anisotropy in single clusters," *Phys. Rev. B,* vol. 69, p. 024401, 2004.
74. W. Wernsdorfer, C. Thirion, M. Jamet, et al., "Quantum phase interference (Berry phase) in single-molecule magnets of [Mn12]2–," *J. Appl. Phys.,* vol. 91, p. 7164, 2002.
75. C. Vouille, A. Thiaville, and J. Miltat, "Thermally activated switching of nanoparticles: a numerical study," *J. Magn. Magn. Mater.,* vol. 272–276, p. E1237, 2004.
76. S. Mørup, "Magnetic hyperfine splitting in mössbauer spectra of microcrystals," *J. Magn. Magn. Mater.,* vol. 37, p. 39, 1983.
77. D. H. Jones and K. K. P. Srivastava, "Many-state relaxation model for the Mössbauer spectra of superparamagnets," *Phys. Rev. B,* vol. 34, p. 7542, 1986.
78. J. L. Dormann, D. Fiorani, and E. Tronc, "Magnetic relaxation in fine-particle systems," *Adv. Chem. Phys.,* vol. 98, p. 283, 1997.
79. J. van Lierop and D. H. Ryan, "Mössbauer spectra of single-domain fine particle systems described using a multiple-level relaxation model for superparamagnets," *Phys. Rev. B,* vol. 63, p. 064406, 2001.
80. P. M. Déjardin and W. T. Coffey, "Mössbauer spectra of noninteracting single-domain ferromagnetic particles including superparamagnetic relaxation: Simplification of the multilevel method," *Hyperfine Interact,* vol. 163, p. 73, 2005.
81. W. Wernsdorfer , E. Bonet Orozco, K. Hasselbach, et al., "Experimental evidence of the Néel-Brown model of magnetization reversal," *Phys. Rev. Lett.,* vol. 78, p. 1791, 1997.
82. W. Wernsdorfer , E. Bonet Orozco, K. Hasselbach, et al., "Macroscopic quantum tunneling of magnetization of single ferrimagnetic nanoparticles of barium ferrite," *Phys. Rev. Lett.,* vol. 79, p. 4014, 1997.
83. W. T. Coffey, D. S. F. Crothers, J. L. Dormann, et al., "Thermally activated relaxation time of a single domain ferromagnetic particle subjected to a uniform field at an oblique angle to the easy axis: Comparison with experimental observations," *Phys. Rev. Lett.,* vol. 80, p. 5655, 1998.
84. H. A. Kramers, "Brownian motion in a field of force and the diffusion model of chemical reactions," *Physica,* vol. 7, p. 284, 1940.
85. P. Hänggi, P. Talkner, and M. Borkovec, "Reaction-rate theory: fifty years after Kramers," *Rev. Mod. Phys.,* vol. 62, p. 251, 1990.

86. T. Dunn, A.L. Chudnovskiy, and A. Kamenev, “Dynamics of nano-magnetic oscillators,” in *Fluctuating Nonlinear Oscillators: From Nanomechanics to Quantum Superconducting Circuits*, M. Dykman, Ed., Oxford, Oxford University Press, 2012.
87. L. Néel, “Propriétées magnétiques des ferrites; Férrimagnétisme et antiferromagnétisme,” *Annales de Physique (Paris)*, vol. 12, p. 137, 1948.
88. N. Gershenfeld, The Physics of Information Technology, Cambridge: Cambridge University Press, 2000.
89. L.D. Landau and E.M. Lifshitz, “On the theory of the dispersion of the magnetic permeability of ferromagnetic bodies,” *Phys. Z. Sowjetunion*, vol. 8, p. 153, 1935.
90. D.A. Smith and F.A. de Rozario, “A classical theory of superparamagnetic relaxation,” *J. Magn. Magn. Mater.*, vol. 3, p. 219, 1976.
91. T. L. Gilbert, “A Lagrangian formulation of the gyromagnetic equation of the magnetic field,” *Phys. Rev.*, vol. 100, p. 1243, 1955 (Abstract only; full report in Armour Research Foundation Project No. A059, Supplementary Report, 1956). Reprinted in: T.L. Gilbert (2004). A phenomenological theory of damping in ferromagnetic materials, IEEE Trans. Magn., 40, p. 1243.
92. W.F. Brown, Jr., “Relaxation behavior of fine magnetic particles,” *J. Appl. Phys. Suppl.*, vol. 30, p. 130S, 1959.
93. S. Dutz and R. Hergt, “Magnetic particle hyperthermia – a promising tumour therapy?,” *Nanotechnology*, vol. 25, p. 452001, 2014.
94. E. A. Périgo, G. Hemery, O. Sandre, et al., “Fundamentals and advances in magnetic hyperthermia,” *Appl. Phys. Rev.*, vol. 2, p. 041302, 2015.
95. M. D. Stiles and J. Miltat, “Spin-transfer Torque and Dynamics,” in *Spin Dynamics in Confined Magnetic Structures*, vol. III, B. Hillebrandsn and A. Thiaville, Eds., Berlin, Springer-Verlag, 2006, p. 225.
96. Y. Suzuki, A. A. Tulapurkar, and C. Chappert, “Spin-Injection Phenomena and Applications,” in *Nanomagnetism and Spintronics*, T. Shinjo, Ed., Amsterdam, Elsevier, 2009, p. 94.
97. W.T. Coffey and Yu.P. Kalmykov, The Langevin Equation, 4th ed., Singapore: World Scientific, 2017.
98. R. Kubo, “Stochastic Liouville equations,” *J. Math. Phys.*, vol. 4, p. 174, 1963.
99. R. Kubo and N. Hashitsume, “Brownian motion of spins,” *Suppl. Progr. Theor. Phys.*, vol. 46, p. 210, 1970.
100. A. Kawabata, “Brownian motion of a classical spin,” *Progr. Theor. Phys.*, vol. 48, p. 2237, 1972.
101. R.L. Stratonovich, Conditional Markov Processes and their Application to the Theory of Optimal Control, New York: Elsevier, 1968.
102. C.W. Gardiner, Handbook of Stochastic Methods, Berlin: Springer, 1985.
103. J.S. Langer, “Statistical theory of the decay of metastable states,” *Ann. Phys.*, vol. 54, p. 258, 1969.
104. D.A. Varshalovich, A.N. Moskalev, and V.K. Khersonskii, Quantum Theory of Angular Momentum, Singapore: World Scientific, 1998.

105. Yu.L. Raikher and M.I. Shliomis, "Dispersion theory of magnetic susceptibility of small ferromagnetic particles," *Sov. Phys. JETP,* vol. 40, p. 526, 1974.
106. I. Klik and L. Gunther, "First-passage-time approach to overbarrier relaxation of magnetization," *J. Stat. Phys.,* vol. 60, p. 473, 1990.
107. I. Klik and L. Gunther, "Thermal relaxation over a barrier in single domain ferromagnetic particles," *J. Appl. Phys.,* vol. 67, p. 4505, 1990.
108. L.J. Geoghegan, W.T. Coffey, and B. Mulligan, "Differential recurrence relations for nonaxially symmetric Fokker-Planck equations," *Adv. Chem. Phys.,* vol. 100, p. 475, 1997.
109. G. Bertotti, C. Serpico, and I.D. Mayergoyz, Nonlinear Magnetization Dynamics in Nanosystems, Amsterdam: Elsevier Science & Technology, 2009.
110. H. Fröhlich, Theory of Dielectrics: Dielectric Constant and Dielectric Loss, 2nd ed., Oxford: Oxford at the Clarendon Press, 1958.
111. P. Debye, Polar Molecules, New York: Chemical Catalog Co., 1929 (reprinted by Dover, New York, 1954).
112. W.F. Brown, Jr., in *Handbuch der Physik,* vol. 17, S. Flügge, Ed., Berlin, Springer-Verlag, 1956.
113. K. Miyazaki and K. Seki, "Brownian motion of spins revisited," *J. Chem. Phys.,* vol. 108, p. 7052, 1998.
114. J. C. Slonczewski, "Current-driven excitation of magnetic multilayers," *J. Magn. Magn. Mater.,* vol. 159, p. L1, 1996.
115. L. Berger, "Emission of spin waves by a magnetic multilayer traversed by a current," *Phys. Rev. B,* vol. 54, p. 9353, 1996.
116. D. C. Ralph and M. D. Stiles, "Spin transfer torques," *J. Magn. Magn. Mater.,* vol. 320, p. 1190, 2008.
117. R. Heindl, W. H. Rippard, S. E. Russek, et al., "Validity of the thermal activation model for spin-transfer torque switching in magnetic tunnel junctions," *J. Appl. Phys.,* vol. 109, p. 073910, 2011.
118. W. H. Rippard, R. Heindl, M.R. Pufall, et al., "Thermal relaxation rates of magnetic nanoparticles in the presence of magnetic fields and spin-transfer effects," *Phys. Rev. B,* vol. 84, p. 064439, 2011.
119. E. B. Myers, F. J. Albert, J. C. Sankey, et al., "Thermally activated magnetic reversal induced by a spin-polarized current," *Phys. Rev. Lett.,* vol. 89, p. 196801, 2002.
120. A. Fabian, C. Terrier, S. Serrano Guisan, et al., "Current-induced two-level fluctuations in pseudo-spin-valve (Co/Cu/Co) nanostructures," *Phys. Rev. Lett.,* vol. 91, p. 257209, 2003.
121. F. B. Mancoff, R. W. Dave, N. D. Rizzo, et al., "Angular dependence of spin-transfer switching in a magnetic nanostructure," *Appl. Phys. Lett.,* vol. 83, p. 1596, 2003.
122. I. N. Krivorotov, N. C. Emley, A. G. F. Garcia, et al., "Temperature dependence of spin-transfer-induced switching of nanomagnets," *Phys. Rev. Lett.,* vol. 93, p. 166603, 2004.

123. R. H. Koch, J. A. Katine, and J. Z. Sun, "Time-resolved reversal of spin-transfer switching in a nanomagnet," *Phys. Rev. Lett.,* vol. 92, p. 088302, 2004.
124. M. L. Schneider, M. R. Pufall, W. H. Rippard, et al., "Thermal effects on the critical current of spin torque switching in spin valve nanopillars," *Appl. Phys. Lett.,* vol. 90, p. 092504, 2007.
125. Z. Li and S. Zhang, "Thermally assisted magnetization reversal in the presence of a spin-transfer torque," *Phys. Rev. B,* vol. 69, p. 134416, 2004.
126. D. M. Apalkov and P. B. Visscher, "Spin-torque switching: Fokker-Planck rate calculation," *Phys. Rev. B,* vol. 72, p. 180405(R), 2005.
127. G. Bertotti, I. D. Mayergoyz, and C. Serpico, "Analysis of random Landau-Lifshitz dynamics by using stochastic processes on graphs," *J. Appl. Phys.,* vol. 99, p. 08F301, 2006.
128. W.H. Butler, T. Mewes, C.K.A. Mewes, et al., "Switching distributions for perpendicular spin-torque devices within the macrospin approximation," *IEEE Trans. Magn.,* vol. 48, p. 4684, 2012.
129. T. Taniguchi, Y. Utsumi, M. Marthaler, et al., "Spin torque switching of an in-plane magnetized system in a thermally activated region," *Phys. Rev. B,* vol. 87, p. 054406, 2013.
130. T. Taniguchi, Y. Utsumi, and H. Imamura, "Thermally activated switching rate of a nanomagnet in the presence of spin torque," *Phys. Rev. B,* vol. 88, p. 214414, 2013.
131. K. A. Newhall and E. Vanden-Eijnden, "Averaged equation for energy diffusion on a graph reveals bifurcation diagram and thermally assisted reversal times in spin-torque driven nanomagnets," *J. Appl. Phys.,* vol. 113, p. 184105, 2013.
132. D. V. Berkov and N. L. Gorn, "Magnetization precession due to a spin-polarized current in a thin nanoelement: Numerical simulation study," *Phys. Rev. B,* vol. 72, p. 094401, 2005.
133. D. V. Berkov and N. L. Gorn, "Non-linear magnetization dynamics in nanodevices induced by a spin-polarized current: micromagnetic simulation," *J. Phys. D: Appl. Phys.,* vol. 41, p. 164013, 2008.
134. K. Itô, "Micromagnetic simulation on dynamics of spin transfer torque magnetization reversal," *IEEE Trans. Magn.,* vol. 41, p. 2630, 2005.
135. T. Taniguchi, M. Shibata, M. Marthaler, et al., "Numerical study on spin torque switching in thermally activated region," *Appl. Phys. Express,* vol. 5, p. 063009, 2012.
136. D. Pinna, A. D. Kent, and D. L. Stein, "Spin-transfer torque magnetization reversal in uniaxial nanomagnets with thermal noise," *J. Appl. Phys.,* vol. 114, p. 033901, 2013.
137. Yu. P. Kalmykov, W. T. Coffey, S. V. Titov, et al., "Spin-torque effects in thermally assisted magnetization reversal: Method of statistical moments," *Phys. Rev. B,* vol. 88, p. 144406, 2013.
138. D. Byrne, W. T. Coffey, Yu. P. Kalmykov, et al., "Spin-transfer torque effects in the dynamic forced response of the magnetization of nanoscale ferromagnets in superimposed ac and dc bias fields in the presence of thermal agitation," *Phys. Rev. B,* vol. 91, p. 174406, 2015.

139. J. Z. Sun, "Spin-current interaction with a monodomain magnetic body: A model study," *Phys. Rev. B,* vol. 62, p. 570, 2000.
140. Yu. P. Kalmykov, S. V. Titov, "Derivation of matrix elements for the system of moment equations governing the kinetics of superparamagnetic particles," *J. Magn. Magn. Mater.,* vol. 210, p. 233, 2000.
141. Yu.P. Kalmykov and S.V. Titov, "Matrix elements for the system of moment equations governing the kinetics of superparamagnetic particles," *Phys. Rev. Lett.,* vol. 82, p. 2967, 1999.
142. R. Bellman, Introduction to Matrix Analysis, New York: McGraw-Hill, 1960.
143. H. Risken, The Fokker–Planck Equation, Berlin (Second Edition, 1989): Springer Verlag, 1984.
144. W.T. Coffey, Yu.P. Kalmykov, and J.T. Waldron, "Relaxational dynamics of a particle in the presence of an external potential: Exact solution in terms of matrix continued fractions," *Physica A,* vol. 208, p. 462, 1994.
145. Yu.P. Kalmykov, "Evaluation of the smallest nonvanishing eigenvalue of the Fokker–Planck equation for the Brownian motion in a potential. II. The matrix continued fraction approach," *Phys. Rev. E,* vol. 62, p. 227, 2000.
146. W.T. Coffey, D.S.F. Crothers, Yu.P. Kalmykov, et al., "Constant-magnetic-field effect in Néel relaxation of single-domain ferromagnetic particles," *Phys. Rev. B,* vol. 51, p. 15947, 1995.
147. Yu.P. Kalmykov and S.V. Titov, "Nonlinear response of superparamagnetic particles to a sudden change of a high constant magnetic field," *Phys. Solid State,* vol. 42, p. 918, 2000.
148. Yu.P. Kalmykov and W.T. Coffey, "Transverse complex magnetic susceptibility of single-domain ferromagnetic particles with uniaxial anisotropy subjected to a longitudinal uniform magnetic field," *Phys. Rev. B,* vol. 56, p. 3325, 1997.
149. W.T. Coffey, Yu.P. Kalmykov, and S.V. Titov, "Nonlinear response of fine superparamagnetic particles to the sudden change of a strong uniform dc magnetic field," *J. Magn. Magn. Mater.,* vol. 241, p. 400, 2002.
150. Yu.P. Kalmykov and S.V. Titov, "Nonlinear response of superparamagnetic particles with a cubic anisotropy to a sudden change of a strong uniform dc magnetic field," *Phys. Solid State,* vol. 44, p. 2276, 2002.
151. Yu.P. Kalmykov and B. Ouari, "Longitudinal complex magnetic susceptibility and relaxation times of superparamagnetic particles with triaxial anisotropy," *Phys. Rev. B,* vol. 71, p. 094410, 2005.
152. B. Ouari and Yu.P. Kalmykov, "Dynamics of the magnetization of single domain particles having triaxial anisotropy subjected to a uniform dc magnetic field," *J. Appl. Phys.,* vol. 100, p. 123912, 2006.
153. P.M. Déjardin and Yu.P. Kalmykov, "Relaxation of the magnetization in uniaxial single-domain ferromagnetic particles driven by a strong ac magnetic field," *J. Appl. Phys.,* vol. 106, p. 123908, 2009.

154. P.M. Déjardin and Yu.P. Kalmykov, "Effect of a dc magnetic field on the magnetization reversal of uniaxial single-domain ferromagnetic particles driven by ac magnetic field," *J. Magn. Magn. Mater.*, vol. 322, p. 3112, 2010.
155. S.V. Titov, P.M. Déjardin, H. El Mrabti, et al., "Nonlinear magnetization relaxation of superparamagnetic nanoparticles in superimposed ac and dc bias magnetic fields," *Phys. Rev. B*, vol. 82, p. 100413(R), 2010.
156. H. El Mrabti, S.V. Titov, P.M. Déjardin, et al., "Nonlinear stationary ac response of the magnetization of uniaxial superparamagnetic nanoparticles," *J. Appl. Phys.*, vol. 110, p. 023901, 2011.
157. B. Ouari, S. Aktau, and Yu.P. Kalmykov, "Reversal time of the magnetization of antiferromagnetic nanoparticles," *Phys. Rev. B*, vol. 81, p. 024412, 2010.
158. B. Ouari, and Yu.P. Kalmykov, "Effect of a dc bias magnetic field on the magnetization relaxation of antiferromagnetic nanoparticles," *Phys. Rev. B*, vol. 83, p. 064406, 2011.
159. H. El Mrabti, P. M. Déjardin, S. V. Titov, et al., "Damping dependence in dynamic magnetic hysteresis of single-domain ferromagnetic particles," *Phys. Rev. B*, vol. 85, p. 094425, 2012.
160. B. Ouari, S. V. Titov, H. El Mrabti, et al., "Nonlinear susceptibility and dynamic hysteresis loops of magnetic nanoparticles with biaxial anisotropy," *J. Appl. Phys.*, vol. 113, p. 053903, 2013.
161. N. Wei, D. Byrne, W.T. Coffey, et al., "Nonlinear frequency dependent effects in the dc magnetization of uniaxial magnetic nanoparticles in superimposed strong alternating current and direct current fields," *J. Appl. Phys.*, vol. 116, p. 173903, 2014.
162. Yu.P. Kalmykov, B. Ouari, and S. V. Titov, "Dynamic magnetic hysteresis and nonlinear susceptibility of antiferromagnetic nanoparticles," *J. Appl. Phys.*, vol. 120, p. 053901, 2016.
163. R. Kubo, M. Toda, and N. Hashitsume, Statistical Physics II. Nonequilibrium Statistical Mechanics, 2nd, Ed., Berlin: Springer, 1991.
164. W.T. Coffey, Yu.P. Kalmykov and E.S. Massawe, "The effective eigenvalue method and its application to stochastic problems in conjunction with the nonlinear Langevin equation," *Adv. Chem. Phys., Part 2,* vol. 85, p. 667, 1993.
165. W.T. Coffey, Yu.P. Kalmykov, and S.V. Titov, "Magnetization reversal time of magnetic nanoparticles at very low damping," *Phys. Rev. B*, vol. 89, p. 054408, 2014.
166. D. Byrne, W.T. Coffey, W. J. Dowling, *et al.*, "On the Kramers very low damping escape rate for point particles and classical spins," *Adv. Chem. Phys.*, vol. 156, p. 393, 2015 ; D. Byrne, W.T. Coffey, Yu.P. Kalmykov, *et al.*, "On a simple derivation of the very low damping escape rate for classical spins by modifying the method of Kramers," *Physica A*, vol. 527, p. 121195, 2019.
167. D. Byrne, W. T. Coffey, W.J. Dowling, *et al.*, "Spin transfer torque and dc bias magnetic field effects on the magnetization reversal time of nanoscale ferromagnets at very low damping: Mean first passage time vs numerical methods," *Phys. Rev. B*, vol. 93, p. 064413, 2016.

168. A. Aharoni, Introduction to the Theory of Ferromagnetism, London: Oxford University Press, 1996.
169. G.D. Billing, and K.V. Mikkelsen, Introduction to Molecular Dynamics and Chemical Kinetics, New York: Wiley, 1996.
170. W.H. de Jeu, Physical Properties of Liquid Crystalline Materials, New York: Gordon and Breach, 1980.
171. G.H. Wannier, Elements of Solid State Theory, Cambridge, UK: Cambridge University Press, 1960.
172. H. Pfeiffer, "Determination of the anisotropy field distribution in particle assemblies taking into account thermal fluctuations," *Phys. Stat. Sol.,* vol. 118, p. 295, 1990.
173. H. Pfeiffer, "Influence of thermal fluctuations on the magnetic properties of particle assemblies," *Phys. Stat. Sol.,* vol. 122, p. 377, 1990.
174. P.M.S. Blackett, Lectures on Rock Magnetism, Jerusalem: Weizmann Science Press, 1956.
175. E.M. Chudnovsky and L. Gunther, "Quantum tunneling of magnetization in small ferromagnetic particles," *Phys. Rev. Lett.,* vol. 60, p. 661, 1988.
176. P.C.E. Stamp, E.M. Chudnovsky, and B. Barbara, "Quantum tunneling of magnetization in solids," *Int. J. Mod. Phys. B,* vol. 6, p. 1355, 1992.
177. E.M. Chudnovsky, "Macroscopic quantum tunneling of the magnetic moment," *J. Appl. Phys.,* vol. 73, p. 6697, 1993.
178. R.L. Stratonovich, "Some Markov methods in the theory of stochastic processes in nonlinear dynamical systems," in *Theory of Continuous Fokker-Planck Systems*, vol. 1, F. Moss and P. McClintock, Eds., Cambridge, Cambridge University Press, 1989, p. 16.
179. W.T. Coffey, "Finite integral representation of characteristic times of orientational relaxation processes: Application to the uniform bias force effect in relaxation in bistable potentials," *Adv. Chem. Phys.,* vol. 103, p. 259, 1998.
180. G. Klein, "Mean First-Passage Times of Brownian Motion and Related Problems," *Proc. R. Soc. London A,* vol. 211, p. 431, 1952.
181. W.T. Coffey, D.A. Garanin, and D.J. McCarthy, "Crossover formulas in the Kramers theory of thermally activated escape rates – Application to spin systems," *Adv. Chem. Phys.,* vol. 117, p. 483, 2001.
182. N.G. van Kampen, Stochastic Processes in Physics and Chemistry, Amsterdam (Third Edition, 2007): North-Holland Physics Publishing, 1981.
183. A. Nitzan, Chemical Dynamics in Condensed Phases: Relaxation, Transfer, and Reactions in Condensed Molecular Systems, London: Oxford University Press, 2006.
184. H. Haken, Laser Theory, Berlin: Springer, 1984.
185. S. Stenholm, Foundations of Laser Spectroscopy, New York: Wiley, 1984.
186. V.I. Mel'nikov, "The Kramers problem: 50 years of development," *Phys. Rep.,* vol. 209, p. 2, 1991.
187. V.I. Mel'nikov and S.V. Meshkov, "Theory of activated rate processes: Exact solution of the Kramers problem," *J. Chem. Phys.,* vol. 85, p. 1018, 1986.

188. H. Grabert, "Escape from a metastable well: The Kramers turnover problem," *Phys. Rev. Lett.*, vol. 61, p. 1683, 1988.
189. E. Pollak, H. Grabert, and P. Hänggi, "Theory of activated rate processes for arbitrary frequency dependent friction: Solution of the turnover problem," *J. Chem. Phys.*, vol. 91, p. 4073, 1989.
190. E.C. Titchmarsh, An Introduction to the Theory of Fourier Integrals, London: Oxford University Press, 1937.
191. P.M. Morse and H. Feshbach, Methods of Theoretical Physics, Part I, New York: McGraw-Hill, 1953.
192. H.C. Brinkman, "Brownian motion in a field of force and the diffusion theory of chemical reactions, II," *Physica (Utrecht)*, vol. 22, p. 149, 1956.
193. R. Landauer and J.A. Swanson, "Frequency factors in the thermally activated process," *Phys. Rev.*, vol. 121, p. 1668, 1961.
194. R. Becker and W. Döring, "Kinetische Behandlung der Keimbildung in übersättigten Dämpfen," *Ann. Phys. (Leipzig)*, vol. 24, p. 719, 1935.
195. B.J. Matkowsky, Z. Schuss and C. Tier, "Uniform expansion of the transition rate in Kramers' problem," *J. Stat. Phys.*, vol. 35, p. 443, 1984.
196. P.M. Déjardin, D.S.F. Crothers, W.T. Coffey, et al., "Interpolation formula between very low and intermediate-to-high damping Kramers escape rates for single-domain ferromagnetic particles," *Phys. Rev. E*, vol. 63, p. 021102, 2001.
197. N.A. Usov, "Low frequency hysteresis loops of superparamagnetic nanoparticles with uniaxial anisotropy," *J. Appl. Phys.*, vol. 107, p. 123909, 2010.
198. W. Wernsdorfer, C. Thirion, M. Jamet, et al., "Temperature dependence of switching fields of single 3 nm cobalt nanoparticles," *J. Appl. Phys.*, vol. 91, p. 7062, 2002.
199. P.M. Déjardin, H. Kachkachi, and Yu.P. Kalmykov, "Thermal and surface anisotropy effects on the magnetization reversal of a nanocluster," *J. Phys. D: Appl. Phys.*, vol. 41, p. 134004, 2008.
200. M.I. Shliomis and V.I. Stepanov, "Theory of the dynamic susceptibility of magnetic fluids," *Adv. Chem. Phys.*, vol. 87, p. 1, 1994.
201. M.I. Shliomis, "Magnetic fluids," *Sov. Phys. Usp.* vol. 17, p. 153, 1974.
202. W.T. Coffey and P.C. Fannin, "Internal and Brownian mode-coupling effects in the theory of magnetic relaxation and ferromagnetic resonance of ferrofluids," *J. Phys. Condens. Matter*, vol. 14, p. 3677, 2002.
203. P.C. Fannin, B.K.P. Scaife, and S.W. Charles, "Relaxation and resonance in ferrofluids," *J. Mag. Magn. Mat.*, vol. 122, p. 159, 1993.
204. P.C. Fannin, "An experimental observation of the dynamic behavior of ferrofluids," *J. Mag. Magn. Mater.*, vol. 136, p. 49, 1993.
205. W. T. Coffey, Y. P. Kalmykov and S. V. Titov, "Itinerant oscillator models of fluids," *Adv. Chem. Phys.*, vol. 126, p. 131, 2003.
206. C. Scherer and H.G. Matuttis, "Rotational dynamics of magnetic particles in suspensions," *Phys. Rev. E*, vol. 63, p. 011504, 2000.

207. Yu.P. Kalmykov, "The relaxation time of the magnetization of uniaxial single-domain ferromagnetic particles in the presence of a uniform magnetic field," *J. Appl. Phys.*, vol. 96, p. 1138, 2004.
208. Yu.L. Raikher and M.I. Shliomis, "The effective field method in the orientational kinetics of magnetic fluids and liquid crystals," *Adv. Chem. Phys.*, vol. 87, p. 595, 1994.
209. P.C. Fannin, "Wideband measurement and analysis techniques for the determination of the frequency dependent complex susceptibility of magnetic fluids," *Adv. Chem. Phys.*, vol. 104, p. 181, 1998.
210. P.C. Fannin, T. Relihan, and S.W. Charles, "Experimental and theoretical profiles of the frequency-dependent complex susceptibility of systems containing nanometer-sized magnetic particles," *Phys. Rev. B*, vol. 55, p. 14423, 1997.
211. B.K.P. Scaife, "On the theory of the complex, frequency-dependent susceptibility of magnetic fluids," *Adv. Chem. Phys.*, vol. 109, p. 1, 1999.
212. R.A. Sack, "Relaxation processes and inertial effects II: Free rotation in space," *Proc. Phys. Soc. B*, vol. 70, p. 414, 1957.
213. C. P. Slichter, Principles of Magnetic Resonance, Berlin: Springer-Verlag, 1990.
214. M.W. Evans, G.J. Evans, W.T. Coffey, et al., Molecular Dynamics, New York: Wiley, 1982.
215. R.C. Tolman, The Principles of Statistical Mechanics, London: Oxford University Press, 1938.
216. R. E. Wyatt, Quantum Dynamics with Trajectories: Introduction to Quantum Hydrodynamics, New York: Springer, 2005.
217. K. Blum, Density Matrix: Theory and Applications, New York: Plenum, 1996.
218. L. D. Landau and E. M. Lifshitz, Mécanique Quantique, Physique Théorique, 3 ed., vol. 3, Moscow: Mir, 1988.
219. W. H. Zurek, "Quantum Darwinism, classical reality, and the randomness of quantum jumps," *Phys. Today*, vol. 67, p. 44, 2014.
220. Yu. P. Kalmykov, S. V. Titov, and W. T. Coffey, "Spin-size effects in stochastic resonance in uniaxial superparamagnets," *Phys. Rev. B*, vol. 81, p. 172411, 2010.
221. Yu. P. Kalmykov, S. V. Titov, and W. T. Coffey, "Classical–quantum crossover in magnetic stochastic resonance in uniaxial superparamagnets," *J. Phys. Condens. Matter*, vol. 22, p. 376001, 2010.
222. N. Bloembergen, E. M. Purcell, and R. V. Pound, "Relaxation effects in nuclear magnetic resonance absorption," *Phys. Rev.*, vol. 73, p. 679, 1948.
223. R. K. Wangsness and F. Bloch, "The dynamical theory of nuclear induction," *Phys. Rev.*, vol. 89, p. 728, 1953.
224. A. G. Redfield, "On the theory of relaxation processes," *IBM J. Res. Dev.*, vol. 1, p. 19, 1957.
225. P. N. Argyres, and P. L. Kelley, "Theory of spin resonance and relaxation," *Phys. Rev.*, vol. 134, p. A98, 1964.

226. D. A. Garanin, "Quantum thermoactivation of nanoscale magnets," *Phys. Rev. E,* vol. 55, p. 2569, 1997.
227. J. L. García-Palacios and D. Zueco, "Solving spin quantum master equations with matrix continued-fraction methods: application to superparamagnets," *J. Phys. A: Math. Gen.,* vol. 39, p. 13243 , 2006.
228. D. Zueco and J. L. García-Palacios, "Longitudinal relaxation and thermoactivation of quantum superparamagnets," *Phys. Rev. B,* vol. 73, p. 104448, 2006.
229. Y. Takahashi and F. Shibata, "Generalized phase space method in spin systems-spin coherent state representation," *J. Stat. Phys.,* vol. 14, p. 49, 1976.
230. Y. Takahashi and F. Shibata, "Spin coherent state representation in non-equilibrium statistical mechanics," *J. Phys. Soc. Jpn.,* vol. 38, p. 656, 1975.
231. F. Shibata, "Theory of nonlinear spin relaxation," *J. Phys. Soc. Jpn.,* vol. 49, p. 15, 1980.
232. A. A. Neufeld, "Quantum-classical approximation beyond Redfield theory," *J. Chem. Phys.,* vol. 119, p. 2488, 2003.
233. D. Kohen, C. C. Marston, and D. J. Tannor, "Phase space approach to theories of quantum dissipation," *J. Chem. Phys.,* vol. 107, p. 5236, 1997.
234. A. Suarez, R. Silbey, and I. Oppenheim, "Memory effects in the relaxation of quantum open systems," *J. Chem. Phys.,* vol. 97, p. 5101, 1992.
235. P. Hubbard, "Quantum-mechanical and semiclassical forms of the density operator theory of relaxation," *Rev. Mod. Phys.,* vol. 33, p. 249, 1961.
236. R. Dümcke and H. Spohn, "The proper form of the generator in the weak coupling limit," *Z. Phys. B,* vol. 34, p. 419, 1979.
237. H. Spohn and J. L. Lebowitz, "Irreversible thermodynamics for quantum systems weakly coupled to thermal reservoirs," *Adv. Chem. Phys.,* vol. 38, p. 109, 1978.
238. J. Łuczka, "On Markovian kinetic equations: Zubarev's nonequilibrium statistical operator approach," *Physica A,* vol. 149, p. 245, 1988.
239. W. T. Coffey and Yu. P. Kalmykov, "Thermal fluctuations of magnetic nanoparticles: Fifty years after Brown," *J. Appl. Phys.,* vol. 112, p. 121301, 2012.
240. J. L. García-Palacios and F. J. Lázaro, "Langevin-dynamics study of the dynamical properties of small magnetic particles," *Phys. Rev. B,* vol. 58, p. 14937, 1998.
241. J. L. García-Palacios, "On the statics and dynamics of magnetoanisotropic nanoparticles," *Adv. Chem. Phys.,* vol. 112, p. 1, 2000.
242. H. Suhl, Relaxation Processes in Micromagnetics, Oxford: Oxford University Press, 2007.
243. F. Shibata and C. Ushiyama, "Rigorous solution to nonlinear spin relaxation process," *J. Phys. Soc. Jpn.,* vol. 62, p. 381, 1993.
244. F. Shibata and M. Asou, "Theory of nonlinear spin relaxation. II," *J. Phys. Soc. Jpn.,* vol. 49, p. 1234, 1980.
245. A. M. Perelomov, Generalized Coherent States and Their Applications, Berlin: Springer, 1986.

246. J. M. Radcliffe, "Some properties of coherent spin states," *J. Phys. A: Math. Gen.*, vol. 4, p. 313, 1971.
247. F. T. Arecchi, E. Courtens, R. Gilmore, and H. Thomas, "Atomic coherent states in quantum optics," *Phys. Rev. A,* vol. 6, p. 2211, 1972.
248. F. A. Berezin, "General concept of quantization," *Commun. Math. Phys.*, vol. 40, p. 153, 1975.
249. G. S. Agarwal, "Relation between atomic coherent-state representation, state multipoles, and generalized phase-space distributions," *Phys. Rev. A,* vol. 24, p. 2889, 1981.
250. G. S. Agarwal, "State reconstruction for a collection of two-level systems," *Phys. Rev. A,* vol. 57, p. 671, 1998.
251. J. P. Dowling, G. S. Agarwal, and W. P. Schleich, "Wigner distribution of a general angular-momentum state: Applications to a collection of two-level atoms," *Phys. Rev. A,* vol. 49, p. 4101, 1994.
252. M. O. Scully and K. Wódkiewicz, "Spin quasi-distribution functions," *Found. Phys.*, vol. 24, p. 85, 1994.
253. C. Brif and A. Mann, "Phase-space formulation of quantum mechanics and quantum-state reconstruction for physical systems with Lie-group symmetries," *Phys. Rev. A,* vol. 59, p. 971, 1999.
254. A. B. Klimov, "Exact evolution equations for SU(2) quasidistribution functions," *J. Math. Phys.*, vol. 43, p. 2202, 2002.
255. A. B. Klimov and S. M. Chumakov, A Group-Theoretical Approach to Quantum Optics, Weinheim: Wiley-VCH, 2009.
256. Yu. P. Kalmykov, W. T. Coffey, and S. V. Titov, "Phase-space formulation of the nonlinear longitudinal relaxation of the magnetization in quantum spin systems," *Phys. Rev. E,* vol. 76, p. 051104, 2007.
257. Yu. P. Kalmykov, B. P. J. Mulligan, S. V. Titov, and W. T. Coffey, "Master equation in phase space for a spin in an arbitrarily directed uniform external field," *J. Stat. Phys.*, vol. 141, p. 589, 2010.
258. Yu. P. Kalmykov, and W. T. Coffey, "Transition state theory for spins: phase-space formulation," *J. Phys. A: Math. Theor.*, vol. 41, p. 185003, 2008.
259. Yu. P. Kalmykov, W. T. Coffey, and S. V. Titov, "Master equation in phase space for a uniaxial spin system," *J. Stat. Phys.*, vol. 131, p. 969, 2008.
260. Yu. P. Kalmykov, W. T. Coffey, and S. V. Titov, "Phase space equilibrium distribution function for spins," *J. Phys. A: Math. Theor.*, vol. 41, p. 105302, 2008.
261. Yu. P. Kalmykov, W. T. Coffey, and S. V. Titov, "Phase-space equilibrium distributions and their applications to spin systems with nonaxially symmetric Hamiltonians," *Phys. Rev. B,* vol. 77, p. 104418, 2008.
262. Yu. P. Kalmykov, W. T. Coffey, and S. V. Titov, "Phase space Langevin equation for spin relaxation in a dc magnetic field," *Europhys. Lett.*, vol. 88, p. 17002, 2009.

263. Yu. P. Kalmykov, S. V. Titov, and W. T. Coffey, "Nonlinear longitudinal relaxation of a quantum superparamagnet with arbitrary spin value S: Phase space and density matrix formulations," *Phys. Rev. B,* vol. 81, p. 094432, 2010.
264. Yu. P. Kalmykov, W. T. Coffey, and S. V. Titov, "Statistical moment equations for stochastic spin dynamics in phase space: A uniaxial paramagnet subjected to a dc bias field of arbitrary orientation," *Phys. Rev. B,* vol. 86, p. 104435, 2012.
265. W. P. Schleich, Quantum Optics in Phase Space, Berlin: Wiley-VCH, 2001.
266. L. M. Narducci, C. M. Bowden, V. Bluemel, and G. P. Carrazana, "Phase-space description of the thermal relaxation of a (2J+1)-level system," *Phys. Rev. A,* vol. 11, p. 280, 1975.
267. S. R. de Groot and L. G. Suttorp, Foundations of Electrodynamics, Amsterdam: North-Holland, 1972, Chapters VI and VII.
268. V. I. Tatarskii, "The Wigner representation of quantum mechanics," *Sov. Phys. Usp.,* vol. 26, p. 311, 1983.
269. M. Hillery, R. F. O'Connell, M. O. Scully, and E. P. Wigner, "Distribution functions in physics: Fundamentals," *Phys. Rep.,* vol. 106, p. 121, 1984.
270. C. K. Zachos, D. B. Fairlie and T. L. Curtright, Eds., Quantum Mechanics in Phase Space, Singapore: World Scientific, 2005.
271. T. L. Curtright, D. B. Fairlie, C. K. Zachos, A Concise Treatise on Quantum Mechanics in Phase Space, Singapore: World Scientific, 2014.
272. W. T. Coffey, Yu. P. Kalmykov, S. V. Titov et. al , "Wigner function approach to the quantum Brownian motion of a particle in a potential," *Phys. Chem. Chem. Phys.,* vol. 9, p. 3361, 2007.
273. R. L. Stratonovich, "On distributions in representation space," *Sov. Phys. JETP,* vol. 4, p. 891, 1957.
274. J. C. Vårilly and J. M. Gracia-Bondía, "The Moyal representation for spin," *Ann. Phys. (NY),* vol. 190, p. 107, 1989.
275. R. R. Puri, Mathematical Methods of Quantum Optics, Berlin: Springer, 2001.
276. J. L. García-Palacios, "Solving quantum master equations in phase space by continued-fraction methods," *Europhys. Lett.,* vol. 65, p. 735, 2004.
277. J. L. García-Palacios and D. Zueco, "The Caldeira–Leggett quantum master equation in Wigner phase space: continued-fraction solution and application to Brownian motion in periodic potentials," *J. Phys. A: Math. Gen.,* vol. 37, p. 10735, 2004.
278. E. P. Wigner, "Über das Überschreiten von Potentialschwellen bei chemischen Reaktionen," *Z. Phys. Chem. Abt. B,* vol. 19, p. 203, 1932.
279. E. P. Wigner, "The transition state method," *Trans. Faraday. Soc.,* vol. 34, p. 29, 1938.
280. R. P. Bell, The Tunnel Effect in Chemistry, New York and London: Chapman and Hall, 1980.
281. M. Razavy, Quantum Theory of Tunneling, Singapore: World Scientific, 2003.
282. W. H. Zurek, "Decoherence, einselection, and the quantum origins of the classical," *Rev. Mod. Phys.,* vol. 75, p. 715, 2003.

283. E. Schrödinger, "Der stetige Übergang von der Mikro- zur Makromechanik," *Naturwissenschaften,* vol. 14, p. 664, 1926.
284. J.R. Klauder and B. Skagerstam, Coherent States, Singapore: World Scientific, 1985.
285. C. W. Gardiner and P. Zoller, Quantum Noise, 2nd ed., Berlin: Springer, 2000.
286. J. E. Moyal, "Quantum mechanics as a statistical theory," *Proc. Camb. Phil. Soc.,* vol. 45, p. 99, 1949.
287. J. Ville, "Théorie et Applications de la Notion dc Signal Analytique," *Câbles et Transmission,* vol. 2, p. 61, 1948.
288. D.E. Newland, An Introduction to random vibrations, spectral and wavelet analysis, 3rd ed., New York: Dover Publications Inc., 2005.
289. H. W. Lee, "Theory and application of the quantum phase-space distribution functions," *Phys. Rep.,* vol. 259, p. 147, 1995.
290. M. Abramowitz and I. Stegun, Eds., Handbook of Mathematical Functions, New York: Dover, 1964.
291. J. P. Dahl and M. Springborg, "The Morse oscillator in position space, momentum space, and phase space," *J. Chem. Phys.,* vol. 88, p. 4535, 1988.
292. A. Franck, A. L. Rivera, and K. B. Wolf, "Wigner function of Morse potential eigenstates," *Phys. Rev. A,* vol. 61, p. 054102, 2000.
293. M. Hug, C. Menke, and W. P. Schleich, "Modified spectral method in phase space: Calculation of the Wigner function. II. Generalizations," *Phys. Rev. A,* vol. 57, p. 3206, 1998.
294. B. Belchev and M. A.Walton, "Solving for the Wigner functions of the Morse potential in deformation quantization," *J. Phys A: Math. Theor.,* vol. 43, p. 225206, 2010.
295. J. J. Włodarz and J. Nowakowski, "Variational calculations of the Wigner distribution function for selected anharmonic oscillators," *Int. J. Quantum Chem.,* vol. 38, p. 373, 1990.
296. A. Ceasu-Velcescu, P. Blaise, and Y. P. Kalmykov, "Phase-space description of a particle in a quartic double-well potential," *Int. J. Mod. Phys. B,* vol. 28, p. 1450164, 2014.
297. N. L. Balazs and A. Voros, "Wigner's function and tunneling," *Ann. Phys.,* vol. 199, p. 123, 1990.
298. V. I. Mel'nikov, "Activated tunneling decay of metastable state: Solution of the Kramers problem," *Physica A,* vol. 130, p. 606, 1985.
299. J. B. Bouchaud, E. Cohen de Lara, and R. Kahn, "Diffusion of H2 and HD in zeolites: Quantum effects at 100 K," *Europhys. Lett.,* vol. 17, p. 583, 1992.
300. V. A. Benderskii, D. E. Makarov, and C. A. Wright, "Chemical dynamics at low temperatures," *Adv. Chem. Phys.,* vol. 88, p. 1, 1994.
301. D. A. Garanin, X. M. Hidalgo, and E. M. Chudnovsky, "Quantum-classical transition of the escape rate of a uniaxial spin system in an arbitrarily directed field," *Phys. Rev. B,* vol. 57, p. 13639, 1998.

302. P. Hänggi and G. L. Ingold, "Fundamental aspects of quantum Brownian motion," *Chaos,* vol. 15, p. 026105, 2005.
303. D. Kohen and D. J. Tannor, "Phase space approach to dissipative molecular dynamics," *Adv. Chem. Phys.,* vol. 111, p. 219, 1994.
304. A. O. Caldeira and A. J. Leggett, "Path integral approach to quantum Brownian motion," *Physica A,* vol. 121, p. 587, 1983.
305. H. Dekker, "Quantization of the linearly damped harmonic oscillator," *Phys. Rev. A,* vol. 16, p. 2116, 1977.
306. Y. Tanimura and P. G. Wolynes, "Quantum and classical Fokker-Planck equations for a Gaussian-Markovian noise bath," *Phys. Rev A,* vol. 43, p. 4131, 1991.
307. Y. Tanimura and P. G. Wolynes, "The interplay of tunneling, resonance, and dissipation in quantum barrier crossing: A numerical study," *J. Chem. Phys.,* vol. 96, p. 8485, 1992.
308. J. J. Halliwell and T. Yu, "Alternative derivation of the Hu-Paz-Zhang master equation for quantum Brownian motion," *Phys. Rev. D,* vol. 53, p. 2012, 1996.
309. D. Cohen, "Quantum dissipation versus classical dissipation for generalized Brownian motion," *Phys. Rev. Lett.,* vol. 78, p. 2878, 1997.
310. L. M. Johansen, "Hydrodynamical quantum state reconstruction," *Phys. Rev. Lett.,* vol. 80, p. 5461, 1998.
311. R. Karrlein and H. Grabert, "Exact time evolution and master equations for the damped harmonic oscillator," *Phys. Rev. E,* vol. 55, p. 153, 1997.
312. G. W. Ford and R. F. O'Connell, "Exact solution of the Hu-Paz-Zhang master equation," *Phys. Rev. D,* vol. 64, p. 105020, 2001.
313. S. Zhang and E. Pollak, "Quantum dynamics for dissipative systems: A numerical study of the Wigner–Fokker–Planck equation," *J. Chem. Phys.,* vol. 118, p. 4357, 2003.
314. G. S. Agarwal, "Brownian motion of a quantum oscillator," *Phys. Rev. A,* vol. 4, p. 739, 1971.
315. E. Pollak, J. Bader, B. J. Berne, et. al, "Theory of correlated hops in surface diffusion," *Phys. Rev. Lett.,* vol. 70, p. 3299, 1993.
316. W. T. Coffey, Yu. P. Kalmykov, S. V. Titov, et. al, "Quantum master equation in phase space: Application to the Brownian motion in a periodic potential," *Europhys. Lett.,* vol. 77, p. 20011, 2007.
317. W. T. Coffey, Yu. P. Kalmykov, S. V. Titov, et. al, "Semiclassical Klein–Kramers and Smoluchowski equations for the Brownian motion of a particle in an external potential," *J. Phys. A: Math. Theor.,* vol. 40, p. F91, 2007.
318. L. Cleary, W. T. Coffey, W. J. Dowling, et.al, "Phase space master equations for quantum Brownian motion in a periodic potential: comparison of various kinetic models," *J. Phys. A: Math. Theor.,* vol. 44, p. 475001, 2011.
319. E. P. Gross and J. L. Lebowitz, "Quantum theory of dielectric relaxation," *Phys. Rev.,* vol. 104, p. 1528, 1956.

320. H. Grabert, "Can quantum Brownian motion be Markovian?," *Chem. Phys.*, vol. 322, p. 160, 2006.
321. U. Weiss, Quantum Dissipative Systems, 2nd ed., Singapore: World Scientific, 1999.
322. D. Zueco and I. Calvo, "Bopp operators and phase-space spin dynamics: application to rotational quantum Brownian motion," *J. Phys. A: Math. Theor.*, vol. 40, p. 4635, 2007.
323. W. T. Coffey, Yu. P. Kalmykov, S. V. Titov, et. al, "Semiclassical master equation in Wigners phase space applied to Brownian motion in a periodic potential," *Phys. Rev. E,* vol. 75, p. 041117, 2007.
324. W. T. Coffey, Yu. P. Kalmykov, and S. V. Titov, "Solution of the master equation for Wigner's quasiprobability distribution in phase space for the Brownian motion of a particle in a double well potential," *J. Chem. Phys.*, vol. 127, p. 074502, 2007.
325. W. T. Coffey, Yu. P. Kalmykov, S. V. Titov, et.al, "Smoluchowski equation approach for quantum Brownian motion in a tilted periodic potential," *Phys. Rev. E,* vol. 78, p. 031114, 2008.
326. L. Cleary, W. T. Coffey, Yu. P. Kalmykov, et. al, "Semiclassical treatment of a Brownian ratchet using the quantum Smoluchowski equation," *Phys. Rev. E,* vol. 80, p. 051106, 2009.
327. W.T. Coffey, Yu. P. Kalmykov, S. V. Titov, et. al, "Nonlinear noninertial response of a quantum Brownian particle in a tilted periodic potential to a strong ac force as applied to a point Josephson junction," *Phys. Rev. B,* vol. 79, p. 054507, 2009.
328. W. T. Coffey, Yu. P. Kalmykov, S. V. Titov, et. al, "Master equation in phase space applied to the quantum Brownian motion in a tilted periodic potential," *J. Phys. A: Math. Theor.*, vol. 45, p. 105002, 2012.
329. D. A. Garanin and E. M. Chudnovsky, "Thermally activated resonant magnetization tunneling in molecular magnets: Mn12Ac and others," *Phys. Rev. B,* vol. 56, p. 11102, 1997.
330. D. A. Garanin and E. M. Chudnovsky, "Quantum-classical escape-rate transition of a biaxial spin system with a longitudinal field: A perturbative approach," *Phys. Rev. B,* vol. 59, p. 3671, 1999.
331. A. Caneschi, D. Gatteschi, et. al, "The molecular approach to nanoscale magnetism," *J. Magn. Magn. Mater.*, vol. 200, p. 182, 1999.
332. M. Oogane, T. Wakitani, S. Yakata, et al., "Magnetic damping in ferromagnetic thin films," *Jpn. J. Appl. Phys.*, vol. 45, p. 3889, 2006.
333. M. C. Hickey and J. S. Moodera, "Origin of intrinsic Gilbert damping," *Phys. Rev. Lett.*, vol. 102, p. 137601, 2009.
334. A. Aharoni, "Effect of a magnetic field on the superparamagnetic relaxation time," *Phys. Rev.*, vol. 177, p. 793, 1969.
335. W. T. Coffey, D. S. F. Crothers, and Yu. P. Kalmykov, "Effect of a uniform bias force on the Brownian movement in double-well potentials," *Phys. Rev. E,* vol. 55, p. 4812, 1997.

336. I. Klik and Y.D. Yao, "Eigenvalue spectrum of Brown's Fokker–Planck equation for coherent rotation of magnetization," *J. Magn. Magn. Mater.,* vol. 182, p. 335, 1998.
337. D.A. Garanin, "Integral relaxation time of single-domain ferromagnetic particles," *Phys. Rev. E,* vol. 54, p. 3250, 1996.
338. Yu.P. Kalmykov, W.T. Coffey, and S.V. Titov, "Analytic calculation of the longitudinal dynamic susceptibility of uniaxial superparamagnetic particles in a strong uniform dc magnetic field," *J. Magn. Magn. Mater.,* vol. 265, p. 44, 2003.
339. Yu.P. Kalmykov and S.V. Titov, "Calculation of longitudinal susceptibility of superparamagnetic particles," *Phys. Solid State,* vol. 45, p. 2140, 2003.
340. W. T. Coffey, "Exact solution for rotational relaxation times for problems governed by axially symmetric Fokker-Planck equations," *J. Mol. Struct.,* vol. 416, p. 221, 1997.
341. W. T. Coffey, D. S. F. Crothers, and S. V. Titov, "Escape times for rigid Brownian rotators in a bistable potential from the time evolution of the Green function and the characteristic time of the probability evolution," *Physica A,* vol. 298, p. 330, 2001.
342. W.T. Coffey, D.S.F. Crothers, J.L. Dormann, et al., "Effect of an oblique magnetic field on the superparamagnetic relaxation time," *Phys. Rev. B,* vol. 52, p. 15951, 1995.
343. W.T. Coffey, D.S. F. Crothers, J.L. Dormann, et al., "Effect of an oblique magnetic field on the superparamagnetic relaxation time. II. Influence of the gyromagnetic term," *Phys. Rev. B,* vol. 58, p. 3249, 1998.
344. E.C. Kennedy, "Relaxation times for single domain ferromagnetic particles," *Adv. Chem. Phys.,* vol. 112, p. 211, 2000.
345. H. Fukushima, Y. Uesaka, Y. Nakatani, et al., "Switching times of a single-domain particle in a field inclined off the easy axis," *J. Appl. Phys.,* vol. 101, p. 013901, 2007.
346. H. Fukushima, Y. Uesaka, Y. Nakatani, et al., "Dependence of prefactor on the angle between an applied field and the easy axis for single-domain particles," *J. Magn. Magn. Mater.,* vol. 323, p. 195, 2010.
347. H.J. Suh, C. Heo, C.Y. You, et al., "Attempt frequency of magnetization in nanomagnets with thin-film geometry," *Phys. Rev. B,* vol. 78, p. 064430, 2008.
348. Yu.P. Kalmykov, W.T. Coffey, U. Atxitia, et al., "Damping dependence of the reversal time of the magnetization of single-domain ferromagnetic particles for the Néel-Brown model: Langevin dynamics simulations versus analytic results," *Phys. Rev. B,* vol. 82, p. 024412, 2010.
349. J. Schratzberger, J. Lee, M. Fuger, et al., "Validation of the transition state theory with Langevin-dynamics simulations," *J. Appl. Phys.,* vol. 108, p. 033915, 2010.
350. A. Prudnikov, Yu. A. Brychkov, and O. I. Marichev, Integrals and Series, More Special Functions, Vols. 3, sec. 1.17.1, New York: Gordon and Breach, 1990.
351. H.B. Braun, "Kramers's rate theory, broken symmetries, and magnetization reversal," *J. Appl. Phys.,* vol. 76, p. 6310, 1994.
352. N. A. Usov and Yu. B. Grebenshchikov, "Superparamagnetic relaxation time of a single-domain particle with a nonaxially symmetric double-well potential," *J. Appl. Phys.,* vol. 105, p. 043904, 2009.

353. E. T. Whittaker and G. N. Watson, A Course of Modern Analysis, 4th ed., London: Cambridge University Press, 1927.
354. Yu.P. Kalmykov, W.T. Coffey, B. Ouari, *et al.*, "Damping dependence of the magnetization relaxation time of single-domain ferromagnetic particles," *J. Magn. Magn. Mater.*, vol. 292, p. 372, 2005.
355. Yu.P. Kalmykov, W.T. Coffey, and S.V. Titov, "Dependence of the magnetization relaxation time of single-domain ferromagnetic particles on damping in the Brown model," *Phys. Solid. State,* vol. 47, p. 272, 2005.
356. T. Taniguchi and H. Imamura, "Current dependence of spin torque switching rate based on Fokker-Planck approach," *J. Appl. Phys.,* vol. 115, p. 17C708, 2014.
357. J. Nogués, J. Sort, V. Langlais, et al., "Exchange bias in nanostructures," *Phys. Rep.*, vol. 422, p. 65, 2005.
358. D.V. Berkov and N.L. Gorn, "Numerical simulation of quasistatic and dynamic remagnetization processes with special applications to thin films and nanoparticles," in *Handbook of Advanced Magnetic Materials*, vol. II, Y. Liu, D. J. Sellmyer and D. Shindo, Eds., New York, Springer, 2006, p. 422.
359. D. A. Garanin, E. C. Kennedy, D. S. F. Crothers, and W. T. Coffey, "Thermally activated escape rates of uniaxial spin systems with transverse field: Uniaxial crossovers," *Phys. Rev. E,* vol. 60, p. 6499, 1999.
360. A. Aharoni, "Relaxation time of superparamagnetic particles with cubic anisotropy," *Phys. Rev. B,* vol. 7, p. 1103, 1973.
361. I. Eisenstein and A. Aharoni, "Asymptotic superparamagnetic time constants for cubic anisotropy. I. Positive anisotropy," *Phys. Rev. B,* vol. 16, p. 1278, 1977.
362. I. Eisenstein and A. Aharoni, "Asymptotic superparamagnetic time constants for cubic anisotropy. II. Negative anisotropy constant," *Phys. Rev. B,* vol. 16, p. 1285, 1977.
363. Yu.P. Kalmykov, "Reversal time of the magnetization of single-domain ferromagnetic particles with cubic anisotropy in the presence of a uniform magnetic field," *J. Appl. Phys.,* vol. 101, p. 093909, 2007.
364. D.A. Garanin and H. Kachkachi, "Surface contribution to the anisotropy of magnetic nanoparticles," *Phys. Rev. Lett.,* vol. 90, p. 065504, 2003.
365. A.J. Newell, "Superparamagnetic relaxation times for mixed anisotropy and high energy barriers with intermediate to high damping: 1. Uniaxial axis in a <001> direction," *Geochem., Geophys., Geosyst.,* vol. 7, p. Q03016, 2006.
366. A.J. Newell, "Superparamagnetic relaxation times for mixed anisotropy and high energy barriers with intermediate to high damping: 2. Uniaxial axis in a <111> direction," *Geochem., Geophys., Geosyst.,* vol. 7, p. Q03015, 2006.
367. P. Talkner and D. Ryter, "Lifetime of a metastable state at low noise," *Phys. Lett. A,* vol. 88, p. 162, 1982.
368. R. Bastardis, P.M. Déjardin, and Yu.P. Kalmykov, "Relaxation time of a Brownian rotator in a potential with nonparabolic barriers," *Physica A,* vol. 387, p. 3432, 2008.

369. W.T. Coffey, P.M. Déjardin, and Yu.P. Kalmykov, "Reversal time of the magnetization of single-domain ferromagnetic particles with mixed uniaxial and cubic anisotropy," *Phys. Rev. B,* vol. 79, p. 054401, 2009.
370. W.T. Coffey, D.S.F. Crothers, Yu.P. Kalmykov, et al., "Exact analytic formula for the correlation time of a single-domain ferromagnetic particle," *Phys. Rev. E,* vol. 49, p. 1869, 1994.
371. P.J. Cregg, D.S.F. Crothers, and A.W. Wickstead, "An approximate formula for the relaxation time of a single domain ferromagnetic particle with uniaxial anisotropy and collinear field," *J. Appl. Phys.,* vol. 4900, p. 76, 1994.
372. V.S. Anishchenko, V. Astakhov, A. Neiman, et al., Nonlinear Dynamics of Chaotic and Stochastic Systems, 2nd edition ed., Berlin: Springer, 2007.
373. A. Moser, K. Takano, D. T. Margulies et al., "Magnetic recording: advancing into the future," *J. Phys. D,* vol. 35, p. R157, 2002.
374. R.F. Fox, "Stochastic resonance in a double well," *Phys. Rev. A,* vol. 39, p. 4148, 1989.
375. J.A. Aseltine, Transform Method in Linear System Analysis, New York: McGraw-Hill, 1958.
376. Yu.L. Raikher and V.I. Stepanov, "Intrinsic magnetic resonance in superparamagnetic systems," *Phys. Rev. B,* vol. 51, p. 16428, 1995.
377. Yu. P. Kalmykov and S. V. Titov, "Calculating coefficients for a system of moment equations used to describe the magnetization kinetics of a superparamagnetic particle in a fluctuating field," *Phys. Solid State,* vol. 41, p. 1854, 1999.
378. J.L. Déjardin and Yu.P. Kalmykov, "Rotational diffusion and dielectric relaxation of linear molecules in a cubic potential," *J. Chem. Phys.,* vol. 111, p. 3644, 1999.
379. E. M. Chudnovsky and J. Tejada, Quantum Tunneling of the Magnetic Moment, Cambridge: Cambridge University Press, 1998.
380. L.D. Landau and E.M. Lifshitz, Mechanics, 3rd edition ed., London: Pergamon Press, 1976.
381. G.T. Landi, "Dynamic symmetry loss of high-frequency hysteresis loops in single-domain particles with uniaxial anisotropy," *J. Magn. Magn. Mater.,* vol. 324, p. 466, 2012.
382. G.T. Landi, "Influence of the magnetization damping on dynamic hysteresis loops in single domain particles," *J. Appl. Phys.,* vol. 111, p. 043901, 2012.
383. W.H. Meiklejohn and C.P. Bean, "New magnetic anisotropy," *Phys. Rev.,* vol. 102, p. 1413, 1956.
384. W.H. Meiklejohn and C.P. Bean, "New magnetic anisotropy," *Phys. Rev.,* vol. 105, p. 904, 1957.
385. V.P. Shilov, Yu.L. Raikher, J.C. Bacri, et al., "Effect of unidirectional anisotropy on the ferromagnetic resonance in ferrite nanoparticles," *Phys. Rev. B,* vol. 60, p. 11902, 1999.
386. K. Rivkin and J. B. Ketterson, "Switching spin valves using rf currents," *Appl. Phys. Lett.,* vol. 88, p. 192515, 2006.

387. P.P.Horley, V. R.Vieira, P.M.Gorley et. al, "Influence of a periodic magnetic field and spin-polarized current on the magnetic dynamics of a monodomain ferromagnet," *Phys. Rev. B,* vol. 78, p. 054417, 2008.
388. Y.-T. Cui, J. C. Sankey, C. Wang et. al, "Resonant spin-transfer-driven switching of magnetic devices assisted by microwave current pulses," *Phys. Rev. B,* vol. 77, p. 214440, 2008.
389. A. N. Slavin and V. S. Tiberkevich, "Nonlinear self-phase-locking effect in an array of current-driven magnetic nanocontacts," *Phys. Rev. B,* vol. 72, p. 092407, 2005.
390. J.-V. Kim, V. S. Tiberkevich, and A.N. Slavin, "Generation linewidth of an auto-oscillator with a nonlinear frequency shift: spin-torque nano-oscillator," *Phys. Rev. Lett.,* vol. 100, p. 017207, 2008.
391. J.-V. Kim, Q. Mistral, C. Chappert et. al, "Line shape distortion in a nonlinear auto-oscillator near generation threshold: application to spin-torque nano-oscillators," *Phys. Rev. Lett.,* vol. 100, p. 167201, 2008.
392. S. Urazhdin, V. Tiberkevich, and A. Slavin, "Parametric excitation of a magnetic nanocontact by a microwave field," *Phys. Rev. Lett.,* vol. 105, p. 237204, 2010.
393. L. Fricke, S. Serrano-Guisan, and H. W. Schumacher, "Parameter dependence of resonant spin torque magnetization reversal," *Physica B: Condensed Matter,* vol. 407, p. 1153, 2012.
394. X. Cheng, C. T. Boone, J. Zhu et. al, "Nonadiabatic stochastic resonance of a nanomagnet excited by spin torque," *Phys. Rev. Lett.,* vol. 105, p. 047202, 2010.
395. O. Prokopenko, G. Melkov, E. Bankowski et. al, "Noise properties of a resonance-type spin-torque microwave detector," *Appl. Phys. Lett.,* vol. 99, p. 032507, 2011.
396. A. Rebei, "Suppression of noise in magnetic layers by a periodic spin torque," *Phys. Rev. E,* vol. 83, p. 061148, 2011.
397. M. d'Aquino, C. Serpico, R. Bonin et.al, "Thermally induced synchronization and stochastic resonance between magnetization regimes in spin-transfer nano-oscillators," *J. Appl. Phys.,* vol. 111, p. 07C915, 2012.
398. W. T. Coffey, P. M. Corcoran, and M. W. Evans, "On the role of inertial effects and dipole-dipole coupling in the theory of the Debye and far-infrared absorption of polar fluids," *Proc. R. Soc. London Ser. A,* vol. 410, p. 61, 1987.
399. Yu.L. Raikher and V.I. Stepanov, "Magnetic relaxation in a suspension of antiferromagnetic nanoparticles," *J. Exp. Theor. Phys.,* vol. 107, p. 435, 2008.
400. Yu. L. Raikher and V. I. Stepanov, "Magneto-orientational behavior of a suspension of antiferromagnetic particles," *J. Phys.: Condens. Matter,* vol. 20, p. 204120, 2008.
401. L. Néel, "Superantiferromagnétisme dans les grains très fins," *Comptes rendus hebdomadaires des séances de l'Académie des sciences,* vol. 253, p. 203, 1961.
402. L. Néel, "Superparamagnétisme de grains très fins antiferromagnétiques," *Comptes rendus hebdomadaires des séances de l'Académie des sciences,* vol. 252, p. 4075, 1961.

403. L. Néel, "Superposition de l'antiferromagnétisme and du superparamagnétisme dans un grain très fin," *Comptes rendus hebdomadaires des séances de l'Académie des sciences,* vol. 253, p. 9, 1961.
404. L. Néel, "Sur le calcul de la susceptibilité additionnelle superantiferro-magnétique des grains très fins et sa variation thermique," *Comptes rendus hebdomadaires des séances de l'Académie des sciences,* vol. 253, p. 1286, 1961.
405. M. S. Seehra and A. Punnoose, "Deviations from the Curie-law variation of magnetic susceptibility in antiferromagnetic nanoparticles," *Phys. Rev. B,* vol. 64, p. 132410, 2001.
406. C. Gilles, P. Bonville, H. Rakoto et al., "Magnetic hysteresis and superantiferromagnetism in ferritin nanoparticles," *J. Magn. Magn. Mater.,* vol. 241, p. 430, 2002.
407. L. Zhang, D. Xue, and C. Gao, "Anomalous magnetic properties of antiferromagnetic CoO nanoparticles," *J. Magn. Magn. Mater.,* vol. 267, p. 111, 2003.
408. N. J. O. Silva, A. Millán, F. Palacio et. al, "Temperature dependence of antiferromagnetic susceptibility in ferritin," *Phys. Rev. B,* vol. 79, p. 104405, 2009.
409. Yu. L. Raikher, V. I. Stepanov, S. V. Stolyar, "Magnetic properties of biomineral particles produced by bacteria Klebsiella oxytoca," *Phys. Solid State,* vol. 52, p. 298, 2010.
410. D. A. Balaev, A. A. Krasikov, A. A. Dubrovskii et.al, "Magnetic properties and the mechanism of formation of the uncompensated magnetic moment of antiferromagnetic ferrihydrite nanoparticles of a bacterial origin," *J. Exp. Theor. Phys.,* vol. 119, p. 479, 2014.
411. J.H.E. Griffiths, "Anomalous high-frequency resistance of ferromagnetic metals," *Nature,* vol. 158, p. 670, 1946.
412. R. Berger, J. Kliava, J.-C. Bissey, and V. Baietto, "Magnetic resonance of superparamagnetic iron-containing nanoparticles in annealed glass," *J. Appl. Phys.,* vol. 87, p. 7389, 2000.
413. R. Berger, J.-C. Bissey, J. Kliava, H. Daubric, and C. Estournès, "Temperature dependence of superparamagnetic resonance of iron oxide nanoparticles," *J. Magn. Magn. Mater.,* vol. 234, p. 535, 2001.
414. R. S. de Biasi and T. C. Devezas, "Anisotropy field of small magnetic particles as measured by resonance," *J. Appl. Phys.,* vol. 49, p. 2466, 1978.
415. E. de Biasi, C. A. Ramos, and R. D. Zysler, "Size and anisotropy determination by ferromagnetic resonance in dispersed magnetic nanoparticle systems," *J. Magn. Magn. Mater.,* vol. 262, p. 235, 2003.
416. E. de Biasi, E. Lima, Jr., C. A. Ramos, A. Butera, and R. D. Zysler, "Effect of thermal fluctuations in FMR experiments in uniaxial magnetic nanoparticles: Blocked vs. superparamagnetic regimes," *J. Magn. Magn. Mater.,* vol. 326, p. 138, 2013.
417. J. Smit and H. G. Beljers, "Ferromagnetic resonance absorption in BaFe12O19 a highly anisotropic crystal," *Philips Res. Rep,* vol. 10, p. 113, 1955.

418. H. Suhl, "Ferromagnetic resonance in nickel ferrite between one and two kilomegacycles," *Phys. Rev.*, vol. 97, p. 555, 1955.
419. G. V. Skrotskii and L. V. Kurbatov, "Phenomenological theory of ferromagnetic resonance," in *Ferromagnetic Resonance*, S. Vonsovskii, Ed., Oxford, Pergamon, 1966, p. 12.
420. N. Noginova, F. Chen, T. Weaver, E. P. Giannelis, A. B. Bourlinos, and V. A. Atsarkin, "Magnetic resonance in nanoparticles: between ferro- and paramagnetism," *J. Phys.: Condens. Matter,* vol. 19, p. 246208, 2007.
421. N. Noginova, T.Weaver, E. P. Giannelis, A. B. Bourlinos, V. A. Atsarkin, and V. V. Demidov, "Observation of multiple quantum transitions in magnetic nanoparticles," *Phys. Rev. B,* vol. 77, p. 014403, 2008.
422. A. Sukhov, K. D. Usadel, and U. Nowak, "Ferromagnetic resonance in an ensemble of nanoparticles with randomly distributed anisotropy axes," *J. Magn. Magn. Mater.*, vol. 320, p. 31, 2008.
423. Yu. L. Raikher and V. I. Stepanov, "The effect of thermal fluctuations on the FMR line shape in dispersed ferromagnets," *Sov. Phys. JETP,* vol. 75, p. 764, 1992.
424. Yu. L. Raikher and V. I. Stepanov, "Ferromagnetic resonance in a suspension of single-domain particles," *Phys. Rev. B,* vol. 50, p. 6250, 1994.
425. I. S. Poperechny and Yu. L. Raikher, "Ferromagnetic resonance in uniaxial superparamagnetic particles," *Phys. Rev. B,* vol. 93, p. 014441, 2016.
426. I. S. Poperechny, Yu. L. Raikher, and V. I. Stepanov, "Ferromagnetic resonance in a dilute suspension of uniaxial superparamagnetic particles," *J. Magn. Magn. Mater.*, vol. 424, p. 185, 2017.
427. Yu.P. Kalmykov, S.V. Titov, W.T. Coffey, and W.J. Dowling, "Finite-barrier correction for the ferromagnetic resonance frequency of nanomagnets with various magnetocrystalline anisotropies," *Phys.Rev. B,* vol. 97, p. 224418, 2018.
428. W. T. Coffey, Yu. P. Kalmykov, S. V. Titov, and D. S. F. Crothers, "High-frequency resonance absorption as evidence for oscillation in a well before escape from a metastable state in the Kramers energy diffusion model," *Recent Advances in Broadband Dielectric Spectroscopy,* p. 151, 2013.
429. A. Lyberatos, E. P. Wohlfarth, and R. W. Chantrell, "Simulated annealing: An application in fine particle magnetism," *IEEE Trans. Magn.*, Vols. MAG-21, p. 1277, 1985.
430. J.-O. Andersson, C. Djurberg, T. Jonsson, et al., "Monte Carlo studies of the dynamics of an interacting monodispersive magnetic-particle system," *Phys. Rev. B,* vol. 56, p. 13983, 1997.
431. D.V. Berkov and N. L. Gorn, "Susceptibility of the disordered system of fine magnetic particles: a Langevin-dynamics study," *J. Phys.: Condens. Matter,* vol. 13, p. 9369, 2001.
432. A. Lyberatos and R. W. Chantrell, "Thermal fluctuations in a pair of magnetostatically coupled particles," *J. Appl. Phys.,* vol. 73, p. 6501, 1993.

433. D. Hinzke and U. Nowak, "Magnetic relaxation in a classical spin chain," *Phys. Rev. B,* vol. 61, p. 6734, 2000.
434. D. Hinzke and U. Nowak, "Magnetization switching in nanowires: Monte Carlo study with fast Fourier transformation for dipolar fields," *J. Magn. Magn. Mater.,* vol. 21, p. 365, 2000.
435. S. I. Denisov and K. N. Trohidou, "Fluctuation theory of magnetic relaxation for two-dimensional ensembles of dipolar interacting nanoparticles," *Phys. Rev. B,* vol. 64, p. 184433, 2001.
436. J. L. Dormann, L. Bessais, and D. Fiorani, "A dynamic study of small interacting particles: superparamagnetic model and spin-glass laws," *J. Phys. C,* vol. 21, p. 2015, 1988.
437. S. Mørup and E. Tronc, "Superparamagnetic relaxation of weakly interacting particles," *Phys. Rev. Lett.,* vol. 72, p. 3278, 1994.
438. P. E. Jonsson and J. L. Garcia-Palacios, "Relaxation time of weakly interacting superparamagnets," *Europhys. Lett.,* vol. 55, p. 418, 2001.
439. D. Rodé, H. N. Bertram, and D. R. Fradkin, "Effective volume of interacting particles," *IEEE Trans. Magn.,* Vols. MAG-23, p. 2224, 1987.
440. W. Chen, S. Zhang, and H. N. Bertram, "Energy barriers for thermal reversal of interacting single domain particles," *J. Appl. Phys.,* vol. 71, p. 5579, 1992.
441. A. Yoshimori and J. Korringa, "A Fokker-Planck Equation for Spin Relaxation," *Phys. Rev.,* vol. 128, p. 1054, 1962.
442. J. Korringa and A. Yoshimori, "Theory of Relaxation of a Two-Spin System," *Phys. Rev.,* vol. 128, p. 1060, 1962.
443. I. Solomon, "Relaxation Processes in a System of Two Spins," *Phys. Rev.,* vol. 99, p. 559, 1955.
444. A. Yu. Zubarev and L. Yu. Iskakova, "Dynamical susceptibility of solid ferrocolloids," *Phys. Rev. E,* vol. 63, p. 061507, 2001.
445. A. O. Ivanov and O. B. Kuznetsova, "Magnetic properties of dense ferrofluids: An influence of interparticle correlations," *Phys. Rev. E,* vol. 64, p. 041405, 2001.
446. A. O. Ivanov, V. S. Zverev, and S. S. Kantorovich, "Revealing the signature of dipolar interactions in dynamic spectra of polydisperse magnetic nanoparticles," *Soft Matter,* vol. 12, p. 3507, 2016.
447. A. O. Ivanov, S. S. Kantorovich, V. S. Zverev, et al., "Temperature-dependent dynamic correlations in suspensions of magnetic nanoparticles in a broad range of concentrations: a combined experimental and theoretical study," *Phys. Chem. Chem. Phys.,* vol. 18, p. 18342, 2016.
448. A. O. Ivanov and V. S. Zverev, "Combined Fokker-Planck-Brown and Yvon approach for describing the dynamic magnetic response of interacting ferroparticles in magnetic fluids," *Magnetohydrodynamics,* vol. 52, p. 43, 2016.
449. B. U. Felderhof and R. B. Jones, "Mean field theory of the nonlinear response of an interacting dipolar system with rotational diffusion to an oscillating field," *J. Phys.: Condens. Matter,* vol. 15, p. 4011, 2003.

450. R. Zwanzig, "Dielectric Relaxation in a High-Temperature Dipole Lattice," *J. Chem. Phys.*, vol. 38, p. 2766, 1963.
451. P.M. Déjardin, S. V. Titov, and Y. Cornaton, "Linear complex susceptibility of long-range interacting dipoles with thermal agitation and weak external ac fields," *Phys. Rev. B,* vol. 99, p. 024304, 2019.
452. S.V. Titov, H. Kachkachi, Yu. P. Kalmykov, and W.T. Coffey , "Magnetization dynamics of two interacting spins in an external magnetic field," *Phys. Rev. B,* vol. 72, p. 134425, 2005.
453. Y. P. Kalmykov, S. V. Titov, D. J. Byrne, W. T. Coffey, M. Zarifakis, and M. H. Al Bayyari, "Dipole-dipole and exchange interaction effects on the magnetization relaxation of two macrospins: Compared", *J. Magn. Magn. Mater.*, vol. 507, p. 166814, 2020.
454. H. Kachkachi, "Effect of exchange interaction on superparamagnetic relaxation," *Europhys. Lett.*, vol. 62, p. 650, 2003.
455. G. Joos, Theoretical Physics, London: Blackie & Son Limited, 1934.
456. A. Abragam, The Principles of Nuclear Magnetism, London: Oxford University Press, 1961.
457. D. Gatteschi, R. Sessoli, and J. Villain, Molecular Nanomagnets, Oxford: Oxford University Press, 2006.
458. R. E. Rosensweig, Ferrohydrodynamics, 2nd ed., New York: Dover, 1997.
459. B.K.P. Scaife. Principles of Dielectrics, 2nd ed., Oxford University Press, London, 1998.
460. F. Bloch, "Nuclear induction," *Phys. Rev.*, vol. 70, p. 460, 1946.
461. F. Bloch, "Dynamical theory of nuclear induction. II," *Phys. Rev.*, vol. 102, p. 104, 1956.
462. F. Bloch, "Generalized theory of relaxation," *Phys. Rev.*, vol. 105, p. 1206, 1956.
463. W. E. Henry, "Spin Paramagnetism of Cr+++, Fe+++, and Gd+++ at Liquid Helium Temperatures and in Strong Magnetic Fields," *Phys. Rev.*, vol. 88, p. 559, 1952.
464. Z. Salman, A. Keren, P. Mendels, et. al, "Dynamics at T0 in half-integer isotropic high-spin molecules," *Phys. Rev. B,* vol. 65, p. 132403, 2002.
465. H. J. Lipkin, N. Meshkov and A. J. Glick, "Validity of many-body approximation methods for a solvable model: (I). Exact solutions and perturbation theory," *Nucl. Phys.*, vol. 62, p. 188, 1965.
466. N. Meshkov, A. J. Glick, and H. J. Lipkin, "Validity of many-body approximation methods for a solvable model:(II). Linearization procedures," *Nucl. Phys.*, vol. 62, p. 199, 1965.
467. R. Ullman, "Rotatory motion of dipolar molecules constrained by a steady electric field," *J. Chem. Phys.*, vol. 56, p. 1869, 1972.
468. M. A. Martsenyuk, Yu. L. Raikher, and M. I. Shliomis, "On the kinetics of magnetization of suspensions of ferromagnetic particles," *Sov. Phys. JETP,* vol. 38, p. 413, 1974.
469. A. Morita, "On nonlinear dielectric relaxation," *J. Phys. D,* vol. 11, p. 1357, 1978.

470. H. Watanabe and A. Morita, "Kerr effect relaxation in high electric fields," *Adv. Chem. Phys.*, vol. 56, p. 255, 1984.
471. J. T. Waldron, Yu. P. Kalmykov, and W. T. Coffey, "Rotational Brownian motion and dielectric relaxation of polar molecules subjected to a constant bias field: Exact solution," *Phys. Rev. E,* vol. 49, p. 3976, 1994.
472. E. Geva, E. Rosenman, and D. Tannor, "On the second-order corrections to the quantum canonical equilibrium density matrix," *J. Chem Phys.*, vol. 113, p. 1380, 2000.
473. F. Shibata, Y. Takahashi, and N. Hashitsume, "A generalized stochastic Liouville equation. Non-Markovian versus memoryless master equations," *J. Stat. Phys.*, vol. 17, p. 171, 1977.
474. F. Luis, V. González, A. Millán, et. al, "Large quantum nonlinear dynamic susceptibility of single-molecule magnets," *Phys. Rev. Lett.*, vol. 92, p. 107201, 2004.
475. R. López-Ruiz, F. Luis, V. González, et. al, "Nonlinear response of single-molecule nanomagnets: Equilibrium and dynamical," *Phys. Rev. B,* vol. 72, p. 224433, 2005.
476. R. López-Ruiz, F. Luis, A. Millán, et. al, "Nonlinear response of single-molecule magnets: Field-tuned quantum-to-classical crossovers," *Phys. Rev. B,* vol. 75, p. 012402, 2007.
477. J. L. Déjardin, and Yu. P. Kalmykov, "Nonlinear dielectric relaxation of polar molecules in a strong ac electric field: Steady state response," *Phys. Rev. E,* vol. 61, p. 1211, 2000.
478. J. L. Déjardin, and Yu. P. Kalmykov, "Steady state response of the nonlinear dielectric relaxation and birefringence in strong superimposed ac and dc bias electric fields: Polar and polarizable molecules," *J. Chem. Phys.*, vol. 112, p. 2916, 2000.
479. R. Löfstedt and S. N. Coppersmith, "Quantum stochastic resonance," *Phys. Rev. Lett.*, vol. 72, p. 1947, 1994.
480. R. Löfstedt and S. N. Coppersmith, "Stochastic resonance: Nonperturbative calculation of power spectra and residence-time distributions," *Phys. Rev. E,* vol. 49, p. 4821, 1994.
481. M. Grifoni, L. Hartmann, S. Berchtold, et. al, "Quantum tunneling and stochastic resonance," *Phys. Rev. E,* vol. 53, p. 5890, 1996.
482. P. Reimann and P. Hänggi, "Quantum features of Brownian motors and stochastic resonance," *Chaos,* vol. 8, p. 629, 1998.
483. I. Goychuk and P. Hänggi, "Quantum stochastic resonance in symmetric systems," *Phys. Rev. E,* vol. 59, p. 5137, 1999.
484. I. Goychuk, J. Casado-Pascual, M. Morillo, et. al, "Quantum stochastic synchronization," *Phys. Rev. Lett.*, vol. 97, p. 210601, 2006.
485. H. Bateman, and A. Erdelyi, Higher Transcendental Functions, vol. 1, New York: McGraw-Hill, 1953.
486. H.S. Wall, Continued fractions, Princeton, USA: Van Nostrand, 1948.
487. I.S. Gradshtein, and I.M. Ryzhik, Tables of integrals, series, and products, Amsterdam: Elsevier, 2007.

488. A. Prudnikov, Yu. A. Brychkov, and O. I. Marichev, Integrals and Series: More Special Functions, vol. 2, New York: Gordon and Breach, 1988.
489. G.B. Arfken, H.J. Weber, and F. E. Harris, Mathematical methods for physicists, Amsterdam: Elsevier, 2012.
490. A. Szabo, “Theory of polarized fluorescent emission in uniaxial liquid crystals,” *J. Chem. Phys.,* vol. 72, p. 4620, 1980.
491. G. Moro and P. L. Nordio, “Diffusive and jump description of hindered motions,” *Mol. Phys.,* vol. 56, p. 255, 1985.
492. A. N. Malakhov and A. L. Pankratov, “Evolution times of probability distributions and averages-exact solutions of the Kramers' problem,” *Adv. Chem. Phys.,* vol. 121, p. 357, 2002.
493. H. Weyl, "Quantenmechanik und Gruppentheorie," *Zeitschrift für Physik*, vol. 46, p. 1, 1927.
494. H. Groenewold, “On the principles of elementary quantum mechanics“, *Physica*, vol. 12 p. 405 (1946),

Index

A

action, 68, 70, 96, 98, 108, 110, 111, 113, 232, 234-236, 238, 240, 244, 246, 247, 253, 256, 258, 264, 267, 272, 273, 281, 283, 420

after-effect function, 131, 296, 513

angular momentum operator, 49, 137, 138, 152, 198, 447, 614, 615, 618, 619, 637

anisotropy constant, 12, 14, 114, 115, 128, 136, 266, 268, 271, 329, 336, 338, 366, 408, 421, 425, 426, 428, 432, 435, 459, 494, 501, 576

antiferromagnetic nanoparticle, 325, 405-408, 424, 425

Arrhenius law (behavior), 92, 121, 124, 242, 251, 286, 302, 326, 337, 342, 344, 389, 463, 499, 601

associated Legendre functions $P_l^m(x)$, 45, 614

asteroid, 114, 118, 120, 121, 470, 471, 476, 500, 501, 597

autocorrelation function (ACF), 131, 133, 174, 301, 314, 334, 341

B

biaxial anisotropy, 13, 15, 41, 77, 117, 236-238, 242, 243, 246, 264, 358, 359, 375, 393, 431, 432, 490

Bloch equation, 21, 22, 536, 537, 561

blocking temperature, 3, 4, 74, 75, 120, 272

Brown's asymptotic formula, 123, 128, 236, 259, 299, 353

Brown's Fokker–Planck equation, 43, 254

Brownian motion, 4, 5, 37, 38, 84, 86, 94, 95, 122, 132, 153, 159, 166-168, 182, 187-189, 191, 192, 214, 217, 219, 221, 358, 426, 439, 472, 510, 519, 524, 525, 575, 595, 596, 599

C

circularly symmetric, 84, 93, 126, 222, 226-228, 235, 236, 252, 253, 259, 270, 287, 291, 318, 336, 421, 460, 473, 505, 506, 544, 603

Clebsch–Gordan coefficients $C^{c\gamma}_{a\alpha b\beta}$, 47, 51, 193, 198, 204, 207, 210, 224, 287, 320, 615, 624, 630, 633, 636, 637

complete elliptic integral of the first kind $K(m)$, 233, 241, 248, 430, 620, 621

complete elliptic integral of the second kind $E(m)$, 248, 620, 621

complete elliptic integral of the third kind $\Pi(v,m)$, 248, 620, 621

complex susceptibility, 126, 131, 133, 135, 291, 296, 300, 314, 318, 322, 334, 445, 452, 457

confluent hypergeometric (Kummer) function $M(a,b,z)$, 290, 605, 612

continued fraction(s), 54-58, 121, 122, 131, 132, 136, 165, 167, 214, 215, 217, 224, 225, 242, 250, 260, 262, 265, 268, 269, 284, 285, 287, 289, 291, 293, 296, 301, 313, 314, 317, 318, 322, 323, 325, 328, 329, 332, 333, 337, 340, 348, 353, 361, 367, 368, 370, 375, 385, 391, 392, 394,

412, 413, 428, 438, 441, 445, 455, 458, 465, 466, 472, 513, 527, 528, 534, 536-538, 546, 547, 561, 565, 566, 585, 594-596, 606, 612, 643
continuity equation, 34, 87, 89, 103
correlation function, 38, 51, 58, 59, 130, 132, 133, 151-156, 158, 160, 288, 289, 293, 294, 313, 340, 444, 452, 461, 473, 532, 536, 543, 577, 587, 641, 643, 644, 646, 647
correlation time, 83, 159, 161, 188, 218, 294, 296-298, 301, 344, 444, 462, 509, 518, 532, 537, 552, 557, 579, 587, 599, 640, 644, 646, 647
cubic anisotropy, 13-15, 35, 77, 78, 117, 133, 212, 224, 225, 266, 268, 270, 271, 273, 329, 334-339, 435, 476, 492

D

Debye equation, 227, 316
Debye relaxation time, 122, 126, 129
Debye theory, 71, 124, 140
depletion effect, 297, 301, 306, 325, 327, 342, 456, 458, 465
depopulation factor, 99, 113, 256, 272, 284
differential-recurrence relations, 30, 46, 58, 127, 136, 164, 222, 224, 288, 289, 346, 359, 393, 394, 408, 441, 448, 460, 510, 562, 583, 593, 602, 633, 643
diffusion coefficient, 29, 32, 33, 67, 188-191, 505, 507, 516, 518, 519, 521, 524, 529, 588
diffusion equation, 36, 95, 98, 441
dipole moment, 8, 11, 124, 133, 139, 140, 345, 442, 444, 510
Dirac-delta function $\delta(t)$, 154, 160, 169, 174, 184, 496, 622
discrete orientation model, 76, 80, 227, 257, 266, 268
distribution function, 26, 59, 62, 66, 70, 73, 81-83, 94-99, 123, 130, 146, 166-172, 178-180, 190, 197, 199, 208, 214, 253, 263, 274, 275, 282, 321, 358, 385, 407, 440, 442, 445, 460, 472, 473, 484, 485, 497, 502, 509, 510, 515, 518, 521, 523, 526-528, 531, 539, 576, 587, 641, 644, 647
drift coefficient, 29, 32
dynamic magnetic hysteresis (DMH), 4, 222, 225, 262, 328, 356, 364-380, 385, 392, 393, 401-405, 415, 421-425, 473, 569-575, 568, 597, 598

E

effective eigenvalue, 130, 132, 136, 291, 314-318
effective relaxation time, 59, 60, 131, 135, 136, 291, 292, 294, 299, 310, 322, 334, 443, 444, 462, 542, 543, 553, 557, 561, 578, 587, 641, 643
eigenvalue, 47, 54, 60, 79, 80, 96, 102, 103, 105, 107, 125, 126, 130, 133, 136, 143, 155, 172, 215, 255, 266, 294, 298, 301, 309, 311, 314-318, 329, 444, 445, 513, 535, 541, 542, 547, 558, 577, 578, 583, 587, 642
eigenvector, 54, 103, 155, 158, 169, 215, 513, 541, 542, 583
energy-controlled diffusion equation, 62, 68-70, 86, 87, 89, 111, 228
ensemble average, 59, 303, 443
error function $\mathrm{erf}(z)$, 103, 276, 607
error function of imaginary argument $\mathrm{erfi}(z)$, 236, 550, 607, 611
escape rate, 5, 25, 36-39, 53, 58, 61, 62, 83, 86, 90-99, 101, 105, 107-110, 112, 113, 121, 122, 168, 182, 185-188, 217, 224-227, 232, 236, 242-244, 248, 250-258, 261, 263-266, 270-275, 279-282, 285-287, 325, 335-337, 390, 396, 405, 419, 438, 441, 457, 465, 466, 470, 472, 494, 496, 498-500, 504, 595, 596, 598, 601
Euler angles, 128, 138, 617, 618, 628

F

ferrofluids, 71, 122-126, 128-136
ferromagnetic resonance (FMR), 33, 36, 105, 124, 126, 130, 132, 134, 225, 289, 313, 314, 327, 345, 355, 356,

362-365, 378, 392, 397, 398, 401-404, 418, 425-428, 430, 431, 435-439, 470, 596, 598
fluctuation–dissipation theorem, 94, 442, 525
Fokker–Planck equation, 26-29, 33, 35-37, 39, 43-47, 53, 62, 66-70, 73, 81, 83, 85, 87, 90, 95, 96, 100-103, 106, 125, 130, 159, 167, 188-192, 214, 217, 218, 221-224, 245, 253-255, 274, 302, 329, 342, 404-408, 421, 440, 441, 445, 472, 504, 505, 509, 510, 515, 518, 521-523, 529, 538, 539, 561, 577, 582, 583, 593-595, 598, 602, 640, 641, 644, 645, 647
Fokker–Planck operator, 36, 46, 47, 60, 61, 82-84, 86, 89, 108, 232, 242, 252, 284, 285, 301, 352, 388, 391, 396, 417, 444, 445, 465, 472, 508, 597, 642, 644

G
gamma function $\Gamma(z)$, 259, 262, 278, 605
Gilbert's equation, 14, 19, 37, 52, 108, 221, 257, 443, 585

H
Hermite polynomials $H_n(z)$, 181, 608

I
integral relaxation time, 59, 292, 299, 301-304, 306, 325, 326, 335, 341, 343, 398, 443-445, 452, 454, 456-458, 461-464, 505, 516, 529, 532, 534, 537, 538, 541, 542, 545, 546, 548-550, 552, 557, 578, 587, 594, 641, 643-645, 648
intermediate-to-high damping (IHD), 37, 38, 95-101, 105-109, 11-113, 133, 252-256, 260, 263, 266, 268, 270, 272-276, 278-282, 328, 337, 338, 343, 345, 441, 445, 446, 452-455, 458, 459

J
Jacobi elliptic functions $\mathrm{sn}(u|m)$, 233, 239, 247, 248, 620, 621
joint probability density function (PDF), 96, 125, 130

K
Kramers theory, 5, 99, 114, 257, 282
Kronecker's delta δ_{mn}, 25, 153, 520, 609
Kubo model, 26, 32, 33, 36

L
Laguerre polynomials $L_n(z)$, 181, 608, 609
Landau–Lifshitz equation, 18, 38, 127, 291, 314, 317, 441, 445
Langer equation, 105, 107, 109, 278, 279
Langevin equation, 24-27, 36, 38-40, 43, 47, 48, 52, 62, 63, 66, 69-71, 73, 76, 90, 95-97, 100, 103, 105, 107, 122, 126, 127, 130, 131, 133, 136, 217, 221-224, 244, 251, 255, 301, 346, 354, 380, 390, 391, 393, 404, 426, 427, 431, 435, 440-442, 465, 466, 472, 519-522, 595, 596, 598
Langevin function $L(\xi)$, 122, 140, 479, 482, 601
Legendre polynomials, $P_n(x)$, 261, 295, 477, 484, 485, 511, 514, 526, 540, 609, 610, 614, 639, 640, 643
linear response theory, 58, 131, 301, 308, 309, 321, 334, 368, 444, 461, 532, 578, 586, 646
longest relaxation time, 36, 61, 86, 294, 297, 301, 337, 453, 553, 555, 559, 561, 588, 643
longitudinal susceptibility, 58, 61, 130, 291, 309, 334, 452, 466, 577
longitudinal response, 126, 134, 225, 226, 293, 313, 318
longitudinal relaxation, 21, 60, 108, 124-126, 130, 159, 289, 293, 317, 336, 345, 354, 358, 362, 363, 418, 419, 473, 506, 509, 516, 518, 519, 523, 527, 536-541, 549, 561-563, 576, 602, 603, 642

M

macroscopic quantum tunneling, 76, 339, 471, 472, 504, 538, 601
magnetic after-effect, 71, 72
magnetic domains, 9, 10, 83
magnetic stochastic resonance, 225, 306, 307, 336, 465, 596, 597, 603
magnetic viscosity, 72, 74, 127
matrix continued fraction, 54-58, 121, 122, 131, 132, 136, 165, 215, 217, 224, 225, 242, 250, 260, 262, 265, 268-270, 284, 285, 287, 289, 293, 296, 301, 313, 314, 317, 318, 322, 323, 325, 328, 329, 333, 337, 340, 348, 353, 361, 367, 368, 375, 385, 391, 392, 394, 397, 412, 413, 428, 438, 441, 445, 455, 458, 465, 513, 565, 566, 594-596, 643
Maxwell–Boltzmann distribution, 94, 96, 97, 168, 179, 182, 190, 505
mean first-passage time (MFPT), 36, 39, 80, 81, 83-87, 89, 90, 110-112, 167, 214, 225-228, 231, 235-237, 244, 250, 299, 302, 420, 421, 472, 505, 594
mixed anisotropy, 121, 212, 270-272, 280, 281, 435, 502
modified Bessel functions of the first kind $I_\nu(z)$, 565, 606, 611, 613
modified Bessel functions of the second kind $K_\nu(z)$, 278, 611, 613
modified spherical Bessel functions $I_{n+1/2}(z)$, 611, 613
multiplicative noise, 29, 31, 63, 519

N

nanomagnet, 3, 5, 24, 27, 38, 58, 113, 114, 128, 221, 222, 225, 245, 252, 254, 256, 263, 268, 270, 282, 287, 301, 312, 318, 319, 334, 336, 339, 345, 355, 358, 375, 380, 384, 391, 392, 399, 401, 404, 425, 426, 428, 465, 466, 469, 472, 501, 538, 561, 562, 569, 572, 574, 575, 578, 582, 592, 594, 595, 598, 603
nanopillar, 53, 237, 380, 390, 391, 404, 432, 598
Néel relaxation, 72, 75, 122-125, 127, 187
Néel relaxation time, 73, 76, 124, 130, 131, 133
noise-induced drift, 29, 48, 443
nonlinear response, 345, 351, 357, 368, 362, 398, 399, 401, 403-406, 408, 414, 416, 423, 425, 441, 444, 461, 513, 528, 537, 561-563, 568, 571, 596, 602, 603, 646, 647

P

P-function, 171,206, 207
paleomagnetism, 4, 74, 272, 469, 472
partition function, 34, 59, 88, 93, 101, 109, 139, 140, 176, 183, 184, 275, 278, 279, 292, 303, 304, 306, 321, 352, 427, 442, 474, 477, 479, 482, 486, 487, 494, 495, 497, 498, 611
phase space, 100, 106, 142, 145, 165, 166-168, 172, 176, 177, 192, 193, 194, 196, 197, 199, 210, 211, 212, 213, 214, 218, 223, 246, 467, 473, 501, 505, 515, 516, 518, 592, 599, 641
phase-space distribution, 167-171, 176, 180, 188, 190, 197, 205, 207, 209, 215, 216, 473-475, 477, 480, 482, 483, 487, 489, 491, 494, 497, 499, 504, 508, 510, 511, 515, 521, 524, 525, 539, 540, 576, 583, 593, 594, 600, 639
phase-space master equation, 190, 214, 215, 220, 473, 509-511, 517, 519, 523, 538, 544, 582, 583, 593, 594, 602
phase-space representation, 166, 172, 193, 208, 213, 216, 218, 219, 475, 500, 502, 504, 507, 508, 512, 515, 516, 561, 582, 591, 594, 568, 631, 632, 636
Poisson bracket, 142

polarization operators, 162-165, 198-200, 203-206, 210, 211, 213-220, 473-475, 478, 510-513, 515, 539, 540, 544, 562, 584, 593, 599, 600, 602, 603, 623, 625-630, 633-636, 641
probability density function, 27, 33, 34, 45, 47, 51, 68, 125, 130, 142, 143, 475, 480, 521

Q
Q-function, 171, 204, 212, 475, 478, 480, 584, 600
quantum oscillator, 191,192
quasiprobability distribution function, 165, 166, 168, 171, 193, 198-200, 205, 219, 473, 480, 599

R
random process, 99,100
reversal time, 4, 36, 54, 57, 61, 62, 77, 83, 86-90, 109-114, 120, 222, 225-230, 236, 243-268, 271, 272, 279-289, 319, 321, 325, 329, 334, 335, 343, 352-355, 363, 386-391, 396, 397, 400, 404-406, 417, 419, 421, 456, 463-466, 469, 473, 476, 494, 499, 502, 504, 542, 553, 567, 568, 572, 574, 578, 582, 586, 587, 589, 594, 596, 598, 642, 643,
rotational Brownian motion, 84, 132, 159, 167, 214, 221, 358, 525

S
separatrix, 42, 88-90, 98, 110, 240, 264, 267, 272-274, 420, 439
sharp values, 28, 30, 48, 51, 222, 287, 443, 447
signal-to-noise ratio (SNR), 4, 262, 307, 308-312, 328, 337-339, 380, 577-581
smallest nonvanishing eigenvalue, 36, 54, 57, 58, 61, 83, 108, 134, 222, 224, 232, 242, 250, 252, 184, 297, 298, 301, 335, 396, 542, 551, 553, 554, 578, 587, 597, 642
spectral density, 157, 160, 174, 188, 307-309
spherical harmonics, $Y_{lm}(\vartheta,\varphi)$, 30, 45, 46, 49-53, 58, 59, 131, 137, 197-200, 203-207, 209, 215, 216, 218-220, 222-224, 287, 320, 325, 329, 359, 392, 408, 442, 446, 447, 460, 473-475, 477, 502, 510, 511, 513, 515, 521, 522, 539, 563, 582, 583, 591, 593, 595, 599, 602, 603, 614, 616, 617, 619, 630, 636, 638, 641, 643
spin operator, 138, 139, 152, 160, 165, 167, 198, 200, 204, 208-213, 216, 220, 474, 478, 480, 481, 483, 489, 491, 515, 523, 528, 539, 545, 584, 586, 593, 594, 599-601, 615, 623, 624, 626, 627, 629, 641
spin-transfer torque (STT), 38-44, 46, 52, 53, 69, 70, 89, 90, 112, 113, 225, 228, 245, 251, 281, 282, 284-287, 358, 380, 381, 390-392, 397-404, 597, 598
stationary process, 94
statistical moment, 30, 37, 45-48, 52-54, 125, 127, 162-165, 197, 204, 214-218, 222-224, 284, 320, 329, 346, 359, 366, 382, 386, 390, 392, 393, 404, 408, 413, 441, 445, 465, 473, 485, 510-512, 522, 582, 583, 591, 593, 595, 597-599, 641-643
steepest descents, 97, 273, 275, 421
stochastic differential equation, 27, 29, 31, 48, 52, 95, 127, 223, 446, 472, 519, 521
stochastic resonance, 4, 221, 222, 225, 262, 289, 290, 306-308, 328, 336, 345, 385, 391, 404, 465, 473, 562, 575, 596-598, 603
Stosszahlansatz, 151, 182, 189
Stratonovich definition, 48,519
STT effect, 38, 42, 44, 69, 90, 225, 228, 245, 251, 281, 282, 285-287, 380, 381, 390-392, 397, 401, 403, 404, 597, 598
Stoner–Wohlfarth orbit, 16, 17, 19, 62, 64-66, 69, 71, 110, 112, 201
superparamagnetic relaxation time, 4, 153
superparamagnetism, 3, 11, 26, 221, 287, 291, 307, 364, 369, 465, 469, 596

switching-field curves (surfaces), 4, 114-122, 261, 271, 404, 465, 470, 473, 476, 499, 500, 501, 503, 596, 598

T

transition probability, 222, 223
transition state theory (TST), 37, 38, 73, 91-98, 109, 110, 182, 183, 185, 186, 227, 255, 262, 282, 456, 466, 471, 476, 494, 496, 498, 499, 502, 504, 596, 601
transverse susceptibility, 130, 136, 291, 313, 438
transverse relaxation time, 21
transverse response, 106, 126, 130, 136, 225, 226, 313, 317, 318
turnover formula, 253, 257, 263, 266, 267, 269, 272, 281, 282, 337
turnover problem, 253
two-mode approximation, 60, 291, 299-301, 368, 559, 560, 570

U

uniaxial anisotropy, 12, 15, 21, 64, 73, 76, 77, 83, 93, 101, 114, 117, 120, 123, 126-128, 130, 132, 133, 186, 228, 230, 232, 238, 240, 242-244, 250, 259, 265, 271, 281, 289, 291, 302, 319, 336, 338, 339, 375, 428, 433, 435, 440, 459, 471, 500, 506, 563, 582, 585
uniaxial nanomagnet, 86, 225, 228, 235, 257, 260, 281, 287, 289, 293, 306, 314, 318, 319, 325, 342, 345-348, 350, 364, 420, 428, 463, 473, 476, 483, 484, 486, 489, 497, 499, 505, 506, 508, 514-517, 534, 538, 539, 544, 549, 552, 557, 560, 561-563, 567, 569, 570, 574, 576, 577, 582, 583, 585, 586, 591, 602, 603

V

very high damping (VHD), 96, 97, 260, 268, 270
very low damping (VLD), 36, 44, 62, 63, 68-70, 86-90, 96-99, 105,106, 109-113, 217, 225, 226, 228-230, 234, 236, 242-245, 248-260, 266-270, 281, 282, 285-287, 337, 388, 418, 426, 466, 595

W

Weyl symbol, 167, 176, 177, 194, 208, 209, 211-214, 216, 218, 220, 478, 485, 497, 499, 501, 503, 508, 509, 515, 544, 556, 586, 593, 594, 600-602, 631, 632, 638-640
white noise, 25, 27, 30, 37, 38, 40, 49, 69, 126, 129, 156, 160, 257, 442, 522
Wigner function, 167, 171, 172, 174-177, 180-182, 184, 192, 207, 219, 473, 480, 483, 484, 489, 490, 492, 502-504, 518, 519, 593, 600-602, 619
Wigner's *D* function, 138, 202, 617-619, 628
Wigner distribution function, 168, 172, 178, 188, 199, 519
Wigner–Stratonovich kernel, 207, 208, 508, 629
Wigner–Stratonovich map, 206, 208, 209, 214, 220, 508, 509, 511, 515, 582, 583, 593, 600, 602, 631, 634, 638

Lightning Source UK Ltd.
Milton Keynes UK
UKHW051152010620
363668UK00006B/113

9 789811 217272